ENVIRONMENTAL SYSTEMS

An introductory text
Second edition

I.D. White
University of Portsmouth

D.N. Mottershead
Edge Hill College of Higher Education

S.J. Harrison
University of Stirling

CHAPMAN & HALL

London · Weinheim · New York · Tokyo · Melbourne · Madras

Published by Chapman & Hall, 2—6 Boundary Row, London SE1 8HN

Chapman & Hall, 2—6 Boundary Row, London SE1 8HN, UK

Chapman & Hall GmbH, Pappelallee 3, 69469 Weinheim, Germany

Chapman & Hall USA, 115 Fifth Avenue, New York, NY 10003, USA

Chapman & Hall Japan, ITP-Japan, Kyowa Building, 3F, 2-2-1 Hirakawacho, Chiyoda-ku, Tokyo 102, Japan

Chapman & Hall Australia, 102 Dodds Street, South Melbourne, Victoria 3205, Australia

Chapman & Hall India, R. Seshadri, 32 Second Main Road, CIT East, Madras 600 035, India

First edition 1984
Fifth impression 1990
Second edition 1992
Reprinted 1993, 1994, 1996

Typeset in 10/11pt Times by Graphicraft Typesetters Ltd, Hong Kong
Printed and bound in Great Britain at the Alden Press, Oxford

ISBN 0 412 47140 X

A catalogue record for this book is available from the British Library

Library of Congress Cataloging-in-Publication Data available

∞ Printed on acid-free text paper, manufactured in accordance with ANSI/NISO Z39.48-1992 (Permanence of Paper).

Contents

Preface

Any book is a compromise between the original aspirations of the author(s) and the realities of publication. For this reason reviewers are often able to take issue not so much with the substantive content of a book, but with pragmatic decisions imposed on authors by this need to compromise. The first edition of this book was no exception, for the breadth of its canvas made compromise inevitable if its focus on functioning environmental systems was not to be lost. Although, hopefully, it has proved to be a successful compromise, several such decisions have been questioned, partly by reviewers, partly by feedback from students and teachers using the book.

In opting to concentrate on the human terrestrial environment in the first edition, the nature and role of the oceans was played down. In this current edition that decision has been reassessed, and a new chapter has been devoted to the hydrosphere, thereby acknowledging the importance of the oceans in an understanding of the energetics and biogeochemistry of the planet.

Similarly, the decision to model the structural and functional organization of the biosphere at the molecular and cellular level, although valid in itself, has also been reassessed, because those readers who constitute the main market for the book are more concerned with the macroscopic properties of the biosphere. In consequence the biosphere and ecosphere chapters have been condensed into one, allowing an expanded treatment of biogeochemical cycling and the perturbations to these cycles occasioned by human activity (with some small loss of the perspective provided by cell biology on the nature of the biosphere).

New chapters dealing with two environments the understanding of which is of profound importance to current environmental problems have been added, mainly in response to reviewers' comments. The first (Aeolian systems) underpins the understanding of desertification and the environmental problems of arid and semi-arid lands. The second (Coastal systems) considers the physical and biological systems that straddle the interface between land and sea.

During the life of any edition of a book, the field it purports to consider will itself evolve and develop, as does its social and cultural context. New research findings enhance understanding or open new avenues of endeavour, which, with inductive reasoning, extend the frontiers of the field. New paradigms emerge, and private and public attitudes and values change. In short, a book is an ephemeral entity, providing at best a synoptic picture: a particular view of the world which is sooner or later out of date and superseded. The first edition of this book, however, was fortunate in being a little ahead of its time in anticipating some of the changes which have occurred since its publication. The preface to the first edition stated that in the environmental sciences the systems philosophy had abandoned its ivory tower to become public property with the rise of environmentalism. Though fundamentally true, that statement was slightly premature for it turned out that the public required time to become aware of what they had acquired, and learn what to do with it. However, the recognition of the depletion of the ozone layer, and the accelerating debate surrounding the threat of global warming changed the status of the environmental and green agenda from parochial and regional issues to global imperatives. Against the background of these changes, for the first edition to have concluded with the hope that politicians and governments would develop an informed picture of their environment through which to formulate and implement policy, seems prophetic indeed. Nevertheless, the first edition has stood in need of revision, first, in recognition of the vastly increased awareness of environmental issues in general, and of the systems approach to environment in particular, and secondly, to catch up with recent developments.

All the chapters have been revised (some extensively, for example that dealing with the glacial system), and updated in line with developments in the appropriate disciplines since the time of the original publication. Recent work has been incorporated within the redrafted text (see, for example, material on the greenhouse gases), and/or appears in new boxed material as appropriate (see new Box on Stream power and the fluvial system). The geophysical and geochemical view

of the planet enshrined in the crustal system at
the heart of the chapter on the lithosphere is now
set in the context of an enriched understanding of
the 'geology' and 'geomorphology' of the other
planets in the solar system.

Among refinements to the ecological chapters
are the addition of material exploring the notion
of ecological strategy, an extended treatment of
ecological change, and a move beyond the focus
on population and community ecology to consider
ecology at a landscape scale. Here ecological
systems have geographical reality, and it is at this
scale that the potentially fertile interface between
ecology and the technologies of remote sensing
and geographical information systems remain to
be fully exploited.

The chapters dealing explicitly with environmental change have been redrafted and revised
to address the contemporary issues of projected
climatic change and its environmental consequences. When the book was originally written,
issues such as the ozone hole and contemporary
climate change were either not fully appreciated
or as yet on the horizon. In these respects the
second edition has been brought right up to date.

Nonetheless, the second edition, like its pre-
decessor, will remain a compromise, flawed by
sins of omission, by bias, and by idiosyncratic
selection. However, our hope is that the book
retains the coherent and consistent commitment
to a comprehensive understanding of our planet
and its surface environments which informed the
first edition. Furthermore, it is our belief that
it remains appropriate that such an understanding should be articulated through a conceptual
framework which regards the Earth as a set of
interconnected and interdependent open systems
susceptible to modelling at a variety of scales.

Part A

A systems framework: to have a picture

The available information would suggest that the appearance of Neanderthal people was dominated by an ugly and repulsive strangeness enhanced by a low forehead, protruding brows, ape neck, and accentuated by low stature and extreme hairiness.

William Golding chose to take such unpromising beings as the central characters of his book, *The inheritors*. Their dialogue, however, is restricted by the intellectual capacity of the extended family to which they belong. They think in pictures; pictures sometimes linked, sometimes isolated . . .

> 'At last Mal . . . began to straighten himself by bearing down on the thorn bush and making his hands walk over each other up the stick. He looked at the water then at each of the people in turn, and they waited.
> "I have a picture."
> He freed a hand and put it flat on his head as if confining the images which flickered there . . . His eyes deep in their hollows turned to the people imploring them to share a picture with him . . .
> Fa put her hand flat on top of her head.
> "I have a picture."
> She scrambled out of the overhang and pointed back towards the forest and the sea.
> "I am by the sea and I have a picture. This is a picture of a picture. I am . . ."
> She screwed up her eyes and scowled ". . . thinking."'

Though the similarities between us and the Neanderthals are slight, thinking is still, as for Mal and Fa, the seeing of pictures, the ability to construct and manipulate mental images. In part these are pictures assembled from empirical facts, observed with our two eyes – seeing is believing. However, the kind of pictures with which William Golding's primitive heroes struggle, and the kind of pictures *we* wish to share with *you*, are seen with another eye . . . the mind's eye. To see with the mind's eye is a much more complex process, for here our pictures may be painted not only with observed visual fact but with words, symbols, ideas, logic and inference, and even with feeling. They may be pictures of pictures, perhaps a collage of different images, some based on our own observation and understanding, some the pictures others have seen and shared with us.

In this book we are going to turn the mind's eye on the world about us, on the natural world at first – our natural environment. Images of this environment and of the landscapes through which it is perceived abound in geography; indeed, to many they *are* geography. However, it is one particular image that we have in our mind's eye – an image of our planet and the landscapes we inhabit at its surface, as ordered and functioning systems. This image is not new, but it has become increasingly significant as a way of looking at the world, and it pervades not only geography but much of the rest of science. It is by no means the only valid image of our environment, but it ha⌐ a strength which has enabled it to 'have illuminated thought, clarified objectives, and cut through the theoretical and technical undergrowth in the third quarter of the 20th century in a most striking manner' (Bennett & Chorley 1978).

So, like Mal, we have a picture and through the images conveyed by this book we hope to share that picture with you, so that you may say with Fa, 'I have a picture . . . it is a picture of a picture'.

I SYSTEMS, HUMANS AND ENVIRONMENT

Why a systems approach?

1.1 Humans, environment and geography

We are animals. We are similar in anatomy and physiology to our mammalian cousins and, like all other animals, we must eat food and breathe air to survive. Nevertheless, we *are* different, but this difference lies in accomplishment and behaviour. We possess not only a genetic inheritance which determines our animal nature, but also a cultural inheritance of knowledge and custom transmitted by language and symbols. As with other animals, however, we occupy a habitat which exists in both space and time and forms our environment. Although this is a physical, chemical and biological environment, it is also a cultural environment with social, political, economic and technological dimensions.

This distinction is reflected in a dichotomy in the study of human–environment relationships, with the natural sciences concerned with the natural environment, and the social sciences and applied sciences concerned with the cultural environment. In reality these two components of our environment are inseparable; the natural environment cannot be fully understood in isolation from us and our interactions with it. But before our cultural relationships with the natural environment can be considered in anything but a superficial way, we *must* understand that environment and, for this reason, this book is concerned with the natural environment – its physical, chemical and biological components, the relationships between these components and those between them and human societies.

How should we begin this formidable task? We could start by taking the environment to pieces to see what it is composed of and how it works, much as we might dismantle a piece of machinery. This process can be called the **analysis** of the environment. It is an example of the **reductionist** approach common to normal science. Here complex phenomena are reduced to their component parts and examined in detail. Indeed different sciences, or different branches of a science, may deal separately with different components. So when we consider how science deals with the environment, we find that oceanography is concerned with the sea, meteorology with the air, hydrology with water, geology with rocks, and biology with organisms. Physics and chemistry too are concerned with the components of the environment, but at a more fundamental level.

The problem with this approach becomes apparent when trying to put the environment back together again: when turning from analysis to **synthesis**. The selected pieces have been looked at in different ways, with different degrees of rigour, and different terminologies have been developed, so we may no longer recognize the pieces as being part of the original whole. The process of reassembling the pieces becomes more difficult because we can no longer remember how they fitted together, or we failed to find out before we dismantled the environment. Also, some of the pieces may have been overlooked in the process of analysis, because none of the specialist sciences regarded them as their concern. The real problem, therefore, is the lack of any true integration. What are the links between a kingfisher, a developing thunder cloud, a breaking wave and a landslide? Much may be known about each of these pieces, but how do they fit together into the environmental jigsaw puzzle? You might ask whether it really matters. Is it necessary to synthesise, to reassemble the pieces, as long as our knowledge of them is pursued in sufficient depth? These questions can perhaps best be answered visually. Fig. 1.1a portrays a section of the envi-

Fig. 1.1 Different perspectives on our environment. (a) A sea loch on the west coast of Scotland. (b) The same landscape type seen from the summit of an adjacent mountain. (c) from a satellite, and finally (d) the planet Earth.

ronment – a landscape in the Scottish Highlands. This is the scale at which our environment is normally observed and most of the components which make up the familiar concept of the natural environment can be seen. There are mountains and hills, valleys, streams, a lake (in fact, a sea loch), plants and animals, and the sky.

The view in Fig. 1.1b is from the top of one of the mountains. The perspective has not really altered very much. Most of the components are still visible, though some are now less clear. The distinction between plants and animals is perhaps more blurred at this distance, but some relationships become more obvious. The spatial relationships between the mountain streams, the river, the loch and the sea can be seen more clearly.

The perspective has altered enormously in Fig. 1.1c. The view here is the one obtained at an altitude of 435 km. Nevertheless, the Highland environment is still visible, but discernible at an entirely different scale. All the components are there; the cows are still grazing by the loch shore. However, they are no longer perceptible as separate entities. It is only the major topographic features that can be distinguished: mountain ridges, valleys, the major drainage network and the sea. Living things are present, but they cannot be clearly discerned.

Finally, in Fig. 1.1d the perspective, at an altitude of 1110 km, has changed so much that only the planet Earth is discernible, though the land, the sea and the atmosphere are still recognizable. All perception of the individual components of the Highland landscape visible in Fig. 1.1a has been lost. Here is visual confirmation that the various components of this original environment, and indeed of the environment as a whole, cannot really be viewed in isolation. There is an integrated unity about the natural environment, which is evident in Fig. 1.1d as the planet Earth – what Kenneth Boulding (1966) has called 'Spaceship Earth'. It is obvious then that to understand the Earth, its environments and the components of these environments, as well as how they function, we must study the interrelationships of all of these components – a synthesis is required. However, in the late 20th century there is a less philosophical, more pressing and practical reason for such a synthesis.

Our finite natural environment represents not only living space and resources for humans and their animal needs but, because people are cultural animals, it must also provide for our developing cultural needs. As our cultural complexity, social organization and technological capacity have increased through time, we have made progressively greater demands on the environment. Also, as cultural development has made us progressively less dependent on our immediate environment, so the nature of our perception of it, and our interactions with it, has changed. We have exploited, modified and damaged or destroyed it to an extent not equalled by any other animal. In the 20th century our demands have begun to outstrip the capacity of the environment to absorb the perturbations which are the result of human activities.

The effects of these perturbations, however, are now felt on a global scale and it is finally being realized that the future equilibrium state of the planet as a whole is in jeopardy. This realization has been hastened by the discovery of the ozone 'hole', by an acknowledgement of climate change and global warming, by desertification and the 'plight' of the rain forest, as well as the more regional impact of acid rain and nuclear disasters of the magnitude of Chernobyl.

This heightened environmental awareness has brought with it an understanding of the need for informed environmental management and for conservation policies at all levels. The key word here is 'informed', for, before we can manage our environment we must understand it. It is this urgent need more than any philosophical consideration that has given added impetus to the integrated study of our natural environment. More than this, it has emphasized the fact that the separation of the natural from the cultural is not only unreal, but also dangerous. Our interaction with the natural environment can only be understood with reference to our perception of it and behavioural response to it, both of which are conditioned by our complex cultural environment.

Indeed, although they will have an important role to play, the solution to the world's environmental problems will lie not with scientists, engineers and technicians. The 'technological fix' is no longer the answer. It is to the development of appropriate political, social and economic structures, and to the will of the peoples of the world that we must look for the future of our environment and of our planet.

1.2 Systems

There are, then, at least two reasons for needing to know more about the physical, chemical and biological environment. As the trustees and potential managers of that environment, and certainly as users of its resources, we need to know how it works. The reductionist approach, fragmented and piecemeal, is at best unsatisfactory and does not impart such understanding. To create a truly comprehensive and integrated picture – our 'picture of a picture' – is none the less a colossal task. What is needed is some sort of framework – a plan, a map, a wiring diagram – to show how to dismantle the environment for analysis and, perhaps more important, to show how to reassemble it so that the results of analysis can be incorporated in an integrated synthesis. Such a framework is provided by the concept of the **system**.

The word 'system' may be very familiar. You, the readers, are probably involved in some part of an education system. The room in which you are sitting probably has a heating system, perhaps even an air-conditioning system. Depending on the time of day, you might be about to close this book and go off to satisfy your digestive system. This might involve travelling home via a public transport system, or, if you have a car or motorcycle you will start it by using the ignition system. What does the word 'system' mean? Why is it used to refer to such dissimilar things? In the first instance, all of these things consist of collections of other things – sets of objects. They may be schools, colleges, universities, classrooms, lecture theatres; or mouth, oesophagus, stomach, colon and rectum; or boiler, pipes, radiators and thermostats; or rails, roads, trains, buses, stations and termini. In all cases, from education to transport systems, we are referring to sets of objects. However, the word 'system' is more than just a collective noun, for it also tells us that the objects are organized in some way, that there are connections and links between the units. So the roads or rails pass through the stations which are arranged along them between the termini, pipes connect the boiler to the radiators, and so on.

The term transport system, however, does not tell us whether we travel in a bus or train, let alone whether it is painted red or green. The term heating system does not specify whether it is gas- or oil-fired, or in fact that it has pipes and radiators: it could be ducted hot air. In other words, the term system is non-specific and the concept can be applied, in these examples, to all such arrangements of units which have a transporting or heating function. The term implies *generality*.

Here another feature of the use of the word 'system' becomes apparent. A system actually functions in some way as a whole. A transport system is not merely a static arrangement of units: it transports people and goods. A heating system circulates hot water or air and warms the room. A common feature of the working of systems is the transfer of some 'material' between the units of the system. This material may be obvious, like the water in the pipes and radiators or the food in the digestive system, or it may be more abstract. In an education system the material may be thought of as a generation of children passing upwards from class to class, or from school to college. Alternatively, one might equally identify the material not as the children but as the transfer of ideas, the flow of knowledge. A political system is concerned not only with the institutions of that system and the transfer of such 'real' materials as people or votes or tax revenues, but also with the transfer of more intangible materials such as directives, decisions, ideas and ideologies.

Such movements of materials require some motivation. In some systems this motivation is again obvious, such as the heat applied to the central heating boiler, or the electricity supply to a hi-fi system. In others it is less obvious: the impetus of the forces of supply and demand on people and goods, for example, or the desire for the acquisition of knowledge and understanding in an education system. In this last example one might be less idealistic and think of the motivation as being provided by the need to obtain qualifications in order to gain employment. But the point remains the same: in all systems there is a driving force that makes them work.

The common characteristics of systems are summarized below.

1. All systems have some structure or organization.
2. They are all to some extent generalizations, abstractions or idealizations of the real word.
3. They all function in some way.
4. There are, therefore, functional as well as structural relationships between the units.

5. Function implies the flow and transfer of some material.
6. Function requires the presence of some driving force, or source of energy.
7. All systems show some degree of integration.

1.3 Environmental systems as energy systems

There is nothing new in the concept of the system used in the broad way encountered so far. Indeed, the antiquity and widespread use of the concept reflect the ability of the human mind to perceive things as a whole, and there is no reason to assume that this ability is the preserve of the late 20th-century. Although all systems have certain characteristics in common – of which organization, generalization and integration are perhaps the most important – this book will be concerned with systems of a particular kind. These are **thermodynamic systems**, and their original description and formalization took place in physics.

Literally, thermodynamics means 'the study of heat as it does work', but this narrow definition is rather misleading. Perhaps thermodynamics could more aptly be termed **energetics**, as it is now concerned not only with heat but also with all other forms of energy. So thermodynamic systems might more profitably be called **energy systems** (Fig. 1.2a). Such an energy system is merely a *defined system of matter, the energy content of that system of matter, and the exchange of energy between that system and its surroundings*. This defined system of matter may be a leaf, a rock-forming mineral, a tree, a length of river channel, a slope segment or a parcel of air. It is merely that part of the physical universe whose properties are under investigation. Although such definitions appear vague, they emphasize two important facts: the first is the importance attached to defining the system; the second is that the definition of the system is somewhat arbitrary (Box 1.1). Energy systems may be isolated, closed or open in terms of the relationships which pertain across the boundary of the system with the surroundings (Fig. 1.2b) (see Box 1.1). However, all environmental systems are open systems exchanging matter and energy with their surroundings, although sometimes it is convenient to treat them as if they were closed systems with only energy exchanges to consider. Furthermore they are dynamic open systems, and as energy or thermodynamic systems they func-

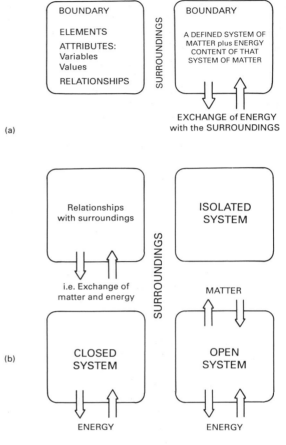

Fig. 1.2 (a) Approaches to the definition of energy systems; (b) Classification of energy systems on the basis of their interactions with their surroundings into isolated, closed and open systems.

tion in accordance with the laws of thermodynamics (Box 1.2, and Chapter 2, Sections 2.3.2., & 2.3.3). Therefore, changes in the constituent elements of the system, in their attributes, and in the relationship between them are important. These can best be thought of as *changes in the state of the system*. Theoretically, the change in state is completely defined when the *initial* and *final* states of the system are specified. In practice, however, the *pathway* of change of state is often of interest, and to specify such a pathway it is necessary to know not only the initial and final states but also the sequence of *intermediate* states in order. The way in which a change in state is effected is called a **process**. When the effect of the operation of a process is to return the system to its initial state, i.e. the process is cyclic and the final state is identical to the initial state, then the process is called a **cycle** (Fig. 1.3).

As open systems are characterized by the continual input, throughput and output of matter and energy (Box 1.2), the maintenance of some structural organization in the face of these throughputs is a critical characteristic of all environmental open systems. A drainage system maintains the organization of the stream and river channel network and of the contributing slopes in spite of the continuous throughput of water. Living organisms, including us, are inconceivable without the maintenance (within very narrow limits) of the extremely complex structural and functional organization of their bodies, though a regular throughput of food materials and energy occurs. In other words, such systems must maintain a more or less stable state defined in terms of their elements, attributes and relationships through time. Although often but wrongly referred to as

---

Box 1.1

DEFINING ENERGY SYSTEMS

An energy system is confined to a definite place in space by the **boundary** of the system, whether this is natural and real like a cell wall or watershed, or arbitrary, though still real, like the walls of a test-tube or vessel in a laboratory, or arbitrary and intangible like the boundary of a cloud. The boundary of a system separates it from the rest of the universe, which is known as the **surroundings** (Fig. 1.2a).

Within its defined boundary the system has three kinds of property. The **elements** of the system are the kinds of substance composing the system. They may be atoms or molecules, or larger bodies of matter – sand grains, raindrops, grass plants, rabbits – but each is a unit which exists in both space and time. Each element has a set of **attributes** or **states**. These elements and their attributes may be perceived by the senses, or made perceptible by measurement or experiment. In the case of such measurable attributes as number, size, pressure, volume, temperature, colour or age, a numerical **value** can be assigned by direct or indirect comparison with a standard. Between two or more elements, or two or more states or attributes, there are **relationships** which serve to define the states of aggregation of the elements, or the **organization of the system**.

The **state of the system** is defined when each of its properties (**variables**), i.e. elements, attributes and relationships, has a definite value.

These definitions apply to all energy systems, but several distinct types of system can be distinguished on the basis of the behaviour of the system boundary (Fig. 1.2b).

In an **isolated system** there is no interaction with the surroundings across the boundary. Such systems are encountered only in the laboratory, but they are important in the development of thermodynamic concepts.

Closed systems are closed with respect to matter, but energy may be transferred between the system and its surroundings. On the Earth, closed systems are rare. It is important, however, to be able to analyse complex environmental systems in terms of simpler component systems, and it is often useful to be able to treat these as closed systems. They are also important because much of thermodynamic theory was developed in relation to closed systems.

Open systems are those in which both matter and energy can cross the boundary of the system and be exchanged with the surroundings. In such systems the transfer of matter itself represents the transfer of energy, as matter possesses energy by virtue of its organization (potential chemical energy, for example). All environmental systems are open systems and are characterized by the maintenance of structure in the face of continued throughputs of both matter and energy.

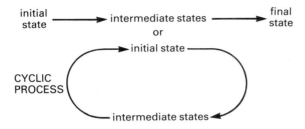

PROCESS ... the mode of operation by means of which a change in the state of the system is effected.

Fig. 1.3 The definition of process.

an **equilibrium state**, the characteristic state of an open system is **stationary** or **steady** (Fig. 1.4a). Open systems do not maintain a true equilibrium because although there may be little or no change in the system variables at what might be called the 'macroscopic' scale, none the less there are exchanges of energy and matter with the surroundings which sustain the 'steady' state of the system. In addition, of course, at the 'microscopic' scale changes do occur: changes such as the continual turnover at the molecular level associated with cell maintenance and repair in living systems, or the continual throughput of water in the channel in a river system.

Box 1.2
THE LAWS OF THERMODYNAMICS

Process implies that work is being done on the system. For work to be done requires the transformation of energy. Environmental systems and our models of them can therefore be seen as energy systems, i.e. a defined system of matter, the energy content of that system of matter, and the exchange of energy (which may involve the exhange of matter) between that system and its surroundings (including with other systems in a cascade).

This view of environmental systems as open energy (or thermodynamic) systems requires us to consider the **laws of thermodynamics** (energetics) which govern the transformation and transfer of energy in real processes. (There are three laws, but only the first two will be presented here).

THE FIRST LAW OF THERMODYNAMICS
The Law of Conservation of Energy, or the Conservation Principle.

ENERGY CANNOT BE CREATED, NEITHER CAN IT BE DESTROYED: ENERGY IS MERELY TRANSFORMED FROM ONE KIND OF ENERGY TO ANOTHER KIND.

This law opens the way for us to account for what happens to energy in the operation of systems, that is, it enables us to look at energy balance sheets, or **energy budgets**.

THE SECOND LAW OF THERMODYNAMICS

NO SPONTANEOUS TRANSFORMATION OF ENERGY IS ONE HUNDRED PERCENT EFFICIENT. IN ALL SPONTANEOUS TRANSFORMATIONS OF ENERGY SOME ENERGY IS DISSIPATED AS HEAT ENERGY AND IS THEREFORE UNAVAILABLE TO DO WORK ON THE SYSTEM.

This law imposes a direction on the operation of real processes in our systems and implies that the transfer of energy is ultimately a one-way flow from high energy levels to low energy levels. The second law prompts us to look for **energy sources** and **energy sinks**, and to look at the efficiency of individual energy transformations, in the operation of systems processes (see also Chapter 2).

a)

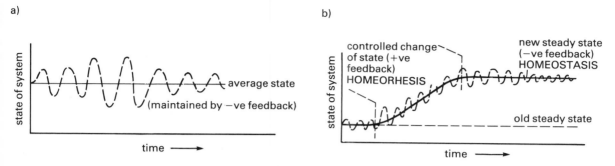

b)

Fig. 1.4 (a) Open system maintaining a steady (average) state through time. Fluctuations about this average state are regulated by negative feedback. (b) A controlled change of state from an 'old' to a 'new' steady state effected by positive feedback.

Although when viewed at the macroscopic scale, the system appears to be maintaining a stable state, the steady state is dynamic, and stability only exists as a statistical abstraction: the **average state**, because of this change at the microscopic scale. The nature of change and equilibrium will be considered further in Chapter 23; for the moment the recognition that open systems maintain a steady state means that they must possess the capacity for **self-regulation** and introduces the concept of **feedback** (Fig. 1.5).

Self-regulation in environmental systems is effected by **negative-feedback** mechanisms or, as they are called in the life sciences, **homeostatic** mechanisms. These are able to damp down change and as such are control mechanisms. Indeed, self-regulation can be profitably termed **homeostasis**.

Negative-feedback mechanisms operate in much the same way as a thermostat in a heating system, in that they are able to sense or interpret the output or effect of some operation, just as the thermostat senses temperature change. They then feed back this information to influence the operation of the process concerned, in the same way as the thermostat switches on or shuts down the source of heat. In terms of the definition of process given earlier, the net effect of the processes involved in negative-feedback is to return the system to the initial state, i.e. deviation from an average or mean condition is not sustained. Therefore, if the system experiences changes in state about this mean condition, these changes are cyclic fluctuations, and over time a steady state is maintained.

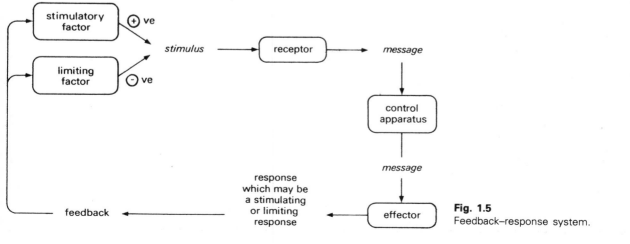

Fig. 1.5
Feedback–response system.

The number, type and degree of sophistication of negative-feedback mechanisms in environmental systems vary considerably. Many physical systems can accommodate quite wide fluctuations about a mean state without jeopardizing their function. The detailed properties of a length of river channel may vary widely about the most effective state, but it will still transfer water and debris. However, self-regulation in living systems has to be far more precise and sophisticated. The temperature of your body, for example, is regulated by a host of interrelated feedback mechanisms, but if it is allowed to fluctuate widely you feel decidedly ill and if control is not established you die. On death your body ceases to function as a living system and the steady state of the system degenerates.

Nevertheless, natural systems do exhibit change. Indeed, one of their most important properties is the tendency to evolve, or develop, such ordered change in systems state through time. This is the case with the extension of an ice sheet, the development of a cyclone or the growth of an organism. These are instances of directional change and they are also regulated by feedback mechanisms. Here they do not stabilize the system, but have a cumulative effect and they reinforce particular directions of change. They are termed **positive-feedback** mechanisms. Although the deviation–amplification character of positive feedback inevitably propels the system further from its initial state, the trajectory of change may be closely regulated. Indeed ordered growth and development in organisms is such a case where **homeorhetic** or positive-feedback mechanisms preserve the order and direction of change in the state of the system (Fig. 1.4b). Just as stabilized behaviour of the system relative to some average state can be termed homeostasis, controlled change and development can be referred to as **homeorhesis**.

In practice the regulation of natural systems states involves the linking of several feedback mechanisms, some positive and some negative, in complex **feedback loops** (Fig. 1.5). Where negative feedback predominates, the overall effect is the maintenance of a steady state. Where positive feedback is dominant and leads to an increase in the order and complexity of the system state, then we speak of the growth or development of the system. Where positive feedback has the cumulative effect of progressively destroying the

organization of the system, it leads to a retrogressive and often irreversible change in state.

Associated with positive feedback is the existence of **thresholds**. These are state variables which, when they assume certain values, are capable of initiating often sudden and sometimes dramatic changes of state. For example, it is possible to define threshold slope angles which, if exceeded, trigger rapid mass-movement processes that result in the complete reorganization of the state of the slope system. In a similar way, in a soil undergoing decalcification by leaching, the moment the calcium content reaches zero the system will be transformed in state and the chemical environment for soil processes will change rapidly. Such thresholds are really only extreme cases, because in all feedback processes certain state variables play similar key roles in controlling the operation of the processes and, in this more general case, we refer to them as **regulators**.

1.4 Systems and models

Much of the discussion of energy systems has been concerned with abstract theoretical systems and with fundamental definitions. The rest of this book will be concerned with systems in the real world. The use of the word 'real' is important, for although these systems may be submitted to precise thermodynamic definition and their functional similarity stressed, they are none the less real. The precipitation input to the denudation system is still the rain you can feel on your face. The throughput of water in the channel system is still the river in which you may swim. The energy store in the biomass of the forest is still the trees you might have climbed as a child. The problem is that the real world is extremely complex and, although the recognition that it consists of thermodynamic or energy systems helps to structure our approach to this complexity, of itself it does nothing to simplify it. The specification of the system, its elements, states, relationships and processes, is, in the last analysis, the specification of the real world in all its complexity. The dilemma is that of being caught between perceiving the system as a whole and the near impossibility of seeing its complete complexity, a dilemma well expressed by Arnold Schultz (1969) when he wrote '...the sheer bulk of information is so overwhelming we can make little use of it. At the other end all the states (of the system) are fused into one grand platitudinous expression and all you can say for it is "There it is"'. Clearly there is a need to simplify, and here the answer to the dilemma lies in the concept of the **model** and the techniques of **modelling**.

1.4.1 Models
The word 'model' is used in everyday speech with three distinct meanings: 'a replica', 'an ideal' and 'to display'. The concept of the model as adopted here combines aspects of all three meanings. In order to simplify environmental systems, models or replicas of them can be constructed. To be useful, these models must idealize the system and they must display or make clear its structure and how it works.

1.4.2 Systems as models
The concept of the system is itself a model. When referring to central heating systems, or transport systems, or thermodynamic systems, what is being presented is an idealized view that is generally applicable to all real situations having the same general character. The London Underground and the 19th-century canal network of Britain, for example, are clearly different and unique entities. To regard both merely as transport systems is to strip away all that is special (but irrelevant) and to idealize their structures and relationships in a generally applicable model – the transport system. In the same way, objects as dissimilar as a bunsen-burner flame and a plant or animal cell can both be modelled as open thermodynamic systems with inputs and outputs of matter and energy. When one begins to regard the natural environment in terms of systems, the process of generalization and idealization has started.

1.4.3 Models of systems
Wishing to simplify the complexity of the real world involves making models of the system itself. Let us for the moment return to upland Britain as portrayed in Fig. 1.1a. Everything visible consists of atoms and their combination in molecules and compounds. These in turn are present as rock-forming minerals, as the water in the loch, as the organic compounds of plants and animals, as the water vapour in the clouds, and so on. Nevertheless, the molecules and compounds cannot be discerned at this scale, but only the physical objects of the rocks outcropping in the distance, the

grass and the cows, the water, the soil, the sky and clouds. Even these components became lost in the broader perception of the land, water, vegetation and atmosphere as the perspective changed, until in the final picture from space all of this matter was perceived simply as the planet Earth. What was happening, apart from changing the perspective, was that the parts of the system were being resolved at different levels until finally there was no resolution of them at all.

The atoms, molecules, cows and rocks, however, are all elements and their states in this system. The third component of a system is the relationship between the elements. So a model of a system needs to incorporate relationships. A topographic map is a model of the terrain it represents. It distinguishes the elements of that terrain by means of symbols for height, rivers, forests, towns, roads, and its power of resolution of these elements varies with scale from, for example, a 1:25 000 sheet map to a school atlas map. However, a map goes further, for it models not just the elements but also their geographical relationships – the distance and direction between any two elements. All modelling can in fact be viewed as a mapping process. The relationships in the system and hence in the model can be of several types: spatial distance, causation, conjunction, succession of events. They can be expressed in words such as 'cow eats grass', or 'river takes soil' through statistical statements of the probability of such events, and through quantitative measurements of, for example, transfers of matter and energy between the elements.

As with the topographic map, some models are static, representing the structure of the system rather than the processes going on within it, and between it and its surroundings. However, to understand the functioning of systems we need dynamic models that are capable of identifying processes and modelling their effect on the system. This is particularly the case if the model is to be used to predict the behaviour of the system; for example, a drainage basin model used to predict flood hazard.

1.4.4 Levels of resolution
In the sections that follow, models of environmental systems are what can be called **homo-morphic** models (*homo* is Greek for *similar*) in that they are imperfect representations of reality. Rarely if at all are the models **isomorphic** (*iso*

is Greek for *same* or *equal*) in the sense that for every element, state, relationship and process there is a corresponding component in the model. The level of resolution of the models will vary with scale, the system being discriminated at some level appropriate to the scale. At the fine end some realism is retained, and at the coarse end there is a gain in generality. However, no matter what level of discrimination is adopted, because the model cannot be truly isomorphic it will be lumping together elements and relationships in single compartments of the model, and a wealth of information inside each compartment is conveniently ignored. This is what Egler (1964) called the 'meat grinder' approach and it introduces another term for these homomorphic models, viz. **compartment** models.

Each compartment of these models is treated as a **black box**, which can be defined as any unit whose function may be evaluated without specifying the contents. At low resolution the models of environmental systems contain a small number of relatively large black box compartments; indeed the whole system may be treated as a black box. As the level of discrimination increases, so these compartments are progressively split into subcompartments, which are in turn treated as black boxes. At intermediate levels of discrimination the

model of the whole system has become a partial view of the system, its structure, relationships and processes, and is called a **grey box** model. Finally, as realism increases towards a truly isomorphic model, the grey box becomes a **white box** with most of the elements, states, relationships and processes of the system identified and incorporated in the model (Fig. 1.6). Even here some black box compartments will remain, for no model is a complete representation of reality.

This approach to modelling has a number of advantages. It provides a hierarchy of models of different levels of discrimination and complexity, which are appropriate to different scales of analysis of the system. It also allows the knowledge and results of specialist work on any part of the system to be coordinated and coupled into the model at the appropriate level in the hierarchy.

1.4.5 Kinds of model

These compartment models may be static or dynamic and they may vary from box-and-arrow (flowchart) models to **quantitative mathematical models** at one end, to scaled hardware models at the other. None is necessarily better than another. However, the most important criterion in assessing the validity of a model is its **ability to predict** the behaviour of the system. In this context it is only

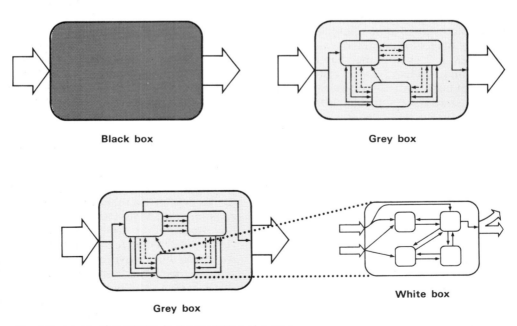

Fig. 1.6 Levels of resolution in the modelling of systems.

dynamic models that have predictive power. Of these it is only some hardware and mathematical models that represent quantitative relationships and predict the behaviour of the system in quantitative terms (Box 1.3) (Fig. 1.7). Indeed Haines-Young and Petch (1986) reject many of the broad characteristics of **conceptual models** presented here as unhelpful. Instead they encourage a specific interpretation of the term related to the way models are actually used in science, as mechanisms by which predictions are made and used to test theory (see also the critique of conceptual models in ecology, Peters, 1991). Models of systems predicated on the assumption that a particular theory will account for the behaviour of the system are used to generate predictions of that behaviour. In accordance with a *critical rationalist methodology*, testing is concerned with the falsification of theory and hence with the failure of the model (and, by inference, the theory on which it is based) to predict the behaviour of the system. Although a valid position to adopt in relation to the concept of the 'model', this view is perhaps too extreme. Indeed conceptual models of all types including the static and non-predictive dynamic models used in this book, are nevertheless very valuable aids to communication and understanding.

Here we are concerned to construct conceptual models of environmental systems as open systems which display their structural organization and explain the way they work, or their function. The process consists in defining the compartments of the model, that is, identifying the boundaries, elements, attributes and structural relationships of the system. These are compartment models, but because they are homomorphic models of reality their description involves the selection of significant elements, properties and appropriate means of describing their structure or morphology (cf. mapping, 1.4.3). Such models are called **morphological models** and vary considerably in sophistication. Statistical or mathematical expressions of concepts such as causation, correlation and probability are used most commonly to describe relationships between variables associated with the properties of the system represented by the compartments of the model.

The relationships between the compartments of such models are also the functional linkages which facilitate the operation of the system. So it is important in functional or dynamic models to identify and depict pathways of 'flux', or transfer of matter and energy. To do so focuses attention on compartments as **stores** of matter and energy, as well as on the inputs to and outputs from such stores. Furthermore the **residence time** of matter and energy in these stores, and the variables that regulate their transfer between stores, become important aspects of these models.

Some of the compartments in such models can be regarded as systems in their own right: that is, they are **subsystems**, which are functionally linked so that the output of one becomes the input to another. Such progressions of matter and energy through several compartments, or perhaps through several subsystems, are known as **cascades**, and they form **cascading models** of open systems.

Even these models, however, are still essentially structural, albeit with the emphasis on functional structure. In order to move towards models which depict the dynamics of the system we must turn to modelling process. Systems respond to processes: processes are the means by which the system transforms and transfers matter and energy, and by which changes of state are effected . . . even if only to maintain the status quo, the steady state.

When we combine the modelling of process with that of the response of the system to process we have models at a higher level, which begin to depict and explain the dynamics of the system and how it works. Such models are **process-response** models and can be conceptualized as combining morphological and cascade models. This is because pertinent variables representing selected properties of the system act as regulators of process, and process by definition involves the transformation and transfer of energy (and, in open systems, matter as well).

1.5 The application of the systems framework

Although the foregoing discussion of systems and models has been conducted largely in the abstract, in the chapters which follow the same conceptual framework will be applied to systems in the real world. We shall begin with a picture of the whole planet (cf. Fig. 1.1d) and regard it at first as a closed system. By adopting a black box approach, we shall concentrate on the transfer of energy across the boundary of the system. Before long, however, it will be evident that we need to 'open up' the black box and examine, at least in a broad

Box 1.3
MATHEMATICAL MODELS

Mathematical models are constructed by translating *conceptual models* into the **formal logic of mathematics**, i.e. the process of translating physical, chemical, and/or biological *concepts* about a system into a set of mathematical relationships. In this process mathematical symbols provide a useful shorthand to describe complex systems, while equations permit formal statements of how system components are likely to interact.

These sets of mathematical relationships are commonly based on four fundamental mathematical elements: the *state* of the system is described by sets of **state variables** representing the condition of the system at any particular time; **transfer functions** are used to represent the functional relationships in terms of the *flows* (transfers) or *interactions* between the components (or compartments) of the system; *inputs* (of matter and/or energy) to the system, or *factors* which can be thought of as affecting the system by influencing the conditions under which it operates, but which are not themselves affected by the components of the system, are modelled by equations known as **forcing functions**; finally, *constants* used in the formulation of the mathematical equations are called **parameters**. The values of these parameters may be determined by the theory on which the model is based, or they may be derived empirically from observation and experience. Models where the values of parameters are derived from underlying theory are termed **fully specified models** (non-empirical models). Those which require calibration, i.e. the estimation of parameter values from empirical observation or experiment are termed **partially specified models** (empirical models). The set of mathematical relationships comprising a model is itself often referred to as a system* (i.e. a system of equations) and the manipulation of this mathematical system can be referred to as systems analysis. The mathematical system is, of course, a model and as such is an imperfect and abstract representation of reality. The degree to which the mathematical statements correspond to the physical, chemical or biological concepts of the system they are intended to represent is a measure of the **realism** of the model.

However, it is the ability to *predict* which makes mathematical models preferable to conceptual models; indeed, mathematical models are more often developed for **prediction** of dynamic change over time rather than for mere description. The ability of the model to predict numerical change, and to mimic the data on which it is based, are measures of the **precision** of the model. The extent to which the model can be applied to different situations, or the number of situations where it holds good, is a measure of the **generality** of the model. Conceptual models represent a number of concepts or ideas about a system, but are not able to test these ideas in any formal way. On the other hand, the validity of a mathematical model can be assessed by comparing its predictions against the empirical observations. The failure of the model to predict change is in itself useful, in that it may point out flaws in the conceptual framework from which the model was developed, thereby contributing to the development of theory.

The equations and the mathematical relationships which define the model can take several forms, and the kinds of mathematical manipulation employed to develop the model can also vary. For example, the symbolism of *matrix algebra* is often used to display system relationships and to manipulate those relationships, while *difference* and *differential equations* are the basis of many models which aim to describe quantitatively the way systems change over time (see the surface-water balance equation in Chapter 12, and the treatment of primary production and population growth in Chapters 19 and 25, respectively). In more advanced models relationships may be modelled *statistically*, and the basis of the logical operation of the model may lie in **probability theory**. Here we may recognize an important, but to some degree arbitrary, distinction which is usually made when considering mathematical models.

Where the model is expressed in such a way that the outcome of the operation of the model (the experiment) is fully determined by the initial values of the state variables, input variables, parameters and factors, it is called a **deterministic model**. In such models the outcome of the operation of the system is fully predicted by the model, so that the precision of the model is only constrained by its realism. Where the model is formulated in such a way that it acknowledges an element of uncertainty, and attempts to include the effects of random variation in forcing functions and parameters, it is called a **probabilistic** or **stochastic** model. Such models, therefore, do not predict the outcome of the operation of the system (the experiment), but instead predict the probability of the outcome (the most probable state, see Chapter 23) with some degree of confidence defined by the model.

There are two reasons for using stochastic models. First, there may be good *a priori* theoretical reasons to believe that the real world system possesses inherent elements of uncertainty, or that its behaviour is, at least in part, random. Secondly, although in theory a system may be capable of being fully specified, and therefore of being described by a deterministic model, in practice that cannot be achieved. This is usually because there is a mismatch between the complexity of the system and the degree of realism of the model we can formulate, i.e. our deterministic model is very coarse. In these situations the uncertainty introduced by lack of realism can be partially overcome by incorporating a random element into the model, i.e. using a stochastic model.

*Reference to mathematical models as systems of *equations* is often confusing, and terms such as dynamical systems, or stochastic systems will be encountered in the literature using the term system in this sense. The same criticism can also be levelled at references to conceptual models, as, for example, cascading systems or process–response systems. It is perhaps better practice to restrict the use of the term system to the defined energy system under consideration, and to deploy the term model for conceptual or mathematical representations of it. (See recommended reading.)

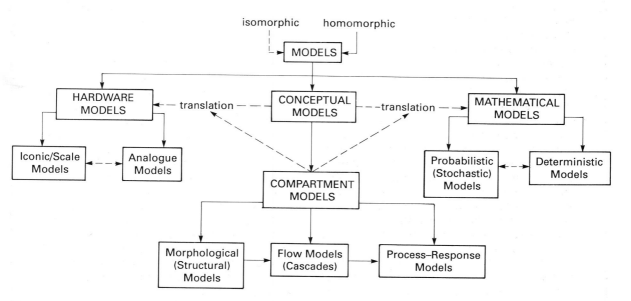

Fig. 1.7 Kinds of model (see text and Box 1.3 for further explanation).

way, the energy exchange between subsystems of the model in order to relate energy input and output by means of an initial appreciation of throughput in the system. Ultimately the closed system model is abandoned and a partial view of the major subsystems of the Earth is presented, where each is modelled as an open system involving transfers of mass as well as energy. For the most part, the models, have a high degree of generalization and the perspective is a global one.

With a change of focus both the resolution of the model and the spatial scale at which it is applied change. It is the scale of the landscape that begins to predominate. In particular, the catchment basin and the ecosystem are functional models applicable at a scale which we all experience and they are concerned with interlinked geographical units having 'both spatial magnitude and location' (Chorley & Kennedy, 1971). The resolution of the models and the level of detail involved increase considerably and represent the white box approach, or near-isomorphic model.

Initially it is the equilibrium relationships and the steady-state characteristics of these environmental open systems that are considered under the prevailing sets of external conditions of existence. Nevertheless, the models presented retain a considerable element of generality in spite of some inclusion of specific spatial or geographical variation.

The developmental or evolutionary tendency of these systems is then considered. Change in the system states is viewed as both an internal readjustment to reach some new steady state, and as a response to changes in input, both of which are natural characteristics of environmental systems.

Here we also consider the interaction of society with the environment, in terms of disturbance by inadvertent human intervention and of deliberate and purposeful regulation of natural systems so that some perceived benefit accrues. Here we meet the wider application of a systems approach, recognizing that the environmental systems discussed in earlier sections are only part of larger *geographical systems* which also have a socioeconomic and cultural dimension.

Further reading

The following serve still as useful introductions to systems ideas:

Beishon, J. and G. Peters (1972) *Systems Behaviour*. Harper & Row, London.

Churchman, C.W. (1968) *The Systems Approach*. Delacorte Press, New York.

Emery, F.E. (1969) *Systems Thinking*. Penguin, London.

Haigh, M. (1985) Geography and general systems theory, philosophical homologies and current practice. *Geoforum*, **16**, 191–203.

More specific application of system thinking to the environment will be found in the following texts, of which Chorley and Kennedy (1971) remains essential reading, while the last (of which the first chapter is most useful at this stage) deals overtly with energetics and thermodynamic systems.

Chorley, R.J. and B.A. Kennedy (1971) *Physical Geography, a Systems Approach*, Prentice-Hall, London.

Gregory, K.J. (1987) *Energetics of Physical Environment: Energetic Approaches to Physical Geography*. John Wiley, Chichester.

Huggett, R.J. (1980) *Systems Analysis in Geography* (Contemporary Problems in Geography). Oxford University Press Oxford.

More demanding texts, but stimulating and challenging are:

Bennett, R.J. and R.J. Chorley (1978) *Environmental Systems: Philosophy, Analysis and Control*. Methuen, London.

Coffey, W. (1981) *Geography, Towards a General Spatial Systems Approach*. Methuen, London.

A very accessible book concerned with the broader philosophy and methodology of science in relation to physical geography is:

Haines-Young, R. and J. Petch (1986) *Physical Geography: Its Nature and Methods*. Harper & Row, London.

Introductions to modelling and to mathematical modelling in particular are:

Kirkby, M.J. (1987) Models in physical geography, part 1.3, in, *Horizons in Physical Geography* (eds M.J. Clark, K.J. Gregory and A.M. Gurnell). Macmillan, Basingstoke, pp. 47–59.

Maynard-Smith J. (1974) *Models in Ecology*. Cambridge University Press, Cambridge.

Thomas, R.W. and R.J. Huggett (1980) *Modelling in Geography: a Mathematical Approach*. Harper & Row, London.

A further introduction (and largely non-mathematical) to modelling in pure and applied contexts is:

Frenkiel, F.N. and D.W. Goodall (eds) (1978) *Simulation Modelling of Environmental Problems*. John Wiley/SCOPE, Chichester.

Research papers demonstrating a range of models and their applications are to be found in:

Woldenberg, M. (ed) (1985) *Models in Geomorphology*. Allen & Unwin, London.

Matter, force and energy

2.1 The nature of matter

In Chapter 1 an energy system was formally defined as a specified system of matter, the energy content of that system, and the exchange of energy between the system and its surroundings. In this chapter we shall explore in more detail the nature of both the matter and energy involved in such systems.

The matter in environmental energy systems can be recognized as the 'real' physical objects of that environment: rocks, soil, water, plants, animals and the gases of the atmosphere. Here, however, matter will be discussed at a fundamental level, for it is only by so doing that the structural unity of these apparently diverse entities can be appreciated.

2.1.1 The structure of the atom

Atomic theory is the key to the current view of the nature of matter. According to this view, matter is composed of very small particles (10^{-7}–10^{-10} m in diameter) – **atoms** – which have a complex internal structure. They consist of a positively charged central **nucleus** around which negatively charged **electrons** orbit (Fig. 2.1). The nucleus has a diameter approximately 10^{-5} that of the whole atom, and in this small volume nearly all of the mass of the atom is concentrated. The nucleus is composed of two different types of particle of almost identical mass: the **protons** are positively charged and the **neutrons** carry no charge, but they are each normally present in approximately equal numbers.

The nucleus of an atom contains a definite number of protons – its **atomic number** – and this defines the charge on the nucleus and ultimately the chemical nature of the atom. The protons and neutrons jointly constitute the nuclear mass, so

that the number of protons plus the number of neutrons gives the **mass number** of an atom. A **chemical element**, which is the *basic unit of matter*, is defined as a substance in which all of the atoms have the *same atomic number* and which cannot be decomposed by chemical reaction into substances of simpler composition. Therefore, although all atoms of an element have the same number of protons and hence the same atomic number and nuclear charge, their masses may not be equal. This is because the number of neutrons

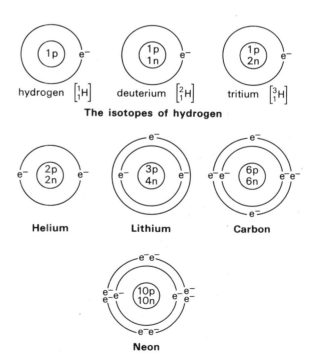

Fig. 2.1 The structures of the hydrogen, helium, lithium, carbon and neon atoms.

present in the nuclei of the atoms of an element may vary.

Groups of atoms which differ in their mass numbers are called **isotopes**. Hydrogen, for example, exists as three **isotopes** with 0, 1 or 2 neutrons in the nucleus respectively, viz:

$$^1_1H, \ ^2_1H, \ ^3_1H.$$

Here the symbol for the element is expanded by a superscript which shows the mass number (protons + neutrons) and a subscript which indicates the atomic number (protons). Most of the chemical elements consists of mixtures of isotopes. Oxygen has three isotopes $^{16}_8O$, $^{17}_8O$ and $^{18}_8O$, as does carbon $^{12}_6C$, $^{13}_6C$, $^{14}_6C$. The separation of isotopes is difficult, for their chemical properties are virtually identical as they have the same nuclear charge and the same number of electrons per neutral atom, and therefore they react in the same way. However, the nuclei of some isotopes may be unstable and liable to disintegrate spontaneously.

The mass of an electron is only 5.45×10^{-4} that of a proton, but it has a charge which is equal, though opposite in sign, to that of a proton. In the electrically neutral atom, therefore, the nucleus attracts a number of negatively charged electrons just equal to the positive nuclear charge, which is in turn determined by the number of protons (atomic number). In the helium atom, for example,

$$2 \text{ protons} + 2 \text{ electrons} = \text{no charge}$$

$$(2+) \quad + \quad (2-) \quad = 0$$

The electrons of an atom occupy certain orbits, or **electron shells**, and there is a definite limit to the number of electrons each shell can contain (Table 2.1). For example, lithium has its third electron placed in a second shell that can contain a maximum of eight electrons (see Fig. 2.1). All electron shells beyond the first can be subdivided. The second shell has s and p subshells, the third has s, p and d subshells, and the fourth, fifth and sixth have s, p, d and f subshells. The distribution of the electrons of an atom between these shells and their subshells is known as the **electron configuration** of the atom (Table 2.2 and Box 2.1).

Table 2.1 The number of electrons in electron shells

shell	1	2	3	4	5	6
electrons (max.)	2	8	8	18	18	32

Table 2.2 Electron configuration of the first 50 elements.

Element	Atomic no.	1s	2s	2p	3s	3p	3d	4s	4p	4d	4f	5s	5p	5d
H	1	1												
He	2	2												
Li	3	2	1											
Be	4	2	2											
B	5	2	2	1										
C	6	2	2	2										
N	7	2	2	3										
O	8	2	2	4										
F	9	2	2	5										
Ne	10	2	2	6										
Na	11	neon			1									
Mg	12	core			2									
Al	13	of 10			2	1								
Si	14	electrons			2	2								
P	15				2	3								
S	16				2	4								
Cl	17				2	5								
A	18				2	6								
K	19							1						
Ca	20							2						
Sc	21						1	2						
Ti	22						2	2						
V	23						3	2						
Cr	24						5	1						
Mn	25						5	2						
Fe	26						6	2						
Co	27		argon core				7	2						
Ni	28		of				8	2						
Cu	29		18 electrons				10	1						
Zn	30						10	2						
Ga	31						10	2	1					
Ge	32						10	2	2					
As	33						10	2	3					
Se	34						10	2	4					
Br	35						10	2	5					
Kr	36						10	2	6					
Rb	37											1		
Sr	38											2		
Y	39									1		2		
Zr	40									2		2		
Cb	41									4		1		
Mo	42									5		1		
Tc	43		krypton core							6		1		
Ru	44		of							7		1		
Rh	45		36 electrons							8		1		
Pd	46									10				
Ag	47									10		1		
Cd	48									10		2		
In	49									10		2	1	
Sn	50									10		2	2	

It is the outer most shell – sometimes called the **valence shell** – that determines the chemical behaviour of an element (with the exception of some elements where an inner subshell is also involved – see Box 2.1) and the number of electrons in the valence shell is important in relation to the type of chemical reaction with which an element may be concerned.

Some elements have their valence shells fully occupied. These are all inert gases (the **noble gases**) and the first five – helium, neon, argon, krypton and xenon – all occur in the atmosphere and make up about 1% (by volume) of air. Their inertness would suggest that they have a very *stable* electron configuration. Elements that do not possess this stable, inert gas configuration may be expected to show a tendency to attain it. One way in which this can be accomplished is for an element to gain or lose electrons so that its valence shell is full and corresponds to that of one of the inert gases. However, electrons added to or removed from a neutral atom give to it a net negative or positive charge respectively. These charged particles are known as **ions** – negatively charged **anions** and positively charged **cations** (Box 2.2).

2.1.2 *Molecules and compounds*

About 92 elements have been recognized in nature, most of which occur in combination with one another. Even elements such as oxygen and carbon, which occur naturally uncombined with other elements, do not do so as individual atoms. A particle of this type, made up of two or more atoms of either the same or different elements, is called a **molecule**. Atoms are held together to form molecules by the sharing of electrons between

---

Box 2.1

ELECTRON CONFIGURATION AND THE PERIODIC TABLE

In 1869, the Russian chemist Mendeleev devised the **periodic table**, in which he arranged the elements by atomic weight (relative atomic mass) in such a way that the similar properties of certain groups of elements became apparent. In modern versions of the periodic table (*Bohr periodic table*, see Table 2.3) the elements are in order of their atomic numbers. The vertical columns – **groups** – contain elements which, though similar, vary in their properties down the columns. For example, *Group IV* elements become more metallic in character. The horizontal rows – **periods** – contain series of elements, the chemical properties of which change in discrete steps. Some elements occupy intermediate positions between *Groups II* and *III*. The most important of these in the natural environment belong to the series known as the **transition elements**.

It is now known that this arrangement reflects the electron configuration of the atoms of the elements. In *Groups I, II, III, IV, V, VI, VII*, and *0*, as the atomic number (number of protons) increases along the periods so the electron shells are progressively filled, from the inner shell outwards. For example, in the second period, carbon, nitrogen, and oxygen appear in sequence. All have two electrons filling the first shell ($1s^2$) but carbon has four ($2s^2 2p^2$), nitrogen five ($2s^2 2p^3$) and oxygen six ($2s^2 2p^4$) electrons in the second shell. The last element in each period (i.e. *Group 0* elements) has no vacancies in its outer shell or subshell. For example, neon in the second period has a full complement of eight electrons ($2s^2 2p^6$) in the second shell.

The elements of all of the groups discussed above are termed **typical elements**, and have in common the fact that the s and p subshells are filled progressively. The remaining elements, the *transition elements*, and those of the *lanthanide series* (the rare earths) and the *actinide series* (including the transuranium elements) differ in that they have incomplete inner shells. They form series where electrons are added to the d (transition elements) and f (lanthanide and actinide series) subshells of the penultimate shell (see Table 2.2).

adjacent atoms in **covalent** and **metallic** bonding, and by the **transfer** of electrons in **ionic** bonding. Molecules showing charge separation (polar molecules) may form **hydrogen bonds** because of the resultant electrostatic attraction.

As with the formation of ions, these bonding mechanisms can be viewed as strategies by which the constituent atoms jointly attain stable, inert gas electron configurations in the compound. In ionic (**electrovalent**) compounds the bonding mechanism arises from *electron transfer*, effectively forming oppositely charged ions that are held together by electrostatic attraction. However, in the compound so formed each ion has the electron

configuration of an inert gas. For example, in sodium chloride (common salt) (see Boxes 2.1 and 2.2 for explanation of notation)

$$Na \quad + \quad Cl \quad \rightarrow \quad Na^+Cl^-$$

1s 2s^2 2p^6 3s 1s 2s^2 2p^6 3s^2 3p^5 1s 2s^2 2p^6 1s 2s^2 2p^6 3s^2 3p^6

neon   argon
configuration   configuration

However, the transfer of electrons in this way cannot be involved in the formations of many compounds. Gaseous oxygen exists in the Earth's atmosphere as molecules (O_2) each containing two atoms of oxygen that cannot reach the stable configuration of an inert gas by electron transfer.

Box 2.2
IONS

The elements of *Group I* of the periodic table (see Table 2.3) have the inert gas structure plus an extra electron. For example, sodium ($1s^2 2s^2 2p^6 3s$) has the same configuration as neon ($1s^2 2s^2 2p^6$) plus one electron in the *s* subshell of the third shell. If it were to *lose* this extra electron it would attain the inert gas configuration. When this happens in a chemical reaction a *positively* charged **ion – a cation –** is formed.

$$Na - e^- \rightarrow Na^+$$

In a similar way, a *Group II* element, such as magnesium ($1s^2 2s^2 2p^6 3s^2$), has to *lose* two electrons from the valence shell to attain the configuration of neon.

$$Mg - 2e^- \rightarrow Mg^{2+}$$

Group VII elements, on the other hand, are one electron short of inert gas configurations and so must *gain* an electron. For chlorine ($1s^2 2s^2 2p^6 3s^2 3p^5$) to reach the configuration of argon ($1s^2 2s^2 2p^6 3s^2 3p^6$) it requires the addition of an electron which forms a *negatively* charged ion – an **anion**.

$$Cl + e^- \rightarrow Cl^-$$

In *Group VI*, oxygen ($1s^2 2s^2 2p^4$) requires two electrons to reach the electron configuration of neon ($1s^2 2s^2 2p^6$).

$$O + 2e^- \rightarrow O^{2-}$$

The charge carried by an ion, which reflects the number of electrons gained or lost by an element in forming it, is known as the **electrovalency** of the element. In the above examples, therefore, sodium has an electrovalency of +1, while in oxygen the electrovalency is –2. However, some elements form ions that may only approximate to the inert gas configuration. This is especially common among the ions of the transition elements and the rare earths, although it is also true of some elements of *Groups V* and *VI*. These elements may, therefore, form more than one type of ion with different electrovalencies (or oxidation states) that represent different approximations to the appropriate inert gas configuration. For example, iron, the fourth most abundant element in the lithosphere, occurs in two electrovalency states – *iron II*, or *ferrous* (Fe^{2+}), and *iron III*, or *ferric* (Fe^{3+}).

However, they can attain such a configuration by *sharing electrons* in pairs, each atom contributing one electron to the pair and each shared pair constituting a covalent bond :

$$\cdot \ddot{O} \cdot + \cdot \ddot{O} \cdot \rightarrow \ddot{O} :: \ddot{O}$$

$$O + O \rightarrow O_2$$

(. represents an electron in the valence shell).

The pairs of electrons between the two atoms are counted towards the electron configuration of both atoms, both having the electron structure of the inert gas neon. Furthermore, the position of the electron pairs is localized and it fixes the positions of the atoms quite rigidly.

When two atoms of hydrogen and one of oxygen combine to form water (H_2O), they do so by strong electron pair or covalent bond. The resultant water molecule is stable and, as a whole, electrically neutral. The electrons, however, have an asymmetrical distribution within the molecule which behaves as a **molecular dipole**, or **polar molecule**, due to the **charge separation**. This partial ionic character of the O–H bond in water lends the hydrogen atom some positive character, permitting electrons from another atom to approach closely to the proton, even though the proton is already bonded. This allows a second, weaker, link to be formed – the **hydrogen bond** (Fig. 2.2).

Hydrogen bonds are found between the atoms of only a few elements, the most common being those in which hydrogen connects two atoms from the group consisting of fluorine, oxygen, nitrogen and, less commonly, chlorine. Hydrogen bonds are always associated with charge separation, but as well as being intermolecular bonds they can also occur between the atoms within a molecule, as in many important organic compounds such as proteins. In fact, hydrogen bonds play a crucial role in biological systems (see 2.1.4 and Chapter 7).

Because of the presence of two unshared pairs of electrons and two protons, the water molecule can form up to four hydrogen bonds with other molecules. These bonds are arranged in the form of a tetrahedron, as in the regular crystal structure of ice (see Box 14.1). In liquid water this arrangement becomes more irregular and the hydrogen bonds are reduced as the water temperature rises. Hydrogen bonds not only form between water molecules but also between them and other

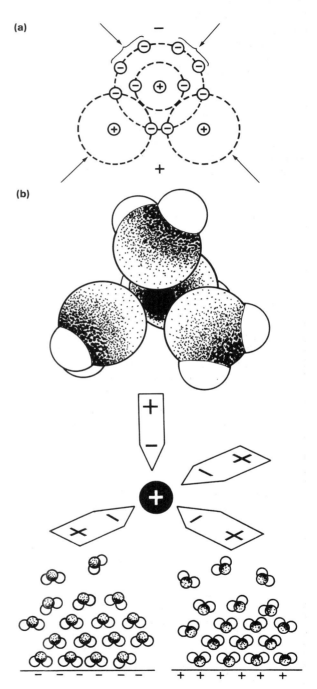

Fig. 2.2 The dipolar water molecule. (a) Charge separation. (b) The formation of hydration shells.

charged particles (ions) or surfaces (soil colloids, for example – see Chapter 22). The polar water molecules orient themselves around such charged particles and surfaces in layers, the thickness of which depends on the intensity of charge. These layers are called **hydration shells** (Fig. 2.2).

In covalent bonding the positions of the electron pairs are localized, but in metallic bonding each atom contributes one or more electrons to a 'sea' of electrons, which tends to move relatively freely throughout the aggregate of atoms. This 'sea' of negatively charged electrons holds the atoms together, although their relative positions are not as rigidly defined as in covalent bonding. It also accounts for many of the properties of metals, such as their malleability and their behaviour as conductors of an electric current.

2.1.3 Chemistry of Life

Life is made up of familiar inorganic, chemical elements. At this level there is nothing inherently unique about the cell or about life. These elements are those abundant in the non-living part of the environment. The elements of the air, rocks and waters of the Earth are those of living cells and organisms. Furthermore, with certain exceptions, they are present overall in much the same proportions. Elements common on the Earth's surface are abundant in living things, while those rare at the surface are rare in organisms. These observations are not merely coincidence, but they provide a first insight into the relationships of living systems with their environment and offer a first clue as to the nature and origin, not only of life, but also of some of the chemical characteristics of that environment (see Chapter 7).

Although estimates of the relative proportions of the elements in the biosphere vary (and these proportions also vary from organism to organism) four elements – hydrogen, carbon, nitrogen and oxygen – are universally present in living systems. They are among the lightest elements and they have atomic numbers (see Table 2.3) of 1, 6, 7 and 8 respectively. Together they account for more than 99% of the atoms of the biosphere. Of those elements with atomic numbers between 9 and 20, sodium (11), magnesium (12), phosphorus (15), sulphur (16), chlorine (17), potassium (19) and calcium (20) are also universally present, but each contributes less than 1% (and most less than 0.1%) of the atoms of the biosphere. Fluorine (9) and silicon (14) are, however, better considered with the remaining 11 elements now known to be present in either plant or animal cells. Eight of these have atomic numbers between 23 and 34 – vanadium (23), chromium (24), manganese (25), iron (26), cobalt (27), copper (29), zinc (30) and selenium (34) – while the remaining three – molybdenum, tin and iodine – have atomic numbers of 42, 50 and 53 respectively. All of these heavier elements, plus fluorine and silicon, are commonly but not universally present in living cells. They are found, however, in such small quantities (usually less than 0.001%) that they are called **trace elements**. The remaining elements of the

Table 2.3 The Bohr periodic table

Group	I	II																	III	IV	V	VI	VII	O
1st period															1 H									2 He
2nd period	3 Li	4 Be																	5 B	6 C	7 N	8 O	9 F	10 Ne
3rd period	11 Na	12 Mg																	13 Al	14 Si	15 P	16 S	17 CL	18 Ar
4th period	19 K	20 Ca	21 Sc										22 Ti	23 V	24 Cr	25 Mn	26 Fe	27 Co	28 Ni	29 Cu	30 Zn	31 Ga	32 Ge	33 As	34 Se	35 Br	36 Kr
5th period	37 Rb	38 Sr	39 Y	rare earths (lanthanides)									40 Zr	41 Nb	42 Mo	43 Tc	44 Ru	45 Rh	46 Pd	47 Ag	48 Cd	49 In	50 Sn	51 Sb	52 Te	53 I	54 Xe
6th period	55 Cs	56 Ba	57 La	58 *Ce*	59 *Pr*	60 *Nd*	61 *Pm*	62 *Sm*	63 *Eu*	64 *Gd*	65 *Tb*	66 *Dy*	67 *Ho*	68 *Er*	69 *Tm*	70 *Yb*	71 *Lu*	72 Hf	73 Ta	74 W	75 Re	76 Os	77 Ir	78 Pt	79 Au	80 Hg	81 Tl	82 Pb	83 Bi	84 Po	85 At	86 Rn
7th period	87 Fr	88 Ra	89 Ac	90 *Th*	91 *Pa*	92 *U*	93 *Np*	94 *Pu*	95 *Am*	96 *Cm*	97 *Bk*	98 *Cf*	99 *Es*	100 *Fm*	101 *Md*	102 *No*	103 *Lw*	104	105													
						actinides													transition elements						

atomic series, mainly the heavy elements, are highly toxic to living systems, as too are the trace elements when present in excess. The overwhelming impression is that living systems are almost exclusively composed of the lighter and more reactive elements, with small but essential traces of heavier elements.

In spite of the broad similarity in elemental composition between life and its inanimate surroundings at the surface of the planet, there remain significant points of contrast. Just as the similarities in composition may suggest clues as to the nature and origin of life, so too do the differences. Silicon, aluminium and iron are, after oxygen, the most abundant elements in the lithosphere, yet they are either absent or are present only in very small quantities in living things. Carbon is scarce in the lithosphere, atmosphere and hydrosphere, yet it is the third most abundant element in living organisms. In an even more striking way phosphorus, the sixth most abundant element in living systems, is rare in the inorganic environment of these systems (Table 2.4). Certainly one implication of this elemental chemistry is that life has, in a sense, been derived from the elements available at the Earth's surface. Nevertheless, the figures in Table 2.4, together with the contrasts in composition alluded to above, indicate that the chemistry of life is not just a reflection of that of the environment.

Hydrogen and oxygen, the most abundant elements, however, are present in roughly the same proportions as in the water molecule. It is not surprising, therefore, that the living cell is be-

tween 60% and 90% water. By weight, vertebrates are 66% and mammals on average 85% water. Even wood is 60% water by weight, while apparently dry seeds are 10% water. Indeed, all living matter is dispersed in water: it is the essential medium of all life on Earth. Its unique properties have already been dealt with, but water also controls the effectiveness of many of the other compounds essential to life through their response to it: whether or not they are soluble, whether they ionize in solution and exist as charged particles, or whether they affect the physical properties of water.

2.1.4 Organic carbon compounds

There are more compounds of carbon than of all of the other 102 elements, and, unlike inorganic compounds, many organic carbon compounds number the atoms in their molecules in hundreds or even thousands. The reason for this behaviour lies in three properties of the carbon atom.

1. The high covalency of carbon (four) permits the attachment of a large number of groups to carbon in a large number of different combinations.
2. The carbon–carbon bond has great strength and permits the formation of chains of carbon atoms of unlimited length.
3. The formation of multiple bonds by carbon further increases the number of organic compounds.

A few of the other elements – for example silicon – possess one or two of these properties, but none has all three. These properties allow carbon to form stable chains containing from two or three to scores of atoms. Practically any number of simple or branched chains may be derived by repetition, while the possibilities are increased by other factors including the introduction of double or even triple bonds between adjacent carbon atoms at certain positions in the structure. Such compounds are referred to as **open-chain compounds**. The carbon bonding can also form closed chains of carbon atoms forming rings – the **cyclic compounds**. Where the ring is composed entirely of carbon atoms (usually five or six), the compound is said to be **monocyclic**, but where besides carbon atoms the rings contain other elements (chiefly nitrogen, oxygen or sulphur) they are **heterocyclic**. Both chain and ring compounds often have attached to the rather inert carbon

Table 2.4 Percentage atomic composition of life, rocks, waters, and atmosphere, for the first ten elements.

Life		Rocks		Waters		Atmosphere	
H	49.8	O	62.5	H	65.4	N	78.3
O	24.9	Si	21.22	O	33.0	O	21.0
C	24.9	Al	6.47	Cl	0.33	Ar	0.93
N	0.27	H	2.92	Na	0.28	C	0.03
Ca	0.073	Na	2.64	Mg	0.03	Ne	0.002
K	0.046	Ca	1.94	S	0.02		
Si	0.033	Fe	1.92	Ca	0.006		
Mg	0.031	Mg	1.84	K	0.006		
P	0.030	K	1.42	C	0.002		
S	0.017	Ti	0.27	B	0.0002		

core of the molecule other groups of atoms such as OH, Cl, NO_2, which are known as functional groups. It is here that some of the other elements listed in Table 2.4 become important, for it is often the functional groups which determine the properties of the compound.

Many of these carbon compounds can exist as **isomers**; that is distinct substances sharing the same molecular formula, but differing, often subtly, in the spatial arrangement of their constituent atoms, especially the functional groups. This may produce asymmetric molecular structures. Many natural organic compounds are asymmetric isomers, but for each kind of compound, nature usually specializes in producing a limited number, often only one, of the many isomers possible. Here we can see not only the further scope for diversity endowed by isomerism, but the selective precision with which it is used in the living cell.

The organic compounds found in the living cell fall mainly into four great classes: **fats**, **carbohydrates**, **proteins** and **nucleic acids** (Fig. 2.3). Structurally the fats are simplest. They are constructed from the hydrocarbon chains of fatty acids (usually three) joined separately to the main part of a molecule of glycerol. They are present principally as energy reserves, but one group of related compounds (the phospholipids) in which one fatty acid is replaced by a phosphoric acid group (H_3PO_4) which in turn is linked to another compound, has a critical role in the formation of biological membranes. Simple sugars consist of pentagonal or hexagonal rings made up of either four or five carbon atoms and one oxygen atom with attached hydroxyl (OH) and CH_2OH side groups. Two such units may form a sugar molecule or they may be lined in chains to form complex sugars or polysaccharides, especially cellulose and starch in plants, and glycogen in animals. Starch, which may be up to 500 or so units long, and glycogen (1000+) serve mainly as energy stores, but cellulose with its much longer polymers (8000 units) is the main structural material in plants and combines in threads and meshes of great complexity.

Even greater complexity is encountered in the nucleic acids, which form large structures built of at least four types of unit, known as nucleotides, each of which consists of a sugar phosphate group (an ester) and a base. These are present in varying proportions and a great variety of sequences. The most important of these molecules are DNA and RNA, and the significance of these information-bearing macromolecules is considered further in Box 2.3.

However, it is with the protein molecules that variety and specificity are seen to be most highly developed. These are the largest and most complex molecules known, and are made up of about 25 different amino acids linked to a carbon backbone. They form chains hundreds to thousands of units long, in different proportions, in all kinds of sequences and with a huge variety of folding and branching.

2.1.5 The states or phases of matter

Matter exists in three **states (phases)** – gas, liquid (together called fluid states) and solid – which are distinguished by the relative motion of the molecules or other units of matter (Box 2.4). In fact, most naturally occurring gases do consist of molecules, but some may be ionized: that is, the molecules may have gained or lost electrons from their component atoms and carry a net charge. Many liquids and all solutions contain ions, while in solids the units may be individual atoms, ions, molecules or arbitrary groups of atoms.

In the gaseous and liquid states, the covalent bonding of the molecules is the only bonding mechanism of significance, but in solids all three types are responsible for the orderly arrangement of atoms. The configuration or crystalline arrangement is known as the lattice structure of the crystal. The topic of crystal structure will be considered again, in relation to rock-forming minerals, in Chapter 5.

The view of the nature of matter which emerges from this discussion implies two important points. It is immaterial whether an atom, a quartz crystal or a tree is under discussion; the nature of a substance depends first on the kind of particles of which it is composed, and secondly on their arrangement or organization. All matter can be considered as a more or less orderly arrangement of particles on various scales from the subatomic to the supramolecular. The nature of these particles and the kind of organization they possess determines the properties and behaviour of that unit of matter.

2.2 Fundamental forces

The fact that matter exists as orderly arrangements of particles on a variety of scales suggests that something is responsible for holding the particles together and maintaining such organization. There must be some force of attraction between the particles. There are four, apparently distinct, categories of natural forces. Two of these classes are relatively familiar and they account for all forces normally experienced. When particles of matter carry opposite electrical charges they exert an attraction on each other. Such a force of attraction is known as an **electrostatic force** and it can be shown to be proportional to the charges carried by the particles and inversely proportional to the square of the distance separating their centres:

$$F_e \propto Q_1Q_2L^{-2}$$
$$\text{or } F_e = EQ_1Q_2L^{-2}$$

where F_e = electrostatic force, Q_1, Q_2 = magnitude of the charges on the two particles, L = length (distance), and E = proportionality constant.

However, when the charges carried by particles

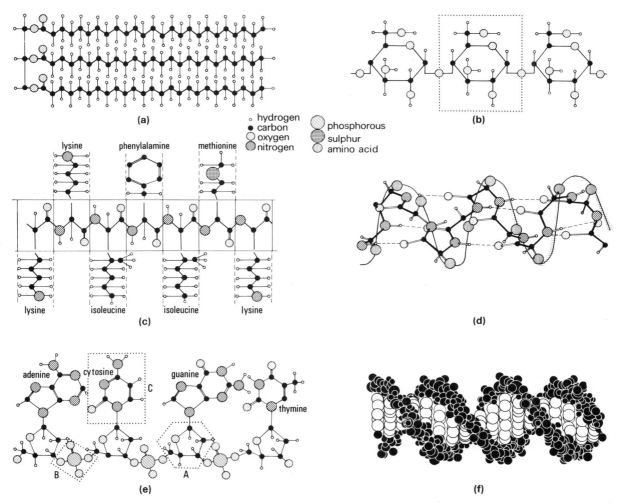

o hydrogen
• carbon
◯ oxygen
⬤ nitrogen

◍ phosphorous
⬤ sulphur
◯ amino acid

Fig. 2.3 Organic macromolecules. (a) Polymerized hydrocarbon chains forming a fat – a triglyceride. (b) Glucose monomers (one is boxed) linked to form a polysaccharide chain. (c) Amino acids linked to form the polypeptide chain of a protein (sequence of seven amino acids from near the *N*-terminal end of cytochrome C from human tissue). (d) Alpha-helix configuration of protein polypeptide chain. (e) One strand (part) of the nucleic acid DNA showing the sugar(A)–phosphate(B) units in ester linkage and the four heterocyclic bases: adenine, cytosine (C), guanine and thymine. (f) Double helix configuration of part of DNA molecule.

Box 2.3
DNA REPLICATION AND TRANSCRIPTION

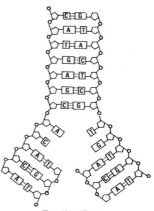

Replication

A Adenine
T Thymine
C Cytosine
G Guanine
U Uracil

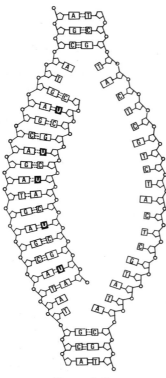

Transcription

The DNA molecule is a double-helix polymer, each strand of which consists of sugars (deoxyribose), phosphate groups and four kinds of heterocyclic base, all linked by strong covalent bonds. The two strands are held together more loosely by hydrogen bonds between pairs of bases. Each base, however, can only legitimately pair with one other. Because of the strict base pairing, the sequence of bases on one strand of the double helix predetermines the sequence on the other. It is this base sequence which holds the information stored in the DNA molecule, each group of three bases being known as a codon. If the two strands of one DNA

molecule are separated, each can serve as a template from which to produce a replica. This **replication** capacity satisfies the need to duplicate information so that it may be passed on to daughter cells during cell division. The other activity of DNA – the control of cellular activity – is, however, accomplished by **transcription**. Here a strand of ribosenucleic acid (RNA) is able to pair and coil round a single strand of DNA to form a double helix, known as a DNA–RNA duplex. In order to pair, the base sequence of the DNA is copied by the RNA, but not exactly as in duplication. It is instead a transcription into a variant of the base code language, for in RNA the base thymine is replaced by uracil. Once synthesized, the RNA uncoils and leaves the DNA, becoming a messenger taking the information beyond the nucleus to the ribosomes. Here this messenger RNA (mRNA) becomes the template for the assembly of amino acids in protein synthesis.

are both of the same type or sign, the electrostatic force is one of *repulsion*, not attraction. This attractive or repulsive force is called **electrostatic** because the charges on the particles can be considered at rest. However, similar attractive and repulsive forces exist between current-carrying conductors and, in this case, because the electric charge is in motion relative to an observer, the forces are termed **electrodynamic**. Both sets of forces are collectively referred to as **electromagnetic** forces. As we have seen, the forces holding an atom together (those binding atoms into molecules, and molecules into liquids and solids) and the contact forces affecting macroscopic bodies, are all electromagnetic.

The second familiar force is **gravitational attraction**. Every particle of matter exerts this force on every other particle, and it is directly proportional to the masses of the particles and inversely proportional to the square of the distances separating them:

$$F_g \propto M_1 M_2 L^{-2}$$
$$\text{or } F_g = G M_1 M_2 L^{-2}$$

where F_g = force (gravitational), M_1, M_2 = masses of two bodies, L = length (i.e. distance) separating them, and G = proportionality constant. This force acts as if the masses of the two bodies were concentrated at their centres, and of course is most familiar as the gravitational attraction of the Earth, which we call **gravity**.

Box 2.4
THE STATES OF MATTER

In the gaseous state the molecules are in constant motion and are continually in collision with each other. This movement is sufficient to overcome the main forces of intermolecular attraction which would lead to cohesion between the molecules. A gas can expand to fill the volume available to it and, conversely, it is susceptible to compression because the volume of the molecules is normally small compared with the total volume of the gas.

water vapour (gas phase)

In a liquid the molecules are in motion, but not sufficiently for them to get away from each other.

liquid water (liquid phase)   ice (solid phase)

They remain attracted to each other, moving past and round each other in a fluid manner. Liquids differ from gases in that their volumes normally remain constant. Their shape, however, is determined by the shape of the 'container', while they characteristically retain a free upper surface.

The molecules in a solid, although capable of vibration, are no longer moving as in a liquid, but take up fixed positions in relation to each other. Solids maintain a particular external shape, resisting change in shape and volume.

The remaining two categories of force are experienced only at the subatomic level. It will be recalled that the protons in the nucleus of the atom are positively charged. We would therefore expect that they would be driven apart by electrostatic repulsion. It is feasible that they are held firmly together in the nucleus by the gravitational attraction of their masses. In fact, it can be shown that the electrostatic repulsive force is very much greater (10^{36} times as great, approximately) than the gravitational attraction. Clearly some new force appears to be involved. This is the **nuclear (or strong) force**, and it is some hundred times stronger than the electrostatic repulsion, although its effect is limited to a short range. The fourth category, the **weak force**, need not concern us in the context of this book except in so far as it is responsible for radioactive decay.

2.2.1 Defining force

So far we have considered force as a quantity acting on particles of matter, but we have not defined it. Before doing so, however, it is important to note that force is a **vector** quantity; that is, a force has not only *magnitude* but also *direction*. If we return to gravitational force, considered above, and imagine a particle of matter released in a situation where it would have no significant gravitational interaction with any other body, it would fall towards the centre of the Earth. The force therefore operates in a specific direction. The distance the particle falls in a unit of time defines its **velocity**:

$$\text{velocity} = \text{distance per unit time}$$
$$v = LT^{-1} = \text{m s}^{-1}$$

where v = velocity, L = length (distance) and T = time. The product of its mass and its velocity defines the **momentum** of the particle:

$$\text{momentum} = \text{mass} \times \text{velocity}$$
$$p = MLT^{-1} = \text{kg m s}^{-1}$$

where p = momentum, M = mass, L = length (distance) and T = time. Because the force of gravitational attraction is inversely proportional to the square of the distance separating two bodies – in this case the particle of matter and the Earth – the smaller the distance becomes, the greater the force of attraction between them, and the

velocity of fall progressively increases. This rate of change of velocity resulting from the application of a force is called the **acceleration** of the particle, in this earthly example the *acceleration due to Earth's gravity* (9.81 m s^{-2}):

$$\text{acceleration} = \text{velocity change per unit time}$$
$$a = LT^{-1}\,T^{-1}$$
$$= LT^{-2} = \text{m s}^{-2}.$$

The momentum of the particle, however, is a function of its velocity and, as the velocity changes, so too will the momentum. As the rate of change of velocity is defined by the acceleration of the particle, so the rate of change of its momentum will be the product of the mass of particle and its acceleration. This rate of change of momentum is used to define the magnitude of the force, in this case the gravitational force:

$$\text{rate of change of momentum} = \text{mass} \times \text{acceleration}$$
$$= \text{force}$$

so,

$$\text{force} = \text{mass} \times \text{acceleration}$$
$$F_g = MLT^{-2} = \text{kg m s}^{-2} = \text{N (newton)}.$$

On Earth this is called the **force of gravity**, and the force of the Earth's attraction for an object is called the **weight** of the object. Thus a mass of 1 kg has an earthly weight of $1 \times 9.81 = 9.81$ N. On the Moon, however, the same object would have a different weight because the acceleration would now be the acceleration due to the gravitational attraction of the Moon, not the Earth. However, the *mass* of the object would remain constant.

2.2.2 Force and motion

From this initial consideration of force it should be apparent that the principal effect of a force is to accelerate bodies of matter. If such a body is at rest, it will remain at rest unless acted on by a force. If the body is already in motion, it will continue to move in a straight line at constant velocity unless it is acted on by a force. These assumptions together represent a statement of **Newton's First Law of Motion**. In 1687, Isaac Newton proposed three basic laws defining the relationships between forces and the motions that forces produce. These laws describe our observations of motion, and they also predict such motions. They became the basis of **Newtonian mechanics**, a system of mechanics applicable at velocities that are considerably less than the velocity of light.

In the natural environment all objects are being acted on by forces all the time, the same fundamental forces discussed in Section 2.2. If these forces are not balanced, the motion of the body is changed in a particular way. The effect of a given force on the motion depends on the mass of that body, for this determines the body's resistance to a change in its motion. If you think of applying your muscle power to two boulders of different mass, this would appear to be self-evident. Furthermore, the effect of the force you apply is to accelerate the boulders from rest, and as long as their mass does not change, for each boulder the acceleration you achieve is proportional to the force you apply, and in the same direction as the force. This is the recognition of **Newton's Second Law**, which relates the force acting on the mass of a body to the change in velocity, or the acceleration of the body. Stated formally the second law reads: the acceleration of a body is directly proportional to the **resultant force** acting on it and inversely proportional to the mass of the body:

$a \propto F/m$ (see definition of force, Section 2.2.1)

The term resultant force is used because under most real situations several component forces will be acting on a body, each of different magnitude and acting in a different direction or sense. It is therefore the net effect of these components – the resultant force – that appears to be responsible for the acceleration of the body.

Newton's Third Law is derived from the everyday experience that, when a force is applied (as for example pushing against a boulder) the boulder appears to be acting back with an equal force. Your subjective impression in this situation is that action is met by reaction. Furthermore, action appears to equal reaction, for if you push harder the reaction from the boulder will be greater. This assumption is Newton's Third Law and it is stated formally as: when one body exerts a force on another, the second body exerts an equal and opposite force on the first – *action equals reaction*.

One of the consequences of Newton's Third Law of motion concerns the effect of action–reaction forces on particles of matter involved in collisions. This is a common occurrence during environmental processes when either one or both of two bodies are in motion. During any small time interval in the collision, the faster body (x) will be acting on the slower body (y) with a force which tends to accelerate it. Over the same short time interval the slower (or stationary) body (y) will be reacting with a force that tends to slow down the faster-moving body (x), so the effect is to increase the velocity of y and decrease the velocity of x. As force is the product of the mass of the body and its acceleration (change in velocity), we can define the force of x acting on y ($_xF_y$) as

$$_xF_y = m_y\, a_y$$

and the force of y acting on x as

$$_yF_x = m_x\, a_x.$$

But Newton's Third Law tells us that these forces are equal, so

$$m_y\, a_y = m_x\, a_x.$$

However, the product ma is the rate of change of momentum, so because the forces, though equal, are opposite in sense (i.e. y has speeded up and x slowed down) the momentum gained by y equals that lost by x. Therefore, the total momentum of x and y has been conserved through the whole collision sequence. This is the **Law of Conservation of Momentum** – in this particular case of linear momentum (Fig. 2.4).

We have been concerned so far mainly with forces acting in what a physicist would term a static frame of reference. However, the Earth and its atmosphere rotate, and therefore it is often necessary to be able to account for the behaviour of objects located in a rotating frame. For example, to an observer an athlete throwing the hammer clearly has to pull, or exert an inward force, on the wire as he rotates in the throwing circle to keep the hammer head moving in a circular path. This is the **centripetal force** and is defined as

$$\text{centripetal force} = mv^2/r$$

where r is the radius of the circle. However, the athlete himself, who would be in a rotating frame

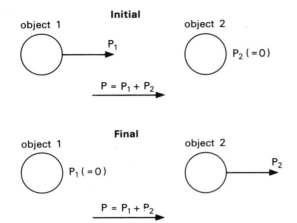

Fig. 2.4 The conservation of linear momentum.

of reference, experiences the mass of the hammer head trying to accelerate outwards in a radial direction as if acted on by some invisible force. This force is the **centrifugal force** and so the athlete applies a force through the hammer wire just sufficient for the radial movement of the head to cease, and from his point of view there is no apparent acceleration of the hammer. In other words, the centripetal force is equal but opposite to the centrifugal force, which must also equal mv^2/r. These forces are important in understanding the motion of the Earth's atmosphere and of the circulation cells within it (see Chapters 4 and 9). In these situations it is necessary to redefine velocity and momentum in terms of the period of rotation, and the radius of the circle as the angular velocity and angular momentum (see Box 4.6).

2.2.3 Applied force: stress and resistance

The concept of **stress** is fundamental to the understanding of the many mechanical processes operating within the natural environment, particularly in denudation systems. Stress in this context is largely concerned with mechanical forces. The basic principles involved underlie studies of fluid, soil and rock mechanics, which embrace the displacement of particles or masses of soil and rock, often associated with moving water.

Stress is defined as force applied per unit area:

$$\text{Stress} = \frac{\text{force}}{\text{area}} = \frac{\text{kg m}^2\text{s}^{-2}}{\text{m}^2} = \frac{\text{N}}{\text{m}^2} = \text{N m}^{-2}.$$

Thus, a block of rock with a weight (mass × gravitational acceleration) of 1000 kg and a basal area of 2 m², resting on a horizontal surface exerts a **normal** (or vertical) **stress** (or pressure) of 500 kg m⁻² or 500 N m⁻² on the surface beneath. Providing that the underlying surface does not deform or yield, then a stress or **resistance** is set up within it, of equal magnitude to the applied stress. In this case the surface has a resistance, or **strength**, at least equal to the applied stress (Newton's Third Law of Motion).

Normal stress and **pressure** are the same thing, and therefore the pressure exerted by the atmosphere on the Earth's surface – **atmospheric pressure** – can be defined as the downward force of the atmosphere per unit area of surface. This force is equal to the product of the mass of the atmosphere above the area of surface and the acceleration due to gravity (g). For a fluid, however, the mass term can be replaced by the product of the density (ρ) and the height, or depth (h) of the fluid, so that pressure is defined as:

$$P = \rho h g.$$

This is known as the **hydrostatic equation**, and it can be modified to apply to the atmosphere. These relationships will be explored further in Chapter 4, particularly in Boxes 4.1 and 4.2.

Different types of stress can be identified in terms of the relationship between the direction of applied force and the plane or body on which it is acting (Fig. 2.5). Where the applied stress is normal to the surface, as in the example cited above of the block of rock on a horizontal plane, it is termed a **compressive stress**. On a plane oblique to the direction of applied stress, as on a slope, the stress becomes a **shear stress**. In this case, the normal stress is attenuated by the obliquity of the plane concerned, and there is a tendency for translational movement (shearing) to take place along this plane. A stress exerted within a mass and tending to prise it apart, is a **tensile stress**.

Examples of all of these types of applied force exist in denudation systems. Tensile stresses, for example, are involved in many processes of rock weathering. Most important, however, are the shear stresses, since these exist whenever gravitational forces operate on gradients (in slope systems, fluvial systems and glacial systems) to

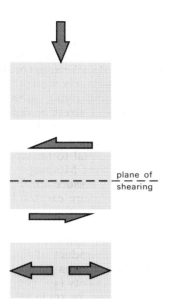

Fig. 2.5 Applied force.

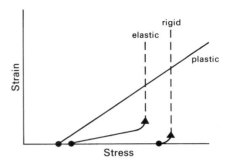

Fig. 2.6 Stress–strain relationships.

provide a gravitational component of shear stress applied to the sloping plane.

The effect of applied stress is to cause a displacement or deformation of the mass to which it is applied. The deformation, or **strain**, is measured as a linear or volumetric proportion of the dimension of the initial body. Different types of behaviour of solid materials under applied stress can be identified (Fig. 2.6). In a rigid solid, insignificant strain takes place at low to moderate stress levels until a critical stress value is reached, at which **failure** takes place in the form of **brittle fracture**, and strain increases very rapidly. An elastic material, on the other hand, will show considerable strain before failure takes place. If the applied stress is less than that required to cause failure, the body will rebound to its original form once the stress is removed. A plastic material will show little strain until a critical threshold stress level is reached, and beyond that will deform in direct proportion to the stress applied. These three modes of behaviour represent idealized and arbitrary states, and most materials exhibit a combination of them. Indeed, a particular material may behave in different ways under different conditions, e.g. according to the rate at which stress is applied, or under different moisture conditions.

Most materials in the natural environment exhibit mainly rigid or plastic behaviour. For ex-

ample, an indurated rock will tend to possess the properties of a rigid solid, and a moist clay will tend to behave in a plastic manner.

The concept of **strength** is intimately related to that of failure. The strength of a material can be defined as the resistance of that material to failure, whether by brittle fracture or plastic deformation. Strength is therefore equal to the force applied at failure and is measured in terms of applied force per unit area ($N\,m^{-2}$).

The behaviour of Earth surface materials under different stress conditions is fundamental to the operation of denudation systems. Properties of these materials relevant to their behaviour under applied stresses, and the nature of the applied stresses, are further considered in Chapters 10–17, where the applications of these basic concepts are taken up.

2.3 Work and energy

Every particle of matter is acted on by forces all the time. Some of these forces may be external to it, such as gravitational and some electromagnetic forces, but others exist inside the particle and result in the aggregation of its constituent parts. When these internal forces are balanced, the body of matter is stable. Equally, when the external forces acting on a body exactly balance each other, the motion of the body will not change. If at rest it will remain at rest, and if moving it will continue to move in a straight line with the same speed, unless acted on by a force (see Newton's Laws of Motion, 2.2.2).

If a force is applied to a body of matter, a certain amount of work will be done against the resistance to the force. The effect of the force is to move its point of application through a distance, but the work done varies not only with the resistance overcome, but with the distance through which it is overcome. In other words, the work performed on the body is expressed through the distance moved. So the work done by a force (or against the resistance of a force) is the product of the force and the distance:

$$\text{work} = \text{force} \times \text{distance}$$

But, as a force has been defined as mass times acceleration:

$$\text{work} = \text{mass} \times \text{acceleration} \times \text{distance}$$
$$W = MLT^{-2}\,L$$
$$= ML^2T^{-2} = \text{kg m}^2\text{ s}^{-2} = \text{Nm} \quad \text{(newton metre)}.$$

To this point the existence of these forces has been assumed, and they have been regarded as being freely available to do work. It is at this stage, however, that the concept of energy must be introduced.

Energy is often defined as the capacity to do work. It is perhaps better thought of as that 'quantity' which diminishes when work is done. This immediately implies that the unit of energy should be the same as that work:

$$\text{energy} = \text{force} \times \text{distance}$$
$$= ML^2T^{-2}$$
$$= \text{kg m}^2\text{ s}^{-2} = \text{Nm} = \text{J (joule)}.$$

The newton metre and the joule are therefore equivalent units of energy. The newton metre is applied to energy in its mechanical form, while the joule is used for heat and potential energy.

All particles or bodies of matter, therefore, possess this 'quantity' or capacity. Just as the existence of forces is related to the relative positions and relative motions of such particles and bodies, so energy can be understood in the same terms.

2.3.1   Potential energy and kinetic energy

Any body of matter at rest will possess an amount of potential energy by virtue of both the configuration of its constituent particles and its position relative to other bodies of matter. Such potential energy can be viewed simply as stored energy (or fuel) which is potentially available to be transferred or to do work. An important example of potential energy is the potential gravitational energy possessed by a body relative to the gravitational attraction between its mass and that of another body, such as the Earth.

$$PE_g = \text{mass} \times \text{acceleration due to gravity}$$
$$(9.81 \text{ m s}^{-1}) \times \text{height}$$
$$= ML^2T^{-2}$$
$$= \text{kg m}^2\text{ s}^{-2} \quad \text{Nm} = \text{J.}$$

Bodies of matter also possess potential chemical energy. It is, in fact, electromagnetic potential energy and it reflects the charges carried by the constituent particles of matter and their relative positions. This is then the binding energy which produces the internal forces that hold the nucleus and electrons together in the atom and the constituent atoms together in molecules. In the same way potential nuclear energy is responsible for the nuclear forces that bind the protons and neutrons together in the atomic nucleus.

This concept of stored energy or potential energy has been applied to bodies of matter at rest. What notion of energy can be applied to bodies in motion? Suppose a force of F newtons acts on a body initially at rest and having a mass m kilograms. If v is the velocity acquired by the mass in t seconds under the action of the forces, F is defined as mvt^{-1}. In displacing the body through d metres, the force F does, by definition, Fd joules of work. But d is $\frac{1}{2}vt$, the average velocity ($\frac{1}{2}v$) multiplied by the time (t). The product Fd is, therefore, $\frac{1}{2}mv^2$ ($MLT^{-2}\,\frac{1}{2}LT^{-1}\,T = \frac{1}{2}ML^2\,T^{-2}$). This is the work done by the force in accelerating the body from rest, and it is now the energy resident in the moving mass. This energy of motion is called the kinetic energy (KE) and, at low velocities

$$KE = \tfrac{1}{2}mv^2 = \tfrac{1}{2}ML^2\,T^{-2} = \text{kg m}^2\text{ s}^{-2} \text{ (joules)}.$$

All familiar forms of energy – light, mechanical, radiant, heat or thermal (Box 2.5), nuclear, chemical and electrical (actually some of these forms are identical) – can be regarded either as forms of potential or kinetic energy, or as some combination of the two.

2.3.2 Energy transfer, entropy and the laws of thermodynamics

Energy is associated with the organization of matter, the relative positions and relative motions of particles of matter. When a change in the state of a system of matter occurs through the operation of a process, some reorganization of the particles of matter occurs, as, for example, during a chemical reaction, or the precipitation of a raindrop. Such changes of state therefore involve a redistribution of energy between the particles of matter that compose the system, or between the system and its surroundings. This redistribution of energy involves both transfers and transformations of energy.

In a chemical reaction, for example, the potential chemical energy of the reactants (bonding energy plus thermal energy) is transformed to kinetic energy and transferred to the molecules in motion during the reaction (Box 2.6). Some of this kinetic energy is transformed to heat during the collision between these molecules, and when the reaction stops, some again becomes potential chemical energy – of the reaction products this time. The kinetic energy of the falling raindrop will be transferred as mechanical energy to the soil on impact, and will perform work by dislodging soil particles. Some will, however, be transformed to heat through friction. Also, as the droplet falls, some of the kinetic energy will be transformed to heat due to the frictional drag of the air.

What these examples suggest is that there is no change in the absolute quantity of energy involved. Energy may be transformed in kind, and transferred both within the system and between the system and surroundings, but when both system

---

Box 2.5
HEAT AND TEMPERATURE

The heat energy content of a body is due directly to the velocity of vibration of the molecules of which it is formed. Temperature is a measure of the amount of this heat energy which a body contains.

Faster vibrating molecules transmit some of their kinetic energy of motion to adjacent slower-moving molecules, a process which is referred to as conduction. In this way, heat energy is transmitted through a body.

In order to measure the temperature of a body, heat energy is exchanged with an indicator, usually another substance which undergoes a known mode of (mechanical) deformation in response to such exchanges. Most substances experience expansion or contraction in response to an input or output of heat energy, according to their coefficient of thermal expansion, which is the deformation produced per unit volume (or length) of the material per unit change in temperature.

If we insert a thin-walled glass bulb full of a fluid such as mercury into a medium such as water, air or soil, kinetic energy will be exchanged between the molecules of the medium and those of the glass and the mercury. If the mercury takes up energy from the medium, it will respond by expanding its volume. This will continue until there exists a dynamic equilibrium between the kinetic energy of the molecules of the substances on either side of the thin wall of the glass bulb.

Should there be a decrease in the kinetic energy of the molecules of the medium, then energy will be transferred away from the mercury, resulting in a contraction in its volume. This change in volume is, therefore, an expression of the change in heat energy content of the medium or a change in its temperature. This is the basic principle behind the measurement of changes in temperature using thermometers.

Box 2.6
CHEMICAL REACTIONS

Chemical reactions can be studied from various points of view. Attention can be directed to the bulk changes of the chemical substances involved. This approach is termed the **stoichiometry** of the reaction and expresses the *active masses* of the reactants and products in terms of gram molecular weights (**moles**). As the **law of conservation of mass** requires, the mass of reactants and of products must balance as there should be no loss of mass in a chemical reaction. For example,

$$2Na^+OH^- + (H^+)_2SO_4^{2-} \rightarrow 2H_2O(1) + (Na^+)_2SO_4^{2-}$$

80 g	98 g	36 g	142 g
sodium hydroxide	sulphuric acid	water (liquid)	sodium sulphate

Alternatively, attention can be centred on the mechanism by which the reaction works, and by which the chemicals come together and how the products are formed. Here the recognition of intermediate products which may be formed momentarily becomes important. So, for example, the stoichiometric equation for the reaction between hydrogen and chlorine molecules, $H_2 + Cl_2 \rightarrow 2HCl$, becomes broken down into stages and indicates that the reaction in this case is a chain mechanism where the stages can be repeated many times, viz.

$$Cl_2 \rightarrow Cl + Cl,$$
$$Cl + H_2 \rightarrow HCl + H,$$
$$H + Cl_2 \rightarrow HCl + Cl.$$

However, chemical reactions occur when the chemical bonds of the reacting substances are broken and reformed to produce the reaction products. The breaking and making of bonds involves the transfer of energy. The likelihood of a reaction occurring and the speed with which it occurs will depend on the energetics of the reaction. This is a third approach, and one that is particularly important in understanding the energy transfers in natural systems. Reactions involving covalently bonded molecules concern collisions between them. In practice this means two-body collisions, for the probability of more than two molecules colliding instantly is extremely small. The occurrence and speed of a reaction will depend not only on the fact that molecular collisions occur, but also on two other factors. First, their **energy of collision** must be sufficient for specific bonds to be broken, and secondly, they must collide in such a way (the **collision geometry**) that these specific bonds will in fact be broken. Together collision energy and collision geometry affect the **activation energy** of the reaction – the minimum energy necessary to break bonds and form products. We can graph the energy changes involved in a reaction in the following way:

Here, as the reactants approach each other their potential energy increases to a maximum at the point of collision, when an **activated complex** of the reactants is formed. The height of the activation energy here shows us the ease or difficulty of a chemical reaction, and if the energy and geometry of collision are favourable the activated complex will break up to form the products. If it does not, it will reform the original reactants. In this diagram the energy of the products is lower than that of the reactants. This energy difference is called the **energy of reaction** or the **enthalpy** change (ΔH) and, because here energy has been lost during the reaction, ΔH is negative and the reaction is **exothermic**. If the energy level of the products had been higher than that of the reactants ΔH would be positive and the reaction would be termed **endothermic**. All chemical reactions can be classified as to whether they absorb energy (usually heat) from, or evolve energy (again, usually heat) to, the surrounding medium. Outward signs of changes in heat energy in exothermic and endothermic reactions are not the only factor to consider in reaction energetics. The concepts of **free energy** (F, or G) and **entropy** (S) must also be introduced (see Box 2.7).

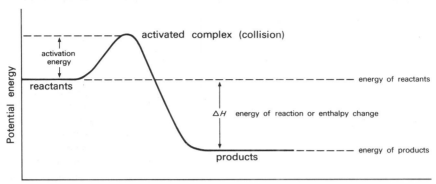

Certain substances – **catalysts** – are able to increase the speed of reactions by influencing the collision geometry of the molecules involved, hence increasing the number of favourable collisions and lowering the activation energy. Catalytic reactions are especially important in living systems where a host of highly specific proteins – **enzymes** – are the catalysts involved in biochemical reactions (see Chapter 7). (Largely after Ashby *et al.*, 1971, see Further reading.)

and surroundings are considered it is found that energy has been neither created nor destroyed. This is the **First Law of Thermodynamics**, and is known as the **Law of Conservation of Energy**. There is no exception known: it is a generalization from experience, not derivable from any other principle, and it states that the total energy of the universe remains constant.

The **Second Law of Thermodynamics** is concerned with the direction of naturally occurring, or real, processes. In combination with the First Law it allows not only the prediction of this direction but also the equilibrium state that will result. To choose a familiar example, if the system consists of a petrol tank and a motor mounted on wheels, the Second Law allows us to predict that the natural sequence of events is: consumption of petrol, production of carbon dioxide and water, and forward motion of the whole device. From the Second Law, the maximum possible efficiency of the conversion of the chemical energy of the petrol into mechanical energy can be calculated. The Second Law also predicts that one cannot manufacture petrol by feeding carbon dioxide and water into the exhaust and pushing the contraption along the road, not even if it is pushed along backwards!

Natural processes are irreversible. They proceed with an increase in the disorder (randomness) of matter and energy. Order or structure in the arrangement of matter disintegrates and concentrations disperse. Energy is degraded from high levels to low. No spontaneous transformation of energy during the operation of a natural process is 100% efficient. Some is dissipated as heat energy (the random motion of matter) and is unavailable to do work, as we have seen in earlier examples. The energy available to be used is continually being diminished through the operation of natural processes. Energy that becomes unavailable to do work is related to the increase in the **entropy** (S) of the system (Box 2.7). Entropy is a measure of the disorder of the system, but it can never be absolutely quantified. However, entropy always increases as a result of the operation of any natural process, so $\Delta S \geq 0$. Therefore, change in entropy (ΔS) can be quantified, though not in absolute terms. The Second Law tells us that all physical and chemical processes proceed towards maximum entropy. At this point there is thermodynamic equilibrium.

2.3.3 Energetics of systems
It is the central premise of this book that energy and its transformations may be viewed as the best way both to systematize and to synthesize the facts and theories of physical geography and environmental science. Therefore, thermodynamic or

energy systems become the most fundamental way of analysing all environmental processes, just as they have long been accepted and used as the basic approach to the analysis of physical, chemical, and more recently, biological processes.

These thermodynamic principles may appear to be abstract and formidable, but the approach and working philosophy of thermodynamics is really quite simple. Furthermore, it is necessary to master only a few such principles in order to examine in a broad way the nature of environmental energy systems and the energy transformations associated with them. The most fundamental of these principles are the First and Second Laws of Thermodynamics, which we have encountered already.

It follows from the First Law that when an energy system changes from an initial state to a final state, it may either receive energy from its surroundings or give it up to the surroundings. The difference in energy content between the initial and final states must be balanced by an equal but inverse change in the energy content of the surroundings. In this case the system is deemed to be in equilibrium when the initial and final states are specified; that is, the process of change has not yet either started nor ceased entirely. Classical equilibrium thermodynamics is not concerned with the time taken to bring the system from the initial to the final state, nor with the rate of change. Equally it is not concerned with the pathways or processes by which physical or chemical change occurs, but merely with the energy difference between the initial and final equilibrium states of the system. As an analogy we can say that if a man travels from London to Edinburgh his change in location is completely specified by stating his initial latitude and longitude and his final latitude and longitude. It does not matter how long the journey took, nor what route he followed.

This is the method of approach of thermo-

Box 2.7

FREE ENERGY AND ENTROPY CHANGE IN CHEMICAL REACTIONS

The enthalpy change ΔH (Box 2.6) does not tell us the amount of work that can be obtained from a chemical reaction. This is given by the **free energy change** of the reaction, ΔF, or ΔG (after Willard Gibbs), where the free energy is the 'useful energy' capable of doing work on the system. Free energy change is related to **enthalpy** change and **entropy** change by the following equation:

$$\Delta G = \Delta H - T\Delta S$$

where T = absolute temperature (degrees Kelvin) ($°K = °C + 273$), and ΔS = the change in the entropy (randomness) of the system.

Therefore, the change in the free energy of the system is equal to the change in the heat energy of the system minus the amount of energy used to change the order or randomness of the system. As an expression of the **Second Law of Thermodynamics** it tells us the direction that the reaction will follow spontaneously, i.e. it will proceed in the direction that results in a more disordered system. For the entropy of the system to increase in this way there must be a decline in the free energy, and ΔG must be negative. Such a spontaneous reaction is known as an **exergonic** reaction (cf. exothermic, see Box 2.6). In this case, the free energies of formation of the products must be less than those of the reactants in the following equation:

$$\Delta G = \underset{\text{(products)}}{\Sigma \Delta G_f} - \underset{\text{(reactants)}}{\Sigma \Delta G_f}.$$

Where the free energy change of the reaction, ΔG, is positive the reaction will not proceed spontaneously and will require an input of energy to start it, i.e. it is **endergonic**. Note, however, that in indicating whether or not the reaction will proceed spontaneously these thermodynamic considerations do not take into account the activation energy (see Box 2.6).

dynamics to the analysis of physical or chemical changes, but the determination of the total energy content of a system in either the initial or final state is in practice formidable. Even in the laboratory it is only possible with very simple systems of gases, so in complex environmental energy systems it is virtually impossible. However, it is the *changes* in the energy content which are our primary concern, and such changes are more easily visualized and measured. If we know the type and magnitude of the energy exchanged with the world outside the system (its surroundings) as it proceeds from an initial to a final state, then we can undertake a thermodynamic analysis of the process. This in practice becomes the method of thermodynamics.

Our reasoning so far has been based on the First Law alone, but the Second Law tells us that no spontaneous physical or chemical change in a system is completely efficient. Such processes have a direction. They lead towards system states where the elements and properties of the system are distributed at random. The energy content of the system is progressively degraded to a random state (which was the way we defined entropy), and is unavailable to do work. So the Second Law tells us that all physical and chemical processes proceed in such a way that the entropy of the system is maximized, and at this point we have equilibrium.

However, these principles were developed to apply to closed systems which do not exchange matter with their surroundings. We are concerned with open systems, which do exchange matter with their surroundings. Such systems do not attain a thermodynamic equilibrium of maximum entropy, but are maintained in a dynamic steady state by a throughput of matter and energy. Nevertheless, they are irreversible systems and, although they may appear to violate the Second Law in that their entropy may remain constant or even decrease as they become more ordered, actually they do not. The entropy of the system or any part of the universe may decrease as long as there is a corresponding and simultaneous increase in the entropy of some other part – total entropy increases.

$$\Delta S_{total} = \Delta S_{system} + \Delta S_{surroundings}.$$

An open system, by definition, maintains a steady state by the transfer of matter and energy by real or irreversible processes with the surroundings. Therefore, time and rate become critical variables in the energetics of entropy production, which as a consequence is itself time-dependent. The most probable state of an open system is one in which the production of entropy per unit of energy flow through the system is at a minimum. Such a state is most likely to persist.

Further reading

Almost any first-year university text in chemistry, physics, and biology will provide further support and background. Those listed below are merely a selection of helpful texts.

Ashby, J.F., D.I. Edwards, P.J. Lumb and J.L. Tring (1971) *Principles of Biological Chemistry*. Blackwell, Oxford.

Carson, M.A. (1971) *The Mechanics of Erosion*. Pion, London.

Coxon, J.M., Fergusson, J.E. and L. Philips (1980) *First Year Chemistry*. Edward Arnold, London.

Davidson, D.A. (1978) *Science for Physical Geographers*. Edward Arnold, London.

Duncan, G. (1975) *Physics for Biologists*. Blackwell, Oxford.

Glymer, R.G. (1973) *Chemistry: an Ecological Approach*. Harper & Row, New York.

O'Neill, P. (1985) *Environmental Chemistry*. Chapman & Hall, London.

Raiswell, R.W., Brimblecombe, P., Dent, D.L. and P.S. Liss (1984) *Environmental Chemistry*. Edward Arnold, London.

Soper, R. (ed) (1984) *Biological Science*. Parts 1 & 2. Cambridge University Press, Cambridge.

Villee, C.A. *et al.* (1989) *Biology* (2nd edn). Saunders, Philadelphia.

Watson, J.D. (1970) *The Double Helix*. Penguin, London.

Part B

A systems model:
a partial view

As we drew back from the landscape seen in Fig. 1.1a and the cattle by the loch shore, the hillslopes and mountain summits receded and our eyes remained fixed on the surface of the planet. As the perspective on our Earthly environment changed we recognized the value of modelling its component systems at different spatial scales such that the properties of each could be discriminated at a level appropriate to a particular scale. In this way we could arrive at a nested hierarchy of systems models, applicable to the entire planet at one end of the scale continuum and to the sand grains on the loch shore at the other. Before we embark on the task of constructing such a hierarchy, let us pause and instead of looking inwards at the Earth and its surface environment which are our home, shift our gaze toward the immensity of space. As we do so it becomes clear that our hierarchy of systems of matter does not stop at the planetary level.

The Earth and its solitary Moon are part of the solar system. The Earth is one of nine planets moving round the central star, the Sun. The solar system is an orderly, harmonious system of matter, but it too is only one of countless star systems with their satellite planets which compose our galaxy. This galaxy, however, is itself but one of an estimated hundred thousand million (10^{11}) galaxies within the maximum radius of the observable universe. Current theories suggest that the solar system originated some 4.7×10^9 years ago from a nebular disc of hot gaseous material. As the disc cooled various minerals and mineral compounds began to condense and aggregate to form small masses of solid matter: the planetesimals. These in turn coalesced, the larger, with the greater gravitational attraction, pulling in the smaller, eventually forming the planets all rotating in approximately the same plane, and in the same direction round the Sun. Density and compositional differences explained by temperature differences in relation to their distance from the Sun divide the planets into two groups. The denser terrestrial planets including the Earth are composed of metallic and rock forming minerals (compounds of oxygen, silicon, and other metallic elements) of high melting point – the first to have condensed, and lie nearer to the Sun. The second group, the giant planets, are largely condensed or frozen gases (water, methane, ammonia) and are more distant, lying in the colder reaches of the solar system. The planet Earth viewed in this context becomes an insignificant speck in a system of matter and energy, the scale of which is beyond comprehension. Within our solar system, however, the Earth is far from insignificant. It occupies a unique position among the planets and its distance from the Sun, almost midway through the zone where water can exist as a liquid, its rate of rotation, its receipt of solar radiation, its surface temperature and the presence of an atmosphere, all combine to create conditions favourable to the development of life, while the existence of life has itself transformed the surface of the planet. Around some distant star in some far-off galaxy another planet may be orbiting under a similar chance combination of conditions. However, as to whether life has evolved elsewhere, we have no way of knowing.

As we turn away from the rest of the universe and fix our gaze once more on our Earth and begin the task of building systems models of the planet, we would do well to remember this larger perspective. In the first place it sets not only the spatial but also the temporal scales of our models in a context. Secondly, many of the properties of the Earth are, as we have seen, related to its place in the solar system, and to its interactions with other bodies in that system, while other properties are inherited from the time of its origin and that of the solar system. The regular rotation of the Earth about its axis and its motion relative to the Sun imposes rhythmical changes in exposure to light and dark, day and night, summer and winter. Both the Sun and the Moon impose tidal rhythms on the atmosphere, on the oceans, and even on the exposed crust as the Earth rotates. The spin of the Earth distorts patterns of circulation in the gaseous atmosphere and the waters of the oceans, while in combination with the liquid metallic core it sets up a strong magnetic field. In attempting to model the planet as an energy or thermodynamic system in Chapter 3, however, it is the continuous output of energy from the Sun which is the single most important characteristic of the solar system, and it represents the major energy source for our planet. Many of the attributes of the Earth's surface environments and of the living systems which inhabit them reflect the magnitude of this energy input and mirror both the solar and lunar rhythms they experience.

In Chapter 3 it will become apparent that to understand the energy relationship of the Earth even at the global scale it is necessary to recognize the existence of a number of major subsystems and to abandon the black box model of the planet with which the chapter commences. Only by identifying the major stores of matter and energy within these subsystems and by considering the pathways and magnitude of transfers between them can we begin to understand the planetary energy balance. We shall therefore arrive at a model of the Earth which gives us a broad, but partial view of the structure of the system and of its functional relationships, particularly in terms of the energy cascade through the system.

In Section III this model will be refined even further. The atmosphere, Earth surface, and Earth's interior – the subsystems of Chapter 3 – will be redefined and each will be treated as a system in its own right. Chapter 4 will consider the atmosphere and Chapter 5 the Earth's interior, but the Earth-surface subsystem of Chapter 3 will be divided into the physicochemical systems operating at the surface of the planet, the denudation system and the living systems which inhabit this surface. Since the denudation system has strong functional links with, and is partly the surface expression of, the geophysical and geochemical activities of the crustal system, they are both treated under the heading of the lithosphere in Chapter 5. In Chapter 6 the critical role of the oceans in the energy fluxes and cycles of mass which occur at the Earth's surface is recognized overtly. As far as we are aware, the existence of life is probably unique to the planet Earth, at least in our solar system, and therefore because of this, and because of the far-reaching effect that life has had on the history of our planet, the living systems of the biosphere are considered in Chapter 7. The evolution and survival of life on the Earth are inconceivable without interactions with the non-living environment and, for this reason, Chapter 7 is entitled The ecosphere, for it is concerned not only with the organisms of the biosphere but with these interactions with the environment.

The focus in Section III is still at the global scale and remains a partial view of the systems concerned. It is, however, a partial view, not of the entire planet, as in Chapter 3, but of each of the major subsystems. The models developed, therefore, are largely at an intermediate level of sophistication, but they nevertheless increase our understanding of the Earth as the level of resolution has improved considerably over that employed in Chapter 3. There are, however, variations in the scale at which these models will be developed. Chapters 4, 5 and 6 are mainly concerned with broad global models of storage and transfer of matter and energy in the physicochemical systems of the atmosphere and lithosphere. Even so, the scale at which particular processes, such as the change of state of water in the atmosphere, or the crystallization of igneous rock in the crust, are considered may be at the molecular scale, or at that of the individual crystal. In the first part of Chapter 7 the decision is taken to model the organization and basic functional activity of living systems, first of all at the molecular and cellular scales. The temporal scale will also, of course, vary from the almost instantaneous operation of the biochemistry of the living cell to the millions of years over which movements of crustal plates must be considered. Nevertheless, in all of these cases the understanding which emerges is generalized and the perspective of the models as a whole remains a global one.

II THE PLANET EARTH

Energy relationships

3.1 The closed-system model

One of the advantages of building systems models is that they simplify the complexity of the real world. In this chapter we shall model the Earth as a closed system: that is, a system of matter whose only interaction with its surroundings is the transfer of energy across its boundary (see Chapter 1). Such a model of the planet is obviously simplified. Meteorites, for example, can penetrate the Earth's atmosphere from space, illustrating that matter also crosses the boundary of this system. So do space vehicles, projected into space perhaps to be lost forever, or perhaps to return as debris after re-entry. Nevertheless, at the planetary scale the closed system model is a useful one for it focuses attention on the energy input and output across the boundary of the system, on the sources of that energy, and on the net energy balance of the whole system. Critical to this approach, however, is the definition of the system. For our present purposes the boundary of the planet Earth will be taken as the outer surface of the atmosphere and our model of a closed-system Earth will be treated initially as a black box (Fig. 3.1).

3.1.1 Energy input: solar radiation

Energy crosses the boundary of the system in several forms, but the most significant input is the radiant energy received from the Sun (Table 3.1). The Earth also receives electromagnetic energy from other bodies in space and it experiences the gravitational energy associated with their masses. Table 3.1 makes clear that these other inputs. although not unimportant, are nevertheless of a different order of magnitude. So the main input of energy to the black-box closed-system model of the planet is in the form of electromagnetic radiation from the Sun (Box 3.1). Its source is the enormous energy locked up in the nuclei of hydrogen atoms, part of which is released by nuclear fusion in the immensely high temperatures of the Sun (Box 3.2). The uncontrolled release of energy by the same process has given humans the awesome capacity for self-destruction which is inherent in the hydrogen (or fusion) bomb. Paradoxically, it is also this same process of nuclear fusion which may one day be harnessed for the peaceful generation of power without the attendant dangers of nuclear waste associated with present nuclear fission technologies.

If we assume the Sun to have an estimated sur-

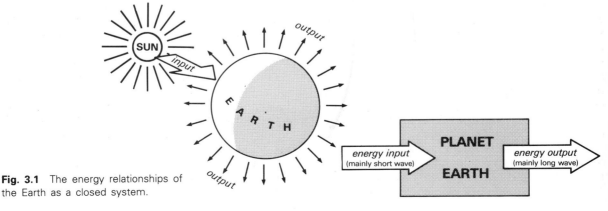

Fig. 3.1 The energy relationships of the Earth as a closed system.

Table 3.1 Energy sources and energy stores (after Campbell 1977, Sellers 1965)

	Energy × 10^{20} J
total annual receipt of solar energy by the Earth	54 385
energy released by 1976 Chinese earthquake	5006
combustive energy stored in Earth's coal reserves	1952
combustive energy stored in Earth's oil reserves	179
combustive energy stored in Earth's natural gas reserves	134
latent heat absorbed by global snow/ice melt in Spring	15
North Sea oil reserves (presently known)	3
annual USA consumption of energy (1970)	0.75
annual UK consumption of energy (1972)	0.09
heat flux from Earth's interior	0.027
total radiation from the Moon	0.006
total energy content of annual grain crop in Britain	0.006
energy released in Krakatoa eruption 1883	1.49×10^{-3}
dissipation of mechanical energy of meteorites	0.89×10^{-5}
total radiation from stars	0.60×10^{-5}

face temperature of 6000 K and to be an ideal, or full, radiator, then we can derive some of the characteristics of the radiant energy emitted by the Sun from three relationships (Box 3.3). The first of these tells us that the amount, or intensity, of solar radiation is relatively high, for it is a power function of the Sun's surface temperature. The second shows indirectly a relationship between the wavelength distribution of solar radiation and the Sun's surface temperature. This is because wave propagation is a function of molecular oscillation and this increases with temperature, and a hot body such as the Sun has a high frequency of wave propagation and hence a shorter wavelength emission than does a cold body. Indeed, the peak of the wavelength distribution – the wavelength of maximum emission – is short, centring round the visible part of the electromagnetic spectrum, for the third law shows that it is inversely proportional to the surface temperature of the Sun. The complete solar radiation spectrum is seen in Fig.

3.2 to consist of ultraviolet (short wavelength, 0.2–0.4 μm) which constitutes on average 7% of the total emission; the visible wavebands (0.4–0.7 μm) which constitute 50%; and infrared radiation (long wavelength 0.7–0.4 μm) which constitutes 43%. Because of the dominance of short wavelengths, solar radiation is typically referred to as shortwave radiation.

Only 0.002% of the total radiation emitted by the Sun forms the input to the system and is received by the Earth. The average amount received per unit area (over a plane at right-angles to the solar beam) per unit time at the outermost boundary of the atmosphere is referred to as the **solar constant**, and is usually given the value 1370 W m^{-2}. However, the notion of the solar constant belies the fact that the effective input to the system varies considerably over space and time. In the first instance, there is a small degree of variation in the amount and nature of radiation from the Sun's surface. The range of variation is relatively small, fluctuating within 1–2% of the solar constant. Unfortunately this falls within the range of error in radiation-measuring techniques employed at the outer boundary of the atmosphere. It has frequently been suggested that a potential cause of changes in the value of the solar constant is the occurrence of darker areas on the Sun's surface, referred to as sunspots.

The effective input to the planet will also be conditioned by the movement of the Earth itself and its motion relative to the Sun. The distance between the two varies, as the Earth pursues its elliptical path round the Sun, and is greatest in early July, at 152 × 10^6 km (aphelion), and least in early January, at 147 × 10^6 km (perihelion). Because of this, the Earth receives slightly more solar radiation in January than in July. The boundary of our model of the Earth as a closed system, however, has been taken to be the outer surface of the atmosphere. As the solar beam can be assumed to consist of parallel rays, the intensity of radiation received on such a horizontal surface is directly related to the angle of incidence of the rays upon it (Box 3.4). This angle of incidence is determined by the angle of latitude, the tilt of the Earth's axis with respect to its orbital plane, and the rotation of the Earth about its own axis.

Using these relationships, it is possible to calculate the solar radiation received at any point and at any time for a planet Earth which is pursuing a known path round the Sun and whose bounding surface is assumed to be completely uniform (Fig. 3.3). Such calculations show that seasonal variation in radiation receipt on a horizontal surface is greatest in polar latitudes and smallest in tropical latitudes (Fig. 3.3). This implies that, although there is the expected latitudinal variation in energy received, the exact nature of this variation will differ according to the time of year, and there is no common seasonal cycle. So, returning to our black box model of the Earth as a closed system (see Fig. 3.1), it is evident that

---

Box 3.1
ELECTROMAGNETIC RADIATION

There are two ways of visualizing electromagnetic radiation. The first is due to Max Planck (a German physicist) who in 1900 suggested that radiation is emitted as a stream of pulses of energy, referred to as quanta – a view supported by the work of Einstein in 1905. The fundamental equation relating the energy content of a single indivisible quantum (E) is given by

$$E = h\nu$$

where h is Planck's constant (6.60×10^{-34} J s^{-1}) and ν (nu) is the frequency of radiation quanta.

The adoption of this quantum approach allows the calculation of direct measures of the energy of radiation, which is essential when considering energy conversion within the system.

The alternative is to view radiation as a sine-wave form. Here, a critical attribute of electromagnetic radiation is its wavelength and the range of wavelength values is represented in the electromagnetic spectrum from the short waves of gamma rays to long radio waves. The waveform of electromagnetic radiation is perhaps more commonly adopted in the environmental sciences and particularly in meteorology, as it lends itself more readily to treatments of transmission, reflection and refraction of radiation.

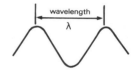

Box 3.2
SOLAR ENERGY

The radiant solar energy received by the Earth arises from nuclear energy. In the immensely high temperatures of the Sun (surface temperature of 6000 K) a part of the enormous energy locked up in the nucleus of the hydrogen atom is released by nuclear fusion. In this process four hydrogen nuclei (four protons) are fused to form a helium nucleus (two protons + two neutrons):

$$4{}^{1}_{1}\mathrm{H} \rightarrow {}^{4}_{2}\mathrm{He} + {}^{0}_{1}\mathrm{e} + h\nu_0.$$

The mass of the helium nucleus is about 0.7% *less* than the sum of the four hydrogen nuclei. This lost mass is converted to a quantum of energy (see Box 2.1) in the form of gamma radiation. This is represented above by the term $h\nu$, in which h is Planck's constant and ν (nu) is the frequency of gamma radiation. After a complex series of reactions, the gamma radiation is emitted again in the form of photons or quanta of light energy.

the input of shortwave solar radiation across the boundary of the system varies in a complex manner, both spatially and through time.

3.1.2 Energy output and the planetary radiation balance

The output of energy from the system is also in the form of electromagnetic radiation, for the planet and its atmosphere not only absorb radiation but themselves act as radiators, emitting radiation. However, applying the basic radiation laws (see Box 3.3) to the Earth, whose surface temperatures and those of its atmosphere are generally below 300K, reveals that this radiation is of low intensity and its entire spectrum lies in the infrared waveband. Consequently, the emission of radiant energy across the boundary of the system is characteristically longwave radiation, in contrast to the shortwave input of solar radiation. However, the radiation laws (see Box 3.3) apply to ideal radiators or perfect **black bodies** which totally absorb all incident radiation. The system in our model, the planet Earth, does not behave as a black body but as an imperfect absorber and emitter of radiation. Such a **grey body**, placed in a stream of radiant energy, will absorb some of it, often differentially in various parts of the spectrum, and transmit and/or reflect the rest. Longwave radiation, therefore, forms only part of the total radiation output of the system, for a proportion of the initial input of shortwave radiation is lost by reflection and is not absorbed.

With the development of satellites it has been possible to measure the radiation ouput near to the outer boundary of the Earth's atmosphere quite accurately (Fig. 3.4). Not only is this radiation loss a combination of short- and longwave

---

Box 3.3
RADIATION EMISSION

Stefan–Boltzmann's Law

The intensity of radiation emitted by a body (I) is proportional to the fourth power of its absolute temperature (T):

$$I = \sigma T^4$$

where σ = Stefan–Boltzmann constant = 5.57×10^{-8} W m^{-2} K^{-4} and K = degrees absolute or degrees Kelvin = °C + 273.16.

Wavelength–frequency relationship

The radiation emitted by a body has a wavelength distribution which is related to its surface temperature, as wave propagation is a function of molecular oscillation which increases with temperature rise. The equation linking wavelength and frequency is

$$\lambda = c/v$$

where λ = wavelength (cm), c = velocity of light (cm s^{-1}) and v = frequency (cycles s^{-1}).

As frequency (v) increases with rise in temperature, so wavelength (λ) decreases.

Wien's Displacement Law

The wavelength of maximum emission (λ_m) is inversely proportional to the absolute temperature (T) of the radiating body:

$$\lambda_m = w/T$$

where w = Wien's constant = 2.897 μm K, μm = 10^{-6} m.

Solids, liquids and gases are all capable of emitting radiation over a range of wavelengths, but these two laws relate to physical ideals known as full radiators, or black bodies. These bodies emit radiation at the maximum possible intensity for every wavelength and absorb all incident radiation. For convenience, it is usually assumed that both the Sun and the Earth radiate as black bodies, although at certain wavelengths they are imperfect radiators, or grey bodies.

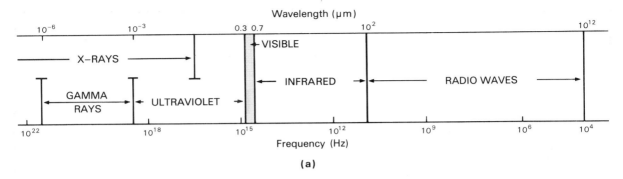

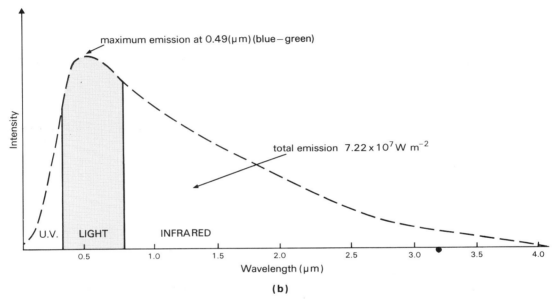

Fig. 3.2 (a) Electromagnetic radiation spectrum. (b) Solar radiation spectrum.

radiation, but it varies in a broadly latitudinal fashion (though this breaks up into a cellular pattern) with a tendency towards outward flux at high latitudes and a net input at low latitudes. However, for the system as a whole – that is, the entire planet and its atmospheric envelope – there is a long-term balance between incoming and outgoing radiation. If this were not the case the total energy content of the system would increase or decrease through time and the Earth would experience a progressive warming or cooling. We should, therefore, be able to represent this balance as

net radiation = incoming solar radiation (mainly shortwave) minus outgoing radiation (mainly longwave)

= 0.

This would be true if we were dealing with finite quantities of radiant energy exchanged over protracted periods of time. In terms of the operation of the system, however, it is more pertinent to consider the **rate of flow** of radiant energy, and here we need to consider not only inputs and

Box 3.4

LAMBERT'S COSINE LAW

The radiant intensity emitted in any direction from a unit radiating surface varies as the cosine of the angle between the normal to the surface and the direction of the radiation beam.

A corollary relates solar radiation received upon a surface (I) to the angle between the radiation beam and the normal to that surface by

$$I = I_0 \cos \alpha$$

where I_0 represents the intensity of the solar beam.

The angular elevation of the Sun above the

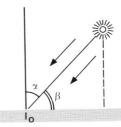

horizon (β), referred to as the solar elevation, can be introduced:

$$I = I_0 \sin \beta.$$

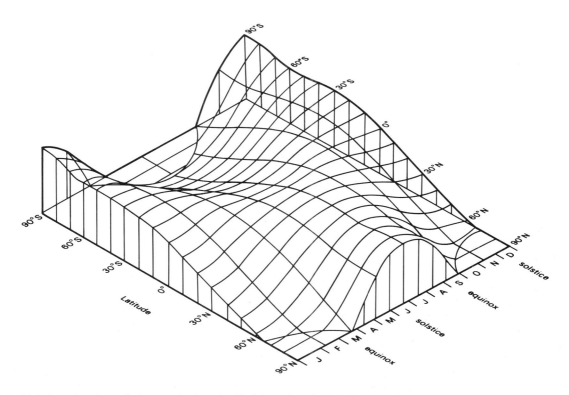

Fig. 3.3 Variations in solar radiation received at the Earth's surface in the absence of an atmosphere.

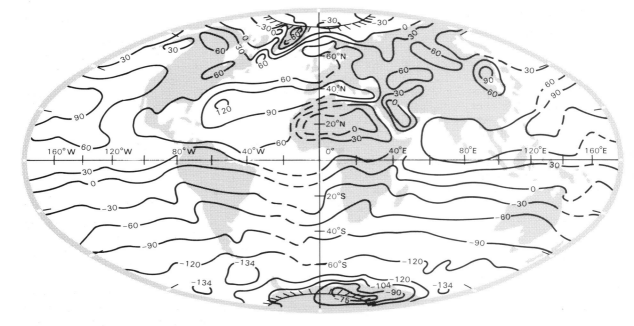

Fig. 3.4 Net radiation (W m^{-2}) balance at the top of the atmosphere (from Nimbus II evidence on 1–15 June 1966, after Barrett, 1974).

outputs, but also the throughput of energy. This is the energy which, at any point in time, is being transferred or stored within the Earth or its atmosphere. This is because within any closed system the balance of energy exchange is not only a function of relative surface temperatures and absorbance and emittance characteristics of all parts of the system, but it also depends on modes of energy transfer other than radiation. These are conduction and convection as well as energy-dependent changes of state, such as the evaporation of water, involving latent heat exchange. So, although a long-term energy balance may exist for the whole system, there will still be energy imbalances, transfers and changes in energy storage within the system.

Therefore, to understand more fully the energy balance of the Earth we need to abandon the black box approach and improve the fidelity of our model in order to follow the way energy is transferred or cascaded through the major subsystems of the model, each of which can be considered to be an energy store. Figure 3.5, therefore, contains separate compartments for the atmosphere, for Earth surface systems and for the Earth's interior. Each of these compartments can be considered as

an open system in its own right, with transfers of matter as well as energy across its boundaries. In this chapter, however, we shall continue to concentrate on energy exchanges and confine our model to the planetary scale.

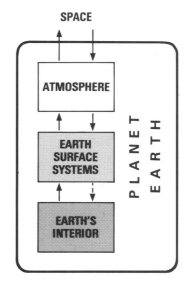

Fig. 3.5 Subsystems of the model.

3.2 The atmospheric system

3.2.1 Exchange and throughput of solar radiation

The input of shortwave solar radiation at the outer surface of the atmosphere has already been considered. During its passage through space solar radiation loses little energy, but on entering the atmosphere the solar beam encounters molecules of gases, liquids and solids, all of which are able partly to absorb or reflect it.

The gases of the atmosphere and suspended liquid and solid matter (see Chapter 4) absorb selectively from the solar beam. Ozone (O_3) and water vapour (H_2O), for example, are major absorbers of radiation but affect different parts of the solar spectrum (Table 3.2). Ozone, which occurs in the upper atmosphere, absorbs a large proportion of ultraviolet radiation in the waveband $0.23-0.32\ \mu m$. This absorption of lethal ultraviolet radiation has had profound effects on the development of life on Earth (see Chapter 6).

Water vapour, which is at its greatest concentration near to the Earth's surface, absorbs in the infrared sector of the solar spectrum in a series of wavebands of high absorption. Between these are bands of low absorption which are referred to as **radiation windows**.

In addition to ozone and water vapour, both oxygen and carbon dioxide absorb radiation. Carbon dioxide, however, has little effect on the solar beam as it absorbs mainly in the infrared range between wavelengths $12\ \mu m$ and $18\ \mu m$. Further absorption also takes place due to the presence of suspended liquid and solid particles. Ice crystals and water droplets suspended in clouds also absorb in the infrared range at wavelengths 3, 6 and $12\ \mu m$.

Table 3.2 The absorption of radiation by the principal gases of the atmosphere

Gas	Wavelengths of greatest absorption
nitrogen (N_2)	no absorption
oxygen (O_2)	$0.69\ \mu m$ and $0.76\ \mu m$ (visible – red)
carbon dioxide (CO_2)	$12\ \mu m$–$18\ \mu m$ (infrared)
ozone (O_3)	$0.23\ \mu m$–$0.32\ \mu m$ (ultraviolet)
water vapour (H_2O)	$5\ \mu m$–$8\ \mu m$
	$11\ \mu m$–$80\ \mu m$ (infrared)
also	
liquid water (clouds)	$3\ \mu m$, $6\ \mu m$, $12\ \mu m$, to $18\ \mu m$ (infrared)

The total effect of all these agents of absorption can be seen in Fig. 3.6. High values, approaching 1 on the curve, indicate a high proportion of absorption. There is considerable loss by absorption in the short- and longwave parts of the solar spectrum, with relatively little absorption of visible radiation, while selective absorption by water vapour, for example, produces a discontinuous solar spectrum at the Earth's surface.

In the atmosphere, gases and suspended matter can both disperse incident solar radiation. A unidirectional solar beam is partially transformed into a multidirectional scatter of radiation, some of which ultimately passes back out of the atmosphere into space. Scattering by gases is related to the wavelength (λ) of the radiation by the expression

$$\text{degree of scattering} = \frac{\text{a constant}}{\lambda^4}$$

In the visible part of the solar spectrum, blue light is scattered to a greater extent than other wave-

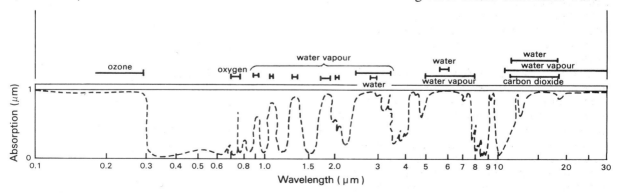

Fig. 3.6 Absorption spectrum of the atmosphere. Total absorption = 1, no absorption = 0 (after Fleagle and Businger, 1963).

lengths, resulting in the predominantly blue colour of the sky.

Scattering by materials suspended in the atmosphere is more correctly termed **diffuse reflection**. Multiple reflection of that part of the solar beam incident upon particle surfaces inevitably results in a scattering effect. The amount of scattering that takes place depends on the size of the particles, particle density in the air, and the distance radiation travels through the atmospheric layer containing the particles. The scattering of solar radiation by the dust from Mount St Helens (USA) which erupted in May 1980 reduced local daytime surface temperatures by 8.0°C (Mass and Robock, 1982). In a similar way, dense Saharan dust storms reduce solar radiation transmission by as much as 30%, resulting in a fall in daytime air temperatures of up to 6.0°C (Brinkman and McGregor, 1983).

The intervention of a band of cloud across the solar beam effectively reduces the amount of radiation reaching the Earth's surface. The upper surfaces of clouds are good reflectors of radiation, the amount reflected being dependent upon cloud type, cover and thickness. A light cover of cirrus cloud is a relatively ineffective reflector, whereas dense stratiform and cumuliform cloud may reflect more than 50% of the solar radiation incident upon them. Reflection from heavy storm clouds may be as high as 90%. Some of the interrelationships between cloud cover and solar radiation are examined further in Fig. 3.7.

The atmosphere directly absorbs a small proportion, some 17 units on average (K_A), of the solar radiation input (100 units), mainly in the short and long wavelengths (see Fig. 3.6). The proportion is small because its major constituents are not efficient absorbers in the wavelengths covered by the solar spectrum. However, this absorption of solar radiation contributes to an increase in the internal energy store of the atmosphere. Of the remaining solar radiation, 29 units are lost as output to space by reflection. This is made up of 6 units lost by scattering $(K\uparrow_{Aa})$ and 23 units lost through cloud reflection $(K\uparrow_{Ac})$. A further 54 units of radiant energy are output to Earth surface systems and can be divided into 36 units $(K\downarrow_D)$ of direct radiation and 18 units $(K\downarrow_d)$ of diffuse radiation, again as a result of scattering. This throughput of solar radiation forms part of what is known as the solar energy cascade (Fig. 3.8). Solar radiation, which is predominantly shortwave, is not the only radiation input to the atmospheric system, for it also receives longwave terrestrial radiation from Earth surface systems.

3.2.2 Exchange and throughput of longwave radiation

Gases such as carbon dioxide and water vapour, together with suspended water droplets and ice crystals in clouds, are more effective absorbers in the infrared part of the electromagnetic spectrum (see Table 3.2), particularly for wavelengths greater than 8 μm. As the spectrum of longwave terrestrial radiation from the Earth's surface extends over wavelengths 3 μm to 30 μm, with a maximum emission at 10 μm, a significant proportion will therefore be absorbed in the atmosphere.

In fact only about 7% of terrestrial radiation passes directly into space, the remainder being absorbed. That which is transmitted through the atmosphere does so within a narrow range of wavelengths where little absorption takes place – these wavelengths are referred to as radiation windows. An extremely important window occurs in the absorption spectrum of water vapour between wavelengths 8.5 μm and 11 μm, within which falls the waveband of maximum emission in the terrestrial radiation spectrum. The atmosphere therefore allows most solar radiation to pass through it, but inhibits the passage of terrestrial radiation. This is commonly referred to as the **greenhouse effect**. As we shall see in Chapter 24, this can be altered considerably by, for example, adding to the amount of carbon dioxide and other **greenhouse gases** in the atmosphere.

The absorbed radiation is re-emitted, as the atmosphere acts as a radiator. Because of relatively low temperatures, generally less than 300 K, atmospheric radiation is longwave and of low intensity. Some of the emission passes into space, but approximately 60% returns to the Earth's surface systems as counter-radiation. As most of the carbon dioxide, water vapour and clouds are in the lowest 10 km of the atmosphere, it is at this level that the greatest absorption and emission take place. Nearly all counter-radiation emanates from the lowest 4 km (Fig. 3.9).

3.2.3 The radiation balance of the atmosphere

The atmosphere receives shortwave radiation from the Sun and by reflection from the Earth's surface. Of this it absorbs only a small proportion (K_A) (see

Fig. 3.8). It also receives longwave terrestrial radiation, a large proportion of which it absorbs ($L\!\uparrow_{EA}$), the remainder passing directly to space through radiation windows. The atmosphere radiates at long wavelengths to space ($L\!\uparrow_A$) and the ground surface ($L\!\downarrow$). The majority of the latter takes part in the complex radiant energy exchanges between surface and atmosphere. These input and outputs of energy are represented in Fig. 3.10.

The equation for the balance of radiation (Q^*_A) may be written as:

$$Q^*_A = K_A + L\!\uparrow_{EA} - L\!\uparrow_A - L\!\downarrow.$$

For the whole atmosphere, assuming a solar constant of 100 units, 17 units are absorbed from the solar beam and 91 from terrestrial radiation, 57 units are radiated into space and 78 units back

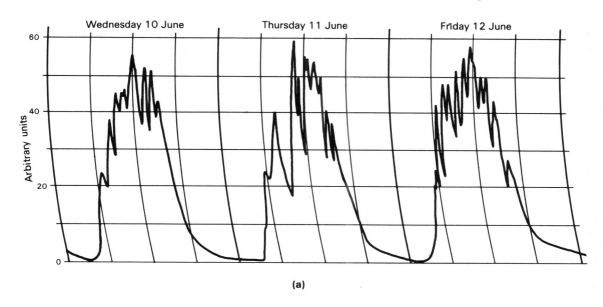

(a)

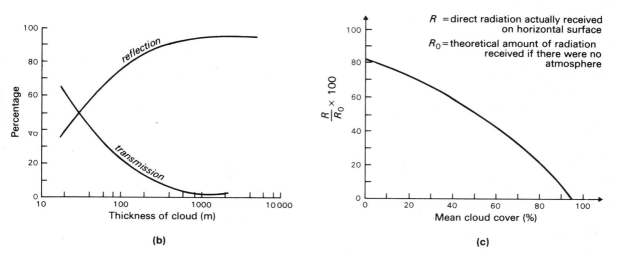

(b)   **(c)**

Fig. 3.7 The effects of cloud cover: (a) on variations in solar radiation received on a horizontal surface, Portsmouth 1976; (b) on reflectance and transmission in clouds of different thickness (Hewson and Longley, 1944); and (c) on radiation receipt (Black *et al.*, 1954).

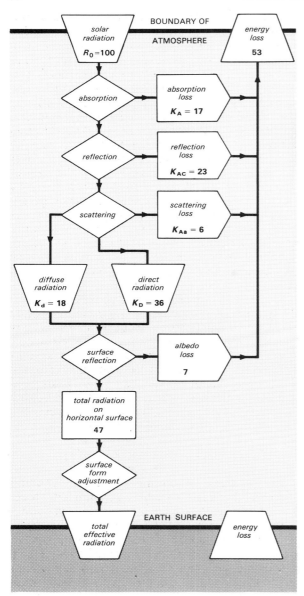

Fig. 3.8 The solar energy cascade.

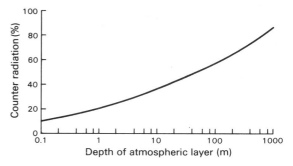

Fig. 3.9 The origin of counter-radiation arriving at the Earth's surface.

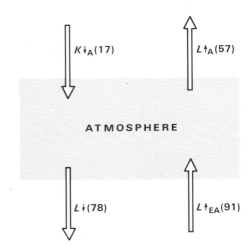

Fig. 3.10 Balance of radiation in the atmosphere.

varies little with latitude, as is indicated in Fig. 3.11. This is largely due to the localized effects of radiation exchange with the ground surface, which is primarily controlled by variation in carbon dioxide, atmospheric water vapour content and cloud cover. The radiation balance of the atmosphere does not appear to explain the marked latitudinal distribution of balance for the whole Earth–atmosphere system.

3.3 Earth-surface systems

3.3.1 Radiation exchange and throughput

The input of shortwave solar radiation transferred from the atmosphere to Earth surface systems consists of both direct and diffuse radiation $(K\downarrow_D + K\downarrow_d)$, where

to the ground surface. The atmosphere thus receives 108 units while losing 135, which represents a net radiation balance of −27 units. In reality this energy loss does not take place because other heat transfer processes are at work maintaining an energy balance, as we shall see later.

The net radiation balance of the atmosphere

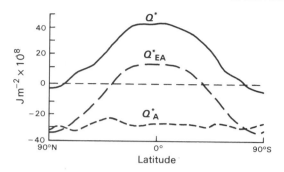

Fig. 3.11 Average latitudinal distribution of the radiation balance of the Earth–atmosphere system (Q^*_{EA}), of the Earth's surface (Q^*), and of the atmosphere (Q^*_A) (after Sellers, 1965).

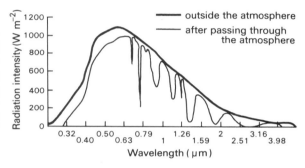

Fig. 3.12 The solar spectrum before and after passing through the atmosphere (after Lamb 1972).

$$(K{\downarrow}_D + K{\downarrow}_d) = K{\downarrow} = K{\downarrow}_0 - K_A - K{\uparrow}_{Aa} - K{\uparrow}_{Ac}.$$

The spectrum of $K{\downarrow}$ differs significantly from that of $K{\downarrow}_0$ (Fig. 3.12). Not only is there an obvious general reduction in radiant intensity, but in certain wavebands this has been severely reduced, mainly through absorption in the atmosphere. That part of the solar beam which experiences scattering, but which subsequently retains a generally earthward direction of motion, reaches the Earth's surface as diffuse radiation. The proportion of diffuse radiation varies considerably, but for the whole Earth surface it constitutes some 33% of total radiation reaching horizontal surfaces. At the Earth's surface the direct solar beam produces zones of full illumination and full shade. Illumination of the shaded zones is, therefore, dependent on diffuse radiation, and its significance can be appreciated by considering the important role it plays in light penetration into the photosynthesizing canopy of vegetation (see Chapter 17).

The distribution of average annual solar radiation falling on horizontal surfaces shows that this does not actually fall into a convenient latitudinal pattern, but tends to develop cellular patterns (Fig. 3.13). The role of cloudiness and atmospheric humidity in the development of this pattern can readily be illustrated by examining the distribution of solar radiation over Africa. Relatively low values over the cloudy and humid Zaire Basin, in equatorial latitudes, contrast with the much higher values around the tropics over the Sahara and Kalahari deserts. In the shorter term, the lifting of dust into the atmosphere by desert dust storms, or by soil erosion arising from poor land management, greatly influences the transmission of solar energy through the atmosphere. Increasing amounts of dust in the atmosphere, derived from a variety of sources, may be instrumental in producing long-term changes in climate (Budyko et al., 1988).

The Earth's surface returns some of the solar radiation incident upon it by reflection back through the atmosphere. In terms of the energy cascade, an average of 13% is reflected, which represents 7 units of radiation lost from the surface. Effective solar radiation input on a horizontal surface is, therefore, reduced to 47 units. If r represents the proportion of radiation reflected, then available net shortwave solar radiation at the Earth's surface (K^*) may be written as:

$$K^* = K{\downarrow} - K{\downarrow}.r \text{ (or } K{\downarrow} - K{\uparrow}\text{).}$$

The amount of longwave counter radiation reaching the Earth's surface depends ultimately upon the absorptive properties of the lower atmosphere. Thus areas of the Earth's surface above which the atmosphere is usually humid and cloudy, such as equatorial regions, receive considerable amounts of counter-radiation. Over the whole of the Earth's surface, counter-radiation constitutes, on average, about 62% of total radiant energy received. The net balance of longwave radiation from the ground surface may be expressed in simple terms in the equation

$$L^* = L{\uparrow} - L{\downarrow}$$

where $L{\uparrow}$ represents the longwave terrestrial radiation emitted from the ground surface, and $L{\downarrow}$ that returned by counter-radiation. This absorption of terrestrial radiation by the atmosphere is

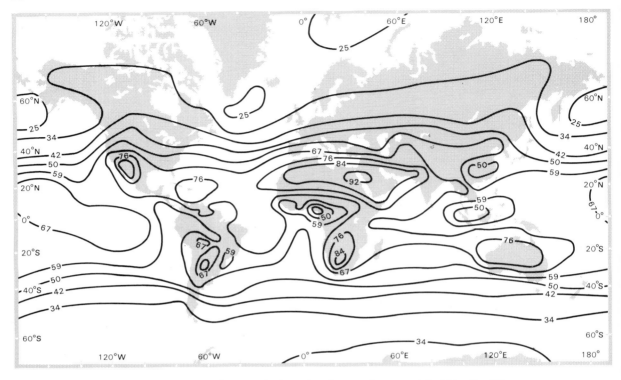

Fig. 3.13 The average annual solar radiation on a horizontal surface at the ground, in MJ m^{-2} × 10^2.

of particular significance in that it creates an insulating blanket (the greenhouse effect) above the Earth's surface, effectively inhibiting heat loss. Without such insulation, the temperature of the surface would be more than 30°C lower than it is at present.

Earth surface systems also receive energy from the Earth's interior as geothermal heat flow to the surface, and a proportion of this is output to the atmosphere as longwave radiation. From the point of view of the atmosphere, it is quantitatively so insignificant as to be lost in the total flux of longwave terrestrial radiation.

The average temperature of the Earth's surface ranges from about 255 K in high latitudes to 300 K in lower latitudes. If we assume an average surface temperature of 288 K and also that the Earth radiates as a black body, then the characteristics of Earth-surface or terrestrial radiation may be determined from the basic radiation laws in Box 3.3. Emission from the Earth's surface is 383 Wm^{-2} and the wavelength of maximum emission

is 10 μm. Radiation of this wavelength falls within the infrared sector of the electromagnetic spectrum. Indeed, the entire spectrum of terrestrial radiation lies within the infrared and is consequently referred to as longwave radiation. This is true for the full range of Earth-surface temperatures (Fig. 3.14).

The changes in surface temperature between polar and tropical regions result in variations in emission of longwave radiation from the Earth's surface. Average rates of emission are less than 150 Wm^{-2} in extremely high latitudes, while values in excess of 400 Wm-2 are typical of low latitudes.

The Earth's surface is, however, an imperfect emitter and absorber of radiation. **Emissivity** and **absorptivity** (Box 3.5), instead of having the black body value of 1, have grey body values of less than 1. For example, the ocean surfaces have an emissivity of between 0.92 and 0.96, while for most land surfaces this may fall to below 0.90 (Table 3.3).

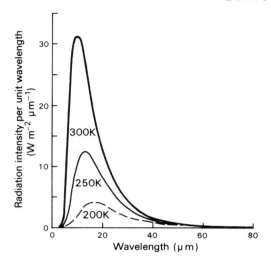

Fig. 3.14 The spectra of black body radiation at temperatures of 200 K, 250 K, and 300 K (after Neiburger *et al.*, 1971).

Table 3.3 Infrared emissivities

ice	0.96
water	0.92–0.96
moist ground	0.95–0.98
desert	0.90–0.91
pine forest	0.90
leaves 0.8 m^{-1}	0.05–0.53
10.0 m^{-1}	0.97–0.98

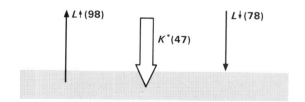

Fig. 3.15 Balance of radiation at the Earth's surface.

3.3.2 The radiation balance of the Earth's surface

The Earth's surface receives shortwave radiation from the Sun (some of which is lost through reflection) and longwave counter-radiation from the atmosphere, while itself emitting longwave radiation. These flows are represented in Fig. 3.15. The equation for the net radiation balance (Q^*) may be derived from the short- and longwave radiation balances:

$$Q^* = K^* + L^*$$

For the whole of the Earth, the surface receives 125 units of radiation, while losing only 98. The balance of +27 units implies that the surface is gaining in energy. In reality, this does not take place, so, by inference, there must be a mechanism for dissipating this excess energy. The equality of atmospheric deficit and Earth-surface surplus of radiant energy suggests that the mechanism is one which transfers energy from the surface into the atmosphere above it.

Unlike that for the atmosphere, the radiation balance of the Earth's surface varies considerably, as indicated in Fig. 3.11. There is a general trend towards a positive balance or surplus in low latitudes, and a negative balance or deficit in high latitudes. Factors such as cloudiness and atmospheric humidity, which affect the transmission of both solar and terrestrial radiation, and the radiative properties of the surface, create the spatial variation in radiation balance over the Earth's surface. Fig. 3.16 is based largely on estimates by Budyko (1958), and it shows that this spatial variation is not entirely latitudinal. Furthermore,

---

Box 3.5
ABSORPTIVITY AND EMISSIVITY

Absorptivity is the fraction of incident radiation absorbed.
Emissivity is the ratio of actual radiation emitted to a theoretical maximum.

Kirchhoff's Law states absorptivity is always equal to emissivity. For a black body:

absorptivity = emissivity = 1.

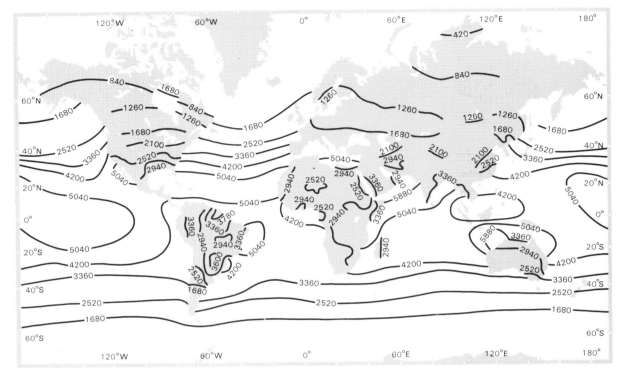

Fig. 3.16 The radiation balance of the Earth's surface in MJ m⁻² a⁻¹ (Budyko, 1958).

it shows that the net radiation balance of ocean surfaces is generally greater than that of land surfaces at the same latitude. The low reflection and high absorption of solar radiation at ocean surfaces largely accounts for this disparity.

During the year, net radiation changes in value. Solar radiation reaches its maximum intensity at the summer solstice, while net loss of radiation from the surface reaches its maximum during August (Fig. 3.17). Between **y** and **x**, surface losses exceed gains, so the net radiation balance is negative – producing cooling – while between **x** and **y** it is positive – producing warming. In a broadly similar manner, changes in the net radiation balance also occur on a diurnal time scale.

We can see from these hypothetical cases that the radiation balance of the Earth's surface is an important factor in determining the distribution of surface temperature. However, as we have already seen, in the long term there is a net radiation imbalance in the atmosphere and upon the Earth's surface. The long-term net radiation balance of the combined Earth–atmosphere sys-

tem, however, was shown earlier in this chapter to equal zero. Indeed, if we were able to measure accurately the radiation arriving at the top of the atmosphere, from the Sun and from the solids and fluids of the Earth and its atmosphere, we should expect to receive the following:

Incoming (from space	*Outgoing (from Earth) and atmosphere)*
$K\downarrow_0$ solar radiation	$K\uparrow_{Ac}$ radiation reflected from clouds
	$K\uparrow$ radiation reflected from the Earth's surface
	$K\uparrow_{Aa}$ radiation scattered by the atmosphere
	$L\uparrow$ terrestrial radiation
	$L\uparrow_A$ atmospheric radiation

Expressing this in the form of an equation:

$$Q^* = K\downarrow_0 - (K\uparrow_{Ac} + K\uparrow + K\uparrow_{Aa} + L\uparrow + L\uparrow_A).$$

net radiation at
outer boundary of
the atmosphere

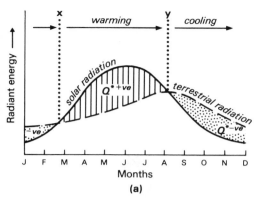

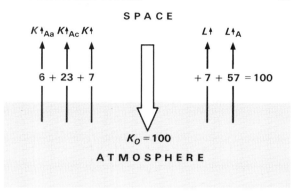

Fig. 3.18 Balance of radiation at the outer boundary of the atmosphere.

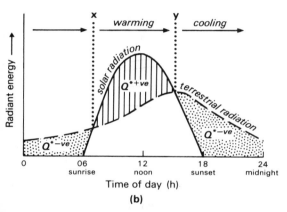

Fig. 3.17 Hypothetical (a) annual and (b) diurnal variation in incoming solar radiation and outgoing terrestrial radiation at the Earth's surface.

Over the whole of the atmosphere boundary, the long-term average value of this balance is zero, as indicated in Fig. 3.18. Therefore, the existence of a long-term net radiation imbalance in the atmosphere and at the Earth's surface implies that energy is being transferred by processes other than those of electromagnetic radiation, in order to maintain the overall net balance. We must therefore consider alternative forms of energy exchange in the form of an energy balance.

3.4 Energy balance of the Earth–atmosphere system

If we consider the simplest of situations – a completely dry, uniform solid surface – solar and atmospheric radiation is absorbed into it, some of which is then reradiated. The heat energy which passes into the surface (Q_G) does so by **conduction**,

while it is also transferred into the atmosphere by conduction and **convection** (Box 3.6). The atmosphere is an extremely poor conductor of heat but, being a fluid, may be set into convectional motion. Heat energy is conducted into a thin air layer in contact with the surface and then redistributed by free or forced convection. Thus some of the available heat energy at the surface is utilized in heating the atmosphere in this manner (Q_H). If we have $Q*$ units in our net radiation balance, we can express its expenditure as a simple equation:

$$Q* = Q_H + Q_G.$$

The signs of Q_H and Q_G may be either positive or negative, in that heat may also pass from atmosphere to ground surface or from subsurface to surface.

For surfaces that contain water, there is the added complication of heat gain or loss as phase changes take place (see Chapter 4). For example, in evaporation, when water changes from liquid to gas, heat energy is consumed (Q_E). So the simple equation becomes

$$Q* = Q_H + Q_G + Q_E.$$

The distribution of heat energy along these pathways varies according to the type of surface. On dry desert surfaces, for example, little energy will normally be consumed in evaporation (Fig. 3.19), while most energy is used to heat the atmosphere. In direct contrast, in a moist environment a large proportion of energy is consumed in evaporation at the expense of heating the atmosphere. The

Box 3.6
CONDUCTION AND CONVECTION

Thermal conduction is the process by which heat energy is transferred from molecule to molecule, and which involves no movement of mass.

The rate of conduction of heat through still air is low and, in terms of large-scale heat transfers within the free atmosphere, it can be regarded as being of negligible significance.

Convection is the transfer of heat within a fluid medium, involving the transfer of mass. This is of considerable significance in the fluid medium of the atmosphere. It may take the form of free or forced convection. In free convection, the movement of air is in the form of density currents as it is heated from below. Heated air rises and is replaced by cooler air.

heating

In forced convection, air movement is in the form of mechanical turbulence in the air flowing across a surface.

basic energy balance equation may be modified to apply to most of the major surface types occurring over the Earth (Box 3.7). However, it must be borne in mind that these are simplifications of what is often a considerably more complex energy balance.

For land surfaces we can ignore, for the sake of simplicity, lateral heat transfer beneath the surface. Not only are rock and soil poor conductors of heat, but temperature gradients within them are also relatively modest. Under a fluid surface, however, a lateral transport of heat energy does take place. The fluid moves in response to free convection or may be driven by the movement of the atmosphere across its surface. The basic energy balance equation for the oceans may be modified to include a lateral movement of heat energy (Q_F), thus:

$$Q^* = Q_H + Q_G + Q_E + Q_F.$$

Such transfers occur on a large scale in the oceans in the form of ocean currents. Fig. 3.20 illustrates the complex nature of the heat energy balance in ocean waters. Very large amounts of heat energy are supplied beneath the surface, to the extent of exceeding the contribution made by net radiant energy.

A summary of global variation in the elements of the energy balance appears in Table 3.4. The significance of subsurface energy flow in the oceans can be seen. Latitudes greater than 30° experience a net gain of energy in this way, while those in lower latitudes are losing energy. The oceans therefore play an important role in the redistribution of heat energy in the Earth–atmosphere system.

So far we have not looked at energy flow in the other fluid – the atmosphere. This will be considered in more detail in Chapter 4, but at this stage it is worth noting that the atmosphere also redistributes heat. We have seen that over the global surface there is a net surplus of radiant energy in low latitudes and a deficit in high latitudes. This imbalance is redressed by heat transfer in a moving atmosphere, through a complex circulatory system.

3.5 The Earth's interior systems

At this point we have considered only two of the three subsystems depicted in Fig. 3.5, the atmosphere and Earth-surface systems. The third subsystem is the Earth's interior, and here the principal processes of energy transfer are conduction and convection. As we have seen, the

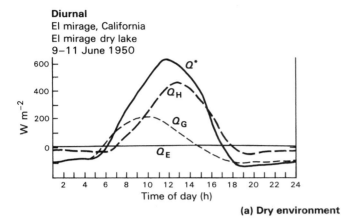

Diurnal
El mirage, California
El mirage dry lake
9–11 June 1950

Annual
Yuma, Arizona (32.7°N)

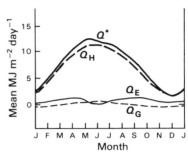

(a) Dry environment

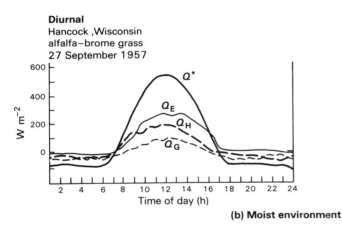

Diurnal
Hancock, Wisconsin
alfalfa–brome grass
27 September 1957

Annual
Madison, Wisconsin (43.1°N)

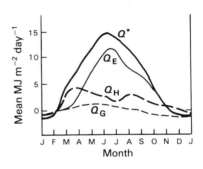

(b) Moist environment

Fig. 3.19 Diurnal and annual variation of heat-energy balances for (a) dry and (b) moist surfaces. $Q*$ = radiation balance, Q_H = heat transferred into air, Q_E = heat utilized in evapotranspiration, Q_G = heat transferred into subsurface.

Box 3.7
ENERGY BALANCES

Vegetated surfaces

The energy balance equation has to include heat consumed in moisture loss from plant surfaces, transpiration (Q_T) (see Chapter 4) and energy consumed in photosynthesis (Q_P):

$$Q_Y^* = Q_E + Q_T + Q_H + Q_G + Q_P + C.$$

C is a complex term as it includes the absorption of heat into the plant itself.

Ice surfaces

The equation has to take into account the melting of ice, which requires heat energy (Q_M). Freezing actually releases heat energy ($-Q_M$):

$$Q_{ice}^* = Q_E + Q_G + Q_H \pm Q_M.$$

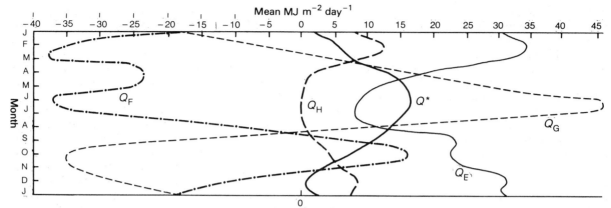

Fig. 3.20 Average annual variation in surface heat energy. Q_F = horizontal heat transfer, Q^* = radiation balance, Q_G = heat transferred into the subsurface, Q_H = heat transferred into lower air layers, Q_E = heat utilized in evaporation.

Table 3.4 Mean latitudinal values of the components of the energy balance equation for the Earth's surface, in MJ m^{-2} (after data from Sellers 1965). NB Data have been converted to SI form and rounded to the nearest whole number. In some cases, therefore, the energy balance equation may not balance

Latitude zone	Oceans				Land			Globe			
	Q	Q_E	Q_H	Q_F	Q^*	Q_E	Q_H	Q^*	Q_E	Q_H	Q_F
80–90°N								−38	13	42	8
70–80								4	38	−4	−29
60–70	97	139	69	−109	84	59	25	88	84	42	−38
50–60	122	164	67	−109	126	80	46	126	118	59	−50
40–50	214	223	59	−67	189	101	88	202	160	71	−29
30–40	349	361	55	−67	252	97	155	307	248	101	−42
20–30	475	493	38	−4	290	84	206	403	307	101	−4
10–20	500	488	25	59	298	122	176	445	340	67	38
0–10	483	336	17	130	302	202	101	441	302	46	92
0–90°N								302	231	67	4
0–10	483	353	17	113	302	210	92	441	319	42	80
10–20	475	492	21	17	307	172	134	437	378	46	13
20–30	424	419	29	−25	294	118	176	395	349	67	−21
30–40	344	336	34	−25	260	118	143	336	311	46	−21
40–50	239	231	38	−29	172	88	84	235	223	42	−29
50–60	118	130	36	−55	130	84	46	118	130	46	−59
60–70								55	42	46	−34
70–80								−8	13	−17	−4
80–90								−46	0	−46	0
0–90°S								302	260	46	−4
globe	344	310	34	0	206	105	101	302	247	55	0

energy exchanges of surface and atmospheric systems are dominated by the massive flux of solar energy. However, this has little direct effect on the Earth's interior for it does not penetrate significantly beneath the surface. Diurnal temperature variations seldom exceed 1°C at a depth of 1 m, while seasonal temperature changes penetrate to 30 m at the most, and therefore processes beneath the immediate surface layer are controlled by energy transfers which reflect changes in the storage of the Earth's internal energy.

3.5.1 Sources of the Earth's internal energy

Although current theories (Box 3.8) hold that the original Earth-forming materials were fairly cool, various processes associated with formation of the Earth can be inferred to have contributed to heat generation. Small bodies arriving at the Earth's surface would possess kinetic energy, and this would be dissipated as heat energy around the point of impact. Although much of this heat would be radiated back to space, a proportion would be transferred to the Earth's interior. As the Earth grew, due to the continuing addition of such material, its interior would become progressively more compacted and dense, and internal temperatures would rise as a result of the increased pressure. If 0.1% of the kinetic energy became trapped, it can be calculated theoretically that the internal temperature would be raised by 30°C, and compression could account for an increase of up to 900°C at the centre. These processes can account for only a fraction of the Earth's internal heat. It is likely, however, that they have contributed in a small way to the Earth's present internal temperatures. Since heat transfer from the interior to the surface is a very slow process, it is possible that a small proportion of the Earth's present heat output may have derived from these events very early in its history.

As the denser elements, originally scattered throughout the homogeneous Earth, concentrated out and settled towards the Earth's centre, large amounts of gravitational (potential) energy must have been released. This would have been in the form of heat, some of which would have contributed to raising the temperature of the core (Box 3.8), but much of which would be available to raise the Earth's temperature. Calculations based on the volume of the core and the density of minerals within it, show that the amount of heat released would be sufficient to raise the tempera-

ture of the Earth by 1500°C. Clearly this is likely to have been an important heat source. It is not yet known when core formation took place, whether early or late in the Earth's history; therefore it is not possible to say how much of the heat from this source has already been lost by radiation, or how much remains.

When a radioactive isotope decays, there is a loss of nuclear energy which is dissipated as heat in the vicinity of the decaying isotope. There are four major radioactive isotopes which produce heat in significant quantities at the present time. These are shown in Table 3.5. The **half-life** is the length of time required for half of the original parent isotope to decay. Only isotopes with a half-life of 10^9–10^{10} years are important in the present context. Short-lived isotopes are very low in abundance and longer-lived isotopes produce too little heat per unit of time to be significant. The quantity of heat released by these isotopes can be determined experimentally in the laboratory, and their abundance in the field can be measured in terms of their concentration in different types of rock. It is possible, therefore, to calculate their potential as heat producers. Such calculations reveal that radioactive decay in this way could easily account for the current observed terrestrial heat flow. In addition it seems likely that other, short-lived, isotopes formerly existed within the Earth. These too would have decayed to produce heat and may have contributed to present Earth temperatures, even though the parent isotopes, having decayed, no longer exist.

Although there are other possible contributors, these are the probable major sources of the Earth's internal heat energy. As yet, we are uncertain of their exact contributions to terrestrial energy, for many uncertainties and assumptions exist about conditions and processes operating during the Earth's early history. Additionally, of course, our knowledge concerning the Earth's interior is largely inferential.

Table 3.5 The major radioactive elements

		Half-life (years)	Heat production ($J g^{-1} yr^{-1}$)
uranium	^{238}U	4.5×10^9	2.97
	^{235}U	0.71×10^9	18.00
thorium	^{232}Th	13.9×10^9	0.84
potassium	^{32}K	1.3×10^9	0.88

Box 3.8

THE EVOLUTION OF THE EARTH AND ITS ATMOSPHERE

The process of accretion which led to the formation of the Earth implies that initially the planet was homogeneous throughout its mass, as similar planetary material was continuously added. In contrast, at the present time, it consists of a series of three **concentric shells** of differing chemical composition, density and material state. The body of the planet consists of a partially liquid **core,** a plastic **mantle** and a solid **crust.** It is surrounded by a blanket of gaseous material – the atmosphere.

The process by which the chemical separation of these various layers took place is known as **differentiation**. It is thought to have occurred at an early stage in the Earth's history and has had profound consequences in respect of materials and conditions at the planetary surface.

At the time of its formation, the Earth is thought to have been relatively cool, in the range 600–1000°C. The radioactive minerals, distributed throughout its mass, would have produced heat as they decayed, more rapidly than it could be conducted to the surface. In this way internal temperatures were raised sufficiently to reach the melting point of iron – a major turning point in Earth history. The melting iron, of high density, gravitated towards the centre to form the core, displacing other material outwards. In so doing, it released large amounts of potential energy which, transformed to heat energy, would have raised internal temperatures further, probably causing the melting of a significant part of the Earths body. Thus less dense molten material would have floated upwards to the surface, there to cool and solidify to form a crust of less dense elements, overlying a mantle of intermediate characteristics.

An important part of the differentiation process was the release of gases from the interior to form the atmosphere. This process of **outgassing** is thought to have been associated with volcanic activity. The time of differentiation appears to have been one of considerable internal turmoil in the history of the planet.

Observations of the chemical composition of gases and lavas of contemporary volcanic eruptions suggest that the original atmosphere consisted of water vapour, hydrogen, chlorine, carbon, oxygen and nitrogen. The Earths primitive atmosphere consisted of a number of gases, in which carbon monoxide, carbon dioxide, methane, hydrogen chloride, and ammonia were present with very little free oxygen – very different from present day conditions. The atmosphere itself has subsequently evolved to attain its present composition. (See Chapters 4 and 7.) The retention of the gaseous atmosphere is a consequence of the gravitational attraction created by the Earths mass although the lighter hydrogen tends to drift off into space. It is a matter of considerable fortune that the Earths distance from the Sun is such that surface temperatures throughout geological time span the range within which water exists as a fluid. As a consequence atmospheric water vapour was able to condense to form the large volumes of water present in the hydrosphere today. Conditions in the primitive atmosphere were such that amino acids formed, leading to the development of organic compounds and hence the evolution of living systems (See Ch. 7). This development in turn had a profound influence on atmospheric evolution. Living creatures accumulated carbon from the atmosphere, some of which was subsequently to be incorporated into carbonate rocks (coal, limestone) in the lithosphere as they died. Living organisms are also oxygen producers during respiration. In these ways they exchanged oxygen for the carbon in the primitive atmosphere, to create the current balance of atmospheric elements.

The contemporary state of the Earth/atmosphere system is therefore a consequence of a particular path of planetary evolution, unique and distinct from paths followed by other planets in the solar system (see Press and Siever, 1986, Ch. 1).

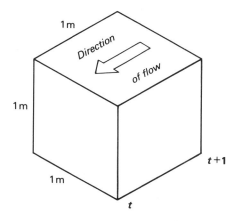

$$Q = \frac{K.A[(t+1)-t]}{L}$$

$$\therefore K = \frac{Q.L}{A[(t+1)-t]}$$

where Q = rate of heat flow (J s^{-1})
   A = cross section area
   L = length of gradient
   K = thermal conductivity of material
   t = temperature (°C)

Fig. 3.21 Thermal conductivity.

3.5.2 Energy transfer from the Earth's interior to the surface

Much of the Earth's internal energy is in the form of heat and is transferred to the surface by conduction, but also in part by convection (see Box 3.6). It seems likely that slow upward movements occur with at least part of the mantle convecting heat to the base of the crust, through which it is transferred mainly by conduction. Studies of heat flow did not begin to take place until 1950, and our current knowledge of the subject is still far from complete. Nevertheless, by indicating the method by which it can be measured and the results obtained, we can go some way towards an understanding of the magnitude and distribution of terrestrial energy inputs to surface systems.

Heat flow per unit area of the Earth's surface at a particular point is defined as follows:

$$q = Kv$$

where q = heat flow, 10^{-6} J m^{-2} s^{-1}, K = thermal conductivity, 10^{-6} J m^{-1} s^{-1} °C^{-1} and v = vertical temperature gradient, °C m^{-1}.

This requires the assessment of the vertical temperature gradient within the crust, obtained by measuring the difference in temperature between two points of differing depth within a borehole in the crust. The average temperature gradient in the upper part of the crust is around 3°C km^{-1}, although significant variations do occur. Also required is the thermal conductivity of the crustal material; that is, the quantity of heat (in joules) flowing across an area of 1 square metre in 1 second in a material where the temperature gradient is 1°C per metre (Fig. 3.21). Since the crust consists of solid materials, this transfer of heat takes place by conduction. Thus heat flows along the geothermal gradient (Fig. 3.22) from the warm interior out towards the cool surface. Heat flow is measured in heat flow units (HFU) where

$$1 \text{ HFU} = 4.19 \times 10^{-6} \text{ J cm}^{-2} \text{ s}^{-1}.$$

The mean world heat transfer to the surface is 1.5 HFU, an amount sufficient to melt approximately 5 mm thickness of ice per year, and representing around 1×10^{-3} of the amount of incoming solar radiation. This mean value masks considerable spatial variations in heat flow, as shown in Table 3.6.

The values of heat flow for crust of continental and oceanic provinces are of the same order of magnitude. Yet the composition of the crust varies considerably between the continental and oceanic realms. Continental crust contains a relative abundance of radiogenic materials, which are sufficient to account for heat flow observed there. Oceanic crust contains only sparse concentrations of such minerals and the magnitude of heat flow in such regions must be accounted for by presence of hot mantle material not far beneath the crust. This in turn leads to a consideration of processes within the mantle and, as will be demonstrated in Chapter 5, large-scale slow circulatory movements occur there, bringing hot mantle material up from greater depths. In this way heat transfer

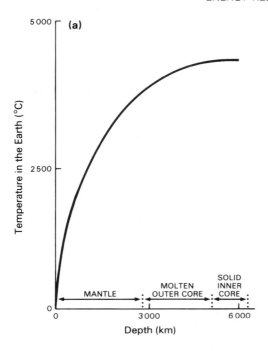

Fig. 3.22 (a) Geothermal gradient. (b) A volcanic cone: this represents an accumulation of lava and ash ejected at the Earth's surface; steam and gases are still issuing from the vent, indicating contemporary transfer of heat from within the crust (Mt. Ngauruhoe, New Zealand).

Table 3.6 Spatial variation in heat flow, shown as mean values for different orogenic regions (after Sass 1971) (1 HFU = 4.19×10^{-6} J cm^{-2} s^{-1})

	HFU
ocean basins	1.27
mid-ocean ridges	1.91
Precambrian shields	0.98
Palaeozoic orogenic belts	1.44
Tertiary orogenic belts	1.77

Table 3.7 Relative magnitudes of solar and terrestrial energy

	Joules/year
solar energy received	5.6×10^{24}
geothermal energy loss by radiation	1.3×10^{21}
volcanic outflows	5.3×10^{18}
earthquakes	1.0×10^{18}

by convection takes place through the mantle towards sections of the oceanic crust, through which it is then conducted.

Small quantities of terrestrial heat also arrive at the surface directly by convection. This results from volcanic and geothermal processes, as a result of which hot materials (lava, steam) from the Earth's interior are ejected directly at the surface. The magnitude of such losses is shown in Table 3.7.

In the wake of the energy crises of the 1970s renewed interest has been shown by the developed nations in the potential of geothermal energy (particularly geothermal waters and steam) (Fig. 3.23). So far, the primary use is for the generation of electricity. World electrical generation from geothermal energy in 1971 was approximately 800 MW, or about 0.08% of the total world electrical capacity from all modes of generation. Geothermal resources have other uses, although to date they have been minor. They include local space heating and horticulture. Much of Reykjavik (Iceland), Rotorua (New Zealand), Boise (Idaho, USA), Klamath Falls (Oregon, USA) and several towns in Hungary and the Soviet Union are heated by geothermal water (Muffler and White, 1975).

Fig. 3.24 represents, in a simplified manner, the cascade of energy within the Earth, a flux of energy which represents a long-term change in the internal potential energy store of the system and a long-term net negative energy balance. The implications of this transfer of energy to the crust and surface systems, where further energy conversions take place, will be considered briefly in the following section and in more detail in Chapters 5 and 10.

3.6 Energy conversion

The Earth–atmosphere system uses heat energy which it receives from the Sun and, to a lesser extent, from within the Earth. This heat energy provides an input to both Earth-surface and atmospheric systems, thereby increasing their own internal energy. In the case of the atmosphere, this

Fig. 3.23 Tapping of subterranean geothermal heat for energy use (Wairakei, New Zealand).

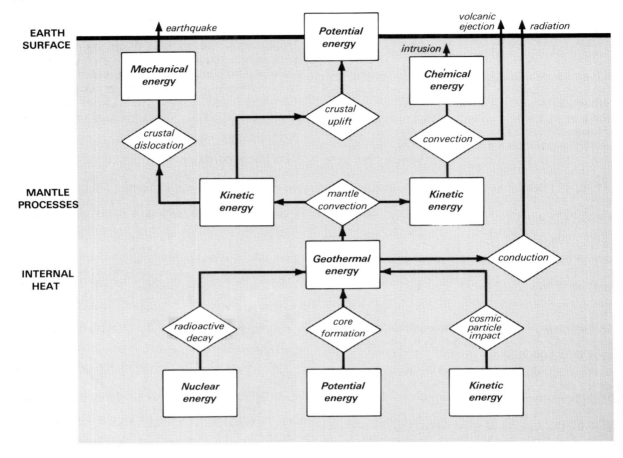

Fig. 3.24 Geothermal energy cascade.

may be seen as increasing its potential energy. The conversion of this potential energy into kinetic energy takes place as the internal energy imbalances of the atmosphere are redressed. In Chapters 4 and 8 we shall see that this conversion is essential to the maintenance of circulation systems and to power the hydrological cascade transferring water from the atmosphere to the Earth's surface and back again. A fraction of the solar radiation input is converted directly to chemical energy by the photochemical reaction of photosynthesis in green plants. This fraction is very small, however, reaching perhaps 2–5% in efficient communities but only a fraction of 1% on a global basis. Nevertheless, the consequences of photosynthesis for the organisms of the biosphere (Chapter 6) are enormous, and the influence of the surface energy balance on water loss and leaf temperature

of plants partially determines the overall productivity of the biosphere.

Although operating over an infinitely longer timescale, the mobility of the Earth's mantle provides yet another input of potential energy to Earth-surface systems. These slow large-scale mantle currents, representing a conversion from heat to kinetic energy, are responsible for both vertical and lateral displacement of the overlying crust and the conversion of kinetic to mechanical energy, which may be released in earthquakes (Fig. 3.24). Vertical displacements of the crust, however, represent a conversion of convective kinetic energy to the potential gravitational energy associated with relief. This potential energy is again converted to kinetic energy which, in combination with the kinetic energy of water moving through the hydrological cascade, is transformed to mechanical energy and heat as surface materials are transferred to lower elevations by the processes of denudation (see Chapters 5 and 10).

In accordance with the Second Law of Thermodynamics, none of the energy transformations referred to above is 100% efficient. At each conversion some energy is dissipated as heat, ultimately as longwave radiation output to space. The continued operation of the subsystems of our model (see Fig. 3.5) therefore depends on a continued flux of energy from the Sun and from the Earth's interior.

The net loss of geothermal energy to space represents an irreversible depletion of the Earth's energy store, which is itself a legacy of the events that caused the formation of the Earth. Ultimately this store of energy within the planet must become exhausted. The present state of the crustal system, therefore, and the processes operating within it, cannot be regarded as permanent. It is but one phase in the geological evolution of the Earth. In the long term, the depletion of the Earth's internal energy will result in a running-down of the system – a cooling down as the heat energy declines, and ultimately solidification. The timescale over which this can be expected to occur, however, need not be a source of anxiety for the Earth's present inhabitants!

The same is also true of the external source of energy, the Sun. It too can be regarded as a finite store of energy which is becoming progressively depleted through time. At its present rate of energy conversion, however, the mass of the Sun (which is being transformed to energy by nuclear fusion) will diminish by only one-millionth of its mass in 15 million years. We are probably safe, therefore, in making the assumption, as we have done in this chapter with our closed-system model, that the flux of energy across the boundary of the system and within it maintains a long-term average steady state.

Further reading

Good general accounts of the Earth's external energy relationships are given in the following:

Atkinson, B.W. (1987) *Atmospheric Energetics*, in *Energetics of Physical Environment: Energetic Approaches to Physical Geography*, (ed K.J. Gregory) John Wiley, Chichester.

Campbell, I.M. (1977) *Energy and the Atmosphere*. John Wiley, London.

Eliassen, A. and K. Pedersen (1977) *Meteorology: an Introductory Course, Vol. 1, Physical Processes and Motion*. Scandinavian University Books, Oslo.

Gates, D.M. (1962) *Energy Exchange in the Biosphere*. Harper & Row, New York.

Oke, T.R. (1978) *Boundary Layer Climatology*. Methuen, London.

The internal energy of the Earth is covered by the following texts:

Clark, S.P. (1971) *Structure of the Earth*. Prentice-Hall, Englewood Cliffs.

Gaskell, T.F. (1967) *The Earth's Mantle*. Academic Press, London.

Smith, P. (1973) *Topics in Geophysics*. Open University Press, Milton Keynes.

III GLOBAL SYSTEMS

The atmosphere

4

The Earth's atmosphere, a complex fluid system of gases and suspended particles, did not have its origins in the beginnings of the planet. The atmosphere of today has been derived from the Earth itself by chemical and biochemical reactions (see Chapter 7). Although this fluid system forms a gaseous envelope around the Earth, its boundaries are not easily defined. They can be arbitrarily delimited as the Earth/atmosphere interface and atmosphere/space interface, but it must be recognized that these are oversimplifications. For example, there is no outer edge to the atmosphere, but a zone of transition where the Earth and solar atmospheres merge. Similarly, at the Earth/atmosphere interface the atmosphere penetrates into the voids and pore spaces between the particles of soil and regolith and is continuous with the so-called soil atmosphere (see Chapter 20).

4.1 Structure of the atmospheric system

Just five gases – nitrogen, oxygen, argon, carbon dioxide – and water vapour together make up 99.9% of the total volume of the atmosphere. Together with suspended particles such as water droplets, dust and soot, and the minor gases, these represent the elements of the system, Omitting water vapour (which fluctuates considerably in quantity) and the suspended particles, the gases are present in dry air in the proportions indicated in Table 4.1. As most of these constituent elements of the system are in a gaseous phase, they are present for the most part as separate molecules, atoms or even ions, and are in a state of constant motion. The organization of these units into more complex compounds and into larger aggregates of matter to form observable structures, such as in the lithosphere and biosphere, is therefore, largely lacking in the atmosphere. Structure exists, but it has to be seen partly in

terms of the differential distribution of the constituents within the system and partly in terms of the variations in the measurable attributes of the system, such as temperature and pressure.

The proportions of the various gases change very slowly over time, but they do show distinct vertical distributions through the atmosphere. In the lowest 11 km the proportions change little due to the turbulent mixing of the atmosphere up to this height. Above this, layering is more in evidence, with gases and suspended particles organized into bands. Before examining this vertical structure we shall consider briefly the distribution of the more important gases in the atmosphere.

Nitrogen. Nitrogen occupies over three-quarters of the atmosphere and occurs throughout a deep layer extending over 100 km from the Earth's surface. The greatest concentrations of molecular nitrogen (N_2) occur in the lowest 50 km, while atomic nitrogen (N) is the more prevalent between 50 and 100 km.

The chemistry of nitrogen is rather complex. To simplify, a distinction is usually made between the relatively inert forms of nitrogen (molecular nitrogen (N_2) and nitrous oxide (N_2O), occurring mainly in the atmosphere, and the more reactive compounds which are susceptible to incorporation

Table 4.1 Constituents of dry air

	Mol. weight	Volume %	Mass %
nitrogen	28.01	78.09	75.51
oxygen	32.00	20.95	23.15
argon	39.94	0.93	1.23
carbon dioxide	44.01	0.03	0.05

in the various reservoirs at the Earth's surface. These compounds, for example nitric acid (HNO_3) are referred to collectively as fixed nitrogen, and processes that convert inert molecular nitrogen in the atmosphere to the more reactive forms are referred to as nitrogen fixation. The reverse process is described by the term denitrification. The basic problem in understanding the biogeochemistry of terrestrial nitrogen is to understand how, where, and at what rate these two processes occur, and whether there are feedback mechanisms that regulate them (see Chapter 7).

Atomic nitrogen is subjected to shortwave cosmic radiation in the upper atmosphere which produces an unstable isotope of carbon (^{14}C) referred to as radiocarbon. This combines with oxygen in the lower atmosphere to produce radiocarbon dioxide which is assimilated by living systems at the Earth's surface. There is normally one molecule of radiocarbon dioxide to 10^{12} molecules of normal (^{12}C) carbon dioxide. The unstable ^{14}C isotope gradually reverts back to nitrogen, which passes back into the atmosphere. This radiocarbon decay has provided man with a means of dating organic deposits at the Earth's surface. Although it occupies such a large volume, nitrogen has little effect on the global radiation balance and, although essential for living systems, it cannot be assimilated directly by either plants or animals.

Oxygen. Oxygen occurs throughout the lowest 120 km of the atmosphere and occupies a little more than one-fifth of its total volume. Below 60 km it exists mainly as molecular oxygen (O_2), while above this the dissociated atomic oxygen (O) is the more prevalent. The latter is brought about by the effects of cosmic radiation on the oxygen molecule at high elevations in the atmosphere. Oxygen, which exists in its gaseous state in the atmosphere, represents only a part of the total stored in the Earth–atmosphere system. Animals and plants store oxygen as a component of organic molecules during their lives, while in the rocks of the lithosphere it is bound into chemical compounds such as oxides and carbonates.

Ozone is present in only very small quantities in the atmosphere but its impact on the Earth–atmosphere system is considerable, particularly in the absorption of shortwave radiation. It is produced as a result of the photodissociation of molecular oxygen into atomic oxygen by radiation of less than 0.24 μm wavelength. The dissociated atoms recombine with molecular oxygen to produce ozone:

$$O_2 + hv \rightarrow O + O$$

and:

$$O + O_2 + M \rightarrow O_3 + M$$

where M is any other molecule present in the atmosphere. It is required to take up the energy released in the reaction. Maximum production of ozone occurs between 30 and 40 km above the Earth's surface, but maximum concentrations occur some 10 km lower than this. At extremely high elevations (greater than 60 km) there are insufficient oxygen molecules to maintain a high rate of production of ozone, while below 10 km there is insufficient shortwave radiation as this has already been absorbed at high elevations. Ozone is also unstable and readily breaks down. There are several reactions which result in the destruction of ozone, but the most efficient make use of the oxides of nitrogen

$$NO + O_3 \rightarrow NO_2 + O_2$$

followed by

$$NO_2 + O \rightarrow NO + O_2$$

These appear to be the only reactions that are able to account for the equilibrium concentration of ozone. The level of maximum concentration thus represents a balance between the processes of creation and destruction of ozone and its mixing into the atmosphere.

The nitric oxide in the stratosphere is believed to originate by a reaction between metastable oxygen atoms and nitrous oxide (N_2O) which has diffused up from the troposphere, having been formed by denitrification at the Earth's surface. These observations indicate how complex are the reaction chains which control the chemistry of our environment. They also justify the concern over the still inadequately understood effects of adding trace pollutants to the stratosphere which enhance the destruction rate of ozone, thereby increasing the influx of harmful ultraviolet radiation to the Earth's surface (see Chapter 24).

Carbon dioxide. Carbon dioxide forms a very small proportion of the atmosphere, but it is essential to life on Earth. Because of its close association with the biosphere it is most prevalent in the lowest 50 km of the atmosphere, and particularly the lowest 2 km. Carbon dioxide is withdrawn from the atmosphere during photosynthesis and returned during respiration and by the decay (oxidation) of organic carbon compounds (see Chapter 7). The residence time of carbon in the surface organic reservoir is of the order of 20 years, and to the extent that the organic reservoir remains fixed in size the cycle maintains a natural balance (see Chapter 7). However, although such combustion sources as forest fires are natural, the burning of carbon compounds which would otherwise have been effectively immobilized in the lithosphere as fossil fuels has had measurable effects on atmospheric carbon dioxide concentrations (see Chapter 24).

Water vapour. Water vapour is the gaseous phase of water and is derived from the diffusion of water molecules from their liquid state at the Earth's surface. Because of this, it is concentrated in the lowest 10 km of the atmosphere, with approximately 90% of it occurring below 6 km. It is one of the most variable constituents of the atmosphere, contributing between 0.5% and 4.0% of the volume of moist air. Fig. 4.1 shows a typical distribution of water vapour through the lowest 8 km of the atmosphere.

Non-gaseous particles. In addition to the major gases, the atmosphere contains materials of non-gaseous nature which are held in suspension or maintained aloft by turbulent mixing. The most important of these are the water droplets and ice crystals that occur in clouds. These are important in that they affect both shortwave (reflection) and longwave (absorption) radiation. The amount of cloud in the atmosphere is highly variable, but there are areas such as equatorial and middle latitudes which experience greater cloud cover than elsewhere (Fig. 4.2).

About 90% of the remaining particles are natural in origin, such as dust from volcanic eruptions, smoke from forest fires, sea-spray (bubble bursting) and pollen (Table 4.2). These are increasingly augmented by the products of human activity, such as dust from soil erosion in areas of poor land management and smoke from the incomplete combustion of fossil fuels.

All particles are carried into the atmosphere by turbulent mixing and may remain there for considerable periods of time, although the heavier particles (>100 μm radius) tend to fall back to the ground through the effects of gravity. About 80% of particles rise no further than 1 km from the surface, but small quantities of fine particles with radii of the order of 10 μm or less may rise to heights or between 10 and 15 km, where they may stay for several years.

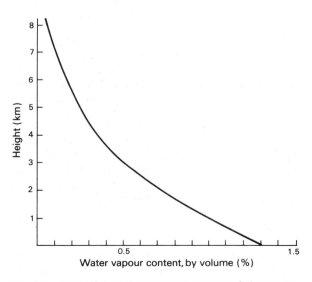

Fig. 4.1 Decrease in water vapour content of the atmosphere with height.

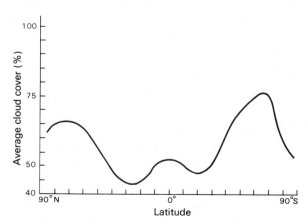

Fig. 4.2 Latitudinal variation in average percentage cloud cover (after data from Sellers, 1965).

Table 4.2 Sources of materials in the atmosphere (from Smith 1975, after Varney and McCormac 1971)

Type	Source
particles	volcanoes
	combustion
	wind action
	industry
	sea spray
	forest fires
hydrocarbons	internal combustion engine
	bacteria
	plants
sulphur compounds	bacteria
(SO_2, H_2S, H_2SO_4)	burning fossil fuels
	volcanoes
	sea spray
nitrogen compounds	bacteria
	combustion

4.1.1 Systems structure: a model of the vertical structure of the atmosphere

It has already been stressed that in the fluid system of the atmosphere structure has to be perceived in terms of the distribution of the properties of the system, particularly those attributes that can be treated as measurable variables. Certainly the preceding discussion has suggested that the atmosphere is not homogeneous, and the distribution of the constituent elements indicates that it has a layered structure. This vertical component of the structure of the atmosphere can be related primarily to the distribution of radiant energy absorption and can be described in terms of the variable of temperature.

Below 60 km there are two main zones of absorption: at the Earth's surface and in the ozone layer. The absorbed energy is redistributed by re-radiation, conduction and convection. There are, therefore, two temperature maxima: at the Earth's surface and at an elevation of c. 50 km. Above each of these maxima there is mainly convectional mixing. Temperatures in these mixing layers decrease with height above the heat source. The lower of these two zones (Fig. 4.3) is referred to as the **troposphere**, and the upper the **mesosphere**. These are separated by a layer of little mixing in which the atmosphere tends towards a layered structure referred to as the **stratosphere**. Between the troposphere and the stratosphere is the **tropopause** which marks the approximate upper limit of mixing in the lower atmosphere. The average height of this is usually given as 11 km, but this varies over the Earth. In tropical latitudes its average height is 16 km and in polar latitudes it is only 10 km. There is one further zone of heating, above the mesosphere and more than 90 km from the Earth's surface, where shortwave ultraviolet radiation is absorbed by any oxygen molecules present at this height. This is referred to as the **thermosphere**. Within this layer, ionization occurs which produces charged ions and free electrons. Beyond the thermosphere, at a height of approximately 700 km, lies the **exosphere** where the atmosphere has an extremely low density. At this level there are increasing numbers of ionized particles which are concentrated into bands referred to as the **Van Allen belts**.

This model of vertical structure can be simplified to provide a model of the atmosphere as two concentric shells, the boundaries of which are defined by the **stratopause**, at approximately 50 km above the Earth's surface, and a hypothetical outer limit of the atmosphere, at approximately 80 000 km. Below the stratopause, in the stratosphere and troposphere, there is 99% of the total mass of the atmosphere and it is at this level that atmospheric circulatory systems operate.

Beyond the stratopause a layer of nearly 80 000 km thick contains only 1% of total atmospheric mass and experiences ionization by high-energy, short-wavelength solar radiation.

4.1.2 Atmospheric pressure: the refinement of the model of vertical structure

Atmospheric pressure is defined as the force exerted by the atmosphere per unit area of surface. If a fluid is at rest, because of the random motion of its molecules it exerts uniform pressure in all directions. The pressure such a fluid would exert under the influence of gravity is given by the **hydrostatic equation** (Box 4.1). This defines pressure as the product of the density (ρ) of the fluid, its depth (or height above the surface, h) and the acceleration due to gravity (g):

$$P = \rho h g. \tag{4.1}$$

The unit of pressure is the newton per square metre ($N\,m^{-2}$) or the pascal (Pa). In the case of atmospheric pressure the unit still commonly used is the millibar, which is equal to $100\,N\,m^{-2}$.

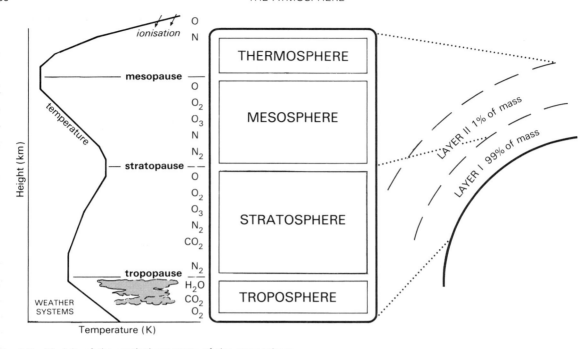

Fig. 4.3 Models of the vertical structure of the atmosphere.

If we assume that the density (ρ) of the atmosphere does not vary with height above the Earth's surface, we can infer that pressure decreases uniformly through the atmosphere (Box 4.2). However, observation has shown that atmospheric pressure does not decrease uniformly with elevation (Fig. 4.4). In the lowest 5 km, decrease in pressure is nearly uniform at a rate of approximately 100 mb km^{-1}. Above this it decreases progressively less rapidly, being about 270 mb at 10 km, 125 mb at 15 km and 56 mb at 20 km. The curve approximates to an exponential change in pressure with height. The reasons for this are that the density of the atmosphere is not constant with height. The atmosphere is readily compressed by the overlying air, such that its density is greatest at the Earth's surface and then decreases rapidly (Fig. 4.4). The relationship between pressure (P), density (ρ) and temperature (T) for dry air is given in the gas equation

$$P = R\rho T \qquad (4.2)$$

where R is the gas constant and is equal to 287 J kg^{-1} K^{-1}.

Using this equation we can modify the hydro-static equation to provide a simple illustration of the effects of decreasing density on atmospheric pressure (Box 4.3). The relationship between pressure and height may be exponential rather than linear, which approaches the observed relationship.

4.1.3 Systems structure: the horizontal component

The model of the vertical structure of the atmosphere developed so far has implied the existence of concentric shells around the Earth: shells which differ in their relative composition and in the distribution of their properties, particularly in the measurable attributes of temperature, density, volume and pressure. The fact that the vertical model varies both in different localities and with time has already been hinted at. The property which best serves to illustrate the horizontal dimensions of this variation in atmospheric properties is again atmospheric pressure.

The horizontal distribution of atmospheric pressure at the Earth's surface is represented in the form of isobars (lines of equal pressure) of sea-level pressure. Pressures recorded over land

Box 4.1

THE HYDROSTATIC EQUATION

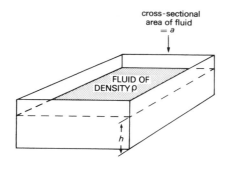

cross-sectional area of fluid = a

FLUID OF DENSITY ρ

h

The pressure exerted by a fluid at rest upon the surface beneath it is given by

$$P = \frac{\text{weight of fluid}}{\text{cross-sectional area}}$$

$$= mg/a \qquad (A)$$

where m = mass of fluid, a = cross-sectional area and g = acceleration due to gravity.

Mass can be expressed in terms of the density (ρ) and the volume (v) of the fluid as:

$$m = \rho v. \qquad (B)$$

The volume occupied by the fluid is given by the product of its cross-sectional area and its depth (h) thus:

$$v = ah. \qquad (C)$$

Substituting (C) into (B) gives:

$$m = \rho ah. \qquad (D)$$

Substituting (D) into (A) gives:

$$P = \rho ahg/a$$

$$P = \rho hg.$$

This is referred to as the **hydrostatic equation**.

surfaces are corrected to sea level using Eqn 4.13 in Box 4.3. On maps of surface pressure there are areas of relatively higher or relatively lower pressure which are referred to as high or low pressures, although there is no strictly quantitative definition of either and they are in no sense absolute terms. The location of areas of high or low pressure varies over time and space, but for convenience two types of pressure are usually identified. The first are the semipermanent areas, which appear on maps showing seasonal average pressure, whose locations are moderately predictable. The second are the ephemeral and mobile pressure cells which appear only on daily pressure charts. The former provide a key to the large-scale circulation of the atmosphere, while the latter (see Chapter 9) are more readily associated with smaller-scale movements of air.

If the Earth's surface were completely uniform, it would be possible to identify, in both hemispheres, a simple zonal pressure pattern of equa-

torial low (0°–15°), subtropical high (15°–40°), middle latitude low (40°–65°) and polar high (65°–90°). However, as the Earth's surface is an assemblage of diverse land surfaces and large expanses of ocean, atmospheric pressure does not follow a convenient latitudinal zonation.

In the distribution of sea-level pressure in July (Fig. 4.5a) values are generally low throughout the equatorial regions. To the north and south of this are the subtropical high pressure areas which, in the southern hemisphere, form a nearly continuous zonal belt. However, in the northern hemisphere, because of the presence of large areas of land surface, high pressure exists over the Pacific and Atlantic oceans, while over much of Asia and North America there are continental low pressure areas. In the middle latitudes pressure is, on average, low. However, the use of averages conceals the great temporal variation of pressure that occurs in these disturbed latitudes. In polar latitudes pressure is not consistently high, but averages tend

Box 4.2

VERTICAL PRESSURE CHANGES IN THE ATMOSPHERE: I

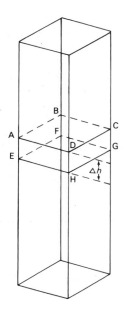

Consider a column of air of unit cross-sectional area and of density ρ.

Consider the air pressure (P_1) on surface EFGH, some height above the base of the column. This will be equal to the weight of the column of air above it (which we can call W).

If we now consider the air pressure on surface ABCD (P_2) this will be equal to the weight of the column of air above EFGH minus the weight of thickness Δh of the column.

The weight of this small section of the column is equal to $\rho g \Delta h$ (see Box 4.1). Thus:

$$P_2 = W - \rho g \Delta h = P_1 - \rho g \Delta h.$$

Therefore

$$P_2 - P_1 = -\rho g \Delta h$$

which may be rewritten as

$$\Delta P = -\rho g \Delta h$$

where ΔP represents a small incremental change in pressure.

Dividing throughout by Δh:

$$\Delta P / \Delta h = -\rho g$$

which, in the form of a differential equation, is

$$dp/dh = -\rho g.$$

If ρ remains constant, the rate of change of pressure upwards from the base of the column is uniform and decreases as height increases.

---

to indicate slightly higher pressure than in middle latitudes.

In January the major pressure areas move southwards, towards the southern hemisphere (Fig. 4.5b). The major changes from July are in the northern hemisphere, the pressure in the southern hemisphere remaining in a zonal pattern. The northern subtropical high pressure is close to exhibiting a zonal distribution, but in the middle latitudes continental high pressure over North America and Asia contrasts with the oceanic low pressures over the Pacific and the Atlantic.

The distribution of sea-level pressure is, there-fore, not necessarily one of zonality, although in the southern hemisphere (81% of which is ocean) this is closely approached. In the northern hemisphere, however, zonality is disrupted by the complex arrangement of land and sea, only 61% being ocean surface. As continental pressure centres are relatively shallow – usually less than 2 km – it is possible to filter them out of the global pressure patterns by considering pressure distribution at some height above the Earth's surface. Such distributions are not represented in terms of isobaric maps but as contoured isobaric surfaces. Contours are drawn based on the heights at which

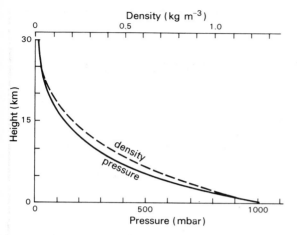

Fig. 4.4 Variation of pressure and density with height in a standard atmosphere.

certain values of atmospheric pressure are reached. The data from these are readily available from soundings taken through the atmosphere. Pressure surfaces may, for example, be drawn for 1000 mb (surface), 700 mb (about 3 km), 500 mb (5–6 km) and 300 mb (9–10 km).

Employing such a device, it is possible to re-examine the complex pressure pattern of the northern hemisphere using either 700 mb or 500 mb surfaces, as these should lie above the continental pressure areas which extend only to the 850 mb level. The 500 mb surface in Fig. 4.6 contrasts with the distribution of sea-level pressure and indicates a meridional pressure gradient southwards from the North Pole, shown by increasing contour values. This pattern, the circumpolar vortex, has a considerable bearing on the circulation of the atmosphere (Section 4.3).

Box 4.3
VERTICAL PRESSURE CHANGES IN THE ATMOSPHERE: II

From Box 4.2 we have:

$$dP = -dH\rho g. \qquad (A)$$

The basic gas equation for an ideal gas is

$$P = R\rho T. \qquad (B)$$

From (B):

$$\rho = P/RT.$$

Substituting this in (A):

$$dP = -dH \frac{P}{RT} g$$

$$\frac{dP}{P} = \frac{-g}{RT} dH. \qquad (C)$$

If we assume that for relatively small changes in height H, changes in T are relatively insignificant.

That is, T is not considered to be a function of H.

Then, by integrating equation (C) to solve for P:

$$\log P = \frac{g}{RT} H = C_1 \qquad (D)$$

where C_1 = constant.

Therefore, there is a logarithmic relationship between P and H. To determine C_1 we require a boundary condition. If $H = 0$ (i.e. sea level) when $P = P_0$ (which we shall call sea-level pressure) substituting this in equation (D)

$$\log P_0 = 0 + C_1$$

$$C_1 = \log P_0$$

Thus equation (D) becomes:

$$\log P = \log P_0 - \frac{g}{RT} H. \qquad (4.13)$$

(a) July

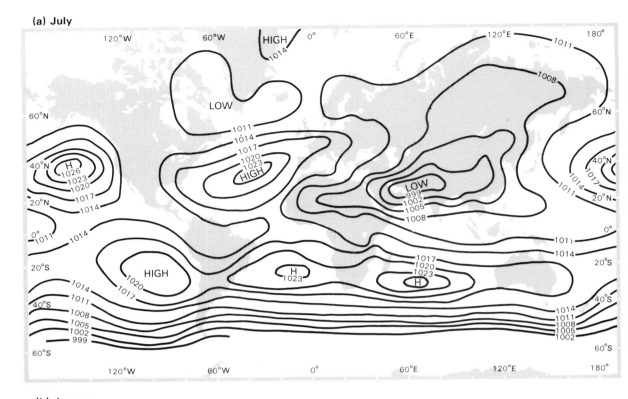

(b) January

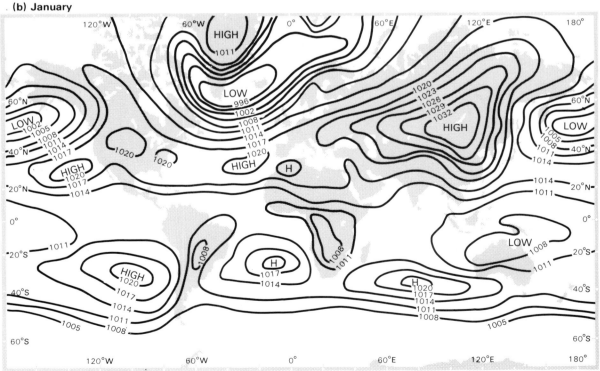

Fig. 4.5 Mean sea-level atmospheric pressure (millibars).

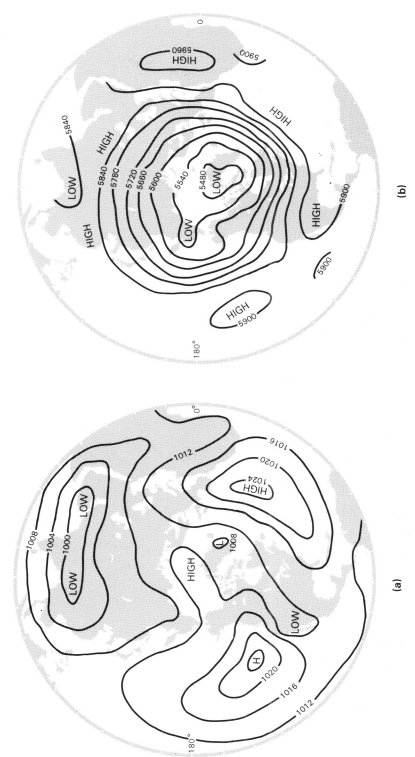

Fig. 4.6 Distribution of atmospheric pressure over the northern hemisphere in July (after Byers 1974). (a) Average sea-level atmospheric pressure; (b) Average height (m) of the 500 mb isobaric surface (after Neiburger *et al.*, 1971).

4.2   The transfer of energy and mass in the atmospheric system

In Chapter 3 we saw that there is an uneven distribution of net radiation over the Earth's surface (Section 3.3.2), some of which was seen to be redistributed by heat energy transfer in the oceans. However, there still remains a large surplus of heat energy in tropical latitudes which is redistributed polewards by the atmosphere (Fig. 4.7). This energy is transferred as sensible heat in the poleward movement of warm air, as the kinetic energy of motion and in the form of the latent heat of vaporization in water vapour (Section 4.4). The

function of mass transfer, or atmospheric circulation, is to carry out this transfer of energy. However, before attempting to produce a functional model of the way in which this is achieved we will begin by considering the forces which act on the atmosphere in the horizontal plane, and the pattern of airflow over the Earth's surface which results.

4.2.1   Forces acting on the atmosphere in the horizontal plane

In accordance with Newton's First Law of Motion (see Chapter 2) the atmosphere will remain in a state of rest or uniform motion unless a horizontal

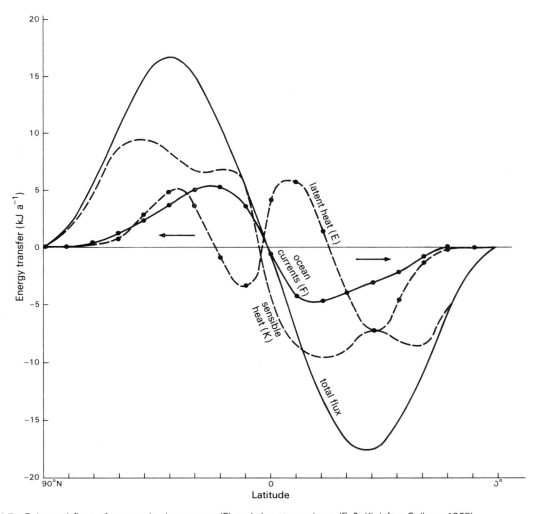

Fig. 4.7   Poleward flow of energy in the oceans (F) and the atmosphere (E & K) (after Sellers, 1965).

force is applied to it. The nature of the motion that ensues when such forces are applied is related very closely to their magnitude and direction of operation. The atmosphere, in reality, is rarely in a state of rest, but moves over the Earth's surface in response to the application of a number of forces, principal of which are **pressure gradient force, Coriolis force,** and **frictional force**.

Pressure gradient force. A difference in atmospheric pressure between two points on the same horizontal plane implies that a greater force is being exerted on one than on the other. In order to compensate for this imbalance, air is accelerated from the higher to the lower pressure, in accordance with Newton's Second Law. However, the same law states that this acceleration is directly proportional to the magnitude of the force. Magnitude, in this case, is expressed in terms of the difference in pressure between the two points, or more specifically, in terms of the pressure gradient between them.

Thus the pressure gradient force operates from high to low pressure at right-angles to the isobars and is given by

$$F_P = \frac{1}{\rho}\frac{dP}{dx} \qquad (4.3)$$

where F_P = pressure gradient force; ρ = density of the atmosphere; dP/dx = rate of change of pressure (P) over distance ($\dot{x}$) in the direction high to low pressure. In simplest terms this is indicated by the closeness of isobars on the pressure map.

If this were the only force in operation, air would flow directly from high to low pressure, and equalization of pressure differences would be extremely rapid. The fact that neither takes place indicates the existence of other forces operating upon the air, modifying the effects of the pressure gradient force.

Coriolis force. The rotation of the Earth about its own axis introduces another force affecting the movement of air over its surface, This is referred to as the Coriolis force, named after a 19th-century French mathematician. For objects moving across the Earth's surface, it operates at right-angles to the direction of motion. It therefore introduces a sideways deflection, which in the northern hemisphere is to the right and in the southern hemisphere is to the left, irrespective of

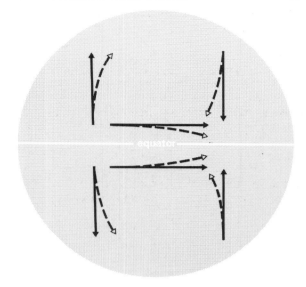

Fig. 4.8 Coriolis deflection of airflow over the Earth's surface.

the initial direction of motion (Fig. 4.8). The acceleration produced by this Coriolis force is, however, only apparent and it arises because air motion must be related to a moving surface and not a stationary one (Box 4.4).

If we were able to view the North and South Poles from the upper atmosphere, the Earth would appear as a disc, rotating anticlockwise around the former and clockwise around the latter. Applying our observations from the disc, Coriolis deflection will be to the right around the North Pole and to the left around the South Pole. However, these polar discs are not flat but hemispherical surfaces. The value of Coriolis acceleration on these hemispheres decreases outwards from the poles, and Eqn 4.14 (Box 4.4) is rewritten as

$$a = 2v\omega \sin \phi \qquad (4.4)$$

where ϕ is angle of latitude. For the Earth $\omega = 7.29 \times 10^{-5}$ radians s^{-1}. As we normally refer to force operating upon unit mass

$$\text{Coriolis force } (F_C) = 2v\omega \sin \phi. \qquad (4.5)$$

At the poles, $\phi = 90°$, so $F_C = 2v\omega$, its maximum value. At the Equator, $\phi = 0°$, so $F_C = 0$, its minimum value.

There are now two forces operating in the atmo-

Box 4.4
CORIOLIS DEFLECTION

There are two large rotating discs, one rotating anticlockwise about A, the other clockwise about B. If an object is projected towards X from either A or B, it will not reach its destination. By the time it reaches the outer rim of the disc, X will have moved to a new position X′ because

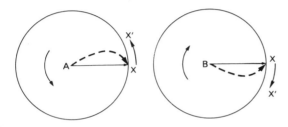

of the rotation of the disc. The object's path relative to a fixed point on the surface of the disc will be a curve, as shown by the dashed line in the figure. The deflection appears to be the result of a sideways acceleration of the object. In the case of A, the apparent deflection is to the right, and of disc B, to the left.

The value of the apparent acceleration experienced by the object is given by

$$\text{acceleration } (a) = 2v\omega \qquad (4.14)$$

where v = velocity at which the object is moving and ω = rate of spin (angular velocity) of the disc.

Box 4.5
THE GEOSTROPHIC WIND

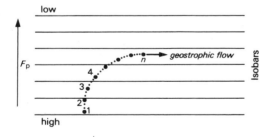

Consider a parcel of air at rest at point 1, which lies on a pressure gradient in the northern hemisphere. As velocity is zero, there is no Coriolis force but it is acted upon by a pressure gradient force. Therefore, it is accelerated according to Newton's First Law of Motion. After travelling the infinitely small distance to point 2, it has acquired velocity and there is, therefore,

a Coriolis force acting to the right of its direction of motion. The air parcel is thereby deflected from its path normal to the isobars. By point 3, it has been accelerated further by the two forces applied to it and it has an even greater velocity. Coriolis deflection is greater so the parcel experiences further deviation from its original path. By point 4, the deviation is greater again, and so on to a point n. Coriolis force thus increases as long as the parcel experiences a resultant accelerating force. By point n, however, the pressure gradient and Coriolis forces are acting in opposite directions to each other and, if they are assumed to be equal, the resultant force on the parcel of air is zero. Therefore, according to Newton's First Law, the parcel will continue in uniform motion in a straight line which, in this case, is parallel to the isobars. This is the **geostrophic wind**.

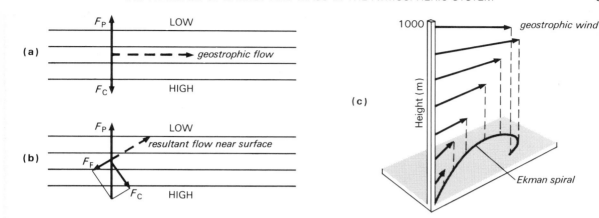

Fig. 4.9 Forces acting on air: (a) at 1000 m and (b) 10 m above the Earth's surface (F_P = pressure gradient force. F_C = Coriolis force, F_F = friction force). (c) The effects of friction on horizontal airflow in the lower atmosphere.

sphere, F_P and F_C. If these are equal, the net result of their action is to produce airflow which is parallel to the isobars and referred to as the **geostrophic wind** (Box 4.5, Fig. 4.9a). Should the isobars be curved, airflow remains parallel to them, held in a curved path by a centripetal force acting towards the pressure centre, and is known as the **gradient wind**. This will be discussed further in Chapter 9.

If $F_C = F_P$, then, from Eqns 4.3 and 4.5:

$$2\omega v_g \sin \phi = \frac{1}{\rho} \frac{dP}{dx}$$

where v_g = geostrophic wind velocity;

$$v_g = \frac{1}{\rho} \frac{dP}{dx} \frac{1}{2\omega \sin \phi}. \qquad (4.6)$$

Frictional force. The Earth exerts a retarding effect on air motion across it, which is due to the transfer of energy into the surface. Such frictional resistance reduces wind velocity in the lowest 1000 m of the atmosphere. The greatest reduction occurs in the air layers in immediate contact with the surface, where wind speeds will tend towards zero. Away from the surface, frictional effects decrease and there is an approximately exponential increase in wind speed with height (see Chapter 8).

As there is a reduction in wind speed, this must also (from Eqn 4.5) reduce the Coriolis force to a value less than that of the pressure gradient force.

The resultant airflow will, therefore, no longer be geostrophic and will be directed towards the lower pressure (Fig. 4.9b). As velocity increases away from the surface, the Coriolis force also increases and thus winds will move progressively toward geostrophic velocity and direction. This is represented in Fig. 4.9c and is referred to as an Ekman spiral.

The frictional drag exerted by the surface is affected by its aerodynamic roughness and is not constant. Over oceans, for example, frictional drag produces a 10°–20° change in wind direction near to the surface and a 40% reduction in velocity, in relation to the geostrophic wind. The greater frictional drag over land surfaces produces changes in direction between 25° and 35°, and a velocity reduction of about 60%.

The effects of these three forces is to produce patterns of airflow related to areas of high or low pressure (Fig. 4.10). Thus air converges on centres of low pressure and diverges from centres of high pressure. However, it must be remembered that the atmosphere is a complex fluid moving across a surface of highly variable physical characteristics. The forces operating within it do not, therefore, produce well ordered patterns of airflow but rather a diversity of airflows which, on first sight, appear to possess no obvious spatial organization. However, it is possible to identify movements which are spatially organized rather than random. These are divided for convenience into **primary, secondary** and **tertiary circulation systems**, according to the space and time scales over which they

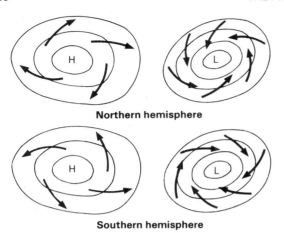

Northern hemisphere

Southern hemisphere

Fig. 4.10 Airflow associated with areas of high and low pressure.

Table 4.3 Characteristic dimensions of atmosphere motion systems (from Barry 1970, Smagorinsky 1979). (Note figures are only accurate to an order of magnitude)

	Spatial scale (km)		Timescale s
	Vertical	Horizontal	
Primary jet streams tropospheric long waves global surface winds	10^1	5×10^3	$7 \times 10^6 – 10^7$
Secondary extratropical cyclones tropical cyclones anticyclones	10^1	$2 \times 10^3 – 5 \times 10^2$	3×10^5
Tertiary squall lines thunderstorms sea breezes mountain and valley winds	$10^0 – 10^1$	$10^0 – 10^2$	$10^2 – 10^4$

operate (Table 4.3). The primary system is the movement of the whole atmosphere and the poleward transport of energy from tropical latitudes. Secondary systems operate within this primary system and are associated with the ephemeral pressure cells referred to on page 81, while tertiary systems operate on the smallest spatial scale and may be referred to as being local circulations. Both secondary and tertiary systems will be considered in detail in Chapter 9. We shall first consider more fully the primary circulation system of the atmosphere.

4.2.2 *Airflow over the Earth's surface*

A simplified view of the distribution of airflow over the surface of the Earth would identify four main wind zones in each hemisphere for an ideal global surface. Around the equator lies an area of slack winds which is associated with a broad area of low pressure, the **intertropical convergence zone** (ITCZ). Converging on this are the **trade winds** from northeast and southeast, renowned for their consistent speed and direction. To the north and south of these, beyond latitude 40°, lie the **westerlies**, associated with weather disturbances. In the polar regions are the **polar easterlies**, highly variable winds but generally northeasterly or southeasterly (northern and southern hemispheres) in direction.

The movement of air across the Earth's surface is by no means this simple, as is evident from Fig. 4.11. However, it is still possible to identify a number of global wind zones (Table 4.4). This latitudinal zonation of surface airflow is based on variations in easterly and westerly components (Fig. 4.12). Any model of the primary circulation system of the atmosphere must be capable of accommodating this observed pattern of surface air movement as well as achieving the equator to pole redistribution of energy.

4.3 The primary circulation system: approximations to a successful model

The fact that the atmosphere operates as a large heat engine transferring heat from a source (the tropics) to a sink (the poles) should, therefore, make it possible to represent its circulation as a simple source–sink–source movement of air, as suggested by Hadley in 1735 (Fig. 4.13). In this single-cell model, warm air rises in the tropics, travels northwards in the upper atmosphere and returns southwards as cooled air across the Earth's surface. These flows are adjusted to the Earth's rotating motion. As the trade winds blow against the direction of rotation of the Earth, they exert a frictional drag which must be counteracted if rotation is not to be decelerated. The westerly winds of the middle latitudes provide such a compensating drag in the direction of rotation.

This simple model may be explained by referring to Fig. 4.14: in (a) there is a uniform decrease in pressure above A, B and C which lie on a uniform horizontal plane. If the surface is heated at B and cooled at A and C, the air column above

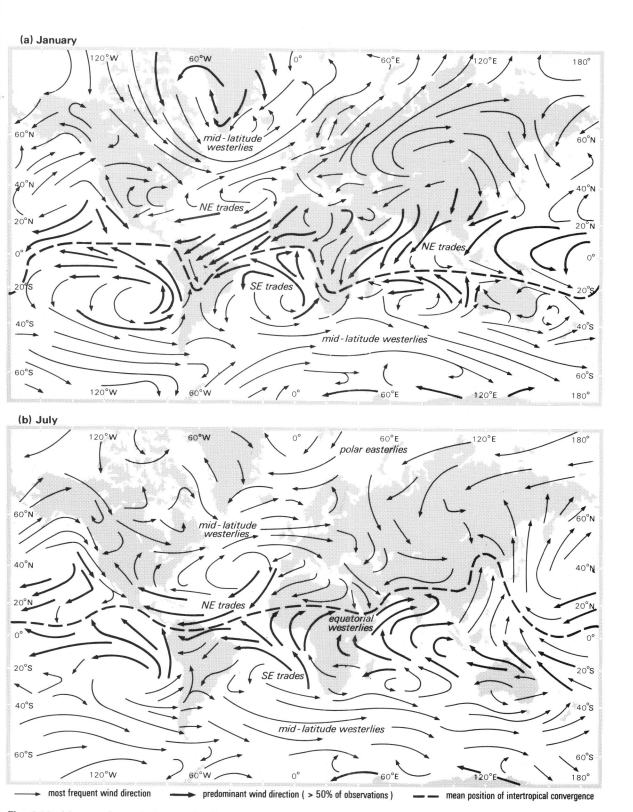

Fig. 4.11 Mean surface winds over the Earth, 1900–1950 (after Lamb, 1972).

Table 4.4 Characteristics of major global winds

	Latitudinal zone	Associated sea-level pressure conditions	Surfaces over which most distinctly developed	Average direction near to surface	Average velocity near to surface	Seasonal variation	Constancy of velocity and direction
doldrums	equatorial – may be displaced 20° from the equator, particularly in the northern hemisphere	generally low pressure with slack pressure gradients	over ocean surfaces as a discontinuous zone	highly variable	less than 3 m s^{-1}	contiguity of the doldrum zone varies, most extensively developed in March and April; disjointed in August	highly variable
equatorial westerlies	equatorial, extending to 28°, particularly over the Indian subcontinent	low, particularly monsoonal (summer) pressure conditions	primarily oceanic but of major importance over land areas of West Africa and India	SW in northern hemisphere; NW in southern hemisphere	less than 6 m s^{-1}	best developed in the summer hemisphere	locally very consistent in speed and direction
trade winds	40° to equator, reaching greatest velocity 5–20°	subsiding air associated with subtropical anticyclone	core areas located in the eastern parts of major oceans but also blowing over subtropical land surfaces	NE in northern hemisphere; SE in southern hemisphere	5–8 m s^{-1}	core areas most extensive in winter hemisphere, but greatest velocities in summer	remarkably consistent; core areas over 70% of recorded wind direction from east; over 50% in most other trade-wind areas
middle latitude westerlies	40°–65°; greatest velocities 40°–50°	variable, but generally low, steep meridional pressure gradients	over oceans disrupted by large land masses in northern hemisphere	SW to W in northern hemisphere; W to NW in southern hemisphere	to 10 m s^{-1} highest velocities in southern hemisphere	strongest in summer when meridional pressure gradient at its steepest	consistent over southern hemisphere oceans where 75% of winds are between S and SW and N and NW; in northern hemisphere less than 50% over ocean areas down to less than 25% over continental interiors
polar easterlies	poleward 65°	variable, occur in the zone between polar high pressure mid-latitude low pressure; well developed when strong, anticyclonic conditions prevail over polar region	subpolar oceans and continental margins	variable, generally from the easterly quadrant – depending on synoptic conditions	variable, little information		variable, dependent very much on synoptic conditions in either middle latitude or over polar ice caps; modified by local katabatic flow in Antarctica

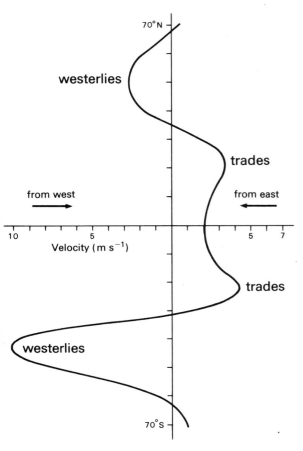

Fig. 4.12 The major zonal winds of the world (after Riehl, 1965).

The heating of air from below, in causing the vertical expansion of the overlying air column, is effectively increasing its potential energy, which is then converted into kinetic energy as motion begins. The simple primary circulation system is represented in Fig. 4.15. A state of dynamic equilibrium is reached in this circulation such that the motion of the air is just adequate to maintain a mean heat balance in the atmosphere. Should the meridional temperature gradient change, negative feedback operates to restore equilibrium. Thus an increased temperature gradient speeds up the circulation of the air and hence the redistribution of heat energy, restoring dynamic equilibrium. A decrease in temperature gradient slows down the circulation and thereby limits the redistribution of heat energy, again restoring equilibrium.

Unfortunately this single-cell model does not fully explain the surface distribution of winds. The main flaw lies in its consideration of the effects of the Earth's rotation. As air moves polewards the radius of its rotation about the Earth's axis decreases. In order to conserve angular momentum the air would have to experience a progressive increase in eastward velocity relative to the Earth's surface (Box 4.6). If there were only one cell, poleward-moving air would be flowing at such extreme velocity by latitude 30° that it would become turbulent, the meridional flow breaking down into eddies. The **Hadley cell** must, therefore, be limited to low latitudes, and alternative mechanisms must be found to achieve the necessary poleward transport of angular momentum.

An alternative to the single-cell model view of the circulation of the atmosphere is the three-cell model where, in place of a single convectional cell, there are three interlocking cells between the tropics and the poles (see Fig. 4.13). The first of these is the low-latitude Hadley cell operating between equatorial regions and 30° north and south, which is a convectional or thermally direct cell. Warm air rises in the intertropical convergence zone, travels polewards and subsides in the subtropical anticyclones, whence it returns westwards to low latitudes. Winds on the surface would be the easterly trades and, in the upper atmosphere, the westerly antitrades. The second cell is a **middle-latitude cell** operating between latitudes 30° and 60°. Air diverges from the subtropics and flows polewards and eastwards across the surface, eventually converging with cold polar air at the

the former expands, while above the latter it contracts. The isobaric surfaces become distended (as in (b)) producing a pressure gradient at the upper levels. The air begins to move in the direction of this pressure gradient, creating divergent airflow at B^1 and convergent airflow at A^1 and C^1(c). The divergence at B^1 induces convergence at B, while convergence at A^1 and C^1 induces divergence at A and C(d). We have, therefore, a convectional circulation of air initiated by surface temperature differences between points.

This theory can be applied to the atmosphere above the curved surface of the Earth. Heating at the equator and cooling at the poles should produce two convectional cells, as indicated in Fig. 4.13. This single-cell circulation is accelerated by a heat energy imbalance and decelerated by friction.

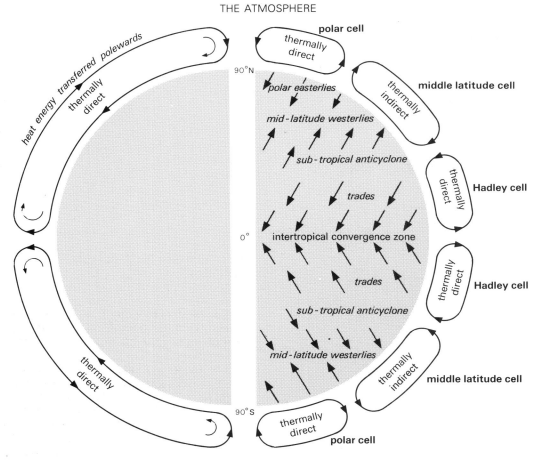

Fig. 4.13 Single-cell and three-cell models of the primary circulation of the atmosphere.

polar front where it rises to the upper troposphere. From here it returns equatorwards aloft. As warm air is effectively sinking at 30° and cold air rising at 60°, this cell is thermally indirect. The third cell is the **polar cell** which is thermally direct, relatively warm air rising at latitude 60° and cold air sinking over the poles. The upper airflow is westerly, while the surface flow is easterly.

The operation of this three-cell model appears to explain the global distribution of surface airflow in that the Hadley cells provide the two trade winds and the intertropical convergence zone, the middle-latitude cell the westerlies and the polar front, and the polar cell the polar easterlies. Surface winds accord with the observed distribution of sea-level atmospheric pressure and the operation of pressure gradient, Coriolis and frictional forces.

In terms of the operation of the primary circulation system, energy transfer is not in a simple single circuit (Fig. 4.15) but is a more complex series of transfers. The two thermally direct cells may be represented as simple energy circuits, but the middle-latitude cell is thermally indirect and must be driven by the other two cells. Thus the main input here is of kinetic energy transferred from the Hadley and polar cells.

During the 20th century, as more information has been gained about the winds in the upper troposphere, the limitations of the three-cell model have become apparent. Observed airflow in the upper troposphere does not match that of the model, particularly in the low and middle latitudes. Above the surface westerlies in middle latitudes, winds also blow from a westerly direction, and not easterly as the three-cell model suggests.

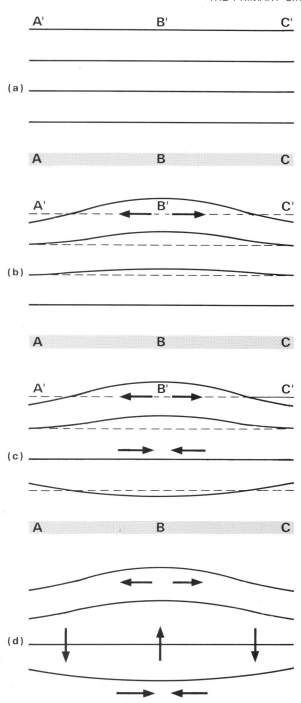

Fig. 4.14 The development of convectional cells.

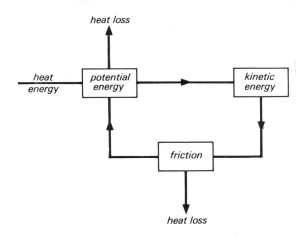

Fig. 4.15 System model of single-cell circulation of atmosphere.

Above the **trade** winds the antitrades do show the required reversal of direction but are often extremely weak, clearly not matching their surface counterparts.

In order for the three-cell model to work, it requires that the middle-latitude cell be driven by the polar and Hadley cells. Not only is the polar cell too weak to fulfil this function, but meridional pressure gradients are at their steepest in both the lower and upper troposphere in middle latitudes. The middle-latitude cell is thus, in contradiction of the three-cell model, the strongest element of the atmospheric circulation and is fundamental to the poleward redistribution of energy.

Examination of the isobaric surfaces in the upper troposphere has revealed a pressure distribution in the form of a circumpolar vortex (Section 4.1.3), in which pressure decreases away from the poles. Pressure reaches a maximum at latitude 20°, whence there is a small decrease into equatorial latitudes. Geostrophic winds blow as circumpolar westerlies between the poles and latitude 20°, and as an easterly flow over equatorial latitudes.

Within the circumpolar westerlies there are two well defined bands of extremely strong winds referred to as the **jet streams**. One of these, the subtropical jet stream, has a circumpolar path between latitudes 20° and 35°. The other, the polar front jet stream, has a path between latitudes 35° and 65° (Fig. 4.16). The subtropical jet stream blows between 12 km and 15 km above the Earth's

Box 4.6
ANGULAR MOMENTUM

Momentum is a property which a body possesses by virtue of its motion and it is given by

momentum = mass × velocity (see Chapter 2).

The law of conservation of momentum states that momentum can be neither created nor destroyed.

Angular momentum is a property which a body possesses by virtue of its motion around a central point.

If there is no relative motion between the atmosphere and the Earth, they are both in solid rotation. In this case, the angular momentum of unit mass of the atmosphere near to the Earth's surface relative to this axis is given by

angular momentum = $\Omega\, r^2 \cos^2 \phi$

where Ω is the angular velocity of the Earth's rotation, r is the radius of the Earth, and ϕ is

the angle of latitude. There is, however, atmospheric motion relative to the Earth. If air is moving zonally with velocity u, then it possesses angular momentum relative to the Earth given by

angular momentum = $ru \cos \phi$.

Therefore, if angular momentum is to be conserved by air moving zonally, there must be an increase in velocity if there is a poleward component of motion.

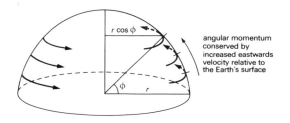

angular momentum conserved by increased eastwards velocity relative to the Earth's surface

surface and, within its narrow core, average wind speed is in excess of 65 m s⁻¹. During winter it is well defined, but in summer it is considerably weakened and may become discontinuous.

As air moves polewards in the Hadley cells, it acquires an increased eastward velocity of the order of 60 m s⁻¹ by latitude 30° in conserving angular momentum, relative to the Earth's surface. As already outlined (Box 4.6), the poleward transfer of angular momentum by increasing velocity of airflow in this way must be limited to lower latitudes, the high-velocity subtropical jet stream marking this limit. As the subtropical jet stream lies above high surface pressure, the air within it subsides towards the surface, where its energy is either dissipated by friction or transferred polewards or equatorwards. The polar front jet stream is located where meridional temperature gradients are steep in higher latitudes, generating a strong upper westerly flow. Mean wind speeds in the jet stream core, which lies between 10 and 12 km

above the ground surface, are 25 m s⁻¹. It tends to be discontinuous and follows a meandering, highly variable path.

Thus a more realistic representation of the atmospheric circulation in its meridional plane is shown in Fig. 4.16. In the northern hemisphere the major feature of the circulation is the Hadley cell, while north of this the thermally indirect cell of the middle latitudes is now much smaller and is confined to the middle and upper troposphere. However, this circulation in itself does not entirely explain how poleward transport of angular momentum and heat energy is achieved beyond latitude 30°. Beyond this, the atmosphere develops eddying motions. It is these which effect the poleward transport of heat and momentum. By using a laboratory model, Hide (1969) has provided an indication of what form this eddying motion may take (Box 4.7). For slow rotation, the movement of the fluid in the dishpan followed a single convectional cell, simulating a Hadley cell operating

Box 4.7
THE DISHPAN EXPERIMENT

Two concentric cylinders enclose a fluid (usually water or glycerol) in which there are small polystyrene granules. The outer cylinder is warmed and the inner cooled, thereby establishing a temperature gradient across the fluid which is analogous to that which exists in the atmosphere between tropical and polar latitudes in one hemisphere.

is symmetrical about its central axis. In contrast, at a rotation of 4 rad s^{-1} there is a well developed wave motion in the fluid.

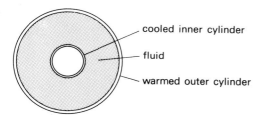

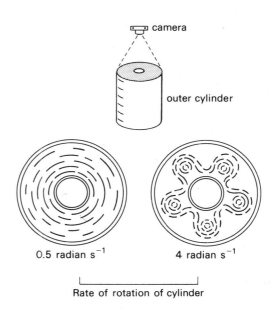

0.5 radian s^{-1} 4 radian s^{-1}

Rate of rotation of cylinder

The movement of the fluid is recorded by photographing the paths taken by the polystyrene granules as the temperature gradient is kept constant and the cylinders are rotated at various speeds.

At low rates of rotation, in the region of 0.5 rad s^{-1}, the movement of the fluid in the chamber

---

between equator and poles. However, as the rotational velocity was increased, thereby changing the Coriolis parameter, the convectional cirulation was unable to maintain the flow of heat and momentum from the warm rim to the cool centre of the dish. In place of the convectional cell, a wave motion was initiated which effectively carried out these transfers.

The isobaric surfaces for the upper troposphere reveal a number of identifiable waves in the upper westerlies, referred to as **Rossby waves**. The number of these waves and the degree of their latitudinal oscillation vary considerably over time. They can develop and decay over relatively short periods of time – as little as a month or two (Fig. 4.17). However, it is possible to identify average positions of the waves over extended periods of

time, and this wave pattern is essential to the atmospheric circulation. Although net meridional airflow is zero in these waves, they fulfil the functions of transferring heat energy and angular momentum away from the subtropics (Box 4.8).

The primary circulation system of the atmosphere is thus considerably more complex than is implied by the single- and three-cell models. However, whatever the model, it can never fully represent what is, in reality, an extremely efficient yet mechanically complex heat engine. The atmosphere manages to maintain a global heat energy balance by effectively redistributing the surplus from lower latitudes. That it does so above a surface whose physical characteristics are highly variable over both time and space adds further to the intricacy of its mechanism.

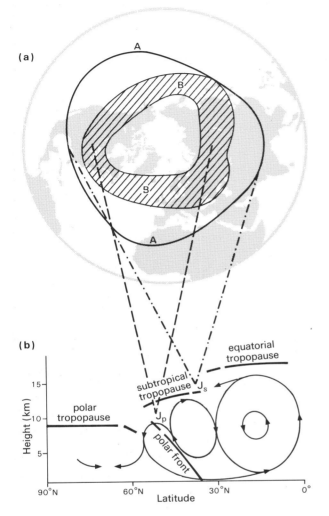

(a)

(b)

Fig. 4.16 The jet streams of the northern hemisphere. (a) Mean winter position of the axis of the subtropical jet stream, J_s (A) and the area of activity of the polar front jet stream, J_p (B) (after Riehl 1965). (b) Mean meridional movement of air between equator and pole (after Palmen, 1951).

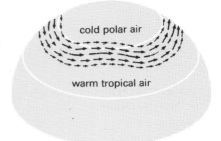

(a) Jet stream begins to undulate

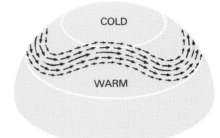

(b) Rossby waves begin to form

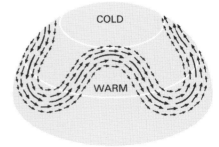

(c) Waves strongly developed

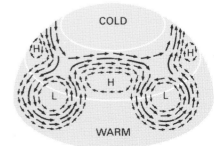

(d) Cells of cold and warm air bodies formed

Fig. 4.17 Waves in the upper westerlies (Strahler and Strahler, 1973).

We have so far considered only the movements of dry air. The presence of water vapour in the air involves additional exchanges of heat energy, in that its extraction from the air releases heat and its addition consumes heat. Thus, in transporting water vapour the atmosphere is also transporting a potential source of heat energy. Water in the Earth–atmosphere system as well as the hydrosphere will be considered in Chapter 6.

Box 4.8
HEAT AND MOMENTUM TRANSFER IN WAVE MOTION

Consider air moving through the wave below, in the northern hemisphere.

(a) If air temperatures at latitude points x_1, x_2, x_3 are greater than at y_1, y_2, y_3, there must be a net loss of heat in the poleward movement of the air.
(b) If we consider the meridional and zonal components of air motion in the wave at points x_1 and y_1, then along the rising limb of the wave air has both northward and eastward components. Along the falling limb, the air has southward and eastward components. We usually find that the northward component along the rising limb and the southward component along the falling limb are roughly equal, meaning that there is no net meridional flow of air, i.e. there is no loss of momentum in this plane. However, if we were to compare the eastward, or zonal, components we would find a considerable reduction in velocity, and hence in momentum. In travelling through the wave the air will, therefore, have lost eastward momentum, effectively transferring this in a poleward direction.

Further reading

Atkinson, B.W. (ed) (1981) *Dynamical Meteorology: an Introductory Selection*. Methuen, London.

Eliassen, A. and K. Pedersen (1977) *Meteorology: an Introductory Course. Vol. 1, Physical Processes and Motion*. Scandinavian University Books, Oslo.

Hanwell, J. (1980) *Atmospheric Processes*. Allen & Unwin, London.

Lockwood, J.G. (1979) *Causes of Climate*. Edward Arnold, London.

Lutgens, F.K. and E.J. Tarbuck (1982) *The Atmosphere*. Prentice-Hall, Englewood Cliffs.

Neiburger, M., Edinger, J.D. and W.D. Bonner (1982) *Understanding our Atmospheric Environment*. Freeman, San Francisco.

Thrush, B.A. (1977) The chemistry of the stratosphere and its pollution. *Endeavour*, **1**, 3–6.

Walker, J.C.G. (1977) *Evolution of the Atmosphere*. Macmillan, New York.

Warneck, P. (1988) *Chemistry of the Natural Atmosphere*. Academic Press, London.

The Lithosphere

In Chapter 3 we considered the gross structure of the Earth, albeit briefly, in order to understand the sources of the planet's internal energy. This structure of concentric shells, differing in chemical and mineralogical composition and in physical properties, evolved from the original uniform aggregation of particles by the process of differentiation under the influence of gravity. The least dense materials became segregated in the thin outer shell of the crust, while the denser iron and nickel settled towards the centre of the Earth to form the core. Table 5.1 illustrates the composition of the Earth and the depth of the layers within it. On the basis of mineralogy, a distinction is made between crust, mantle and core, each separated by a major boundary and each further subdivided. The crust, it should be noted, is very thin in comparison with the other layers, and constitutes only 1.55% of the Earth's total volume. It is this thin layer, and primarily its surface characteristics, with which this book is largely concerned. The bulk of the Earth's mass is composed of mantle, of a density between that of crust and core.

Within the mantle, at a depth of about 50 km, geologists recognize a discontinuity separating the rigid crust and upper mantle from the lower mantle, which displays the dual properties of a viscous fluid and an elastic solid. This chapter is largely concerned with the upper zone (or **lithosphere**) but consideration will also be given to the plastic zone of the mantle which extends down to about 250 km and is termed the **asthenosphere**. Below this is the **mesosphere**. It is important to note that these major subdivisions are based on the mode of behaviour of the materials involved and are quite distinct from the mineralogical subdivision of crust, mantle and core (Fig. 5.1).

Table 5.1 Structure and composition of the Earth

Layer	Depth to boundary (km)	Volume (%)	Composition	Major subdivisions
crust (continental) (oceanic)	av. 33 10–11 Mohorovičić discontinuity	1.55	mainly granitic basaltic	lithosphere
upper mantle	ca 50		peridotite	
viscous layer	ca 250		partly fused peridotite	asthenosphere
lower mantle	2900	82.25	high density peridotite	
outer core	Gutenberg discontinuity 5000		iron–nickel, liquid	mesosphere
inner core	6371 (centre)	16.20	iron–nickel, solid	

5.1 The crustal system

The lithosphere can be regarded as a system whose concentric boundaries are the outer surface of the solid Earth and the discontinuity within the mantle. The outer boundary forms a complex interface with the atmosphere and hydrosphere and is also the environment in which life has evolved. The inner boundary is adjacent to rock, which is near its melting point and is capable of motion relative to the lithosphere above. For the remainder of the book we shall refer to the system just defined as the **crustal system**. It is an open system and hence there are exchanges of matter as well as energy across both its outer and inner boundaries. At its outer boundary the crustal system is responsible for the structure and distribution of the continents and ocean basins and for the major relief units within them. As a prelude, therefore, to a closer examination of the systems operating at the Earth's surface, we shall consider the transfer of matter and energy within the crustal system and across its boundaries. In particular we shall consider on a broad scale the functional relationships between it and the denudation systems at the surface of the planet.

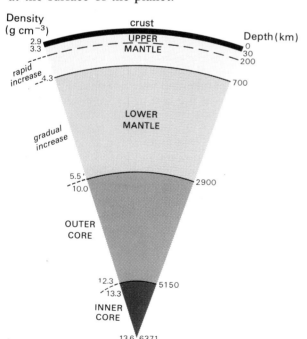

Fig. 5.1 A schematic section through the crust, mantle and core.

5.2 System structure

5.2.1 Chemistry and mineralogy of the lithosphere

The elements of the crustal system are the chemical elements from the atoms of which the minerals and rocks of the lithosphere are composed. Table 5.2 shows the elemental composition of the crust for the eight most abundant elements as a percentage by weight, atomic composition and volume. Oxygen is by far the most abundant. For a volatile element such as oxygen to be held in the solid crust, in quantities which greatly exceed its presence as free molecules in the atmosphere, it must be firmly bonded to other constituents of the lithosphere. Indeed, the atoms of oxygen and those of the other elements are combined, largely by covalent and ionic bonds (see Chapter 2), into rock-forming minerals. These in turn combine in characteristic mineral assemblages to form the major rock types. The second most abundant element in the lithosphere is silicon, so it is not surprising that most of the oxygen is in combination with it to form silicate minerals. The basic unit of these minerals is the silica tetrahedron (Box 5.1).

The subsequent combination of these units in silicate minerals represents a compromise between the geometrical constraints imposed by the tetrahedral structure and those associated with the net charge on the unit. There are two such compromises: either the oxygens of adjacent units are shared – hence reducing the negative charge deficit

Table 5.2 Elemental composition in percentages of the Earth's crust (after Mason 1952)

Element	Weight (%)	Volume (%)	Atomic composition (%)
oxygen	46.6	93.77	62.5
silicon	27.7	0.86	21.22
aluminium	8.13	0.47	6.47
iron	5.00	0.43	1.92
magnesium	2.09	0.29	1.84
calcium	3.63	1.03	1.94
sodium	2.83	1.32	2.64
potassium	2.59	1.82	1.42
others*	1.43	0.01	0.05

* In this category is included a wide variety of elements of very low abundance. These are termed **trace elements**, and are often of great significance in the biosphere

Box 5.1
SILICATE MINERAL STRUCTURES

Most of the common rock-forming minerals of the Earth's crust are silicates, formed by the combination of silicon and oxygen with one or more of the abundant metals.

The fundamental unit of silicate minerals is the silica tetrahedron, in which a central silicon atom is linked to four oxygen atoms. The small silicon ion (Si^{4+}) and four large oxygen ions (O^{2-}) pack together to form a pyramid (SiO_4). This forms a strongly bonded structure and the bonds are usually considered to be covalent.

The silica tetrahedron is a complex ion in which the association of four oxygen ions, each with two units of negative charge, and one silicon ion with four positive units, leaves the resultant tetrahedral ion with a net negative charge of four units. To form an electrically neutral unit, either the tetrahedral ion must be bonded with additional positive ions (for instance magnesium, Mg^{2+}, or iron, Fe^{2+} in the case of the mineral olivine, $(Mg, Fe)_2SiO_4$), or it must share the oxygen ions at the corners with adjacent tetrahedra.

Adjacent silica tetrahedra can be linked in different ways by the sharing of two or more oxygen atoms (Fig. 5.2). Various degrees of oxygen sharing occur, the result of which is the grouping of the tetrahedra into chains (single or double), sheets or three-dimensional lattices. In combination with different mineral elements these varied structures give rise to a wide variety of rock-forming minerals.

Inosilicates

In the **inosilicate** group of minerals, the silica tetrahedra are linked into chains. In the **pyroxenes**, single chains are formed by the sharing of two oxygens between adjacent tetrahedra. The chains are of indefinite length with an Si:O ratio of 1:3. This leaves a double negative charge for each tetrahedron, which is satisfied by a single divalent ion such as Mg^{2+}, Al^{2+} or Fe^{2+}. In the **amphibole** group of minerals the lattices of the tetrahedra are linked into double chains, again of indefinite length. This gives an Si:O ratio of 4:11 and the electrical balance is formed by the addition of metallic cations, together with hydroxyl (OH^-) groups.

Phyllosilicates

In the **phyllosilicate** group, the tetrahedra are linked to form sheets of indefinite lateral extent, with an Si:O ratio of 2:5. This commonly results in a hexagonal pattern of atomic structure in the plane of the sheet. Cations and water are accommodated between the sheets. This group contains a large number of commonly occurring minerals – the clay minerals – and mica and possesses a very strong cleavage parallel to the plane of the sheets.

Tectosilicates

The greatest degree of oxygen sharing is attained in the **tectosilicates**, in which every oxygen is shared between adjacent tetrahedral groups to form a three-dimensional lattice. The Si:O ratio is thus 1:2 and at its simplest this forms the mineral quartz (SiO_2). Structurally similar is the abundant feldspar group of minerals, in which a proportion of the silicons are replaced by metallic cations. Thus, in the alkali feldspars one silicon in four is substituted by aluminium. This causes a deficiency of one positive unit of charge per four tetrahedra. The electrical balance is achieved by a single K^+ ion in the case of the mineral **orthoclase** ($KAlSi_3O_8$) and Na^+ in the case of **albite** ($NaAlSi_3O_8$) which are accommodated within the lattice structure. Tectosilicate minerals tend to be hard and do not cleave easily.

– or a neutral mineral is formed by using the positive charges of other metallic cations to balance the negative charge on the tetrahedra. In practice, both occur together in almost all silicate minerals. The degrees to which tetrahedra are linked in a mineral structure form a sequence from unlinked tetrahedra, through single chains, double chains and sheets, to continuous three-dimensional networks (Box 5.1, Fig. 5.2). The more the tetrahedra are linked in a mineral structure, the greater is the reduction in the negative charge imbalance and the smaller the number of additional cations necessary to produce a neutral mineral (Fig. 5.2).

The manner in which these mineral structures are formed and their relation to the processes operating in the crustal system will be discussed in this chapter. For the moment, suffice it to say that they occur in particular combinations and amounts in response to the conditions under which they formed, and it is these mineral assemblages that we recognize as rock types. Of the rocks composing the lithosphere, over 95% by volume are of igneous origin (Box 5.2) and consist of primary silicate minerals (Table 5.3). Metamorphic and sedimentary rocks which make up the remaining 5%, however, can also be thought of as having been derived from these same primary silicate minerals of igneous rock. Igneous rocks are dominant in terms of volume, but the sedimentary rocks (shale, sandstone and limestone) are exposed over 70% of the land surface area of the crust, where they occur as a thin veneer, with igneous rocks occupying only 18% of the area (Table 5.4). In addition to their chemical and mineralogical properties, igneous, metamorphic and sedimentary rocks all possess important physical and mechanical attributes, which also govern their response to processes operating both within the crustal system and at its interface with the atmosphere. However, a consideration of such properties will be deferred until later in the chapter. Partly on the basis of rock type and mineralogy, the lithosphere is divided into three gross structural units: the **continental crust**, the **oceanic crust** and the **upper mantle**.

5.2.2 Gross structure of the system: major components

Continental crust. The continental crust is composed of granitic rocks rich in silicon and aluminium and with a mean density of 2.8 (Box 5.2).

These rocks form a discontinuous outer shell to the planet and they underlie the continents. Each continent has a core of ancient Precambrian crystalline rocks of metamorphic or igneous origin, which are structurally very stable and are termed **cratons**. This long-term stability is a property which has recently prompted their exploration as possible sites for the disposal of nuclear waste. The rocks of these core (or shield) areas are the oldest exposed at the surface, often with ages in excess of 2000 Ma (Box 5.3). In North America this ancient core is represented by the Laurentian Shield, in Europe by the Fenno-Scandian Shield, and much of central Africa consists of three major shields. Sometimes these crystalline shields have a veneer of derived sedimentary rocks. This mantle of younger sediments is, however, largely undeformed since they rest on a stable crustal base and are hence protected from lateral pressure. This is not true of sediments deposited at the margins of cratons.

The cratons are rimmed and separated by more mobile portions of crust. These zones (up to 300 km across) are known as **orogens**. They are more readily deformed by crustal pressure and are the sites of seismic activity and crustal folding – termed **orogeny**. Here sedimentary rocks are deformed into fold mountains and are metamorphosed and intruded by igneous rocks. These areas of complex geological structure generally form strongly linear patterns in their surface expression and, although old fold mountains may have been reduced to low elevations, the zones of more recent activity stand out as areas of high relief (>4000 m relative relief). Indeed, the surface of the continents which, remember, is part of the outer boundary of the crustal system, is highly varied in topography. The mean land elevation of 870 m above sea level (asl) (Fig. 5.3) is strongly influenced by these orogenic zones, for 70% of the continental surface lies below 1000 m asl. The highly skewed distribution of land surface elevation in Fig. 5.3 highlights the existence of steep gradients in certain continental areas.

Oceanic crust. The oceanic crust differs petrologically from the continental crust above it. It is basaltic in composition, consisting of more basic minerals, and has a mean density of 3.0. It is structurally simple and nowhere are oceanic basalts older than the Mesozoic (225–65 Ma) (see Box 5.3). This relative youth of the oceanic crust

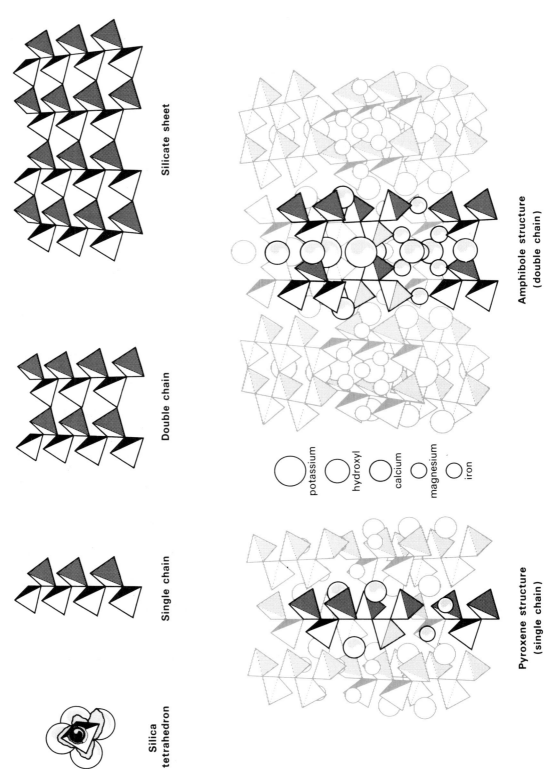

Silica tetrahedron

Single chain

Double chain

Silicate sheet

potassium

hydroxyl

calcium

magnesium

iron

**Pyroxene structure
(single chain)**

**Amphibole structure
(double chain)**

Fig. 5.2 The structure of silicate minerals, showing the silicate framework built up from silica tetrahedra and the inclusion of non-framework ions (see text).

Box 5.2
MAJOR ROCK TYPES

The rocks of the Earth's crust can be conveniently grouped into three major classes, according to their mode of origin.

Igneous rocks are formed by the cooling and crystallization of molten rock (magma) or mineral fluids derived from the mantle. They are formed either as intrusive rocks, where the magma is injected into existing crustal material, or as extrusive rocks, which are formed by ejection at the Earth's surface, as for example volcanic material. Intrusive rocks form at depth and they crystallize slowly, forming large crystals and a coarse-grained rock. The extrusive rocks, cooling rapidly at the Earth's surface, are generally fine-textured.

With few exceptions, igneous rocks are composed of interlocking crystals of primary rock-forming minerals. The mineral composition may vary considerably depending on the particular minerals present, and a distinction is frequently made between acidic and basic rocks.

The mode of formation of igneous rocks means that there are few voids within the rock, and this low porosity is often associated with a high mechanical strength.

Sedimentary rocks are formed by the denudation of existing rocks, broken down by weathering processes and transported in the form of detrital grains. The sediments are transported across land masses and the bulk of them is carried to the oceans, where they accumulate on the continental margins. More restricted and isolated sediment traps exist on the continental surfaces, for example lakes and river valleys.

The detrital grains of which sedimentary rocks are formed are frequently composed of resistant minerals, such ás quartz. Other sedimentary rocks are formed of the accumulation of organic materials, for example chalk and limestone, which consist of the calcareous shells and skeletons of marine creatures.

Sediments are transformed into sedimentary rocks by the process of **lithification**, the compression and compaction of the sediments associated with the extrusion of water. The detrital grains may subsequently become cemented by deposition of salts from percolating waters to form an indurated rock.

Sedimentary rocks are typically stratified, variations in composition and texture resulting from changes in depositional conditions.

Metamorphic rocks are formed by alteration of previously existing rocks, themselves of igneous, sedimentary or even metamorphic origin. Conditions of high temperature and/or high pressure may develop in the crust and these cause mechanical deformation or, more commonly, chemical recombination of the elements in the rock-forming minerals. Metamorphic rocks exist as many types, depending on the original minerals present and the type of metamorphism (whether thermal or dynamic) or a combination of both. A great variety of both mineralogy and texture is possible within the metamorphic group. Typical examples are slates and schists, representing lower and higher grades of metamorphism respectively.

Table 5.3 Minerals exposed on the land surfaces of the Earth

Mineral	Exposed area (%)
feldspars	30
quartz	28
clay minerals and mica	18
calcite	9
iron oxides	4
others	11

Table 5.4 Rock types exposed on the land surfaces of the Earth

Rock type	Exposed area (%)
shale	52
sandstone	15
granite	15
limestone	7
basalt	3
others	8

is emphasized by the fact that no sediments older than Jurassic rest upon it. It forms the outer boundary of the system only on the floor of the ocean basins, which by implication are considerably younger features than the continents.

The topography of the floor of the ocean basins is generally much more subdued than that of the continents and does not have the local variations of relief in relation to area which the continents possess. Submarine elevations fall off quite sharply around the margins of the continents, but 85% (by area) of the ocean floor lies between 3 km and 6 km below sea level. There are, however, two features of ocean floor topography that are

quite distinctive – the mid-ocean ridges and the deep submarine trenches. The mid-ocean ridges form an interconnecting worldwide chain over 60 000 km in length, found in all major ocean basins, although not always in a central position (Fig. 5.4). Between 500 and 1000 km in breadth and reaching heights of up to 3000 m, with a central trough or rift along the centre line, they are less pronounced features than most of the orogenic zones of the continents. One characteristic feature of these ridges is that they are associated with anomalously high geothermal heat flows. Also strongly linear features, but of more restricted area (1% of the ocean floor) are the deep ocean

---

Box 5.3
THE GEOLOGICAL TIMESCALE

Era	Period	Age of base (10^6 a)
Cenozoic	Quaternary	2
	Tertiary	65
Mesozoic	Cretaceous	135
	Jurassic	200
	Triassic	240
Palaeozoic	Permian	280
	Carboniferous	370
	Devonian	415
	Silurian	445
	Ordovician	515
	Cambrian	600
Precambrian	Proterozoic	2500
	Archaean	3900

Sedimentary rocks occupy the bulk of the exposed rocks in the geological succession from the Cambrian period onwards. The Archaean basement rocks are generally metamorphic rocks of ancient and complex history. The oldest known crustal rocks, from Greenland, have an age of 3.9×10^9 a. The age of the Earth is estimated at 4.5×10^9 a. The majority of the world's present major zones of high relief date from orogenic constructional episodes during the Tertiary period, and surface erosion and deposition during the Quaternary.

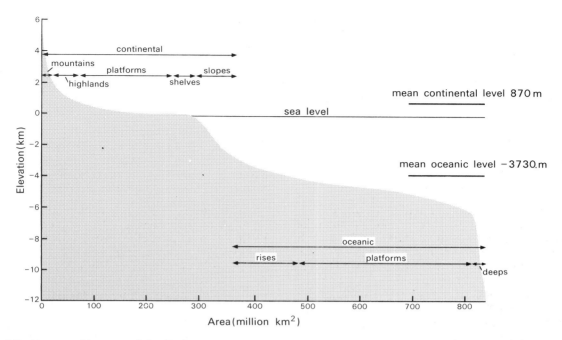

Fig. 5.3 Hypsographic curve of the Earth.

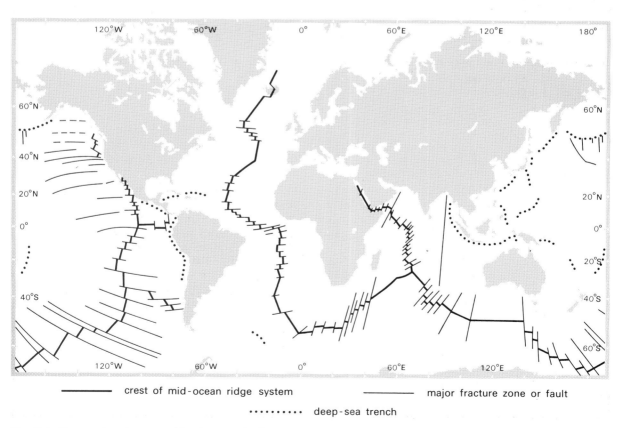

——————— crest of mid-ocean ridge system   ——————— major fracture zone or fault

· · · · · · · · · · · deep-sea trench

Fig. 5.4 Structural and topographic elements in the ocean basins (from *Continental drift* by J. Tuzo Wilson, © 1963 by Scientific American, Inc., all rights reserved).

trenches. Located close to the margins of the ocean basins, particularly around the Pacific ocean, they attain depths of 10 000–15 000 m. Here the trenches often contrast strongly with the emergent tips of submarine mountains, with which they are associated and which protrude above sea level as island arcs. That ocean trenches are zones of crustal instability is indicated by their coincidence with plots of deep earthquake epicentres. Perhaps the most remarkable feature of the ocean basins is not the mid-oceanic ridges, or the deep trenches, but the bilateral symmetry which they show in relation to these features, and particularly to the ridges (Table 5.5). This is best exemplified by the Atlantic ocean basin (Fig. 5.5).

Now that we have discussed the structure of the crustal system at scales appropriate to both the individual rock-forming mineral and the continental land mass, we are in a position to look at the working or function of the system.

5.3   The crustal system: transfer of matter and energy

5.3.1   A geophysical model
The gross structural units described in the preceding section are so organized as to combine to form a large-scale functioning system which is responsible for the formation and destruction of crustal

Table 5.5   Type of symmetry about mid ocean ridges

age of oceanic crust rocks	become progressively older away from ridges, particularly well exemplified by accessible rocks of oceanic islands
age of ocean floor sediments	become progressively older away from ridges; repeats and confirms pattern of age symmetry displayed by crustal material
magnetic anomalies	variations in Earth's magnetic field arranged in bands parallel to each other and to ridge axis
magnetic polarity reversals	patterns of normal and reversed polarity, as preserved in the palaeomagnetism of oceanic crustal rocks; show bilateral symmetry about ridges
continental margins	topographic and stratigraphic congruence of some opposed continental margins

material and for the fragmentation and distribution of the continents. The processes responsible for the transfer of crustal material on such a large scale operate but slowly on the human timescale. Consequently, they require sophisticated scientific observation for their detection and it is only as a result of recent suboceanic research that they have begun to be appreciated. Even now, processes deep within and beneath the crust (especially the continental crust) are in part the subject of speculation, but logic suggests that they must take place in order to account for the directly observed events, while the ancient rocks themselves provide evidence for deep-seated processes.

The key to the operation of the crustal system is the recognition of the process of sea-floor spreading. New basaltic material from the mantle is intruded into the crust along the central rift of a mid-ocean ridge. There it solidifies to form new crustal rock. As this new material is added to the crust, existing crust is pushed aside laterally in both directions in order to accommodate it. Thus the high heat flow of the ocean ridge is explained by the proximity or presence of hot mantle material beneath the ridge. The pattern of progressively older crustal material (and sedimentary material overlying it) away from the ridge axis is explained by the lateral translation of crustal material in that direction. The mirror image pattern of both magnetic anomalies and magnetic reversals is due to the same process, as the rocks record magnetic conditions at the time of their formation and continue to display them as they are transferred laterally away from the ridge axis in opposite directions.

This continuous formation of new crustal material at the mid-ocean ridges, and the transfer away from them by sea-floor spreading, implies that all the ocean floor is in motion. It also explains the relative youth of the oceanic crust. It is possible to ascertain the speed of sea-floor spreading by dating rocks at varying distances from the ridge axes. Fig. 5.6 indicates that in the ease of the North Atlantic the average spreading rate is about 2 cm a^{-1}. The Pacific is more active, with a spreading rate commonly of 4.5 cm a^{-1}. Overall, the creation of oceanic crust falls within the range 1–8 cm a^{-1}.

Since the Earth is not expanding, the creation of new crustal material must be balanced by the destruction of crustal material elsewhere at the same rate. This occurs beneath the deep ocean trenches, where ocean crust slides down along a shear zone (the **Benioff zone**) inclined at approximately 45° beneath the adjacent section of crust. The shearing of the descending section of crustal plate creates friction and sets up earthquake epicentres, often as deep as 700 km. The seismic instability of such zones is reflected in volcanic and geothermal activity at the surface above.

Thus, distributed across the surface of the Earth

are zones where new crustal material is rising to the surface, and other zones where it is descending towards the mantle, where it is consumed. The lateral movement of crustal material across the surface by sea-floor spreading is presumed to be balanced by a counter movement in the opposite direction at depth within the mantle. Therefore, on the basis of these observable movements at the surface, it seems probable that

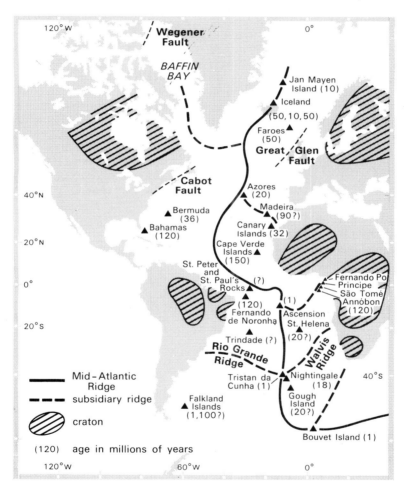

Fig. 5.5 Symmetry of the Atlantic ocean basin and age of Atlantic islands (from *Continental drift* by J. Tuzo Wilson, © 1963 by Scientific American, Inc., all rights reserved).

there is a convective movement of material within the mantle which has widespread and large-scale repercussions on the crustal surface. A series of convective cells, driven by terrestrial energy, appears to exist within the asthenosphere, which over a long period of time behaves in a plastic or semifluid manner (Fig. 5.7). The pattern of convective cells is irregular and there is no reason to assume that it has been constant throughout time. It seems that old rises and sinks can die out, while new ones become active. One can draw an analogy with a pan of gently simmering, viscous fluid – for example custard – in which the pattern of convective cells and rises and sinks changes through time. For instance, a new rise appears to be developing beneath East Africa, causing arching up of the continental crust, which has fractured to form the East African rift system. This feature is analogous to a mid-ocean ridge in the oceanic crust and appears to herald the break-up of the African continent.

Through the operation of the crustal system, then, there is a circulation of oceanic crustal material from the mantle to the surface and back again. Lateral movements of oceanic crustal material carry sections of continental crust across the surface of the globe.

The surface of the Earth can be conveniently thought of as a series of plates, of which there are six major ones and several minor ones (Fig. 5.8). Most of the major plates carry a continent, though this is not an essential feature. The plate margins are either rises where crustal material is generated (**constructive margins**), sinks where crustal material is consumed (**destructive margins**), or **transform faults** where two plates slide laterally past one another. Events occurring at plate margins have a very significant effect on the form of the Earth's surface topography. Constructive margins beneath the oceans create the mid-ocean ridges, described in detail earlier. A constructive margin developing beneath a continent causes the rifting and eventual fragmentation of that continent. At destructive margins, where there is a convergent movement of surface crust, significant major relief forms are developed. Three types of destructive margin can be identified. Where two oceanic plates converge, one may slide down beneath the other, giving rise to deep-seated volcanic activity. Such eruptions may cause volcanic accumulations which break the

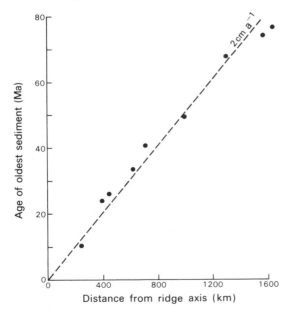

Fig. 5.6 Age of oldest sediment against distance from ridge axis compared with a spreading rate of 2 cm a⁻¹ (after Maxwell *et al.*, 1970).

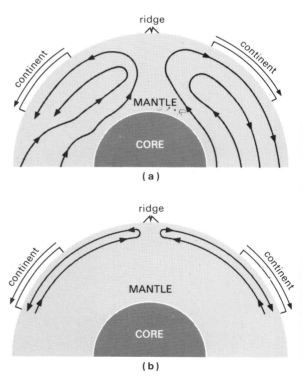

Fig. 5.7 Convective cells: (a) in the mantle and (b) confined to the asthenosphere.

ocean surface to form island arcs, common around the Pacific plate, as for example the Aleutian Islands. At a continental/oceanic plate boundary, a greater amount of sediment may be available as a result of denudation of the adjacent continent. Thus, a major mountain chain may develop as the sediment is piled up against the advancing continental plate, as in the examples of the Andes and Cordillera of South and North America respectively. The addition of a major mass of rock material to the continental crust may cause a depression of the latter by isostasy into the lower crust beneath (Box 5.4). The third type of destructive margin – continent to continent – also results in major mountain chains. The movement of India northwards into continental Asia resulted in the massive uplift of former marine sediments to form the Himalayas.

The operation of the crustal system, therefore, has enormous implications for the distribution of continents and relief across the Earth's surface. Lateral transfer of continents has moved them to their present positions and indeed is continuing

to move them yet further. Continents may become fragmented and rejoined as the pattern of convective cells and the position of rises and sinks change through time. There have been times during the geological record when all the continental crust coalesced to form one or two supercontinents. At other times, such as the present, continental crust is fragmented and the continents become scattered (Box 5.5). In addition to these major lateral movements, the operation of the crustal system also effects vertical movements of crustal material. Such transfers of mass involve changes in the potential gravitational energy of these materials. For example, orogenic activity transfers crustal rocks and sediments to higher elevations, thereby increasing their potential energy. These vertical movements, however, are complicated by isostatic reactions, for the fact that the continents float on the denser lower crust permits them to rise or fall according to whether mass is added or removed from them. These isostatic movements also involve major transfers of gravitational potential energy (Box 5.4).

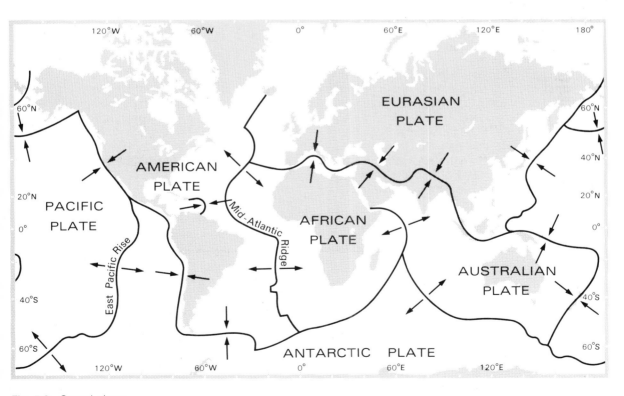

Fig. 5.8 Crustal plates.

Box 5.4
ISOSTASY

The fragments of continental crust float upon the denser oceanic crust at a level determined by the density difference between the two. This is known as isostatic equilibrium.

A simple analogy would be a cube of ice of density 0.9, floating nine-tenths submerged in water of density 1.0. Its isostatic equilibrium would be upset if light downward pressure were exerted, in effect increasing the load by pushing it down. When the pressure is removed, the ice cube returns to its equilibrium level.

So it is with continents. Any increase in mass, by mountain building, local accumulation of sediment or development of an ice sheet, imposes an increased load and the continent subsides. Processes of denudation, or the decay of an ice sheet, will decrease the load and cause isostatic rebound, as the lightened crust regains a new equilibrium level. Whereas in the example of the ice cube rebound is almost instantaneous, with crustal material it takes place much more slowly. In several regions from which glaciers have decayed within the past 10 000 years, contemporary isostatic rebound is measured at a rate of several millimetres per year. Likewise, the continuing process of erosion of the continental surface is counterbalanced by slow and continuous isostatic rebound.

Isostatic adjustment also implies some rearrangement of material at depth beneath the continental crust. This may be accounted for by a lateral movement of crustal material at depth (a), counterbalancing the erosion and deposition

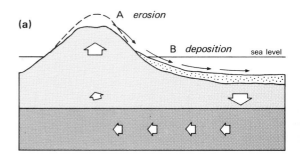

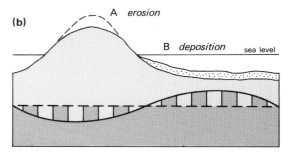

of material at the surface. Alternatively, it may be accommodated by a phase change in the rock-forming minerals in the mantle. It is known that under increasing pressure olivine, which forms some 70% of the upper mantle, becomes altered to a more dense form without change in chemical composition. Thus, an increase or decrease in pressure at depth could be accommodated by a phase change and a vertical movement of a phase boundary (b).

5.3.2 *The denudation system: the exchange of matter and energy at the lithosphere/ atmosphere interface*

Interaction between the crustal system of the lithosphere and the atmosphere and biosphere takes place where continental crust is exposed above sea level. At the land/air interface crustal material becomes exposed to inputs of solar radiant energy, precipitation and atmospheric gases. These inputs are often modified by, or operate through, the effects of the living systems of the biosphere (see Chapters 6 and 7). Under the influence of these inputs, crustal rocks are broken down by weathering processes and are transferred downslope by erosion processes. The effect is the denudation of the continental landscapes. These interactions and processes occurring at the Earth's surface operate within, and define, the denudation system. It is an

Box 5.5
THE CONGREGATION AND DISPERSAL OF THE CONTINENTS

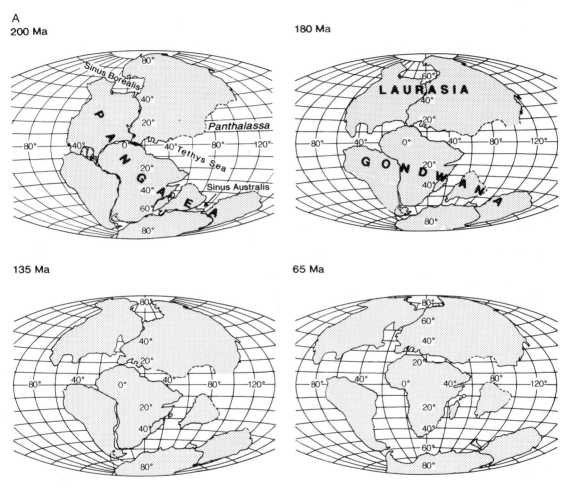

A
200 Ma

180 Ma

135 Ma

65 Ma

The continuous renewal of crustal material, and the lateral motions of the crust itself, have enormous implications for the distribution and location of the continents, both at present and in times past. All the major continents, in their geological record of sedimentary rocks, carry evidence of a range of past climatic conditions, implying that they have in the past occupied various different positions in relation to the equator. Belts of folded sedimentary rocks on the continents mark sutures, sites of past colli-

sion, of continental fragments which have become welded together. The distribution of folded sedimentary belts of different ages permits reconstructions of past continental dispositions.

The past 200 M years have been dominated by the breakup of a single supercontinent, and the dispersal of its fragments around the globe. At the close of the Palaeozoic era all the continental crust was clumped into one mass named *Pangaea*. According to the model of Dietz and

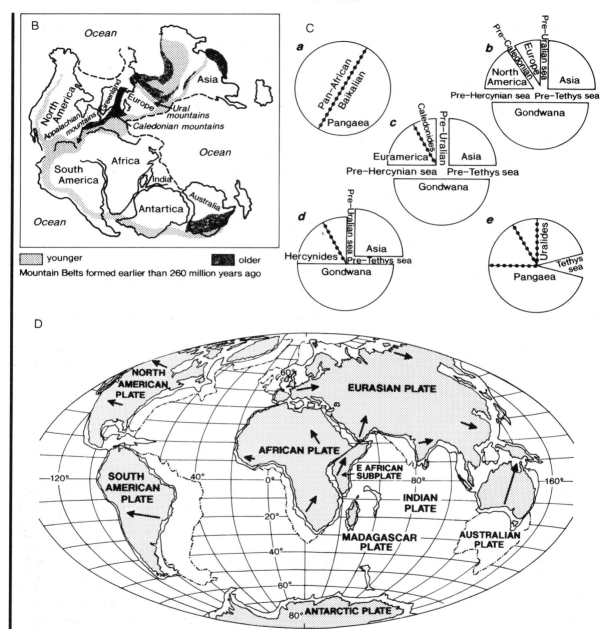

B

Mountain Belts formed earlier than 260 million years ago

younger / older

Holden, by the end of the Triassic period 180 M years ago, this supercontinent had fragmented into two masses – a northern group of continents, named *Laurasia*, and a southern group, *Gondwana*. The dispersal of the continents has continued subsequently (A), a process in which the opening of the Atlantic rift (Fig. 5.5) has played a prominent part. India has become reunited with the Asian crustal mass, compressing the intervening body of massive sediments into sedimentary rocks to form the young fold mountain belt of the Himalaya. This geologically recent phase of continental dispersion is of enormous significance in relation to current biogeographical distributions (Chapter 7, Section 7.1.1).

Looking further back in time, the distribution of folded sedimentary rocks in the reconstruction of Pangaea (B), implies that the supercontinent itself was formed by the congregation of previous crustal fragments. Each belt of folding represents the closure of an intervening sea and the folding and thrusting up of sediments during continental collision to form a suture (C). During Ordovician times, the Iapetus Ocean opened up a rift in the combined Euramerican continent. Thus marine fossils in southern Scotland have their counterparts in Newfoundland and are typologically distinct from those in nearby northern Scotland, which were then on the opposite margin of the Iapetus. This sea closed during the Caledonian orogeny, forming folded sedimentary ranges as northern

Europe and America reunited, subsequently to rift apart again in the Mesozoic, rupturing the previous suture.

An interesting exercise is to extrapolate the effects of contemporary lateral crustal motions in order to predict future continental distribution (D). The map of the world 50 M years hence reveals a continuing widening of the Atlantic rift, with westward drift of the Americas complemented by the eastward drift of Europe and Africa. Australia has moved markedly equatorwards to abut against southeast Asia. Africa is beginning to rift apart, with a fragment of east Africa moving off into the Indian Ocean. Southern California has moved northward along the San Andreas fault to Alaska.

The diagrams referenced in this box are collected together in Wilson (1976) – see Further Reading at the end of this chapter.

open system, therefore, which receives inputs of matter and energy from the crustal system and from the atmosphere and biosphere, and its output can be viewed as the products of rock breakdown and erosion delivered as sediments or solutes to the oceans. In this chapter we shall concentrate on the external relationships of the denudation system, while its internal operation will be discussed in detail in Chapters 10–17.

The elevation of crustal material above sea level (the base level of land-surface erosion) imparts potential energy to rock at the surface, related to its height above sea level. Repeated detailed geodetic surveys of the land surface may reveal significant changes in elevation over quite a short timescale. Fig. 5.9 shows contemporary vertical crustal movement on a continental scale in northern and eastern Europe. Both positive and negative movements are shown to be taking place at rates up to 12 mm a^{-1}. Table 5.6 lists measured rates of change of continental elevations caused by various large-scale geological processes.

In this way, crustal uplift can impart potential mechanical energy to the land mass in calculable amounts. Assuming a mean rock density of 2.8, the increase in potential energy of a 1 m^3 mass of rock elevated through 1 mm is as follows:

$$\Delta E_p = mg\,\Delta h$$

where m (mass) = 2.8×10^3 kg; g (gravitation acceleration) = 9.81 m s^{-2}; Δh = elevation = 0.001 m.

Therefore

$$\Delta E_p = 2.8 \times 10^3 \times 9.81 \times 1 \times 10^{-3} \text{ kg m}^2 \text{ s}^{-2} \text{ (J)}$$
$$= 27.47 \text{ J}.$$

Such uplift, however, does not simply affect a cube, but a column of rock 1 m^2 in area through the crust from the land surface down to base level. Total increase in potential energy per square metre will, therefore, depend on the total thickness of rock being uplifted. Within a given region, the total potential energy of uplift must be related to the area of crust being uplifted. Clearly, when the length of the column and the total area are taken into account, the increase in potential mechanical energy associated with a crustal uplift of as little as 1 mm a^{-1} is very considerable.

In addition to potential gravitational energy imparted to rock masses by uplift, the rocks themselves and their constituent minerals represent further inputs of potential energy to the denudation system. The internal mechanical forces associated with the packing of the individual mineral grains within the rock represent a potential strain

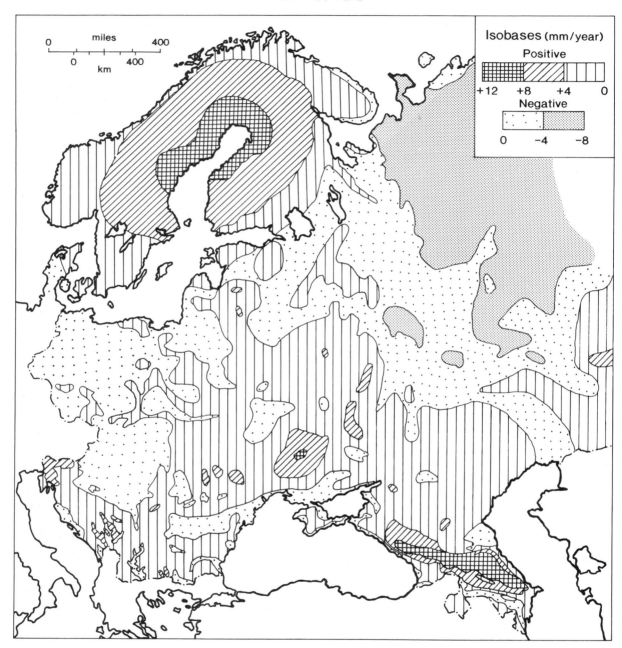

Fig. 5.9 Contemporary crustal deformation in eastern and northern Europe. Significant areas of uplift are evident in the Caucasus region due to orogeny, and around the Gulf of Bothnia, caused by postglacial isostatic rebound. The intervening area occupied by the European platform demonstrates gentle flexuring (after Lilienberg *et al.*, 1975).

Table 5.6 Representative rates of relative land and sea-level changes (after Carson and Kirkby 1972, from various sources)

	mm a^{-1}
glacial eustatic	up to 25
current eustatic rise	1.2
orogenic uplift	
California	3.9–12.6
Japan	0.8–7.5
Persian Gulf	3.0–9.9
epeirogenic uplift	0.1–3.6
isostatic	
Fennoscandiá	10.8
Southern Ontario	4.8

energy inherited from the confining pressure experienced during magmatic crystallization, metamorphism or lithification. This energy is released in the denudation system with the expansion, fracture and disaggregation of the rock under conditions of low pressure and temperature at the Earth's surface. Furthermore, the organization of the constituent atoms of the rock-forming minerals into complex structural configurations (see Box 5.1) represents a further input to the denudation system, this time of potential chemical energy. The most significant component of this chemical energy is the bond energy associated with the interatomic and intermolecular bonds in the lattice structure of rock-forming minerals. This energy is inherited from that involved in the chemical reactions under which the minerals originally formed in the crust. In the denudation system this potential chemical energy is released, at least in part, during the chemical breakdown of mineral structures as they undergo weathering reactions at the surface of the Earth.

Rock materials, however, represent an exchange of matter between the crustal system and the denudation system, and the potential energy content of these rock materials is only one aspect of their properties. Their elemental chemical composition, chemical structure and physical and mechanical properties are also important in influencing their behaviour in the denudation system. Like the forms of potential energy these rocks represent, many of their properties are also a legacy of the processes they experienced in the crustal system.

The frequency and distribution of fractures and fissures in the rock mass – referred to as its pattern of jointing – may have resulted from contraction during the cooling of magmatic igneous rocks; in sedimentary rocks, joints may have developed along bedding planes, or as a result of compaction and lithification, normal to the bedding. Superimposed on all these joint patterns, crustal movements may create a variety of further fractures as a result of tension, compression, wrenching or torsional forces. The physical properties of a rock will also reflect its origin and antecedent history. There is, for example, a clear difference between igneous rocks, whose formation involves the growth of crystals until they are often in intimate contact, and sedimentary rocks, where loose detrital grains accumulate by accretion. Unconsolidated sediments will incorporate a much higher proportion of voids, for example, than those which have undergone lithification.

These then are the principal inputs of matter and energy to the denudation system from the crustal system. The prime source of energy input from the atmospheric system is solar energy. The solar energy cascade was considered in Chapter 3 (see Fig. 3.8). Of the total radiant energy arriving at the Earth's surface, some is employed in sustaining the biosphere, while the remainder is used in heating the ground surface and causing evaporation of soil and surface moisture. Fig. 3.16 shows Budyko's (1958) estimate of solar energy received at the Earth's surface. Clearly, higher values occur in the tropics than in lower latitudes. The highest values, however, are found in the desert regions, where the filtering effect of cloud and vegetation is minimized. Of the energy actually incident at the ground surface, between 15% (in deserts) and 50% (in humid regions) is required to evaporate surface moisture. The remainder raises the ground temperature.

The major material input from the atmosphere is water. The mean annual world precipitation is estimated at 857 mm. This generates a total volume of runoff from the continents of 37×10^3 km^3 with a total mass of 37×10^{15} kg. If, for the sake of simplicity, we assume that this mass falls from the mean continental altitude of 870 m, the potential energy associated with precipitation is over 35×10^{18} kgf m, or 34×10^{19} J. As this massive volume of water drains off across the continental surfaces, much is lost directly to the atmosphere by evaporation, but of the remainder the potential energy is converted to kinetic energy. Thus, only a portion of the potential energy is ultimately employed in the mechanical transfer of debris. It is shown in Chapter 15 that a reasonable average rate of erosion of the Earth's continental surface is 0.05 mm a^{-1}. This involves the removal of some 20×10^{12} kg of rock debris per year. Assuming that it descends from the mean continental elevation, the potential energy thus released from the land masses is about 18×10^{15} J, less than one-thousandth of the potential energy available from precipitation.

Within the framework of these massive energy transfers, the interaction between lithosphere, biosphere and atmosphere results in the processes of continental denudation. The presence of heat energy and moisture causes rock breakdown by weathering into transportable debris. Running water and gravitational energy combine to transfer clastic material downslope and downvalley towards the oceans, where it is deposited as sediment, mainly along the continental margins. This interaction, the denudation system, can therefore be seen as a conflict between the constructive forces of the lithosphere and the destructive forces of the atmosphere. The uniqueness of continental denudation on Earth is illustrated by comparison with other planetary bodies (Box 5.6).

5.3.3 A geochemical model: the cycling of rock-forming materials

Implicit in earlier sections of this chapter is the concept that rock-forming materials are circulated both within the crustal system and between it and the denudation system. This is the **rock cycle** first described by James Hutton, the father of modern geology, in the 18th century. Indeed, massive amounts of energy are employed in the operation of the crustal and denudation processes that accomplish this circulation.

In Fig. 5.10 the primary source of rock-forming materials can be seen to be the upper mantle, from which molten magma rises to form primary igneous rocks. These basaltic magmas from the mantle are complex mixtures of many elements, but because they are dominated by silicon and oxygen, they are often also referred to as silicate magmas. At certain temperatures and pressures in the crust these magmas start to crystallize. Crystallization proceeds along reaction series (Bowen, 1928; Fig. 5.11) with the earlier-formed minerals either changing gradually as the magma cools (continuous series), or being dissolved and

reconstituted as different minerals at lower temperatures (discontinuous series). Alternatively, the earlier-formed and denser minerals may sink through the magma, or may be left behind as it migrates upwards (fractionation) and give rise to more basic rock types such as peridotite and gabbro. The composition of the molten fraction is now different, being relatively richer in such elements as silicon, aluminium and potassium, so that the later-formed minerals and the rocks they produce will be less dense and more acidic, such as granite. Finally, residual watery silicate fluids remain and cool and solidify in fissures and cavities.

Where magmatic injection, differentiation and solidification occur at depth in the crust, intrusive igneous rocks result, but magma may be ejected at the surface and cool rapidly in contact with air or water and form extrusive igneous rocks (Fig. 5.12). The mineral composition of these extrusive rocks will reflect the composition of the extruded magma and this will depend on the degree of magmatic fractionation that has occurred before extrusion.

These primary igneous rocks are relatively heterogeneous in the elemental composition, though magmatic fractionation may concentrate certain elements. Intrusive igneous rocks are exposed to denudation at the surface, either after the erosion of overlying rocks or by a combination of uplift and exhumation. Extrusive rocks are, by definition, exposed to denudation processes almost immediately. Under the influence of energy and material inputs from the atmosphere and biosphere, both types of igneous rock undergo weathering reactions at the surface. Interestingly enough, Goldich (1938) maintained that the susceptibility of the primary minerals of these rocks to such reactions is in the reverse order to that of the Bowen series (Fig. 5.11 and Chapter 11).

The products of weathering are the more resistant of the original minerals, secondary minerals derived from them, and soluble and volatile products. Particular elements may be concentrated preferentially in residual weathering products, whether of primary or secondary origin, and accumulate in the weathered mantle or **regolith**. Erosion and transport, however, ultimately deliver rock debris and soluble products to the oceans, along whose margins they may be deposited or precipitated to form sediments. Other minor sediment traps (environments where sediment may

Box 5.6

LANDFORMS AND DENUDATION: THE EARTH COMPARED WITH PLANETARY MORPHOLOGY OF THE MOON AND MARS

The conditions and features of the Earth surface environment which we have described thus far are probably very familiar to most readers. They contrast, however, with other bodies in the solar system, which can provide a broader perspective on terrestrial landforms and denudation systems.

On the global scale, the forces which condition the nature of terrestrial landforms lie in the lateral and vertical mobility of the crust, interacting with denudation processes driven by the atmosphere. There is a continuous process of production and consumption of crustal material by the conveyor belt of plate tectonics, by which crustal material is exchanged with the mantle. As a consequence, most of the Earth's surface is younger than 200 million years. The Earth possesses a substantial circulating atmosphere, whose pressures and temperatures are such that water readily crosses the threshold between the liquid and gaseous states. The Earth's mass is such that its gravitational energy ensures that raindrops have significant impact energy, and that water drains readily to lower elevations, applying mechanical energy to terrestrial surfaces. The surfaces of Earth's continents are thus subject to active denudation as a consequence of the particular environmental conditions which prevail. The continuing renewal of the continents is a function of the present stage in the Earth's planetary evolution.

On other planetary bodies however, conditions are very different, both in respect of the nature of the crust and surface temperature and atmosphere. Two examples, the Moon and Mars, will demonstrate some of the variety which occurs.

It is now known, as a result of recent lunar exploration, that most lunar surface rocks are very old, in the range $3.1–4.5 \times 10^9$ years. The heavily cratered highland areas consist of older granitic rocks, whereas the darker lowlands ('maria') are formed of younger basaltic rocks.

It is evident that the Moon's crust formed shortly after its formation by accretion, the result of solidification of a fluid granitic magma generated by the differentiation of its small iron core. Beneath this crust existed a zone of fluid denser basalts.

The absence of any significant atmosphere or surface water meant that Earthly processes of denudation could not take place, leaving the Moon's surface to be shaped by meteorite bombardment. This process has continuously affected all planetary bodies in the solar system. (The effect of meteorite impact on Earth is limited by the fact that many meteorites may oxidize or burn up passing through the Earth's dense atmosphere, thus reducing the number of impacts. Furthermore, the processes of denudation serve to obliterate the geomorphological traces of such events in a short space of geological time.)

The lunar highlands have thus been exposed to bombardment for over 4000 million years, and are intensely pitted with impact craters and mantled with a regolith of debris thrown out by the impacts. The lunar lowlands, formed around 3500 million years ago, represent sites of major impacts, at a time when still-molten basalt lay beneath the crust and was able to spill through the punctured crust and consequently solidify. These portions of the lunar surface, being younger, are considerably less cratered.

The internal evolution of the Moon took place very early in its history. Its crust, therefore, is an ancient creation, and the landforms it now bears have been shaped over a long period by agents external to the lunar system.

Mars is an interesting planet in that it possesses a mix of lunar and terrestrial surface forms. It has been orbited by spacecraft and scientific instruments have been landed on its surface. It is half the diameter of Earth, with one-eighth of Earth's mass, 40% of its gravity, and possesses less than 1% of its atmosphere. Mars is characterized by extreme temperatures. The low atmospheric pressure means that water cannot exist in the fluid form on the Martian surface, and that any atmospheric disturbance results in very strong winds.

The Martian surface possesses an abundance of terrain cratered to a similar intensity to that of the Moon, thus suggesting a similar age. Large lowlands of impact are also present. Major volcanoes exist, though it is not clear that any are still active. The shield volcano Olympus Mons is 600 km wide, 27 km high and possesses a summit caldera 80 km wide – a giant by Earthly standards. With a rigid crust the source of magma would be in a constant position and would permit the long-continued growth of such a massive form, in contrast to Earthly volcanoes which move away from the magma source due to the lateral progress of plate tectonics. The presence of rift valleys on Mars suggests crustal fracturing and extension as a result of tensile stresses. These features are indicative of a rigid and immobile crust of considerable antiquity.

There are also present on Mars landforms associated with denudation processes. Fretted escarpments, large-scale channels with fluid bedforms, landslides and dunefields are all indicative of surface sediment transport, many of them by fluids. Yet there is no large body of fluid evident at the present time. Mars does possess two polar icecaps composed of frozen CO_2 and water, but these do not contain a suf-ficient volume to account for the denudational forms. It is possible that there is a large subsurface store of water, perhaps in the form of permafrost, which may be released by periodic large impacts or climatic changes.

Thus on Mars the pattern of planetary evolution has created a strong and rigid crust. The materials associated with agents of denudation have been present on the Martian surface in the past. Denudation processes are currently slow, permitting the survival of ancient crustal terrain in parts; yet there have also been episodes of denudation which have left their traces, and which also survive in this currently denudationally inactive environment.

A major control on the rate of planetary evolution may be that of size. Smaller planets, with a higher surface area to volume ratio, will cool more quickly, and hence develop a rigid crust earlier in their life. Earth's major landforms, then, can be seen as a function of a planet of a particular size, at a particular stage of its geological evolution. As it too cools down, crustal mobility can be expected to decline and cease, as the Earth's store of internal heat runs down. Yet the supply of solar energy will continue, as will the atmospheric circulation. What, then, will Earthscapes be like in the planetary future?

See Press and Siever (1986), Ch. 22, and Lowman and Garvin (1986).

---

collect) exist at a variety of scales on the way. Soluble products may also be reprecipitated before reaching the oceans. Meander cores, river flood plains, lake basins, glacier forelands and desert basins are all terrestrial environments where significant volumes of deposition may occur. Eventually most sediments become lithified due to compaction by overburden and by internal redistribution of minerals by percolating solutions to produce sedimentary rocks. Transport processes, the environment of sedimentation and redistribution during lithification, all tend to act selectively, and therefore enhance the likelihood that the elements of the original igneous rock will become further segregated into distinct facies of sedimentary rock.

Sediments and sedimentary rocks may in their turn be exposed to denudation by exhumation and uplift to complete one possible geochemical cycle. Alternatively, sedimentary and igneous rocks may undergo metamorphism. Here rocks experience crushing, recrystallization and/or recombination under conditions of high temperature or pressure (or both) within the crust and are converted to metamorphic rocks. Indeed, existing metamorphic rocks may themselves be converted into new types as a result of the same processes. Again, metamorphism often has the effect of segregating the elements present in the original rock, sometimes to form relatively rare minerals.

Metamorphic rocks, of course, like their sedimentary and igneous counterparts, may also be exposed at the surface and be subject to denudation, and hence complete a second type of cycle.

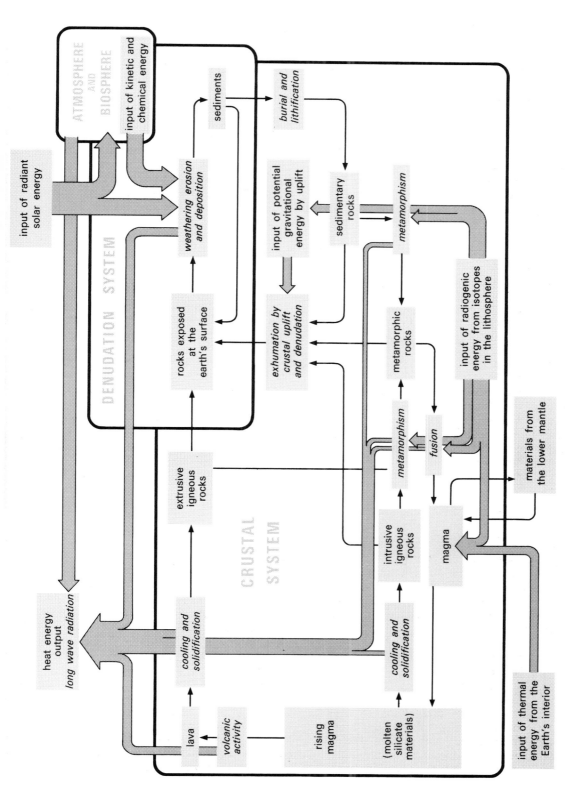

Fig. 5.10 A geochemical model of the lithosphere. Stippled pipes represent energy flow and arrowed lines the pathways of material circulation.

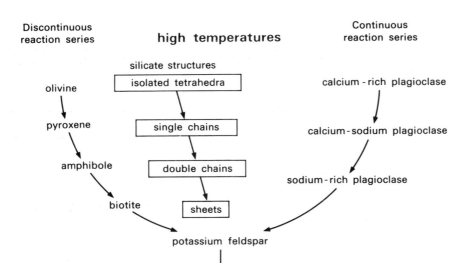

Fig. 5.11 The Bowen reaction series.

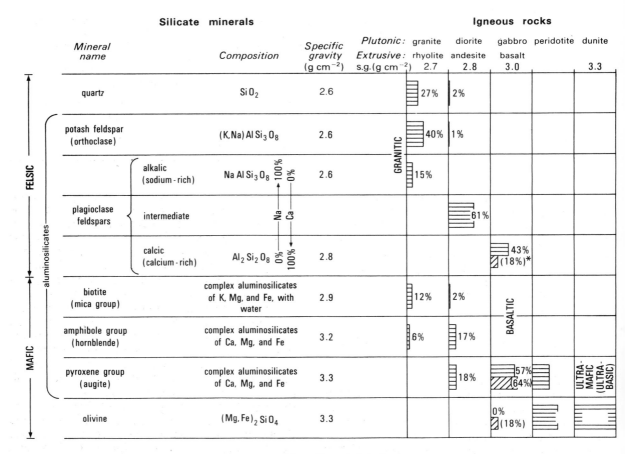

Fig. 5.12 The mineralogy of igneous rocks (adapted from *Planet Earth* by A.N. Strahler, © 1972 by Arthur N. Strahler).

Table 5.7 The exploitation of mineral resources depends on their occurrence as concentrations which render them economically viable. Such concentrations reflect both crustal and denudational processes that promote the segregation (fractionation) of materials involved in geochemical cycles

Process		Examples
Mineral deposits originating in the crustal system		
magmatic segregation	(a) settling of early formed minerals to base of magma	magnetite, chromite, platinum (South Africa)
	(b) settling of immiscible sulphide or oxide melts, sometimes with injection along fissures	copper–nickel (Norway and Canada); magnetite (Sweden)
contact metasomatic	wall rocks of intrusion replaced by minerals derived from magma	magnetite and copper (Utah and Arizona)
deposit from watery silicate fluids	filling fissures in wall or outer part of intrusion (pegmatites)	mica (New Mexico)
deposited from hot watery fluids	filling fissures in, and replacing, wall and outer part of intrusion (hydrothermal)	lead–copper–zinc (Cornwall and northern Pennines)
Mineral deposits originating in the denudation system		
sedimentary	(a) evaporation of saline waters	salt and potash (Northumberland)
	(b) precipitation from solution	iron (Northamptonshire)
	(c) deposition and sorting of detrital grains	placer gold (Australia, California and Alaska); titanium (India and Australia); diamond (Namibia)
residual	weathering leaving concentrations of insoluble elements in residual material	aluminium (bauxite) (USA, France, Jamaica and Guyana)
secondary enrichment	precipitation at depth from groundwater in a mineral deposit	copper (Arizona and Miami)

A third type of geochemical cycle is completed when rocks of sedimentary, igneous and metamorphic origin are carried down the Benioff zone towards the mantle. The fractionation of their constituent minerals is lost as they undergo a phase change, fuse or melt, and are absorbed back into the largely homogeneous magma in the upper mantle, perhaps to be intruded into the crust again at a later date.

5.4 The implications of the geochemical model

It is evident from the geochemical model developed above, and from Fig. 5.10, that the elements of the crust are involved in several alternative cycles between a number of stores in the lithosphere. It is important, however, to realize that the scales of these cycles, both spatial and temporal, may vary considerably. The cycle within the denudation system from weathered rock to unconsolidated sediment and once more to weathering at the surface may be relatively rapid on a geological timescale. Over the northern hemisphere, for example, many superficial deposits of Quaternary age – glacial tills, fluvioglacial, solifluction and interglacial terrace deposits – are all undergoing contemporary weathering and erosion. The cycle, therefore, is being completed in between 10^3 and 2×10^6 years. The same point is true for some of the unconsolidated sediments of arid and semi-arid areas, and for large areas of recent alluvium. At least with some of these sediments, such as solifluction or glacial lake deposits, the spatial scale of this pathway of circulation may be very localized. In contrast, the route via lithification and consolidated sedimentary rock usually takes considerably longer ($>10^7$ a) and operates on a larger spatial scale. Finally, the sub-duction and fusion of rock-forming materials into the mantle, and the upwelling of magma to form igneous rocks, is a circulation route which operates over a very long timescale and at the spatial scale of the crustal plate and convection cell.

The conclusion is that all of the routes taken as elements circulate within the crustal system are relatively slow and they involve considerable residence times in the various compartments or stores in the model. However, two alternative routes remain by which elements circulate, and they are much more rapid in their operation. Both are coupled to the cycles in the denudation and crustal systems which we have been considering. The first of these alternative routes is through the atmosphere and hydrological cascade to the denudation system, and the second is via the living systems of the biosphere as the mineral nutrients essential to life. In the latter case, the model becomes one of biogeochemical cycles. The atmospheric route has already been encountered in Chapter 4, and the biospheric pathway will be discussed in Chapters 6 and 7.

There is a further implication of the geochemical model which has profound significance, both for the organisms of the biosphere and for humans with their developed technology. The elements of the crust and the minerals and rocks they form are critical resources, both as the nutrient elements on which life depends and as the raw materials of human technologies. Magmatic differentiation, metamorphism and denudation have all been shown to act to fractionate and concentrate rock-forming materials. From the point of view of the organisms of the biosphere, this means that essential mineral nutrients are not uniformly available across the Earth's surface. The normal situation is rather that the rocks, from which nutrients are released by weathering, display either deficiencies or excesses of some essential elements. It is also the concentrations of certain elements in particular minerals and rock types which form economically viable ore deposits, and their discovery and exploitation depends on an understanding of the processes operating in the crustal system (Table 5.7). Although the existence of some ore bodies is predictable, many of the large concentrations of mineral resources represent freak geological events. This fact, coupled with the immense timescale involved in many crustal processes, underlines why mineral resources are considered to be non-renewable.

Further reading

Much of this chapter is concerned with the science of geology and the reader is referred to any of the large number of student texts covering the field. At an introductory level an excellent overview is provided by:

Press, F. and R. Siever (1978) *The Earth*. Freeman, San Francisco.

A geomorphological perspective on crustal and tectonic processes and landforms is provided by Part One of:

Summerfield, M.A. (1991) *Global Geomorphology*. Longman, London.

A basic source on minerals and mineralogy is:

Read, H.H. (1970) *Rutley's Elements of Mineralogy*, (26th edn) Allen & Unwin, London.

The nature of rocks themselves is covered by:

Nockolds, S.R, R.W.O'B. Knox and G.A. Chinner (1978) *Petrology for Students*. Cambridge University Press, Cambridge.
Brownlow, A.H. (1978) *Geochemistry*. Prentice-Hall, Englewood Cliffs.

The recent revolution in geophysics and global geology and the development of the unifying theory of plate tectonics is covered by the following, of which the article by Oxburgh and the book by Wylie are by far the most accessible:

Cocks, L.R.M. (ed) (1981) *The Evolving Earth* (Published for the British Museum [Natural History]). Cambridge University Press, Cambridge.
Fifield, R. (ed) (1985) *The Making of the Earth* (New Scientist Guides). Blackwell, Oxford (Reprints of New Scientist reports covering the period of development of the modern theory of plate tectonics and related areas).
Gass, I.G., P.J. Smith and R.C.L. Wilson (1973) *Understanding the Earth: a Reader in the Earth Sciences*. (Published for the Open University) Artemis Press, Horsham, Sussex.
Oxburgh, E.R. (1974) *The Plain Man's Guide to Plate Tectonics*. Proceedings of the Geologists Association.
Wilson, J. Tuzo (1976) *Continents Adrift and Continents Aground*. (Readings from Scientific American) Freeman, San Francisco.
Wyllie, P.J. (1976) *The Way the Earth Works: an Introduction to the New Global Geology and its Revolutionary Development*. John Wiley, New York.

The following texts treat specific phenomena and landforms associated with crustal activity:

Ollier, C.D. (1988) *Volcanoes*. Blackwell, Oxford.
Ollier, C.D. (1981) *Tectonics and Landforms*. Longman, London.
Weyman, D. (1981) *Tectonic Processes*. Allen & Unwin, London.

Lastly, the cycling of crustal materials is well treated by:

Garrels, R.M. and F.T. MacKenzie (1971) *Evolution of Sedimentary Rocks*. Norton, New York. (especially Chapters 4, 5 and 10).

The hydrosphere

6.1 Water in the Earth–atmosphere system

6.1.1 The nature of water

The water molecule is stable and forms only weak bonds with neighbouring molecules of other substances (Box 6.1). Water exists in three phases – gas (water vapour), liquid and solid (ice) – and it is present in the Earth–atmosphere system in all three phases. Most water vapour is found within the atmosphere; the oceans store most of the water in its liquid phase. At the poles, in permafrost regions, and in high alpine regions there are stores of water in the form of ice.

Water possesses physical properties that are unique when compared with substances of similar molecular mass. It has, for example, the highest specific heat of any known substance, which means that temperature changes take place very slowly within it. It also has a high viscosity and a higher surface tension than most common liquids. Its boiling point at standard atmospheric pressure (1013.25 mb) is 100°C and its melting point 0°C (Table 6.1). In the solid and liquid phases, the physical properties of water vary with its temperature. Of these, the variation in density differs markedly from the behaviour of most liquids, which usually reach their maximum density at their freezing point. A maximum density at 4°C means that freezing to great depths in lake and ocean waters will be suppressed by water cooled below 4°C, which floats to the surface.

In the gaseous phase, water vapour constitutes less than 4% of the total atmospheric volume (see Chapter 4, Section 4.1). The amount present may be expressed in terms of absolute humidity,

---

Box 6.1
THE WATER MOLECULE (AFTER SUTCLIFFE 1968)

Hydrogen and oxygen atoms combine to form water, the electrons being shared between them as in the diagram. There is electrostatic attraction between molecules because of the asymmetry of the distribution of electrons which leaves one side of each molecule with a positive charge. The water molecule can form four such hydrogen bonds, which are relatively weak. In a solid state, as ice, the tetrahedral arrangement of this bonding produces a tetrahedral crystalline structure. In a fluid state, increases in temperature weaken the hydrogen bonding.

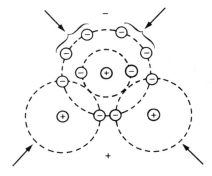

Table 6.1 Physical constants of pure water (Sutcliffe 1968)

specific heat (15°C)	4.18 J g^{-1} deg^{-1}
latent heat of melting	334.4 J g^{-1}
latent heat of vaporization (15°C)	2462 J g^{-1}
surface tension	7340 mN m^{-2} cm^{-1}
tensile strength	1418.5 kN m^{-2} cm^{-2}
melting point (1013 mb)	0°C
boiling point (1013 mb)	100°C

specific humidity or mixing ratio (Box 6.2). Water vapour exerts a partial pressure in the atmosphere which is referred to as its vapour pressure (e), which normally varies between 5 and 30 mb. When the atmosphere lies above a liquid water surface, water molecules are constantly being exchanged between them. If the atmosphere is dry, the rate of uptake of molecules is greater than the rate of return to the surface. When a point of equilibrium is reached where the number of molecules leaving the surface is equal to the number arriving, the vapour pressure of the air has reached saturation with respect to water. Subsequent additions of water molecules to the air are balanced by deposition on to the surface.

The value of vapour pressure at which saturation occurs is dependent on air temperature, as indicated in the **saturation vapour pressure curve** (Fig. 6.1). If we have a parcel of dry air at point X with temperature T_a and vapour pressure e_a, we can derive measures of its humidity. If its temperature remains constant and more water molecules are added, saturation is reached at Y where saturation vapour pressure is e_s. Saturation deficit and relative humidity for air at temperature T_a may be derived as indicated in Box 4.10. If vapour pressure remains constant but temperature is reduced, saturation is eventually reached at temperature T_d, which is referred to as the dew-point. The difference between T_a and T_d may also be used as an indication of the humidity of the air.

Should the surface underlying the air be ice, then the saturation vapour pressure is slightly less than that over a water surface at the same temperature, as shown in Fig. 6.1. If air lies over both water and ice simultaneously, and is cooled from temperature T_b to T_i, then with respect to the water surface, the air is unsaturated and will accept more water molecules from it. However, with respect to the ice surface it is at its dew point and it must deposit any further water molecules that it receives. Thus there is a simultaneous withdrawal of water molecules from the liquid surface and deposition on the ice surface. This process is

---

Box 6.2
WATER VAPOUR IN THE ATMOSPHERE

Absolute humidity (χ)
Mass of water vapour in a given volume of air at a given temperature:

$$\chi = m_v/v$$

where m_v = mass of vapour and v = volume of air in which it is contained.

Specific humidity (q)
The proportion by mass of water vapour in moist air:

$$q = \frac{m_v}{(m_v + m_a)}$$

where m_a = mass of moist air.

Saturation deficit
Saturation deficit at temperature T_a is given by:

$$S = (e_s - e_a)$$

where e_a is vapour pressure, and e_s is saturation vapour pressure at temperature T_a.

Relative humidity

$$RH = \left[\frac{e_a}{e_s} \times 100 \right] \%.$$

of particular importance in the development of precipitation from clouds (Section 6.15).

The processes which bring about such changes in phase, the operation of these processes and the transport of water in vapour or liquid form within the Earth-atmosphere system are commonly represented in simplified form as a hydrological cycle (Fig. 6.2b).

6.1.2 The movement of water in the Earth–atmosphere system

As no water is exchanged between the atmosphere and space, we can regard its movement as taking place within the Earth–atmosphere system as a closed system. The atmosphere alone must, however, be viewed as an open system, because the movement of water across the Earth/atmosphere interface represents a transfer of both matter and energy across the boundary of the

system. The movement of water across this boundary and within the system is initiated and maintained by a flow of energy through the system (Fig. 6.2b). Within this hydrological system there are a number of locations at which water is stored. The major stores are the atmosphere, the land, the oceans and the polar ice caps. The distribution of water between these is indicated in Fig. 6.3. The smallest store is the atmosphere, which contains 0.001% of the total water in the Earth–atmosphere system, while the greatest is the ocean store, which has 97.6%. Transfers between stores are carried out by the processes of evaporation, condensation, precipitation, runoff and freezing and melting. The greatest exchanges are those between ocean and atmosphere: 86% of evaporation takes place over oceans, while 78% of precipitation occurs over them (Baumgartner and Reichel, 1975).

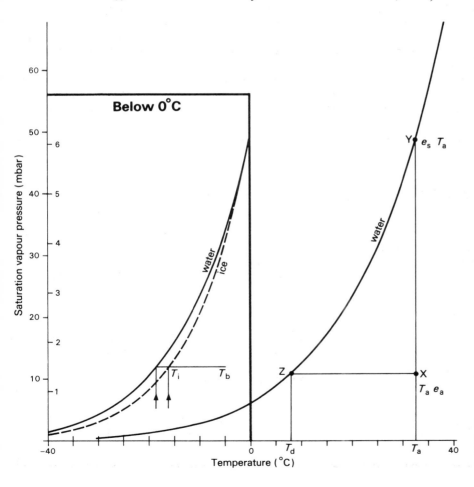

Fig. 6.1 Saturation vapour pressure curves with respect to water and ice surfaces.

6.1.3 Evaporation

Evaporation is the process by which liquid water is changed into its gaseous phase (water vapour). If heat energy is supplied to a water surface, it has the effect of weakening the bonding between water molecules and increasing their kinetic energy. The faster-moving molecules thus have a greater capacity to break away from the water surface and enter the air above. This transfer is partly offset by water molecules returning to the surface, and the net loss represents the rate of evaporation from the water surface.

The change in phase from liquid to gas requires an input of heat energy which is referred to as the latent heat of vaporization. At 0°C, 2501 J are required to evaporate 1 g of water, while at 40°C this is reduced to 2406 J. If heat energy ceases to be supplied from an external source, the energy for evaporation will be drawn from the remaining water, thus having the effect of reducing its temperature and suppressing further evaporation.

There are a number of factors that affect the rate of evaporation, the most important of which are the heat energy supply, the humidity of the

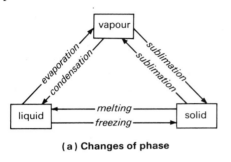

(a) Changes of phase

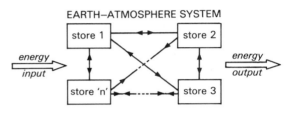

(b) The movement of water in a closed system

Fig. 6.2 The movement of water in the Earth–atmosphere system.

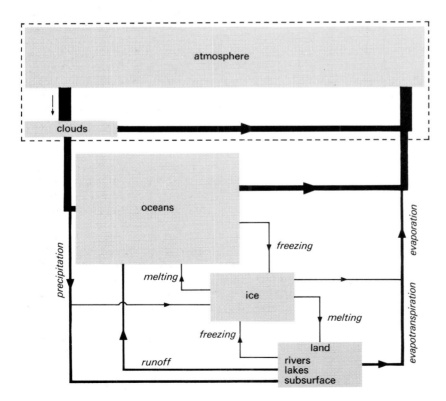

Fig. 6.3 The hydrological system.

air, the characteristics of the airflow across the surface and the nature of the evaporating surface.

The rate at which water molecules diffuse into the lower layers of the atmosphere is directly governed by the heat energy made available to them. As a component of the heat balance equation (see Chapter 3), evaporation depends not upon solar energy input but on net radiation. For open water and wet soils, most of this net radiation is consumed in evaporation. The spatial and temporal variations of net radiation are thus imposed upon evaporation, producing, for example, summer and early-afternoon maxima.

The balance between water molecules leaving the surface and those returning to it depends on the number of them in the overlying air. If the air is relatively dry, their numbers will be small in relation to those leaving the surface and thus the rate of evaporation will be high. If, however, the air approaches saturation, it has a higher population of water molecules, from which those returning to the surface will be only slightly fewer than those leaving, and consequently the rate of evaporation will be low. The rate of evaporation is related to the difference between saturation vapour pressure (e_s) at surface temperature and the vapour pressure of the air above (e_d):

$$\text{evaporation } (E) = \text{constant} \times (e_s - e_d). \qquad (6.1)$$

Molecules will continue to leave the surface while e_a is less than e_s. However, the effect of this is to increase e_a until it eventually reaches e_s. At this point the rate of evaporation will be zero. So, unless there is a mechanism for mixing the air and thereby redistributing water molecules, evaporation will always exhibit decrease with time.

The mixing of the lower atmosphere by vertical and horizontal motion effectively results in the replacement of air before saturation is reached. Thus evaporation is maintained at a higher rate than if there were no such replacement, as e_a no longer increases towards e_s. Increases in horizontal wind speed cause increases in the rate of evaporation, but only up to a point where further increases have little effect. This maximum is determined by the heat energy available for evaporation and by the humidity of the air. Eqn 6.1 may be rewritten to take into account mean wind speed ($\bar{u}$):

$$E = Bf(\bar{u}) (e_s - e_a) \qquad (6.2)$$

where B is a constant, and $f(\bar{u})$ expresses a function of wind speed.

Where a water surface is exposed to the atmosphere evaporation may take place, whether this surface be of an ocean or a glass of water. However, the rate of evaporation is determined by the rate at which molecules lost to the atmosphere can be replaced from the subsurface store. In the case of an open water surface there is effectively an unlimited supply of water freely available for evaporation. However, in a soil there is usually a limited supply, thereby suppressing rates of loss. Should the soil be saturated, for example after a long period of rain, then evaporation may be considerably higher than from an equal area of open water, because of the water film held on soil particles near to the surface. The quality of the water also affects the rate of evaporation, in that it is reduced by the presence of impurities such as dissolved salts. The greatest reduction is from the salt water of, for example, the oceans where there is, as a rough approximation, a 1% reduction in evaporation for a 1% increase in salinity.

If there is vegetation growing upon the surface, this provides an extra pathway for water molecules transferring from the ground surface into the atmosphere. Water vapour is diffused through pores (stomata) on leaf surfaces and is transferred into the atmosphere (see Fig. 19.3). This is associated with a suction pressure inside the plant causing it to withdraw water from the soil through its root system. This transfer of water into the atmosphere is referred to as **transpiration** and it accounts for a large proportion of moisture losses from vegetated surfaces (see Chapter 19).

The rate of transpiration loss is governed by two groups of factors: those already discussed – the extrinsic factors which affect rates of evaporation from water surfaces – and the factors intrinsic to the plant. For example, transpiration loss from most plants will be large when the atmosphere is relatively dry. A good example of this is the effect of the dry mistral wind of the Rhône valley in France, which causes harmful increases in transpiration loss from crops. In most plants, leaf stomata open under daylight conditions and close at night, thereby introducing a clear diurnal variation in transpiration losses. However, should the plant be unable to withdraw adequate water through its roots to match transpiration loss, internal stresses may be set up. In response to this there may be a partial or complete closing of the stomata in order to limit transpiration.

Over vegetated surfaces with a dense plant cover, water losses to the atmosphere are largely accounted for by transpiration. In a dense forest, for example, over 60% of water loss will be achieved through transpiration. If evaporation of precipitation intercepted by the trees is included, over 80% of water transferred to the atmosphere may be due entirely to the presence of a vegetation cover. In semi-arid areas where there is virtually no surface water, transpiration will account for the entire surface-to-atmosphere water transfer.

Most surfaces are neither absolutely bare nor completely vegetated, but possess elements of both. Such is the complexity of these surfaces that separation of evaporation and transpiration is impracticable and the two are combined in the one term, **evapotranspiration**. As we have already seen, a shortage of soil moisture limits the rate at which evaporation and transpiration can take place. Under such 'limiting' conditions, actual evapotranspiration is occurring. If, on the other hand, no such limiting condition prevails, then evapotranspiration can take place at its maximum rate within the constraints placed upon it by the availability of heat energy, atmospheric humidity and wind speed. Under these 'non-limiting' conditions, we have potential evapotranspiration. This concept is used extensively in irrigation studies, for it represents the worst possible situation with regard to water loss from the ground surface.

Over the Earth's surface, actual evapotranspiration is greater over oceans than over land surfaces, with maxima occurring not at the equator but over the tropical oceans between latitudes 10° and 40° (Fig. 6.4). The meridional cross-section of average rates of evapotranspiration illustrates the operation of the various factors that control it. Taking account only of heat balance we should expect a clear low-latitude maximum and high-latitude minimum. However, this simple distribution is modified in the zones of persistent winds, for example the subtropical trade winds. These winds are warm and dry and they clearly increase evapotranspiration, while in humid equatorial regions low wind speeds and high atmospheric humidity tend to limit evapotranspiration which is, to some extent, offset by the high rates of transpiration from equatorial forests.

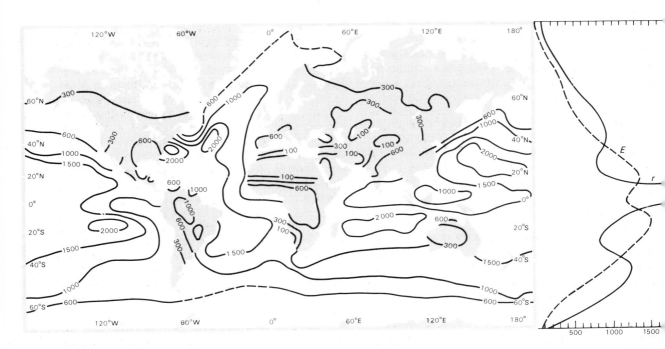

Fig. 6.4 Global distribution of annual evaporation (*E*) and evapotranspiration (*r*); values in millimetres (after Barry, 1970).

6.1.4 Condensation

When the number of water molecules returning to a surface exceeds those leaving, there is a net deposition of water from the air, which is referred to as **condensation**. The implication of this statement is that a necessary condition for condensation is saturation, since a condition of non-saturation produces a net loss of water molecules from the underlying water surface, as we have already seen. Therefore in order for air at point X in Fig. 6.1 to produce condensation, it must first achieve a state of saturation. For simplicity we can consider the two alternatives represented by paths XY and XZ. By physically adding more water vapour to the air, while keeping its temperature constant, saturation is eventually achieved at Y, at which point any further addition will initiate condensation. By cooling the air to its dew-point, saturation is achieved at Z and condensation is initiated by further cooling. In the atmosphere, the latter is responsible for the majority of condensation forms.

If heat energy is withdrawn from the water vapour molecules, the subsequent decrease in their kinetic energy and strengthening of intermolecular bonding prevent them remaining in the gaseous state. They return to liquid water and in so doing release the latent heat of vaporization they gained during evaporation (Section 6.1.3).

If moist air is in contact with a water surface, the molecules of the cooling air are readily absorbed into it. In the free atmosphere, however, there are apparently no such surfaces over which condensation takes place. If pure air, devoid of any suspended foreign matter, is cooled to its dewpoint, no condensation occurs and it may continue to take up water molecules to a state of supersaturation. Under laboratory conditions it has been possible to reach relative humidities of over 400% before condensation into water droplets takes place.

In reality, the Earth's atmosphere contains a number of impurities resulting from natural events or from human activities (see Chapter 4, Table 4.2). These provide surfaces upon which condensation may take place. The relative humidity at which this occurs is largely dependent upon the nature and number of the particles, which are referred to as condensation nuclei. Condensation on to these results in the rapid growth of water droplets. The radii of nuclei range in size from less than 10^{-3} μm to more than 10 μm. These two extremes in the range of sizes are the least effective in initiating condensation, the former producing unstable droplets readily evaporated back into the atmosphere and the latter falling quickly to the ground under the influence of gravity. The most effective particles are in the range 10^{-1} μm–1.0 μm, some of which may also be **hygroscopic** (attracting water molecules to them), and they include substances such as sodium chloride (sea salt) and ammonium sulphate.

The presence of hygroscopic nuclei may initiate condensation in free air well before it reaches its saturation vapour pressure, at relative humidities as low as 80%. Other less hygroscopic or non-hygroscopic materials, such as terrestrial dust, are less effective, but when present in large numbers they encourage condensation in slightly supersaturated atmospheres.

Condensation takes place as a result of the cooling of the atmosphere in the presence of a receiving surface, whether this be a terrestrial surface or a suspended condensation nucleus. The condensation forms produced may be classified according to the nature of this cooling process – the most important of which are **contact cooling**, **radiation cooling**, **advection cooling** and **dynamic cooling**.

Contact cooling. When terrestrial surfaces lose heat rapidly by radiation, this produces a lowering of their temperatures. Heat is then conducted from the air to the surface. As air is an extremely poor conductor of heat, this cooling does not penetrate far above the surface, particularly if there is little or no air movement. Should the temperature of the surface fall below the dew-point of this thin layer of cooled air, direct condensation of water droplets on to it occurs in the form of dew. If the dew-point temperature is less than 0°C, deposition occurs in the form of ice crystals of hoar frost. If, however, there is a small amount of turbulence in the air layers near to the ground, there will be limited mixing and a cooling to greater heights. In this case, reduction of air temperature below its dew point produces a radiation, or ground, fog which will persist until it is thoroughly mixed with the drier air that overlies it, or until temperatures rise above dew-point through solar heating.

The downslope drainage of air cooled by contact may lead to its accumulation in lower-lying areas, especially where there is standing water.

Such movement may create deep fogs, commonly referred to as valley fogs.

Radiation cooling. The atmosphere also experiences direct loss of heat by radiation. However, because of the slow rate of cooling that arises from this, it is rarely the sole cause of condensation. It can, however, enhance the cooling of the air already in contact with cold ground surfaces.

Advection cooling. Cooling may result from a horizontal mixing of air, which is referred to as **advection**. Condensation forms can be produced when two streams of air mix together if both have vapour pressures approaching saturation and if there is a relatively large temperature difference between them. If, for example, warm moist air is transported over a cool moist surface, it mixes with the shallow layer of cooled air associated with the surface. When this has taken place, the mixture may have a saturation vapour pressure, determined by its temperature, which is less than its actual vapour pressure and it is therefore supersaturated. This excess is condensed in the form of **advection fog**. These fogs may be associated with relatively turbulent airflow and may therefore be deep. The cold surface may be either sea or land, although most advection fogs are associated with sea surfaces. For example, warm moist air passing over a cool ocean current will produce advection fog. An excellent example of this occurs off the coast of Newfoundland, where warm air associated with the Gulf Stream mixes with the cold air over the Labrador Current. Warm tropical air approaching the British Isles from the south crosses cool ocean waters, particularly in spring, and it may also produce thick advection fog.

One further form of advection fog occurs where cold air flows over warm waters. Water evaporating from the surface condenses in the cold air above and produces a fog which resembles smoke or steam rising. This usually occurs where there are large temperature differences between surface and air, such as in the polar latitudes where cold air flowing from the ice caps over warmer seas produces **sea smoke**. In terms of the basic condensation process, it is produced by an addition of water vapour to the air rather than a cooling process.

Dynamic cooling. If air is forced to rise, it is subjected to **adiabatic cooling** (Box 6.3). The form of the condensation resulting from this depends upon the magnitude and rapidity of the enforced rise and the stability of the air. A gradual rising of relatively stable air will produce less spectacular condensation forms than a violent uplift of unstable air. Uplift may be the result of the passage of air over mountain ranges, of localized convection or of frontal contact between masses of air of differing temperatures at fronts (Fig. 6.5). In all cases the air rises and cools at the dry adiabatic lapse rate until it reaches saturation. Above this the air continues to cool at the saturated adiabatic lapse rate, while condensation takes place in the form of clouds. The shape and depth of these clouds vary from shallow layered forms (stratus) characteristic of stability, to clouds with great vertical extent (cumulus) that are associated more with unstable than stable air. The types of cloud are listed and described in Table 6.2.

6.1.5 Precipitation

Water that falls from clouds to the ground in either solid or liquid form is referred to as **precipitation**. Although this is derived from condensed water vapour, it is not true to say that condensation automatically leads to precipitation.

There are initially two forces operating on a cloud droplet, these being a gravitational attraction towards the Earth and the frictional force between the droplet and the air through which it is moving. These are represented by G and F in Fig. 6.6a. When these two forces balance each other, in accordance with Newton's First Law of Motion, the droplet falls towards Earth at a constant velocity – its **terminal velocity** – which is directly related to droplet size. The radius of cloud droplets ranges from 1 μm to 20 μm which, in still air, would have terminal velocities between 0.0001 and 0.05 m s^{-1}. However, the air is not still, for there are updraughts in clouds which may reach speeds of 9 m s^{-1}, which represent a further force (D) acting upon the cloud droplets.

The terminal velocities of cloud droplets are so small in relation to those of the updraughts that they are held within the cloud mass and are not precipitated. Precipitation is not usually initiated until drops have reached a radius of approximately 1000 μm, although fine drops of radius as low as 200 μm may be precipitated in particularly calm conditions. The average cloud droplet (of radius 10 μm) must, therefore, increase its volume a millionfold to reach the average precipitation drop

Box 6.3
ADIABATIC TEMPERATURE CHANGES

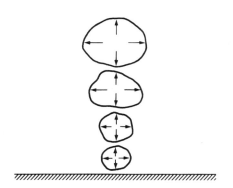

Consider a parcel of air at the ground surface which is forced to rise. As it does so, it expands in response to decreasing pressure. In so doing, the parcel does work and therefore expends energy. If the parcel is a closed system and there is no heat exchanged between the parcel and its surroundings, then the energy for work is drawn from inside it, causing its temperature to decrease. This temperature decrease is referred to as **adiabatic**.

In the atmosphere, when the air is unsaturated, the rate of temperature decrease is referred to as the dry adiabatic lapse rate and has a value of 9.8°C km⁻¹. If the air reaches saturation point, further cooling results in condensation which releases latent heat and so offsets the rate of temperature decrease. Thus an average value of the saturated adiabatic lapse rate is 6.5°C km⁻¹, but its value varies because of the relationship between saturation vapour pressure and temperature.

Stability and instability
Consider two separate vertical temperature profiles, xy and ST, referred to as environmental lapse rates. Consider also two points on these profiles, A and B. If a parcel of air at A is forced to rise it will cool at the dry adiabatic lapse rate.

At A' it will be warmer than its surroundings and will, therefore, continue to rise. If the same parcel is forced downwards, it warms at the dry adiabatic lapse rate. At A" it will be cooler than its surroundings and will continue to fall. Thus, parcel A shows a disinclination to return to its former position once given an initial upward or downward displacement, and it is therefore unstable.

In contrast, if the air at B is forced to rise, by B' it is cooler than its surroundings and it will fall back to B. If forced downwards, by B" it is warmer than its surroundings and will rise back to B. The parcel B thus shows an inclination to

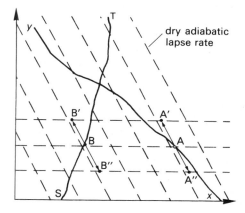

return to its former position given an initial displacement, and it is therefore stable.

In the atmosphere, stability and equilibrium are not synonymous. The atmosphere may be in a state of either stable or unstable dynamic equilibrium, its stability determining the disruption caused by an external impetus. An unstable atmosphere may continue in a state of disequilibrium once displaced, while a stable atmosphere may readily return to equilibrium.

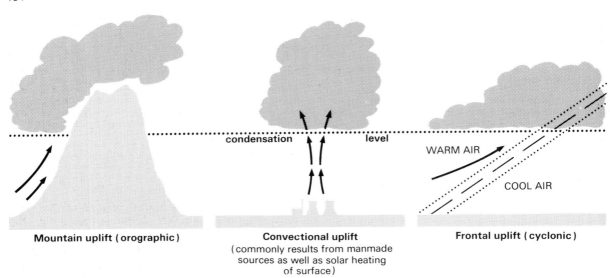

Mountain uplift (orographic)   Convectional uplift   Frontal uplift (cyclonic)
                                (commonly results from manmade
                                sources as well as solar heating
                                of surface)

Fig. 6.5 A simplified representation of dynamic cooling.

Table 6.2 Cloud types and their main characteristics. The height and temperature values given refer to the British Isles (from the *International cloud atlas*, WMO 1956)

Cloud genera	Abbreviation	Characteristics	Height of base (km)	Temperatures at base (°C)
cirrus	Ci	white filaments – fibrous appearance or silky sheen, or both	5–13	−20 to −60
cirrocumulus	Cc	thin white patch or sheet cloud without shading; composed of very small elements	5–13	−20 to −60
cirrostratus	Cs	transparent whitish cloud veil of fibrous or smooth appearance	5–13	−20 to −60
altocumulus	Ac	white and/or grey, patch, sheet or layer of cloud; composed of elements	2–7	+10 to −30
altostratus	As	greyish cloud sheet; layered appearance – totally or partly covering the sky	2–7	+10 to −30
nimbostratus	Ns	grey layer cloud associated with continual falling rain or snow; thick enough to blot out the sun	1–3	+10 to −15
stratocumulus	Sc	grey or whitish, patch, sheet or layer; formed of rounded masses	0.5–2	+15 to −5
stratus	St	grey cloud layer with uniform base which may give drizzle	0–0.5	+20 to −5
cumulus	Cu	detached cloud, dense with sharp outline; vertically developed in rising towers resembling a cauliflower	0.5–2	+15 to −5
cumulonimbus	Cb	heavy and dense cloud with considerable vertical extent; precipitation falling	0.5–2	+15 to −5

of radius 1000 μm. There have been a number of explanations offered as to why this growth should take place, of which two are discussed here.

Ice crystal process. If ice crystals and super-cooled water occur together in a cloud at temperatures between –10 and –25°C, water vapour is transferred directly from the water surface and is deposited as ice on the crystal surface (Section 6.1.1). These crystals may be the result of the freezing of supercooled water droplets at upper levels in cumulus clouds or they may have fallen from the higher cirrus clouds, which are composed almost entirely of ice crystals. Ice crystal growth

takes place, which increases their earthward velocity. The growing crystal may grow further as it falls and collides or coalesces with other crystals. If the crystal melts, it continues its fall as a water droplet and will grow as a result of coalescence with much smaller water droplets in its path (Fig. 6.6b). On reaching a certain critical size, the droplet becomes unstable and divides into many medium-sized droplets which will collide and coalesce in turn, thereby providing an efficient precipitation process operative over short intervals of time.

Collision and coalescence. Although the ice crystal process explains precipitation from clouds that have temperatures below –10°C, it fails to explain why precipitation should occur from clouds whose temperatures may be as high as 5°C, such as those in tropical latitudes. The movement of air within the clouds produces water droplets of different sizes through chance collisions and coalescence. As terminal velocities are directly related to droplet size there will be relative motion between the droplets, in which case the process of coalescence illustrated in Fig. 6.6 will take place and cause the growth of some droplets at the expense of smaller ones. Such growth may eventually produce droplets large enough to fall to the ground.

The main forms of precipitation are listed in Table 6.3. A broader classification can also be developed from the nature of the condensation process from which precipitation is derived. Thus we have **orographic** (mountain), **convectional** and **cyclonic** precipitation.

Over the Earth's surface there is considerable variation in precipitation, which may be attributed to a number of factors operating over a range of spatial scales. On a global scale, the two areas associated with convergent airflow at the surface – the intertropical convergence and the middle latitudes – are both areas of relatively higher precipitation, arising from the consequent uplift of air and its adiabatic cooling. In contrast, the subtropical and polar regions have surface airflow of a divergent type and hence have much lower precipitation due to the subsidence of air. This elementary zonation is apparent in the distribution of mean annual precipitation, especially over ocean areas (Fig. 6.7).

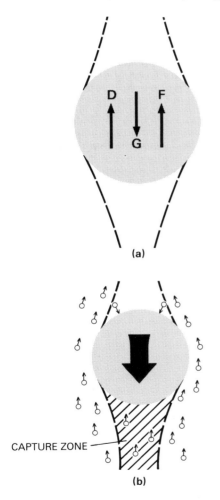

Fig. 6.6 (a) Forces acting on a cloud droplet (see text). (b) Collision and coalescence between a large falling drop and smaller cloud droplets.

The mean atmospheric circulation is, however, only one of a number of controls on the global distribution of precipitation. We must also consider, for example, the thermal structure of the atmosphere – whether it is stable or unstable. Precipitation also depends upon the number and type of condensation nuclei in the air and the presence of conditions conducive to the growth of cloud droplets.

The land masses, particularly those in the northern hemisphere, exert a marked control on the distribution of precipitation. As most atmospheric water vapour is derived from the oceans, the continental interiors may experience much drier atmospheric conditions than do locations near to coasts. This is reflected in the interior of Eurasia, for example, which derives much of its moisture from the Atlantic ocean, conveyed inland by the middle-latitude westerlies. As this air travels eastwards over the continent it loses more water to the surface than it gains through evapotranspiration. There is thus an eastward decrease in precipitation, especially in winter when frozen surfaces give up little water to the atmosphere and the divergent airflow of the Siberian high pressure restricts the penetration of moist oceanic air.

One further modification to the global distribution of precipitation results from airflow over major mountain ranges. The intensification of both condensation and precipitation processes as air is forced to rise over, for example, the Rockies or the Andes, produces higher precipitation totals.

6.1.6 The surface water balance

Of the precipitation that falls upon the Earth's surface, some is returned to the atmosphere by evaporation and transpiration while the remainder is stored on or within the surface and is available for use in surface systems. We can express this distribution in terms of the **surface water balance equation**:

$$(\text{precipitation}) \quad P = E + T + \Delta S + \Delta G + R \tag{6.3}$$

where E is evaporation; T is transpiration; ΔS is the change in soil water storage – water is held between the particles of the soil by retentive forces – some of the incident liquid precipitation will infiltrate into available spaces in the soil; ΔG is the change in groundwater storage – some water will percolate from the soil into the deeper groundwater reservoir; R represents overland flow across the surface, initially in rills but ultimately in streams and rivers. The structure of this balance varies over both time and space according to the prevailing moisture conditions in both the atmosphere and the surface. However, on a global scale, we can for simplicity consider the two basic surface forms of open water and land.

The water balance equation (6.3) may be rewritten for an open water surface, such as a large pan, in the simple form

$$E = P + \Delta V. \tag{6.4}$$

Table 6.3 Precipitation types and their characteristics

Type	Characteristics	Cloud type from which derived	Measurement
rain	waterdrop radius greater than 250 μm	Ns, As, Sc, Ac	rain gauge
drizzle	fine waterdrop radius less than 250 μm	St, Sc	rain gauge
snow	loose aggregates of ice crystals	Ns, As, Sc, Cb	snow gauge; snow run; photogrammetric
sleet	partly melted snowflakes or rain and snow falling together	as above	rain gauge
hail	pieces of ice of radius 2500–25 000 μm; concentric shells of ice	Cb	rain gauge; hail gauge
direct precipitation	directly on to surfaces under low-lying cloud – referred to as horizontal interception	St	interception gauges (experimental)

Fig. 6.7 Global distribution of average annual precipitation totals, in millimetres (after Lamb, 1972).

We may assume that there is no runoff from a water surface ($R = 0$) and that there is no transpiration loss ($T = 0$). ΔV represents a change in the volume of water contained within the reservoir and is the equivalent of storage changes under land surfaces ($\Delta V = \Delta S + \Delta G$). We can use this equation to calculate open water evaporation by recording changes in the level of water in an open pan exposed to the atmosphere (Box 6.4).

In the case of the oceans we can ignore the R term, but there is an extra inflow of water in the form of runoff from the surrounding land masses by way of rivers and of flow from one ocean area to another. By considering all the oceans as one surface we can effectively ignore interocean flow of water, and changes in storage over time are negligible. The surface water balance equation therefore becomes

$$P = E - \Delta F \qquad (6.5)$$

where ΔF is the flow of water from land to ocean. The average precipitation depth over the oceans amounts to 1066 mm or 385.0×10^{-12} m^3 of water, while evaporation amounts to 1176 mm or 424.7×10^{-12} m^3 (Baumgartner and Reichel, 1975). Over the ocean surfaces there is a net loss of 39.7×10^{12} m^3 of water returned through ΔF.

The water balance equation for a land surface must remain as written, as all components are present. However, for practical purposes, evaporation and transpiration are considered together as evapotranspiration. The equation is therefore

$$P = E_t + \Delta S + \Delta G + R. \qquad (6.6)$$

Indirect measurements may be made of both potential and actual evapotranspiration using this equation. If we assume that our surface is perfectly horizontal, then there is no gravitational acceleration of water drops along the surface, only into it. Thus R is zero. If soil moisture levels are maintained such that evapotranspiration takes place at its potential maximum rate, then, effectively $\Delta S = 0$. In this case the evapotranspirometer (Box 6.4) may be used to determine potential evapotranspiration. If irrigation of the soil tanks is not carried out, and a method of monitoring soil moisture changes (ΔS) is incorporated, actual evapotranspiration may be determined. Over the global land surface, the average depth of precipitation amounts to 746 mm, or a total volume of

111.1×10^{12} m^3, while evapotranspiration amounts to 480 mm, or 71.4×10^{12} m^3, which means that there is a surplus of 39.7×10^{12} m^3 at the surface. As storage factors ΔS and ΔG may be regarded as negligible in the long-term balance, this surplus is discharged in the form of runoff.

Within these broad global balances there are considerable differences in the structure of water balances, between oceans and between continents (Table 6.4). Of particular note is the high proportion of precipitation consumed in evapotranspiration over Africa (84%) and Australia (94%), which are predominantly tropical continents, compared to the 17% over Antarctica.

6.1.7 The atmospheric water balance

Our discussion of water balances would be incomplete without a brief reconsideration of the atmosphere, which is receiving water from the surface beneath it and returning it through precipitation. The horizontal motion of the atmosphere transports water vapour so that the surfaces from which it is derived are not necessarily those to which it returns as precipitation. This is best illustrated by reference to the data in Table 6.4. Over the oceans, the atmosphere receives 424×10^{12} m^3 of water, yet it releases only 385×10^{12} m^3, while over the land it receives only 71.4×10^{12} m^3 and releases 111.1×10^{12} m^3. The implication is that an exchange of water exists between the two, part of which will be through the flow of water vapour from over the oceans to over the land, through movement of the atmosphere.

This transport of water vapour is of significance in the distribution of heat energy in the atmospheric circulation. Water vapour transported in the atmosphere and subsequently condensed has, in releasing its latent heat of vaporization, acted as a vehicle for heat-energy transfer. The meridional transfer of latent heat (see Chapter 4, Fig. 4.7) is equatorwards between latitudes 30°N and 30°S, where the trade winds blow across the ocean surfaces. In the middle latitudes, between 30° and 65°, there is a net poleward transport. This form of heat transfer is clearly greater in the southern hemisphere where there is a larger proportion of ocean surfaces acting as sources of water vapour.

6.2 Chemistry of atmospheric water

Evaporation from the land and ocean surfaces provides an essentially pure source of water to

Box 6.4
METHODS OF MEASURING EVAPORATION (a) AND POTENTIAL EVAPOTRANSPIRATION (b) BASED UPON THE SURFACE WATER BALANCE EQUATION

(a)   Standard evaporation pan

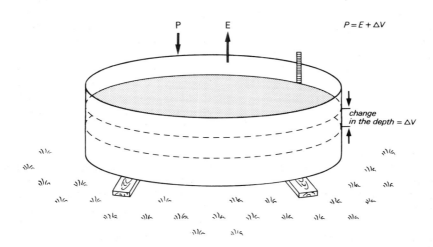

P   E $P = E + \Delta V$

change
in the depth $= \Delta V$

(b)   Evapotranspirometer

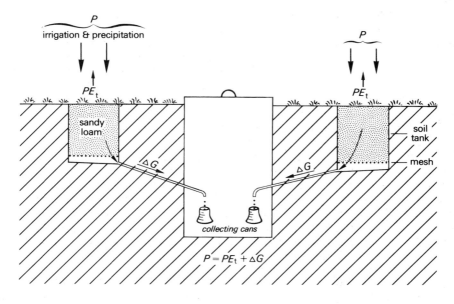

P
irrigation & precipitation

P

PE_t PE_t

sandy
loam

soil
tank

mesh

ΔG ΔG

collecting cans

$P = PE_t + \Delta G$

the atmosphere. This water contains small amounts of dissolved gases through equilibration with the more soluble constituents of the atmospheric reservoir. As a result the water in atmospheric water droplets and ice crystals is a weakly acidic solution of carbon dioxide (CO_2), sulphur dioxide (SO_2), and oxides of nitrogen (NO_x). They also carry particles such as fine dust and sea spray (Table 4.2) many of which act as **hygroscopic nuclei** (Section 6.1.4). This is because there is another route by which relatively small amounts of water (0.15% of that contained in the atmosphere) may be supplied to the atmosphere involving **bubble-bursting** at the ocean surface. Ultimately this is a more important pathway in determining the composition of rainwater than is evaporation. Bubble-bursting ejects tiny droplets of liquid seawater into the atmosphere where they readily undergo evaporation to form microparticles of sea salts. Some of these particles may fall out back into the sea or on to the land, but many form condensation nuclei around which raindrops can form. Rainwater, therefore, contains small amounts of dissolved salts representative of those found in seawater: Since Na^+ and Cl^- are the dominant ions in seawater, it follows that these are the most abundant ions in rainwater, though there is a spatial variability in the amounts, with a fairly rapid falloff over land to a more or less constant background concentration in the interior of land masses.

The deposition onto the Earth's surface of these particles and ions from a saturated atmosphere occurs through the processes of **rainout** and **washout**. Rainout occurs when material is incorporated into ice crystals or water droplets which are subsequently precipitated from the atmosphere. This may occur as free-falling precipitation or, alternatively, by the interception of cloudwater droplets by, for example, the leaves and branches of a hillside forest. Washout is the sweeping of material from the atmosphere beneath a cloud by precipitation falling through the air in which it is suspended. This can be seen during rainstorms when concentrations of particulate pollutants in the lower atmosphere have a tendency to decrease very rapidly, causing a marked improvement in visibility. One of the more complex forms of washout occurs on hillslopes, where precipitation from an upper level of cloud falls through a lower level of orographic cloud (Fig. 6.8), thereby enhancing the particulate and dissolved ion concentrations in the precipitation reaching the ground surface. This is referred to as a **feeder-seeder** mechanism.

Rainout and washout, which are collectively referred to as 'wet deposition', can deposit moderately large amounts of particulates. One of the best examples of this occurs when dust in tropical continental air (see Fig. 8.11) from North Africa is washed out as 'red rain' over western Europe.

Atmospheric water in equilibrium with CO_2 is weakly acidic (see Section 11.1.4) with a pH of

Table 6.4 Water balance of the Earth's surface (Baumgartner and Reichel 1975)

	Water volume (10^3 km³)			Proportion (%) of precipitation	
	Precipitation	Evaporation	Runoff	Evaporation	Runoff
Europe	6.6	3.8	2.8	57	43
Asia	30.7	18.5	12.2	60	40
Africa	20.7	17.3	3.4	84	16
Australia (without islands)	3.4	3.2	0.2	94	6
North America	15.6	9.7	5.9	62	38
South America	28.0	16.9	11.1	60	40
Antarctica	2.4	0.4	2.0	17	83
Arctic Ocean	0.8	0.4	0.4	55	45
Atlantic Ocean	74.6	111.1	−36.5	149	−49
Indian Ocean	81.0	100.5	−19.5	124	−24
Pacific Ocean	228.5	212.6	15.9	93	7
all land	111.1	71.4	39.7	64	36
all ocean	385.0	424.7	−39.7	110	−10

5.6 (Box 11.2). However, the presence of hydrogen ions (H^+) together with sulphate (SO_4^{2-}) and nitrate (NO_3^-) ions tends to be associated with pH values reduced to below 5.0, and in extreme cases to below 4.0. When precipitated from the atmosphere this water is now known by the more familiar term *acid rain*, which has been associated with deleterious changes in some ecosystems.

Atmospheric sulphate ions are derived from both natural sources such as the oceans and anthropogenic sources such as fossil fuel combustion and metal smelting, in approximately equal proportions as a global average. However, in heavily populated and industrialized areas of the world the latter sources increase to more than 90%. Atmospheric sulphur dioxide (SO_2) is oxidized in the presence of hydrogen peroxide (H_2O_2) to sulphuric acid (Box 6.5). NO_3^- ions are similarly derived from natural sources, including biochemical processes in soils, in addition to anthropogenic sources such as fossil fuel combustion. The photochemical oxidation of oxides of nitrogen (NO_x) results in the formation of nitric acid (Box 6.5).

The net result of these two processes is to reduce the pH of precipitation, an effect which is most marked where anthropogenic emissions of SO_2 and NO_x are greatest. Fig. 6.9 shows the low

pH values of the industrial eastern United States. In Britain, anthropogenic emissions of SO_2 are currently decreasing from a maximum in the early 1970s, while NO_x emissions continue to increase (Table 6.5).

Low pHs also occur in the vicinity of hill areas. This is due in part to the feeder–seeder mechan-

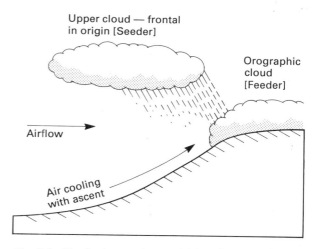

Fig. 6.8 The feeder–seeder mechanism for increasing ion concentrations in rainfall.

Box 6.5
LIQUID PHASE REACTIONS IN CLOUDS AND PRECIPITATION

The chemical transformations which take place in the presence of atmospheric water (liquid phase) are complex. The following simplified reactions have been based on Wellburn (1988) and Mason (1990).

Sulphur dioxide
Sulphur dioxide to bisulphite (HSO_3^-)

$$2SO_2 + 2H_2O \longrightarrow SO_3^{2-} + HSO_3^- + 3H^+$$

Bisulphite to bisulphate (HSO_4)

with hydrogen peroxide

$$HSO_3^- + H_2O_2 \longrightarrow HSO_4^- + H_2O$$

with ozone

$$HSO_3^- + O_3 \longrightarrow HSO_4^- + O_2$$

Sulphite and hydrogen ions

$$HSO_4^- \rightleftharpoons H^+ + SO_4^{2-}$$

Nitrogen oxides
From nitrogen dioxide

$$O_3 + NO_2 \longrightarrow NO_3 + O_2$$

$$NO_3 + NO_2 \longrightarrow N_2O_5$$

$$N_2O_5 + H_2O \longrightarrow 2HONO_2 \text{ (nitric acid)}$$

Nitrate and hydrogen ions

$$HONO_2 \rightleftharpoons H^+ + NO_3^-$$

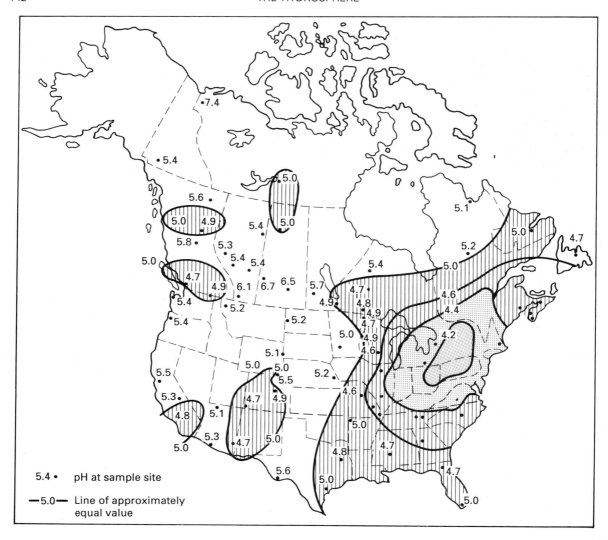

Fig. 6.9 The distribution of pH in wet deposition in North America in 1982 (from Elsom, 1987 after Miller, 1984).

Table 6.5 Annual emissions of SO_2 and NO_x for the United Kingdom in megatonnes per year (after Mason, 1990)

Year	SO_2	NO_x
1900	1.40	0.21
1950	2.30	0.30
1960	2.80	0.41
1970	3.00	0.50
1980	2.33	0.54
1984	1.77	0.63
1987	1.93	0.74

Table 6.6 Dimensions of the major ocean basins (after Briggs and Smithson, 1985)

Ocean	Area km × 10⁶	%	Mean depth m
Total ocean area	361	100.0	3650
Pacific	165	45.7	4270
Atlantic	81	22.4	3930
Indian	75	20.8	3930
Arctic	14	3.9	1250
Minor seas	26	7.2	–

ism in Fig. 6.8, but also to the fact that hilltops are more frequently enveloped in cloud. The rough surfaces intercept cloud droplets which tend to have a higher concentration of ions than precipitation. This is referred to as **occult deposition**. Low pH values may thus occur some considerable distance from sources of SO_2 and NO_x which are readily transported in the atmosphere. For example, the lowest precipitation pH values in both Scotland and Norway have been found to be associated with air which originates in industrial eastern Europe.

6.3 The oceans

So far we have considered only one of the principal fluids of the Earth–atmosphere system. The oceans, as can be seen in Fig. 4.7, also act to distribute heat away from tropical latitudes. Indeed, at 20° N and S the oceans contribute as much as 40% of the total poleward flow of energy. Approximately 71% of the Earth's surface is covered with water, of which the principal oceans are the Pacific (33%), Atlantic (16%), Indian (14%) and Arctic (3%) (Table 6.6). The mean depth of the oceans is approximately 4000 m, but there is considerable variation about this value. In the ocean trenches, such as the Mindanao Trench off the east coast of the Philippines, the depth falls to more than 9000 m, but over 75% of the ocean area depth is between 3000 and 6000 m. Around the edges of the oceans there is a shallow ledge of varying width – the continental shelf – over which water depth decreases towards the coasts. In comparison to the atmosphere, the oceans are shallower and are bounded by land within ocean basins, which determines the nature of the circulatory motion which develops within them.

6.3.1 Physical Properties of Seawater

Salinity. Seawater carries within it both dissolved and suspended matter derived from sources such as river inflow and coastal erosion (see Chapter 17). Of the dissolved matter, the ionic composition (Table 6.7) indicates that the dominant ions are those of sodium (30%) and chloride (55%), which comprise sodium chloride, or common salt. Together with other solutes these form the **salinity** of seawater. The average salinity of the world's oceans is 35 g kg^{-1} or parts per thousand (‰), which normally varies between approximately

Table 6.7 Mean ionic composition (by weight) of seawater (from Beer, 1983)

Ion		% of seawater
chloride	Cl$^-$	55.04
sodium	Na$^+$	30.62
sulphate	SO$_4^{--}$	7.68
magnesium	Mg^{++}	3.69
calcium	Ca^{++}	1.15
potassium	K$^+$	1.10
bicarbonate	HCO$_3^-$	0.41

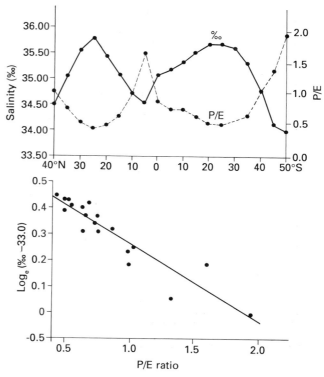

Fig. 6.10 Relationship between salinity (‰) and P/E ratio (after data from Sverdrup *et al.*, 1942).

33‰ and 37‰ in the open sea according to geographical location and season.

One of the principal factors affecting salinity is the precipitation–evaporation ratio. Fresh water inputs from the atmosphere will tend to reduce salinity, while evaporation of surface waters will tend to concentrate the solution. Thus in Fig. 6.10 there is a clear correlation between the P/E ratio and salinity, with extremes of both occurring in the trade-wind latitudes (see Table 4.4) where there is strong surface evaporation. This contrasts with the lower salinity around the equator where there are greater amounts of rainfall. There is, therefore, a **salinity gradient** in surface waters between these latitudes.

Salinity tends to increase with depth, but where there is surface concentration due to evaporation there may be a general decrease down through the upper 800–1000 m, then an increase to 2000 m or more, where values become relatively stable.

Salinity is also affected by freshwater inputs from rivers and melting ice. Coastal salinity values are much reduced by river discharge and there is a well marked landward salinity gradient through estuarine systems (Fig. 6.11).

Temperature.   The very high thermal capacity of seawater (see Table 8.29) means that large inputs of heat are required in order to bring about only moderate increases in temperature. Solar radiation passes into the water surface and is progressively absorbed as depth increases. In clear water a little less than half of the solar energy is absorbed within one metre of the surface, and most has been absorbed by 100 m (Table 6.8). The rate of decrease is expressed in the form of Beer's Law (Box 6.6). In turbid coastal waters, in which there is a greater amount of suspended sediment, the extinction coefficient is considerably larger so solar radiation may penetrate to only a metre or so.

The absorption that takes place is wavelength-selective, with ultraviolet and infrared wavebands being almost totally absorbed within the first metre below the surface (Fig. 6.12). The radiation which is transmitted to greater depths is almost wholly in the blue wavelengths in the visible spectrum, which, when back-scattered, gives the characteris-

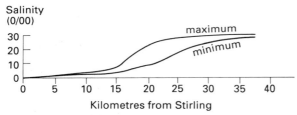

Fig. 6.11   Changes in salinity in the Forth Estuary, Scotland (from McLusky *et al.*, 1980).

Table 6.8   Radiant energy penetration into seawater (%) (after data from Pickard and Emery, 1982)

depth (m)	0	1	2	10	50	100	Extinction coefficient
clear ocean water	100	45	39	22	5	0.5	0.047
turbid coastal water	100	18	8	0	0	0	0.631

Box 6.6

Beer's Law relating to the transmission of electromagnetic radiation through translucent media:

$$I_x = I_o\, e^{-ax}$$

where flux density at distance $x(I_x)$ is related to the incident flux density I_o and the *extinction coefficient a*.

Rewritten as $\log_e I = \log_e I_o - ax$

'a' can be determined from the gradient of a straight line plot of $\log_e I$ against x:

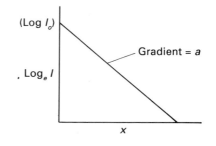

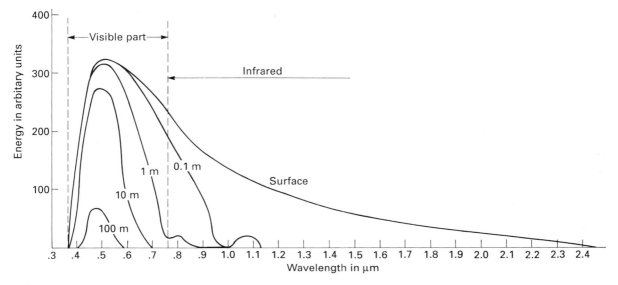

Fig. 6.12 Spectrum of transmission of solar radiation through seawater (after Neumann and Pierson, 1966).

tic colour to seawater. In the presence of phyto-plankton, scattering becomes more prominent in the green wavelengths, while other algae can cause a shift to red–yellow wavelengths.

Latitudinal variation in solar energy inputs (see Fig. 3.3) is reinforced by increases in reflection coefficients as solar elevation decreases (Table 8.1) towards higher latitudes. There is, therefore, a well marked latitudinal zonation in ocean surface temperature (Fig. 6.13).

The surface heat energy balance also varies with latitude and is dominated by latent heat flux (Q_E) which consumes more than 90% of available heat energy between latitudes 10° and 30° (Table 3.4). Sensible heat flux (Q_H) into the atmosphere is generally small in tropical latitudes (less than 4%), but it increases to more than 66% in temperate latitudes as the contrast between warmer ocean and colder atmosphere becomes more marked. Sensible heat is also transferred advectionally in the movement of the oceans, which provides a large input to waters in higher latitudes (Table 3.4) thereby supplementing the decreasing radiant energy inputs.

Inputs of geothermal heat to the oceans are, in global terms, very small but in some locations, such as over the Mid-Atlantic Ridge and in the Indian Ocean, they may result in warming. Average geothermal heat fluxes to the oceans are approximately 0.055 Wm⁻² (Section 3.5.2) but may rise to 0.227 Wm⁻² over localized sources.

With the exception of geothermal energy, oceans are heated from above, in contrast to the atmosphere, which is mainly heated from below. Left to the process of molecular diffusion alone the result would be a shallow surface layer through which there would be a very steep vertical gradient of change. However, observations of temperature

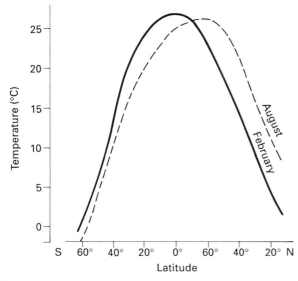

Fig. 6.13 Meridional change in sea-surface temperature in the Atlantic Ocean (after data from Pickard and Emery, 1982).

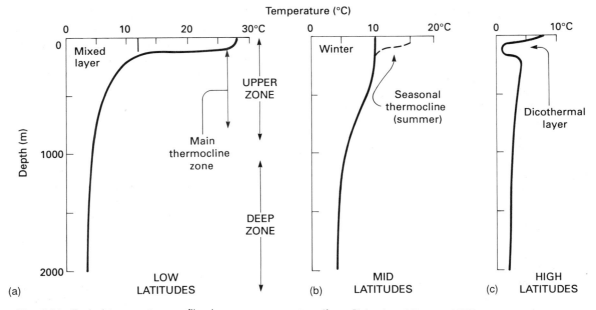

Fig. 6.14 Typical temperature profiles in open ocean waters (from Pickard and Emery, 1982).

in the uppermost 200 m of the oceans reveals a layer in which the vertical temperature distribution tends towards the isothermal. This is a result of mixing of the water by the action of winds blowing across the surface (Section 6.3.5).

Beneath this **mixing layer** the temperature falls sharply across a narrow zone referred to as the **thermocline**, which separates a warmer surface layer from cooler deep waters. In low latitudes strong surface heating accompanied by only moderate surface mixing produces a sharply defined thermocline (Fig. 6.14a). A similar feature develops during the summer in temperate latitudes but during the winter, reduced surface heating and stronger surface winds combine to deepen the mixing layer and reduce the temperature change across the thermocline (Fig. 6.14b). In high latitudes the much reduced solar heating means that water temperatures are low and tend to decrease steadily downwards. Ice melt during the summer creates a very shallow warm surface layer which overlies a layer of cold water (Fig. 6.14c). In all these cases, a profile of decreasing temperature with depth indicates a dynamically stable system in which there is little potential for free convective mixing (buoyant mixing).

Density. The density of ocean waters is expressed as the σ (sigma) value where $\sigma = \rho - 1000$, and ρ is the

true density in kg m^{-3}. The density of surface waters varies typically within the range $\sigma = 22.00-28.00$. Because water is a compressible fluid its density increases with depth such that it may exceed 50.00 at depths greater that 5000 m. However, it is common practice to use σ, which is density at atmospheric pressure. There is a relationship between density, temperature and salinity. This is illustrated in Fig. 6.15, in which the lowest densities are clearly associated with the

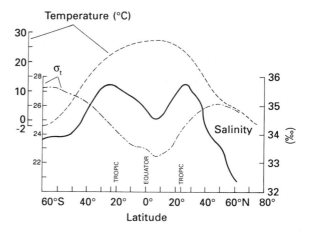

Fig. 6.15 Latitudinal variation in mean temperature, salinity, and density (from Pickard and Emery, 1982).

warmest temperatures and the lowest salinities. Thus the processes which control temperature and salinity also determine density.

6.3.2 Movement of the Oceans

There are circulation systems in the oceans which operate over a range of spatial and temporal scales, from the large-scale motion of water in ocean currents down to tidal eddies along coasts. In order to explain how ocean movement is initiated, Newton's First Law of Motion (Section 2.2.2) requires us first to identify the nature of the applied forces. These are:

1. The Sun and Moon exert a gravitational pull on the earth which results in **tidal forces**.
2. Variation in density within the oceans generates deep thermohaline circulations through **free convective forces**.
3. The movement of the atmosphere across the ocean surface causes **wind stress**, which transfers momentum into the surface water.
4. All bodies in motion across the Earth's surface are subject to **Coriolis force**, which results in a deflection of water flow to the right in the northern hemisphere and the left in the south (Box 4.4). As this is related directly to velocity, the greatest Coriolis deflections are experienced in near-surface waters.

As we have already seen, these forces act in confined ocean basins on a water body which is dynamically stable.

6.3.3 Tidal motion

The gravitational force which bodies such as the Sun and Moon exert on the earth results in a measurable and periodic upwards **tidal distension** of the ocean surface. The magnitude of the gravitational force between two bodies is related directly to their masses and inversely to the square of the distance between them (see Section 2.2). Although the sun is much further away from the Earth, its greater mass means that it exerts a

stronger gravitational force at the Earth's centre than does the Moon (Table 6.9). However, distension of the ocean surface is a result of the difference between this force and the gravitational force acting at the Earth's surface. Thus the magnitude of the distending, or *tidal*, force is given by

$$F_m = \pm \frac{2\,G \times M \times r}{R^3}$$

where F_m is the force exerted per unit mass, G is the universal constant of gravitation ($6.664 \times 10^{-11}\ kg^{-1}\ m^3\ s^{-2}$), M is the mass of the extraterrestrial body (Sun or Moon), r is the radius of the Earth (approximately 6371 km) and R is the distance from the Earth's centre to the centre of the body. The tidal force exerted by the Moon thus becomes stronger than that exerted by the Sun (Table 6.9). Whereas the principal action of these forces is along the line joining the centres of the respective bodies, over the spherical surface of the Earth there must also be a component of the forces which acts in a direction tangential to the surface.

If the whole of the Earth's surface were to be covered in water the net effect of the application of such forces would be the building up of an upward bulge of water, one facing and one diametrically opposite the Sun or Moon, while at 90° to the line joining centres there would be two depressions of the water surface (Fig. 6.16). Because the Earth rotates on its axis relative to the Sun (24 hr) and Moon (24 hr 50 min), the bulge in the water surface (the **tidal wave**) travels westwards, in a direction counter to that of rotation, and the tidal waves produced by the two bodies will move in and out of phase with each other.

In the simple model in Fig. 6.17 location A experiences semidiurnal tides which are almost equal in magnitude, while C will experience only one tide per day. Location B, on the other hand, will experience one stronger and one weaker tidal wave per day. This difference is further depend-

Table 6.9 Gravitational forces exerted by the Sun and Moon

	Mass (kg)	Distance between Earth and body (km)	Force per unit mass (dyn/g)	Tidal force (dyn/g)
Sun	1.971×10^{29}	149.5×10^6	5.876×10^{-2}	52.92×10^{-12}
Moon	7.347×10^{22}	384.4×10^3	3.317×10^{-3}	116.04×10^{-12}

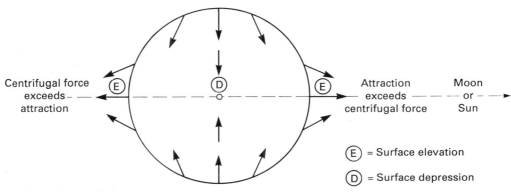

Fig. 6.16 Action of tidal forces on the earth's surface.

ent on the angle of **declination** 'a' of the Sun or Moon relative to the Earth.

In the example in Fig. 6.18 another fluctuation in tidal range is also apparent, which has a periodicity of nearly 15 days. This arises because the gravitational effects of the Sun and Moon move in and out of phase with each other. When in phase (Fig. 6.19) the alignment of forces produces a combined tidal bulge and corresponding depression, but when out of phase, the forces act at 90° to each other (Fig. 6.19) so the respective bulges and depressions interfere. These two conditions are referred to as **spring** and **neap** tides.

The Earth's water surface is, of course, not continuous but separated into a number of ocean basins of varying size and depth (Table 6.6). Such basins possess their own natural periodicity, or **resonance** (Box 6.7), rather like water slopping about in a domestic basin. When the resonant frequency is of the order of 12–13 hours, as in the Bay of Fundy in the Maritime Provinces of Canada, the effect of a semidiurnal tide will be to produce tides of exceptionally large range. The highest tidal ranges tend to be experienced in such embayments, whereas in the open ocean, tidal ranges are at their smallest. The effect of land masses is to produce complex tidal interactions. The North Sea, for example, experiences its own tidal regime but is also affected by those from the Atlantic from the north and the English Channel from the south.

Because of the Earth's rotation, tides are affected by Coriolis force, which imparts a component at 90° to the direction of motion of the tidal wave. Thus, in a large water body, tides tend to vary around a nodal point (**amphidromic** point) where there is little or no tidal variation (Fig. 6.20). Tidal range increases outwards from these points while the timing of the tidal peak, indicated by the **cotidal lines**, rotates anticlockwise. Where tidal mechanisms are complex, such as in the North Sea, there will be several such amphidromic points. The effect of Coriolis force in estuaries is to deflect tidal flow to the right or left, depending on the hemisphere, which results in a cross-channel slope on the water surface and a tidal circulation.

The passage of tidal waves results in the horizontal motion of the water. In the open sea the mean **tidal current** velocity (V) is given by

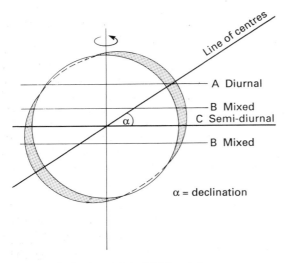

Fig. 6.17 A simple model of tidal variation.

$$V \times d = A \ (g \times d)^{0.5}$$

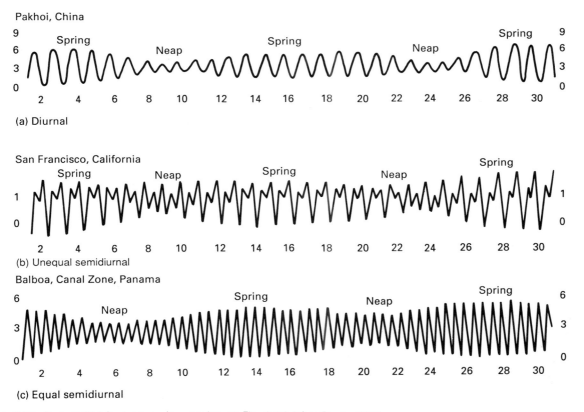

Pakhoi, China
(a) Diurnal

San Francisco, California
(b) Unequal semidiurnal

Balboa, Canal Zone, Panama
(c) Equal semidiurnal

Fig. 6.18 Typical tidal fluctuations (a – c relate to Fig. 6.17) (after Davis, 1972).

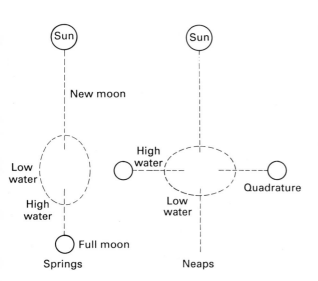

Fig. 6.19 Interaction of lunar and solar tidal forces.

where d is the depth of the water, A is the tidal amplitude, and g is acceleration due to gravity. The range of values of V for the open ocean is 0.02–0.04 m s^{-1} for a mean ocean depth of 4000 m. However, as depth decreases across the continental shelf, there is a marked increase in tidal current velocity (Fig. 6.21) to more than 0.5 m s^{-1}. In coastal waters, factors such as bottom topography and friction, channel configuration and locally modified tidal amplitudes influence the tidal current, which may exceed 1.0 m s^{-1} at the water surface. The net distances gained by water as a result of tidal flow are, however, relatively small due to the reversal of flow direction during ascent and descent through a tidal wave.

6.3.4 Thermohaline motion
The variation in density in ocean waters is generally small and results from differences in temperature or salinity. Despite the fact that water depths are small relative to surface area and the

---

Box 6.7

RESONANCE OF OCEAN BASINS

The period of the resonant wave (T) is given by

$$T = \frac{4 \cdot L}{\sqrt{g \cdot d}}$$

where L = length of the water body, d = depth of the water body and g = acceleration due to the earth's gravity.

(a)   **(b)**

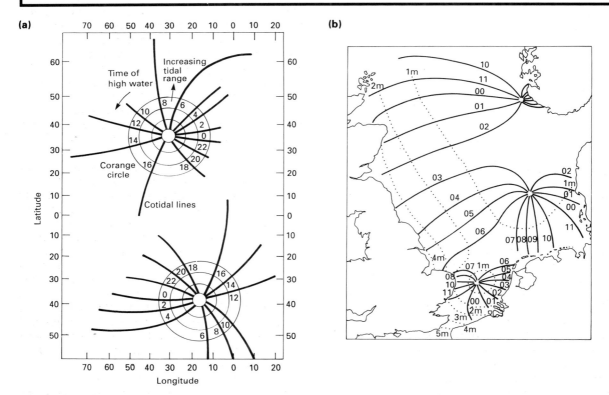

Fig. 6.20   (a) A simple model of amphidromic points in an ocean (from Russell and Macmillan, 1952). (b) Amphidromic points in the North Sea (after Harvey, 1979).

vertical distribution of density through the water column is one which provides dynamic stability, thermohaline circulations are of major significance in the large-scale movement of the oceans, where massive deep meridional motion results from the density contrasts between cold polar and warm tropical waters.

Considering temperature as the primary control, density-driven meridional circulation can perhaps be seen as two single convectional cells (Fig. 6.22a), in a similar manner to the single Hadley cells in the atmosphere (Fig. 4.13). However, the effect of precipitation and evaporation is to modify the density distribution in lower latitudes such that the density gradient effectively operates towards equatorial latitudes. The resulting thermohaline circulation is, therefore, complex (Fig. 6.22b). In the Atlantic, for example, the densest water is Antarctic bottom water, which moves northwards under the slightly less dense

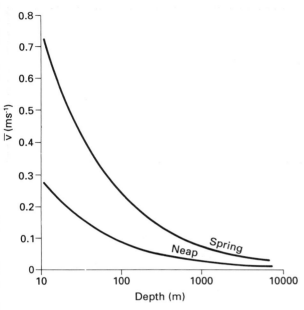

Fig. 6.21 Relationship between tidal current velocity water depth for equilibrium spring and neap tides.

Atlantic deep water. The deep convergence of these waters does not take place under the equator but meanders between 40° and 20°N.

6.3.5 Wind-driven motion

As the atmosphere moves across the oceans there is frictional resistance between the two fluids. This applies a horizontal stress to the water surface which initiates motion in the surface layer. There is, in effect, a direct transfer of momentum between the fluids. The stress exerted by the air is proportional to the square of the surface wind speed (u). The motion which is initiated at the surface extends down through the water to a depth of only 100 m or so. It is subject to Coriolis deflection, so the surface water moves obliquely to the wind, at 45° to the right in the northern hemisphere, and 45° to the left in the southern. Beneath the surface, momentum is carried downwards by internal friction within the water and there is a progressive displacement from the surface flow direction, eventually reaching a point where water is flowing in the opposite direction. This velocity–direction–depth relationship is expressed in the Ekman spiral (see also Fig. 4.9). Over the whole near-surface layer set in motion by the wind, net transport is 90° to the right of the wind direction in the northern hemisphere.

In the oceans, the primary circulatory system of the atmosphere sets water in motion in the form of surface **ocean currents** (Fig. 6.23). These are best developed where the winds are strongest or most consistent in direction, including the circumpolar westerlies, particularly in the southern hemisphere, and the trade winds (see Table 4.4). Either side of the equator there are large anticyclonic ocean circulations, referred to as **gyres**, which have a warm polewards arm in the west and a cool equatorwards arm in the east. The former is the stronger of the two, so the circulations are asymmetric. These large gyres are driven primarily by the trade winds which are consistent in terms of both direction and speed. Between the large gyres the convergence of the wind systems from the two hemispheres sets up two smaller narrow gyres, between which there is a westerly current, the equatorial counter current. Polewards, the great subtropical gyres link into the circumpolar westerly motion of the middle latitudes.

In the absence of major continental land masses in the southern hemisphere, the westerly winds acquire their greatest strength and the Antarctic circumpolar current sweeps unhindered around the earth. In contrast, in the northern oceans there are strong poleward currents of warm water in both the Atlantic (North Atlantic drift) and Pacific (North Pacific drift) oceans which are accompanied by relatively weaker cool northerly polar currents such as the Labrador (Atlantic) and Oyashio (Pacific). Currents in polar regions can be complex, due in large part to the ephemeral nature of wind systems in these latitudes, but also to the effect of seasonal freezing and melting of ocean surfaces.

The linear velocity of ocean currents is relatively slow, but measured in terms of volume flow, we can see from Table 6.10 that very large quantities of water are in motion, ranging from 10×10^6 to 2000×10^6 m^3 s^{-1}. For comparison, the mean discharge from the Tay river system in Scotland, the largest in the United Kingdom, is only 155 m^3 s^{-1}. Because of the very high thermal capacity of seawater, these currents are carrying very large amounts of stored heat away from the tropical region to be released into the atmosphere as sensible and latent heat. The oceans, therefore, operate alongside the atmosphere in redistributing energy in the Earth–atmosphere system.

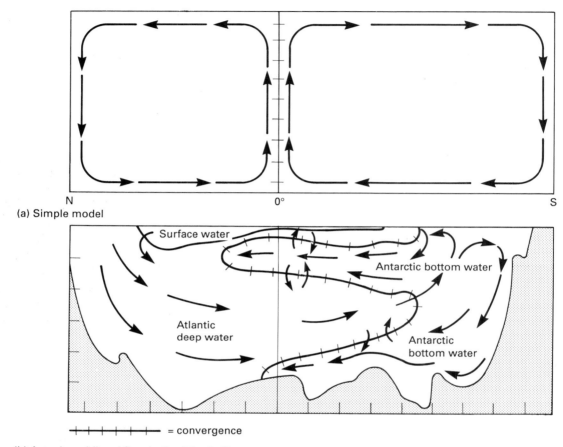

(a) Simple model

= convergence

(b) Actual meridional flow in the Atlantic Ocean

Fig. 6.22 Thermohaline circulation of the oceans. (a) Simple model. (b) Actual meridional flow in the Atlantic Ocean.

Table 6.10 Volume flow in the major ocean currents (from Beer, 1983)

Current	Flow ($m^3s^{-1} \times 10^6$)
Antarctic circumpolar	200
Gulf Stream	100
Kuroshio	65
Agulhas	40
Pacific North Equatorial	30
Equatorial countercurrent	25
East Australian	20
Peru	18
Beguela	15
Flinders	15
California	12
West Australian	10
Brazil	10
Pacific South Equatorial	10

6.3.6 The Ocean System

The oceans are an open system. They receive inputs of water through precipitation, river discharge and, to a lesser extent, direct groundwater discharge; of solid and gaseous matter in river sediments and coastal erosion and deposition and solution from the atmosphere; and of organic matter from the lifecycle of marine organisms. Water is lost principally though evaporation into the atmosphere and groundwater seepage; solid and organic matter accumulates on the ocean floor or is deposited along coasts; and gaseous matter is emitted directly into the atmosphere. Energy exchanges are considerably more difficult to define and are bound into a complex coupling with atmospheric and terrestrial systems.

Adopting the simple form of energy system

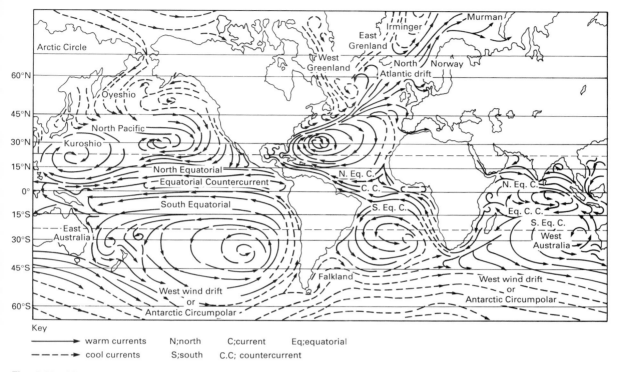

Fig. 6.23 Major surface ocean currents of the world (from Tolmazin, 1985).

which we used for the atmosphere, we can identify energy inputs into potential and kinetic energy
stores in ocean circulations (Fig. 6.24). Potential
energy is gained principally from solar and
geothermal inputs which directly influence water
density, from tides, and from the rotational energy of the earth. The conversion of potential to
kinetic energy is enhanced by the direct uptake
of energy of motion from the atmosphere. The
circulation is retarded through internal friction and
through external friction at the ocean floor and
particularly along coasts. Apart from friction, some
of which will be returned to the potential store as
heat, energy is lost through radiative transfer and
sensible and latent heat exchanges with the atmosphere. Thermal coupling with the atmospheric
system is particularly complex. Air motion generates currents, which in turn distribute heat through
the surface layer of the oceans. At the same
time, the transfer of sensible heat and latent
heat from the oceans to the lower atmosphere
operates directly on atmospheric circulation
systems.

6.4 Chemistry of the land-based hydrosphere

The chemistry of surface water, soil water and the
fresh water of streams, rivers and lakes can be
understood as the progressive alteration of the
relative composition and relative concentration of
what was originally rainwater. This alteration is
effected by exchanges, both additions and losses,
with the soil and regolith, involving weathering
reactions, and with plants by nutrient uptake and
decomposition of dead organic matter. Effective
concentrations are altered by dilution and concentration, occasioned by precipitation and
evaporation respectively. However, the most significant of these processes is probably the addition of dissolved and suspended matter originally
derived from the rocks of the lithosphere by
weathering.

The traditional focus when considering the
processes of weathering and their outcomes is
the solid weathering products which contribute to
the **regolith** and to the solid phase of soils. The

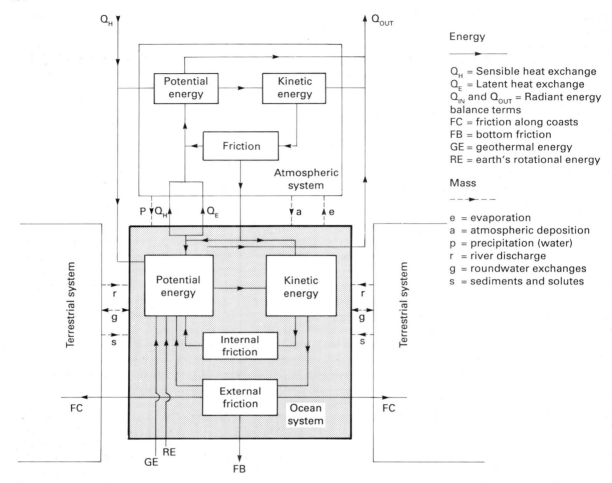

Fig. 6.24 A systems model of the oceans.

residence time for these solid particles, or rather their chemical constituents, in soil and regolith may be many thousands of years. So, from the perspective of global cycles of materials both of these compartments act as **reservoirs**, or **sinks**, albeit short-term sinks when viewed on a geological timescale, where elements are in storage.

It is rather the dissolved products of weathering and the suspended sediment load of terrestrial waters (often secondary minerals derived from regoliths) which contribute to the flux of elements between the lithospheric and hydrospheric reservoirs involved in the global circulation of materials that are important in understanding such circulations. Of course the proportion of dissolved

products from any weathering environment depends on the minerals present in the bedrock and on their solubility.

Waters from calcareous catchments are distinguished by the predominance of Ca^{2+} and HCO_3^- over other dissolved species, by high **total dissolved solids** (TDS) and low particulate suspended residues from weathering. On the other hand, waters draining igneous catchments have low TDS values, a more varied chemistry, and contain secondary clay minerals in suspension. Of course, one would expect such contrasts on the basis of a knowledge of weathering reactions (see Chapter 11).

None the less, the composition of real river

waters cannot be explained solely as a result of bedrock composition. That we need to understand the controls on the composition of river waters becomes apparent when we consider that it is the solute load of streams and rivers which represents the flux of elements through the land-based part of the **hydrological cascade**. This flux is also at least partially responsible for the maintenance of the steady-state composition of the oceans. It also forms an important part of the circulation pathway of elements which lack a gaseous or volatile state, and which are involved in the so-called **sedimentary** cycles (see Chapter 7). The reasons for the observed differences in river water composition were discussed in a now classic paper published in 1970 (Gibbs, 1970). In investigating global water chemistry Gibbs examined the behaviour of cationic and anionic ratios which incorporated aspects of the chemistry of both rainwater (see Section 6.2) and dissolved weathering products, as a function of total dissolved solids (TDS), for rivers (and some lakes and oceans). The ratios used were $Na^+/Na^+ + Ca^{2+}$, and $Cl^-/Cl^- + HCO_3^-$ and the data when plotted were found to fall within an envelope with two diverging arms which represented two series of transitions in water chemistry, and allowed three mechanisms to be identified as controlling world water chemistry (Fig. 6.25). Similar conclusions were found to result from the use of an anionic ratio.

6.4.1 Controls on water chemistry

Where water chemistry is dominated by rainwater, sodium will be the dominant cation because rainwater is (as seen in Section 6.2) greatly diluted seawater in terms of its composition, so the ratio $Na^+/Na^+ + Ca^{2+}$ will be close to 1. Furthermore, the TDS value will be low, as the salts derived from seawater are greatly diluted with pure water from evaporation. Any river where water chemistry is dominated by rainwater will fall on the bottom right of the graph (Fig. 6.25). The typical rivers in this category are those of tropical Africa and South America, draining large basins of low relative relief, well weathered substrate and high rainfall. The supposition is that under such conditions the contribution of rock weathering to the dissolved load of the river will be small, and its chemistry will be dominated by that of precipitation. This state of affairs, therefore, is known as **precipitation dominance**. However, a cautionary remark is

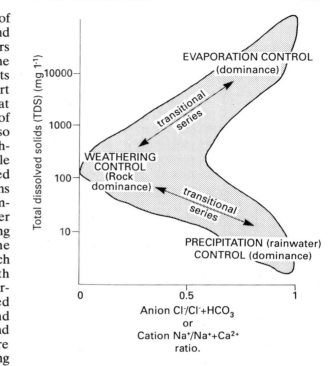

Fig. 6.25 Processes controlling the chemistry of global surface waters (after Gibbs, 1970).

pertinent here, for the low contribution of dissolved products of weathering to the solute loads of rivers in such environments is also under the control of ecological factors. This is because the large biomass of the native forest ecosystems of the lowland tropics, with their huge and rapid within-system nutrient cycles, effectively immobilizes large quantities of elements which would otherwise be available for denudation.

River waters, the chemistry of which is conditioned by the weathering reactions in their drainage basins, are said to exhibit **rock dominance** or to be under **weathering control.** They tend to show intermediate values for TDS, and occupy the first half of the cation ratio axis (Fig. 6.25). The value of the cation ratio, however, will depend on the nature of the minerals undergoing weathering in the catchment rivers draining igneous rocks having a higher value than those with a significant proportion of carbonate rocks in their catchment. Precipitation and rock dominance are not mutually exclusive, and many rivers exhibit properties between the two extreme compositions. Accordingly, rock and precipitation dominance controls are best seen as the two end-members of a continuous series. Indeed the role of each can change spatially within a catchment, and the overall effect of one may come to dominate the river's chemistry even though spatially it is restricted in its influence to the minority of the catchment area. Such a case is the Amazon, where 85% of the solute load of the river is derived from weathering in the small part of its catchment in the Andes, where rock dominance is the main factor.

The remaining important mechanism controlling the chemistry of rivers is associated with the evaporation and crystallization which occurs in some catchments in hot and arid environments. Here, evapotranspiration greatly exceeds precipitation, and the evaporation of water from rock-dominated rivers increases the TDS, and as evaporation continues $CaCO_3$ will eventually precipitate, thereby increasing the value of the cationic ratio. As a consequence such rivers comprise a series which forms the upper arm of the graph (Fig. 6.25) with a positive slope as both TDS and cationic ratio increase towards the upper end member of the series represented by the composition of seawater. This situation is termed **evaporation control** or **dominance**.

6.4.2   Some implications of global water chemistry

These consideration arising from Gibbs' work have considerable significance for the understanding of global biogeochemical cycling, which will be developed further in Chapter 7. This significance is perhaps twofold. First, it becomes apparent that the flux of elements as the solute load of streams and rivers linking their release from the lithospheric store by weathering to the oceanic reservoir will vary considerably in different environments according to the particular dominance of these control mechanisms. Secondly, it would appear that variations in relief, vegetation and actual bedrock composition (as opposed to its relative role) are all of secondary importance compared to the interplay between atmospheric precipitation, rock dominance (through weathering) and evaporation/crystallization in controlling world water chemistry, at least or the global scale. These secondary factors become of paramount significance, however, when considering the water chemistry within any one catchment.

So under precipitation control, although the total flux will depend on total discharge, the concentrations of elements will be low, and for any given discharge the flux will be smaller than for a rock-dominated river of the same discharge. The implication is that either elements are not being released by weathering, and therefore remain immobilized in rocks, or if released they are following alternative pathways. Here the contenders are uptake by plant roots and incorporation in the biomass of living organisms, or reprecipitation, resynthesis or immobilization within the regolith or soil.

Similarly, under arid conditions, although the concentrations of TDS may be high, total discharges are likely to be low because of evaporative losses, so that total element flux is also likely to be small. Furthermore, some of the elements released by rock weathering will be immobilized as precipitated solids, further reducing the total flux.

Further reading

The first two texts (the first in particular being easy to read) present the case for a unified approach to the ocean and atmosphere as a single system, while the third is an overview of the hydrological cycle:

Perry, A.N. and J.M. Walker (1977) *The Ocean–Atmosphere System*. Longman, London.
Harvey, J.C. (1976) *Atmosphere and Oceans: our Fluid Environments*. Artemis Press, London.
Berner, E.K. and R.A. Berner (1988) *The Global Water Cycle*. Prentice-Hall, Englewood Cliffs.

The next group of references is concerned with water in the atmosphere, concluding with a more chemical perspective:

Baumgartner, A. and E. Reidell (1975) *The World Water Balance*. Elsevier Amsterdam.
Mason, B.J. (1975) *Clouds, Rain, and Rainmaking*. Cambridge University Press, Cambridge.
Miller, D.H. (1977) *Water at the Earth's Surface: an Introduction to Ecosystem Hydrodynamics*. Academic Press, New York.
Ministry of Agriculture, Fisheries and Food (1967) *Potential transpiration*. Technical Bulletin No 16. HMSO, London.
Sumner, G. (1988) *Precipitation: Process and Analysis*. John Wiley, New York.
Warneck, P. (1988) *Chemistry of the Natural Atmosphere*. Academic Press, London.

The following suggested reading covers the oceans as physical systems, in terms of ocean chemistry, and, although not treated explicitly in this chapter, as ecological systems, with some references on marine ecology included:

Stowe, K.S. (1984) *Principles of Ocean Science*, (2nd edn). John Wiley, Chichester.

Thurman, H.V. (1987) *Essentials of Oceanography.* Merrill, Columbus.

Gross, M. Grant (1989) *Oceanography: a View of the Earth,* (5th edn). Prentice-Hall, Englewood Cliffs.

Broecker, W.S. (1974) *Chemical Oceanography.* Harcourt Brace Jovanovich, New York.

Holland, H.D. (1978) *The Chemistry of the Atmosphere and Oceans.* John Wiley, New York.

MacIntyre, F. (1970) Why the sea is salt. *Scientific American,* **223**, 104–115.

Riley, J.P. and R. Chester (1971) *Introduction to Marine Chemistry.* Academic Press, London.

Barnes, R.S.K. and R.N. Hughes (1988) *An Introduction to Marine Ecology,* (2nd edn) Blackwell, Oxford.

Berger, W.H., V.H. Smetack and G. Wefer (eds) (1989) *Productivity of the Oceans: Present and Past.* Wiley, New York.

Cushing, D.H. and J.J. Walsh (eds) (1976) *Ecology of the Seas.* Blackwell, Oxford.

Turekian, K.K. (1976) *Oceans.* Prentice-Hall, Englewood Cliffs.

The following sources provide insight into various aspects of the land-based hydrosphere:

Degens, E.T., S. Kempe and J.E. Richey (eds) (1990) *Biogeochemistry of Major World Rivers.* John Wiley, New York.

Drever, J.I (1988) *The Geochemistry of Natural Waters.* Prentice-Hall, Englewood Cliffs.

Gibbs, R.J. (1970) Mechanisms controlling world water chemistry. *Science,* **170**, 1088–1090.

Raiswell, R.W., P. Brimblecombe, D.L. Dent and P.S. Liss (1984) *Environmental Chemistry.* (especially Chapters 2 & 3) Edward Arnold, London.

Ward, R.C. (1975) *Principles of Hydrology,* (2nd edn). McGraw-Hill, Maidenhead.

Lastly are some texts which examine the relationships between plants and water:

Bannister, P. (1976) Water relations of plants, in (P. Bannister) *Introduction to Physiological Plant Ecology.* Blackwell, Oxford.

Fitter, A.H. and R.K.M. Hay (1987) *Environmental Physiology of Plants,* (2nd edn) Academic Press, London.

Grace, J. (1983) *Plant–Atmosphere Relationships.* (Outline Studies in Ecology Series) Chapman & Hall, London.

Meidner, H. and D.W. Sheriff (1976) *Water and Plants.* Blackie, London.

The ecosphere

7.1 The biosphere and ecosphere

At the top of the lithosphere, throughout the hydrosphere and into the lower atmosphere, lies a **transition zone**, which both contains and is created by an enigmatic arrangement of matter which we know as **life**. The existence of a global veneer of life (not forgetting the dead and decaying remains of that which once was alive) is undoubtedly the most profound feature of the surface of the planet.

Life far outweighs its small relative mass in the significance of its effect on the nature of the lithosphere, hydrosphere and atmosphere. In this chapter this living veneer will be termed the **biosphere**, and the biosphere together with the transition zone which supports it and with which it interacts will be called the **ecosphere** (Cole, 1958; Hutchinson, 1970). Hence the ecosphere model includes the physical systems already discussed, *viz.* the upper part of the lithosphere, a major part of the atmosphere and most of the hydrosphere (Fig. 7.1), insofar as they have functional links involving transfers of energy and matter with the living material of the biosphere.

7.2 Structural organization of the biosphere

The biosphere at the global level can be treated as one large living system. We shall not define the boundaries of this system too closely, for that would involve a precise definition of what life is and what it is not. Admittedly most of us would claim to be able to decide intuitively if an object is alive or not, but that is not quite the same thing, and this is not the place to embark on a philosophical discussion as to the nature of life. So we shall make the *a priori* assumption that a distinction exists between life and non-life and that our

'biospheric system' contains all living material on the planet. What then are the elements composing this system, the biosphere? In what ways are these systems elements aggregated and organized? What are their properties and what are their links with each other and with the other systems that form the compartments of the ecosphere model?

These are awesome questions when one reflects that estimates of the number of different kinds of organisms constituting the biosphere range between two and four million, and that the notion of 'life' encompasses a vast spectrum of phenomena associated with all of these organisms. Fortunately, there are certain characteristics to be found in the simplest of them which also prove essential for the most complex. The simplest organisms consist of a single cell and even the most complex consist of comparatively few cell types. The **cell** has therefore come to be regarded as the simplest independent structure that possesses all of the necessary properties of life. In attempting to answer these questions, therefore, it is appropriate to model the biosphere in terms of the structural and functional organization of the living cell. To do this involves a consideration of the molecular basis of cellular activity but, fortunately, cells contain relatively few types of molecule and, although these include the most complex molecular structures known, many are universal in their occurrence in the biosphere. The generality of a biosphere model at the cellular level should, therefore, be assured.

7.2.1 The chemistry of life
Fig. 7.2 makes it plain that the organism and the cell are merely two of a number of levels of organization at which we can examine the structure of living systems. If we adopt both a **reductionist approach**, and opt to model the biosphere at the **cellular level**, then we can reduce the cell to its

constituent parts at lower levels of organization. In this way we can begin to understand the living cell, and perhaps the nature of life itself, in terms of its elemental and molecular chemistry and its supramolecular structures. In systems terms we are modelling the cell through its component subsystems. There is nothing unique in the chemical elements, the atoms of which form the building blocks of living systems (Table 7.1). However, the chemistry of life is not just a reflection of the chemistry of the environment. Living systems have concentrated some elements, such as carbon, in proportions far beyond any obvious source at the planet's surface, but have rejected other naturally occurring elements, such as silicon, even when they have been readily available (see Chapter 2, Section 2.1.4). Now, not only does this selection raise questions concerning the reasons why certain elements have been incorporated in living systems at the expense of others, but also prompts the enquiry 'What effect has this differential selection had on the elemental composition of the atmosphere, hydrosphere and lithosphere?' (see Section 7.3).

There are limits, however, to the usefulness of these elemental inventories in providing answers to such questions, and to what they can tell us about the structure of the biosphere and the nature of life. We need to know how the atoms of the elements are organized, what molecules and compounds they form and what structures these units build to produce the living cell.

Equally, when we turn to the molecular chemistry of life we find that living systems can be described in terms of their constituent chemicals, and are subject to the same physical and chemical laws that govern non-living systems.

The constituent chemical elements of the biosphere can be visualized as being combined in living systems into both relatively simple inorganic molecules (often ionized in solution) and organic carbon compounds. The vast majority of naturally occurring compounds containing carbon are truly organic (see Chapter 2, Section 2.1.4), and they attain a level of complexity and sophistication in the arrangement and organization of their molecules not experienced in non-living systems. However, this complexity is achieved by carbon in combination with only some half-dozen elements – mainly hydrogen, oxygen and nitrogen. The most important of these organic carbon compounds in the living cell are the fatty acids, simple sugars (monosaccharides), mononucleotides, amino acids and heterocyclic bases (Chapter 2, Fig. 2.3). These in turn form the precursors of very complex molecules indeed: the lipids (fats), the polysaccharides (complex sugars, starch, cellulose), nucleic acids (RNA, DNA) and proteins (Box 2.3). These large molecules are known to chemists as high polymers. The artificial synthesis of such compounds is the basis of the synthetic fibre and plastics industries, but in nature there seems to be an almost infinite variety of compounds formed in this way.

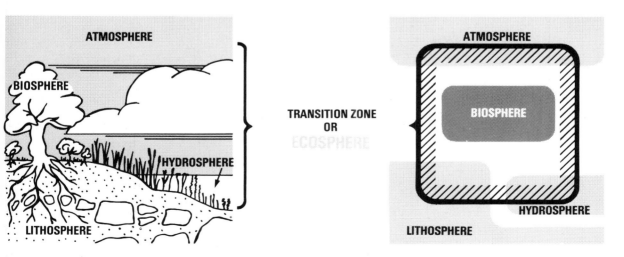

Fig. 7.1 Defining the biosphere and the ecosphere.

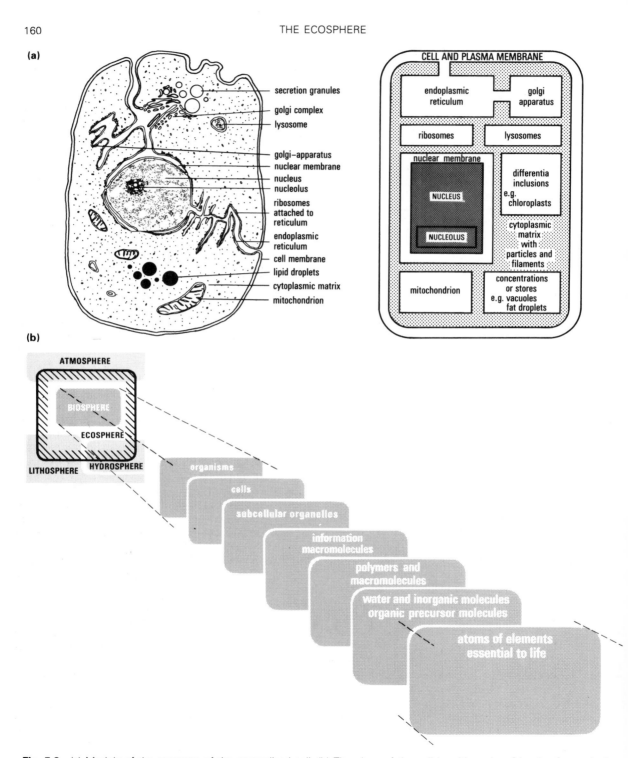

Fig. 7.2 (a) Models of the structure of the generalized cell. (b) The place of the cell in a hierarchy of levels of organization of living systems.

Table 7.1 Percentage atomic composition of the biosphere, lithosphere, hydrosphere and atmosphere, for the first ten elements

Biosphere		Lithosphere		Hydrosphere		Atmosphere	
H	49.8	O	62.5	H	65.4	N	78.3
O	24.9	Si	21.22	O	33.0	O	21.0
C	24.9	Al	6.47	Cl	0.33	Ar	0.93
N	0.27	H	2.92	Na	0.28	C	0.03
Ca	0.073	Na	2.64	Mg	0.03	Ne	0.002
K	0.046	Ca	1.94	S	0.02		
Si	0.033	Fe	1.92	Ca	0.006		
Mg	0.031	Mg	1.84	K	0.006		
P	0.030	K	1.42	C	0.002		
S	0.017	Ti	0.27	B	0.0002		

The search for structure at the molecular level in the cell is, however, not yet finished. There are in fact three levels of structure. The first we have just dealt with (primary). This level involves the arrangement of the monomers (sugars, amino acids or nucleotides) to form the polymer chain. The secondary structure is the way in which the chain itself is coiled or folded (in the proteins, for example, this often involves bonds between the sulphur atoms of some of the constituent amino acids). The tertiary level describes the way in which several polymers may come together, again with a definite three-dimensional arrangement. Because of the carbon bond angle of 110°, these arrangements are often helical or spiral in configuration. Giant molecules formed in this way are known as **macromolecules** and they have high molecular weights: haemoglobin, the protein in red blood cells, has a molecular weight of 68 000 (see Chapter 2, Section 2.1.4).

Among these macromolecules of the cell there is an important distinction between the proteins and nucleic acids and the simpler polymers. This is not because the types of chemical bond linking the building blocks together are particularly complicated, but because these precursor molecules are more diverse and because they occur in the structure of the molecules in an exact and specific order, or sequence. These sequences mean that such macromolecules carry an enormous amount of information, a property of profound significance in the functional activity of the cell. It can be said that if the word 'life' has meaning, it is at the level of these information macromolecules that it begins to take effect (Box 7.1).

7.2.2 Structure at the cellular level

Of course, cells differ considerably in size, shape and many other characteristics, even within the same organism. This specialization of cells, however, merely represents the extreme development of a particular property or function, which in principle all cells possess or have possessed at some time. In other words, at some stage in their development all cells have many features in common. A model of the cell as a living system is shown in Fig. 7.2, but this diagram also serves to make clear the existence of a level of organization between the molecular and the cellular. There are a number of subcellular structures known as organelles, but here we can make a more basic subdivision of the cell. The first of these subdivisions is the cytoplasm, which is a viscous fluid, largely water, but with inorganic ions, simple organic molecules and macromolecules dispersed through it and forming a colloidal suspension. It has an enigmatic physical condition best described by the physical state known as 'liquid crystal', as it displays properties of both fluid and crystalline states. This cytoplasmic matrix contains the subcellular organelles, one category of which – the cell membrane – encloses it and penetrates deeply and intimately through it. All of these organelles have definite structures built up of combinations of complex structural macromolecules and are best thought of as biochemical compartments within the cell. The second subdivision – the nucleus – is also really an organelle and it too is a biochemical compartment. Its importance, however, lies in the fact that it is the prime site in the cell of the information macromolecules of the nucleic acids.

The hierarchical approach adopted here has shown us that, in the structure of the living cell, a level of organization is reached which is unmatched in the non-living world. If the *a priori* assumption that there is a distinction between life and non-life is true, it is not one of kind but one of level. It is in the level of complexity and precision in the internal organization of the cell and the organism that the distinction must lie. But this is not the complete story, for the model of living systems developed so far is a static one. The precise and definite structures made up of a multitude of complex chemicals have equally precise functions. The nature of life cannot be fully appreciated until we have considered the functional organization and activity of the cell.

7.3 Functional organization and activity of the cell

Function, of course, implies the performance of work and hence the utilization of energy. This has led many authors to draw an analogy between the living cell and a manufacturing plant. Although perhaps overworked, this analogy remains useful as there are indeed parallels in the functional organization and activities of the cell and those of a factory. Like a factory, the cell needs raw materials from which to manufacture its products. It requires an energy supply to power the production line and to transport both materials and products, as well as to dispose of waste. However, in a factory the whole process works only because it has been designed to do so. By analogy, the cell must also possess such design specifications and blueprints.

Three kinds of work take place in this factory. Chemical work is performed both for cell maintenance and during active growth, when the cell makes and assembles all of the complex components required for repair or to make a new cell from comparatively simple substances. These processes are collectively called **biosynthesis**. Secondly, there is the work of transport, where materials are moved within the cell or across its bounding membrane. Such work often involves changes in the relative concentration of materials and may take place against gradients of concentration of electrical potential. Finally, there is mechanical work, most obvious in the muscle cells of animals but performed in all cells and associated with contractile filaments. The performance of all these kinds of work involves energetically uphill, or **endergonic**, processes (see Chapter 2), and therefore they all require an investment of energy.

Box 7.1

ORGANIC MACROMOLECULES

The organic compounds found in the living cell fall mainly into four great classes: fats, carbohydrates, proteins and nucleic acids. Structurally the fats are simplest. They are constructed from the hydrocarbon chains of fatty acids (usually three) joined separately to the main part of a molecule of glycerol. They are present principally as energy reserves, but one group of related compounds (the phospholipids) in which one fatty acid is replaced by a phosphoric acid group (H_3PO_4) which in turn is linked to another compound, has a critical role in the formation of biological membranes. Simple sugars consist of pentagonal or hexagonal rings made up of either four or five carbon atoms and one oxygen atom with attached hydroxyl (OH) and CH_2OH side groups. Two such units may form a sugar molecule or they may be linked in chains to form complex sugars or polysaccharides, especially cellulose and starch in plants and glycogen in animals. Starch, which may be up to 500 or so units long, and glycogen (1000+) serve mainly as energy stores, but cellulose with its much longer polymers (8000 units) is the main structural material in plants and combines in threads and meshes of great complexity.

Even greater complexity is encountered in the nucleic acids which form large structures built of at least four types of units, known as nucleotides, each of which consists of a sugar phosphate group (an ester) and a base. These are present in varying proportions and a great variety of sequences. The most important of these molecules are DNA and RNA, and the significance of these information-bearing macromolecules is considered further in Box 2.3.

However, it is with the protein molecules that variety and specificity are seen to be most highly developed. These are the largest and most complex molecules known, and they are made up of about 25 different amino acids linked to a carbon backbone. They form chains hundreds to thousands of units long, in different proportions, in all kinds of sequences and with a huge variety of folding and branching.

7.3.1 The transfer of matter and energy in the cell

In our cellular factory the necessary energy is stored in what can be thought of as molecular batteries, which can be moved around the cell to the places where the energy is needed. Here they are coupled to the mechanism concerned in the processes of cellular activity, and the energy they carry is discharged. These spent batteries are then recharged using the chemical energy of fuel molecules. These are relatively complex molecules, particularly carbohydrates (starch, glycogen) and the fats or lipids, reserves of which are often stored in specialized regions of the cell.

This charge–discharge cycle is illustrated in Fig. 7.3 and, in the cell, some of the energy of the fuel molecule released during its breakdown or oxidation is conserved when the compound adenosine diphosphate (ADP) is converted to adenosine triphosphate (ATP) in coupled reactions. It is these compounds that act as the mainline energy store and transfer system of the cell – in other words, as our cellular batteries. ADP represents the exhausted, discharged state and ATP the fully charged state of the battery (Box 7.2).

The charging of the batteries is accomplished by the harnessing of chemical energy from fuel molecules by their oxidation during the process of cellular **respiration**. The total output from this **oxidative respiratory system** for the complete oxidation of one molecule of glucose is 38 molecules of ATP. The batteries have been recharged! Apart from the fuel molecule, the process requires one atom of oxygen for each pair of electrons which pass along the respiratory chain. At the same time, one molecule of water is produced by this chain, while two molecules of carbon dioxide are released. Finally, as the conservation of the energy released is not 100% efficient, some is lost as heat, in accordance with the Second Law of Thermodynamics. This is known as respiratory or catabolic heat loss. The respiratory oxidative systems of cells are located in the mitochondria, which have therefore been called the powerhouses of the cell.

The discharge of the batteries in cellular work can be illustrated by the chemical work of biosynthesis. The phosphate group transfer potential energy of ATP molecules is either donated directly, or through intermediate phosphate compounds, to the synthesis of component macromolecules from both inorganic and organic precursor molecules at precise locations in the cell, often particular organelles. Both the process of synthesis and the rate of formation are controlled and regulated by an army of specific enzymes. Only in this way can thousands of individual chemical reactions go on in the cell at the same

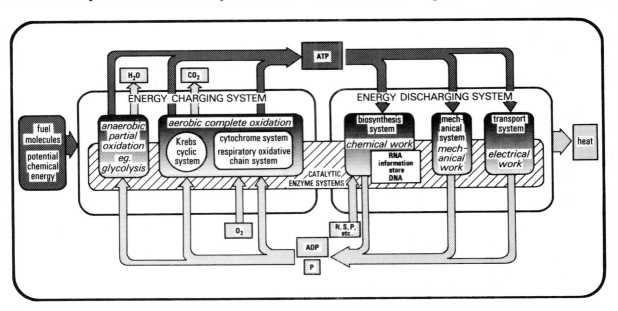

Fig. 7.3 The ADP–ATP energy transfer system of the cell (developed from an idea in Lehninger, 1965).

time, at room temperature and in a medium of liquid water. These enzymes, themselves proteins, are the mechanism by which biological information is expressed in the cell. In terms of our factory analogy, they are the shop-floor directives, standing orders and production targets by means of which the objectives of the plant manager and production manager are realized. The information contained in master plans and production schedules is represented in the cell by the information coded in the molecules of deoxyribose nucleic acid (DNA) in the chromosomes of the nucleus. These are the control systems which direct all of the functional activities of the cell (Box 2.3). Although the factory analogy has proved useful, it sometimes seems inadequate in the face of such a complex, precise and sophisticated array of mechanisms.

7.3.2 *The steady state of the living cell*

This complex synthesis of cell parts would be more acceptable if it took place once only – if the appropriate molecules were assembled and put together to produce the appropriate structure, and that was that. Resynthesis or replacement would

---

Box 7.2
ATP-ADP ENERGY TRANSFER SYSTEM

The phosphate compounds of adenosine are all mononucleotides. The structure of these molecules is given in the adjacent diagram (see Fig. 2.3 for key to symbols). The heterocyclic aromatic ring of adenine is linked to the five-carbon sugar D-ribose, to which are attached one, two or three phosphate groups by ester linkage at the 5' position. It is adenosine triphosphate (ATP) which forms the main energy transfer system of the cell. Adenosine triphosphate is unique in that it is intermediate in position in the energy-level scale of cellular phosphate compounds and, therefore, acts as a go-between for phosphate transfer from high-energy to low-energy forms. The energy level of ATP, however, is not just due to the bond energy of the terminal phosphate group. Because at pH 7 each of the three phosphate groups is completely ionized, four negative charges exist in close proximity, and therefore repel each other. The transfer of the terminal phosphate (hydrolysis) relieves some of this electrostatic stress. The like charges are thus separated as ADP^{---} and HPO_4^{--} which have little tendency to approach each other and undergo stabilization. The energy level, in terms of electron configuration, of these products once separated is considerably lower than when part of ATP. It is this difference in energy content between the initial reactants and the final products which is measured as the free energy of hydrolysis of ATP. At standard conditions of pH 7 and 25°C and with one molal concentration of reactant and products, the free energy change:

$$\Delta G = -2.929 \times 10^4 \text{ J/mole.}$$

In the cell there is reason to believe it is as high as -12×10^3 cal/mole (-5.016×10^4 J/mole).

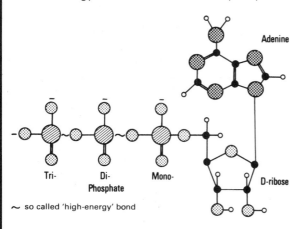

Adenine

Tri-   Di-   Mono-
Phosphate

D-ribose

~ so called 'high-energy' bond

then be necessary only after the structure in question had worn out. Experiments with radioactive tracers, however, have proved that the cell is constantly being rebuilt over and over again. In this continuous state of flux the molecule which survives for more than a few days is exceptional. Clearly, one major function of the cell is to re-create itself constantly, but its ability to synthesize its component parts is not only harnessed to this dynamic turnover at the molecular level. Biosynthesis can be cumulative, giving cells the capacity of growth and thereby increasing their structural complexity. Cells, however, do not grow indefinitely – they divide. The resultant cells then grow, perhaps developing specialized structures and functions and/or dividing further. It is true to say that all organisms – indeed the entire biosphere – have resulted from endless cycles of regulated cell growth and division.

If we look at these functional activities of the cell in thermodynamic terms, then we are dealing with systems that possess relatively little *entropy* (see Chapter 2) compared with the universe around them: they are highly ordered systems. In fact, when cells are dividing and growing rapidly they actually decrease their entropy. Furthermore, our knowledge of evolution implies that, in general, species have evolved from lower primitive to higher, more complex forms. Over time, the biosphere as a whole has progressively decreased its internal entropy as it has increased in diversity. The laws of thermodynamics, however, would suggest just the opposite, for the Second Law leads us to expect the entropy of the universe to increase, as no spontaneous energy transformation is 100% efficient.

There are two answers to this apparent paradox of life. First, one of the secrets of life is the way that the cell is able to build complex structures and to carry out equally complex functional activities by diverting part of the natural downhill thermodynamic trend of energy transfer to uphill energy-demanding processes. It does this, as we have seen, by means of a vast array of coupled chemical reactions, where each stage of each reaction is catalysed by specific enzymes and energy is conserved and redirected in a highly efficient manner. This does not mean, however, that the cell escapes the implications of the Second Law.

Here we must turn to the second secret of life. Although the entropy of the cell (the system) decreases, it does so at the expense of the surroundings, which gain in entropy. The result is that the total entropy of the system and surroundings increases in accordance with the Second Law, even though the system itself, the cell (or biosphere), is becoming more ordered and hence decreasing its internal entropy. This gain in entropy by the surroundings occurs, of course, because cells are taking low-entropy fuel or food molecules representing high-grade potential chemical energy from their surroundings (their environment). However, they return to it simple inorganic molecules (CO_2, H_2O) and low-grade heat energy, which have higher entropy.

Cells and all living systems are, then, open systems. They maintain their internal structures in the face of continual throughputs of matter and energy. This internal structure is, however, constantly being broken down and resynthesized and is, therefore, maintained in a steady state. Only by the continuous self-adjustment of the steady state, facilitated by enzyme systems, can the cell, and indeed all living systems, keep the production of entropy or the tendency towards disorder at a minimum. The steady state is the orderly state of an open system.

The model of the cell used so far is adequate for most cells and for a large proportion of the organisms of the biosphere composed of such cells, which live by breaking down molecules that they or other cells or organisms have manufactured, using energy from earlier generations of fuel molecules. But there is a catch in the use of this model, for it implies that the biosphere is in this sense devouring itself: it is living on capital. Such a closed-system model is untenable on thermodynamic grounds. The dissipation as heat of some proportion of the available energy during each transfer in respiration, biosynthesis or any other kind of cellular work would inevitably mean a running-down and eventual exhaustion of the fuel, in a way analogous to human experience with coal and oil reserves. If the cell and the organism are truly an open system maintaining a steady state, then there must be an external energy source somewhere which represents a continuous and renewable energy input to the system.

There is another snag in the use of the model of what is known as the heterotrophic cell, which relies on preformed organic molecules for chemical energy. Such cells also depend on these food molecules for materials which they cannot obtain as simple inorganic substances for their environment, and cannot synthesize themselves. For example, heterotrophic cells cannot synthesize all of the 20 essential amino acids of protein. The cells of the human body, for example, can only manufacture 12. Also, for each heterotrophic organism there are certain essential organic substances required in small quantities which the organism cannot obtain from the environment, and either cannot synthesize at all or cannot do so in sufficient quantities to meet its requirements. These are known as vitamins and they contribute to enzyme systems. Again, somewhere there must be cells which can produce these compounds and make them available to heterotrophic cells. The conclusion is, therefore, that our model of the generalized cell is incomplete, and either needs refinement or we need a model of a different kind of cell.

7.3.3 The autotrophic cell model

Autotrophic cells are those which can produce organic fuel molecules (usually carbohydrates) from inorganic molecules using some external source of energy. There are two such sources utilized by living systems. First, there are those cells and organisms that can use the bond energy of inorganic compounds to reduce carbon dioxide to form organic carbon compounds. These are known as **chemotrophs** and the process as **chemosynthesis**. Secondly, there are cells and organisms that are capable of using light energy for the process of **photosynthesis**.

At the cellular level, photosynthesis can be viewed as a reversal of the pathway by which carbohydrates are decomposed to water and carbon dioxide during respiration (Fig. 7.4). Just as the outcome of respiration is the conversion of ADP to ATP (oxidative phosphorylation, see Fig. 7.3) which is then available for cellular work, so in photosynthesis ADP is converted to ATP which in the autotrophic cell donates energy to the building of carbohydrates and other organic molecules, starting from carbon dioxide and water. The critical difference is that the energy that converts ADP to ATP in photosynthesis is entirely derived from absorbed light energy, and the process is therefore known as **photophosphorylation**.

The trapping and utilization of this light energy takes place in most autotrophic cells in the specialized organelles – the chloroplasts – which contain the pigment chlorophyll (Box 7.3). Because chlorophyll absorbs the red and blue

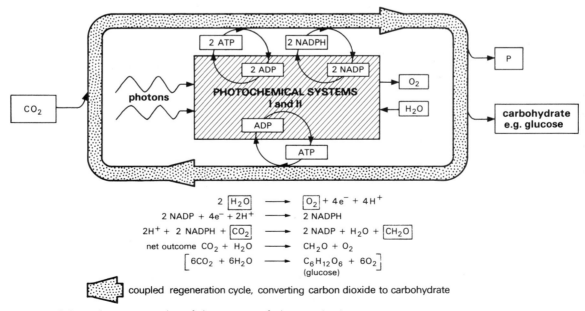

$$2 \; \boxed{H_2O} \longrightarrow \boxed{O_2} + 4e^- + 4H^+$$
$$2 \; NADP + 4e^- + 2H^+ \longrightarrow 2 \; NADPH$$
$$2H^+ + 2 \; NADPH + \boxed{CO_2} \longrightarrow 2 \; NADP + H_2O + \boxed{CH_2O}$$
$$\text{net outcome} \quad CO_2 + H_2O \longrightarrow CH_2O + O_2$$
$$\left[6CO_2 + 6H_2O \longrightarrow C_6H_{12}O_6 + 6O_2 \right]$$
$$\text{(glucose)}$$

coupled regeneration cycle, converting carbon dioxide to carbohydrate

Fig. 7.4 Schematic representation of the process of photosynthesis.

Box 7.3
CHLOROPHYLL

There are several kinds of chlorophyll, but in the higher green plants chlorophyll *a* and chlorophyll *b* are the most important. They are similar in molecular structure, both consisting of four pyrrole units arranged to form a porphyrin ring with a magnesium atom at its centre and with a long hydrocarbon side chain. The only difference is the replacement of the *methyl* group of

chlorophyll *a* by a *formyl* group on chlorophyll *b*. The pair of bonds holding the magnesium atom in the centre of the ring alternate between the four available nitrogens, a phenomenon known as **resonance**.

There are other photosynthetic pigments – the carotenoids in all photosynthetic plants and the phycobilins in blue-green algae and red algae. Chlorophylls absorb mainly red and blue–violet wavelengths, reflecting green light. Carotenoids absorb strongly in the blue–violet range and are therefore yellow–orange, red or brown pigments. Often masked by chlorophyll, they are responsible for the autumn colours of leaves when chlorophyll is lost, as well as for the colours of many non-photosynthesizing tissues such as flowers and fruits and, of course, the orange–red of the carrot. Some phycobilins absorb strongly in the green wavelengths, some in the orange and red, and hence appear red and blue respectively and are responsible for the colour of red and blue–green algae.

These light-absorbing pigments are distributed in two photochemical, or pigment, systems – PSI and PSII – in both of which chlorophyll *a* is the primary pigment in what are called the trapping centres of the system. In PSI, only chlorophyll *a* pigments are utilized, but in PSII chlorophyll *b* and other accessory pigments gather absorbed light energy. In both PSI and PSII this energy is passed to modified molecules of chlorophyll *a* which are termed photoreactive or trapping centres. Each system has a variant of chlorophyll *a* with a different absorption peak (PSI 700 nm and PSII 680 nm).

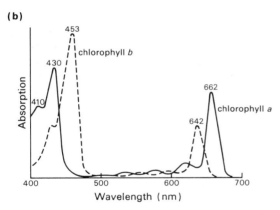

(a)

phytyl chain

- ● carbon
- • hydrogen

(b)

Absorption vs Wavelength (nm)

453 — chlorophyll *b*
430
410
662 — chlorophyll *a*
642

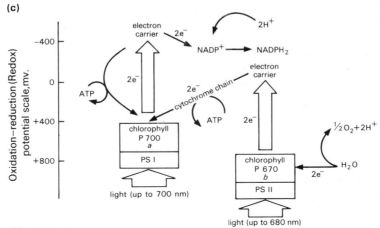

(c)

PS = photosystem

wavelengths of the visible part of the spectrum, it appears green. The action of the absorbed light is to raise the energy level of the electrons in chlorophyll, and their energy is in turn redirected through phosphate compounds as chemical energy to further chemical reactions. When a photon of light representing a quantum of energy falls on the chlorophyll molecule, it excites one of the electrons sufficiently for it to be donated from the chlorophyll to a series of electron acceptors, or enzyme catalysts. The electron passes along this chain which, like the respiratory chain (see Fig. 7.3), contains cytochromes, generating ATP before being returned to the chlorophyll molecule. This is the so-called light phase of the photosynthetic process and, although many autotrophic cells are aerobic, it does not require the presence of oxygen. The synthesis of organic carbon compounds – mainly, but not exclusively, carbohydrates from carbon dioxide and water – is termed the dark phase, and involves the biochemical reduction of these inorganic molecules using the energy of intermediate phosphorous compounds, which in turn have derived their energy from the ATP molecules produced by the light phase. This biochemical reduction liberates molecular oxygen from the water molecules.

7.4 The organism, the population and the community

So far in this chapter the functional organization and activity of the living systems of the biosphere have been discussed in terms of models constructed at the cellular level. Such models may have seemed at first sight to be far removed from the global perspectives of physical systems presented in the other chapters in this section. The universality of these cellular models is really quite remarkable, in spite of the astounding richness and diversity of life forms on this planet. It is precisely this generality that makes an understanding of the fundamental organization and functioning of the cell a necessary prerequisite for a consideration of living systems at the level of the organism, the population and the community.

It is at the level of the individual organism that we encounter *discrete* living systems in direct contact with their non-living environment. In single-celled creatures such as the protozoa or the unicellular algae, the individual may differ little from the generalized model of the cell.

Most organisms however, are multicellular. In such cases, cells specialized both structurally and functionally form equally specialized tissues, organs and organ systems, that together constitute and function as discrete organisms – a crocus, or a giant redwood, a housefly, or a horse – each of which exchanges matter and energy with its inorganic environment – to which it is functionally linked. These exchanges may be the heat transfer with which an animal living on the surface of the Earth is involved (illustrated in Fig. 7.5), or the transfer of chemical energy as food from grass to sheep, or the uptake of water and nutrients from the soil by the tree in the same figure. Clearly, however, it is with the organism that we begin to appreciate more fully the interaction of living systems with the global physical systems of the ecosphere model (see Fig. 7.1).

7.4.1 Species and populations

The Earth is inhabited by an enormous number of different kinds of organism, and each is in some sense unique. These are known as **species** and they are populations of organisms that show an overall resemblance, are distinct and reproductively isolated from any other group, and have a degree of constancy through time. At a conservative estimate there are about 400 000 species of plant

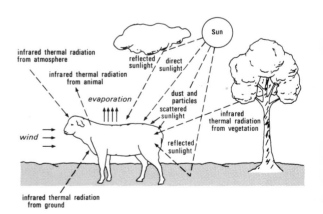

Fig. 7.5 The energy budget of an animal at the Earth's surface (Gates, 1962).

living today. Among the animals, more than a million species are animals without backbones (invertebrates) and 850 000 of these are insects. The mammals are represented by about 4000 species, the birds by approximately 8000 species and the fishes by more than 20 000 species, and there are about 6000 species of reptiles and amphibians. All of these species are the result of a long-continued process of evolution and speciation. Each of these contemporary species of organism, therefore, has arisen by an interplay between changes that have occurred to the genetic information (genotype) that it carries encoded in the DNA molecules in the nuclei of its cells (see Box 2.3) and the selective pressures exerted by its physicochemical and biotic environment.

Because the genetic information controls, through transcription, the metabolic activity, growth and development of each individual, changes in genotype give rise to parallel changes in the morphology or anatomy of the organism, or to changes in the way that it functions – physiology – or to changes in its behaviour, or to some combination of all three. The result of these changes is the occurrence of hereditable variation in the population of that species – variation that arises at random but that survives in the population only if it has no fatal or deleterious effect on the organisms concerned. It may, however, endow some individuals with a selective advantage in their interactions both with the inorganic environment and with other members of the same species or other species. This process – the selection of adaptive variation in the population – is known as the environmental sieving of genotypes. The adaptive significance of such variation may be apparent immediately in the generation in which it appears and in relation to the existing environment. It may be, however, that the variation, though reappearing from one generation to the next, only becomes of adaptive significance when the environmental conditions change or the species experiences a new environment as it spreads by dispersal and migration (Fig. 7.6). This is Darwin's process of natural selection, and it ultimately leads to speciation, either abruptly or gradually, as new species arise from old and successful variants either replace ancestral populations or become reproductively segregated and geographically isolated from them as they spread out, colonizing new environments. This process has been repeated over millions of years and has led, through the **adaptive radiation** of evolutionary change, to the great diversity of organisms that exist today and to those that, though now extinct, existed in the past (Fig. 7.7).

Each of these species exists in a whole complex of environmental interactions. It exists because it can survive and compete successfully under the particular conditions of existence to which it is adapted, or of which it is tolerant, and the presence of a species population in any location will reflect the often long and complex history of evolution, dispersal and migration of that species. It is this history that, at least partially, explains the present distribution of plant and animal species

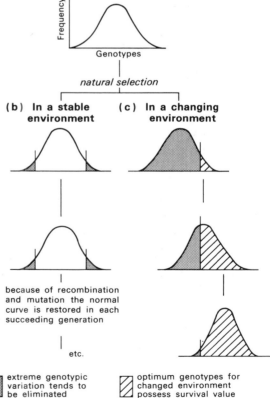

Fig. 7.6 Selection of genotypes in a stable and changing environment (Heywood, 1967).

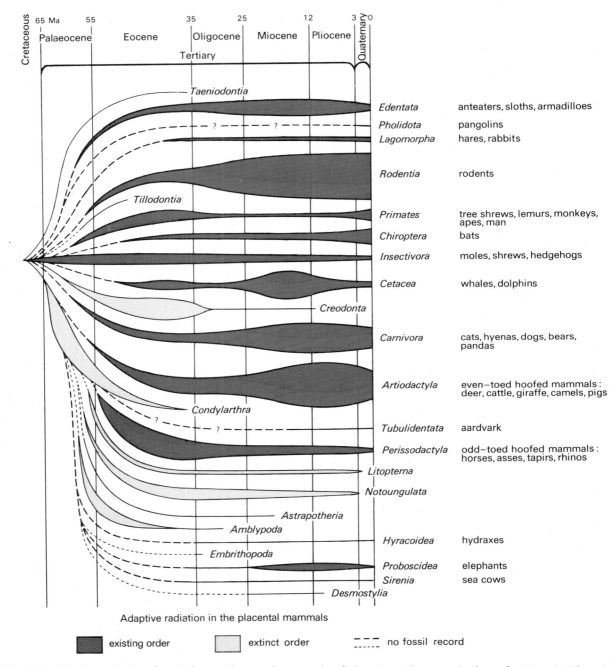

Adaptive radiation in the placental mammals

existing order   extinct order   - - - no fossil record

Fig. 7.7 Adaptive radiation of evolutionary change: the example of the placental mammals (from *Continental drift and evolution* by B. Kurtén, © 1969 by *Scientific American*, Inc., all rights reserved).

(Fig. 7.8). Some are young species with restricted distributions. Some are very ancient species whose area has become drastically reduced to a single relict distribution. However, both are described as narrowly **endemic**. Other species may be present in almost all of the world's land masses, and hence are known as **cosmopolitan**, while others may have markedly discontinuous (**disjunct**) distributions. Some of these distribution patterns reflect not only the evolution and migration of the species, but also changes in the distribution of land and sea, changes in climate, the appearance of mountain ranges and the extension and retreat of continental ice sheets. In other words, distribution reflects the history of the environment as well as that of the species, and in many ways it embodies the response of the species to environmental change through time (see Chapter 25).

The pattern of floral and faunal regions of the world can be seen in Fig. 7.8 to reflect in part the changes in the relative positions of the continents through geological time. The greater floral and faunal affinities in the northern hemisphere are associated with the fact that dispersal and migration routes have existed between the northern land masses until relatively recently. In contrast, evolution and speciation has proceeded in isolation for longer on the land masses of the southern hemisphere, resulting in distinctive floral and faunal regions with affinities often only apparent at higher taxonomic levels. Nevertheless, problematic disjunct distributions in the southern hemisphere can also be explained with reference to the timing of the fragmentation of an earlier supercontinent (Chapter 5). Thus the marsupials which reached Australia via Antarctica from South America are absent from New Zealand, Africa and India, all of which had separated before their arrival. Some northern hemisphere disjunctions such as the **amphi-atlantic** distribution may also reflect the fragmentation of a once more continuous distribution by continental drift, but many are associated with the effects of late Tertiary and Quaternary climatic change on plant and animal distributions which had been circumboreal in the early and mid-Tertiary (e.g. the redwoods). In such cases recent divergent evolution of these isolated descendants of common ancestors has resulted in species with endemic distributions today (e.g. the plane tree *Platanus* spp. and the tulip tree, *Liriodendron* spp.).

7.4.2 Communities, ecosystems and the ecosphere

Wherever they occur, species of plants and animals rarely do so as pure populations. Normally, populations of different species grow and live together as members of plant and animal communities. The variety and diversity of such communities is almost as great as that of their component species, and furthermore, the community concept can be applied at a range of scales from that of a small pond to the thousands of square kilometres of the Amazonian rainforest. Like the organism, the community as a whole is adapted to and reflects the conditions imposed by the external environment in which it exists and with which it interacts. This adaptation is partly the sum total of the individual adaptive strategies (see Chapter 19) displayed by the species which make it up, but it is also adaptation at a higher level than the organism. This community-wide adaptation to environment involves the structural and functional organization of the entire community.

If we now extend the model of the biosphere developed in Fig. 7.2b to incorporate the organism, the population and the community, it becomes clear that they are merely higher levels of integration in the model (Fig. 7.9). Just as organelles function as integrated units in a model at the cellular level, or as cells are functional units in the organization of the multicellular organism, so individual organisms and species populations are functional units in a model at the level of the community. The biosphere is, in its turn, composed of all of the different community types that occur at the Earth's surface. But from the level of the organism upwards these models involve direct interactions with the non-living environment. At the global level this is recognized by the ecosphere model. At the community level these environmental or ecological interactions are incorporated by modelling the community as an ecological system, or **ecosystem**. By grouping ecosystems which are broadly similar in their structure (particularly of the vegetation) (see Box 18.3) and which occur in similar environments, we can define the major broad subdivisions of the ecosphere. These are known as **biomes** and biome-types, and in the northern hemisphere we can talk of a broad-leaved deciduous or summer forest biome, or a tundra biome (Fig. 7.10).

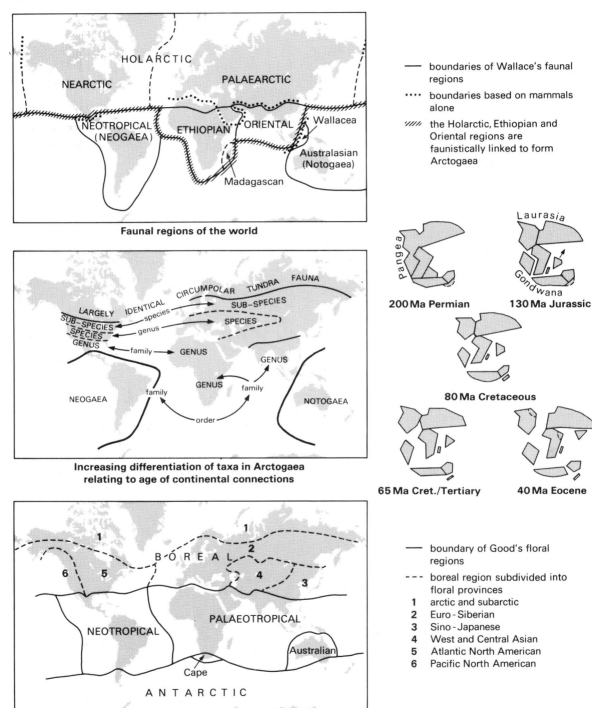

Faunal regions of the world

boundaries of Wallace's faunal regions

•••• boundaries based on mammals alone

///// the Holarctic, Ethiopian and Oriental regions are faunistically linked to form Arctogaea

Increasing differentiation of taxa in Arctogaea relating to age of continental connections

200 Ma Permian

130 Ma Jurassic

80 Ma Cretaceous

65 Ma Cret./Tertiary

40 Ma Eocene

boundary of Good's floral regions

- - - boreal region subdivided into floral provinces

1 arctic and subarctic
2 Euro-Siberian
3 Sino-Japanese
4 West and Central Asian
5 Atlantic North American
6 Pacific North American

Floral regions of the world

Fig. 7.8 Types of plant and animal distribution pattern (compiled from various sources). See text for explanation.

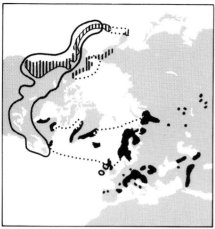

||||| circumboreal – *Picoides tridactylus* (three–toed woodpecker)

≡ Europe – E. Asiatic disjunction – *Cyanopica cyanus* (azure–winged magpie)

■ E. N. American – Asia Minor disjunction – *Platanus* (plane)
Platanus occidentalis endemic to N.E. America, *Pl. orientalis* to S.E. Europe and Asia Minor

||||| *Luzula piperi* and ||||| *L. wahlenbergii* (woodrushes) – transBeringian distribution

■ *Potentilla crantzii* (alpine cinquefoil) – amphi–Atlantic disjunction

□ *Spiranthes romanzoffiana* (drooping ladies' tresses) – amphi–Atlantic disjunction

≡ *Symphonia* – American – African tropical discontinuity at generic level

□ *Ancystrocladus* – Africa – Asia tropical discontinuity at generic level

■ *Buddleia* – pantropical discontinuous distribution

— *Palmae* – pantropical distribution at family level

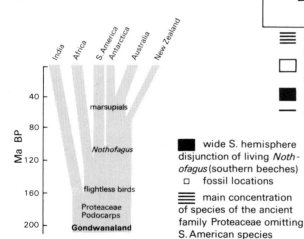

Time of arrival of certain taxa in relation to the fragmentation of Gondwanaland

■ wide S. hemisphere disjunction of living *Nothofagus* (southern beeches)
□ fossil locations

≡ main concentration of species of the ancient family Proteaceae omitting S. American species

▲ present and △ fossil distribution of the genus *Dachrydium* of the southern conifers, the Podocarpaceae

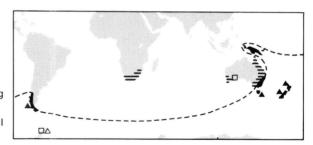

||||| N. E. America – E. Asia disjunction of the genus *Liriodendron*
L. tulipifera endemic to N.E. America
L. chinense endemic to China

■ *Sequoia sempeviriens* and *Sequoiadendron gigantea* (redwoods) both endemic to W. N. America
○ fossil redwood locations

■ *Metasequoia glyptostroboides* (dawn redwood) – endemic to Yangtse Valley

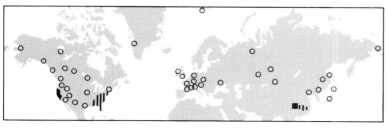

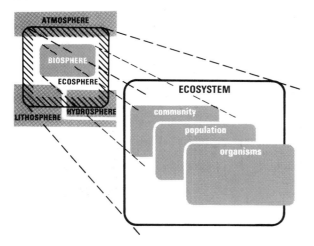

Fig. 7.9 Nested hierarchy model of the ecosphere.

7.5 A functional model of the ecosphere

In spite of the vast diversity that we see displayed in the real world at the levels of the organism and of the community, it is still possible to present a generalized model of the functional activity of the ecosphere. This is possible because such activity, as in the case of the cell discussed in Section 7.3, remains the exchange of matter and energy between the system and its surroundings, and the pathways and processes of energy transfer within the living system, whether that system is an organism, a complex community of organisms, or the entire biosphere. The distinction between the autotrophic and heterotrophic functions of the cell will also, of course, hold true for the organism. Plants for the most part are autotrophic, and animals are heterotrophic systems. Not all of the cells of plants, however, are autotrophic. The light energy fixed during photosynthesis and transformed to the chemical energy of the synthesized carbohydrate molecules may be transferred by respiration to ATP, and hence to biosynthesis and other activities in the autotrophic cells of these plants. Alternatively, the products of photosynthesis may be translocated within the plant and the energy they represent may be utilized by non-photosynthesizing heterotrophic cells elsewhere in the plant. Heterotrophic

organisms – animals – will obtain the products of both photosynthesis and subsequent plant biosynthesis, and hence the chemical energy represented by these compounds, when they consume parts of the plant as food, either directly or indirectly.

The functional organization that emerges at the organism and community level is, therefore, hierarchical. It reflects the energy flux from radiant light energy via photosynthesis in the presence of chlorophyll in green plants to chemical energy subsequently ingested and assimilated by herbivorous animals whose food the plants represent, and then to carnivorous animals, again as chemical energy when they devour their prey. Because communities, and indeed the whole of the biosphere, are composed of both autotrophic and heterotrophic organisms, the functional organization of the ecosphere must also reflect this energy flux.

7.5.1 The trophic model

This hierarchical organization is the food chain. Its formulation as a series of thermodynamically valid steps in the energy flow through the biosphere was first developed in a now classic paper by Lindeman in 1942 (Box 7.4). In this model, organisms are lumped together into compartments depending on their place in the food chain or on their energy source. These compartments are called **trophic levels** (Fig. 7.11). In accordance with the Second Law of Thermodynamics, as the original radiant energy input is passed from one trophic level to the next, some of the energy is dissipated at each step as unavailable heat energy. This is the expression of the catabolic or respiratory heat loss associated with cellular activity, but manifested here at the level of the organism (Fig. 7.12).

Fig. 7.11, however, also emphasizes the fact that the biosphere contains not only living but also dead organic matter. Furthermore, there is a compartment in the model which we have not considered when taking the simple food-chain view point. Assigned to this compartment are organisms whose energy source is not the chemical energy stored in the tissues of living plants and animals, but that which remains after their death, or is excreted or shed by them while still alive. These are the **decomposers** that feed on dead and

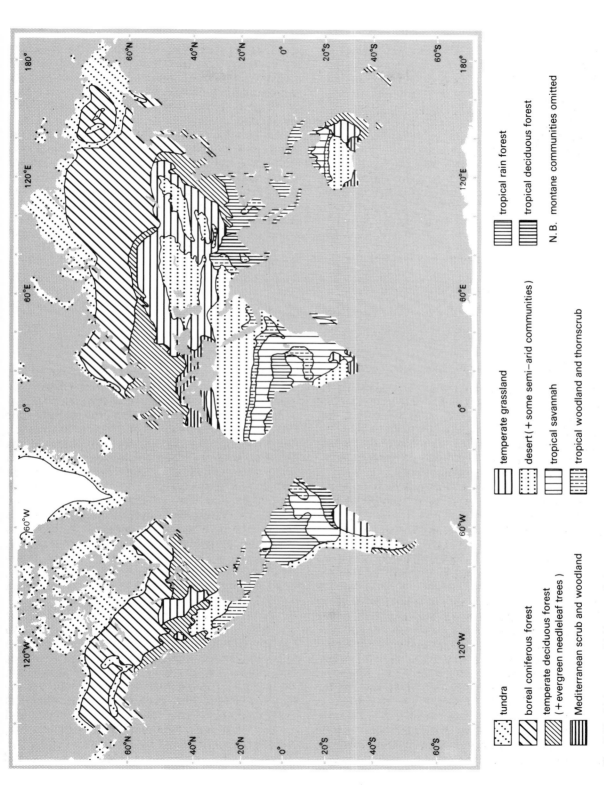

tundra

boreal coniferous forest

temperate deciduous forest
(+ evergreen needleleaf trees)

Mediterranean scrub and woodland

temperate grassland

desert (+ some semi–arid communities)

tropical savannah

tropical woodland and thornscrub

tropical rain forest

tropical deciduous forest

N. B. montane communities omitted

Fig. 7.10 Major terrestrial biomes of the world.

decaying organic matter. The most important in the chemical breakdown of dead organic matter are the bacteria and fungi, both of which belong to groups that on evolutionary grounds are best regarded as being distinct from the plant and animal kingdoms. It will become apparent in Chapter 21 that this decomposer trophic level is an oversimplification, for some organisms are in a specialized category of herbivore – **the detritivores** – while others are carnivorous. Nevertheless, the inclusion of the decomposers emphasizes the fact that the energy flow through the biosphere can take one of two routes: a fairly direct pathway through the so-called grazing food chain, and a more indirect pathway, often involving a timelag, through the detrital food chain. Whichever pathway is followed, however, the ultimate output is unavailable heat energy.

In common with most hierarchies, the character of the individual is lost in this trophic model of the biosphere. All organisms, from algae to forest trees, are lumped together in a trophic level, in this case as autotrophic. Each organism is significant only in so far as it is a contribution to the total mass of organic matter (**biomass**) and total energy content of that trophic level per unit area of the Earth's surface at a moment in time. This is because from the functional point of view each

---

Box 7.4

TROPHIC LEVELS AND PYRAMIDS OF BIOMASS

Raymond Lindeman was a young American ecologist who spent 5 years studying Cedar Bog Lake in Minnesota, as a PhD student at the University of Minnesota. After completing his thesis he went to Yale University where he developed part of his thesis into the paper (Lindeman, 1942) which was destined to set a new paradigm for ecology.

This paper was published posthumously, for while at Yale he died of a liver disease at the early age of 27. In the 'classic' 1942 paper he advanced the notion that ecosystems could be seen as being composed of **trophic** (or feeding) levels arranged in a hierarchy, and that ecosystems could be studied by charting the flow of food between these trophic levels, and that these flows of food could be expressed in units of energy. Ideally ecologists could (according to Lindeman's model) find the weight or biomass, and by conversion, the energy content of plants and animals in each of the trophic levels. They could also observe the flow of energy between trophic levels by measuring the transfers of mass (organic matter) that passes from one level to the next. Lindeman hypothesized that the progressive reduction of available energy for each successive trophic level was the explanation of the **pyramid of biomass**. Furthermore, based on the evidence at his disposal in 1941, he suggested that the amount of biomass supported by a unit of energy (1 g of food) entering a trophic level increased up the hierarchy, that is, there was an improvement in the efficiency of food use at successively higher trophic levels.

The elegant simplicity and apparent power of the concept of the so-called trophic dynamic view of the ecosystem meant that it came to almost totally supersede earlier approaches (particularly the pyramid of numbers and size view of trophic relationships expounded by the British animal ecologist Charles Elton). Indeed, the attraction of measuring the contents of levels, and transfers between levels in energy units, proved overwhelming, and was to form the basis of the International Biological Programme (IBP) which spanned the years 1964–1974. As Steve Cousins (1985) has expressed it: 'no one wanted to go around counting organisms of different sizes now that energetics and the second law of thermodynamics had entered ecology'. Whether this was an entirely valid point of view will be addressed in Chapter 20 when we shall consider if Lindeman's work is in need of some re-evaluation.

level in this trophic model represents a temporary store of potential chemical energy.

Ignoring the diurnal cycle for the moment, we can regard the input of radiant energy as a continuous process. The same is true for the metabolic activities of both plants and animals. The energy transfers involved in photosynthesis – the ingestion and assimilation of food, maintenance, activity, growth and reproduction – are taking place all the time, so there will be a continuous loss of energy as heat. Over a period of time (such as a year) a balance will exist for each trophic level between the gains (or inputs) of energy and the losses (or outputs) of energy. This balance is the net accumulation or deposit of energy over that period of time (**production**, P) and it is manifest as a change in the biomass of the trophic level in question (ΔB) (as will become apparent in Chapter 19, P and ΔB have a more complex relationship than implied here). For the autotrophic level this is referred to as the **net primary production** (P_n) but for heterotrophic levels, although termed secondary production, it is more strictly called **conversion**.

Production expressed as a rate of change in the energy content of a trophic level is termed **productivity** and is expressed as mass or its energy equivalent per unit area of the Earth's surface per unit time ($kg\ m^{-2}\ a^{-1}$ or $kJ\ m^{-2}\ a^{-1}$). The biomass of a trophic level then, whether expressed in units of mass or energy, is a measure of its 'bulk' in terms of the amount of organic matter present at that point in time. It is of course this 'bulk' that is often impressive to an observer. Who would not be impressed by the bulk of organic matter represented by the plants of an undisturbed forest? However, the net productivity of a trophic level is a measure of its efficiency, of the net rate of accumulation of this biomass. Of course, this rate may bear little or no relation to the biomass actually present at one moment in time (Macfadyen, 1964; see Chapter 19).

As the higher trophic levels in our model (see Fig 7.11) all depend on the autotrophs for their energy supply, the net primary productivity and its integral net primary production of the biosphere are important parameters. Estimates of the total net primary production of the land and ocean areas

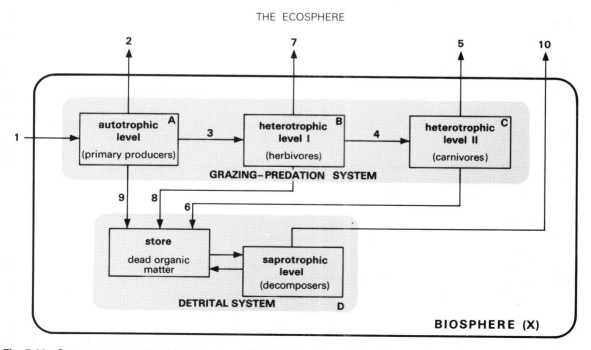

Fig. 7.11 Compartment model of the biosphere (X), which has been divided into compartments A, B, C, D (trophic levels). The energy transfers into and out of these compartments are represented by arrows numbered from 1 to 10.

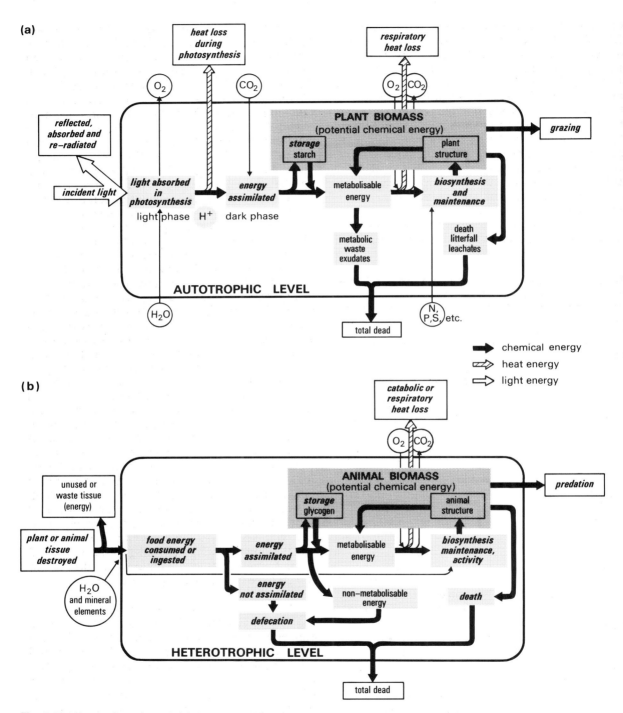

Fig. 7.12 Energy flow through (a) the autotrophic and (b) the heterotrophic levels.

of the Earth are given in Table 7.2 (see also Fig. 7.13). If we take Lieth's (1971 and 1973) estimates of 100.2×10^{12} kg a^{-1} for the continents and 55×10^{12} kg a^{-1} for the oceans, these figures represent 178.24×10^{23} J a^{-1} fixed in net primary production on land and 109.2×10^{23} J a^{-1} in the oceans. The total net primary production of the globe is therefore 287.44×10^{23} J a^{-1}. This quantity of energy, fixed annually, is the equivalent of the total generating capacity of approximately 300 million 1000 mW power stations and compares with 2.2×10^{18} J released by one of the largest series of volcanic eruptions this century, at Kamchatka in the Soviet Union in 1955–56.

Photosynthesis not only requires light, it is also dependent on the availability of water and it is sensitive to temperature. As all three factors are far from constant over the Earth (see Chapter 4), it is not surprising that the global figures quoted above conceal a good deal of variation (Fig. 7.14). The nature of this will be discussed further in Chapter 19, but it is worth noting here that even when such variation is taken into account considerable dangers remain in the use of global estimates of production. These have been lucidly reviewed by Newbould (1971), who counsels caution in what he calls the 'numbers game'. Even so, the estimates used in Table 7.2 are probably sufficiently valid at the level of generalization of the ecosphere model used in this chapter.

However, one further point needs to be made about these figures. The production figures in Table 7.2 are expressed as weight of 'dry matter'. This is the weight after drying in an oven at 105°C until no further loss of weight occurs. In addition to organic compounds, dry matter contains some inorganic material. Ignoring this we have the organic weight, which for most plants is between 75% and 95% dry weight, and of this only 45–48% is carbon. The remainder is mainly oxygen and hydrogen, but as we saw at the beginning of the chapter, it also contains small but significant quantities of other elements. Here, then, we will shift our perspective on the stored chemical energy, which the biomass of the trophic levels of the biosphere represents, to the composition of those same chemical compounds. By doing so we can recognize that the net production transferred from one trophic level to the next is also a transfer of materials. When a herbivore grazes a plant, it gains not only energy but also the elements it needs to build new cells and for growth, elements that were first fixed in organic compounds by green plants using the energy of sunlight. But where did these elements come from and where do they go to after the death of the plants or animals?

Table 7.2 Comparison of the various production estimates for the Earth (after Box, 1975)

Area (10⁶ km²)	NPP est. (10¹² kg yr⁻¹)	Method
Land		
140	96 (38.4 × 10¹² kg C)	planimetering Lieth's (1964) productivity map and checking it against annual, global CO_2 fluctuation
149	109.0	sum of estimates by means for major vegetation types
149	100.2	sum of estimates by means for major vegetation types
149	116.8	sum of estimates by means for major vegetation types
140.2	104.9	evaluation of Innsbruck productivity map (Lieth, 1972)
140.2	124.5	evaluation of Miami model (Lieth, 1972)
140.3	118.7	evaluation of Montreal model (Lieth and Box, 1973)
149	121.7	sum of estimates by means for major vegetation types
Ocean		
332	46–51 (23 × 10¹² kg C)	sum of estimates by means for major zones
361	55.0	evaluation of major zones
361	43.8	evaluation of oceans Productivity map (Lieth and Box, 1972)
361	55.0	sum of estimates of major zones

7.5.2 Transfer of matter in the ecosphere

Fig. 7.15 shows the major pathways taken by the elements necessary for life as they are exchanged between the living systems of the biosphere and their non-living environment (see also Fig. 5.10). Although at first sight a complex diagram, it is still basically the simple compartment model of Figure 7.1, for the extent of the interaction of the biosphere with the atmosphere, hydrosphere and lithosphere delimited by these pathways helps to define the boundary of the ecosphere.

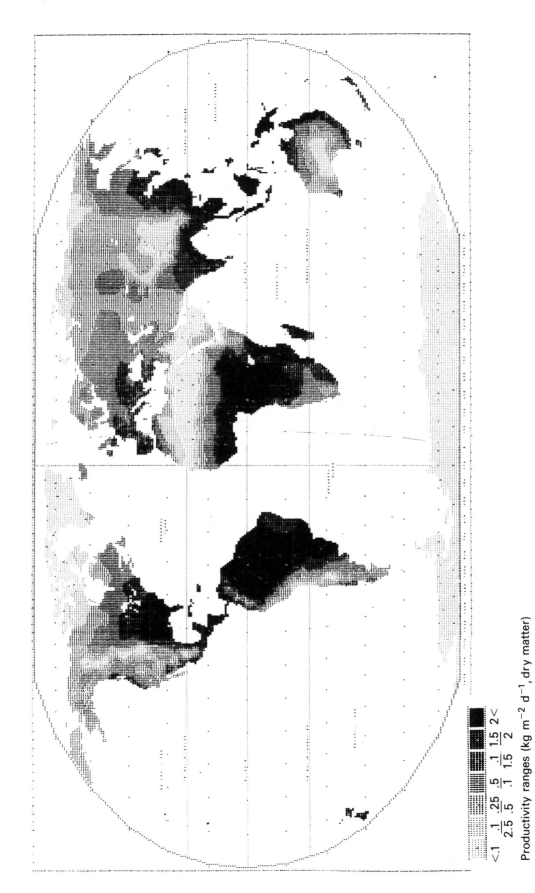

Productivity ranges (kg m^{-2} d^{-1}, dry matter)

Fig. 7.13 Computer-generated productivity map of the land areas of the world (after Lieth, 1975).

If the three kinds of pathway identified in Fig. 7.15 are traced through the diagram, it will be found that they pass one into the other. Therefore, at least some of the elements travelling these routes are involved in enormous global cycles through the lithosphere, atmosphere and hydrosphere. Many of these pathways and the cycles they form have already been considered implicitly in Chapters 4 and 5. However, those chemical elements necessary to the functioning of living systems can be seen to take a detour, to follow an alternative pathway through the biosphere. Here too, all three kinds of pathway

converge to follow a common route as organic molecules through living systems.

The continuity of the movement of elements along these pathways is, however, broken by processes, the operation of which involves transfers of energy, as was stressed in Chapter 2. Again, many of these processes have been examined in Chapters 4 and 5 and are therefore already familiar. Some of the processes carry a positive and some a negative sign. These signs refer to the potential energy level of the materials involved in the process. If, after completion of the process, the material is at a lower energy level, then the sign is negative. If, on the other hand, the operation of the process increases the energy level, the sign is positive. The energy level in question refers in some cases to the potential chemical energy of the molecules of which the element is a part, as for example the decline in potential chemical energy from primary rock-forming minerals to weathering products during the process of weathering. In other cases there may be no change in chemical energy, but merely a change in potential energy by virtue of a change in relative position – as when rocks are uplifted.

It is by means of these processes that the one way flow of energy required by the laws of thermodynamics is coupled into and powers the closed circulation of matter within and between the atmospheric, hydrospheric, lithospheric and biospheric systems. Here the distinction made in Fig. 7.15 between the positive, energy-demanding (endergonic) processes and the negative, energy-yielding (exergonic) processes is critical. Without the uphill processes the cycles would not exist; the pathways would not close to turn full circle. From the point of view of the ecosphere, however, these processes have a further significance in that they function as rate-limiting mechanisms. This means that the rate at which they take place determines the speed with which materials flow along the pathways. For example, the rate at which weathering processes occur (see Chapter 11) determines the rate at which elements are made available for uptake by plant roots. Now some of these processes – evaporation for example – take place rapidly, while others, such as the subduction of ocean sediments into the mantle, take a long time. The speed of transfer between the different compartments in Fig. 7.15 will, therefore, vary considerably. Again, from the ecosphere's point of view, some of these stores may act as sinks for

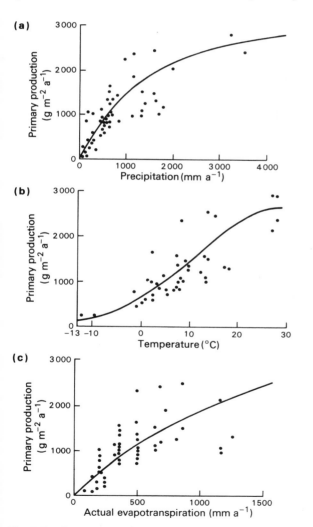

Fig. 7.14 Relationship of productivity to (a) precipitation, (b) temperature and (c) evapotranspiration.

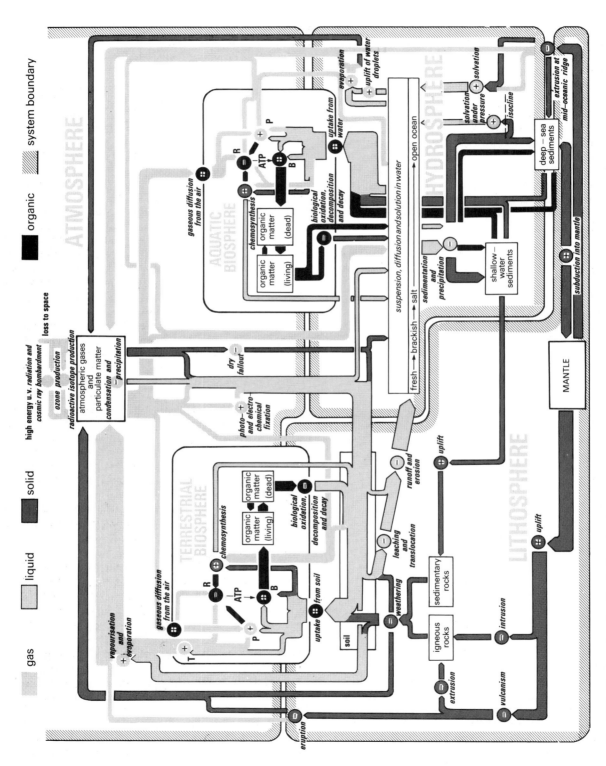

Fig. 7.15 The transfer of matter in the ecosphere.

some elements and effectively immobilize them. This is the case if the residence time of an element in a compartment is long and its cycling is slow. Here, however, our viewpoint is relative. This statement would be true at one scale for a critical element such as phosphorus, immobilized in undecomposed soil organic matter for a matter of years, and at another for those elements bound up in deep-sea sediments perhaps for millions of years.

For matter moving through these pathways to be raised to a higher energy level requires an input of energy, of course. In the atmosphere this input is radiant solar energy, as for example in the vaporization of water during evaporation (see Chapter 4). In the lithosphere the energy source is both terrestrial energy in the crustal system and solar energy in the denudation system (see Chapter 5). An input of energy is also required for the transfer of elements at a relatively low energy level (oxidized) in the environment to a high energy level (reduced) in the cells of the living systems of the biosphere.

7.5.3 The energetics of matter transfer through the ecosphere

As we have seen, the organisms of the biosphere are, structurally at least, mainly carbon, hydrogen and oxygen. The dominant process that transfers these elements from the environment to living systems is, of course, the photoreduction of carbon dioxide to carbohydrate with the liberation of molecular oxygen from water during photosynthesis. From autotrophic organisms carbon, hydrogen and oxygen are passed to heterotrophs along the food chain. However, light is not the only source of energy necessary for photosynthesis. The carbon dioxide enters the leaves in response to diffusion gradients which require the expenditure of metabolic energy to establish and maintain, while water molecules are lifted from the roots by the transpiration stream which is maintained by radiant energy.

The other elements essential to life enter the high-energy route through living systems largely as ions in solution in water. They may be passively transported by the transpiration stream, but to enter plant cells, whether in response to diffusion gradients or by 'active ion pumps', involves the expenditure of energy ultimately derived from respiratory oxidation. These essential elements are only then available for the synthesis of organic compounds using ATP energy during biosynthesis. Like carbon, hydrogen and oxygen these elements are passed on to heterotrophs as the constituent atoms of the organic molecules of plant tissue consumed as food by animals (Chapters 19 and 20).

Living plants and animals return to their environment some of the elements they have drawn from it. Carbon dioxide is released to the atmosphere during respiration, and oxygen during photosynthesis. Some bacteria, using nitrates or sulphates as a substitute for oxygen in respiration, reduce them to nitrogen gas (dinitrogen) and hydrogen sulphide respectively, both of which are released to the atmosphere. Yet other inorganic and organic compounds are excreted by living organisms as waste products. On death, however, the life-sustaining input of energy and matter stops and the spontaneous, if sometimes slow, breakdown of the complex molecules of the organism and its constituent cells begins. These processes of decay and decomposition are oxidation reactions, and they proceed with a net loss of free energy and release the constituent elements of the organic molecules back into the environment as simple inorganic compounds. However, much of this dead organic matter forms the food supply of the decomposers. The elements it contains may recirculate through this compartment of the ecosphere many times before being released. Even then the more resistant organic residues may accumulate, as, for example, soil humus or peat in terrestrial environments, or as organic sediments in aquatic environments, so immobilizing the elements they contain.

For all of the elements essential to living systems on land, including hydrogen and oxygen combined as water, the soil is the immediate store on which the biosphere draws and to which it returns these elements. The only exceptions are photosynthetic carbon dioxide and respiratory oxygen, which are both obtained from and returned to the lower atmosphere. In aquatic environments the biosphere draws all elements from those dissolved, suspended or diffused in water, even carbon dioxide for photosynthesis and oxygen for respiration, and returns them to the water.

If the rate of uptake from the soil or water is balanced by the rate of release of elements by decomposition, then the circulation rate is limited by the rate of energy flow from photosynthesis through autotrophic to heterotrophic organisms, i.e. the biosphere energy flux. On the other hand, this flow of energy, and hence the productivity of the biosphere, may itself be **rate-limited** if the rate of return of essential elements through decomposition is inadequate. For many elements this is indeed the case, for not only may they be immobilized in undecomposed organic matter, but even when released they may be converted to a largely unavailable inorganic form. For example, in well aerated and alkaline calcareous soils the availability of copper, manganese and iron decreases, as they are readily oxidized or precipitated in such environments and are not easily absorbed by plant roots.

These pathways and processes of transfer of elements between the living systems of the biosphere and their immediate environment can be thought of as the ecosphere loop of the circulation of materials depicted in Fig. 7.15. This diagram, however, also shows us that the ecosphere loop is linked to the pathways of transfer of matter through the atmosphere, hydrosphere and lithosphere. Elements are lost from the soil, mainly via the denudation system (see Chapters 5 and 10), ultimately to sediments, and in aquatic environments by precipitation and deposition, again ultimately to sediments. There are also elements lost from the ecosphere loop in a volatile or gaseous state. Carbon dioxide released during respiration by autotrophic, heterotrophic and decomposer organisms is returned to the atmosphere. The same is true for the nitrogen gas (dinitrogen) and hydrogen sulphide released by nitrate- and sulphate-reducing bacteria respectively, as well as for the oxygen evolved during photosynthesis, and in marine environments dimethyl sulphide (DMS) from phytoplankton.

The pathways by which the losses of these volatile elements, and of those which follow the denudation system, are made good differ fundamentally not only in the routes taken but also in the timescale involved. For these oxidized gaseous elements to be returned to the biosphere requires the investment of energy to raise them to a higher energy level or to a chemically reduced form. In the case of nitrogen, ultraviolet radiation or lightning may provide sufficient energy, if only for a fraction of a second, for nitrogen to combine with oxygen or with the hydrogen of water. The oxides and hydride of nitrogen so formed dissolve in rainwater to reach the soil or the oceans as nitrate and ammonium ions available for ab-

sorption by plants. The main route, however, is the biological fixation of gaseous nitrogen by both free-living and symbiotic bacteria and algae (Box 7.5), which ultimately make it available to higher plants as nitrate (Fig. 7.16).

Oxygen is coupled to respiratory oxidation as an electron acceptor and is reduced to water. In the case of sulphur, it enters the cells of plants as an inorganic sulphate ion in a similar way to nitrate, but it would not be recycled through the atmosphere without the bacterial reduction to sulphides under anaerobic conditions. When not precipitated as iron sulphide, sulphur escapes to the atmosphere as H_2S and DMS, which is then reoxidized to sulphate. Finally carbon as CO_2 is photochemically reduced to carbohydrate during photosynthesis, which is where we started. The cycling of these gaseous elements between the atmosphere and the ecosphere depends, therefore, on four reduction reactions. All require an investment of energy which, in green plants and autotrophic bacteria and algae, is derived from an external source, but in the case of the heterotrophic bacteria involved in nitrogen and sulphur reduction must be supplied from food energy.

All of the remaining elements essential for life are lost from the ecosphere by leaching and they enter the denudation system, eventually reaching the ocean sink. Their recycling only becomes possible over a long geological timescale, following sedimentation, lithification, uplift, erosion and weathering. The only way that this route can be short-circuited is for mineral dust from the land surface, or droplets of seawater as spray, to enter the atmosphere, where they are mixed and circulated. The elements contained eventually fall out, or are washed out by precipitation, thereby completing the cycle. However, they only enter the biosphere when taken up by plants or ingested by animals and incorporated in organic compounds using energy originally fixed during photosynthesis (Fig. 7.16).

7.6 Biogeochemical Cycles

In this section we will turn from the insights provided by a focus on the energetics of the transfer processes involved in the circulation of matter in the ecosphere. Instead we will adopt one which

Box 7.5
NITROGEN FIXATION

Due in part to the very stable triple bond ($N \equiv N$) that links the two atoms of nitrogen in the dinitrogen molecule (N_2) of atmospheric nitrogen gas, it is normally very unreactive. However, biological nitrogen fixation succeeds in making the molecule reactive at normal temperatures and pressures. As with other biochemical reactions, this is only possible because the reaction is catalysed by an enzyme – in this case the enzyme **nitrogenase**. Although the mechanisms are as yet not fully understood, nitrogenase is known to be a complex of two proteins, both of which have attached iron and sulphur atoms. The larger of the two proteins also has two atoms of molybdenum per molecule. In the nitrogen fixation process an activated complex of the two proteins is formed with the nitrogen molecule bound to the molybdenum of the larger protein and ATP (see Box 7.2), as a mono-magnesium salt is bound to the smaller protein. Electron transfer takes place between the iron atoms of the two proteins (a redox reaction) assisted by the energy from the ATP–Mg which in turn is converted to ADP and Mg^{++}. The result of these electron transfers is the splitting of the molecular bond in dinitrogen and the release of the reduced product of the enzyme reaction as ammonia (NH_3).

It is now known that a very large number of organisms contain nitrogenase and are capable of nitrogen fixation, but they are all primitive groups – bacteria and blue-green algae. They exist both as free-living organisms and in symbiotic relationships with higher plants (in the case of the legumes and *Alnus*-type associations of bacteria) and some lesser plants (as with the lichens where the blue-green algae *Nostoc* and *Calothix* grow within a fungal matrix). None of the 4×10^{12} kg of nitrogen gas in the atmosphere, or of the 2×10^{14} kg bound in rocks, is available to plants until it is fixed by these organisms – with the exception of the application of industrially produced nitrogen fertilizer – but even under agriculture the contribution of nitrogen-fixing microorganisms is enormous.

Estimated N-input into plant crops in 1971–72 ($\times 10^9$ kg N) (Postgate, 1978) is as follows:

Source of N	UK	USA	Australia	India
fixation in legumes	0.4	8.6	12.6	0.9
fixation by free-living microorganisms	<0.04	1.4	1.0	0.7
take up from N fertilizer	0.6	4.9	0.1	1.2

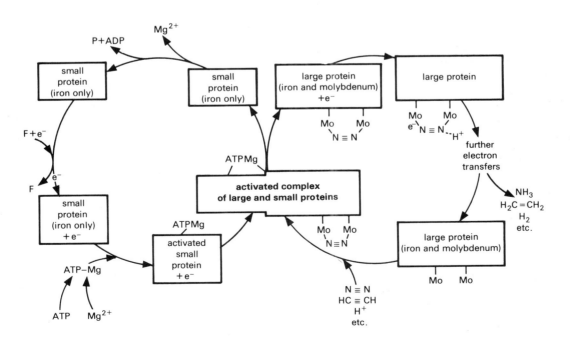

concentrates on the structure of these cycles in terms of the main compartments, stores or reservoirs involved. In so doing we shall assess their relative importance, the residence time of elements in them, and the relationships between them.

7.6.1   The Phosphorus Cycle

The **geochemical** cycle of phosphorus (see Fig. 7.16) is essentially that of the rock cycle (Chapter 5): one of denudation, sedimentation, uplift, and once again denudation. It is the quintessential **sedimentary cycle** with a volatile species (phosphene) of the element which is insignificant. As such it will serve as an example for many other elements, such as calcium and potassium, which follow similar pathways of circulation. The reservoirs feeding the phosphorus cycle, therefore, are to be found principally in the rocks of the Earth's crust, in ocean sediments, in soils, and in the waters of rivers, lakes and the oceans. Furthermore, it differs from other cycles that we shall examine in that microbial action is not significant. However, within this geochemical cycle of sedimentation, uplift and erosion, are nested three subsidiary cycles. Two of these are inorganic, the first being transfers between deep and surface waters in the oceans. This involves the long-term circulation of deep-water columns, perhaps taking thousands of years to complete. The second is a small but not insignificant pathway which short-circuits the rock cycle. It starts with bubble-bursting and sea-spray over the oceans (and dust over the land) which carries phosphorus into the atmosphere, from where it returns via precipitation and dry fallout back to the land surface (and the ocean surface). The final subsidiary cycle links phosphorus circulation to the biosphere and involves the incorporation of phosphorus in organic compounds in terrestrial and marine biomass, so that the circulation of the element becomes overall a **biogeochemical** cycle.

On land, therefore, the geochemical part of the cycle is linked to terrestrial ecosystems by phosphate uptake from the soil and the biosynthesis of organic compounds containing phosphorus. Most of this phosphorus comes from the weathering of phosphate minerals such as apatite, plus, as we have just seen, a small input from precipitation and dry fallout from the atmosphere. We already know that phosphorus is a critical element for living systems (ADP and ATP for example),

and as we shall see in Chapter 19 it is actively conserved within living biomass during the life of organisms, and is involved in very tight nutrient cycling between living and dead biomass. Indeed, the residence time of phosphorus in terrestrial biomass can lie between 1 and 100 years before it is finally leached or translocated to drainage waters. Reserves of mineral phosphorus in the soil can be considerable, but only a fraction of it is immediately available as orthophosphate ions. Here, symbiotic or mutualist relationships between the roots of most higher plants and fungi, **mycorrhizae**, considerably improve nutrient (phosphorus) uptake, especially on poor soils with an inherent phosphorus deficiency.

Once lost to the soil and terrestrial nutrient circulation, phosphorus passes relatively rapidly through the terrestrial hydrosphere, with residence times in aquatic ecosystems of the order weeks to years only. In these aquatic ecosystems phosphorus and other elements are passed rapidly oceanwards in downstream nutrient spirals. Indeed, in common with many other elements which have sedimentary cycles, much of the phosphorus flux through streams and rivers takes place in particulate form as suspended sediment load.

In the oceans surface water phosphorus concentrations are relatively low, and are rate-limiting for marine productivity, with the turnover between the water and marine organisms being very rapid. Phosphorus in surface waters, however, is linked to circulations with the deep ocean, which is itself a significant reservoir of phosphorus. So an atom of phosphorus may have a residence time in surface waters (where it may be utilized over and over again by the marine ecosystem) of a few thousand years, but its total residence time in the oceans may be much longer; about 100 000 years. During this time it may be transferred between the surface and deep water up to 100 times before immobilization in ocean sediments for perhaps ten million years removes it from this subsidiary cycle. Subsequently the circulation of phosphorus becomes closed only when uplift and erosion once again releases phosphorus at the land surface to undergo weathering, after perhaps hundreds of millions of years.

7.6.2   The Nitrogen Cycle

Although the vast reserve of **dinitrogen** gas (N_2) which the atmosphere contains is virtually inert,

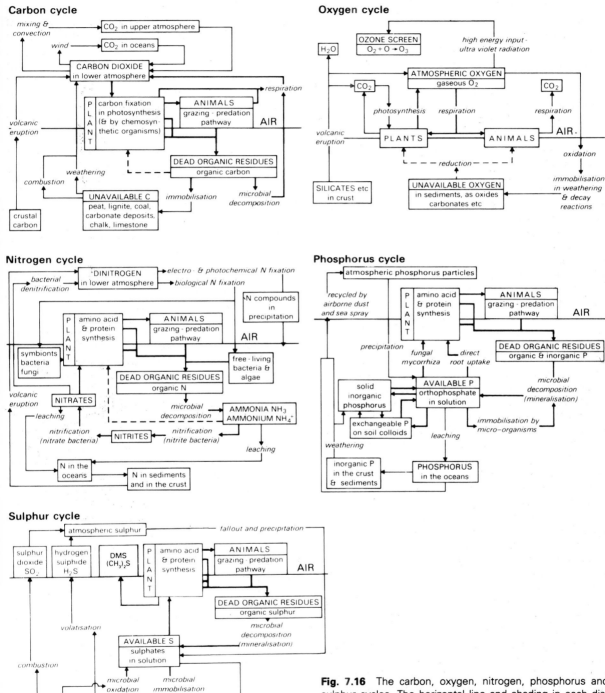

Fig. 7.16 The carbon, oxygen, nitrogen, phosphorus and sulphur cycles. The horizontal line and shading in each diagram represents both the soil, in terrestrial environments and the water/ocean surface in aquatic environments as appropriate.

it is the principal reservoir of the nitrogen cycle. This reservoir is linked to the biosphere by **nitrogen fixation** and **denitrification**, which maintain a steady-state concentration of dinitrogen in the atmosphere with a turnover time of 10^7 years. The cycle, therefore, is essentially a gaseous one, but just for a moment we shall regard it as a purely geochemical cycle. The only pathway of transfer in such a case would be very slow moving. This would involve photoelectrochemical fixation of atmospheric nitrogen and the rainout of nitrogen compounds, which would also include dust particles and volatilized ammonia from the atmosphere. From this point leaching and transportation would culminate in nitrogen compounds bound in sediments and rocks. The cycle back to the atmosphere would only be completed by the volatilization of nitrogen compounds, particularly the oxides and hydrides, and their emission into the atmosphere through volcanic activity and crustal degassing. In fact without organic denitrification all the nitrogen would effectively be sequestered in the oceans and in organic sediments.

In reality, of course, photoelectrochemical fixation accounts for only 3–4% of the nitrogen fixed from the atmosphere. The other pathways of nitrogen fixation, as we already know, involve the organisms of the biosphere, and so in a very real sense the nitrogen cycle is a biogeochemical cycle. The biosphere part of the cycle is linked to the geochemical cycle described above both by nitrogen fixation – the vast majority effected by microbial activity – and by denitrification, again mediated by microorganisms. On land, something of the order of 7 kg N ha^{-1} an^{-1} is fixed by symbiotic organisms, and 3 kg N ha^{-1} an^{-1} by free living organisms. However the total of these two figures underestimates the amount available to plants each year; indeed, it only represents 12% of available nitrogen. The remainder comes from efficient recycling between living and dead biomass, or reflects nutrient conservation strategies within ecosystems. In any event it underlines the view that nitrogen, a critical element in living systems (e.g. as a constituent of amino acids and proteins), is under tightly controlled circulation in most ecosystems (see Chapter 19).

In fact, there are relatively small amounts of nitrogen in terrestrial biomass and soil organic matter (high C:N ratio) and equally very small amounts of inorganic nitrogen reside in the soil as ammonium or nitrate ions. However, as we have just implied, the flux through these compartments is relatively large and very rapid. It is a flux which is dependent on the activity of the bacteria and fungi of decay, on nitrite bacteria converting ammonium compounds to nitrites, and on nitrate bacteria converting nitrites (which in any significant concentration are toxic to plants) to nitrates, and finally on uptake by plant roots. The efficiency of this pathway is such that under undisturbed conditions little nitrate is leached and lost to drainage and stream waters, though some is returned to the atmosphere as nitrous oxide by bacteria denitrification. The small loss to rivers in turn can inhibit and limit the level of nutrition of aquatic ecosystems (see also phosphorus).

In total, however, the input of nitrogen to the oceans from terrestrial runoff is highly significant, accounting for perhaps one-third of the total (precipitation input and biological fixation accounts for the remainder). In the oceans nitrogen has a mean residence time of 8000 years, and like other elements is involved in vast, slowly circulating columns of water which carry elements between the surface and the deep and back, with a periodicity of between 200 and 400 years. While in the surface waters, nitrogen is cycled rapidly through marine ecosystems a large number of times. However, in contrast to some other elements, nitrogen for the most part is not immobilized and incorporated in deep ocean sediments. On the contrary, its major pathway of return to the atmosphere is not the rock cycle route, but by microbial denitrification.

Denitrification, in both the terrestrial and marine biospheres is, under natural conditions, the principal source of nitrous oxide (N_2O) to the atmosphere. Here some nitrous oxide diffuses up into the stratosphere, where it becomes unstable, forming nitric oxide (NO_2), which has an important role in catalysing the destruction of ozone (see Chapter 4).

7.6.3 The Sulphur Cycle

The cycling of sulphur is in many ways a hybrid between the gaseous cycle exemplified by nitrogen and the sedimentary cycle of elements such as phosphorus. Certainly the largest reservoir is in the crust, in minerals such as pyrite (iron II sulphide), and gypsum (calcium sulphate) in evaporite deposits derived originally from ocean water, and as a precipitate in saline soils (see Chapter 22), thereby resembling the phosphorus cycle. Additionally, oxidative weathering, bubble-bursting and sea-spray, and volcanism and crustal degassing transfer large amounts of sulphate aerosols and ions to the oceans. Half of this input comes from precipitation and dry fallout, and half from runoff and denudation. The mean residence time of sulphur in the oceans is very long – about 3 Ma – but sulphur is lost continuously to the comparative sink of ocean-floor sediments by the conversion of hydrogen sulphide (H_2S) to iron II (ferrous) sulphide and its precipitation.

However, in other ways the sulphur cycle has strong similarities with that of nitrogen. Like nitrogen, the largest annual transfers are through the atmosphere, though not mediated through the mechanisms referred to above, but by the production of reduced gases by microbial activity. These gases return sulphur to the atmosphere in significant quantities, closing the cycle and thereby short-circuiting the longer-term sedimentary route and providing a relatively rapid turnover of sulphur. Hardly any of the gaseous states of sulphur is long-lived, so sulphur compounds in the atmosphere have short mean residence times because of oxidation to sulphate, but the annual flux through the atmosphere rivals that of nitrogen. However, in contrast to both the nitrogen and the phosphorus cycles only a small part of the global flux of sulphur is reduced and assimilated as organic sulphur compounds (again as with nitrogen, some amino acids and proteins will serve as examples) in biospheric cycling.

So, just as with the nitrogen cycle, the global sulphur cycle is driven by microbial transformations. In anaerobic environments sulphate forms the substrate for microbial sulphate reduction, which may lead to either the release of gases, particularly hydrogen sulphide (H_2S) in freshwater wetlands and in anaerobic and hydromorphic soils, or the precipitation of sulphides in the presence of iron. The latter occurs in both of these terrestrial situations when iron is present in excess, and commonly in marine sediments where the release of H_2S is of minor importance. These same anaerobic environments also support a variety of primitive (in the sense of early) photosynthesis, where hydrogen is donated from sulphur compounds rather than from water:

$$CO_2 + 2H_2S \rightarrow CH_2O + 2S + H_2O$$

In contrast, in aerobic environments reduced sulphur compounds undergo microbial oxidation, sometimes coupled to the reduction of carbon dioxide in chemosynthetic reactions based on sulphur.

From the oceans, sulphur is returned to the atmosphere mainly by the biogenic emissions of dimethylsulphide (DMS). DMS ((CH_3)$_2$S) is produced by marine phytoplankton in the upper 50 m of the oceans (Fig. 7.17). Although concentrations are low it is ubiquitous in the surface waters, and over the enormous surface area of the oceans the small flux to the atmosphere per unit area adds up to about a third to a half of all the sulphur entering the atmosphere. However, DMS is rapidly oxidized in the atmosphere and only has a mean residence time of a day or so. It is returned as sulphate aerosol particles (sulphuric acid) in precipitation. Under natural conditions the increase in rainwater acidity which results is slight and spread widely over the Earth's surface. These aerosols also have a significant role in relation to atmospheric energy exchanges and global climate (see Chapters 3 and 4), reflecting and scattering shortwave radiation, and over remote open oceans forming the main source of condensation nuclei for cloud droplet and precipitation (see Chapters 4 and 6).

Also entering the atmosphere from the surface waters of the oceans is carbonyl sulphide (COS), which is formed photochemically by the action of sunlight in the photic zone on various organic compounds (Fig. 17.17). Unlike DMS, COS is practically inert in the lower atmosphere (troposphere) but oxidizes in the stratosphere under the influence of high-energy ultraviolet radiation. However, like DMS, COS also gives rise to sulphate aerosols, creating a haze high in the stratosphere which reflects incoming solar radiation back to space.

7.6.4 The Carbon and Oxygen Cycles

As we are already aware, the driving forces behind the global carbon cycle are the opposing processes of photosynthesis and respiration. As the first of these is also responsible for the oxygen content of the atmosphere, the carbon cycle is closely associated with that of oxygen, as indeed it is through oxidation and reduction reactions with the cycles of many other elements. The carbon cycle is predominantly gaseous, with the main fluxes occurring as carbon dioxide (CO_2). The largest of these fluxes are between the atmosphere and the hydrosphere and biosphere, particularly the terrestrial biosphere. Even under natural conditions the atmospheric reservoir of carbon dioxide (mean residence time about 3 years, close to the mixing time for the atmosphere) would have fluctuated in response to both the seasonality of photosynthetic requirements, and of oceanic uptake, but overall the atmospheric concentration of c. 350 ppm would have maintained a steady state. Historically the lithosphere was unimportant and played a minor role in relation to the main fluxes of carbon dioxide.

On land, carbon dioxide is fixed and stored in plant and animal tissue and in soil organic matter. The overall residence time in the biosphere seems to be of the order of 20 years. In living biomass this figure would be nearer 6 years, and in dead organic matter perhaps 25 years, so the average residence time reflects both the average lifetime of vegetation and the time taken for microbial decomposition (see Chapter 21). Like the terrestrial biosphere, the well mixed surface waters of the oceans maintain a balance with the carbon dioxide content of the atmosphere by the processes of dissolution in seawater and by degassing, but mixing with deeper waters is very slow and this influences the mean residence times for carbon in the oceans. In the surface ocean, where most of the exchange with the atmosphere occurs, residence time is of the order of 6 years. Turnover in the ocean as a whole is much slower – c. 350 years – reflecting the slow circulation and mixing with deeper waters.

Although it is by far the most important, carbon dioxide is not the only chemical species involved in carbon fluxes, and here we will briefly consider methane (CH_4). **Methanogenesis** from wetlands and hydromorphic soils is the main source of methane; its atmospheric concentrations are low, only 1.7 ppm, and it has an atmospheric residence time of about 10 years. The most important current concern in regard to methane is that it appears to be increasing slowly in concentration, and as a greenhouse gas has occasioned some study (see Chapters 4 and 24).

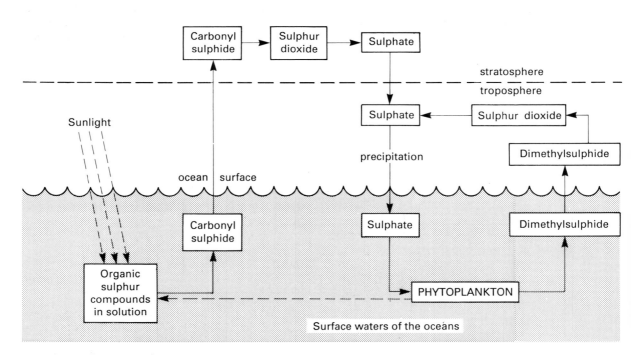

Fig. 7.17   Biogenic Sulphur Compounds emitted from the oceans (derived from Andreae, 1987)

Oxygen first appeared in the Earth's atmosphere as a result of autotrophic photosynthesis (Section 7.7) and accumulated there when production exceeded that needed to oxidize the exposed rocks of the lithosphere. Of course the current atmospheric reservoir bears no relation to the total produced over geological time, and like carbon dioxide maintains a steady-state concentration with the biosphere between production by photosynthesis and consumption in respiration. However, in both the carbon and oxygen cycles the exchanges between the atmosphere and the terrestrial biosphere on the one hand, and the oceans on the other, are apparently uncoupled (Fig. 7.18). In the oxygen cycle, that part in the oceans is well buffered, for increases are compensated for by the expansion of the depth of aerobic respiration in marine sediments, thereby consuming the increase. In the carbon cycle increases in carbon dioxide may be photosynthetically withdrawn from seawater and accumulate as carbonate and organic sediments or buffered by the dissolution of carbonate sediments. In both cases the feedback mechanism which connects the independent terrestrial and oceanic circulations involves the burial and subsequent uplift of sediments and their oxidation by weathering reactions at the land surface. Both the carbon and oxygen cycles are characterized, therefore, by large annual fluxes

superimposed on slow fluxes which maintain a steady state over long periods (Walker, 1984; Schlesinger, 1991)

Oxygen is also coupled to many other elemental cycles. For example, 14% of the annual consumption of oxygen is used to oxidize ammonium in nitrification, oxygen that might otherwise be available for the oxidation of organic carbon. Methanogenesis is responsible for 4% of oxygen consumption each year, while sulphate reduction to iron sulphides in the oceans releases oxygen, regulating the oxygen content of the oceans, and by interaction with the atmosphere affects atmospheric oxygen concentrations.

7.6.5 *Perturbations to biogeochemical cycles*

All the biogeochemical cycles considered above have, in the 20th century, come under the influence of anthropogenic interference and modification. The perturbations to these cycles occasioned by humans will be considered more fully in subsequent chapters, most notably in Chapters 24 and 26. However, to conclude this section we shall look briefly at the main perturbations to the five cycles we have just discussed.

Perhaps the most significant effect of human activities on the phosphorus cycle has been the inflation of the quantities of phosphorus being transferred through the denudation system from

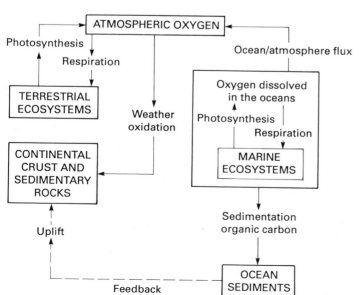

Fig. 7.18 Oxygen regulation: The feedback between the marine and terrestrial biosphere is the 'flow' of organic carbon in sediments which on exposure to the atmosphere on uplift represents an oxygen demand.

land to ocean. Human activity now accounts for approximately two-thirds of the phosphorus reaching the oceans as the suspended and solute loads of rivers. This increase can be traced to several sources. In the first instance, biomass removal and accelerated erosion, runoff and discharge is one factor. Next, runoff from agricultural land carries with it excess phosphate fertilizer. This fertilizer is originally mined from the lithospheric reservoir or transferred from the ocean reservoir via a marine foodchain comprising fish and fish-eating birds, and finally their phosphate- and nitrate-rich droppings subsequently mined for fertilizer. Indeed, fishing transfers a significant amount of phosphate from oceans to land, and again it ultimately manifests itself as increased phosphorus in fresh waters. Domestic and industrial detergents also follows the same route, via industrial effluent and domestic sewage respectively. The increased input to the oceans is, of course, insignificant when set against the magnitude of the oceanic reservoir, but its effect on the streams, lakes and estuaries through which it is transferred can be catastrophic and lead to eutrophication problems.

Perturbations to the nitrogen cycle are similar to those affecting phosphorus, at least as far as that part of the cycle which follows the denudation pathway is concerned. Here, too, nitrates contribute to eutrophication problems associated with runoff from agricultural land using nitrate fertilizers, and from the treatment and discharge of sewage effluent, while biomass removal accelerates the loss of fixed nitrogen from the biosphere. However, the nitrogen cycle differs from that of phosphorus in terms of the transfers between the Earth's surface and the atmosphere. Both nitrogen fixation and denitrification are influenced by human activity. **Leguminous** crops with **symbiotic bacteria** increase artificial fixation, so that together with industrial fixation processes and the fixation of nitrogen as oxides as a byproduct of the internal combustion engine, it is now of the same order of magnitude as natural fixation.

Denitrification is also increased under productive fertilized agriculture, and paradoxically by deforestation and vegetation clearance. The small increase in nitrous oxide (N_2O) in the atmosphere which appears to be resulting, and its significance for the ozone shield have already been mentioned. Finally the oxides NO and NO_2 emitted by the combustion of fossil fuels and from automobiles contribute to the 'acid rain' problem as nitric acid in rainwater (about 30–40% responsible) afflicting the industrialized world and those unfortunate enough to be downwind.

Acid rain is also the main impact of perturbations to the sulphur cycle, by the burning of fossil fuels which contain small but significant amounts of sulphur (about 3% on average in both coal and oil). As sulphate aerosols which dissolve in raindrops as sulphuric acid, sulphur emissions are washed out of the atmosphere in a short time and are the main culprit in acid rain. As we have seen, the natural release of sulphur, principally as DMS, over the oceans and artificial emissions are roughly equal. The anthropogenic emissions however, are highly concentrated in the industrialized areas, particularly of the developed world. In these areas the carbon dioxide equilibration pH of rainwater of 5.6 can be reduced as low as 2.1 (Mooney *et al.*, 1987).

With the carbon cycle, of course, the principal perturbation is the increase in carbon dioxide concentrations in the atmosphere resulting from the combustion of fossil fuels, thereby bringing the immobilized carbon reservoir of the lithosphere into play in global carbon cycling. The alleged impact of this state of affairs is global warming and global climate change, topics which will both be discussed fully in Chapter 24.

7.7 The ecosphere and the evolution of life

We are now in a position to address again those questions raised earlier in the Chapter about the effects that the evolution of life has had on the development of our planet. There is ample evidence to suggest that life evolved in an environment very different from that which exists today. The atmosphere had a secondary origin from the gases evolved from the Earth's interior by volcanism, and it lacked free oxygen. Furthermore, as the ozone shield did not exist, destructive ultraviolet radiation was received at the Earth's surface. In this atmosphere, and particularly in the sterile oxygen-free seas, molecules would have been continually circulated and mixed, and collisions between them may well have facilitated the production of the precursors of 'organic molecules'. That such molecules could form without life as we know it has been demonstrated by Miller (1953 and 1957) in what is now a classic experiment, and repeated many times since with many combinations of conditions (Chang *et al.*, 1983). This involved circulating methane (CH_4), ammonia (NH_3) and water vapour – all of which were present in the primitive atmosphere – over an electric spark for a week. The spark was to simulate the energy input of electrical discharges in the atmosphere. At the end of the period, glycine and alanine (the two simplest amino acids) and other 'organic' compounds had been synthesized.

At some stage the spontaneous but ordered arrangement of these precursor molecules reached a complexity which could be considered to be the first living organism. From that moment on, there began an interaction between life and its inorganic environment which was to change that environment completely. This interaction was to have three major steps.

The first organisms were undoubtedly heterotrophs with no option but to live on the 'organic' molecules from which they themselves had evolved. As we have seen when considering the heterotrophic cell model, such a strategy implies living on capital, while in the absence of oxygen the only process by which the energy of the food molecules could be released was **fermentation**. Fermentation is a partial oxidation, essentially similar to the first stage of the respiration process outlined in Section 7.3.1. Compared with total oxidation, however, fermentation is extremely inefficient (Fig. 7.19), while the products are waste which must be disposed of.

Fortunately, these primitive forms of life developed the capacity of manufacturing their own organic food molecules before they exhausted the capital on which they had been living, i.e. pre-existing 'organic' molecules in the primeval seas. To do this they utilized carbon dioxide (one of the waste products of fermentation), which had been accumulating in the atmosphere and oceans. Initially they derived the energy necessary for the synthesis of organic carbon compounds from the bond energy of inorganic compounds, but about 3×10^9 years ago the first organisms appeared which were capable of utilizing light energy from the Sun. Photosynthesis had arrived.

At first, however, the impact of photosynthesis

was limited. These early autotrophs were not equipped to deal with free oxygen and they may have acquired electrons for photosynthesis from substances other than water. Even when water was split during photosynthesis to release molecular oxygen, it has been suggested that the early photosynthetic organisms used inorganic compounds as oxygen acceptors. The potential destructive effects of free oxygen were therefore avoided by these anaerobic organisms. Eventually, protective enzymes were evolved and life became not only autotrophic but for the first time aerobic as well, and oxygen began to enter the atmosphere in gradually increasing quantities (Table 7.3).

The presence of oxygen had two important effects. First of all the early photosynthesizers were limited by fermentation in the energy yield they could win from the fuel molecules they produced. The presence of free oxygen allowed the development of complete oxidative respiration, which, by releasing a far greater amount of energy from the fuel molecules, enabled the biosphere to invest the surplus in an accelerated process of evolution, diversification and accumulation of biomass. Here the second effect of free oxygen becomes important, for ultraviolet radiation caused some of the molecular oxygen to dissociate and the subsequent recombination of the highly reactive atomic oxygen formed ozone. So the developing ozone screen high in the atmosphere (Chapters 3 and 4) absorbed the lethal ultraviolet waves of solar radiation, and this allowed life to emerge from

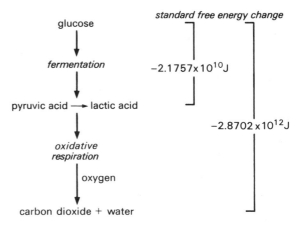

Fig. 7.19 The energy yield from fermentation and for oxidative respiration (after Lehninger, 1965).

Table 7.3 The evolution of the atmosphere and the fossil record

Period	Age in years	Solar radiation reaching Earth	Terrestrial atmosphere	Metabolic systems	Structural fossils

Solar radiation reaching Earth: visible light; longwave ultraviolet; shortwave ultraviolet

Terrestrial atmosphere components: H_2, NH_3, CH_4, N_2, H_2O, CO_2, O_2

Period	Age in years
birth of the Earth	5.0×10^9
	4.5×10^9
early Precambrian	3.5×10^9
	3.3×10^9
	3.2×10^9
	3.1×10^9
	2.8×10^9
	2.7×10^9
	2.3×10^9
	2.2×10^9
middle Precambrian	2.0×10^9
	1.9×10^9
	1.6×10^9
	1.3×10^9
late Precambrian	1.1×10^9
	1.0×10^9
	5.8×10^8
Cambrian	4.0×10^8
Palaeozoic and Devonian younger	
Carboniferous	3.4×10^8

Metabolic systems:
- abiogenic production of organic compounds
- origin of life?
- anaerobic heterotrophs
- anaerobic heterotrophs using solar energy to produce ATP
- chemical remains of bacterial-type systems
- water-dissociating organisms anaerobic photoautotrophs using solar energy to produce ATP
- microbial photoautotrophic aerophiles producing more ATP
- O_2 gradually becoming more tolerable constituent of atmosphere
- aerobic respiration established
- organisms with modern type of metabolism and sexual reproduction
- O_2 production by photosynthesis at maximum

O_2 levels: 1% PAL, 10% PAL, 20% PAL

Structural fossils:
- Prokaryotic organisms
- Bacteria
- Blue-green algae – coccoid
- Stromatolites
- Blue-green algae – filamentous
- Eukaryotic organisms
- Metazoans
- Invertebrates
- Higher algae – marine evolution
- Vascular plants
- Terrestrial animals

– – – possible occurrence; ——— probable occurrence; ——— established occurrence; PAL, present atmospheric level.

the protective sediments and waters of the oceans and to colonize the land surface.

The changes in the chemistry of the atmosphere and oceans which accompanied these milestones in evolution had profound effects on the lithosphere. The presence of free oxygen in the surface waters of the Earth changed the nature of sedimentary rocks by altering the sedimentation environment and affecting the solubility of many inorganic compounds; iron-rich sediments, for example, first appeared 1.8×10^9 years ago. The immobilization of carbon in organic compounds and in insoluble carbonates by the photosynthetic withdrawal of carbon from atmospheric carbon dioxide and from soluble bicarbonates not only reduced the carbon dioxide content of the atmosphere but produced most of the limestones of the stratigraphic column. Today the carbon dioxide concentration of the atmosphere (if we ignore the burning of fossil fuels) is in equilibrium with the photosynthetic system of the biosphere. In organic compounds the immobilization of elements other than carbon has also affected the processes of erosion and denudation, and the sediment loads delivered to the oceans are very different from those which would arrive in the absence of terrestrial organisms. Finally, weathering and soil formation are both processes whose character has been determined by the biosphere.

The organisms of the biosphere, therefore, inhabit an environment which they themselves have largely produced. Why this should have come about can be understood if we return to the simple systems model of Fig. 7.1. The appearance and evolution of life can be regarded as an apparent reversal of the trend to maximum entropy of the universe. As we have seen, however, for the central compartment in our model – the biosphere – to have reduced its internal entropy throughout evolution there must have been a corresponding gain in entropy by the surroundings. In our model the surroundings can be thought of as that part of the atmosphere, hydrosphere and lithosphere included in the ecosphere, with which the organisms of the biosphere exchange matter and energy. Therefore, the changes discussed above can all be seen as an expression of a progressive gain in entropy by the surroundings of the biosphere.

Further reading

The scope of this chapter is very wide indeed, and in consequence only a limited range of suggested reading can be provided for each major topic. There are many introductory biology and biochemistry texts which cover the subjects treated in the first part of the chapter, e.g.:

Moore, D.M. (1982) *Green Planet: the Story of Plant Life on Earth*. Cambridge University Press, Cambridge.

Rose, S. (1970) *The Chemistry of Life*. Penguin, London.

Soper, R. (ed) (1984) *Biological Science. Part 1. Organisms, Energy and Environment*, and *Part 2. Systems, Maintenance and Change*. Cambridge University Press, Cambridge.

Villee, C.A. *et al.* (1989) *Biology*, (2nd edn). Saunders, Philadelphia.

Williams, V.R., Mattice, W.L., & H.B. Williams (1978) Basic *Physical Chemistry for the Life Sciences*. Freeman, N.Y.

The biogeography of species populations is covered by:

Cox, B.A. and P.D. Moore (1985) *Biogeography: an Ecological and Evolutionary Approach*, (4th edn) Blackwell, Oxford.

Hengeveld, R. (1990) *Dynamic Biogeography. (Cambridge Studies in Ecology)*. Cambridge University Press, Cambridge.

Myers, A.A. and P.S. Giller (eds) (1988) *Analytical Biogeography: an Integrated Approach to the Study of Animal and Plant Distributions*. Chapman & Hall, London.

Vincent, P. (1990) *The Biogeography of the British Isles*. Routledge, London.

The ecology of individuals, populations and communities is the subject matter of an increasing number of broad texts in ecology, several of which are cited at the end of Chapter 18. Only two of those texts are referred to here:

Begon, M., J.L., Harper and C.R. Townsend (2nd edn) (1990) *Ecology: Individuals, Populations, and Communities*. Blackwell, Oxford.

Ricklefs, R.E. (1990) *Ecology* (3rd edn) Freeman, New York.

Biogeochemistry is a topic of increasing importance with some good recent tests now available, of which Schlesinger is perhaps the best:

Drever, J.I (1988) *The Geochemistry of Natural Waters*. Prentice-Hall, Englewood Cliffs.

Garrels, R.M., F.T. MacKenzie and C. Hunt. (1975) *Chemical Cycles and the Global Environment*. Kaufman, California.

Fergusson, J.E. (1982) *Inorganic Chemistry and the Earth*, Pergamon, Oxford.

Holland, H.D. (1978) *The Chemistry of the Atmosphere and Oceans*. John Wiley, New York.

Schlesinger, W.H. (1991) *Biogeochemistry: an Analysis of Global Change*. Academic Press, San Diego.

Woodwell, G.M. (1978) The carbon dioxide question. *Scientific American*, **238**, 38–43

Part C

Open system model refined: environmental systems

A number of important conceptual ideas have emerged from the pursual of a systems approach to the natural environment in the first three sections of this book. First of all we have realized that all natural systems are open systems, but more than this we have seen that they exist as components of cascades through which matter and energy flow: the output of one system forms the input to the next. The solar radiant energy which is output from the atmosphere to the Earth's surface is the energy input to the autotrophic cells of the biosphere. Part of the biomass of producer organisms is the energy and nutrient input to the consumer organisms of the trophic model of the biosphere. The precipitation that falls from the atmosphere is the input of water to the denudation system.

Within any component system of such cascades, however, part of the flow of matter and energy through the system is transferred temporarily to stores. Energy and chemical elements flowing through the biosphere are diverted temporarily to the biomass store of the organisms of the different trophic levels. The flow of geothermal heat energy and geochemical elements through the crustal system is diverted to the store of potential and chemical energy and rock-forming materials in the uplifted relief of the Earth's surface. Some energy or matter will entirely bypass these stores, but this, together with the transfer of matter and energy into and out of storage and between stores, represents the throughput of the component systems of the cascade.

We have also seen that the destinies of energy and matter as they cascade through these systems are fundamentally different. As each energy transfer is completed, part of the original input of energy is dissipated, in accordance with the Second Law of Thermodynamics, as unavailable heat energy ultimately radiated to the energy sink of space. Energy flow is, therefore, a one-way process. When we consider the transfer of matter, we have found that ultimately it turns full circle and the cascades of matter in our global models have become cycles. The completion of these cycles, however, only occurs because they are interlocked with and driven by the cascade of energy.

In Part C we are going to increase the resolution and fidelity of our models of natural systems considerably, and alter both our spatial and temporal perspectives. From broad global vistas, with one eye on geological time, we shall move to the scale of the local landscape, which is both familiar and accessible. We shall wade in streams, walk through woodlands and dig through leaf litter to the soil beneath. We shall don our raincoats as a frontal depression approaches and we shall feel a sea breeze on our faces. For the most part, we shall be concerned with much shorter spans of time – with the day, the year and perhaps the century. Nevertheless, we shall still be concerned with open systems functioning as parts of energy and mass cascades. Although it is possible to consider each type of cascade separately as the solar energy cascade, the hydrological cascade and the debris cascade, for example, we shall define the component systems of these cascades and examine the ways in which the throughputs of matter and energy are interlocked as they cascade through and between these systems. As our models of these systems increase in their level of resolution we shall identify nested hierarchies of systems operating at different spatial and temporal scales. For example, the denudation system is modelled in terms of the catchment or drainage basin, but within it we shall recognize weathering, slope and channel subsystems, and within them further subsystems such as the soil aggregate breakdown/raindrop impact system on a slope, or the chemical reaction system in rock weathering. Primary productivity of the ecosystem will be considered at the scale of the individual leaf, the plant and the entire vegetation canopy, while in the atmosphere we shall recognize secondary and tertiary circulation systems within the general pattern of global circulation.

IV ATMOSPHERIC SYSTEMS

The atmosphere and the Earth's surface

8.1 Introduction

The lowest 11 km of the atmosphere – the **troposphere** – are characterized by turbulent mixing of constituents and exchange of gases, liquids and solids, as well as heat energy, between the air and the Earth's surface. Water molecules, for example, are transferred from surface to air and air to surface. Similarly, materials such as dust and soot are derived from the surface, to which they eventually return. Heat energy transferred from the surface into the air (see Chapter 3) is convectionally mixed through the troposphere. A net transfer back to the surface tends to engender stability in the lower troposphere, which inhibits convectional mixing.

The continuous exchange of mass and energy between the Earth's surface and the atmosphere (and within the atmosphere) represents the functioning of circulatory systems. In these, the movement of mass is initiated and sustained by inputs of energy. While these inputs are ultimately derived from the solar energy cascade, it is possible to be a little more specific as to their origins within the Earth–atmosphere system.

An uptake of heat energy from the Earth's surface in the form of sensible heat or latent heat (see Chapter 4) increases the potential energy of the atmosphere which, in a circulation system, is converted into the kinetic energy of motion. The motion of large volumes of the lower atmosphere is a result of differences in potential energy arising from simple contrasts in sensible heat content, expressed in terms of horizontal temperature gradients, or latent heat content, or a combination of the two.

In examining the general circulation of the atmosphere (see Chapter 4), we made broad generalizations concerning latitudinal variation in potential energy. If, however, we are to consider circulation systems which operate over smaller temporal and spatial scales, we must first look a little more closely at the characteristics of the Earth's surface which affect the exchanges of mass and energy between it and the atmosphere. The most important of these are its physical properties, its roughness and topography.

8.2 The physical properties of a surface

Shortwave radiation incident upon a surface may be absorbed, reflected or transmitted. The proportion reflected depends upon the shortwave reflectivity (the **albedo**) of the surface and upon the wavelength and angle of incidence of solar radiation.

The albedo values in Table 8.1 are for wavelengths in the solar spectrum and are generalized for a range of solar elevations. Snow and peat are at opposite ends of the range of albedo values for natural surfaces. Freshly fallen snow may reflect 95% of incident shortwave radiation, while for the moorland peat this is only 10%. Weller and Holmgren (1974) have provided a useful illustration of this contrast over tundra surfaces in Alaska. Because of its dark colour, the tundra surface has a relatively low albedo (between 15 and 20%), but after snowfall this rises rapidly to over 80% (Fig. 8.1).

For other surfaces, less drastic but no less significant fluctuations in albedo may occur, some of which are related to moisture content. Many soils tend to have lower albedo values when moist, and over plant surfaces there may be changes in albedo relating to the annual growth cycle.

In the case of water, both **reflection** and **refrac-**

Table 8.1 Albedo of various surfaces (wavelengths less than 4.0 μm) from Monteith, 1973; Sellers, 1965; Lockwood, 1974; Oke, 1978)

Surface	albedo
Snow	
fresh	0.80–0.95
old compacted/dirty	0.42–0.70
Ice	
glacier	0.20–0.40
Water	
calm, clear seawater	
solar elevation 60°	0.03
30°	0.06
10°	0.29
Soils	
dry, wind-blown sand	0.35–0.45
wet, wind-blown sand	0.20–0.30
silty loam (dry)	0.15–0.60
silty loam (wet)	0.07–0.28
peat	0.05–0.15
Plants	
short grass (0.02 m)	0.26
long grass (1.0 m)	0.16
heather	0.10
deciduous forest (in leaf)	0.20
deciduous forest (bare)	0.15
pine forest	0.14
field crops	0.15–0.30
sugar beet (spring)	0.17
sugar beet (early summer)	0.14
sugar beet (midsummer)	0.26
Man-made	
asphalt	0.05–0.20
concrete	0.10–0.35
brick	0.20–0.40

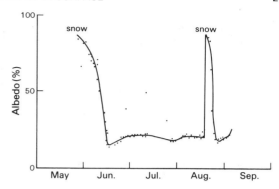

Fig. 8.1 The albedo of the tundra surface (Weller and Holmgren, 1974).

tion take place at its surface, with radiation penetrating into the water body. The amount of shortwave radiation reflected depends largely upon the angle at which it strikes the surface. When solar elevation is low, reflection is of the order of 29%, while for high solar elevation, reflection is relatively small at 3% (Table 8.1). In the latter case, much of the radiation is transmitted into the water and is progressively absorbed by it. In clear waters, by a depth of only 5 m, 70% of radiation has been absorbed, mainly in the infrared wavelengths. In turbid waters 70% has been absorbed by a depth of less than 1 m.

For all surfaces the absorbed solar radiation is redistributed by radiation, conduction and convection and the relative importance of these processes varies according to the physical character of the surface. In the absorption and emission of long-wave radiation, for example, we have already seen that natural surfaces act as grey rather than black bodies (see Chapter 3).

The rate at which heat is transferred downwards into the subsurface is directly related to the nature and efficiency of the distribution mechanisms. In solids, heat is redistributed by conduction, and the rate at which this takes place depends upon **thermal conductivity** (Box 8.1). Most naturally occurring materials, such as soil and rock, have low conductivities and, in relation to commonly used metals (Table 8.2), can be regarded as insulators rather than conductors. The penetration of heat is, therefore, to relatively shallow depths. For example, on a clear summer's day, a typical diurnal range of temperature at a depth of 0.01 m in a sandy soil may be of the order of 33°C, while at a depth of 0.03 m it is only 17°C and at 0.3 m it is less than 1°C. The greatest rates of heating and cooling are experienced in the upper layers of the soil.

The rate at which temperature changes take place in a material depends upon its **thermal capacity** (Box 8.1). For example, it requires less than 1.2×10^3 J of heat energy to raise the temperature of 1 m³ of still air by 1°C, while to achieve a similar increase in soil temperature requires 2.5×10^6 J. Conversely, when air is cooled by 1°C it releases less heat than soil cooled by the same amount. Thermal capacity affects the amount of heat energy transferred into and out of the subsurface store.

A vegetation canopy presents to the atmosphere not a simple solid/fluid interface but an ill-defined transition zone between free air and air contained in pockets among the foliage of the plants. Both heat and moisture are transferred across this zone by the movement of air. The effect a vegetation cover has on soil surface temperatures beneath it and air temperatures within it is determined largely by the type of plant and the number of individual plants growing within a certain area. Should the plant cover density be relatively sparse, the soil surface is not shielded from direct radiant energy exchange with the free atmosphere. If, however, dense foliage does shield the surface, intracanopy radiant energy exchanges are largely responsible for canopy and subcanopy temperature variations.

In dense woodland as little as 10% or less of solar radiation may reach the underlying soil surface, mainly as diffuse radiation and in selected

Table 8.2 Thermal properties of selected materials

	Thermal conductivity $W\ m^{-1}\ {}^{\circ}C^{-1}$	Thermal capacity $J\ m^{-3}\ {}^{\circ}C^{-1} \times 10^{6}$
still clear water	0.57	4.18
pure ice	2.24	1.93
still air	0.025	0.0012
fresh snow	0.08	0.21
moist sand	2.20	2.96
dry sand	0.30	1.08
moist peat	0.50	4.02
dry peat	0.06	0.58
iron	87.9	3.47
granite	4.61	2.18

Box 8.1
HEAT PARAMETERS

Thermal capacity is the amount of heat required to raise the temperature of unit volume of a substance by one degree. It is given by

$$s = \rho \text{ (density)} \times c \text{ (specific heat).}$$

Units are joules per cubic metre per degree Celsius ($J\ m^{-3}\ {}^{\circ}C^{-1}$).

Thermal conductivity determines the rate at which heat flows through a substance, and is defined as the rate of flow of heat through a unit area of plate of unit thickness when the temperature difference between the faces is unity.

Under steady-state conditions, ∂Q units of heat flow through a plate of thickness ∂x, whose faces differ in temperature by $\partial \theta$, in time ∂t. The rate of flow of heat ($\partial Q / \partial t$) is related to the temperature gradient ($\partial \theta / \partial x$) and thermal conductivity (k) by

$$\frac{\partial Q}{\partial t} = -k \frac{\partial \theta}{\partial x}.$$

Units of k are watts per metre per degree Celsius ($W\ m^{-1}\ {}^{\circ}C^{-1}$).

Thermal diffusivity: a major problem in soils is that steady-state conditions are rarely achieved. An alternative parameter, thermal diffusivity (α), is used which is given by

$$\alpha = k/s.$$

Units are metres squared per second ($m^2\ s^{-1}$). Thermal diffusivity, for a homogeneous medium, defines the rate at which temperature changes ($\partial \theta / \partial t$) take place:

$$\frac{\partial \theta}{\partial t} = \frac{\partial^2 \theta}{\partial x^2} \alpha.$$

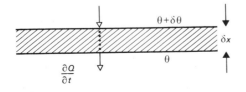

wavebands transmitted through leaves. Most incoming solar radiation is absorbed in the canopy (Fig. 8.2). The absorption of this radiation by the leaves and twigs of the woodland canopy raises their own surface temperature. The longwave radiation emitted from these surfaces is absorbed in the air above the trees and by the air pockets within the foliage. The whole of the canopy space then acts as an emitter of longwave radiation, some of which is transmitted to the trunk space and the soil. Heating of this lower zone within the vegetation is, therefore, not by direct absorption of solar radiation.

Fig. 8.3 illustrates the effect of these radiant energy exchanges upon net radiation at various levels within a woodland. Above the canopy there is a large diurnal variation with $Q*$ reaching a relatively large positive value around midday and falling below zero for much of the hours of darkness. At the woodland floor, however, $Q*$ varies little over the 24-hour period and has a large longwave component (Table 8.3). Of the net radiation available at the top of the canopy, most is transferred as either latent heat or sensible heat into the air above. Only a small amount is exchanged with the canopy heat store (Fig. 8.3).

A marked contrast in thermal properties exists between soil and water surfaces and between the land and the oceans (see Table 8.2). The difference in thermal conductivity is small, but in thermal capacity it is relatively large, that of the oceans being approximately twice that of the land. Cooling by 1°C releases 4.18×10^6 J m^{-3} of heat energy from water, compared to 2.5×10^6 J m^{-3} from an average soil.

Because it is a fluid, water also redistributes

heat by convection. As heating is from above, most of this heat redistribution is by forced convection by mechanical mixing in the upper layers of the water body (see Chapter 6, Section 6.3). At temperatures below 4°C, where water reaches its maximum density, cooler yet less dense water rises to the surface in free convectional mixing. By these mechanisms, heat is transferred away from the

(a)

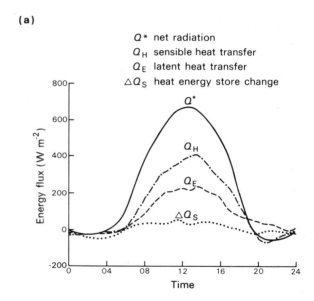

(b)

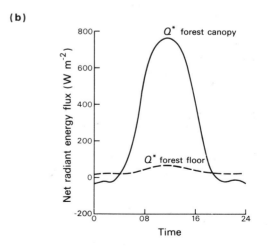

Fig. 8.3 Net radiation and heat-energy balances in a forest canopy. (a) Diurnal energy balance of a Scots and Corsican Pine forest at Thetford, England (after Oke, 1978). (b) Diurnal variation in net radiation in a spruce forest (after Lee, 1978).

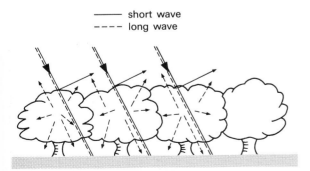

Fig. 8.2 Model of radiant energy exchanges in a deciduous woodland.

Table 8.3 Radiant energy balance above and within a forest

| | Above | | Forest floor | |
	Daily total (MJ m^{-2})	Relative to net radiation flux	Daily total (MJ m^{-2})	Relative to net radiation flux
K↓	18.2	3.19	1.0	3.33
K↑	1.8	0.32	0.1	0.33
K*	16.4	2.87	0.9	3.00
L↓	24.6	4.32	29.5	98.33
L↑	35.3	6.19	30.1	100.33
L*	−10.7	−1.87	−0.6	−2.00
Q*	5.7	1.0	0.3	1.0

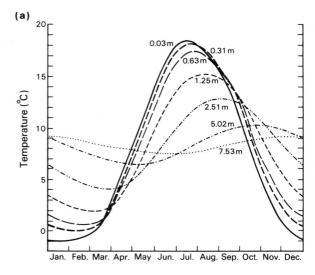

(a)

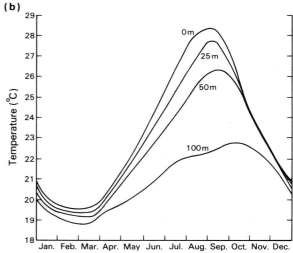

(b)

Fig. 8.4 (a) Annual variation in Earth temperature at Königsberg (after Schmidt and Leyst, from Geiger, 1965). (b) Annual variation in sea temperature off the south coast of Japan (after Sverdrup, 1945, from Harvey, 1976).

surface and redistributed through the water body. The significance of this is illustrated by the annual temperature regimes beneath ocean and ground surfaces (Fig. 8.4). While annual variation in terrestrial temperature is negligible at a depth of only 10 m, in oceans there is little difference between temperatures near to the surface and those at a depth of 25 m.

8.3 Surface roughness

The relationship between mean wind speed ($\bar{u}$) and height (z) above a surface approximates to the logarithmic form

$$\bar{u} = \text{constant} \times \log\ (z/z_o). \qquad (8.1)$$

In this equation z_o is the roughness of the surface, referred to as its roughness length or roughness parameter. The values of z_o for various surfaces are given in Table 8.4. A particularly significant contrast in roughness is that between a water surface (z_o is approximately 0.5×10^{-5} m) and the land surface (z_o is approximately 0.1×10^{-2} m), the latter exerting considerably greater frictional drag on air flowing across it (Fig. 8.5). In the case of the sea surface, the friction layer – defined by the height at which gradient wind velocities are attained – is 270 m deep over the water surface. Over the much rougher rural land surface, this increases to 400 m, or 520 m over urban areas.

Because of the penetration of airflow into a vegetation canopy, wind velocities do not tend towards a zero value at the canopy surface. The form of the velocity profile depends on the morphology of the canopy surface and, in particular,

its permeability to airflow. For example, in the case of a stand of conifers (Fig. 8.6) the foliage of the canopy presents resistance to airflow, while in the more open trunk space it is possible for airflow to penetrate at this level.

Air flowing across a vegetated surface causes some disruption of the canopy. Sway of trees, for example, while absorbing momentum from the air, also allows turbulent downdraughts to penetrate through the canopy. Grass surfaces, on the other

Table 8.4 Roughness length of natural surfaces

Surface	Roughness length, $z_0(m)$
still water	0.1×10^{-5}
ice, mudflats	0.1×10^{-4}
fresh snow	0.1×10^{-2}
sand	0.3×10^{-4}
soils	$0.1 \times 10^{-3} - 0.1 \times 10^{-2}$
short grass (less than 0.01 m)	0.1×10^{-2}
tall grass (up to 0.1 m)	0.2×10^{-1}
forest	4.0

hand, often become aerodynamically less resistant at moderate wind velocities as the sward bends before the wind.

8.4 Topography

The slope angle, aspect and elevation of surfaces affect the amount of direct solar radiation which is incident upon them. The effect of slope may be related to Lambert's Cosine Law (see Box 3.4) by which, for direct solar radiation, the angle between the solar beam and the sloping surface determines

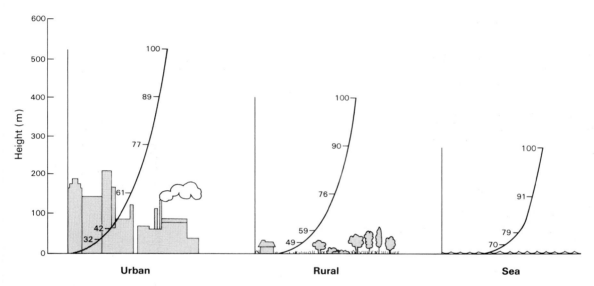

Fig. 8.5 Typical wind velocity profiles over three contrasting surfaces. Numbers refer to mean horizontal wind speed expressed as a percentage of the gradient wind speed (after Leniham and Fletcher, 1978).

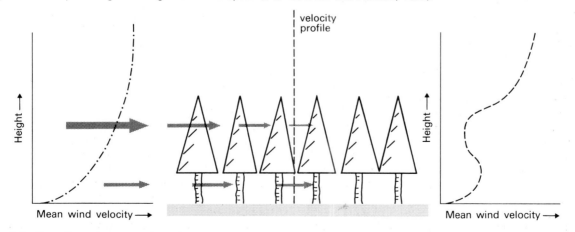

Fig. 8.6 The effect of a stand of conifers on the profile of mean wind velocity.

its intensity. Thus, in Fig. 8.7 the intensity of radiation incident upon slope A is greater than that on slope B. The orientation of such a slope, referred to as its aspect, also affects the intensity of solar radiation upon it. For example, in the northern hemisphere, a slope of southerly aspect will receive solar radiation when solar elevation is greatest. Conversely, a slope of northerly aspect will receive a low intensity, if indeed it receives any at all.

The simple example in Fig. 8.7 illustrates how slope and aspect combine to determine the intensity of direct solar radiation falling upon sloping surfaces. Over subdued relief, aspect and slope combine to produce greater solar heating of slopes X than of slopes Y. If slope angles are increased, X still receives direct solar radiation and is referred to as the **adret** slope. Y, however, receives no direct solar radiation and is referred to as the **ubac** slope. An illustration of this shading effect may be seen in many of the north-facing mountain corries of Scotland and Wales. Only small amounts of direct solar radiation are received on the steep back wall of corries and this results in delayed snow melt, which, in cooler climatic epochs, has been critical in the development of these features. This topographic variation in the solar heating of surfaces produces complex patterns of soil and air temperatures and rates of evaporation. Jackson (1967) has illustrated how monthly potential evapotranspiration varies according to aspect, slope, surface albedo and season (Fig. 8.8). A reduction in available heat energy on cooler south-facing slopes in the southern hemisphere inhibits both evaporation and transpiration. This is particularly noticeable at times of relatively lower solar elevation in June and September.

The relationship between surface elevation and direct solar radiation is complex and depends largely upon atmospheric transparency. Under cloudless skies the effect of increases in surface elevation is to increase direct solar radiation received, because of the shortened path through the atmosphere. High alpine areas may receive as much as 90% of potential solar radiation, whereas areas at sea level in similar latitudes may receive only 54%. However, under atmospheric conditions conducive to orographic cloud formation, the relationship changes. The increasing cloud amounts characteristic of the maritime uplands of Britain produce a decrease in solar radiation received as elevation increases. Harding (1979) has estimated that in Britain the average rate of decrease in total solar radiation received at the surface decreases at the rate of 2.5 MJ m^{-2} day^{-1} for every 1000 m increase in altitude.

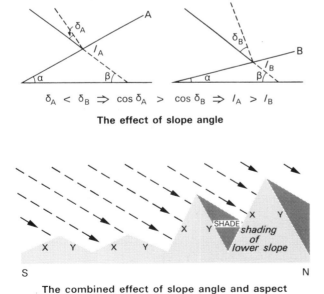

$$\delta_A < \delta_B \Rightarrow \cos \delta_A > \cos \delta_B \Rightarrow I_A > I_B$$

The effect of slope angle

The combined effect of slope angle and aspect

Fig. 8.7 The effect of topography on the intensity of direct solar radiation arriving at a surface.

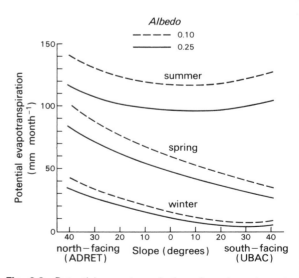

Fig. 8.8 Potential evapotranspiration of north- and south-facing slopes of two different albedo values; New Zealand (from Jackson, 1967).

8.5 Surface–air interaction

The operation of the factors discussed above creates energy imbalances over much smaller distances and time intervals than those already discussed in Chapter 4. Along coasts, for example, there are two fundamentally different materials lying next to each other, which gives rise to localized differences in surface heat energy balance. However, there is a complex relationship between the distribution of heat energy in the lower atmosphere and these contrasts in surface energy balance which depend upon the nature of both horizontal and vertical air movements.

Over relatively short distances of the order of 1 km, air blowing from over a smooth, warm and dry surface to a moist vegetated surface undergoes adjustment to these new conditions (Fig. 8.9). This adjustment is not instantaneous but occurs gradually with increasing distance from the point at which the surface character changed; it is referred to as the **fetch** of airflow.

Along coasts, air blowing inland from the sea possesses distinctly maritime properties and has a mean wind profile adjusted to the low roughness parameter of the water surface. As it moves inland there is progressive readjustment to the new surface. This gradual change is evident in the increasingly oceanic character of climate away from coasts. For example, on a relatively large scale, under a middle-latitude wind system there is a clear west-to-east change in air temperature regimes (Fig. 8.10). Equability of temperature variation, typical of sea surfaces, is evident on the coast of Ireland, while extreme temperature variation, typical of land surfaces, is characteristic of central and eastern Europe.

Over the surface of the Earth there are extensive masses of air which have acquired temperature and moisture characteristics closely related to the surface beneath them. These air masses are associated with extensive source regions. There are two principal groups of air masses associated with high latitudes (polar and arctic air) and with low latitudes (tropical air). These may be further subdivided into maritime or continental, according to whether their source areas are over oceans or land masses. In the middle latitudes, movement of these air masses from their source areas generates the characteristic variability of weather. In addition to the fundamental contrasts in source area characteristics, there are the modifications to which the air masses are subjected in moving over the Earth's surface. Both direction of movement and seasonal timing are important in this respect. This can be illustrated best by considering examples of the principal air masses which affect the British Isles (Fig. 8.11).

Polar maritime air, which is the most frequently occurring air mass, has its source area over the north Atlantic ocean, off the coast of Greenland. In travelling southeastwards towards Britain it

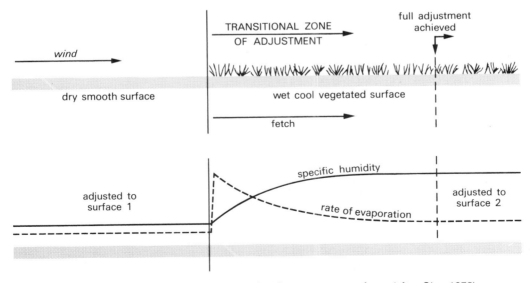

Fig. 8.9 Adjustment of air after crossing a boundary between two surfaces (after Oke, 1978).

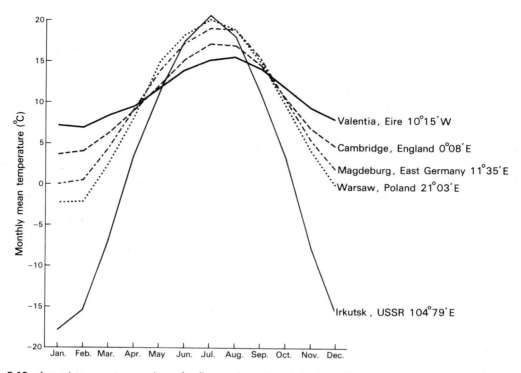

Fig. 8.10 Annual temperature regimes for five stations along latitude 52°N with mean monthly temperatures corrected to sea level (Meteorological Office, 1972, and Lydolph, 1977).

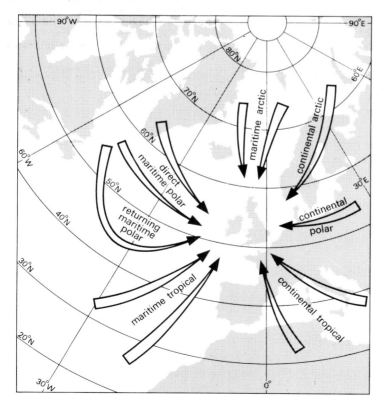

Fig. 8.11 The principal air masses affecting the British Isles (from Barry and Chorley, 1971).

passes over relatively warmer ocean surfaces and is gradually heated from below. This causes it to become unstable, so that when it reaches Britain it is frequently associated with deep cumulus cloud and a showery type of precipitation. If, however, the air initially travels southwards from its source and subsequently approaches Britain from the southwest, it has been slightly recooled from below in returning northwards and is, therefore, more stable. This returning polar air is usually associated with higher air temperatures than is the direct polar maritime, and it provides more continuous precipitation.

Tropical continental air is the least frequent air mass to affect Britain and has its source region over North Africa, whence it travels northwestwards. It is initially hot, dry and unstable, but on travelling northwards it may be cooled from below and become more stable. In summer, this cooling may be negligible as the air passes over the warm land surfaces of western Europe, and during this season it may be sufficiently unstable to give thunderstorms in Britain. In winter, however, a much greater degree of cooling from below makes the lower air very stable and, upon reaching Britain, tropical continental air is associated with stratus cloud and generally dull weather.

The movement of air masses towards Britain and in all other latitudes not only affects the heat and moisture balances of the atmosphere and surface but is also associated with the development of atmospheric circulations within the primary circulation system. For example, where polar air meets tropical air at the polar front (see Chapter 9), the steep temperature gradient between them gives rise to the development of cyclonic storms. Where tropical maritime air moves over warm seas around the western flanks of the subtropical high pressure areas, storms develop in the form of the tropical cyclone.

Spatial variation in the characteristics of the Earth's surface generates energy imbalances over a range of space and time scales, from the larger-scale development and modification of air masses to the smaller-scale differences between land and sea. The air motion resulting from these imbalances may be considered in terms of secondary and tertiary circulation systems respectively.

Further reading

Arya, S.P. (1988) *Introduction to Micrometeorology.* Academic Press, New York.

Bannister, P.J. (1976) *Introduction to Physiological Plant Ecology.* Blackwell, Oxford.

Geiger, R. (1965) *The Climate Near the Ground.* Harvard University Press, Cambridge Mass.

Grace, J. (1983) *Plant–Atmosphere Relationships. (Outline Series in Ecology)* Chapman & Hall, London.

Lockwood, J.G. (1979) *The Causes of Climate.* Edward Arnold, London.

Monteith, J.L. (1973) *Principles of Environmental Physics.* Edward Arnold, London.

Oke, T.R. (1978) *Boundary Layer Climates.* Methuen, London.

Rosenberg, N.J. (1974) *Microclimate: the Biological Environment.* John Wiley, New York.

Secondary and tertiary circulation systems

9.1 Secondary circulation systems

9.1.1 The pressure cell

Over the surface of the Earth there are cells of high and low atmospheric pressure which are both ephemeral and mobile, being apparent on the daily weather map but frequently undetectable in the distribution of seasonal means of atmospheric pressure. These are identified as areas of closed isobars in the form of low-pressure cells (cyclones) and high-pressure cells, or anticyclones. Both these are seen on a typical weather map for the eastern Atlantic ocean and western Europe (Fig. 9.1).

Air converging upon a low-pressure cell rises and diverges aloft, while air diverging from a high-pressure cell is subsident with convergence aloft. In the former, air is cooled as it rises and generates cloud, in contrast to the warming and drying of subsiding air in an anticyclone. This contrast, together with the steeper pressure gradients into the low-pressure cells, means that in terms of weather events, cyclones are the more active features.

9.1.2 The extratropical cyclone

In the middle latitudes there are mobile cyclones (depressions) which move eastwards under the influence of a westerly component of atmospheric circulation and which possess an organized pattern of air movement. These features were identified by Abercromby in 1883 (Fig. 9.2), who noted that they contained areas of greater cloudiness and precipitation, and well defined zones where both air temperature and wind direction changed rapidly over short distances. Work by Norwegian meteorologists in the early 20th century identified these zones, or discontinuities, as the leading edge (warm front) and trailing edge (cold front) of a

sector of warm air. These fronts are the principal cloud and precipitation zones of the depression (Fig. 9.2). A satellite photograph of a well developed cyclone in the north Atlantic (Fig. 9.3) clearly shows its associated spiralling bands of cloud and the denser cloud along the frontal zones.

At both fronts air is being lifted aloft over cooler air and is thereby subjected to adiabatic cooling. At the warm frontal zone warmer air overrides the cooler air in advance of the depression. The weather conditions associated with its passage across the surface are mainly of increasing amounts of cloud, precipitation and wind speed in advance of its arrival (Box 9.1). In its wake, cloudiness may persist but precipitation becomes intermittent and wind speeds decrease.

The cold frontal zone is located where advancing polar air undercuts the air of the warm sector. The enforced uplift of air is rapid, resulting in deep cumulus cloud and intense precipitation. After its passage, precipitation may continue in the form of heavy showers. The depression is therefore asymmetric, the distribution of heat and moisture and the nature of airflow depending upon position relative to its centre, as suggested by Abercromby. Speed of eastward movement of the whole system is variable, each depression having its own individual character, but an average speed would be in the region of 11.5 m s^{-1}.

The Norwegian model of cyclone development was based upon airmass interaction along the polar front, which most commonly separates polar maritime and tropical maritime airmasses (Fig. 9.4a). Along this zone of contact a small wave or perturbation may develop, as warm air begins to move polewards into the cooler air (Fig. 9.4b). A distinct warm sector develops between warm and cold fronts. As the former moves forwards at a

(a)

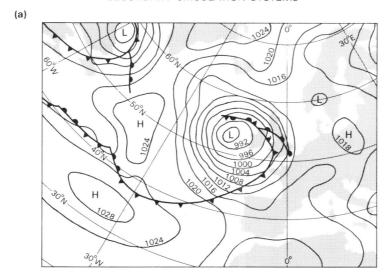

(b)

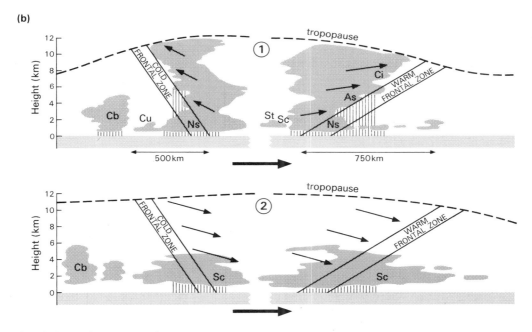

Fig. 9.1 A typical weather map: (a) for the North Atlantic Ocean and western Europe showing an extratropical cyclone approaching the British Isles (weather log 12 GMT, 24 August 1977) (reproduced by permission of the Meteorological Office); (b) cross-sections through the extratropical cyclone where warm air is (1) rising and (2) falling relative to the frontal zone. (See Table 6.2 for cloud abbreviations.)

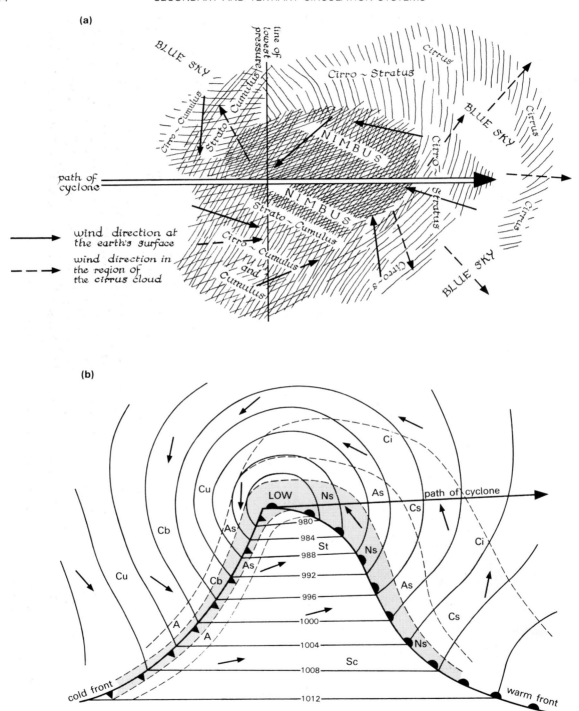

Fig. 9.2 Representations of cloud distribution in an extratropical cyclone by (a) Abercromby (1883) and (b) Norwegian meteorologists (1914 onwards).

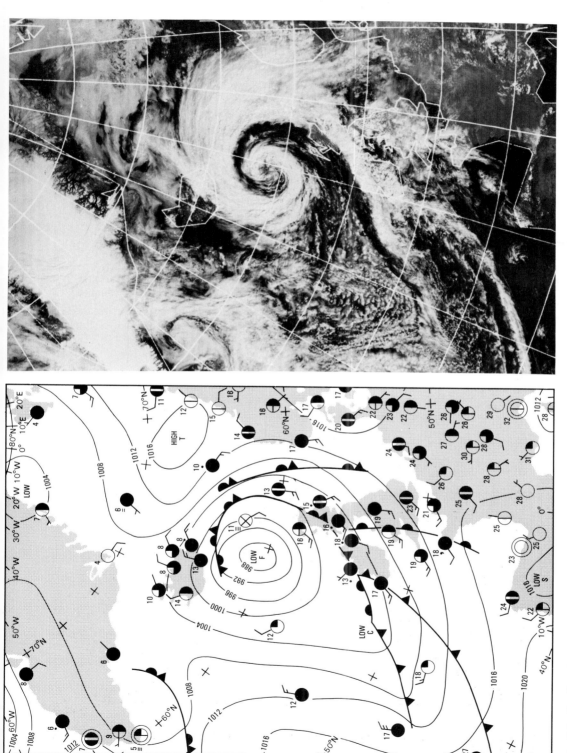

Fig. 9.3 Satellite photograph and daily weather map for the north Atlantic ocean and western Europe, 6 August 1979 (reproduced by permission, University of Dundee and the Meteorological Office).

Box 9.1

CHANGES ASSOCIATED WITH THE PASSAGE OF WARM AND COLD FRONTS (from Meteorological Office, Aviation Meteorology)

Element	In advance	At the passage	In the rear
warm front			
pressure	steady fall	fall arrested	little change or slow fall
wind	backing and increasing	veer and decrease	steady direction
temperature	steady, or slow rise	rise	little change
dew-point	rise in precipitation	rise	steady
relative humidity	rise in precipitation	may rise further if not already saturated	little change; may be saturated
cloud	Ci, Cs, As, Ns in succession; St fra, Cu fra below As and Ns	low Ns and St fra	St or Sc may persist; perhaps some Ci
weather	continuous rain (or snow)	precipitation almost or completely stops	dry, or intermittent slight precipitation
visibility	good, except in rain (or snow)	poor, often mist or fog	usually moderate or poor; mist or fog may persist
cold front			
pressure	fall	sudden rise	rise continues more slowly
wind	backing and increasing, becoming squally	sudden veer, perhaps squall	backing a little after squall, then fairly steady or veering further in later squalls
temperature	steady, but fall in prefrontal rain	sudden fall	little change, variable in showers
dew-point	little change	sudden fall	little change
relative humidity	may rise in prefrontal precipitation	remains high in precipitation	rapid fall as rain (or snow) ceases; variable in showers
cloud	St or Sc, Ac, As then Cb	Cb with St fra, Cu fra or very low Ns	lifting rapidly, followed for a short period by As, Ac and later further Cu or Cb
weather	usually some rain perhaps thunder	heavy rain (or snow) perhaps thunder and hail	heavy rain (or snow) usually for short period, but sometimes more persistent; then fine but followed by further showers
visibility	moderate or poor, perhaps fog	temporary deterioration followed by rapid improvement	very good

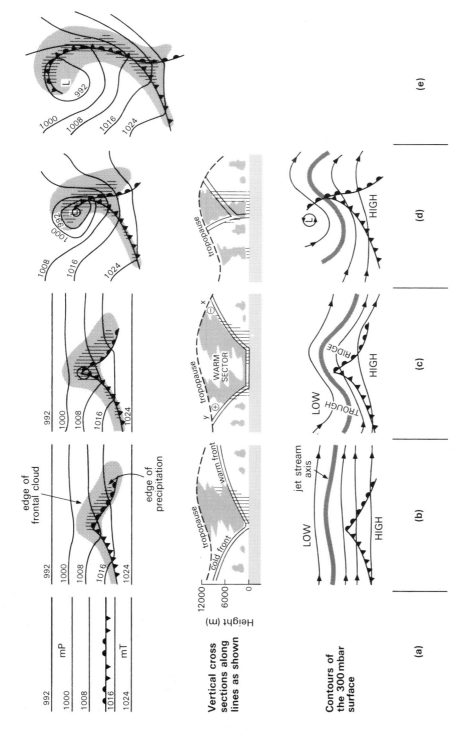

Fig. 9.4 Stages in the development of an extratropical cyclone (isobars at 8-mb intervals; from Pedgley, 1962).

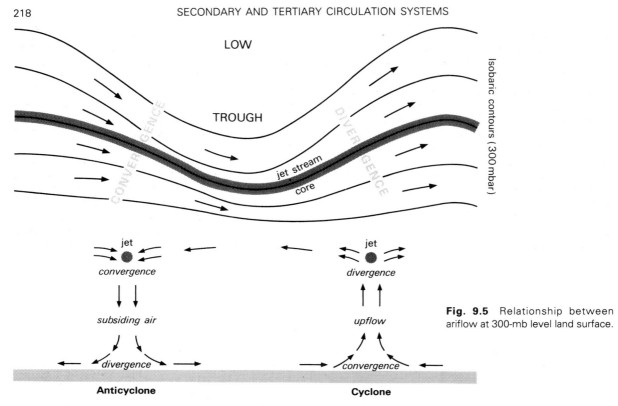

Fig. 9.5 Relationship between airflow at 300-mb level land surface.

slower speed than the latter, the polar air is, in effect, executing a pincer movement on the air of the warm sector which is therefore being laterally constricted. Thus, there is a net convergence of air in the lower troposphere which is associated with decreasing atmospheric pressure at the apex of the wave. The depression develops as this low pressure deepens (Fig. 9.4c).

The meeting at the surface of cold and warm fronts marks the maximum development of the depression when pressure at its centre has reached a minimum (Fig. 9.4d). After this point, decay, or **occlusion**, begins to occur as the warm sector is lifted clear of the surface. With continued lifting its remaining heat energy is dissipated and the depression weakens, although precipitation may continue to fall from the occluded front (Fig. 9.4e). Eventually the perturbation of the polar front has been removed and only a weak cyclonic cell of cool air in the lower troposphere remains. In so doing, the atmosphere has released the considerable amount of heat energy contained within the sector of warm air.

This polar front model closely matches the observed growth and decay of a typical middle-latitude depression. However, inspection of weather charts for the north Atlantic does not always reveal a clear developmental pattern of depressions and associated fronts. Waves on the polar front may develop and decay rapidly in the western Atlantic, or may be still intensifying when they reach Britain. The development of waves along the trailing cold front of an Atlantic depression over the Celtic sea gives rise to secondary depressions. These often intensify and produce deteriorating weather conditions as they move eastwards.

The increasing number of meteorological observations made in the middle and upper troposphere, particularly during the past 20 years, have revealed a close relationship between the upper westerly flow and the formation of extratropical cyclones. The contours of the 300-mb surface above the cyclone in Fig. 9.5 indicate a wave pattern in the upper westerly airflow. Indeed, if we were to look at the circumpolar vortex in the northern hemisphere (Fig. 9.5), we would find that as it moves northwards through the waves, cyclone development (**cyclogenesis**) commonly occurs. On the western limb of a trough in one such wave (Fig.

9.5) airflow is convergent, indicated by a closer proximity of the isobaric contours. In order to compensate for the net inflow of air at this level (some 9000 m above sea level), underlying subsiding flow towards the Earth's surface develops. The resulting inflow of air into the lower troposphere gives rise to divergent anticyclonic flow across the surface. On the eastern limb of the trough, airflow is divergent, indicated by a spreading out of the isobaric contours. Divergence of airflow at this level results in an upward replacement flow from the lower troposphere. This upflow of air from the Earth's surface is replaced by a convergent cyclonic inflow across the surface.

Thus, to the west of wave troughs convergence in the upper troposphere results in anticyclone development (**anticyclogenesis**) in the lower troposphere, while to the east divergence results in cyclogenesis (Fig. 9.4). The latter affects very markedly the development of extratropical cyclones along the polar front over both the Atlantic and Pacific oceans, particularly when there are well developed troughs in the upper westerlies.

As a circulation system, the middle-latitude depression results from a combination of strong horizontal temperature gradients across the polar front and the external impetus provided by the upper westerly airflow. Its operation relies on the continued conversion of potential energy into kinetic energy (Fig. 9.6). Potential energy is

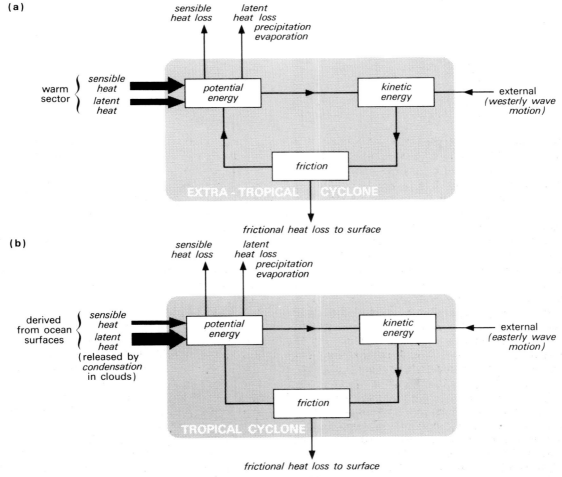

Fig. 9.6 Cyclones as energy systems. (a) Extratropical cyclone. (b) Tropical cyclone.

derived from the temperature contrast between the warm and cold air masses, and its conversion into kinetic energy. Heat energy is also gained from the surface in the form of latent heat, this being released by condensation of water vapour in the warm sector. Although some of this energy will be returned into the atmosphere by evaporation, a net loss of water by precipitation will ensure a net gain of energy through condensation.

The kinetic energy of air motion is dissipated through the effects of friction, both within the air itself as turbulent friction or at the Earth's surface. Kinetic energy is transferred directly to the surface (for example, in ocean waves) or is converted into heat energy. In the case of the latter it is returned to the potential energy store or is lost from the circulation system.

The vigour of the depression measured in terms of the velocity of both horizontal and vertical air motion ultimately depends on the balance between the rate at which energy is converted from potential into kinetic, and the rate of frictional energy losses. In the depression's occluded state, the supply of potential energy has been exhausted and kinetic energy is rapidly depleted. The depression thus weakens and pressure begins to rise at its centre.

9.1.3 The tropical cyclone

In tropical latitudes, cyclones occasionally take the form of the rotating storm known as the hurricane or typhoon. These are variously described as severe, violent, destructive or devastating. Wind speeds frequently exceed 50 m s^{-1}, causing considerable damage, and associated with the hurricane there is a storm surge at sea which frequently causes heavy flooding along low coasts. Such a surge was responsible for as many as 250 000 deaths in Bangladesh when a typhoon struck the Bay of Bengal overnight on the 29/30th April 1991. Waves of 7.0 m combined with mean wind speeds in excess of 65 m s^{-1} to devastate island and coastal communities.

The hurricane appears on the weather map as a series of concentric and closely spaced isobars (Fig. 9.7) at the centre of which sea-level atmospheric pressure usually falls to 950 mb and, in exceptional cases, to 900 mb. The whole system has a diameter of 200–1000 km and is at its most active over a period of between 3 and 14 days.

Within the eye, air is subsiding, maintaining relatively high temperatures and low relative humidity through adiabatic warming (Fig. 9.7). The

skies here are relatively clear, there being only a thin veil of high-altitude cirrus cloud. Around this there is a towering wall of cumulonimbus cloud rising as high as 15 000 m, from which falls most of the cyclone's precipitation. Precipitation intensities greater than 50 mm h^{-1} have been recorded in this zone of the cyclone. The zonation of clouds within the tropical cyclone is apparent in satellite photographs of hurricanes. Both the central eye and the surrounding ring of cumulus are particularly obvious.

The tropical hurricane may be compared to the middle-latitude depression, in that it is an area of low pressure which gives rise to airflow spiralling into its centre and within which precipitation occurs. However, in most other respects it is strikingly different (Table 9.1). In addition to having only half the overall diameter and developing twice the wind speed of the middle-latitude depression, the hurricane possesses no

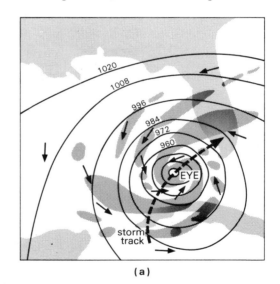

(a)

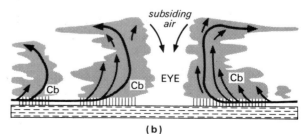

(b)

Fig. 9.7   (a) Tropical cyclone approaching Florida. (b) Vertical section.

Table 9.1 Middle-latitude depression and tropical hurricane: a comparison of scale (partly from Eliassen and Pedersen, 1977)

System	Horizontal scale (diameter) (km)	Vertical scale (km)	Wind speed (mean) (m s^{-1})	Average lifetime (days)
tropical hurricane	200–1000	15	30	7
middle-latitude depression	1000–3000	10	10–20	7

discontinuities. The weather pattern is more symmetrically organized around its central eye.

The main hurricane areas are at latitudes greater than 5° over the tropical oceans, along the western flanks of the subtropical high-pressure areas, as indicated in Table 9.2. Most occur in the northern hemisphere, only the southwestern Indian ocean and the offshore waters of northern Australia being greatly affected in the southern hemisphere. Hurricanes usually occur during the late summer and autumn when sea temperatures are high, and they usually develop from less violent tropical disturbances which form in tropical maritime air at low latitudes. Many of these small tropical disturbances from which hurricanes develop are formed within westward-moving waves in the tropical easterly airflow. To the east of a

wave trough there is divergence in the middle or upper tropospheric flow, which encourages convergence at sea level and the development of deep cumulus cloud. Many such tropical lows form, some of which develop into tropical storms of moderate intensity with wind speeds in excess of 17 m s^{-1}, but only a few develop further to form hurricanes.

From a vertical section (Fig 9.7) we can see that within the hurricane there is a strong convergence and uplift of air near the sea surface, the latter developing to 15 000 m. A vital feature which differentiates the hurricane from the tropical storm, is well developed divergence in the upper troposphere, beneath which there is a warm core of subsiding air. Air over the ocean surface takes up both heat and moisture, which are then carried aloft in the strong vertical currents. Condensation and precipitation at upper levels cause the release of massive amounts of latent heat, which maintains a supply of warm air for the core. An adequate uptake of heat and moisture is, therefore, essential and it is supplied by warm ocean surfaces. A critical threshold ocean temperature of 27°C is commonly accepted, and the world distribution of hurricanes is closely related to this isotherm.

As we have already seen in Chapter 4, the Coriolis force has a value of zero at the equator but it increases rapidly as the sine of latitude. The absence of a strong Coriolis force inhibits the development of a cyclonic vortex and it is not

Table 9.2 Areas of the world most subject to hurricane acitivity (after Trewartha, 1961; Barry and Chorley, 1976)

Area		Hurricane season	Mean annual frequency
I	south and southwest Atlantic ocean		
	Cape Verde Islands	August and September	
	east and north of West Indies	June to October	
	north Caribbean	May to November	4.6 (1901–63)
	southwestern Caribbean	June and October	
	Gulf of Mexico	June to October	
II	North Pacific ocean (west coast of Mexico)	June to November	2.2 (1910–40)
III	southwestern North Pacific ocean (including China Sea, Japan)	May to December	19.4 (1924–53)
IV	north Indian ocean		
	Bay of Bengal	April to December	4.7 (1890–1950)
	Arabian Sea	April to June	
		September to December	0.7 (1881–1937)
V	south Indian ocean (eastward from Madagascar)	November to April	4.7 (1848–1935)
VI	south Pacific ocean (eastward from Australia)	December to April	4.0 (1940–56)

until latitude 5° that it becomes sufficiently large for hurricane development. In migrating northwards in the northern hemisphere, air moving within the hurricane experiences an increasing Coriolis force and may, therefore, be expected to intensify. However, as the ocean surface gradually becomes cooler, the supply of sensible heat and moisture decreases, and consequently there is less energy being released in the upper troposphere and the hurricane cannot be sustained. Should this hurricane move over land surfaces, the supply of sensible heat may be maintained, but the supply of moisture, and hence latent heat, is restricted. The total energy input is therefore inadequate to sustain the hurricane. There is also a considerable increase in energy loss due to greater frictional drag over the aerodynamically rougher land surface. Thus, the hurricane's energy supply is reduced in leaving tropical latitudes, with the result that there is a gradual reduction in energy exchange within it, bringing a decrease in wind speed and precipitation intensity. Hurricanes which have followed the warm Gulf Stream of the western Atlantic eventually develop middle-latitude characteristics as they meet cold polar maritime air. In many cases these become deep depressions, which bring stormy weather to Britain.

As a circulation system, the tropical hurricane is maintained by large inputs of heat energy derived from the condensation of water vapour. The uptake of sensible and latent heat from the oceans provides potential energy, which is converted into kinetic energy. As is the case with the middle-latitude depression, the operation of the system relies upon this conversion (see Fig. 9.5). The brake on the system is similarly provided by friction within the atmosphere and at the surface.

While the system is located over warm ocean areas there is a surplus of potential energy uptake over frictional loss and this maintains an intensifying circulation. Reduction of potential energy and increase in frictional energy losses slow down the circulation. Once energy uptake is exceeded by energy loss, the hurricane begins to decay unless a fresh input of potential energy becomes available.

Although wind speeds and destructive capacity are greater in a hurricane than in a middle-latitude depression, its total kinetic energy is considerably less. The kinetic energy of a developing hurricane is of the order of 10^{16} J, increasing to 10^{18} J at maturity. In comparison, a relatively intense middle-latitude depression develops 10^{19} J. Related to these energy exchanges in the Earth–atmosphere system, the energy of a devastating earthquake, at point 8 or more on the Richter scale, is 10^{18} J and an average earthquake, at point 6, 2×10^{13} J. The problem in developing such a comparison, however, lies in the concentration of energy. Clearly, the hurricane at any one time is a more locally concentrated phenomenon than the more diffuse middle-latitude depression.

9.1.4   The anticyclone

No discussion of circulation systems would be complete without some consideration of the anticyclone, which is the less energetic counterpart of the cyclone. It is associated with relatively shallow pressure gradients and consequently much lighter winds. Air in the high-pressure centre is subsiding and hence it experiences increasing atmospheric pressure as it moves towards the Earth's surface, which results in adiabatic warming (see Box 6.3). As the air temperature increases, it is less likely that the air will become saturated unless extra water vapour is added, and any suspended water droplets are likely to be evaporated. We may, therefore, expect anticyclones to bring cloudless weather with little likelihood of cloud or precipitation. Indeed, we immediately associate high atmospheric pressure during the middle-latitude summer with clear skies and hot weather. The winter anticyclone with its attendant clear skies brings rapid nocturnal radiative cooling of ground surfaces and a high risk of frost.

There are, however, occasions when anticyclones bring long periods of dull cloudy weather, often referred to as anticyclonic gloom. The occurrence of shallow stratus and stratocumulus clouds is closely related to the marked inversion of temperature in the lower atmosphere which is a distinctive feature of the anticyclone. This layer, in which air temperatures increase rather than decrease with height above the Earth's surface, may be in immediate contact with the surface or may be elevated at some distance from it. The cold winter anticyclone of Asia is an example of the former, the warm Azores anticyclone of the latter.

Subsiding air begins to slow down as it approaches the Earth's surface, and air in the lower 1000 m or so may not experience subsidence. The subsiding air diverges above a layer of relatively cooler air in contact with the surface, in which an input of extra moisture may encourage condensation to take place. Radiation fogs are a result of cooling of the ground surface (usually under anticyclonic conditions) and subsequently of the air above it, which reduces the temperature of moist air to its dew-point. The resulting fog thickens as the temperature inversion in the lower atmosphere deepens, and its depth of development is limited to the vertical extent of the inversion layer. In other situations, turbulent mixing of the lower atmosphere may take place beneath an inversion layer not in contact with the surface. Where turbulent uplift of air is localized, as in the case of thermal convection, adiabatic cooling may be sufficient to initiate condensation. This results in isolated cumuliform cloud capping the thermal upcurrents, which are limited in their vertical development by the inversion layer. A moist lower atmosphere which has experienced more extensive vertical mixing may produce more widespread stratus or stratocumulus cloud, which may occasionally produce low-intensity precipitation.

In the Earth–atmosphere system there are a number of readily identifiable types of anticyclone which, while sharing the common factor of subsiding air, have fundamentally different origins. Some are mobile features, such as those of the middle latitudes, while others move little, either from day to day or month to month, such as the subtropical anticyclones over the Atlantic and Pacific oceans. Some have a deep vertical development extending through the troposphere; others extend vertically for only 1 or 2 km.

During winter intense radiational cooling over the interior of high-latitude continents remote from oceanic sources of heat energy results in the cooling of a shallow layer of the lower atmosphere. The cold air has a relatively high density, which results in an increase in the pressure which the atmosphere exerts on the surface beneath it. The resulting anticyclone is a shallow feature extending vertically for less than 2 km.

Although winter anticyclones are a feature of synoptic charts for both North America and Asia, the one which develops over the latter is the more persistent. The 0°C isotherm of mean January temperature encloses most of Soviet Asia, with mean temperatures in many easterly locations falling below −40°C. The Asian anticyclone has its centre in the vicinity of Mongolia and it extends northeastwards in a ridge across eastern Siberia

(Fig. 9.8). There is a well developed temperature inversion layer in the lower atmosphere which extends upwards to the 850-mb level (Fig. 9.9). Surface winds indicate a divergence of airflow from the high-pressure centre, and to the west westerly winds are clearly deflected northwards into the Arctic ocean. However, at the 800-mb level there is little trace of any deflection from a circumpolar westerly airflow. While high pressure dominates the synoptic charts, temperatures remain low and precipitation is infrequent, although low stratus cloud does bring long periods of poor visibility. Occasional incursions of cyclonic activity bring most of the winter precipitation.

In direct contrast to these cold anticyclones, those which occur in subtropical latitudes exhibit deep development through the troposphere and are detectable on pressure surfaces from sea level up to the tropopause. They also vary in position by as little as 5° of latitude during the year. As areas of subsidence and surface divergence, they are associated with some of the Earth's most arid areas, such as the Sahara Desert. A characteristic of the lower atmosphere in these warm anticyclones is a marked inversion of temperature. Along the eastern flank of the anticyclone this inversion is relatively near to the surface, but away from this area it is an elevated feature and it lies over a turbulently mixed air layer. The Cape Verde Islands lie to the south of the main axis of subtropical high pressure in the north Atlantic, and the temperature inversion here lies between the 925-mb and 825-mb levels (Fig. 9.10).

9.2 Tertiary circulation systems

9.2.1 Scale

Within the secondary circulation systems are yet smaller-scale circulations of air which operate over relatively short distances, usually less than 160 km, and over short periods of time. They develop limited kinetic energy, of the order of 10^{13} J.

Such 'local' or tertiary circulations may be thermally direct, that is, convectional cells which arise from differential heating of the Earth's surface. In these case, energy transfers may be likened to those of the single-cell model of the primary atmospheric circulation systems (Chapter 4). They may also operate as modifications of established larger-scale airflows, which are referred to here as 'regional' winds. The topography of the ground surface greatly modifies the characteristics of the air flowing across it. In this case, kinetic energy is derived directly from the regional wind. In the succeeding sections we shall consider examples of both types of tertiary circulation.

9.2.2 The sea breeze

Characteristic of many coastal sites is a diurnal change in wind speed and direction which results from local circulations of air. Analysis of hourly mean wind speeds over a period from 1 June 1975 to 19 August 1975 at Portsmouth, England, has shown that there was a marked diurnal change from a minimum 2.9 m s⁻¹ in the early morning to a maximum of 5.4 m s⁻¹ in the early afternoon. Wind roses for these two times show a clear difference in dominant wind direction (Fig. 9.11). In the morning there is a high frequency of northerly winds which are offshore along the Hampshire

coast. In the afternoon these have been replaced by dominant southerly, or onshore, winds. There is, therefore, evidence for the presence of weak offshore winds in the early morning and strong onshore winds in the mid-afternoon.

Peters (1938) identified this sea breeze in south Hampshire, detecting its onset by noting rapid changes in wind direction associated with decreases in air temperature and increases in relative humidity. On this evidence he identified a sea breeze season extending from March to September. A more extensive study by Simpson (1964) revealed a considerable inland penetration of sea breezes, of the order of 40 km. Using a rise in dew-point temperature as the indicator of sea breeze arrival at a point, he determined that the breeze is established during the late morning in the immediate coastal zone, but then develops

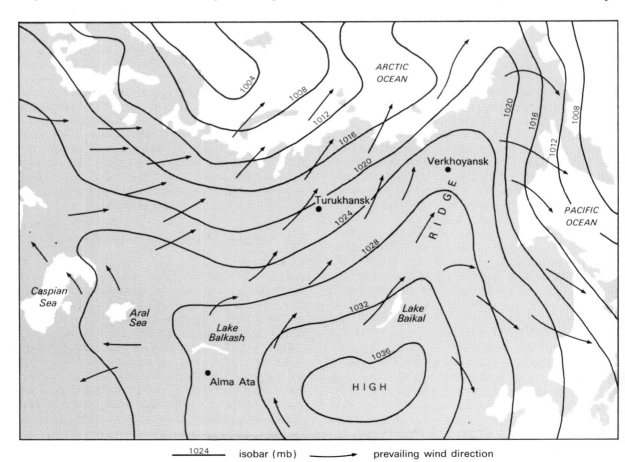

Fig. 9.8 Mean sea-level atmospheric pressure and prevailing wind direction over Asia in January (after Borisov, 1965).

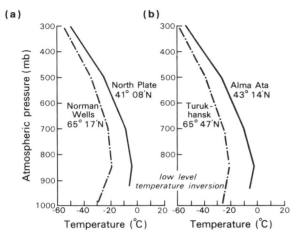

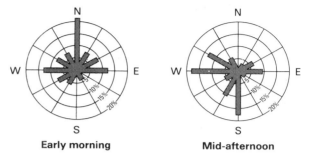

Fig. 9.11 Wind roses for Portsmouth (Lion Terrace) in the summer of 1975.

Fig. 9.9 Vertical profiles of temperature in the winter anticyclones of (a) North America and (b) Asia (data after Crowe, 1971).

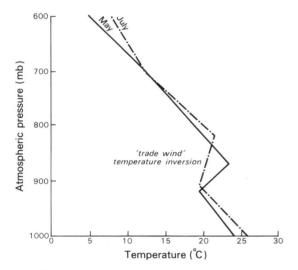

Fig. 9.10 Vertical temperature profiles in the subtropical anticyclones of the north Atlantic ocean; Sal, Cape Verde Island (16°44′N) (data after Crowe, 1971).

landwards. The incursion of sea air over the land surface may do so in the form of a weak cold front, at which cumulus cloud develops.

Broadly similar sea breeze characteristics have been found for many coastal areas in both temperate and tropical latitudes. A typical sea breeze develops during the late morning at the coast, then, gradually extends its influence inland, when geostrophic wind velocities are relatively low.

Wind speed increases rapidly during the late morning and into the early afternoon when it reaches its maximum, which in Britain would be of the order of 4 m s⁻¹. The onshore flow of air extends upwards from the surface to a height of 750 m. In addition to this, there is a weaker offshore return flow aloft which reaches its maximum velocity some 2000 m above the surface.

A simplified view of the sea breeze as a localized circulation of air is as a thermally direct cell. During the daylight hours land surface temperatures increase more rapidly than those of the sea surface, due to a difference in heat distribution mechanisms. This differential heating creates a single thermally direct cell. Air rises above the warmer land surface and is replaced at lower levels by an inflow of air from the sea. To compensate for this, there is a return circulation aloft and a sinking of air over the sea (Fig. 9.12). On this basis, the greatest differences in surface temperature should give rise to the strongest circulation.

The pressure gradient force in the simple model operates directly from sea to land, and thus wind direction may be mistakenly assumed to be normal to the coast. All moving bodies are, however, subjected to a Coriolis force which deflects their motion away from the direction of the pressure gradient force. A sea breeze blowing northwards across a coast in the northern hemisphere will, for example, develop an eastward component of motion. Defant (1951) has illustrated the combined effect of Coriolis force and friction upon sea breezes along the Massachusetts coast. Using both theoretical calculation and actual observation, he has shown that the sea breeze is deflected as much as 45° away from normal to the coastline.

Sea breezes are most frequently developed

under warm coastal conditions during the middle-latitude summer and in tropical latitudes, which tends to support a simple interpretation based upon temperature differences. However, observation has shown that the breeze may not develop despite there being relatively large temperature differences between land and sea, or may develop when there are much smaller differences. Watts (1955) determined that along the Sussex coast, at Thorney Island, the timing of the onset of the sea breeze was not related to the magnitude of temperature differences alone but also to the stability of the atmosphere. Under unstable atmospheric conditions the onset was rapid, whereas under stable conditions only gradual changes in wind direction were observed. The inference is that sea breezes are most readily developed where there are atmospheric conditions conducive to the rapid vertical expansion of air over the land surface.

As the air is heated over the land surface under 'suitable' atmospheric conditions, there is rapid vertical expansion of the overlying air and, at a height of about 2 km, pressure begins to rise. This results in air flowing seawards at this level (Fig. 9.12). This divergence aloft causes pressure at the land surface to fall as air rises to take its place, while the arrival of air above the sea surface causes a convergence aloft. Air subsides beneath this convergence, causing an increase in atmospheric pressure at sea level. The development of a surface pressure gradient gives rise to the landward air flow of the sea breeze.

In terms of energy flow within a circulation system, the sea breeze derives potential energy from the greater release of heat into the lower atmosphere over land surfaces than over the sea surfaces. This is converted into kinetic energy which is partly dissipated by friction. As sea breezes may develop when there is a regional wind blowing, this may add a direct input of kinetic energy.

A land breeze which blows in the opposite direction to the sea breeze is often similarly repres-

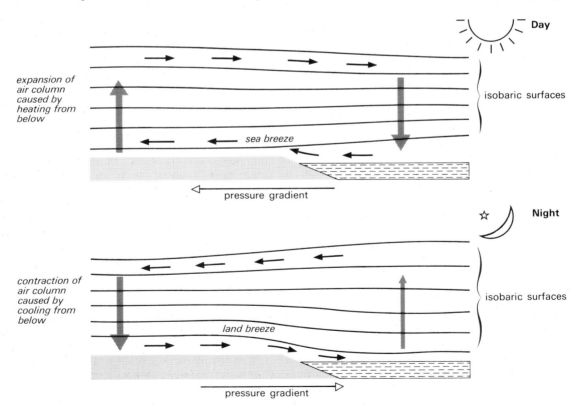

Fig. 9.12 Simple model of the circulation of air over coasts.

ented as a thermally direct cell. Cooler land temperatures at night cause the air to subside and diverge at the surface, while over the sea surface there is convergence and uplift (Fig. 9.12). The land breeze begins to blow offshore before midnight and develops to reach a maximum velocity immediately after sunrise. The low velocity of the land breeze, usually less than 2 m s^{-1}, may be a result of the low magnitude of vertical uplift in this circulation system and the slow progress of subsidence over the land. In many cases, however, there is no obvious return circulation aloft, this being weak and readily obscured by the regional wind. Alternatively, the land breeze may be interpreted as a gravitational downslope movement of cold dense air, related as much to inland topography as to temperature differences between land and sea surfaces. This type of airflow will be considered in more detail in the following section.

9.2.3 Slope winds

Over sloping terrain, local winds are generated as an upslope (**anabatic**) flow or a downslope (**katabatic**) flow of air. If we consider a uniform slope receiving direct solar radiation (Fig. 9.13), its surface is heated, which in turn heats the overlying air. The result of this is the vertical expansion of the air and the creation of an upslope pressure gradient. The resulting upslope flow is compensated by a downward sinking of air over the foot of the slope and a flow away from the slope summit, completing a circulation. Over a slope receiving solar radiation directly after sunrise, the unstable anabatic wind begins to blow within an hour of initial slope heating, and continues to intensify until maximum slope temperatures are reached in the early afternoon.

Nocturnal cooling of the same slope causes a vertical contraction of the air above it (Fig. 9.13) thereby developing a downslope pressure gradient along which there is a katabatic flow of air. This is compensated by a net inflow of air at the top of the slope and a rising of air above the slope foot, thereby completing a circulation. The stable katabatic wind begins to blow about one hour after sunset and it gradually intensifies as cooling proceeds, to reach a maximum velocity around dawn.

Both upslope and downslope flows are best developed under clear skies, commonly under anticyclonic conditions. Over extensive slopes anabatic winds may reach maximum speeds of about 4 m s^{-1} compared to the 2 m s^{-1} of the katabatic wind. The former develop to depths of 200 m or more, while the latter are more closely confined to a layer up to 150 m thick. Within these moving layers of air, maximum velocity is usually attained a little way from the surface, due largely to the retarding effects of friction (Fig. 9.13).

The development of these slope winds is closely related to the radiation balance of the slope surface which, as can be inferred from Chapter 8, Section 8.4, is extremely complex. Aspect, slope angle and degree of shading by nearby topographic features will affect the character of airflow over slopes.

As a circulation system, the anabatic wind is thermally direct (Fig. 9.14). Potential energy is derived from the heating of the slopes and the raising of the temperature of air in contact with them. As a result there is a temperature difference between the air layer in contact with the slope and air in the free atmosphere at the same elevation above the slope foot. In the case of katabatic winds, the potential energy is derived from a reversal of this temperature difference as the slopes are cooled. This potential energy is converted into kinetic energy which is partly dissipated by friction, particularly that between slope surface and the air.

If we consider two slopes forming a symmetrical valley, each will develop a thermal slope wind (Fig. 9.15). Assuming that the slopes are equally heated, the daytime movement of air, particularly under clear skies and relatively calm weather conditions, will be upslope (Fig. 9.15a–d). This creates divergence in the centre of the valley, where air subsides to take its place (Fig. 9.15e,f). If we also assume that the slopes cool at similar rates, then the resulting downslope flow of air should create convergence in the valley floor (Fig. 9.15g, h). However, as the air is usually cold and dense, uplift above the valley floor is restricted. The cold air may remain in the valley, accumulating there as a cold pool in which temperatures may fall below 0°C.

If the radiation balances of the two slopes are dissimilar, the two-cell circulation becomes asymmetric. Valleys that trend east–west have a warm and a cold slope. In this case, the cold slope develops a weak circulation during the day and valley circulation is dominated by that above the warm slope.

The slope of the valley floor itself, from mountain to plain, superimposes another circu-

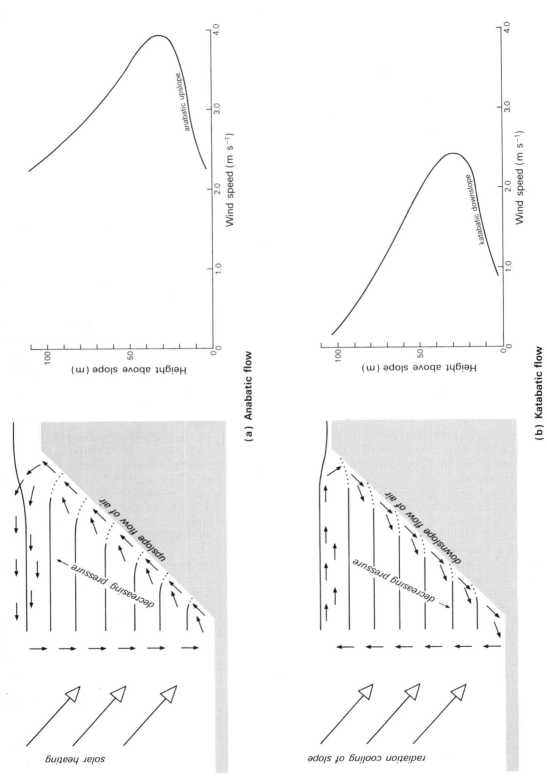

Fig. 9.13 Development of anabatic and katabatic airflow over a uniform slope.

(a) **Anabatic flow**

(b) **Katabatic flow**

solar heating

upslope flow of air

decreasing pressure

anabatic upslope

Wind speed (m s^{-1})

Height above slope (m)

radiation cooling of slope

downslope flow of air

decreasing pressure

katabatic downslope

Wind speed (m s^{-1})

Height above slope (m)

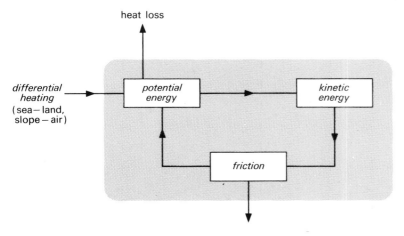

Fig. 9.14 Tertiary circulation of a thermally direct type as a simple energy system.

lation upon that within valleys. The temperature contrasts between the air over mountain slopes and over the distant plains generate a larger-scale circulation of air which uses the valleys as natural channels of flow. These are referred to as mountain-valley winds. The flow of air within these valleys will be a complex combination of mountain to plain (and vice versa) and valley-side flows, which have been represented in simplified form by Defant (1951) (Fig. 9.15).

9.2.4 The Föhn wind

Air which is forced to flow over hills or a mountain range does not necessarily generate localized circulations, but may be modified to produce distinctive changes in the characteristics of flow. One such modification is the Föhn wind, which blows as a warm, dry and blustery wind down lee slopes. The name Föhn was originally applied in the Austrian Tyrol. There is also a variety of local names, including the Chinook of the Rockies and the Zonda of the Andes.

The onset of the wind is usually accompanied by extremely rapid increases in temperature. Changes of 21°C in 4 minutes have been recorded in Canada at the onset of the Chinook. On a smaller scale, Lockwood (1962) has investigated the occurrence of Föhn winds in Britain which produce unseasonal high temperatures. Weather records for Kinloss in Scotland contain evidence of a southerly Föhn wind, most probably emanating from the Cairngorms, its arrival being marked by a distinct decrease in relative humidity and an increase in air temperature (Fig. 9.16).

One explanation of the Föhn effect is outlined in Fig. 9.17a. Conditionally unstable air at temperature T_1 is forced to rise over the mountain ridge and it cools at the dry adiabatic lapse rate (AB). After cooling to condensation level, further uplift causes cooling to continue at the slower saturated adiabatic lapse rate (BC). This produces orographic cloud, which presents a cumulus wall to the lee side – known as the Föhn wall. If precipitation falls from these clouds, the air experiences a net loss of water and a net gain of sensible heat through the release of latent heat of condensation. As water has been lost there is little consumed in re-evaporation once the air begins to travel down the lee side of the ridge and to warm adiabatically. In its descent, the air is therefore warmed at the dry adiabatic lapse rate (CD) to temperature T_2, which is greater than T_1. The unmodified airflow contains both potential (internal heat and gravitation) and kinetic energy. As the air is forced to rise over the mountain barrier, kinetic energy is converted into potential. As the air flows down the lee slope, this is converted back into kinetic. The outflow of energy in the modified flow will be equal to that of the unmodified flow, with the exception of energy lost through friction.

As the temperature increase is directly attributed to a net gain from latent heat release, this model of the Föhn requires precipitation to fall from the orographic cloud. In this respect, theory does not match observation, as Föhn winds often occur without orographic loss of moisture. Therefore, more recent theories have tended towards the alternative view that the Föhn wind is generated by the forced descent of upper air rather than the ascent and descent of lower air (Fig. 9.17b). This relatively dry air may be forced to

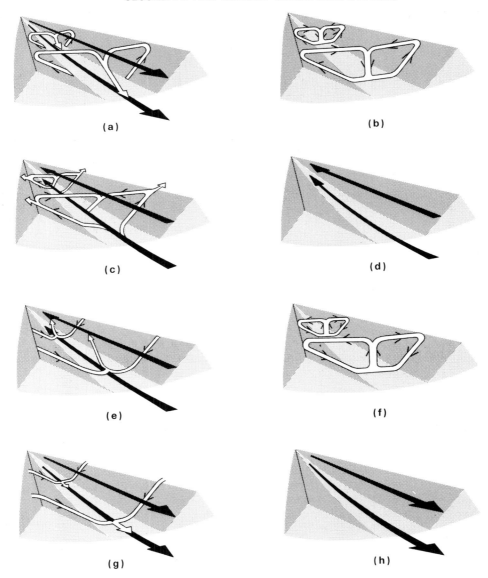

Fig. 9.15 Schematic illustration of the normal diurnal variation of air currents in a valley (after Defant, 1951). (a) Sunrise: onset of upslope winds (white arrows), continuation of mountain wind (black arrows), valley cold, plains warm. (b) Morning (c. 0900 h): strong slope winds, transition from mountain to valley wind, valley temperature same as the plains. (c) Noon and early afternoon: diminishing slope winds, fully developed valley wind, valley warmer than plains. (d) Late afternoon: slope winds have ceased, valley wind continues, valley continues warmer than plains. (e) Evening: onset of downslope winds, diminishing valley wind, valley only slightly warmer than plains. (f) Early night: well developed downslope winds, transition from valley to mountain wind, valley and plains at the same temperature. (g) Middle of night: downslope winds continue, mountain wind fully developed, valley colder than plains. (h) Late night to morning: downslope winds have ceased, mountain wind fills valley, valley colder than plains.

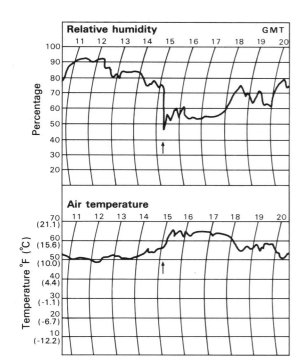

Fig. 9.16 Extracts from humidity and air temperature traces (Kinloss, Morayshire, Scotland, 12th March 1957) showing the onset of a Föhn wind (from Lockwood, 1962).

descend and warm adiabatically with the formation of a lee wave downwind of the mountain barrier. This alternative thus involves no release of latent heat. Descending air converts potential energy to kinetic and, as warming is adiabatic, no internal heat is either lost or gained by the air.

9.3 Linkages between atmospheric systems

The movement of air within the Earth–atmosphere system, represented here as secondary and tertiary circulation systems, arises most frequently from spatial and temporal variation in heat energy and moisture supply from the underlying water or ground surface. The range of scales of such

Fig. 9.17 Alternative models of the Föhn wind.

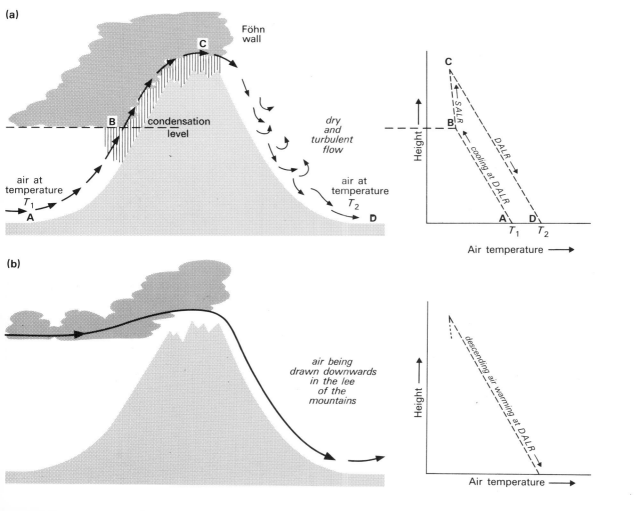

variation results in both air movements of short duration operating over small distances, such as the sea breeze, and movements of longer lifespan such as the extratropical cyclone which influences weather patterns over extensive ocean and land surfaces. However, irrespective of scale, there is a basic similarity in all these systems in that they all involve the conversion of potential into kinetic energy and its dissipation by both internal and external friction. Essential to their continued operation is the maintenance of inputs of adequate potential energy. However, we have also seen that, while inputs of potential energy are frequently a result of interaction between the atmosphere and the underlying surface, the operation of circulation systems is also conditioned by the larger scales of atmospheric motion within which they develop. For example, the development of the extratropical cyclone is very closely related to the waves of planetary scale in the upper westerlies of middle latitudes. Similarly, the smaller-scale sea breeze circulation requires certain conditions of atmospheric stability and relatively low gradient wind velocity in order to develop. We have, therefore, a series of mutually dependent – rather than self-contained – atmospheric systems.

Further reading

Atkinson, B.W. (ed) (1981a) *Dynamical Meteorology: an Introductory Selection*. Methuen, London.

Atkinson, B.W. (1981b) *Meso-scale Atmospheric Circulations*. Academic Press, London.

Barry, R.G. and R.J. Chorley (1976) *Atmosphere, Weather and Climate*, (3rd edn). Methuen, London.

Chandler, T.J. and S. Gregory (eds) (1976) *The Climate of the British Isles*. Longman, London.

MacIntosh, D.H. and A.S. Thom (1972) *Essentials of Meteorology*. Wykeham, London.

Riehl, H. (1979) *Climate and Weather in the Tropics*. Academic Press, New York.

DENUDATION SYSTEMS

The catchment basin system

Fig. 10.1 shows a scene typical of upland landscapes in a humid temperate region. In it we can discern a number of features, some of which we can conceptualize as systems and some we can regard as elements (or components) of these systems. The overcast sky represents the atmospheric system containing water vapour which we considered in Chapters 8 and 9. However, we can see that the form of the land surface is composed of a number of elements. In the foreground a river of significant dimensions sweeps by and an accumulation of fine sediment is visible on the inside of the meander bend, in addition to some large boulders scattered in the channel. The river is flanked by slopes of very low gradient, where again large boulders can be seen on the ground contained within the meander bend. Rising above the valley floor are slopes of moderate gradient, whose complex three-dimensional form can be described in terms of both vertical profile and horizontal curvature. Small outcrops of solid bedrock are visible towards the top of the cliff at the left-hand side of the scene, but otherwise the slopes are mantled by weathered bedrock and soil, not visible here because of the sward of vegetation. The relationships between these elements – the arrangement of the slopes composing the land surface, the river, the solid bedrock and the sediments – represent the denudation system. The land surface is mantled by an almost continuous cover of vegetation, for the most part grassland and heath, but with significant areas of woodland. This of course represents the ecosystem.

In this chapter we shall be concerned with the relationships at the interface between these major systems – the atmospheric system, the denudation system, and the ecosystem. At this interface vast quantities of water are held in, and transferred between, various stores (see Chapter 4). Here

the rocks of the Earth's crust and their mineral constituents are broken down physically and chemically and are transferred to lower elevations. All of these processes, involving both water and debris, take place in the denudation system. Indeed, it is the operation of the denudation system on relief created by the crustal system that determines the morphology of landscape and provides the functional environment that sustains life. However, at the scale at which we perceive landscape – the scale represented by Fig. 10.1 – it is more appropriate to model the operation of the denudation system as the **catchment basis system**. Here, therefore, we shall consider in a broad way the organization of the catchment basin system, the elements of which it is composed, and their functional relationships. In Chapters 11–16 individual subsystems of the model are isolated and considered in detail. However, the inputs of matter and energy to the catchment basin system show significant spatial variations and it is these that in turn are largely responsible for spatial variations in the state of the denudation system, and hence in the form of the world's landscapes. In Chapter 17, therefore, we shall examine such spatial variation.

10.1 The structure of the catchment basin system: functional organization

The basic functional unit through which the denudation system operates is the drainage or catchment basin. For the purposes of this chapter, the catchment basin can be regarded as a system in its own right. As such, it has a well defined boundary and its component elements show clear relationships, both structurally in terms of its morphology, and functionally by virtue of the flow of matter and energy through the system. The

Fig. 10.1 A typical fluvial landscape in an upland temperate region: West Dart River, Devon, England.

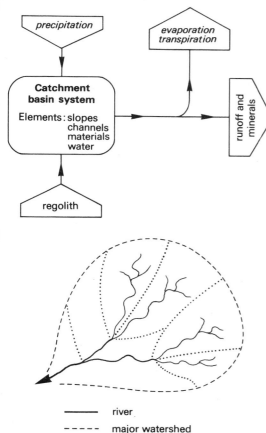

- ——— river
- — — — major watershed
- ············ minor internal watershed

Fig. 10.2 (a) Simplified schematic model of the catchment basin system. (b) Delimitation of catchment watersheds.

inputs and outputs across the system boundary can also be clearly distinguished (Fig. 10.2a).

The boundary of the catchment basin system can in part be regarded as the ground surface and the surface of water bodies within the catchment. Its areal limits are set by a major watershed, while minor watersheds define subcatchments within it (Fig. 10.2b). Its lower boundary within the lithosphere is more difficult to define. For the moment this can be thought of as a surface defining both the lower limit of the weathering system and of the active transfer of water within the system.

The major elements in the catchment system are channels, slopes, bedrock and regolith, and water. These are related in an organization of slopes and converging channels which facilitates the basic function of the system: to evacuate runoff and debris from the catchment basin. If the simple model of the catchment basin system in Fig. 10.2a is expanded and refined, these elements can be considered as composing several functional units or subsystems. Each is related to the others and all are an integral part of the function of the basin as a whole.

In Fig. 10.3a the slope system, channel system and weathering system have been separated, and the relationships between them indicated in a more sophisticated model. The slope system is functionally linked to the weathering system by the transfer of materials, both directly as weathering products from the regolith (Fig. 10.3b) and indirectly through the soil and the ecosystem (see Chapter 21) in the form of litter and decaying organic matter. Precipitation forms a link between the slope system and the atmospheric system, as also do the dissolved materials in that precipitation and the dry fallout from the atmosphere. In addition, some elements from the atmosphere are fixed in organic compounds by the organisms of the ecosystem, and these form inputs to the slope system when the organisms die and decay. Water output occurs as evaporation and transpiration back to the atmosphere, as surface runoff into channels and as infiltration entering the soil and regolith as either soil water or ground water. Surface runoff will remove material as solid sediment particles, and also in solution.

The outputs from the slope system form the major link with the channel system. Overland flow of water down hillslopes during storms feeds directly into the river channel, and the bulk of the percolating soil water and ground water finds its way into the channel, sustaining channel flow between rainstorms. The third water input to the channel (volumetrically the least important) is direct precipitation onto the river channel surface. Output of water from the channel is partly by direct evaporation from the water surface, although this is normally limited except under high temperatures and very broad channel sections, as for example where the river passes through a lake (Box 10.1). The bulk of the water output from the channel is via runoff which contain sediments and solutes. This channel runoff normally flows to the ocean, where it can be regarded as input to the ocean store, which contains 97% of the Earth's water.

The weathering, slope and channel systems, therefore, form a cascading system organized so that the outputs of one are the inputs of another, and together they form the catchment basin system. This model of the denudation system can be applied even to deserts and to landscapes experiencing active glaciation, as well as to fluvial landscapes. These can be regarded as special cases which differ only in the inputs to the system and in the pathways and rates of matter and energy flow through the system. The subsystems of this model are treated in more detail in Chapters 11–16.

10.2 System structure: spatial organization

So far, the structure of the denudation system has been treated in terms of its functional organization. Its structure, however, can equally be described in terms of its spatial organization. Although it often parallels functional organization, this is a particularly useful approach for it enables the structure of the system to be interpreted in terms of the surface morphology or form of the catchment basin.

If we turn again to the simple model of the catchment basin system (see Fig. 10.2), both slopes and channels can be regarded not as subsystems but as elements in the system. These elements can be described in terms of their attributes and, in particular, their spatial relationships. The quantitative treatment of these attributes is known as **morphometric analysis**. Measures have been devised which analyse the linear aspects of the catchment, its areal properties and its relief characteristics. Each of these will be examined in turn, but initially we shall consider the basic morphological element of catchment basins – slopes.

According to the simple classification scheme of Young (1972), based on surface gradient and slope plan form, any catchment basin consists of no more than five fundamental types of slope. Young classifies these into two groups: flats (or nearly so) and valley slopes (Fig. 10.4). Flats are found in two locations: as interfluve remnants they can represent undissected portions of an original surface into which the catchment basin is incised, and can contribute to catchment processes by groundwater flow; flats also occur on the valley floor, where they normally represent the flood plain of the river. In terms of the mass movement of materials by slope processes, flats are insignificant, since the effects of gravitational energy are nullified by such low gradients.

Valley slopes are more important in terms of erosional processes, for on them the movement of material by gravitational processes is facilitated. Slope gradient is important in this respect, since

(a)

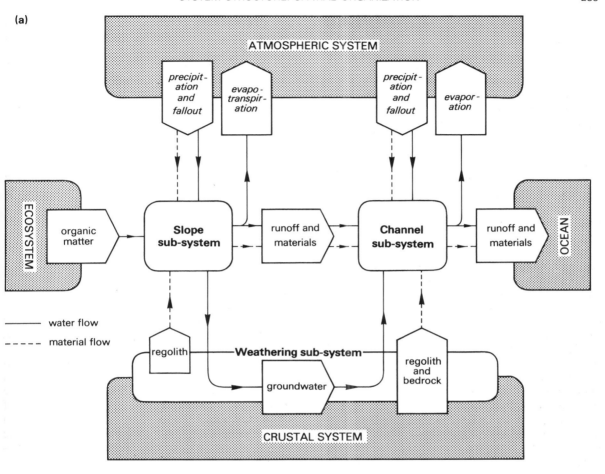

(b)

Fig. 10.3 (a) A schematic model of the functional organization of the catchment basin system. The intimate functional relationship between the slope and channel systems is clearly to be seen in the accompanying photography (b) of the Colorado River, Arizona, USA, where rock debris is fed directly into the river channel from the steep rock walls and talus slopes.

Box 10.1
Lake catchment systems

Many catchment basin systems contain natural or artificial lakes. Since the rivers flowing into them suffer a very significant decrease in velocity, lakes act as sediment traps. The sediment which accumulates represents the output of the catchment draining to the lake, and as such the lake acts as a sensor of environmental denudation.

Lake sediments accumulate, usually in a stratified manner, as mineral sediments representing the mechanical erosion of the upstream catchment are delivered by rivers as suspended load. The accumulating sediment also incorporates minerals in solution, indicative of the chemical quality of the inflowing water. Organic remains in the form of pollen and plant fossils will represent the vegetation throughout the catchment, while the lake ecosystem itself will be represented by fossil remains of diatoms and other lacustrine organisms. Additional to these are particulate airfall deposits and minerals washed out of the atmosphere during precipitation. The range of materials present, then, forms a wide-ranging inventory of ecological, erosional and chemical conditions within the catchment area upstream of the lake.

In recent years the value of the record stored by lake sediments has been increasingly recognized. Cores of lake sediments can be dated by a variety of stratigraphic (varve, pollen) and radiometric (^{14}C, ^{210}Pb, ^{137}Cs) means, as a consequence of which a chronology of events within the catchment can be built up. In this way, observations of contemporary conditions and monitoring of present processes can be linked to the history of the catchment measured in decades, recording the period of human industrial impact, to millennia recording climatic and denudational changes of postglacial time.

Some of the major events recorded by lake sediment archives are as follows:

- The mineral sediment flux into the lake, and the water quality indicated by individual elements present within it, can provide evidence of the nature and volume of denudation through time, and provide an indication of changing environmental conditions.
- Plant remains provide evidence of changing vegetation and therefore climatic conditions, human impact and consequent implications for the water balance and denudation processes.
- Acidification is indicated by diatom spectra.
- Eutrophication in diatoms, caused by the influx of phosphates due to human impact, in the form of sewage and detergents.
- Pollution, input of magnetic minerals and heavy metals indicative of industrialization.

See: Engstrom and Wright, (1984), Oldfield (1987), O'Sullivan et al., (1982).

---

gravitational force is proportional to the sine of slope angle (see Chapter 12). Also important is the form of the slope in plan. Three variations are possible: valley-side slopes, valley-head slopes and spur-end slopes. Valley-side slopes are straight in plan. They can be considered in profile as a simple linear form, since lateral movement across them can be neglected. Valley-head slopes are concave in plan and they converge towards the base. The opposite is the case on spur-end slopes, convex in plan, where divergence occurs downslope. The

plan of the slope in respect of divergence or convergence towards the base has important implications for the balance of materials and processes on the slope, and it leads to differing basal conditions.

This arrangement of morphological components holds true for all landscapes on Earth. Even plainlands are rarely entirely flat, but consist of gently shelving slopes separated by divides. According to Young, only 7% of the Great Plains of the USA is truly flat land, and of the Mato

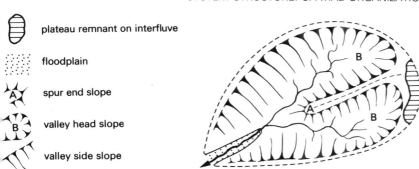

plateau remnant on interfluve

floodplain

spur end slope

valley head slope

valley side slope

Fig. 10.4 Morphological components of the catchment basin (after Young, 1972).

Grosso plateau of Brazil only 5% is truly flat. At the other end of the scale are mountain landscapes which consist entirely of steep valley-side slopes separated by sharp crested divides. In such landscapes, plateau remnants have been entirely removed and flood plains have not developed. The five basic morphological components, therefore, represent the possible variety of slope types, though they may not all be present within an individual catchment basin system.

The spatial organization of the system – that is, the form of the landscape – can be described and measured in various ways. Any landscape possesses linear, areal and relief properties, and we shall examine each of these in turn.

10.2.1 Linear properties
The linear properties of the system primarily concern the distribution in plan of channel and valley networks. The network consists of a series of channels converging towards a single outlet. The organization of the network can be analysed by allocating each channel segment a rank according to its position in the network (Box 10.2).

Data for the network of the River Wallington in Hampshire (England) are set out in Table 10.1, and the network is shown in Fig. 10.5. Analysis of the network shows that it is a fifth-order basin. It is shown to have an internal geometry with consistent relationships between stream order, stream number and stream length, and as such it is characteristic of stream networks in general (Fig. 10.6).

This kind of analysis can be applied also to valley networks, which are generally more extensive than the networks of perennial streams. This is particularly the case in regions of highly permeable rocks, seasonal climates, or where a former climatic regime created a more humid

environment with greater surface runoff. Accordingly, it is sometimes more valuable to analyse the valley network as indicated by contour lines. In this way, for example, the linear patterns of chalk and limestone terrains may be assessed.

10.2.2 Areal properties
Descriptive measures of the areal attributes and relationships of catchment basins are numerous, and only a representative selection of the more significant measures will be discussed here. Gardiner (1974) offers a comprehensive review.

One of the most fundamental catchment characteristics is that of drainage density. This is defined as the total length of drainage channel divided by drainage area, thus:

$$Dd = \frac{L}{A}$$

where Dd = drainage density, L = sum of total stream lengths, and A = catchment basin area; and is expressed as a ratio in kilometres per square kilometre. Using the data for the Wallington system presented above, drainage density is shown to be 1.22 km km^{-2}. Comparative values for other areas are discussed in Chapter 17.

The value derived for drainage density will depend closely on the origin of the data used. Often based on cartographic data, it is common to use the mapped stream network for morphometric analysis. Clearly, the correspondence between this network and functioning channels in the field will depend on both the scale and the cartographic conventions of the map employed. For instance, the Mississippi system on an atlas map at a scale of 1:12 500 000 may be shown as a fourth-order basin, whereas Leopold, Wolman and Miller (1964) estimate that its network of perennial

Box 10.2
DRAINAGE BASIN MORPHOMETRY

Various methods of network description have been put forward, but the one most widely used is that of A.N. Strahler (1952). In this method, headwater streams which receive no tributaries are designated as first-order. Where two first-order streams are confluent they form a second-order stream. The confluence of two second-order streams forms a third-order stream and so on, so that a stream of given order can receive streams of any lower order without its own order being raised. Stream order is raised only when two streams of the same order converge. Finally, the basin order is defined by the highest-order stream within it. Applying this form of analysis to the Wallington system it is shown to be a fifth-order basin. Within it are subcatchments of successively lower orders forming a nested hierarchy.

Analysis of the network can be carried out by tabulating the number of streams in each order. The number of streams decreases progressively through the higher orders in a geometric manner. The rate of change of stream number with stream order is described by the bifurcation ratio (R_b). This is calculated by dividing the number of streams in a given order by the number in the next (higher) order. The mean value of the ratios so calculated is the mean bifurcation ratio ($\overline{R}_b$) and it is characteristic of the network. The bifurcation ratio normally lies in the range 2–5. There is thus a simple geometric relationship between stream number and stream order, of the form

$$N_u = \overline{R}_b{}^{(s-u)}$$

where N_u = number of streams of order u, $\overline{R}_b$ = mean bifurcation ratio and s = basin order. For the River Wallington, this equation becomes:

$$N_u = 2.97^{(5-u)}.$$

These data are shown graphically in Fig. 10.6.

The values of mean stream length also show a progressive change with stream order. In the same way as the bifurcation ratio describes the rate of change between orders, so the ratio of mean stream lengths can be calculated and a characteristic value derived for the basin. The relationship between mean stream length and stream order shows an increase in the direction of higher orders. It is described by the following equation:

$$\overline{L}_u = \overline{L}_1 . R_L{}^{(u-1)}$$

where $\overline{L}_u$ = mean length of streams of order u, $\overline{L}_1$ = mean length of first-order stream and R_1 = ratio of mean stream lengths. For the River Wallington, this relationship becomes:

$$\overline{L}_u = 0.38 \times 1.97^{(u-1)} \text{ (see Fig. 10.6).}$$

Thus the simple geometric relationship between stream order and stream number is repeated with stream length. Similar geometric relationships exist between stream order and basin area (increasing with stream order), and stream order and stream gradient (decreasing with stream order).

The major advantage of the Strahler method is that it is easy to apply and it has also gained widespread acceptance. Within any given network, streams and basins of each successive order can be found up to the largest value present, and this facilitates morphometric comparisons of populations of basin systems. There are, however, some limitations to this method which prompted Shreve (1966) to put forward an alternative method of network analysis.

Table 10.1 River Wallington: network characteristics

Stream order	No. of streams	Bifurcation ratio	Total stream length (km)	Mean stream length (km)	Ratio of stream lengths
1	69	3.6	26.3	0.38	2.8
2	19		20.5	1.08	
		3.8			1.4
3	5	2.5	9.8	1.96	1.7
4	2	2.0	6.5	3.25	2.0
5	1		6.5	6.50	
		$\bar{R}_b = 2.97$	$L = 69.6$		$R_L = 1.97$

channels forms a 12th-order system. By omitting headwater tributaries, stream order is lowered throughout the system and drainage density is correspondingly reduced. Within the British Isles, Gardiner (1974) has shown that the 1:25 000 Second Series maps of the Ordnance Survey offers the most reliable cartographic source of morphometric data.

The most accurate method of determining drainage density is the field mapping of actually functioning channels. This underlines the fact that drainage density and basin order are not constant values for a given system, but they vary through time. The amount of water within a catchment basin, recorded in groundwater levels and extent of the drainage net, will vary according to variations in precipitation input. These variations will occur throughout individual storms and also from season to season. Drainage density and basin order are therefore dynamic variables and are related to basin discharge. Hanwell and Newson (1970) show the variations in drainage density in a small catchment (1.18 km^2) at different levels of flow between extreme flood and drought (Fig. 10.7). There is a twenty-fold variation in drainage density. As a result of this variation, the basin ranges from second-order at low flows to fourth-order at high flows.

Simple descriptive measures of basin shape are manifold, but at the same time are beset by problems of definition. Gardiner (1974) recommends the **elongation ratio**, defined as

$$E = A/L$$

where E = elongation ratio, A = basin area and L = basin length. This yields values ranging from zero for a circular basin to values approaching 1 for more elongated basins. Basin shape is important in influencing the travel time of runoff from various parts of the basin to its outlet.

Basin relief can be expressed by a variety of indices. A simple measure is the **relief ratio**:

$$R_h = H/L$$

where R_h = relief ratio, H = vertical difference between highest and lowest points in the basin, and L = maximum basin length. As such, it is a generalized measure of gradient and therefore of potential energy of the basin. It is thus a fundamental influence on denudation processes in the drainage basin, both on slopes and in river channels.

10.2.3 Relief properties
More detailed information concerning the distribution of altitude within the drainage basin can be gained from the hypsometric curve, in which altitude is plotted against area. This is derived by measuring the area between successive contours throughout the altitudinal range of the catchment. The data can be plotted in dimensionless form, where percentage height is plotted cumulatively against percentage area (Fig. 10.8a). The area beneath the curve, expressed as a proportion of the total volume, represents that volume of land remaining, assuming that the catchment basin has been eroded from a rectangular block of relief equal to the highest point on the watershed. This technique has been widely used as an index of the proportion of the basin removed by erosion.

A more flexible way of handling the same data is to plot relief against percentage area as a frequency distribution of altitude (Fig. 10.8b). From such data, mean relief, standard deviation and skewness can be calculated, giving a better statistical description of the distribution of relief within the catchment system. A negatively skewed distribution, for instance, would indicate a plateau with narrow incised valleys, and positive skewness would result from a broad lowland area with isolated residual hills.

To be able to assign numerical values to the spatial attributes and relationships of the system and its elements allows us not only to specify the state of the system's surface morphology but also to make meaningful comparisons between

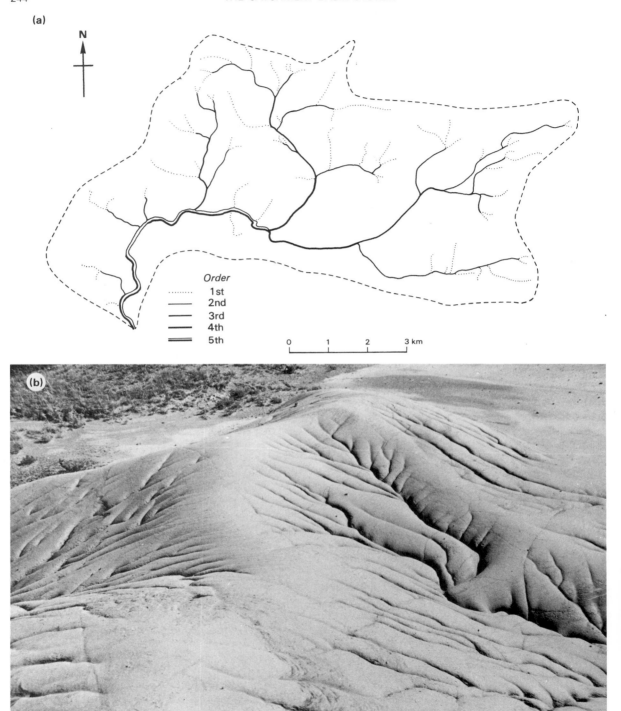

Fig. 10.5 (a) The drainage net of the River Wallington (Hampshire, England) showing stream orders according to the Strahler system. (b) Small rills can be seen forming an integrated drainage network (Alberta, Canada).

(a)

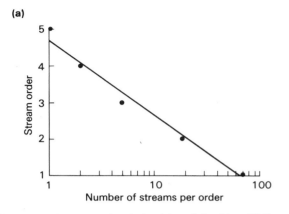

(b)

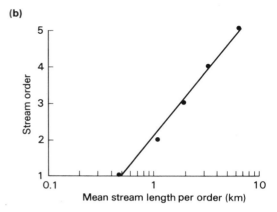

Fig. 10.6 Morphometric relationships of the River Wallington (Hampshire, England).

different systems on the basis of morphometric parameters.

As a result of the operation of the catchment system, its geometrically distinctive spatial organization will change with time and, as it changes, so the morphology of landscape will evolve. The state of the system's spatial organization at any moment in time will, therefore, reflect the inputs to the system, and the relative roles and rates of the processes operating within it. As the latter are to a large extent conditioned by the former, the spatial organization of the system will be determined mainly by the major environmental inputs – precipitation, lithology, relief, vegetation cover and time.

10.3 System structure: material components

The mineral components of the denudation systems are the rocks and materials derived from them by weathering processes. Their mineralogical characteristics have been outlined in Chapter 5. We are here concerned with their physical and mechanical characteristics, for it is these that govern their behaviour in relation to the applied forces of denudation processes.

The physical nature of the mineral components can vary widely, from solid rock to loose particulate material. Solid (indurated) rock behaves mainly in a rigid manner, whereas soil, or soil–rock mixtures, tend to display plastic or elastic behaviour. There is, then, a fundamental distinction based on the degree of induration, but we

can nevertheless derive descriptive parameters that can be used to characterize the wide variety of states of mineral material in the system.

10.3.1 Basic properties
All rocks and soils can be considered to be composed of minerals and voids. Thus:

$$V = V_s + V_v$$

where V = total volume of the mass, V_s = volume of solids and V_v = volume of voids. The voids may be occupied either by air or by water. Thus:

$$V_v = V_w + V_a$$

where V_w = volume of water and V_a = volume of air. Thus we can define the material as comprising three elements: solids, air and water. The relative proportions of these elements permit the derivation of basic descriptive parameters.

1. Voids ratio (e) is defined as the ratio of voids to solids:

$$e = V_v/V_s.$$

2. Porosity (n) is defined as the ratio of voids to total volume:

$$n = V_v/V.$$

There is wide variation in porosity, particularly between consolidated and unconsolidated materials (Table 10.2).

3. Moisture content (m) is defined as the ratio of weight of water to weight of solids, expressed as a percentage:

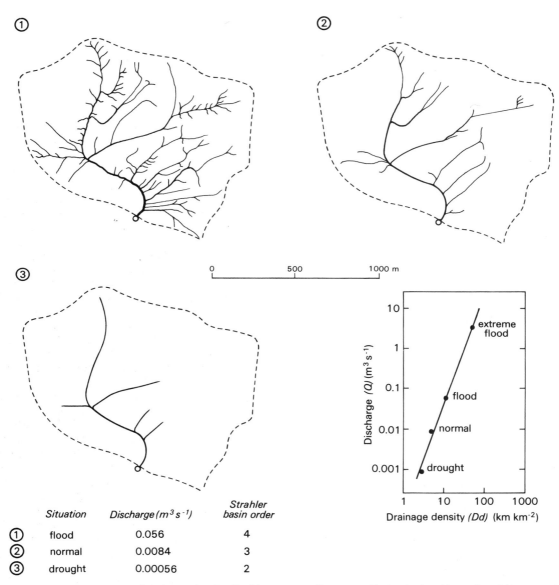

Situation	Discharge ($m^3 s^{-1}$)	Strahler basin order
① flood	0.056	4
② normal	0.0084	3
③ drought	0.00056	2

Fig. 10.7 Seasonal variation of drainage density (Swildon stream, Somerset, England: after Hanwell and Newson, 1970).

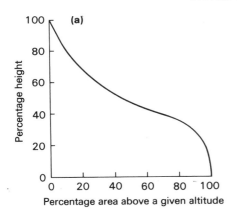

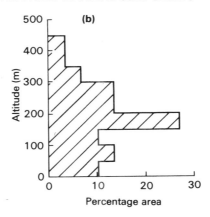

Fig. 10.8 (a) Hypsometric integral. b) Altitude frequency distribution.

Table 10.2 Average porosities for different materials (after Leopold *et al.*, 1964 from various sources)

	Porosity (%)
Consolidated rocks	
granite	1
basalt	1
limestone	10
sandstone	18
shale	18
Unconsolidated materials	
gravel	25
sand	35
silt	40
clay	45

$$ m = \frac{W_w}{W_s} \times 100 $$

where W_w = weight of water and W_s = weight of solids.

These basic descriptive properties are used in relation to both indurated and unconsolidated materials. Voids ratio and porosity are fundamental in two ways: they determine the amount of water a given material is capable of absorbing and they are an indirect measure of the extent to which adjacent grains or crystals are in contact, which contributes in part to the strength of the material. Furthermore, moisture content closely affects the way in which materials, especially unconsolidated ones, behave under stress.

10.3.2 Strength properties
It is essential to be able to determine strength properties if the way materials in the field behave

under stress is to be understood. Engineering practice has developed a number of standardized procedures which permit strength to be determined in relation to the various types of applied stress (see Chapter 2). Since the most common stresses in the natural environment are shear stresses, we shall confine discussion to these alone, while bearing in mind that it is possible also to test for compressive and tensile strength (Duncan, 1969). Strength properties are applicable to both consolidated and unconsolidated materials.

Values of shear strength can be determined under controlled conditions in the laboratory (Box 10.3). These also permit the components of strength to be isolated, in addition to other factors which influence strength properties.

Rigid materials (indurated rock) will yield under stress by brittle fracture and, after fracture has taken place the shear strength of the body of material can be considered to be insignificant. However, unconsolidated materials exhibit shear strength after failure has taken place. In these materials, therefore, we can recognize two strength values; maximum shear strength (τ_{max}), and residual shear strength (τ_{res}). The latter value is of importance in considering unconsolidated material on slopes across which shearing has already taken place. In this case stability is related to residual (not maximum) shear strength.

The effect of normal stress is important. Box 10.3 shows that strength increases directly with normal stress. Thus, material at depth beneath the surface will be further consolidated by the pressure of overburden exerting a normal stress, thereby increasing the shear strength at depth. The effect of moisture is to decrease shear strength, as shown by Terzaghi's equation (Box 10.3). Thus moist materials will have a lower shear strength than dry ones.

Representative values of strength for a variety of indurated rocks are presented in Table 10.3. These data should be treated with some degree of caution, since they are drawn from a variety of sources and are probably derived from a variety of test conditions, which may influence the values obtained. They do, however, demonstrate the range and order of magnitude of rock strength, and show the variation between different types of rock. They show also the different degrees of resistance possessed by rock in relation to the type of stress. All of the materials illustrated are most

Table 10.3 Strength values for different rock types (in MN m^{-2}) (after Billings, 1954, from various sources)

	Compressive		Shear		Tensile	
	Mean	Range	Median	Range	Median	Range
granite	145	36–372	22	15–29	4	3–5
sandstone	73	11–247	10	5–15	2	1–3
limestone	94	6–353	15	10–20	4.5	3–6
marble	100	30–257	20	15–25	6	3–9
serpentine	121	62–121	25	18–33	8.5	6–11

BOX 10.3
SHEAR STRENGTH

The most basic method of testing shear strength is the simple shearbox, consisting of two separate halves, the lower one of which is fixed, while the upper half is free to slide over it under an applied force. The applied force (stress) is measured through a proving ring, and strain is measured as the distance travelled during shear.

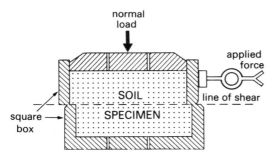

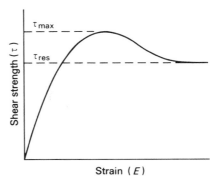

The behaviour of material under shear can be described by plotting stress against strain. For an unconsolidated material of mixed calibre the following can be expected to occur.

At low stress, little strain takes place until the point of failure. This defines the **maximum shear strength** (τ_{max}). Thereafter, the shearing force required to cause further strain diminishes to a steady level, defining the **residual shear strength** (τ_{res}).

The effect of consolidation of the material on its shear strength can be examined by varying the load on the sample, normal to the shear plane and repeating the test. The relationship between τ_{max} and normal load permits the identification of the two major components of shear strength – cohesion and friction – since the effect of increasing the load is to increase interparticle contact, thereby increasing friction. Shear strength at zero load is assumed to be due to cohesion alone.

For most mixed materials, the shear strength: normal load relationship is shown in Figure (a). Strength has a finite value at zero load and increases linearly as load increases. This rate of

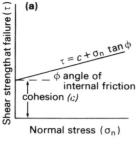

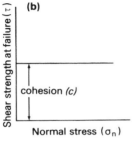

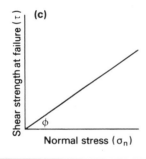

increase in strength with load is due to the increase in friction. This can be expressed by Coulomb's equation (1776), as follows:

$$\tau = c + \sigma_n \tan \phi$$

where τ = shear strength (N m^{-2}), c = cohesion (N m^{-2}), σ_n = load normal to shear plane (N m^{-2}) and ϕ = angle of shearing resistance.

The two components of shear strength are, therefore, cohesion (c) and friction ($\sigma_n \tan \phi$). A continuum exists between frictionless materials, such as some clays, where loading does not increase the strength ($\phi = 0$) and all the strength is due to cohesion, and cohesionless materials, such as loose sand ($c = 0$), in which strength is assumed to be due entirely to friction.

A further refinement of the shear strength equation can be made by taking into account the effect of pore water. This exerts a force against adjacent particles (pore water pressure) which operates against the normal force, thereby reducing the friction component of strength. This is expressed in the equation

$$\tau = c + (\sigma_n - u) \tan \phi$$

where u = pore-water pressure.

Thus, with positive pore-water pressure (saturated soil) shear strength is reduced, with negative pore-water pressure shear strength increases.

---

resistant to compressive stresses, moderately resistant to shear stresses and least resistant to tensile stresses, with an order of magnitude difference between the values of each.

It should be borne in mind that these values relate to small samples of intact rock and are not representative of the rock mass as a whole. The overall strength of a mass of rock in the field is dependent on the presence, abundance and orientation of planes of fracture or discontinuities within it, and it will be lower than the value obtained for a small sample (Box 10.4).

Strength values for unconsolidated materials are, of course, much lower. Data quoted by Carson and Kirkby (1972) for the shear strength of residual soils and taluvial material range from 0–85 kN m^{-2}, several orders of magnitude less than the values of solid rock. Under zero normal load

(Box 10.3) many unconsolidated materials may have no measurable shear strength, i.e. they possess no cohesion. The difference in strength between solid rock and soil is therefore shown to be very great indeed.

10.3.3 Consistency properties

Properties of consistency are a measure of the physical reaction of unconsolidated materials to moisture content. Known as **Atterberg limits**, arbitrarily defined tests indicate the moisture content at which the material changes from the solid to the plastic state (**plastic limit:** *PL*) and from the plastic to the liquid state (**liquid limit:** *LL*). The plastic limit is defined as the minimum moisture content at which the soil can be rolled to a thread 3 mm in diameter without breaking. The liquid limit is the minimum moisture content

Box 10.4
ROCK MASS STRENGTH

Values of rock strength determined in the laboratory on small intact samples (Table 10.3) reasonably reflect the behaviour of rock in relation to small-scale processes, such as forces of mechanical weathering. Large masses of rock however are dissected by systematic fractures or discontinuities (Chapter 5, Section 5.3.2) which have a major influence on the behaviour of a rock mass in relation to mechanical stresses. There is therefore a need to assess rock mass strength, in manner which incorporates the effects of discontinuities. M.J. Selby (1982) has devised a scheme for doing this in relation to rock exposed on hillslopes, and which can readily be employed in field situations.

Rock mass strength is assessed, for an area of slope of some 10 m², employing the following seven parameters, among which the abundance and nature of discontinuities play a dominant role.

- Intact rock strength – resistance to a blow from the pick end of a geological hammer, or value obtained by Schmidt hammer (a device for assessing rock hardness).
- Weathering – visible evidence of alteration by weathering processes.
- Joint spacing – linear distance between joints.
- Joint orientation – dip of joints in relation to

the rock face, i.e. favourably or unfavourably aligned in relation to shear stresses.
- Joint width – linear width across joints.
- Joint continuity and infill – assessment of the degree of friction between opposed joint faces and the cohesion supplied by any infill.
- Outflow of groundwater – observation of flow volume.

The parameters are given differential weightings, with maximum values shown in Table 1.

Table 1

	% rating
Intact rock strength	20
Weathering	10
Joint spacing	30
Joint orientation	20
Joint width	7
Joint continuity and infill	7
Outflow of groundwater	6
	100

Each of the parameters is rated on a five-point scale, with a specified numerical value applied to each category of each parameter, to yield a

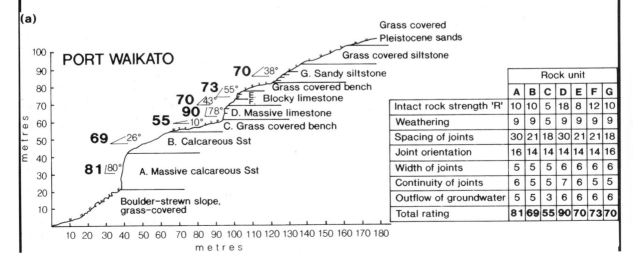

(a)

PORT WAIKATO

70 ∠38° — G. Sandy siltstone

Grass covered
Pleistocene sands
Grass covered siltstone

73 /55° — Grass covered bench
70 /43° — E. Blocky limestone
90 /78° — D. Massive limestone
55 —10° — C. Grass covered bench
69 ∠26° — B. Calcareous Sst
81 /80° — A. Massive calcareous Sst

Boulder-strewn slope, grass-covered

	Rock unit						
	A	B	C	D	E	F	G
Intact rock strength 'R'	10	10	5	18	8	12	10
Weathering	9	9	5	9	9	9	9
Spacing of joints	30	21	18	30	21	21	18
Joint orientation	16	14	14	14	14	14	16
Width of joints	5	5	5	6	6	6	6
Continuity of joints	6	5	5	7	6	5	5
Outflow of groundwater	5	5	3	6	6	6	6
Total rating	81	69	55	90	70	73	70

metres (vertical axis: 10 20 30 40 50 60 70 80 90 100)

10 20 30 40 50 60 70 80 90 100 110 120 130 140 150 160 170 180
metres

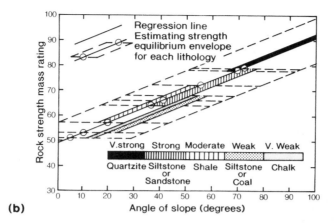

(b)

rock mass strength scale ranging from 100 down to 26.

Rock mass strength can be assessed in this way on slopes formed in rock where there is no significant development of soil or talus. For a large sample of slopes there is a good correlation between rock mass strength value and slope angle, suggesting that the observed slopes have attained an equilibrium in accordance with their properties of resistance. Fig. (a) illustrates

the rock mass strength in relation to slope angle in a single profile; Fig. (b) illustrates the general relationship between rock mass strength and slope angle.

The concept of discontinuous rock mass strength is thought to be particularly significant in relation to the processes of glacial erosion, as studied in Snowdonia by Addison, (1981). (See Chapter 14, Section 14.4.1).

at which the soil flows under its own weight, as defined by a standard test procedure.

The plasticity index (*PI*) is defined as the range of moisture contents at which the soil behaves in a plastic manner, and is defined as

$$PI = LL - PL$$

(all terms expressed as % moisture content).

Atterberg limits vary widely between different soils, depending on the percentage of clay and the type of clay minerals present. Whalley (1976) quotes the following values as typical of London Clay:

PL	30–45%
LL	70–105%
PI	35–65%.

Values are lower with a lower clay content and, for cohesionless materials, *PI* becomes zero.

These material properties have considerable influence on the behaviour of rock and soil under the applied forces of denudation. They are es-

sentially descriptive parameters, yet at the same time they are essential to an understanding of the resistance of materials to denudation and are valuable tools in the analysis of denudation processes.

10.4 System function: energy flow

When we considered the functional organization of the denudation system in the context of its component catchment basins, the links between the subsystems of the model where presented mainly as pathways of material transfer (see Fig. 10.3). These links are also pathways of energy flow, while the processes associated with the operation of the subsystems involve the transformation and transfer of energy. The two fundamental sources for all Earth surface systems are terrestrial energy and solar energy.

Input of the geothermal component of terrestrial energy to denudation systems is so small that its effect on most aspects of systems operation can be regarded as negligible, with the significant exception of the glacial system (see Chapter 14).

The expression of terrestrial energy as inputs of potential energy, however, is of great significance in the operation of the denudation systems. First, the primary minerals of the rocks forming the crust can all be considered as possessing potential chemical energy by virtue of their ordered structure and the atomic and molecular bonds that maintain it. In the weathering system this energy is released as weathering reactions proceed and complex compounds are broken down to produce structurally simpler weathering products. Secondly, and of great significance, is the potential energy of relief allied to gravitational force. Where gradients exist, as on hillslopes and in river channels, the motion of both solid and fluid materials results in the conversion of the potential energy to kinetic energy. It is clear that velocity is the major component of kinetic energy (see Chapter 2). Thus the more rapid movements of materials (e.g. flowing water in channels and rapid mass movements) possess higher levels of kinetic energy. The more slowly moving materials (e.g. soil creep) also involve kinetic energy, although of much less magnitude.

The input of both direct and diffuse solar radiation manifests itself in a number of ways in denudation systems. In the weathering system, absorbed solar radiation raises the temperature of both soil and regolith and, in most cases, increases the rate of chemical reactions. It is responsible for the evaporation of water from land and water surfaces and in part for the transpiration of plants. The latent heat of evaporation means that 2450 J are lost to the denudation system per gramme of water evaporated at 20°C. After condensation in the atmosphere, the gravitational potential energy of each water droplet becomes the kinetic energy of the falling raindrop, defined by its mass and velocity. Accordingly, each rainstorm produces its own specific kinetic energy input, related to its intensity and duration. This kinetic energy is converted to mechanical energy on impact and it performs work on the inorganic and organic particles of the land surface – the work of erosion. Since work is defined as the product of force and distance, the application of this erosional energy results in the movement of materials. These materials, therefore, possess kinetic energy as they move downslope through the slope system, as does the water flowing on the surface and below it. Some of this kinetic energy becomes the input to the next system in the cascade, as runoff and debris are delivered to the

channel system. Some of the incident solar energy, however, is absorbed by the ecosystem, particularly during photosynthesis by plants, and is subsequently stored as the potential chemical energy of organic compounds. This energy subsequently becomes available to the weathering system in the soil, while the vegetation itself has profound effects on slope erosional processes and on the catchment water balance.

Outputs of energy from the denudation system occur mainly in the form of heat energy. Direct heat loss from the ground surface by radiation occurs particularly at night. As we have seen, evaporation involves considerable energy loss as latent heat, while much of the kinetic energy involved in erosional processes becomes frictional heat loss to the atmosphere as the material in motion moves relative to static material. Flows of energy through denudation systems, however, although understood in principle, have as yet undergone little quantitative study.

10.5 System function: material flows

The flow of materials through denudation systems involves both water and mineral elements. The flow of water is fundamental to catchment basin processes, since water either acts directly as the transporting medium or plays an important role in most other modes of movement of mineral materials.

The dynamics of the flow of water through the catchment basin are shown in simplified form in Figs 10.9 and 10.10. The input of precipitation is distributed by a series of transfers through a number of stores to outputs of channel runoff and evaporation. Precipitation falls directly upon vegetation, where present on the ground surface, and on the water surface of channels and lakes. Vegetation controls the interception store and transfers water via evaporation back to the atmosphere, and via stemflow and throughfall to the ground surface beneath. From the surface store, water can be transferred to the channel storage by overland flow, evaporated directly to the atmosphere, or transferred by infiltration to the zone of aeration beneath the ground surface. This zone of aeration may consist of soil, regolith and bedrock, and it represents that part of the subsurface, above the permanent water table, which is intermittently wet; in other words, the pore spaces are occupied sometimes by air and sometimes by water. Evaporation and transpiration via vegetation can transfer water from the zone of aeration back to the atmosphere. Lateral flow within this zone (throughflow) can transfer water to the river channel, and deep percolation can carry water down to the groundwater zone. From here, deep transfer takes place by groundwater flow into river channels.

In terms of one element (water), the change in state of the denudation system per unit time can be specified if the transfers between the stores can be given numerical values and related to inputs and outputs. This can be done by reference to the water balance equation (see Chapter 4):

$$P = R + (E + T) + (\Delta S + \Delta G)$$

where P = precipitation, R = runoff, E = evaporation, T = transpiration, ΔS – change in soil water and ΔG = change in groundwater.

By combining the elements in brackets above, the equation can be simplified to the following form:

$$\text{precipitation} = \text{runoff} + \text{evaporation} + \text{infiltration}.$$

This water balance, or budget, approach can be used at all scales from the global to the local in order to illustrate spatial differences in the operation of the denudation system (Box 10.5).

The movements of mineral elements through denudation systems for the most part follow the same pathways as those of water, which itself acts as the main transporting agent. Until they pass into solution as charged ions, or into suspension as fine particles, the speed of transfer is much slower than with the flow of water. In other words, the residence time of mineral elements in the stores of the systems is much longer. In the weathering system, the products of weathering accumulate as regolith, until surface erosion brings them within reach of mass movement processes which move them downslope. Here fluvial erosion may ultimately transfer them from the slope to the channel subsystem, to become part of the sediment load of the river. Other weathering products pass into solution and follow the flow of water, either as overland flow and throughflow, again downslope to the channel system, or as percolation to the groundwater storage and hence to the channel baseflow. Some mineral elements are held temporarily in the soil in the exchange complex (see Chapter 22), while both these exchangeable elements and those free in the soil water may be taken up by plant roots and

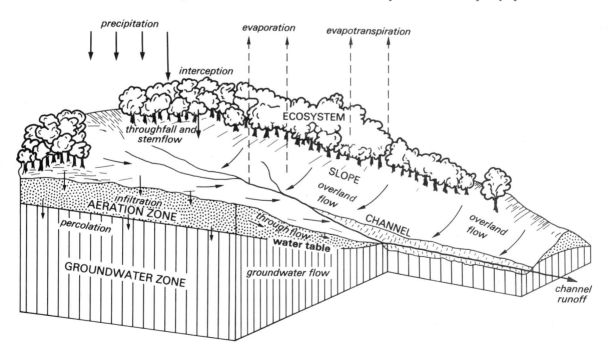

Fig. 10.9 Block diagram of the catchment basin system.

temporarily immobilized in the biomass of the ecosystem (see Chapter 17). These elements may subsequently be returned to the slope system in the stemflow and throughfall components of precipitation, where they join materials already dissolved in rainfall. Alternatively, they may remain in organic compounds until death and litter fall followed by decomposition release them to the slope system once more. The major inputs of materials, therefore, are from the atmosphere, dissolved in precipitation or as dry fallout; from the crust through the weathering system as the weathering front progresses; or directly to the channel subsystem where the erosion of bedrock

occurs. The main output of mineral elements from the catchment basin system is the mineral load in channel runoff, although some elements which can exist in gaseous form may go direct to the atmosphere in a way analogous to water vapour during evaporation.

10.6 Special cases of the model

The basic functional model of the catchment basin system developed so far can be modified to cater for situations where the system operates under more extreme sets of conditions, which radically affect the magnitude of the inputs to the system and the distribution of matter and energy through its components.

In glacial landscapes the precipitation input is largely in the form of snow, and runoff is mainly in the form of slowly moving streams of ice, supplemented by meltwater streams in marginal environments. Thus the glacier takes the place of the river channel and the rate of throughput is much slower than in a normal fluvial basin. Mineral material can enter glacier systems either from above, where slopes overlook the glacier surface, or from below as the glacier erodes material from its margins.

Desert landscapes are characterized by very high values of evaporation loss, often reaching 100%, which result in a zero net surface runoff. Consequently, deserts often contain a series of enclosed catchment basins with centripetal drainage systems. When runoff occurs, it is shortlived and streams ultimately peter out as infiltration and evaporation take their toll. Often, therefore, there is no net outflow of surface runoff.

Limestone and chalk landscapes are characterized by high rates of infiltration. Streams on limestone disappear underground in swallet holes, while percolation is more diffuse in chalk. Thus the amount of surface runoff is strictly limited, as catchment basins in such areas drain underground, and infiltration and ground water play an important role.

All of these variations can be accommodated by the basic functional model. It is merely that climatic and lithological variations involve differences in the relative significance of different processes and the relative importance of different pathways of energy and matter transfer through the denudation system. These variations are no

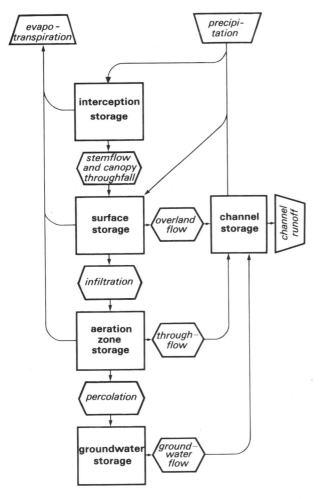

Fig. 10.10 Hydrological processes in the catchment basin system.

Box 10.5

COMPARISON OF CATCHMENT WATER BALANCES

Stoddart (1969) quotes values in mm depth of water for the processes of precipitation, runoff, evapotranspiration and infiltration for the total land surface areas of the world, as follows:

$$P = R + E_T + I$$
$$730 = 171 + 478 + 81$$

where E_T = total evaporation loss and I = infiltration. It should be remembered that these values are extremely generalized and are based on estimates from many regions of the Earth for which few accurate data yet exist. If we assume that the water infiltrating rejoins river channels and reappears as surface runoff, then a total of 252 mm (171 mm + 81 mm), out of the original precipitation of 730 mm, ultimately flows from the Earth's land areas. This represents a total proportion of only 34% of the precipitation value and puts into perspective the relative importance of the paths of water flow shown in Fig. 10.3.

These general figures for water balance processes on the world scale obscure the considerable differences that exist between continents, shown below (from Budyko, 1958):

	Precipitation (mm)	Evaporation (mm)	Runoff (mm)
Europe	600	360	240
S. America	1350	860	490
Australia	470	410	60
Africa	670	510	160

Clearly, these figures also must be regarded as first-order estimates, but they do reveal important differences between the continents. The temperate mid-latitude continent of Europe exhibits moderate values throughout. South America has a high precipitation value, resulting in high runoff. The arid continent of Australia, on the other hand, shows a relatively low rainfall and a high annual evaporation loss due to high temperatures, resulting in a low rate of runoff. The values for Africa show a moderate precipitation and a high value for evaporation, since the continent straddles the tropics and contains large arid regions.

The same model of water balance can be applied at the scale of the individual catchment. Values in millimetres for a small sample of catchment basins in selected areas of Britain are tabulated below:

	Precipitation	Evaporation	Runoff	Runoff (%)
E. Riding[1]	645	458	187	28.9
S. Hampshire[2]	926	439	487	52.6
E. Devon[3]	1033	551	467	45.2

[1]Pegg and Ward (1971); [2]Mottershead and Spraggs (1976); [3]Gregory and Walling (1973)

Hanwell and Newson (1973) present a map of selected catchments throughout Britain showing variations in mean annual runoff as a percentage of mean annual precipitation. Values range from as low as 10–20% in catchments in Suffolk, to 80–90% in the west Highlands of Scotland. Clearly then, great variations exist between different individual catchments in the amount of water flowing along the various pathways of the denudation system. These differences can be related both to the nature of the precipitation input and to the catchment basin characteristics.

more, therefore, than particular cases of a general denudation system model.

10.7 Time, denudation and relief in the catchment basin system

As we have seen, the effect of the operation of the denudation system is to strip rock debris from the surface of the continents and deliver it to the oceans. Monitoring of the denudation processes involved has provided information on the rate at which this takes place, and it is shown to be considerably slower than the rate of uplift by orogenic process. Accordingly, it would appear that land masses are affected by periods of relatively rapid uplift, creating relief, followed by long-continued periods of denudation, which gradually reduce it.

A major effect of the denudation process–response system is that denudation itself affects the relief upon which it is operating. This further modifies the nature and rate of operation of denudation processes. Thus there exists a positive feedback loop between denudation processes and relief (Fig. 10.11). As shown in Fig. 10.12, denudation rate is closely related to relief (Ahnert 1970). Relief in turn is related to gradient, which controls many of the denudation processes, especially the physically based ones. Assuming a constant basin area, the lower the absolute relief in the basin, then the lower its mean gradient will be both in channels and on slopes. Thus potential

energy also is reduced, which in turn leads to a decrease in the rate of operation of denudation processes. This positive feedback loop, then, leads in the direction of progressive change, towards lower relief.

The decline in denudation through time is suggested by the empirical evidence in Fig. 10.12. Using data derived from a sample of mid-latitude humid drainage basins, the rate of denudation is shown to be directly and linearly related to relief. This can be expressed simply as

$$d = 0.0001535\ h$$

where h = mean relief in metres and d = mean denudation rate in mm a^{-1}.

Such a relationship assumes that erosion is distributed evenly throughout the catchment, and is based on the premise that the higher the relief, the greater the tendency for steep gradients and corresponding erosional stresses. Since the rate of denudation is functionally related to relief, and the latter is reduced through time, it is possible to calculate on the basis of the above relationship that after 11 Ma a land mass would be reduced to 10% of its original relief, and after 22 Ma it would be reduced to 1%, as the rate of denudation declines with both time and relief.

Denudation rates, however, refer only to material stripped from the continental surfaces. In considering the destruction of relief we must also take into account isostatic rebound, as the gradu-

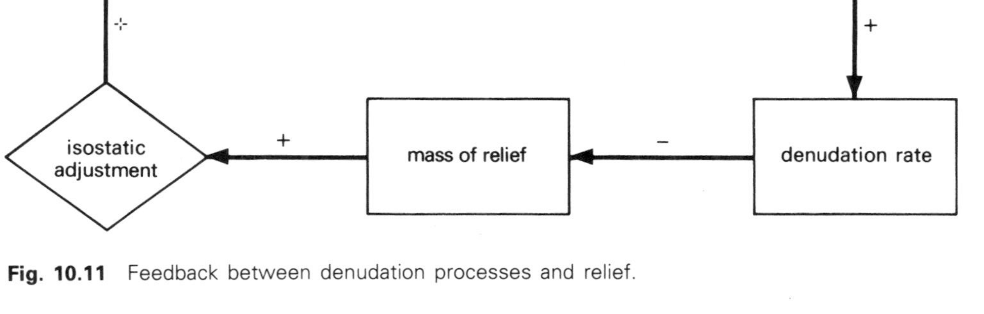

ZONAL AND DYNAMIC CHANGE

Fig. 10.11 Feedback between denudation processes and relief.

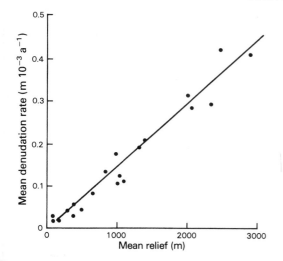

Fig. 10.12 Denudation rate as a function of relief (after Ahnert, 1970).

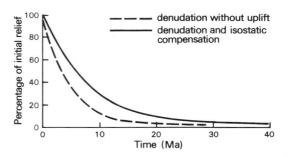

Fig. 10.13 Time needed for relief reduction (after Ahnert, 1970).

ally thinning continent becomes more buoyant on the underlying mantle. Isostatic rebound can be calculated thus:

$$h = Br/A$$

where h = isostatic compensation, B = specific gravity of surface rocks removed, A = specific gravity of material replacing at depth, and r = thickness of surface layer removed. Assuming A = 3.4 and B = 2.6, then h = 0.76 r.

Assuming then that isostatic rebound in response to erosional losses is a continuous and widespread process, then three-quarters of the relief removed within a given period of time is thus replaced. Accordingly, the real length of time taken for the destruction of the relief of a land mass will be longer, and Ahnert calculates that, with isostatic compensation, 18.5 Ma will be required for reduction to 10% of the initial relief, and 37 Ma to 1% (Fig. 10.13).

All this presupposes that external conditions remain essentially constant throughout the operation of the denudation process over periods of millions of years. Yet, as we shall see, considerable variations in system inputs can take place over much shorter timescales.

Further reading

A wide range of general geomorphological texts is available, covering the material presented in this chapter, and also serving

Chapters 11–17. Still a useful introduction to geomorphological systems is:

Bloom, A.L. (1969) *The Surface of the Earth*. Prentice-Hall, Englewood Cliffs.

Good general texts are:

Rice, R.J. (1977) *Fundamentals of Geomorphology*. Longman, London.
Selby, M.J. (1988) *Earth's Changing Surface*. Oxford University Press, Oxford.
Summerfield, M.A. (1991) *Global Geomorphology*. Longman, London.

A modern account of geomorphological processes is:

Derbyshire, E., K.J. Gregory, and J.R. Hails (1979) *Geomorphological Processes*. Dawson, Folkstone.

A succinct and useful introduction to the same subject, stressing the mechanics and dynamics of geomorphological processes is:

Stratham, I. (1977) *Earth Surface Sediment Transport*. Oxford University Press, Oxford.

More advanced general texts include:

Chorley, R.J., S.A. Schumm and D.E. Sugden (1984) *Geomorphology*. Methuen, London.
Embleton, C.E. and J.B. Thornes (eds) (1979) *Processes in Geomorphology*. Edward Arnold, London.

Useful chapters are to be found in:

Cooke, R.U. and J.C. Doornkamp (1974) *Geomorphology in Environmental Management*. Oxford University Press, Oxford.

At the catchment scale the following are useful sources:

Burt, T.P. and D.E. Walling (eds) (1984) *Catchment Experiments in Fluvial Geomorphology*. Geobooks, Norwich.
Smith, D.I. and P. Stopp (1978) *The River Basin*. Cambridge University Press, Cambridge.

The weathering system

Rocks formed within the lithosphere under conditions of high temperature, high pressure or both, are relatively stable physically and chemically until exposed at the Earth's surface (see Chapter 5). Here they encounter entirely different conditions. For example, pressure is reduced, temperature is subject to considerable fluctuation, and both oxygen and water are abundant. Rocks then become altered by weathering processes into forms which are stable under these new conditions. Thus weathering can be defined as the response of materials which were at equilibrium in the lithosphere, to new conditions at or near the contact with the atmosphere and biosphere.

This response is seen on the left of Fig. 11.1. On the right, however, you can see that a similar process is affecting organic compounds which are also being exposed to a parallel change in conditions when, on death or excretion, they cease to be components of living systems. These two directions of alteration and adjustment merge in the soil system, where the processes of weathering and decay, together with the properties of the weathering products and organic residues, interact in soil formation (**pedogenesis**). In a sense, then

the soil can be viewed as a new equilibrium state established by these adjustments in both inorganic and organic materials in response to changes in their conditions of existence.

It should be clear from Fig. 11.1 that it is difficult to separate weathering, decay and soil-forming processes, and to delimit the systems within which they operate. The three clearly overlap and there are intersections between the sets of variables which characterize the systems and interactions in the operation of the processes. In this chapter we consider the weathering system set, as far as possible in isolation. The pedogenesis system set will be treated in part in Chapter 12 and again in Chapter 22. We are concerned here with the overall process of adjustment that the rocks of the lithosphere undergo in order to attain a new equilibrium with the conditions they experience at and near the Earth's surface.

The consequence of the operation of the weathering system is, therefore, the production of new materials, which differ both chemically and mechanically from the original parent rock. Substances are produced which are less massive, less indurated and possess much lower mechan-

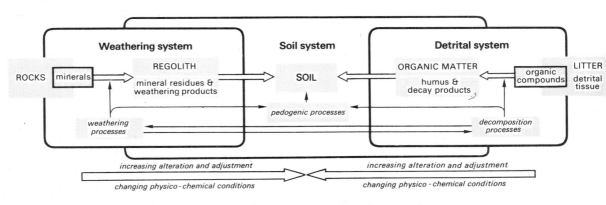

Fig. 11.1 Model of the weathering system and its relations with the soil and detrital systems.

ical strength. As such, they become available for transportation by denudation processes, for their resistance falls to a level at which the energy available for erosion is sufficient to transport them. This material is the solid weathering residue. In addition, soluble substances are produced which are readily removed by circulating water. These minerals are therefore rendered highly mobile and made available to both the ecosystem and the denudation system. As such, they are removed readily, thus being released from storage in the lithosphere and transferred in geochemical or biogeochemical cycles.

11.1 A process–response model of the weathering system

Geographers have traditionally divided weathering into mechanical (physical) processes and chemical processes, with some acknowledgement of the role of the biotic agents in both of these categories. When the fundamentals are considered, however, it becomes apparent that the chemical processes are in fact governed by the basic laws of physics operating at the molecular and submolecular scales. The distinction between chemistry and physics is, therefore, essentially one of scale. Accordingly, we adopt here an original approach to the study of weathering, based on the scale at which breakdown occurs. The model of the weathering system which we shall develop is a process–response model (Fig. 11.2). The response of the system will be viewed as a progression of states, defined in terms of their mineralogical, petrological and mechanical properties, as they replace each other through time. The processes are both those directly responsible for the adjustment and reaction to forces brought to bear on the rock when exposed at or near the surface, and the processes that activate or enable rock breakdown to occur. In this model it is contended that there are two, and only two, **primary mechanisms** of rock breakdown. These are **brittle fracture** and **crystal lattice breakdown**, the former operating at the scale of the rock mass, clast or crystal, and the latter at the molecular and submolecular scales. These primary mechanisms are triggered by environmental inputs of various kinds and promoted by the operation of the second group of processes alluded to above; these we shall term the **activating processes**, which in turn are

promoted by a range of **activating agencies** (Table 11.1).

The nature of the environmental inputs will vary from one place to another across the Earth. Some enacting processes require hot conditions, others cold. Some are restricted to humid environments, others to arid ones. There are surface processes and subsurface processes. Therefore, the combination of different enacting processes in operation at a particular location will in turn reflect the conditions under which the system functions – the **weathering environment**.

Approximately one-third of all minerals making up about 90% of the Earth's crust belong to the silicate group (see Box 5.1). For the purposes of this model, therefore, we shall consider a massive indurated rock composed of primary silicate minerals as the initial state of the system. The first compartment in the model (the initial state), therefore, is the unaltered rock, retaining original structures, composed of primary minerals and with characteristic mechanical properties of permeability, porosity and strength. From the point of view of the model, it is regarded as a system organized hierarchically from the atoms of the constituent elements into the orderly arrangement of these atoms and the molecules they form in crystal lattice structures, to the individual mineral paricles they produce. These mineral particles are organized in turn into a three-dimensional mosaic to form massive rock (see Chapter 5). This structure is maintained by internal mechanical forces due to the packing of the grains and by the chemical bonding forces within the crystal lattices and between the crystals. These forces are the internal or potential chemical energy of the system.

Under the influence of the primary mechanisms of weathering, this initial state is replaced in time by clastic fragments of rock, still composed of primary minerals but with different mechanical properties and with a greater surface area. The next state is one where the minerals of these fragments are altered chemically and the weathered residues of the primary minerals now exist in an environment containing the mobile products of weathering, usually in solution. Also new minerals, not directly derived from primary minerals, may have appeared. The physicochemical properties of the mineral residues and of these new minerals, together with the mechanical characteristics of the regolith they form, may differ radically from those of the original rock and

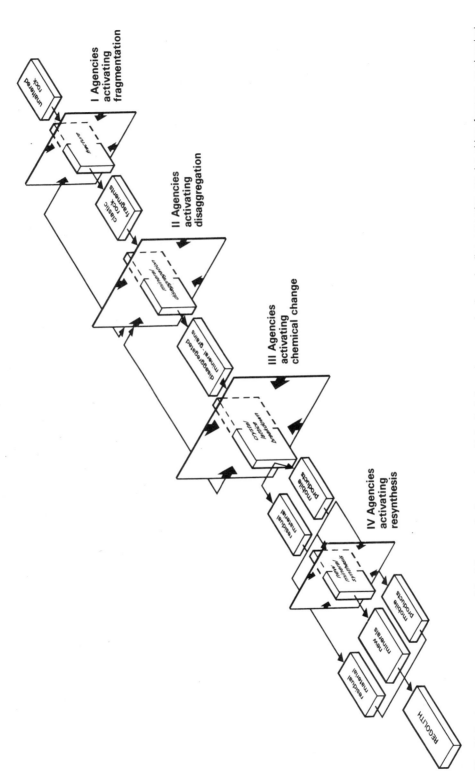

Fig. 11.2 A process–response model of the weathering system. Note that the primary mechanisms in disaggregation can be either fracture or chemical breakdown.

Table 11.1 A classification of weathering processes

Primary mechanisms	Enacting processes	Activating agencies
brittle fracture	strain release (dilatation) applied mechanical stress	unloading thermoclasty haloclasty gelifraction biomechanical forces
crystal lattice breakdown	modification of electrostatic forces, by hydration hydrolysis and redox reactions	exposure of mineral surfaces supply of water supply of hydrogen ions (pH) oxidation/reduction status of weathering environment (Eh) agencies removing weathering products (e.g. by leaching, cheluviation)

primary rock-forming minerals. Particle size will be much reduced, exposed surface-area/volume ratios will have increased enormously, density and mass are reduced and strength parameters will have changed. The smallest (colloidal) particles will exhibit various degrees of mobility and will have marked ion exchange properties.

11.1.1 The primary mechanisms of weathering: brittle fracture

Brittle fracture occurs when the rock experiences tensile and shear stresses. Such stresses will place under tension the bonds between mineral grains and those within crystal lattices. This in turn stretches or deforms these bonds, placing them under strain or giving to them a certain **strain energy**. This strain energy is dissipated as the bonds break and a crack or fissure forms. Although usually regarded as a macroscopic process operating on massive rock, the actual mechanism of fracture is located at the apex of the crack and it operates at the molecular level where the strain is concentrated.

The development of most cracks and fissures in rocks probably takes place along grain boundaries, while the site of initiation and the direction of propagation will be determined by existing points, lines and planes of relatively weak bonding within

the granular mosaic of the rock. Under a maintained stress such cracks can spread very rapidly. Fracture, however, also takes place within the crystal lattices of rock-forming minerals. Such crystals are rarely perfect and strain can be concentrated at defective points in the lattice, or along displaced layers of the lattice known as dislocations. Once initiated, such crystal fracture will be guided preferentially by cleavage planes. Although fracture can overcome the physical and chemical forces holding the lattice together, it does not alter its nature but merely produces smaller fragments of crystal and appears, at its lower limit, to be restricted to the formation of silt size particles. Beyond this threshold of rock dislocation we move into the realm of the second primary mechanism of weathering, the fundamental breakdown of the crystal lattice itself. Lattice breakdown, in contrast, involves basic changes in both the composition of the lattice and in its structural configuration, thereby producing chemical change in the mineral concerned.

11.1.2 Activating agencies of brittle fracture

The immediate processes causing fracture are either the release of internal strain or the application of external stress to the rock mass (Fig. 11.3). The expression of both, however, is controlled by the mechanical properties of the rock in question and of its constituent minerals. Therefore, these mechanical properties – particularly strength parameters – act as regulators which, by assuming threshold values, determine the incidence of fracture under any given stress. As the rock is subject to progressive fracture and fragmentation, so the values of these regulatory parameters change and a negative-feedback loop is established which reduces the effectiveness of fracture until a size threshold is passed, beyond which weathering can proceed only by chemical change.

Internal strain is present in rock masses because of the compressive stress experienced during their formation under conditions of high pressure in the crust, or as a result of the normal compressive stress of overlying rocks. The enacting process of **strain release** is the erosion and removal of overburden, leading to unloading and the reduction of normal stress and the consequent expansion of the rock. Rocks rarely expand uniformly, and the usual result is the production of joint planes (**dilatation joints**), which are subparallel to the topographic surface which has experienced

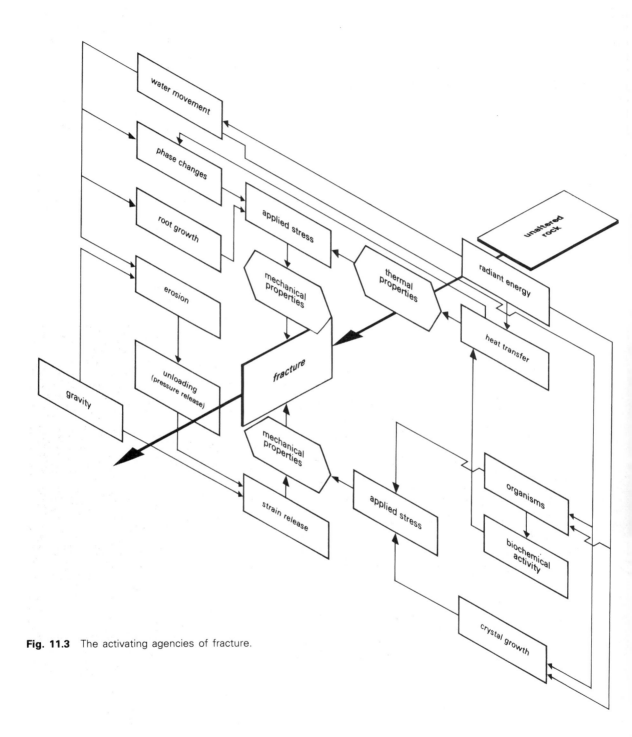

Fig. 11.3 The activating agencies of fracture.

unloading. These joints are often spaced 1–3 m apart and may affect the rock to a depth of 20 m or more and they can result in large sheets or slabs becoming detached from the rock mass. Massive igneous rocks (Fig. 11.4) display dilatation joints most clearly, particularly when forming rock domes or **inselbergs** (residual hills), from which the surrounding rock has been denuded. Such joints are also well displayed in glacial troughs and corries, where again bedrock has responded in this way to the removal of overburden by denudation. Indeed, Whalley (1976) suggests that mountain landscapes may owe far more to strain-release fracture than is usually realized.

The enaction of rock fracture by applied mechanical stress can be divided into two groups of processes. First, there are stresses set up in the rock as a direct or indirect result of heat transfer,

Fig. 11.4 Dilatation jointing. (a) Sheet joints nearly parallel to the land surface above (after Chapman and Rioux, 1958). (b) Large-scale sheet structures and exfoliation processes in massive Colorado Plateau sandstones; arrows show inferred directions of dilatation; X is an exfoliation dome, Y an exfoliation cave (after Bradley, 1963). (c) Dilatation sheets in massive igneous rocks (Yosemite, California, USA).

and secondly, stresses that result from the processes of growth and expansion within pores and fissures of the rock.

Because of the poor thermal conductivity of rock (see Chapter 3), the transfer of heat (derived from solar energy) by conduction from the surface is slight, and steep temperature gradients can exist through only a limited thickness of rock. The surface layer responds to heat by expanding, while adjacent mineral grains within it, having different coefficients of expansion, expand (or strain) at different rates and cause increased mechanical stress. These stresses can eventually produce thermal disintegration (**thermoclasty**), a process which was formerly thought to be a major cause of breakdown on exposed bedrock surfaces in desert regions where high air temperatures are attained. However, laboratory experiments have failed to reproduce thermal disintegration without the presence of moisture. This has led to the conclusion that thermal disintegration is limited in occurrence and is normally limited to situations where moisture has previously weakened intergranular and intercrystalline bonds by chemical change, notably by hydration and some hydrolysis along grain and crystal boundaries, thereby lowering the tensile strength of the rock and emphasizing the feedback relationship between mechanical and chemical processes.

A more indirect effect of heat transfer (in this case the latent heat of fusion), however, is associated with a phase change affecting the water occupying joints, fissures and pores in the rock. Water freezing to form ice crystals undergoes a volume expansion by 9%, and can develop pressures in a confined space theoretically in excess of $200 \, MN \, m^{-2}$. In reality, it is doubtful whether such values are reached, since water within rocks will not be in a totally confined space. Nevertheless, when compared to the tensile strengths of rocks, normally in the range 1–$10 \, MN \, m^{-2}$ (see Table 10.3), it can be seen that there is a sufficient margin for ice crystal growth to be very effective. Rock breakdown by this process (**gelifraction**) will tend to be most effective in conditions with frequent alternations of temperature about 0°C. It is clearly more effective in montane periglacial environments, which may experience frost more than 300 times per year, than in the arctic latitudes where freezing and thawing may be limited to short periods in the spring and autumn. Short-term fluctuations about the freezing point penetrate not more than a few centimetres beneath the ground surface. Effective frost weathering is, therefore, a surface phenomenon. Long-term freezing cycles may also be expected to cause rock fracturing, but they are events of limited frequency and hence may be restricted in their overall effect. Freezing within pore spaces tends to pulverize rock into individual grains and small angular fragments. Chalk, for example, responds in this way. The effect of ice formation in the joints of well consolidated rocks is to prise away joint-bounded blocks.

The growth of ice crystals, of course, can also be thought of as belonging to the group of processes that involve the growth and expansion of material in pores and fissures. The other mechanisms concerned in this second group of processes are the growth of salt crystals and the growth and activity of living organisms (**biomechanical forces**), both of which exert tensile stresses.

Salt crystal growth, normally thought of as a chemical process, can exert mechanical forces within rocks (Winkler and Wilhelm, 1970) – a conversion of chemical to mechanical energy. The effectiveness of salt weathering processes – known as **haloclasty** – in the breakdown of building stone has long been appreciated in the literature of building technology, although its significance in geomorphology had been underestimated in the main until the comprehensive review by Evans (1969).

Salts may be emplaced within pore spaces in rocks from saline solutions, especially groundwater and seawater. Saline groundwater is drawn to the surface by evaporation and capillary action in desert and semi-arid environments, while the inundation of rocks by seawater spray is an everyday occurrence in coastal environments. The evaporation of these saline solutions results in the precipitation of crystalline minerals within the pore spaces of surface rocks and often forms visible efflorescences of salts.

Salt weathering is caused commonly by three types of process. First, as salt crystals grow by crystallization out of saline solutions they exert a **force of crystallization**. The forces developed are related to the density and molar volume of the salt, the degree of supersaturation of the saline solution, and the temperature at which the process takes place, according to the model of Correns (1949). The force of crystallization at 20°C is calculated for several salts and is shown in Table 11.2 in relation to the supersaturation of the saline solution. Extremely modest levels of supersaturation are required to generate a force of 1 MN m^{-2}, and a force of 10 MN m^{-2} is attained by only moderate levels of supersaturation. Halite (NaCl) is shown to be the most effective of the salts tabulated. These levels of tensile stress are sufficient to cause the disaggregation of most rocks. Secondly, anhydrous salts (for example anhydrite, $CaSO_4$), already emplaced within the rock, may absorb water and become hydrated (to form gypsum $CaSO_4.2H_2O$). In order to accommodate the water they modify their lattice structure and expand (Table 11.3). The growth of crystals due to hydration generates **hydration pressure**. This is particularly common in desert environments, where low night temperatures cause high humidity and the formation of dew on exposed rock surfaces. Winkler and Wilhelm calculated hydration pressures for several salts and show that values of 10 MN m^{-2} are easily exceeded in many hydration reactions. Thirdly, differential thermal expansion of entrapped salts may cause rock

breakdown by thermoclasty. Cooke and Smalley (1968) point out that many of the common salts have higher coefficients of expansion than most rocks. Sodium chloride, for example, expands volumetrically three times as rapidly as granite with increasing temperature. In rocks exposed to a high diurnal temperature range this may well lead to damaging levels of tensile stress.

Rock weathering by salts is a topic that lends itself well to experimental investigation in the laboratory. Fig. 11.5a illustrates an attempt to isolate experimentally the elements of the weathering environment responsible for the rapid breakdown of a schist on an exposed coast. The experiment sought to identify whether it was the presence of water alone, or the salts dissolved in seawater, or alternations of wetting and drying which was responsible for rapid weathering. Accordingly, rock samples were immersed in either deionized water or salt water for 1 hour daily and allowed to stand in a 100% humid atmosphere, or to air-dry for the remainder of the time. Over a period of 180 days a clear distinction was revealed by the samples immersed in seawater and air-dried, as weight loss took place as a result of weathering. The treatment to which this particular set of samples was subjected permitted the activity of salt crystallization and it was concluded that this was the dominant weathering process. As a result, flakes of rock became detached and mechanical disaggregation took place. Since weathering took place from one surface only, a rate of weathering of 0.78 mm a^{-1} could be calculated – a clear demonstration of the effectiveness of salt weathering (Mottershead, 1982).

Finally, the mechanical effect of plant growth is also to exert tensile stress on rock and to promote fracture. The most significant effect of these mechanisms is the growth of roots as they penetrate joints in bedrock. Research in the field of agricultural technology has demonstrated the power of root growth. Taylor and Burnett (1964) demonstrate that penetration of roots into soil is sometimes possible with a soil strength of 2–2.5 MN m^{-2}, and always possible when soil strength is less than 1.9 MN m^{-2}. Thus, root growth is possible in materials at the lower end of the scale of the tensile strength of rocks. Taylor and Ratcliff (1964) show that advancing roots and rootlets are capable of exerting a continuous radial pressure of 1.5 MN m^{-2}, probably sufficient to prise apart already existing fractures.

Table 11.2 Levels of supersaturation (in %) of various salts in water required to generate crystallization pressures of 1 MN m^{-2} and 10 MN m^{-2} at 20°C

	1 MN m^{-2}	10 MN m^{-2}
NaCl	1.16	12.18
$MgCl_2$	1.69	18.34
$MgSO_4$	1.87	20.42
$CaSO_4$	1.91	20.78
Na_2SO_4	2.19	24.30
$CaSO_4.2H_2O$	2.28	25.33

Table 11.3 Volume increase during hydration of selected salts (after Goudie, 1977)

Salt	Hydrate	% volume increase
Na_2CO_3	$Na_2CO_3. 10 H_2O$	374.7
Na_2SO_4	$Na_2SO_4. 10 H_2O$	315.4
$CaCl_2$	$CaCl_2. 2 H_2O$	241.1
$MgSO_4$	$MgSO_4. 7 H_2O$	223.2
$MgCl_2$	$MgCl_2. 6 H_2O$	216.3
$CaSO_4$	$CaSO_4. 2 H_2O$	42.3

(a)

Air temperature (°C)

maximum	31.0	33.0	36.0	27.0	24.5	20.5	20.5	18.5	15.0
minimum	17.0	15.0	16.0	13.5	12.0	5.5	4.5	4.5	5.0

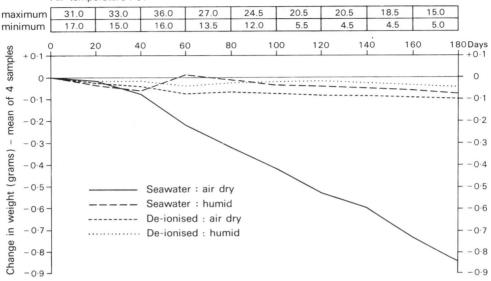

Change in weight (grams) – mean of 4 samples

Seawater : air dry
Seawater : humid
De-ionised : air dry
De-ionised : humid

Fig. 11.5 (a) The weight loss shown by four samples of schist when subjected to different weathering treatments (see text for explanation). (b) Crystallization of salts in brickwork as evidenced by salt efflorescences: the surface of the bricks has been disrupted by flaking while the mortar is largely unaffected and stands proud to produce an unusual form of differential weathering (Hurst Castle, Hampshire, England).

The below-ground biomass of most plant communities is at least equal to, and in some cases more than, that of the visible parts of the plants, so it is clear that plants are capable of expending massive amounts of mechanical energy on the bedrock below (Fig. 11.6).

11.1.3 The primary mechanisms of weathering: lattice breakdown

In order to comprehend the breakdown of the crystal lattices of rock-forming minerals we must consider the primary mechanism at the scale of the component atoms. Lattice breakdown is accomplished by chemical reaction. Such reactions occur when the chemical bonds of the reacting

Fig. 11.6 Tree roots penetrating fractured and partly weathered rock. Clearly, as the roots grow and expand radially, further tensile stress is exerted on the surrounding rock (Mallorca).

substances are broken and reformed to produce the reaction products. The breaking and making of bonds involves the transfer of energy. Reactions which proceed with the aid of a net input of energy are **endothermic** while those that proceed with a net loss of energy are **exothermic**. Most weathering reactions that occur spontaneously are exothermic, although an energy input is still required in many cases to overcome the activation barrier of the reaction (see Box 2.5). Lattice breakdown, therefore, involves the breaking of bonds in the lattice and the formation of new bonds between both the original constituents of the crystal and other reactants, but with a net loss of free energy. Because of the nature of chemical bonds (see Chapter 2), weathering reactions must involve the rearrangement of electrostatic forces within the lattice. This in turn will involve changes in the disposition of the protons and electrons, the fundamental particles responsible for these forces. Before we can see how this is accomplished, we must give more consideration to the nature of the principal reactants.

It will be recalled that in the structure of silicate minerals (see Box 5.1) silica tetrahedra form the lattice framework, which increases in three dimensional complexity from ring silicates such as olivine to the framework silicates such as the feldspars and quartz (note that in some minerals there is some substitution of silicon by aluminium in the tetrahedra). Occupying spaces of various kinds within these structures are various non-framework ions, of which the most important are iron, aluminium, magnesium, potassium, sodium, calcium and hydroxyl ions. These crystal lattices are held together by both ionic and covalent bonds. The silicon and oxygen in the tetrahedral framework of silicate minerals are bound by covalent bonds, many of the non-framework ions by ionic bonds, while elements such as iron and aluminium form intermediate bonds, showing in different proportions tendencies to be ionic or covalent. The strength of the ionic bonds in silicate mineral lattices, however, varies considerably. It is best expressed by the relationship between the charge on the ion – its *valency* – and the number of ions surrounding it, and with which the valency charge is shared – its **coordination number**. Although in theory the configuration of the tetrahedra and/or the inclusion of non-framework ions produces an electrically neutral structure, this is rarely the case in real crystals. Certain sites in the lattice remain vacant or contain impurities, while the faces and edges of crystals, which will rarely be perfect, consist of incomplete units. These facts mean that the lattice as a whole will carry a charge, usually a net negative charge.

It was stressed at the beginning of this chapter that the almost ubiquitous presence of water is one of the properties of the environment within which the weathering system operates. In the present context, water possesses two important attributes. First, because of the charge separation (polarization) of water molecules, they behave as electrical dipoles (polar molecules) (see Chapter 2). Hydrogen bonds formed between these molecules give some crystallinity even to liquid water, though this decreases as temperature rises. Hydrogen bonds also form between water molecules and other charged particles or surfaces, producing layers of oriented water molecules. This is the process of **hydration**, and the thickness of the layers depends on the density of charge carried by the particles or surfaces, which in turn are said to be hydrated. Secondly, in pure water, the molecules dissociate to a small extent into $H^+_{(aq)}$ and $OH^-_{(aq)}$ ions. (aq) stands for aqueous and tells us that these ions are hydrated – they have water molecules attracted to them. However, the hydrogen ion is a single proton and is usually regarded as being attached to one water molecule forming the hydroxonium ion (H_3O^+) or a hydrated proton ($H^+ + H_2O$):

$$H_2O_{(l)} \leftrightharpoons H^+_{(aq)} + OH^-_{(aq)}$$
$$H_2O_{(l)} + H_2O_{(l)} \leftrightharpoons H_3O^+_{(aq)} + OH^-_{(aq)}.$$

Bearing in mind these properties of silicate minerals and water, the presence of a charged mineral surface in contact with water will result in the formation of a layer of oriented water molecules surrounding it. A similar layer of preferentially oriented water molecules will be formed on internal surfaces within the lattice wherever water can penetrate, as for example along cleavage planes. The first result of this process of hydration is that, as a polar solvent such as water begins to surround the lattice, there is a weakening of the interionic (electrostatic) attraction within the lattice. This effect, due to the relative permitivity of the water, aids the detachment of non-framework ions from external and internal surfaces of the lattice. Their removal is

effected as ion–water molecule bonds are established, with a consequent release of energy. This release of energy facilitates the detachment of ions from the lattice and they diffuse as hydrated ions into the free water away from the crystal surface. This process where water is the solvent is also known as hydration, but is more generally termed **solvation**, and the ions are said to be solvated.

The effect of the presence of layers of oriented water molecules does not stop at simple hydration and limited solvation of non-framework ions exposed at mineral surfaces, but leads to the initiation of **hydrolysis**. In this process, hydrogen ions from a variety of sources, including the dissociation of water molecules, are able to penetrate the hydrated lattice as hydroxonium ions. Here the hydrogen protons compete with and tend to replace non-framework ions in the

(a)

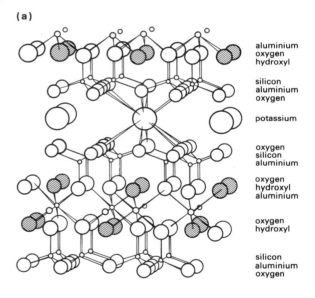

aluminium
oxygen
hydroxyl

silicon
aluminium
oxygen

potassium

oxygen
silicon
aluminium

oxygen
hydroxyl
aluminium

oxygen
hydroxyl

silicon
aluminium
oxygen

(b)

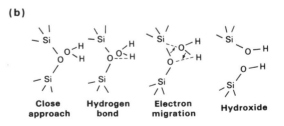

Fig. 11.7 (a) Lattice structure of muscovite; note the potassium ion with a coordination number of twelve. (b) Silicate hydration (after Curtis, 1976).

crystal structure. These displaced ions then pass into solution as hydrated cations. When this happens and the lattice begins to become saturated with hydrogen, it is no longer stable and the mineral may disintegrate, although initially it may retain its silicate structure (Box 11.1).

In both hydration and hydrolysis the electrostatic forces of the lattice are disturbed and rearranged by the hydrogen ion which, remember, is a single proton. In the only other processes of lattice breakdown the electrostatic forces of the lattice are disrupted by the transfer of the electron, the other elemental particle.

Substances that lose electrons are oxidized, while those gaining electrons are said to be reduced. Processes involving the transfer of electrons are known as oxidation–reduction (or **redox**) reactions. The oxidation state of an element describes the charge that an atom of that element appears to have in a compound and this reflects its electron configuration. Now, many elements can exist in more than one oxidation state. For example, iron has oxidation states of +2 and +3 and, in the compounds it forms in many rock-forming minerals such as olivine (where iron is a non-framework ion), it exists in the lower

---

Box 11.1
HYDROLYSIS OF SILICATE MINERALS

Because of their size there is almost a one-to-one relationship between the water molecules and the oxygens of the tetrahedral units of the lattice at the crystal/water interface. The single unsatisfied negative charge of each oxygen will, therefore, be in proximity to the double positive charge of each oriented water molecule. This produces an accumulation of excess positive charge at this interface. Now some of these water molecules will be the 'hydration' water of hydrogen ions, that is they will be hydroxonium ions (H_3O^+) and the extra proton will add further to this excess positive charge. In order to correct this charge imbalance, these hydrogen ions (protons) and those from both additional dissociation of water molecules and from other sources tend to penetrate into the lattice, passing from one water molecule to another, each of which temporarily behaves as a hydroxonium ion.

Within the lattice, hydrogen ions (protons) compete with non-framework ions (such as the potassium ion seen in the muscovite lattice in Fig. 11.7a) to neutralize the charge on the silicate framework. Each valency charge on the non-framework ions, however, is often shared with a relatively large number of coordinating ions (e.g. 12 in the case of potassium in Fig. 11.7a) so that the bonding is weak. Because of its small size, the hydrogen ion can coordinate with

only one or two of these neighbouring oxygens, and hence its single charge is shared to a much smaller extent and the bonds formed are therefore stronger. Not surprisingly, then, in competition hydrogen ions will tend to replace non-framework ions. These diffuse as hydrated cations through the lattice to the free water beyond, ultimately leaving the silicate framework at least partially saturated with hydrogen. However, the mineral is now unstable, for small units of the tetrahedral framework of the lattice individually have their negative charge imbalance neutralized by hydrogen, and they are no longer ionically bonded to other units of the lattice. In addition some of the oxygens of the silica tetrahedra may carry unsatisfied excess negative charges, thereby setting up forces of electrostatic repulsion as like charges repel. Therefore, the overall effect is the disintegration of the lattice structure of the original mineral, though the covalently bonded silicate units may initially remain intact as detached but hydrated chains, sheets or individual tetrahedra. However, these too may be disrupted by the breaking of the O–Si–O bond (the fundamental structural unit of the silicate framework) as the hydrogen bonding of hydration water changes through electron migration ultimately to produce silicic acid which passes into solution (Fig. 11.7b).

oxidation state, as ferrous iron Fe++ (iron II). The transfer of an electron from iron II to an electron acceptor such as oxygen converts (or oxidizes) iron II to iron III (ferric iron), distorting the lattice and leading to the disintegration of the crystal. Here oxidation is an enacting process of lattice breakdown in an iron-bearing silicate mineral and, in other non-silicate minerals such as pyrites (iron sulphide), oxidation is even more effective as an enacting process of weathering. Redox reactions, however, are also important as secondary reactions associated with lattice breakdown. This is because the solubility of such elements as iron varies with the oxidation of their ions. In its ferrous (iron II) state iron is soluble and mobile, but when oxidized to the ferric (iron III) state it is precipitated, and is insoluble. Consequently, ferrous ions released by hydrolysis from silicate structures may be oxidized immediately to the iron III state, and hence rendered immobile. The nature and proportions of the oxidizing and reducing substances in the weathering system or in the soil determine the redox potential (Eh), which is a measure of the tendency of the system to receive or supply electrons.

11.1.4 Activating agencies of lattice breakdown

We have already seen that fracture-activating processes (Fig. 11.3) function as regulatory mechanisms which, together with the physico-chemical properties of the rock and its minerals, control the type and rate of process active in weathering. The same is true of lattice breakdown (Fig. 11.8). It begins as a surface reaction at the mineral/water interface. Therefore, any process that increases the rate at which mineral surfaces are exposed will be an activating agency of lattice breakdown. At the macro-scale, the amount of exposed mineral surface will depend on the mechanical properties of jointing and porosity, while the rate at which surfaces are produced will reflect the effectiveness of fracture and mechanical disintegration, which increases the rock's exposed surface area. At the micro-scale, the ratios of surface-area/volume for particular minerals, the existence of cleavage planes and the voids within the lattice become important and reflect the properties of the crystal itself. Processes which remove the products of lattice breakdown, such as leaching and nutrient uptake by plant roots, will also be important activating agencies, because

the removal of products from a weathered mineral surface will expose fresh surfaces of the lattice to chemical reaction. Removal of soluble products, particularly the hydrated non-framework ions released by weathering, depends on the maintenance of concentration (or diffusion) gradients (actually gradients of electrical potential) between the mineral and the free water. In this context it is essential that the voids and pores in soil and regolith are regularly flushed by percolating water.

The removal of reaction products, particularly in solution by leaching, is important for another reason. In any chemical process with a given active mass of reactants, there is an inherent negative feedback at work which brings the reaction to a point where no further chemical change takes place. This point of chemical equilibrium is defined by the thermodynamic equilibrium constant of the reaction and it is conditioned by the energy transfers involved. To prevent this negative feedback and to stop the reaction reaching equilibrium, there must be a continuous replenishment of reactants as well as a continual removal of reaction products.

We have already considered how the supply of fresh mineral surfaces is maintained, but what of the other reactants? In *hydration* and *hydrolysis* the active agents of lattice breakdown are the water molecule and the hydrogen ion or proton dissociated in water. Processes that ensure the movement of water through the weathering zone will, therefore, activate and promote lattice breakdown. Natural waters (rain, soil and groundwater), however, are not pure; they are aqueous solutions containing molecules and ions of other elements. Furthermore, natural waters are normally acidic (if only slightly), that is, their hydrogen ion concentration is higher than pure water (Box 11.2). But just as the dissociated hydrogen ions in pure water are balanced by hydroxyl ions, so in natural waters the excess of hydrogen ions must be balanced by other anions. The source of these excess hydrogen ions, and their balancing anions, is the dissociation of acids in water. Processes that deliver these acids to percolating water are, therefore, highly significant activating agencies in weathering which, by increasing hydrogen ion concentrations, promote the breakdown of lattice structures by *hydrolysis*.

By far the most important process in this category is the dissolution of carbon dioxide to give carbonic acid, which dissociates as the hydrogen ion (H$^+$) and the bicarbonate ion (HCO$_3^-$):

$$H_2O + CO_2 \leftrightharpoons H_2CO_3 \leftrightharpoons H^+ + HCO_3^-.$$

Some atmospheric carbon dioxide will dissolve directly in rainwater, but the major source of carbonic acid is carbon dioxide photosynthetically withdrawn from the atmosphere and released again by root respiration and by the respiration of detritivores and decomposers during the decomposition of organic matter (see Chapter 19) to become dissolved in percolating waters. In addition, organic compounds released into solution (such as simple carbohydrates) undergo further oxidation and produce carbon dioxide. Other organic compounds – the organic acids – dissociate in water to yield hydrogen ions directly. The decomposition of organic matter is, therefore, an activating agency of the most profound significance, and, as we shall see, its regulatory role does not stop at the provision of hydrogen ions for hydrolysis.

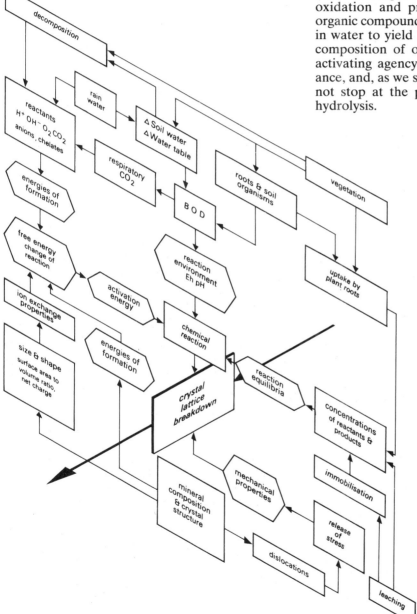

Fig. 11.8 The activating processes of lattice breakdown.

Box 11.2
ACIDITY OF NATURAL WATERS

The acidity of rainwater and stream waters is associated with the presence of hydrogen ions derived from the dissolution of carbon dioxide, sulphur dioxide and the dissociation of organic acids. The presence of aluminium ions also has the effect of increasing the hydrogen ion concentration. In soils, which can be thought of as a solid phase dispersed through a liquid phase – i.e. a soil/water suspension – hydrogen and aluminium ions adsorbed on the solid phase (the exchange complex) and the existence of an equilibrium solution of hydrogen ions in the soil water are responsible for the **reserve** and **active acidity** of the soil, respectively. Therefore, acidity can be measured in terms of the hydrogen ion concentration (strictly activity) expressed as pH.

pH is defined as the negative logarithm of hydrogen ion activity (or effective concentration in g ions l^{-1})

$$pH = -\log_{10} [H^+].$$

In pure water the hydrogen ion and hydroxyl ion activities are equal and have a value of 10^{-7} g ions l^{-1}. Therefore, the pH of pure water is 7. As the hydrogen ion activity increases, the pH value falls. In most natural soils the pH value lies in the range 4–8. However, the chemical definition of pH which refers to simple aqueous solutions is complicated by the presence of charged solid particles in soil water suspensions which give to the soil the capacity to resist changes in its active acidity by ion exchange – the so-called **buffer capacity** of the soil.

The other fundamental enacting process of weathering concerns redox reactions, which are regulated by the presence of oxidizing and reducing agents and by the redox environment (Eh). The principal variable here is the degree of aeration of the weathering zone, so that processes and conditions which maintain the availability of free oxygen will activate oxidation reactions, while conversely, those producing anaerobic conditions will promote reduction reactions. The permeability and drainage relationships of the rock and regolith relative to the water table and the quantity and characteristics of precipitation input will be important regulators of aeration, while the diffusion coefficients of oxygen in air and water will determine the degree of oxygenation of the reaction environment. In addition, the biological oxygen demand (total requirement of respiratory oxygen) will also affect oxygen availability and, even in aerated soils and regoliths, it can produce localized reducing conditions. Conversely, the radial loss of oxygen from roots can initiate localized oxidation, as is evident in the mottling of water-logged soils.

The redox environment is important not only in regulating active weathering reactions but also in controlling the mobility and hence the removal of some weathering products (Fig. 11.9). Such elements as iron and manganese are markedly more mobile in their lower reduced valency state – Fe^{2+} (Fe II, ferrous ion) and Mn^{2+} (Mn II, manganous ion) – and pass into solution under reducing conditions. Iron, in its Fe^{3+} (Fe III, ferric) valency state, together with aluminium, silicon and titanium, precipitates to form insoluble compounds, mainly complexed oxides and hydroxides, in the normal pH and Eh environments of weathering. In the absence of reducing conditions or extremes of pH, the removal of these elements appears to depend on the supply of organic complexing agents released by the decomposition of organic matter. These are known as chelating agents and they consist of organic molecules, particularly of phenolic compounds capable of forming at least two coordinate covalent bonds with a central metallic ion to form a complex chelate. The charge on the complex may be positive, negative or zero, but it carries into solution such metallic ions as iron and aluminium which would not otherwise be soluble. **Chelation** is often cited as an active weathering process, but its significance lies not in its contribution to lattice breakdown but in its ability to remove elements already partially released by hydrolysis, which would otherwise accumulate as insoluble products. As such, it is a leaching, erosion or pedogenic process, and in the weathering system it functions as an activating agency and regulator.

11.1.5 Resynthesis and the formation of new minerals

The immediate result of lattice breakdown is the appearance of stable but highly mobile cations such as calcium, magnesium, potassium and ferrous iron in aqueous solution. Some silicon (as the monomolecular silicic acid, H_4SiO_4) and a limited amount of aluminium will also exist in solution, though most will be present as colloidal particles. The larger of these particles will partially retain the silicate structure of the original lattice, usually as chains of tetrahedra. The chemical environment within the disintegrating lattice will be alkaline because of the high concentration of basic cations, but as they and the silicon and aluminium in solution diffuse into the soil water, the environment becomes more acid. It is at

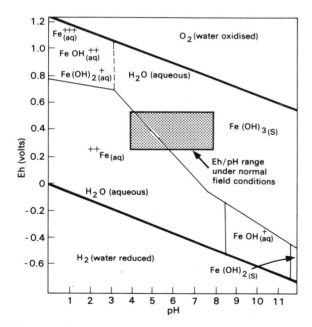

Fig. 11.9 Stability areas of the Fe II–Fe III aqueous system. Note that under normal field conditions insoluble ferric compounds would occur.

this point that the first category of new minerals appears.

As we have seen already, iron and aluminium, and to a lesser extent silicon, are soluble only under reducing conditions and/or extremes of acidity and alkalinity. In the oxidizing and mildly acid environment of aerated weathering zones they are immobile. Iron and aluminium, therefore, are precipitated as complexed oxides and hydroxides (e.g. goethite $FeO.OH$, and gibbsite $Al(OH)_3$) which also usually contain some silicon as an impurity. These form amorphous or microcrystalline particles, normally but not always of colloidal dimensions ($<2\ \mu m$). They tend to carry a net positive charge in acid environments, and may be adsorbed to form coatings on the fragments of the silicate framework of the original mineral.

The second group of new minerals begins to form as a consequence of the same processes. Monomolecular silicon in solution becomes polymerized (see Chapter 7) under the more acid environment and precipitates together with much of the aluminium to give, as we have just seen, mixed aluminium hydroxide and silicic acid particles. At the same time silicon and aluminium are present together in the residual silicate framework fragments. These two entities then form the precursors of a group of new or secondary minerals (crystalline hydrated aluminosilicate minerals) known as clay minerals (Box 11.3). They are sheet silicates (phyllosilicates) with a structure similar to the micas (Fig. 11.10). The mechanisms by which aluminium and silicon in weathering residues become reorganized to form the silica tetrahedral and aluminium octahedral layers of the developing clay minerals is complex and not fully understood. In some cases, however, as in regoliths where the unaltered rock contained a large proportion of sheet silicates such as biotite and muscovite, relatively little modification of the lattice is sufficient to convert these micas to secondary clay minerals. The type of clay mineral produced, however, depends on the ions available in the regolith to be incorporated in the lattice as non-framework ions. As Fig. 11.11 indicates, this depends on the intensity of the two processes which control the rate at which these ions (and the precursors of the aluminosilicate sheets) become available and the rate at which they are removed, i.e. lattice breakdown and leaching, respectively. Indeed, the combination of these processes can lead in time to the conversion of one type of clay mineral to another, e.g. illite to montmorillonite to kaolinite.

The properties of clay minerals differ from those of primary minerals in several important respects. The particles they form are smaller, giving them colloidal properties; their composition is more heterogeneous, even within the same crystal, and at least some have expanding lattices (Box 11.3) giving them enormous surface areas in contact with soil water. Clay particles carry a net negative charge (surface charge), partly originating in unsatisfied charges within the silicate and aluminium sheets (permanent charge) and partly due to the dissociation of hydrogen ions and to other broken bonds (pH-dependent charge) (see Box 11.4). This surface charge is compensated for by a layer of oriented water molecules and adsorbed ions, largely cations (volume charge) (Fig. 11.12). Cations adsorbed on clay minerals in this manner are protected against removal by leaching, and the process of ion adsorption is an important regulator of both leaching loss to the solute load of drainage water and of nutrient availability to plants. Both are controlled by the complex process of cation exchange.

In arid environments a third type of new mineral is formed, in addition to clay minerals, for here leaching does not carry away excess ions. Instead,

---

Box 11.3

CLAY MINERALS

There are many different kinds of clay mineral, but all form very small, flat lamellar crystals with a very large surface-area:volume ratio and resemble the micas, such as muscovite and biotite. They are hydrated aluminosilicate minerals with a lattice structure composed of two basic units: the silica tetrahedron and the aluminium octahedron. In the first the silicon atom is equidistant from, and at the centre of, four oxygen atoms in tetrahedral coordination, and in the second, aluminium atoms are equidistant from, and at the centre of, six oxygen atoms or hydroxy ions in octahedral coordination.

These basic units join by covalent bonding to form tetrahedral and octahedral sheets, respectively. In the lattice structure of clay minerals these sheets are combined in various ways to form layers. In minerals such as kaolinite the two sheets are hydrogen bonded in the ratio 1:1, tetrahedral sheet:octahedral sheet. This structure does not swell on wetting, as hydration water molecules cannot penetrate the lattice layers. In 2:1 lattice clay minerals, such as vermiculite and montmorillonite, a single octahedral sheet lies between two tetrahedral sheets. The result is that the mineral swells on wetting as water of hydration penetrates between successive 2:1 layers, carrying with it

magnesium and calcium, respectively. In the hydrous mica clay minerals such as illite the 2:1 layers are held together by potassium ions with no hydration water in the interlayer space. With loss of this structural potassium, illite weathers to montmorillonite.

One of the most important properties of clay minerals is that they carry a large surface charge. This is a net negative charge and much of it is a **permanent charge** originating within the lattice structure by **isomorphic substitution**. Here atoms or ions of similar size but lower valency replace some of the silicon and aluminium in the tetrahedral and octahedral sheets, without straining their structure, leaving unsatisfied negative valency charges associated with the oxygens and hydroxyls. In addition broken bonds occur, particularly at the imperfect edges of the crystals, but these charged sites (which may be either positive or negative) are pH-dependent. The surface charge of clay minerals means that they can attract and adsorb on their large external and, in the case of expanding 2:1 lattice clays, internal surfaces a swarm of cations. This phenomenon of **cation adsorption** accounts for their importance in **ion exchange** processes in soils and regoliths.

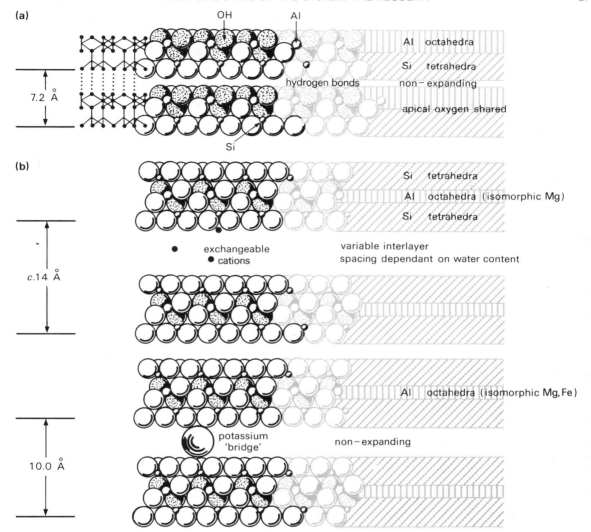

Fig. 11.10 Crystal structure of (a) 1:1 and (b) 2:1 lattice clay minerals.

they are precipitated in the form of salts such as carbonates and sulphates of calcium and sodium as soil water evaporates. The mechanical effects of such salt crystal growth have already been discussed.

11.2 The final state of the system: the regolith

As a response to the primary mechanisms of weathering, the final state of the system will be characterized by the following.

1. A group of minerals which have proved resistant to dissolution, acid hydrolysis and redox reactions. These will exist as grains released by the disaggregation of the rock, but will be little altered chemically and structurally. In 1938, Goldich suggested that the persistence of primary silicate minerals under weathering was the reverse of their order of formation as predicted by the Bowen series (see Chapter 5). This order of persistence is in broad agreement with the energy change undergone during weathering, so that minerals with a high loss of

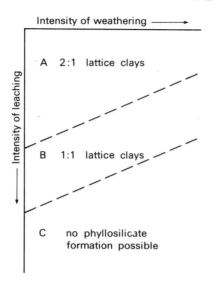

Fig. 11.11 The type of secondary mineral formed in relation to the weathering/leaching ratio.

free energy on breakdown are less stable than those with a small negative free-energy change. Free-energy change correlates with the bond energy and the energy of formation of primary minerals, thereby providing the thermodynamic link with Goldich's empirical weathering sequence (this does not hold, however, for minerals where oxidation is the prime weathering reaction). On both thermodynamic and empirical grounds, quartz (crystalline silica) appears to be the most resistant mineral under most weathering regimes and it tends to dominate this residual category (Fig. 11.13).

2. Resynthesized or secondary minerals, of which the hydrous oxides and hydroxides of iron and aluminium, the clay minerals and, in some environments, accumulations of soluble salts, are the most important. At any particular moment, however, most regoliths will contain also materials which represent transient or intermediate states of the system (see Fig. 11.2).

Box 11.4
POINT OF ZERO CHARGE AND THE ISOELECTRIC POINT

The edge or prism surfaces of sheet silicates consist of aluminium and silica groups as hydroxide and oxides, and involve the breaking of the primary valence bonds of the sheets and the equilibriation of these exposed groups with the surrounding soil solution. Exposed Si^{4+} and Al^{3+} ions conserve their coordination by the adsorption of H^+, or OH^-, or H_2O. In very acidic solutions hydrogen ions are adsorbed on to the oxide surfaces, making the surface positive and allowing anion retention. In alkaline conditions hydrogen ions are removed into solution so that the surface becomes negatively charged, and allowing the surface adsorption of cations. At an intermediate point, the **point of zero charge (PZC)** neither cations nor anions are held significantly by this mechanism.

The point of zero charge is not necessarily at pH 7, and in fact for the sheet silicates the presence of Al^{3+} ions close to the surface facilitates the loss of H^+ ions into solution, and the point of zero charge (PZC) occurs at lower pH values.

For aluminium oxides the PZC is in the region of pH 9, and for silicates the PZC lies between pH 1 and 2. The edges of clay crystallites can be regarded as mixed aluminium silicates, though it is more appropriate to talk of the **isoelectric point** in the case of crystallites, for the charge on their edge surfaces is never actually zero; it is instead the net charge that is zero. The isoelectric points of mixed oxides depend on the ratio of silicon to aluminium, and when it is 50% the isoelectric point is at pH 6.5. Edge surface effects are of greatest significance in the kaolinites, which have low permanent charges (2–5 me 100^{-1} g) and high edge/planar ratios (1:10–1:5). Conversely edge charges are unimportant in the 2:1 clay minerals such as montmorillonite, with high permanent negative charges (40–150 me 100^{-1}g) and which exist as smaller crystals offering smaller edge/planar ratios. The charge characteristics of soil clays may be considerably modified from that of pure clay minerals because of the presence of thin

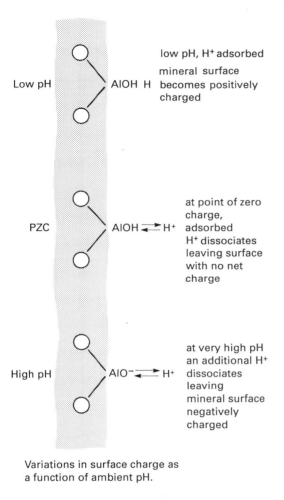

Low pH

AlOH H

low pH, H⁺ adsorbed

mineral surface becomes positively charged

PZC

AlOH ⇌ H⁺

at point of zero charge, adsorbed H⁺ dissociates leaving surface with no net charge

High pH

AlO⁻ ⇌ H⁺

at very high pH an additional H⁺ dissociates leaving mineral surface negatively charged

Variations in surface charge as a function of ambient pH.

films of the sesquioxides, with PZC values in the range 7–9, and hence normally positively charged in the soil, adsorbed on the planar surfaces of the clays.

The implication for soil clays is that over the whole pH range of most natural soils, and especially over the pH range of interest for plant growth the edge surfaces of clays carry both positive and negative charges, and can therefore retain both anions and cations. Free oxides, especially of Fe and Al, also exhibit pH-dependent charges due to the adsorption of potential determining H⁺ ions. Again the pH at which the net charge is zero is the PZC. Other components of the soil, such as humus and iron III oxide, also have equilibria between their surfaces and the soil solution which affect their ability to hold ions. For humus the charge is completely dependent on pH, due to the dissociation of carboxyl and phenolic groups, and from pH 3 upwards the charge is negative and augments the permanent negative charge of the clay minerals. For humus then, the sequence

$$R-CO_2H_2^+ \leftrightarrow R-CO_2\dot{H} + H^+ \leftrightarrow R-CO_2^- + 2H^+$$

is important and, since $R-CO_2^-$ ions are readily formed at the pH values prevalent in soils, humus is important in retaining cations. In principle very acid conditions would eventually encourage the formation of $R-CO_2H_2^+$, and hence anion retention. In soils of high humus content the effect is to considerably increase the pH dependence of the soil's CEC. (See also Chapters 21 and 22.)

Therefore, in addition to the two categories above there may also be a) some minerals present as individual grains or aggregates of grains which are still actively undergoing chemical breakdown; and b) rock fragments (clasts) of various sizes and shapes which have resulted from mechanical fracture but which have not yet undergone significant chemical change.

These four categories differ considerably in particle size. The quartz residue, for example, largely exists as particles in the range 0.02–2.0 mm

(sand-size fraction); the secondary oxides and clays are all smaller than 0.002 mm and most are colloidal; mineral grains actively undergoing weathering, though variable in size, largely fall in the range 0.02–0.002 mm (the silt-size class); and finally, the clastic fragments can vary from 2.00 mm to sizeable stones and small boulders. It is the relative proportions of these size fractions that determine the texture of the regolith. The more advanced weathering is, and the closer it approaches the final state, the more it will be dominated by the quartz and clays and the more bimodal its texture will become. In other words, the nature of the regolith as it approaches the final state of our model is governed by elapsed time.

The changes wrought by the operation of the weathering system include a reduction in induration, or a change from the massive to the clastic state, in addition to the chemical changes. These inevitably cause changes in the mechanical properties as the initial rock changes its state. Dearman *et al.* (1978) describe the changes which affect a granite on weathering. Of the three main constituent minerals, the quartz remains essentially unchanged. The biotite becomes bleached as iron is removed in solution to form chlorite and other clay minerals. Feldspars become altered to kaolinite and associated clay minerals. The rock passes through a series of physical states, from the original sound rock, through stages of discoloration indicating the removal of solubles, to decomposed rock. In the latter state, the original structure is still visible in the form of individual crystals and macrostructures and as joints and dykes. Yet the substance of the material has

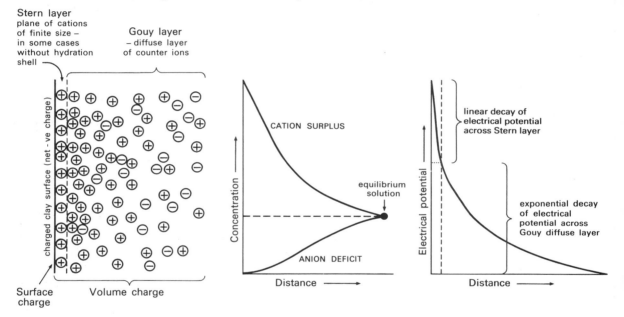

Fig. 11.12   The ionic distribution and electrical potential gradients at a negatively charged planar clay surface.

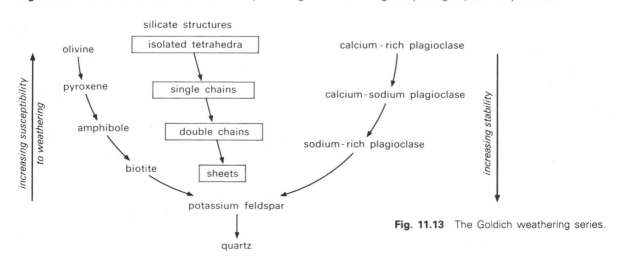

Fig. 11.13   The Goldich weathering series.

altered and what remains are unaltered quartz grains set in a matrix of clay minerals which have replaced, yet retain the outlines of, the original parent minerals. As breakdown takes place and mobile products are removed, so porosity increases and mechanical strength decreases. These various grades of weathering are set out in Table 11.4. The relationships between mechanical strength, weathering grade and porosity are illustrated in Fig. 11.14.

We can, however, substitute space for time and represent the system as a vertical series of horizontal zones forming a profile, from the surface to unweathered bedrock at depth. This, of course, is the traditional approach to the description of

(a)

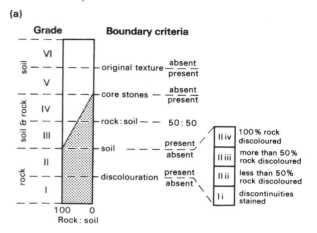

(b)

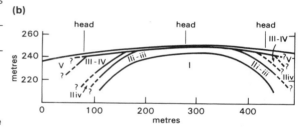

Fig. 11.14 (a) Sequence of weathering grades of granite – an engineering classification based on field properties. (b) Cross-section through a quarry face showing distribution of weathering grades. (After Dearman *et al.*, 1976.)

Table 11.4 (a) Characteristics of granite at different grades of weathering (after Fookes *et al.*, 1971)

Grade	Description	Physical characteristics
VI	soil	completely weathered, original structure and fabric lost
V	completely weathered	completely weathered, friable, original structure and fabric intact
IV	highly weathered	rock discoloured, discontinuities extensive <50% original rock intact
III	moderately weathered	rock discoloured, discontinuities open, >50% original rock intact
II	slightly weathered	discoloration by bleaching or staining, discontinuities opening
I	fresh rock	no sign of weathering

(b) Material properties of granite at different weathering grades (after Dearman *et al.*, 1976)

Grade	Material	Uniaxial compressive strength (kN m^{-2})	Effective porosity (%)
I	fresh granite	246	0.11
IIii	partly discoloured	219	0.57
IIiv	completely discoloured	165	1.52
V	intact soil	3.5	9.98

the weathered mantle or regolith, and in the upper zones it becomes the classic model of the soil in terms of the soil profile and soil horizons (Fig. 11.14 and Chapter 20).

In this model it is assumed that the zones nearer the surface, in closer proximity to inputs from the atmosphere and biosphere, will be in a more advanced state of adjustment. Conversely, zones at depth will show less alteration and resemble the original bedrock most closely. Intermediate zones will, therefore, represent various states of transition. Following this gradient in the states of the regolith are parallel gradients in physico-chemical conditions and properties (including particle size) and in the combination and relative significance of weathering processes. The model is not entirely static, for with time the depth of the most altered zone increases in relation to the rest of the regolith as a weathering front advances down through the profile.

As we have seen, the upper part of this vertical

model is the soil profile, and it is really impossible to separate weathering from pedogenesis, except in the few environments that are totally devoid of life. Under these pedogenic influences, therefore, the regolith will be further modified by the incorporation of organic matter and the activity of soil organisms, by aggregation and the formation of organomineral complexes, and by translocation and horizon development (see Chapter 22).

11.3 The generalization of the model: other rock types

So far, the model of the weathering system has been restricted to igneous rocks. It can be generalized to cope successfully with both sedimentary and metamorphic rocks, by the inclusion of what are essentially feedback loops. One of these loops, which operates within the weathering system itself, has already been considered implicitly. This is the pathway whereby the products of weathering are resynthesized to produce new minerals such as clays, which can then experience renewed or continued weathering *in situ* to produce a series of possible final states of the mineral, depending on the intensity of weathering and/or the length of time over which it is effective. This concept can be extended to incorporate cyclic pathways outside the weathering system itself.

As we have seen in Chapter 10, the products of weathering are ultimately destined to be removed and deposited by the processes of denudation. Mobile products in solution can be transported rapidly and continuously, and their denudation is separated in time from that of the residual products forming the regolith, the rate of removal of which is conditioned by the slower processes of mass movement on slopes. Eventually, after deposition and sedimentation, perhaps followed by lithification, these weathering products, now components of sediments or sedimentary rocks such as mudstones, shales and sandstones, may be exposed to another cycle of weathering after uplift and exhumation. Such a cycle can be modified to include a diversion through living organisms and the genesis of biogenic sediments incorporating siliceous or carbonate skeletal remains. It can also be extended to include metamorphism, not only of sediments but also of the igneous rocks.

In fact, such an approach is the geochemical model developed in Chapter 5 when considering the crustal system and its relationship to the denudation system.

The generalization of the model to other rock types in this way must necessarily take into account the differing mineralogical composition and mechanical structure of those rocks. There is a very wide range of variability in the properties of both sedimentary and metamorphic groups. Factors such as porosity, permeability, the degree of fracturing and induration all influence the access and passage of weathering agents (Fig. 11.15).

Sedimentary rocks generally include a lower proportion of primary rock-forming minerals than the igneous rocks from which they were originally derived. Therefore, the most abundant minerals – quartz and white mica – in sedimentary rocks are also the most resistant. On the other hand, less chemically stable minerals such as feldspar are less abundant. Arenaceous sedimentary rocks are commonly indurated with a cement consisting of a metallic compound (e.g. Fe_3O_4, $CaCO_3$) which is more susceptible to weathering than the resistant detrital grains. The weathering of such rocks commonly involves the chemical breakdown of the cementing material by, for example, redox reactions or solvation, and the release of the constituent detrital grains. In the case of argillaceous sedimentary rocks – mudstones and shales – the resistant detrital grains are usually set in a matrix of finely divided material of secondary origin (resynthesized minerals) with a high proportion of base cations. Such material is of course potentially highly vulnerable to weathering reactions, yet the rate of weathering may be strictly limited by the restricted access of water in materials of such low permeability.

Sedimentary rocks consisting almost entirely of soluble minerals constitute a special case. Limestones are the most important of this group; less commonly exposed is rock salt. Here weathering proceeds by the direct solvation of the bulk of the rock, leaving only a small residue of insoluble impurities.

Metamorphic rocks possess such great variety that it is difficult to make valid generalizations. Suffice it to say that many contain minerals that are structurally similar to the primary minerals of igneous rocks, as well as some relatively rare minerals which are potentially unstable under most weathering regimes.

Fig. 11.15 Lithology and weathering; different rock types react in different ways to weathering processes. (a) Massive limestone – joints have opened up as a result of solution of the calcium carbonate composing the rock (Malham, Yorkshire, England). (b) Weakly cemented sandstone – weathering of the cement produces disintegration of the rock; less resistant lenses in the faced blocks have also been etched out (Taunton, Somerset, England). (c) Regolith formed from massive dolerite – weathering and formation of new minerals have taken place along former joints, leaving massive corestones of unaltered rock set in a matrix of weathered material; the exposed face is several metres high (North Queensferry, Fife, Scotland).

11.4 Exogenous variables and the control of weathering

The process–response model (see Fig. 11.2) of weathering which we have been considering, applies not only to all rock types but also in all environments. It will have become evident, however, that the activation and regulation of weathering processes are ultimately conditioned by the exogenous (input) variables to the model. It is these variables that define the conditions of existence under which the system operates. In the weathering system, as in the denudation system as a whole, these variables are climate, lithology, vegetation and time, and the control they exert at any site will be modified by its position in a landscape, particularly in relation to slope. That the model can accommodate the lithological variable has already been demonstrated, and the effects of climatic and biotic controls at a global scale, are summarized in Fig. 11.16 and Table 11.5. The effects of both vegetation and climate have

an important manifestation through the control they exert on the hydrological relationships of the weathering environment: that is to say the disposition of the locus of weathering in relation to the movement of water. Table 11.6 shows a classification of hydrological zones in relation to conditions that affect weathering.

Weathering of rock in the subaerial zone occurs only where the rock is exposed at the surface, and for any given rock its rate will depend on the nature and quantity of the reactants available. Weathering is probably most intense in the percolation zone, where the renewal of reactants and removal of weathering products is continuously facilitated. This zone is also subject to considerable fluctuations in temperature. In the subaqueous zone weathering is probably limited; since in this zone the voids are permanently waterfilled and conditions are anaerobic and reducing. Furthermore, the groundwater will be highly charged with solutes derived both *in situ* and particularly from above. Under these conditions reaction rates will

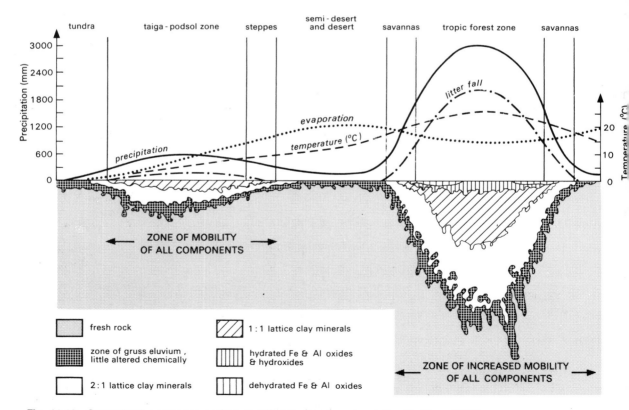

Fig. 11.16 Generalized weathering profiles in different climatic zones (after Strakhov, 1967).

Table 11.5 Geochemical types of rock weathering related to climatic conditions (after Lukashev, 1970)

Type of residual weathering product	Geochemical nature of the process	Weathering environment and solute transfer conditions
skeletal, clastic	formation of mixture of debris; slight removal of solutes	low temperature, slight chemical and biological breakdown of rocks
siallitic–argillaceous (iron-pan type)	SiO_2 and Al_2O_2 hydrate, mixtures formed with accumulation of SiO_2 in podzol horizons and removal of Al_2O_3 and Fe_2O_3 to underlying horizons, leaching of such elements as Cl, Na, Ca, Mg and K	moderate humidity and temperature; active organic and humic acids; downward migration of solutes
siallitic–carbonatic (calcrete type)	silica, iron and aluminium hydrates formed, together with accumulation mainly of calcium carbonate, but also Mg, K and Na carbonates	Mediterranean and related semi-arid, seasonal climates; organic and humic activity; both upward and downward migration of solutes
siallitic–chloride–sulphate type (gypsum type)	formation of hydrated weathering products (siallites); high mobility of SiO_2, accumulation of chloride and sodium, calcium and magnesium sulphates	warm, arid conditions; upward migration of solutes dominant; greatly reduced organic activity
siallitic–ferritic and allitic (ferricrete–bauxite type)	accumulation of iron and aluminium with general loss of silica and more soluble elements	hot, wet climates; widespread leaching and migration of solutes

Table 11.6 Some characteristics of the principal hydrological zones (after Keller 1957)

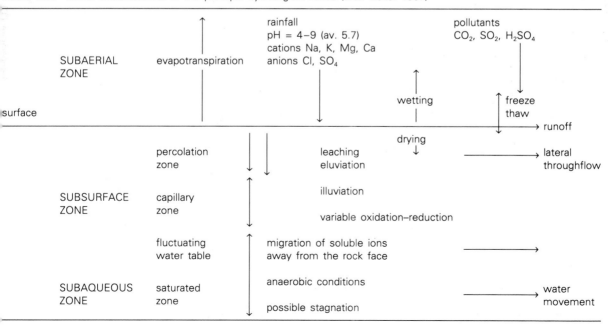

be slow and the slow rate of removal of weathering products by groundwater seepage will limit the rate of weathering.

11.5 Conclusion: further perspectives on weathering

Weathering has traditionally been separated from erosion, but such separation is misleading and has caused some confusion. Weathering is not restricted to material *in situ*; it also affects debris being transported actively. Furthermore, as we have seen, the operation of weathering processes is partially regulated by the removal of weathering products in order to prevent the establishment of equilibrium conditions. Here erosion and transport are seen to be integral processes in the weathering system, and this particularly applies to leaching. Because of its association with soil and the two-dimensional model of the soil profile, it has been the vertical component of movement that has been stressed in relation to the leaching process. However, leaching and erosion by solution are really synonymous. The significance of through-flow in the vadose zone to the solute load of streams is now appreciated, and this is therefore directly related to weathering on the contributing slopes of their catchments.

The debris cascade is only one perspective on the weathering system. The ecologist's view is different. In his or her eyes, weathering, pedo-genesis and ecosystem function are intimately linked and, as we have seen, many of the rate-limiting regulators and thresholds that control the operation of the weathering system are at least indirectly biological. The same is true for the control of the throughput of weathering products on slopes. Indeed, the ecosystem operates to conserve and regulate not only the release of nutrient ions but also their circulation and loss to the denudation system. The functional link between weathering, soil, the ecosystem and denudation clearly falls within the domain of the slope system and the processes operating within it.

Further reading

The original approach adopted in this chapter makes the recommendation of a comprehensive text difficult, although many useful texts exist on different aspects of weathering. A sound introduction, particularily from a geomorphological point of view, is provided by:

Birkeland, P.W. (1984) *Soils and Geomorphology*. Oxford University Press, New York.
Ollier, C.D. (1984) *Weathering*. Longman, London.

Basic texts on the processes of chemical weathering are:

Keller, W.D. (1957) *The Principles of Chemical Weathering*. Lucas, Columbia.
Loughnan, F.C. (1969) *Chemical Weathering of the Silicate Minerals*. Elsevier, New York.

A thermodynamic approach to weathering is provided by:

Curtis, C.D. (1976) Chemistry of rock weathering: fundamental reactions and controls, in *Geomorphology and Climate*, (E. Derbyshire (ed)). John Wiley, Chichester.
Ross, S.M. (1987) Energetics of Soil Processes, *Energetics of Physical Environment: Energetic Approaches to Physical Geography* (ed K.J. Gregory). John Wiley, Chichester.

Other useful sources are:
Mottershead, D.N. (1982) Coastal spray weathering of bedrock in the supratidal zone at East Prawle, Devon. *Field Studies*, **5**, 663–684.
Paton, T.R. (1978) *The Formation of Soil Material*. Allen & Unwin, London.
Trudgill, S.T. (1977) *Soil and Vegetation Systems*. Oxford University Press, Oxford.
Whalley, W.B. and J.P. McGreevy (1985, 1987, & 1988) *Weathering. Progress in Physical Geography*, vols. 9, 11, 12. pp. 559–581, 357–369, 130–143.
Wilson, R.C.L. (1983) *Residual Deposits: Surface-Related Weathering Processes and Materials*. Geol. Soc. Sp. Pub. 11 Blackwell, Oxford.
Winkler, E.M. (1975) *Stone: Properties, Durability in Man's Environment*, (2nd edn). Springer Verlag, Berlin.

The slope system

From the functional point of view the slope system is a cascade, in the literal as well as the conceptual sense. Its inputs are from atmospheric, weathering and biological systems and its outputs principally to the channel system, but also to the atmosphere and to the vegetation growing on the slope (Fig. 12.1). Its functional role is one of throughput: the evacuation of rock and debris prepared by weathering and of water carrying elements in solution and suspension. In many ways the visible form of a slope is a balance between the rates of input of these materials and the rates of throughput and output, as regulated by the interplay of denudation forces and the resistance of the slope materials.

The disposition of slopes and the spatial relationships of individual slope facets are funda-

mental units of all landforms, and hence create the character of the landscapes they form. Important relationships exist between the state of the system, in terms of slope form, and the processes operating in it. There is a complex feedback between the two. On the one hand, slope form (in terms of angle, profile and depth and disposition of regolith) is a consequence of the past operation of processes, yet at the same time it strongly influences the operation of contemporary processes.

12.1 The initial state of the system

The form of a slope at any moment in time is a function of initial form, geological structure and subsequent modification by denudation processes.

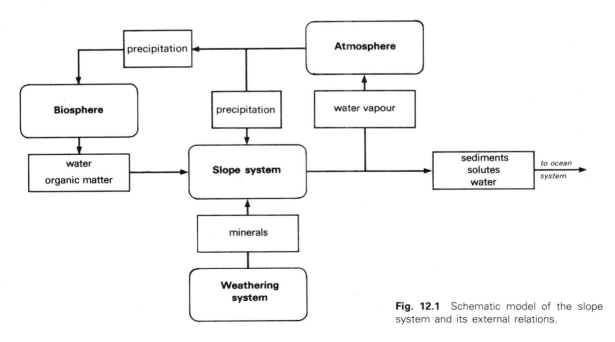

Fig. 12.1 Schematic model of the slope system and its external relations.

Slopes may be initiated in a number of ways. First, uplift of a relief mass may take place as a result of the activity of the crustal system, by means of folding or faulting. Orogenic activity of this nature has been measured at a variety of locations, commonly yielding values of several millimetres per year. Secondly, linear erosion by rivers or glaciers can incise into a landscape to produce valley-side slopes. At Grand Canyon, the Colorado River has incised itself at a mean rate of 0.25 mm a^{-1} over approximately 8 Ma. Thirdly, marine action can create slopes, either by erosion, whereby the margins of a land mass are trimmed back by cliff formation, or by a fall in sea level, which results in the exposure of the slopes of the former sea floor. These various modes of slope initiation are means by which relief is created faster than it is being destroyed. Furthermore, the initiation of slopes is not an instantaneous process but a gradual one, and therefore doubt is cast on models of slopes that assume an instantaneous origin. The number of ways in which slopes can be initiated reveals that a considerable variation may be expected in initial slope form, ranging from the vertical cut of a rapidly incised river to a gently shelving former sea floor.

12.2   The operation of the slope system: the transfer of water

Fundamental to the understanding of slope systems is the recognition of the modes of transfer of water downslope, both at and beneath its surface. This will be considered here, followed by a study of the transfer of minerals in Section 12.3. The retention and movement of water in the slope system, both across the surface and within the soil and regolith, is best understood by reference to the forces acting on the water at any point in the system. The ability of the water to do work under the constraints of these forces is an expression of the potential energy of the water in the system (specific free energy) or the **water potential**, ψ (Box 12.1). By treating the water in the slope system in terms of water potentials, we are able to specify the magnitude and direction of the flow of water, for it will tend to move from regions of high potential to low potential, just as heat is transferred from regions of high to low temperatures.

Soil water potential (Box 12.2) is a composite term whose components are related to the chemical and mechanical forces concerned. The move-

ment of water within the system will depend on the difference in potential between any two points, which in turn is a function of soil moisture conditions on the slope. At any given time these will be determined partly by the duration and intensity of rainfall input, or the time elapsed since the last input, and partly by the characteristics of soil and regolith, which act as regulators of water transfer.

The initial movement of water into the soil is by **infiltration**, though some precipitation may be temporarily held in microtopographical irregularities at the ground surface as depression storage. Infiltration is measured in mm h^{-1} and the **infiltration capacity** is the maximum possible rate of infiltration for a particular soil in a specified condition. During infiltration, three moisture zones can be distinguished within the soil:

1. the thin saturated zone at the surface, whose water content decreases rapidly with depth to pass into
2. the transmission zone, where water content diminishes more gradually, and
3. the wetting zone and wetting front, where it decreases rapidly once more.

Water movement into the soil through these transition zones takes place in response to a gradient of matric potential (ψ_m) with high potentials at the surface and low potentials at the wetting front. Although soil properties control the nature of the matric potentials concerned, normally the surface tension or capillarity component of matric potential is most significant in infiltration. The capillary properties of the soil (**hydraulic conductivity**) depend partly on such physical properties as texture, and on the presence and type of vegetation, because of the effects of roots and organic matter on hydraulic conductivity. The most important regulator of infiltration in a particular soil, however, is its initial or antecedent moisture content, for water movement is proportional to total soil potential differences and these will be less in a wet soil than in a dry soil receiving precipitation. So infiltration capacity may be as high as 50 mm h^{-1} initially for a dry sandy soil, declining to a few millimetres per hour for a wet soil or clay.

Movement of water in soils is a continuous process dependent upon differences in soil water potential (ψ_s). Within the intergranular pore spaces and voids of the soil this movement has both a vertical and a lateral component, and it is characterized by a diffuse or matrix flow. It may take place under saturated or unsaturated conditions. Lateral transfer of water downslope by this pathway is referred to as **matrix throughflow** (Fig. 12.2). In unsaturated throughflow, differences in matric potentials contribute most to the potential gradient on the slope, and control the direction and rate of diffusion, perhaps with a minor contribution from osmotic or solute potential differences (ψ_π). When the matrix is saturated with water, however, the gravity potential differences and the hydrostatic pressure potential become the determinants of hydraulic potential gradients (Box 12.2). Saturated throughflow can take place within the soil where saturated conditions develop as a result of differences in the permeability of soil horizons. These conditions are most typical at the foot of the slope, where a wedge of saturated soil

develops and extends upslope as water accumulates in the soil (Fig. 12.2). Whether saturated or unsaturated, throughflow will be conditioned and regulated by the characteristics of the soil as a matrix for water movement, by the forces of moisture retention and by the antecedent moisture conditions.

Within the slope itself the flow of water is, theoretically at least, described by models of flow through a porous medium, such as Darcy's Law, which states that

$$q = K \times \frac{H}{L}$$

where q = discharge, H/L = hydraulic gradient (height/length), and K = hydraulic conductivity,

---

Box 12.1
WATER POTENTIAL

Water potential (ψ) is a term which expresses the difference between the chemical potential of water at any point in the system (μ_ω) and that of pure free water under standard conditions of temperature and pressure or elevation (μ_ω^0). This difference ($\mu_\omega - \mu_\omega^0$) is an indication of the ability of water in the system to do work, as compared with that of pure free water. Chemical potential, however, is not easily measured in absolute terms, but water potential can readily be determined because

$$\psi = \mu_\omega - \mu_\omega^0 = RT \ln (e/e^0)$$

where R = universal gas constant (J mole^{-1} degree^{-1}), T = absolute temperature ($^\circ$K), e = vapour pressure of water in the system at temperature T, e^0 = vapour pressure of pure water at the same temperature and elevation as water in the system, and $\ln$ = the natural logarithm, $\log_e$. The units of $RT \ln (e/e^0)$ are J mole^{-1}. Pure free water is arbitrarily defined as having zero water potential under standard conditions. When the ratio e/e^0 is less than 1, $\ln e/e^0$ is negative

and water potential is a negative quantity. Water potential is increased by increase in pressure or temperature, but decreased by the presence of solutes, hydrostatic or capillary forces and by electrostatic attraction to charged surfaces. Water potential can be expressed in pressure units by dividing by the partial molar volume of water ($\overline{V}_\omega$):

$$\psi = \frac{\mu_\omega - \mu_\omega^0}{\overline{V}_\omega}$$

$$\psi = \frac{RT \ln (e/e^0)}{\overline{V}_\omega}$$

A conversion table for various expressions of water potential assuming 1 cm^3 of water weighs 1 g (after Bannister, 1976) follows:

Atmospheres (STP)	Bar	N m^{-2}	J g^{-1}	m of water
1	1.013	1.013×10^{-5}	0.1013	10.33
0.987	1	10^5	0.1	10.17
9.87×10^{-6}	10^{-5}	1	10^{-6}	1.017×10^{-4}
9.70×10^{-2}	9.833×10^{-2}	9.833	9.833×10^{-3}	1

Box 12.2
SOIL WATER POTENTIAL

Soil water potential (ψ_s) is a composite quantity to which different types of force, both mechanical and chemical, contribute. It is defined as follows:

$$\psi_s = \psi_m + \psi_\pi + \psi_p + \psi_q$$

where ψ_m = matric potential, ψ_π = osmotic potential, ψ_p = pressure potential, ψ_q = gravitational potential.

The first of these is the surface tension force in the water menisci of the pore spaces between soil particles. There are also forces of adsorption between water molecules and charged particles and surfaces, particularly in the colloidal fraction of the soil. Because both of these forces are associated with the solid phase of the soil (the soil matrix), both inorganic and organic, they are combined to form one component – the matric (or matrix) potential, ψ_m. Which force predominates, however, depends on soil textural properties and soil water content, but usually adsorption forces increase as soil water content decreases. In dry soils, therefore, adsorption forces alone largely determine the matric potential.

The second component contributing to the water potential in soils is the concentration of solutes in the soil solution and this determines the osmotic potential (ψ_π). In moist soils, when the matric potential approaches zero as saturation is approached, the osmotic potential may be the predominant component of ψ_s.

A third component of total soil water potential is a pressure potential (ψ_p) which is directly proportional to the excess hydrostatic pressure exerted over atmospheric pressure by the column of soil water above a point. Because of its relation to height of water column, it can be of some importance at depth, but only below a water table in saturated soils. In contrast to the other components mentioned so far, it is positive (see Box 12.1) and it acts to increase the total soil water potential. A related component is the gravity potential (ψ_q) which is relatively insignificant in dry soil, but of some importance in saturated soils on slopes and below the water table. It is proportional to the density of the water, the height of a point above its base level and to the acceleration due to gravity (see Box 1.4).

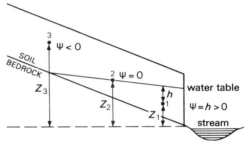

datum $Z = 0$ at base of slope

In slope hydrology it can be said that differences in ψ_m and ψ_π determine the direction and rate of water movement in unsaturated conditions, whereas ψ_p and ψ_q become the dominant controls under saturated conditions. Where the interest is in saturated flow alone, the matrix and osmotic potential components are often ignored, and the sum of pressure and gravitational potentials is referred to as the total hydraulic potential (ϕ). The calculation of ϕ for various points on the slope relative to elevation and water table is shown in the adjacent diagram (after Atkinson, 1978).

Hydraulic potential	=	Gravitational potential	+	Pressure potential
point 1 ϕ_1	=	gz_1	+	h
point 2 ϕ_2	=	gz_2	+	0
point 3 ϕ_3	=	gz_3	+	ψ

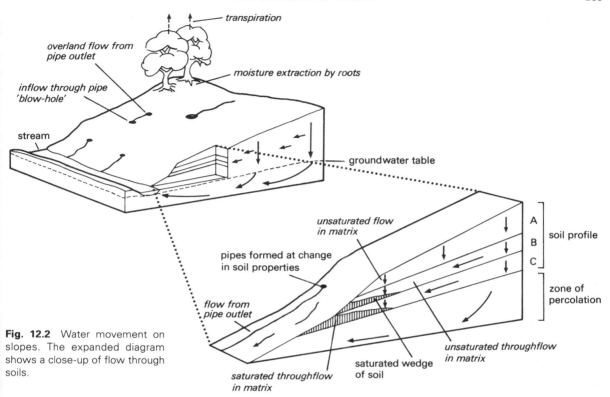

Fig. 12.2 Water movement on slopes. The expanded diagram shows a close-up of flow through soils.

an empirically determined property of the particular medium.

In reality, most soils are structured in some way, and therefore anisotropic to the passage of water through them. Commonly, soils are layered into horizons of differing hydraulic conductivity, as, for example, a porous sandy horizon overlying a zone of silt and clay accumulation (see Fig. 12.2, soil horizons A, B, C). In other cases soils may be separated by desiccation cracks into discrete peds, the intervening fissures permitting rapid transit of water. The peds themselves may possess pores of differing sizes. Trudgill *et al.* (1984) identify micropores as <60 μm, capillary macropores in the range 60–1000 μm, and non-capillary macropores as >1000 μm. Non-capillary macropores permit the rapid movement of water, which may bypass substantial areas of the soil peds. Under these circumstances the diffuse movement of water through the soil matrix is supplemented by a more concentrated throughflow pathway (Fig. 12.2), and Darcy's Law no longer provides a complete explanation of soil water movement. The large

voids may be further enlarged by soil fauna and the growth and decay of roots.

These may concentrate throughflow movement into networks of soil pipes (Fig. 12.3), through which water is transmitted by turbulent flow. Discharge in completely filled pipes varies in response to pressure and gravity potentials and in partially filled pipes (which behave as channels) in response to the slope of the water surface. Pipeflow velocity is usually much more rapid than matrix flow. Weyman (1975) quotes estimates of pipeflow velocity of 50–500 m h^{-1} and matrix flow 0.005–0.3 m h^{-1} (see Table 13.1). The remaining water which has infiltrated passes down to the water table and is stored there as groundwater, ultimately returning to the channel by deep percolation.

The generation of soil throughflow is enhanced by various factors. Towards the base of a slope, runoff will reach a maximum due to the increasing extent of the catchment area above it. Lateral variation in plan curvature of a slope to create spurs and hollows will create a convergence of streamlines within the latter. Anderson and Burt (1978)

(a)

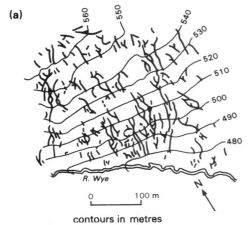

Fig. 12.3 (a) Pipe network, Plynlimon, Wales (after Atkinson, 1978). (b) Soil pipe developed at the interface between organic soil and the underlying mineral soil, exposed in a shallow cutting. Immediately after a heavy storm a strong discharge is issuing from the pipe. A trickle of overland flow is falling from the vegetated surface of the organic soil. (Sutherland, Scotland.)

indicate the magnitude of the effects of slope planform curvature on throughflow discharge. They demonstrate that for a small valley in Somerset, UK, the discharge per unit slope width from hollows is 5–10 times greater than from spurs.

Monitoring of discharge from slope pits dug to intercept flow from differing horizons reveals the relative importance of differing hydrological pathways. Williams *et al.* (1984) have identified four different throughflow pathways on slopes on the granite of Dartmoor (Table 12.1). Considerable variations in both frequency and magnitude of flow are demonstrated in different subsurface pathways, and contrasted with the discharge of overland flow.

After a period of prolonged rainfall, all the pore spaces may become filled with water, thus saturating the soil. At this point the water table has effectively risen to the surface and the infiltration capacity is reduced to zero. The soil is incapable of absorbing any further water, and subsequent rainfall runs off directly across the surface of the slope as **saturated overland flow**. This situation is likely to come about towards the base of a slope, where both local infiltration and throughflow received from higher up the slope contribute to soil moisture.

Under certain circumstances the rate of precipitation may exceed the infiltration capacity of the soil, even when the latter is not saturated. In this case the excess of precipitation runs off downslope as **overland flow**. This appears to be a common process in semi-arid regions, where precipitation intensities are high and infiltration capacity of the

Table 12.1 Frequency of discharge, and discharge for a specific storm, for different hydrological pathways in a granite slope, Dartmoor, UK (after Williams *et al.*, 1984)

	Frequency of flow event (% of time)	Discharge for storm of 26/9/77 (ml)
overland flow	88	>1150
1 eluviated horizon above iron-pan	29	364
2 granite regolith above fragipan	82	>1150
3 saturated regolith above fragipan	44	247
4 saturated fragipan	27	123

sparsely vegetated soils is low. It is further encouraged by the development of a crust (Fig. 12.4) on the soil as the surface layer becomes compacted and the pores blocked as a result of the redistribution of soil particles following raindrop impact (Farres, 1978). This kind of overland flow seems to be absent in temperate environments with only modest precipitation rates and well structured soils, except under certain conditions of cultivation (see Chapter 26).

The flow of water down and through slopes is therefore a process of some complexity. Whereas the driving force is the input of water under gravitational energy, the rates and patterns of flow are very strongly influenced by the form of the slope, soil water potentials and the nature of the hydrological pathways present within the soils.

12.3 The operation of the slope system: the transfer of minerals

The transfer of minerals downslope takes place in a wide variety of ways. Carson and Kirkby (1972) distinguish between **mass movement, particulate movement** and **movement in solution**. This forms a convenient classification for discussion, but it must be recognized that the distinctions are not always clearcut. The difference between particle movement and mass movement is that in the latter the debris moves as a coherent mass, and in the former, particles may move as individual bodies, constantly changing their positions in relation to their neighbours.

12.3.1 Mass movement

Three basic mechanisms of mass movement can be identified: slide, flow and heave (Fig. 12.5). In the case of a slide, the mass moves as a coherent unit with minimal internal dislocation along a discrete plane of failure (the slide plane). A flow, in contrast, suffers internal deformation as it takes place. Velocity decreases downwards through the flow towards its bed, in much the same way as in a river channel (see Chapter 13). Heave processes are characterized by expansion movement normal to the surface and subsequent contraction, causing alternate elevation and lowering of the ground surface.

In practice, however, most processes of mass movement involve a combination of these three basic mechanisms. Accordingly they can be used as a basis for classification (Fig. 12.6). Different types of mass movement are located on the triangular diagram according to the relative proportions of flow, slide and heave involved. An additional feature of this classification is that it shows a gradation from heaves to slides and flows. It shows also a gradation of increasing moisture along the slide–flow axis. Within this framework we shall now consider the major forms of mass movement.

Soil creep. Soil creep is caused by the disturbance and subsequent settlement of soil particles in the regolith cover of a slope (Fig. 12.7). The net result is the slow, continuous downslope movement of the debris sheet *en masse*. The fundamental mechanism of creep is heaving. This

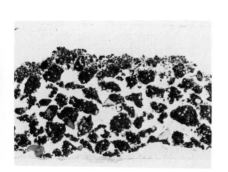

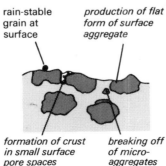

rain-stable grain at surface

production of flat form of surface aggregate

formation of crust in small surface pore spaces

breaking off of micro-aggregates

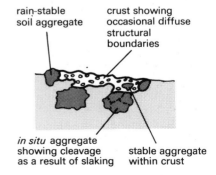

rain-stable soil aggregate

crust showing occasional diffuse structural boundaries

in situ aggregate showing cleavage as a result of slaking

stable aggregate within crust

Fig. 12.4 Crusting at the soil surface produced by raindrop impact (after Farres, 1978).

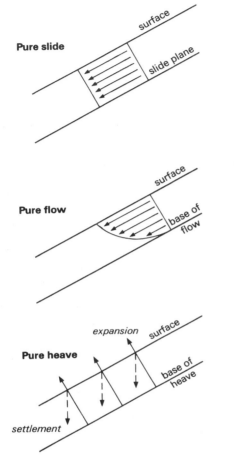

Pure slide

surface

slide plane

Pure flow

surface

base of flow

Pure heave

expansion   surface

base of heave

settlement

Fig. 12.5 Basic mechanisms of mass movement on slopes.

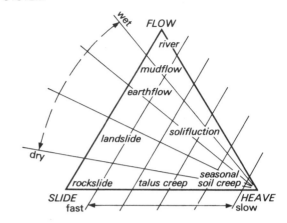

Fig. 12.6 Classification of mass movement processes (after Carson and Kirkby, 1972).

SYSTEM

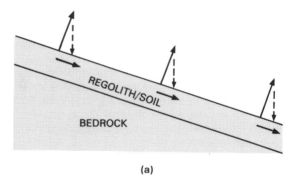

REGOLITH/SOIL

BEDROCK

(a)

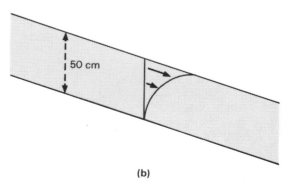

50 cm

(b)

Fig. 12.7 Soil creep: (a) fundamentals of movement, (b) velocity distribution.

raises particles in a direction normal to the slope, and they subsequently tend to settle back in a vertical direction to a position downslope from their original location. The cumulative effect of such repeated disturbance is a ratchet-like downslope movement, affecting the whole of the surface layer of the debris mantle.

The motive forces initiating the heave are many and varied. The alternate wetting and drying of soils, especially those with a significant clay component, can cause repeated expansion and contraction. Freezing and thawing of moist and wet soils occurs seasonally, or intermittently in temperate environments, and causes similar effects. The growth of plant roots and the activities of burrowing animals apply much mechanical energy to the soil layer and contribute considerably to the disturbance of soil particles. It has been estimated, for example, that in humid temperate environments the entire surface layer to a depth of 10 cm is entirely ingested and passed through the gut of earthworms in a period ranging from 18 to 64 years. Paton (1978, see Chapter 8) gives a good review of the role of soil fauna in subsurface soil processes.

Measurement of soil creep over a 12 year period on a 25° slope in humid temperate conditions shows a mean downslope movement by creep of 0.4 mm a^{-1} throughout the top 20 cm of soil (Young, 1978). Rates of movement in the range 0.2–3.0 mm a^{-1} have been reported from a variety of sites in humid temperate environments.

Frost creep and gelifluction. In periglacial environments freeze–thaw processes are much more frequent and soil moisture values are generally higher. Accordingly, heaving of the ground surface may attain values of 5–20 cm. During thaw, the water content of the upper thawed layer of soil, unable to drain through the still-frozen ground below, is close to the liquid limit and tends to flow (*gelifluction*). The combined effects of frost creep and gelifluction under such conditions (they are difficult to separate) have been shown by many studies to attain velocities of several centimetres per year, and they generally affect soils down to a depth of 50 cm.

The consequence of these creep processes is to produce a gradual downslope movement of the surface layers of the debris mantle, often with little visible effect. Where a section is cut into a slope, the dragging-over of steeply dipping strata is sometimes evident. Surface effects include the accumulation of soil upslope of retaining walls or tree trunks, whereas a hollow often develops downslope of a tree trunk. Where the surface of a creep layer is mantled by a turf cover, the latter is sometimes ruptured to form terracettes. Under the more rapid movement typical of gelifluction, terrace and lobe features develop, penned back by a barrier of turf or stones.

Rockslides and landslips. Sliding involves the sudden and rapid downslope movement along a discrete plane of an intact mass of rock on unconsolidated material. In the case of rockslides, this plane may be a major joint or a bedding plane between two adjacent strata. In homogeneous materials such as clays, an arcuate failure plane is liable to develop leading to a rotational slide. Whenever the shear force comes to exceed the resistance to shear along any potential failure plane, sliding will be initiated.

Slides can be thought of as a consequence of inherent and initiating factors. Inherent factors that predispose slopes to instability and failure include the steepness and height of the slope, which impose high shear stresses towards its base (Fig. 12.8). They also include properties of the materials composing the slope itself, such as the low shear strength of the material involved or pre-existing places of weakness. Initiating factors are trigger mechanisms which push the slope beyond the threshold of stability (Box 12.3). These include the excessive loading of the slope after prolonged rainfall, which increases the overburden pressure and hence the shear stress, at the same time reducing the shear strength by raising the pore-water pressure. Long-term weathering at the surface of the slope can cause a reduction in shear strength and consequent failure. The removal of supporting material by erosion at the base of a slope is a further cause of instability. This is exemplified by the active undercutting of slopes by fluvial or marine action.

There are many varieties of slide and slip (Fig. 12.9). Deep planar slides, often more than 10 m in depth, are usually related to inherent structural weakness within the slope. Shallow planar slides are usually caused by loss of strength in the surface

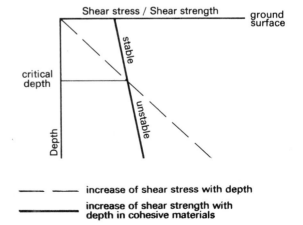

Fig. 12.8 Shear stress and strength in relation to depth within a slope.

Box 12.3

FUNDAMENTAL FORCES ACTING ON THE HILLSLOPE

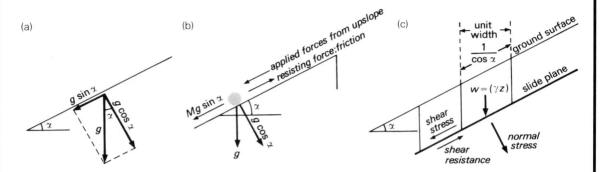

(a)   (b)   (c)

Denudation processes involving the transport of solid material down hillslopes can be resolved in terms of fundamental mechanical forces. The movement of material depends upon the nature and magnitude of these forces. There are forces that tend to promote movement and those that resist movement. Whether or not movement takes place depends upon the balance between these two sets, These fundamental mechanical principles apply both to individual particles resting at the surface and to the movement of masses of material comprising the hillslope.

The basic gravitational force (g), operating vertically downwards, can be resolved into a downslope force (or shear stress), $g \sin \alpha$, and a force normal to the slope, $g \cos \alpha$ where α is the slope gradient (see (a)).

An individual particle
In the case of an individual particle of mass M, resting on a slope of gradient α, the downslope shear stress is given by $Mg \sin \alpha$. This shear stress acts as a driving force and may be augmented by an external agent, for example raindrops, flowing water or wind, and produce a tendency to downslope movement (see (b)).

Resistance to movement is the normal stress (i.e. normal to the slope surface) which causes a tendency for the particle to subside into the slope, or at least retain its position. The resistance to movement will be complemented by the frictional resistance between the particle and

the surface on which it rests. If the particle is partially embedded in the slope, or is angular in form and interlocked with adjacent particles, it will clearly possess a much greater frictional resistance.

A rock or soil mass
Mass movement takes a wide variety of forms, depending on the induration on the material concerned and the presence of discontinuities within it. However, for the purpose of explaining the basic principles we shall consider the simplest case of an inclined failure plane parallel to the slope surface, and ignore side and edge effects (see (c)). This is a simple shallow planar slide. Assuming a slope of unit width, gradient α and horizontal length l, the vertical force will be distributed over an area $l/\cos \alpha$. Vertical stress is therefore

$$\frac{\gamma z}{l/\cos \alpha} = \gamma z \cos \alpha$$

where γ = of soil mass and z = depth of mass
This can be resolved into downslope shear stress:

$$S = \gamma z \cos \alpha \sin \alpha$$

and normal stress $\Theta_n = \gamma z \cos^2 \alpha$. The **driving force** is the shear stress. The **resisting force** is the shear strength, given by the Coulomb equation (Box 10.2) distributed over the area of the failure plane. Substituting for τ in the Coulomb equation gives:

$$S = c + (\gamma z \cos^2 \alpha - u) \tan \phi.$$

The ratio between resisting force and driving force is defined by the **factor of safety** (F). Thus

$$F = \frac{c + (\gamma z \cos^2 \alpha - u) \tan \phi \ \text{(resisting force)}}{\gamma z \cos \alpha \sin \alpha \ \text{(driving force)}}$$

where u = pore water pressure.

When $F > 1$ the slope is stable, but whenever conditions lead to $F < 1$ failure will occur. This equation then reduces to

$$F = \frac{\tan \alpha}{\tan \phi \left(1 - \dfrac{u}{\gamma z \cos^2 \alpha}\right)}$$

from which it is clear that dry soil ($u = 0$) is stable up to the angle of repose ($\alpha = \phi$) moist soil ($u < 0$) to a higher angle but saturated soil ($u > 0$) only to a lower angle.

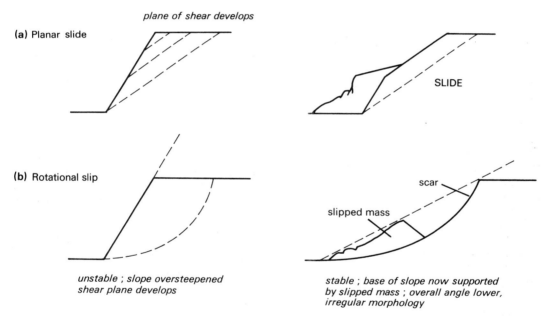

(a) Planar slide

plane of shear develops

SLIDE

(b) Rotational slip

scar

slipped mass

unstable ; slope oversteepened shear plane develops

stable ; base of slope now supported by slipped mass ; overall angle lower, irregular morphology

Fig. 12.9 (a) Planar slide development on a slope containing discontinuities. (b) Rotational slip development in homogeneous material.

layers of the slope, caused by weathering. Deep rotational slips, along an arcuate failure plane, may be simple or multiple in occurrence and they occur in homogeneous materials on slopes upon which the factor of safety has been exceeded.

The morphological effect of rockslides and landslips is to produce an upslope scar, from which the slipped material has been detached, and an accumulation of transported material on the lower slope to form a very irregular topography.

Earthflows and mudflows. Earthflows and mudflows occur when non-indurated materials, usually with a high clay component, attain a high water content. Consequently, the pore-water pressure increases, thereby decreasing cohesion and causing loss of strength. Flows are therefore closely related to rainfall events. The material involved has a high mobility, and velocities often reach several metres per second.

Morphologically there are three components to

an earthflow or mudflow (Fig. 12.10). The source area, from which the flow originates, is usually a hollow in the hillslope where moisture accumulates. A scar is formed in the hillslope by the removal of material, which moves downslope along a narrow flow track. At the base of the slope the material then fans out to form a broad lobate toe, where the flow comes to rest.

Although flows, slides and creep have been discussed as separate processes, in reality individual mass movements in the field usually incorporate a combination of these mechanisms. In a particular instance of slope failure, one mode may be dominant but not exclusively so. Landslides, for example, rarely retain the intact form of the original mass and some deformation usually occurs. In the case of flows, the shear may be limited to the lower portion of the material in transit, which may carry a raft of intact material at the surface of the flow.

12.3.2 Particulate movement

The movement of individual particles through slope systems can be caused by gravitational forces alone, or by the applied forces of falling and running water. In the present context, gravitational forces alone are normally an effective agent only on vertical or near-vertical faces, and they cause **rockfall** or **debris fall**. The applied force of kinetic energy of raindrops falling on a slope produces **rainsplash erosion**, and overland flow produces **surface wash**.

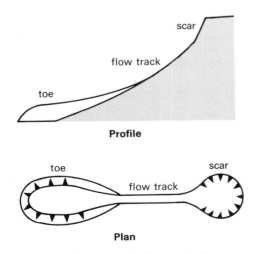

Profile

Plan

Fig. 12.10 Simplified diagram of an earth flow.

Fall processes. Large joint-bounded slabs of rock, or individual small particles on a steep rock face, may become detached by weathering processes. Once detached, they fall freely under the influence of gravity to the slope beneath, accumulating there as a **talus slope**. Larger particles, by virtue of their greater kinetic energy, tend to travel further on landing. All particles tend to roll after falling on to the accumulation slope until they encounter particles of similar size, where they tend to lodge. There is thus a sorting on the talus slope, with finer particles at the top and coarser material distributed downslope. Such slopes are commonly linear in profile, sometimes with a basal concavity. They normally possess gradients of 30–35°, the angle of repose of coarse angular loosely packed debris.

Rainsplash. As a raindrop falls on the ground surface, a proportion of its kinetic energy is transferred to loose sediment particles lying on the surface. These particles are consequently disturbed and, assuming vertical rainfall, tend to be thrown out radially in all directions equally from the point of impact. Where this happens on a slope, the gradient has the effect of lengthening the travel of a particle thrown in the downslope direction, as compared to a particle thrown upslope with a similar trajectory (Fig. 12.11). Thus the effect of a rain-storm operating on a slope is to produce a net downslope transfer of surface sediment as a result of the impact of myriads of individual raindrops. The kinetic energy of a raindrop is related to its terminal velocity and size. Raindrops vary in diameter from 0.2–5.0 mm according to rainfall type, and they possess terminal velocities ranging from 1.5–9.0 m s^{-1}. Thus the applied kinetic energy for a storm of known size and intensity can be calculated and its effects measured in the field. Sand- and silt-size particles are readily removed, whereas clays, which possess cohesion, are more resistant to this process.

For a given energy input, the magnitude of rainsplash erosion is directly proportional to the sine of slope angle, but it is clearly effective only where loose sediment is exposed at the surface and there is no vegetation cover to absorb the raindrop impact.

Surface wash. Overland flow can be generated both as excess of rainfall over infiltration and as a result of soil saturation. Water flowing downslope applies a shear stress to particles in its path (see Box 12.3) proportional to its velocity, which is governed by the Manning equation (see Box 13.3):

$$V = \frac{1.009}{n} R^{2/3} s^{1/2}.$$

In the case of **sheetwash**, flow down a slope plane of great width and minimal depth, the hydraulic radius (see Box 13.2), is effectively equal to the depth of flow, and velocity is closely determined by surface gradient and the roughness (microtopography) of the slope. Since the catchment area increases in the downslope direction, so distance

(a)

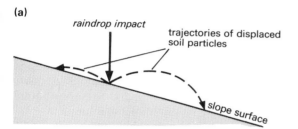

(b)

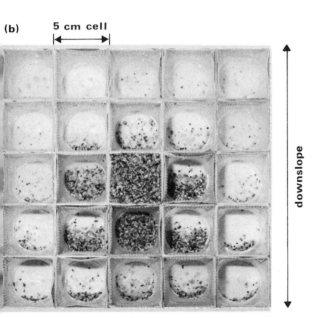

Fig. 12.11 Rainsplash erosion. (a) Schematic diagram of the displacement of particles by rainsplash. (b) The result of a laboratory simulation of rainsplash: particles have been displaced outwards from the central cup, preferentially in the downslope direction.

and depth of flow increase. Thus the flow accelerates to a threshold velocity at which entrainment of soil particles begins. Erosion by sheetwash is, therefore, negligible at the top of a slope, increasing in the downslope direction once the critical shear stress has been attained for the soil particles present.

Surface irregularities and vegetation cause the flowing water to be concentrated into anastomosing threads, which then incise themselves into confined channels to form **rills**. The flow of water is now more efficient due to the increased hydraulic radius, and higher velocities are attained. Rill wash, therefore, is a much more effective erosional process than sheetwash. The laws of hydraulic geometry cause a tendency for rills to develop concave long profiles, just as in river channels (see Chapter 13), and this in turn creates a tendency towards concavity in the lower part of the hillslope profile. Rills are ephemeral features forming in response to individual storms and being obliterated subsequently by collapse, creep, frost processes and agricultural practices between storms.

The effectiveness of overland flow as an agent of downslope transfer of particles in suspension is visibly demonstrated by the sheets or ribbons of dirty sediment-laden water traversing slopes when this process occurs. An informative account of soil erosion by surface wash is contained in Morgan (1979).

12.3.3 Movement in solution

Rainwater cascading through the slope system initially has a high potential (see Box 12.1) and possesses only low concentrations of ions in solution. It is far from being saturated in respect of ionic minerals, and is therefore aggressive towards them. In flowing through the slope system along various pathways – throughfall and stemflow, overland flow, matrix throughflow, pipeflow and groundwater flow – slope water comes into contact with the rock and soil minerals of the slope system.

The primary control on solute removal from slopes will be the rates and patterns of water flow (Section 12.2). The solution of minerals will then depend on the solubility of available minerals, the degree of undersaturation (aggressivity) of the water toward each of them, and the length of time the water remains in contact with the slope minerals (residence time). During the past decade several experimental field investigations have demonstrated the effectiveness of solutional denudation on slopes. One experimental technique is to emplace small tablets of rock within the slope in a range of weathering locations, and after a period of elapsed time, to retrieve them and determine any weight loss. This loss is then interpreted as a measure of solutional denudation. In this way Burt *et al.* (1984) have demonstrated that in a humid temperate valley in Devonian sandstone, solution is more rapid in hillslope hollows than on spurs, and is more rapid upslope than downslope.

Another approach is to dig a pit into the slope and sample hillslope water in order to identify and quantify the uptake of solutes during its passage through the system. Ternan and Wiliams (1979) and Williams *et al.* (1984) have made observations of this kind in the granite catchment of Narrator Brook, Dartmoor, UK, set in a landscape similar to that depicted in Fig. 10.1. These authors investigate the movement of silicon along various hydrological pathways on the experimental slopes. The significance of silicon in this context is twofold. First, since the atmospheric input of silicon to the slope is negligible, any occurrence of this element in slope waters has to be derived from the slope itself. Secondly, the major constituents of granite (quartz, feldspar, mica) are all silicate minerals, and the loss of silicon from the slope must therefore represent the breakdown of these rock-forming minerals, and the removal of the weathering products from the system. Table 12.2 demonstrates values of silicon in the form of silica (SiO_2) sampled from various hydrological pathways. With the exception of the overland flow pathway, with relatively high values of silica derived from the decay of vegetation, silica concentrations increase at successively lower levels within

the slope. The highest concentrations are observed in springs discharging water which has penetrated deeply into the slope, and with presumably a long residence time. This study reveals an interesting positive correlation between the silica concentration in springs and the steepness of the slope from which the spring emerges; this is interpreted as reflecting the deeper penetration of spring water emerging from steep slopes. If it were possible to combine these concentrations with absolute values of discharge for these various pathways, then the total loss of silica could be determined, and the zones of maximum silica loss determined. The loss over the area of catchment as a whole was determined as equivalent to a surface lowering of 0.0035 mm a^{-1} due to silica alone: total solutional loss would, of course, incorporate the whole range of soluble ions.

The solutional denudation of slopes attains its maximum effectiveness on slopes composed exclusively of soluble rocks such as limestone and chalk. In pure limestone the process of aqueous solution is capable of dissolving the entire substance of the rock, to leave solid residue of insignificant proportions. In practice, however, many limestone and chalk slopes possess a mantle of soil, a residue of a former cover rock or a consequence of deposition by some other external agent such as glaciers or wind. Rates of solution are controlled by the discharge of water and the acidity of the immediate weathering environment. Both deductive (Pitty, 1968) and experimental (Trudgill *et al.* 1984) studies have demonstrated that maximum solution takes place in the subcutaneous zone, that is, at the soil/rock interface. Removal of solutes will take place either downslope by soil throughflow, or by percolation of water into pores or fissures within the rock (self drainage), ultimately to emerge via groundwater to springs. In chalk springs, the concentration of solutes may exceed 300 ppm.

Clearly the mode of slope evolution under solutional denudation of this kind will depend strongly on the relative magnitude of vertical and lateral transfers of weathering products, which in turn are dependent on the depth and nature of any soil mantle.

The physical effects of solutional denudation on a humid temperate slope developed on sandstone have been indicated by Young (1978). Markers set into the slope over a 12-year period enabled the nature of movement of slope mate-

Table 12.2 Average concentrations of silica in various hydrological pathways on granite slopes, Narrator Brook, Dartmoor. (adapted from Ternan and Williams (1979), and Williams *et al.*, 1984)

	Parts per million
overland flow	3.62
eluviated horizon overlying iron pan	1.33
granite regolith overlying fragipan (saturated)	1.44
granite regolith overlying fragipan (unsaturated)	1.64
non-indurated fragipan	5.78
spring flow	7.23

rial to be determined. A pattern of movement consistent with downslope soil creep was revealed, with the exception that the markers indicated a subsidence into the slope in the top 20 cm. This is interpreted as representing a settling of the surface layer due to loss of mass, as solubles were removed by throughflow processes. In humid environments this may be an effective mechanism of slope development.

12.4 The balance of slope process and slope form

The operation of slope processes is closely governed by the input and output of material. The input from the weathering system is either shed directly on to the slope from exposed bedrock or accumulated at the base of the regolith as the weathering front penetrates more deeply to produce a thickening of the weathered mantle. Material undergoing transport can be regarded as throughput (the migratory layer) and may continue to undergo weathering and breakdown during transport. If the rate of transport and rate of weathering are in balance, then the debris mantle on the slope will be maintained through time (Fig. 12.12).

Output of mineral material takes place at the foot of the slope by processes of basal removal, where the operative agent is usually the stream

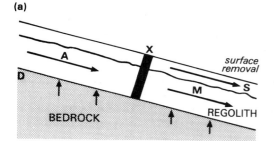

(a)

(b)

Fig. 12.12 (a) The budget of regolith on a slope: S = surface removal of material; M = removal of material by mass movement; A = material input from upslope; D = input of regolith from weathering. Budget in column X is therefore given by D + A = S + M. (b) Resistant outcrop of massive rock (Navajo Sandstone) forming near-vertical slope facet. This is an example of a weathering-limited slope facet from which debris is released by rockfall to the talus slope beneath. (Monument Valley Utah, USA.)

channel. Varying degrees of basal activity can be recognized. At the most continuously active end of the scale is the situation where a perennial stream channel undercuts the base of a slope. This facilitates the continuous removal of slope material by causing collapse of the channel banks and by receiving mineral-laden water flowing directly off the slope. Several cases exist where basal removal is intermittent. In chalk landscapes, channel flow may be seasonal (in winter only) as is also the case with tundra landscapes, where channel flow takes place only in summer. Under arid conditions, stream flow is intermittent and often less than annual. In the same category are slopes bordering floodplains. These suffer direct basal removal only when the outside of a channel meander undercuts them (Fig. 12.13). As the meander belt shifts downstream, an individual slope profile will be abandoned by the channel until the next meander sweeps by. Basal removal is at a minimum on slopes in dry valleys. In this situation, removal takes place by slow downvalley creep along the valley floor. The magnitude and frequency of basal removal can therefore vary widely under different conditions.

We have seen that there are many different kinds of process responsible for the downslope transfer of debris. They can be classified broadly into two groups: the slow continuous processes such as creep and wash, and the rapid intermittent processes of falls, sliding and flowage. The former can be considered as erosional events of low magnitude and high frequency, whereas the latter occur with low frequency and high magnitude.

It is pertinent to consider which of the groups of processes has the greater overall effect in the long term. Data on the rate of operation of these processes are gradually accumulating but are as yet insufficiently complete for large-scale generalizations to be made. One clue may be gleaned from slope profile morphology. The rapid mass movement processes tend to produce dissected slopes with an irregular profile. The continuously operating processes tend to produce slope profiles of a smooth nature, with little irregularity of form. Taken overall, more slopes conform to the latter form, which suggests that creep and wash processes are more dominant on the world scale.

Data determined by field experiment permit comparisons to be made between the relative effectiveness of creep and wash processes in different environments. The ratio of creep to wash in several environments is shown in Table 12.3.

There are clearly enormous variations between these environments, depending on inputs of erosional agents. The input of precipitation is particularly important in determining the nature of erosional process: the total influences the amount of vegetation cover and total available water, and rainfall intensity determines the erosional energy of the individual storm. The humid temperate environment, with abundant vegetation and well distributed rainfall of low intensity favours creep. In semi-arid environments, where the opposite conditions prevail, wash processes prevail.

The analysis of slope form has two aspects. The first is the plan form of a slope, whether it be concave, linear or convex. It is an important influence on the concentration or dispersion of materials moving down it (see Chapter 10). Secondly, and easier to handle, is the linear aspect of the slope's long profile. Reducing the slope to two dimensions,

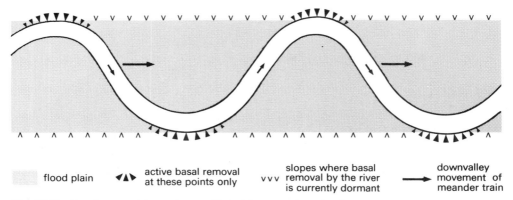

| | flood plain | ◀▲▶ | active basal removal at these points only | v v v | slopes where basal removal by the river is currently dormant | ➝ | downvalley movement of meander train |

Fig. 12.13 Basal removal from slopes affected by meander migration.

Table 12.3 Ratio of creep/wash in different environments

	Creep	:	Wash
humid temperature (UK)	10–20	:	1
savanna	1	:	5
warm temperature (Australia)	1	:	7
semi-arid (New Mexico)	1	:	98

by assuming it to be linear in plan, simplifies the study of both process and consequent form.

Several approaches have been employed in modelling slope form. At the most basic level there is the descriptive model, as exemplified by a simple two-dimensional hillslope long profile, which may be characterized in descriptive terms, e.g. convexo-concave or rectilinear.

At a greater level of sophistication is what we might call the interpretive model, as exemplified by the nine-unit land surface model of Dalrymple *et al.* (1968) (Fig. 12.14). This scheme identifies nine possible components of which a hillslope may be composed, and relates them to the processes that may be expected to dominate on each facet (Table 12.4). As such it is capable of widespread

application under varying conditions of lithology, climate and process.

Individual slopes of greater or lesser complexity can easily be encompassed by the model. For instance, if units 2 and 4 are omitted (Fig. 12.15a), the result is a simple convexo-concave slope profile. Alternatively, a more complex profile in which a series of resistant beds outcrop to form scarps would be represented by repetitions of unit 4 at intervals (Fig. 12.15b), as, for example, in the slopes of Grand Canyon, Arizona.

In theory, therefore, any hillslope profile can be interpreted in terms of this model as a particular combination of the units identified. It should be noted, finally, that units 8 and 9 (channel wall and channel bed) properly belong to the fluvial system in the framework adopted in this text.

On any hill-slope there is a relationship between the weathering rate and the transportation of debris. Clearly this relationship will depend on the properties of the bedrock and the energy available for both weathering and transportation processes. On the one hand, weathering may be relatively slow in relation to transportation processes, such that the latter are able to remove all weathered material as rapidly as it is produced.

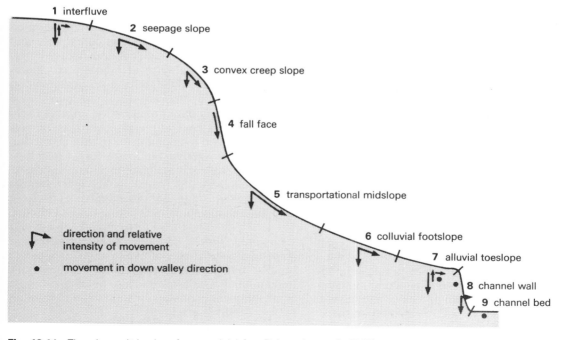

Fig. 12.14 The nine-unit land surface model (after Dalrymple *et al.*, 1968).

Table 12.4 The nine-unit land surface model; slope components and associated processes

Slope facet	Dominant processes
(1) interfluve	pedogenic processes, vertical subsurface water movement
(2) seepage slope	mechanical and chemical eluviation by lateral subsurface water movement
(3) convex creep slope	soil creep
(4) fall face	rockfall, rockslide, chemical and physical weathering
(5) transportational midslope	transportation by mass movement, creep flow and wash
(6) colluvial footslope	deposition of material by mass movement, creep flow and wash; limited transportation by creep, wash and subsurface water
(7) alluvial toeslope	alluvial deposition, subsurface water processes
(8) channel wall	corrosion, slumping, fall
(9) channel bed	transportation of material down valley by surface water processes, chemical and mechanical

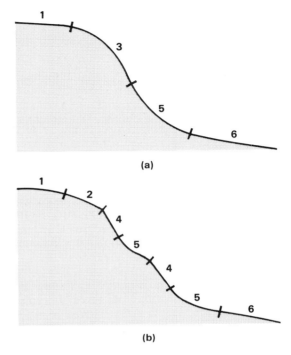

Fig. 12.15 Alternative slope forms derived by combination of units from the nine-unit model: (a) simple slope, (b) complex slope.

Thus the factor limiting the rate of development is weathering, and this is known as the **weathering-limited** case. Alternatively, it may be that transportation processes are insufficiently vigorous to remove all the weathered material, and the weathered mantle consequently becomes progressively deeper. As the solid bed-rock is buried more deeply, so weathering is reduced. The bedrock is insulated from the effects of surface weathering processes, and chemical weathering agents are used up within the weathered mantle. In this case it is the slow transport of debris that inhibits hillslope development, and this is the **transport-limited** case.

Weathering-limited slopes. Weathering-limited slopes tend to develop in regions where climatic inputs favour transportation processes and inhibit weathering processes, or where resistant bedrock exists. It is considered that the weathering-limited condition leads to straight slope profiles. A vertical cliff is the extreme case of a weathering-limited slope, where weathered rock is completely and immediately removed by rock fall, thereby preventing the accumulation of regolith and maintaining the vertical slope form. A similar condition exists on many upland areas formed by resistant rock, where abundant precipitation ensures rapid debris removal by creep and other mass movements, and even overland flow. In this case a thin debris mantle overlies the long straight midslope profile. Soluble rocks such as limestone and chalk, which yield very little clastic debris since weathered material is carried away almost entirely in solution, also support weathering-limited slopes.

Transport-limited slopes. Transport-limited slopes develop in areas of weak rocks that weather readily and produce lowland topography with limited relief. This condition also tends to suppress transportation processes, which will be inhibited by low hillslope gradients.

Let us assume an initial straight slope with limited transport capability. The rate of soil creep will tend to be constant throughout because of the constant gradient. Since an increasing volume of regolith to be removed exists downslope, the limited transport will lead to a thickening of the regolith in that direction. Consequently, weathering is inhibited on the lower slope. The more rapid weathering continuing on the upper section will produce debris available for transport and lead to the development of an upslope convexity (Fig. 12.16). On the convex slope the rate of transport by creep will be proportional to gradient and there will be an increasing transporting capacity in the downslope direction corresponding with the increased amount of debris available. Accordingly, a state of balance is reached in which gradient and creep velocity are closely related to the rate of weathering over the whole slope. A convex segment thus becomes self-perpetuating.

12.5 Complexity of hillslope process–form relationships

Hillslope form and process depend on a number of variable inputs. Variables which have their origin outside the slope are those of temperature, precipitation and vegetation. Initial slope form, geological structure and the nature of the weathering products are the internal variables.

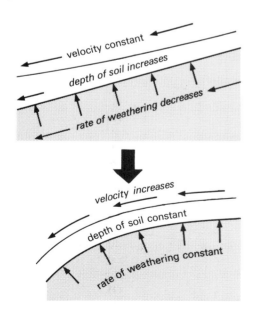

Fig. 12.16 The development of slope convexity as a result of transport-limited conditions.

In the present state of knowledge, all hillslope models are at best partial, in that they usually describe only a limited range of processes, or they rest on assumptions of the initial slope form. In reality we rarely have the opportunity to observe the initial slope. A second problem is that on any slope a number of different processes will be operating in combination during a particular period of time. The effective role played by each constituent process will vary depending on the inputs, and it is very difficult to assess. Models based on the operation of one process alone are, therefore, limited in application. To some extent the transport–weathering model avoids this problem by treating transport and weathering as grey boxes and generalizing the results of the individual processes within them. Although this may simplify the model, it renders it more difficult to verify by field measurement.

A further source of complexity in hillslope process–form relationships is the fact that the form itself may change through time. This in turn affects process, particularly the gradient-dependent processes. Thus there is feedback from form to process (Box 12.4).

12.6 Conclusion

As a consequence of the operation of processes in the slope system, materials are transferred in various forms to the base of a slope, where they form the output of the system. In the common case where the slope cascades into a river channel, this output forms the material input to the river system.

Rock minerals in solution are delivered to the river channel by overland flow and throughflow following rainstorms, and by the more continuous flow of emergent springs. Sediment transferred downslope by creep and overland flow tends to accumulate on the lower slopes and valley floor, to be stored there until incorporated into the river channel by lateral erosion of the channel margin. In the more extreme case of steep high slopes, rockfalls or landslides may deliver material directly to the channel. From the point of delivery at the base of the slope, denudation is then continued by processes within the domain of the fluvial system, and the outputs of minerals from the slope system may be regarded as sources of materials in respect of the fluvial system.

Box 12.4
MATHEMATICAL SIMULATION OF SLOPE DEVELOPMENT

The process–response relationship between slope process and form has increasingly been studied in recent years with the aid of mathematical simulation modelling. This involves developing mathematical models of erosion processes and applying them to the transformation of slope profiles in order to observe the changes they effect. These are, therefore, predictive models of slope development, and in order to be valid should be based on reasonable assumptions and models of process verified by field study.

One such set of models, based on the *continuity equation*, has been developed by Kirkby (1971), and Carson and Kirkby (1972). In its simplest form it is applied to a two-dimensional slope profile, on which the position of any point can be referred to a *y* axis denoting elevation, and an *x* axis denoting horizontal distance from the crest, although equations can be developed to encompass slope curvature in the lateral dimension. In these models it is common practice to use dimensionless axes, expressing *y* and *x* values as proportion of total elevation and slope length, respectively, i.e. relative elevation and relative length.

In its simplest form the continuity equation may express the rate of lowering of a point as

$$M + D = \frac{-dy}{dt}$$

where M = rate of mechanical lowering, D = rate of chemical lowering, y = elevation, and t = time.

Change of soil thickness on a slope can be expressed as

$$\frac{dz}{dt} = \frac{dy}{dt} + W$$

where z = soil thickness, W = rate of lowering of point of weathering, and

$$\frac{dz}{dt} = W - \frac{dS}{dx}$$

where S = rate of debris transport.

Capacity for debris transport, C, which describes the ability of processes to transport debris, can be expressed in the form of the process model

$$C = f(a) \frac{dy^n}{dx}$$

where a = area drained per unit contour length, thus $f(a)$ describes the effect of increasing distance from the divide, and n = exponent of constant value describing the effect of increasing gradient.

From this the following empirical relationship can be derived to describe transport capacity:

$$C \propto a^m \; slope^n$$

Values of m and n have been determined by field observation from different processes, as shown in Table 1.

Table 1

	m	n
soil creep	0	1.0
rain splash	0	1.0–2.0
soil wash	1.3–1.7	1.2–2.0
rivers	2.0–3.0	3.0

The link between process and the resulting form is now given by the following slope model equation, which can be calculated for each point on the profile.

$$y = y_0 \left\{ 1 - \left(\frac{x}{x_1} \right) \left[\frac{1-m}{n} + 1 \right] \right\}$$

where y_0 = height of divide, y = height of the point, x = horizontal distance from divide, and x = total horizontal slope length.

Thus a series of curves, representing characteristic form slope profiles can be derived for a range of mechanical denudation processes (see (a)).

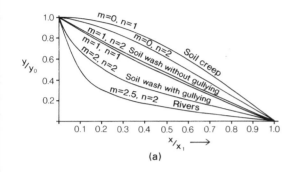

(a)

These basic forms of the slope model are capable of considerable development and elaboration; in due course field and laboratory experimentation will yield more comprehensive and increasingly valid process models, which in combination with the refined slope models should lead to more effective simulations of slope development.

NB: The derivation of the equations is given in Carson and Kirkby (1972), Appendix B.

For a practical illustration of continuity equation-based slope models in action, the reader is referred to the computer programme Slopes (Pethick, 1985).

Further reading

Apart from the appropriate chapters in the textbooks cited under further reading in Chapter 10, a good general account is to be found in the first reference, while the second is a good introduction to techniques of analysis:

Finlayson, B. and I. Statham (1980) *Hillslope Analysis*. Butterworths, London.
Young, A. (1972) *Slopes*. Oliver & Boyd, Edinburgh.

A more advanced and mathematical treatment is given in:

Carson, M.A. and M.J. Kirkby (1972) *Hillslope Form and Process*. Cambridge University Press, Cambridge.

The following provide a technical explanation of many hillslope processes, the last being particularly strong on rock slopes, on slope hydrology, and on soil erosion on slopes:

Brunsden, D. and D.B. Prior (eds) (1984) *Slope Instability*. John Wiley, Chichester.
Burt, T. (1986) Runoff processes and solutional denudation rates on humid temperate hillslope, in *Solute Processes*, ed S.T. Trudgill, Wiley, Chichester.
Kirkby, M.J. (ed) (1978) *Hillslope Hydrology*. John Wiley, Chichester.
Morgan, R.P.C. (1986) *Soil Erosion and Conservation*. Longman, London.
Selby, M.J. (1982) *Hillslope Materials and Processes*. Oxford University Press, Oxford.
Statham, I. (1977) *Earth Surface Sediment Transport*, Oxford University Press, Oxford.

An overtly energetic perspective is provided by:

Brandt, C.J. and J.B. Thornes (1987) *Erosional Energetics*, in *Energetics of Physical Environment: Energetic Approaches to Physical Geography*, (ed. K.J. Gregory) John Wiley, Chichester.

Finally, several interesting case studies can be found in:

Brunsden, D. (ed) (1971) *Slopes: Form and Processes*. Inst. Br. Geog. Sp. Pub. 3.

The fluvial system

Rivers exhibit a wide variety of forms. They may range from small turbulent mountain streams flowing down irregular beds mantled with coarse boulders and plunging over more resistant outcrops of rock as waterfalls, to the broad lowland river flowing in a tranquil manner in a channel incised into alluvium and often meandering in form. Nevertheless, all streams are governed by the laws of hydraulics, which apply to stream channels at all scales from fingertip tributaries to major trunk streams. As it is possible to enter and observe channels (particularly small ones), and because their boundaries are easily defined and channel processes operate over short timescales (facilitating monitoring and measurement), our knowledge of river channels is in many respects more advanced than is the case with other types of system.

Channel flow takes place when the major stores within the catchment (canopy interception, surface depression, soil moisture and groundwater) have been satisfied, or when a critical regulator such as infiltration capacity has been exceeded. Water may, therefore, arrive at the channel by a variety of routes, along which it may have flowed at widely differing rates (Table 13.1). A proportion of water in the channel is derived by direct precipitation on the channel surface, while water falling on slopes may run off directly as overland flow, or more slowly as soil throughflow (see Chapter 12) or by seepage from groundwater. We can therefore regard channel flow as the sum of the outputs of these various stores.

Since these major inputs to the channel operate at different rates and come into play at different times, there are important temporal variations in the magnitude of channel flow, and these are recorded graphically in the **hydrograph** in which discharge is plotted against time. It is convenient to separate the hydrograph into two components: **direct runoff** (water which finds its way rapidly into the channel), and **indirect runoff** (water which has followed a slower route).

13.1 Energy and mass transfer in channel systems

As in the operation of all systems, energy is the fundamental motive force. The stream channel processes are closely related to the magnitude and expenditure of this energy. At each point along the stream channel, the total effective energy will be the sum of the potential gravitational and kinetic components. Total energy can be related to an individual particle of water or, more usefully in geomorphological terms, to the total mass of water flowing in the channel at any one point. In this respect the mass of water is fundamental and, expressed as discharge, it is the major independent controlling variable in the operation of stream channel processes. Velocity is a major component of kinetic energy and is closely related to discharge, though not in a simple way.

The expenditure of energy in the channel takes place in two main ways. More than 95% is used

Table 13.1 Estimated velocities along different flow routes in the catchment (after Weyman, 1975, from various sources)

	Flow routes	Velocities (m h^{-1})
surface	channel flow	300–10 000
	overland flow	50–500
soil flow	pipeflow	50–500
	matrix throughflow	0.005–0.3
groundwater flow	limestone (jointed)	10–500
	sandstone	0.001–10
	shale	10^{-8}–1

in overcoming the frictional drag of the channel margin on the flowing water. The actual amount will vary according to channel size and shape, and the roughness of the bed and banks. Energy used in this way is converted to heat and lost to the surroundings by radiation and conduction, though it is barely measurable. The remainder is converted to mechanical energy and is used in transporting mineral sediments, which form part of the load of the river. Therefore, the proportion of the energy of the river actually used in erosion and transportation is only small.

The inputs and outputs of water to the river channel system are shown in Fig. 13.1. There are four major contributors to the discharge of water in the channel, which is measured in terms of volume per unit time – usually cubic metres per second (cumecs).

During rainfall, a proportion of the rain falls directly on the surface of the river channel. Normally this contributes only a small volume of water to discharge, but it will be larger when the channel has a large effective surface area, as when it incorporates a lake. Water falling on slopes may reach the channel by overland flow, or more slowly by way of throughflow. That proportion of water which percolates to groundwater may return to the channel by seepage from the groundwater store. This is known as an **influent channel** and it is typical of humid environments.

The outputs of water from the channel system are, first, evaporation directly from the channel surface. This clearly becomes more important when the channel has a large surface area exposed to the atmosphere. Secondly, under certain conditions, when the water table is below the level of the channel floor, water may be lost from the channel to groundwater. This condition, defined as an **effluent channel**, is sometimes found in arid environments and in areas of permeable rock such as limestone. The major form of output from the channel system, however, is normally runoff, the process by which water is transferred along the channel system into the ocean store.

The magnitude of runoff varies considerably in both space and time. Within a catchment basin, for example, the discharge through any channel cross-section is closely related to the catchment area drained through that section (Fig. 13.2). More important in the operation of channel dynamics is the temporal variation in discharge. This results from the fact that the four major inputs of water to the channel system are of different magnitudes and they come into play at different times. The inputs consequent upon precipitation (direct run-off) are of high magnitude and are closely

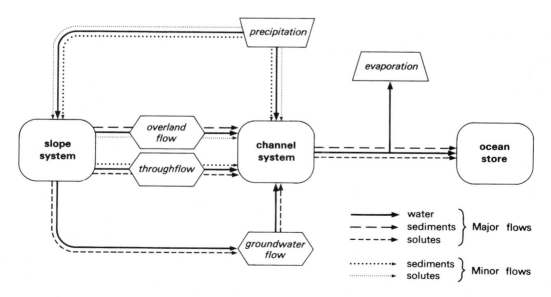

Fig. 13.1 Pathways of water and mineral flow through the catchment basin system. (This is a more detailed model than that in Figure 10.2.)

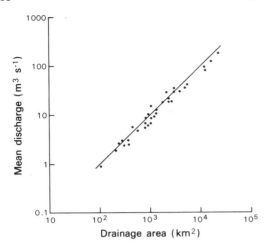

Fig. 13.2 Relationship between discharge and drainage area for gauging stations on the Potomac River. USA (after Hack, 1957).

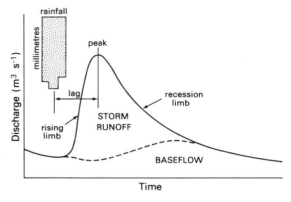

Fig. 13.3 Characteristics of the flood hydrograph.

related to the periodic distribution of rainfall events. Inflow from ground water (indirect runoff) is normally of much lower magnitude, but is distributed continuously through time.

Fig. 13.3 shows a storm hydrograph in relation to the rainfall input. Following rainfall, discharge rises sharply to a peak then falls gradually. The length of time between maximum rainfall intensity and peak discharge is the **lag time**. The section of the hydrograph showing increasing discharge is the **rising limb**, and the **recession curve** describes the decrease in discharge. The area under the peak measures the volume of direct runoff, as opposed to indirect runoff, upon which the former is superimposed. Thus the processes associated with a rainfall event lead to storm runoff, while groundwater seepage contributes to base flow. The magnitude and shape of the hydrograph will vary according to many factors associated with the nature of the rainfall and the characteristics of the catchment.

Fig. 13.3 plots an idealized hydrograph of an individual storm flow. In reality, of course, a hydrograph consists of a series of flood peaks distributed through time and superimposed upon base flow. Fig. 13.4 shows two such hydrographs for a 1-year period. Two adjacent rivers in southern England, of similar catchment area, are contrasted in their responses to the same rainfall inputs. The River Wallington, draining a catchment based mainly on clay, shows a series of sharp flood peaks superimposed upon a base flow which is higher in winter

than in summer, reflecting seasonal differences in groundwater storage level. Because of the impermeable base of the catchment, runoff response to storm rainfall, is immediate and highly peaked. The River Meon, with a chalk catchment which precludes direct runoff due to high infiltration, exhibits a much more subdued regime and is fed mainly by groundwater flow.

The flows of water described in the previous paragraphs are, of course, also associated with flows of materials (in both solid and dissolved forms) and are shown in Fig. 13.7 together with the state of the materials.

Rainwater is not pure water, for it contains a range of ions in solution as well as particulate matter washed out of the atmosphere. Overland flow is capable of transporting both sediment and solutes to the river channel, though the former is dominant. Elements are therefore washed into the channel from adjacent slopes by this process. Throughflow contributes elements in solution derived from regolith and soil. Sometimes deposits of elements transported by throughflow can be seen in the channel banks. A reddish-brown stain, for instance, may mark the redeposition of formerly dissolved iron where the throughflow emerges. Similarly, groundwater seepage contributes solutes to the channel.

Mineral sediment is contributed in considerable measure by erosion of the banks of the channel, which collapse into the stream when undermined. Fresh erosion scars are often seen on the outside of bends in the channel.

Once in the channel, solutes tend to be transported continuously by the stream. Sediments, on the other hand, are largely intermittent

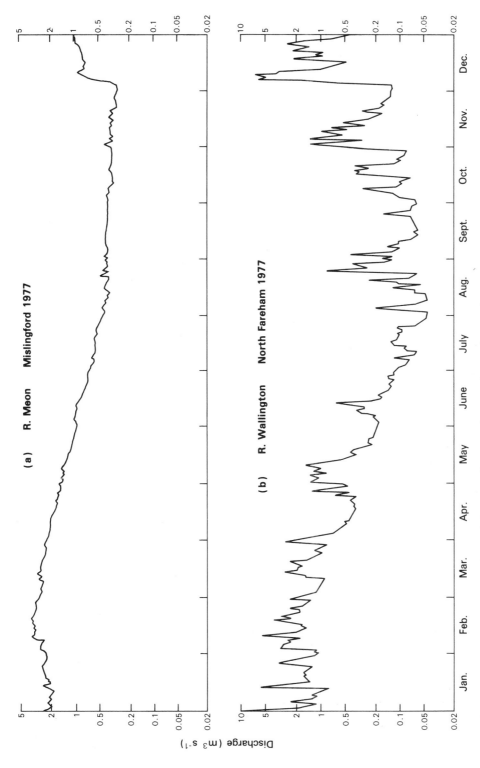

Fig. 13.4 Comparison of the annual hydrograph for two adjacent and contrasting catchments for the same year: (a) the Meon, fed mainly by baseflow from the Chalk, shows mainly seasonal variations; (b) the Wallington, draining a largely clay-based catchment, is much more flashy and responds strongly to individual storms. (Hampshire, England).

in their movement. High concentrations may be moved in times of flood, afterwards being deposited on the channel bed until the next flood. Accordingly, there is a two-way exchange of sediment between the channel bed and the stream. The relationships between erosion, transportation and discharge will be examined in more detail below.

13.2 Channel dynamics: flow of water

In this section we examine channel processes and form in relation to the moving body of water. The channel can be defined by a system of variables (Boxes 13.1 and 13.2). Water is a newtonian fluid which is unable to resist shearing stresses, and it will deform at a rate directly proportional to the applied stress. Another way of expressing this is to say that water will flow under its own weight within the solid boundaries that contain it.

The flow of water can be expressed as a balance between impelling forces and resisting forces (Box 13.3). In this way basic hydraulic relationships can be developed which express the velocity of flow in relation to channel form characteristics. Expressions such as the Manning equation (Box 13.3), however, deal only with the mean velocity, and neglect important variations which occur through the channel cross-section.

There is resistance to flow not only at the channel boundary but also within the fluid mass itself. The adjacent layers of water are able to slip past each other with differing velocities. Accordingly, velocity in a vertical profile increases away from the bed, where the shearing resistance is greatest, and the highest velocities are to be found at the water surface. This is known as **laminar flow** (Fig. 13.5a). When the bed is rough and velocity is high, this simple pattern of parallel streamlines breaks down and is replaced by **turbulent flow**, in which jets of water travel obliquely to the general direction of flow, and eddies develop causing greater mixing of the water and a more even distribution of shear across the channel (Fig. 13.5b).

The relationship in the channel between water and cross-section form is described by **hydraulic**

Box 13.1

RIVER CHANNEL VARIABLES

The geometry of a river channel can be defined in terms of a number of variables.

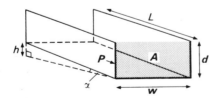

(a) *Channel width (w)*: measured at the water surface.
(b) *Channel depth (d)*: except in the case of the rectangular channel cross-section illustrated, this varies across the section and is normally expressed as mean depth $(\bar{d})$.
(c) *Cross-section area (A)*: calculated as the product of width and mean depth; thus $A = \bar{d}w$.
(d) *Wetted perimeter (P)*: the length of channel margin in the cross-section in contact with flowing water; thus $P = 2d + w$.

(e) *Channel gradient (S)*: expressed as change in elevation per unit length, thus:

$$S = \frac{h}{h \cos \alpha}, \text{ or } \tan \alpha.$$

(Note that for small gradients S approximates to $\sin \alpha$.)

(f) *Velocity of flow (V)*: distance travelled per unit of time (m s^{-1}).
(g) *Discharge (Q)*: volume of water passing through a cross-section per unit time:

$$Q = w\bar{d}v \text{ (m}^3 \text{ s}^{-1}\text{)}.$$

(h) *Channel length (L)*: slope length of channel (see Box 13.3).

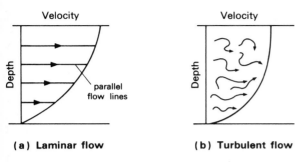

(a) Laminar flow   **(b) Turbulent flow**

Fig. 13.5 Contrast in flow patterns between (a) laminar and (b) turbulent flow.

geometry, and the response of channel-form variables to changes in discharge can be assessed by field investigation. Discharge (the controlling variable) varies through time at any section in the channel and varies also in the downstream direction.

Let us consider the effect of increasing discharge at one point in the channel. The first visible effect is an increase in channel depth as the water level rises, and this is usually accompanied by an increase in width, since most channels are trapezoidal in cross-section. These responses determine an increase in channel cross-section, resulting in greater hydraulic radius and channel efficiency. The gradient of the channel remains essentially

---

Box 13.2
HYDRAULIC RADIUS

Hydraulic radius is a measure of the amount of water in the channel cross-section in contact with the channel margins (the wetted perimeter). As such, it is a measure of frictional retardation and therefore of channel efficiency. It has dimension length and is almost equal to mean water depth.

By taking the case of the simplest channel cross-section, rectangular in form, and substituting different values for width and depth, significant differences in channel efficiency can be demonstrated.

Hydraulic radius $(R) = \dfrac{A}{2d + w}$.

Case 1
With $d = 2$, $w = 4$; $A = 8$ and $R = 1.0$

Case 2
With $d = 1$, $w = 8$; $A = 8$ and $R = 0.8$

Thus a wide, shallow channel is less efficient than a more compact one.

Case 3
With $d = 4$, $w = 8$; $A = 32$ and $R = 2.0$

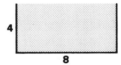

Thus a large channel, of the same proportions but of greater magnitude than Case 1, is more efficient.

A corollary of these relationships is that river channels become more efficient in times of flood, when depth increases to fill the channel and render it greater in area and more compact in form. The channel is most efficient at the bankfull stage, for above this level it spreads out across the floodplain, much increasing its wetted perimeter and thereby decreasing the hydraulic radius.

It follows also that since cross-section area increases downstream with increasing discharge, rivers will tend to become more efficient in the downstream direction.

Box 13.3
FLOW IN OPEN CHANNELS

The flow of a fluid in an open channel can be expressed as a balance between impelling forces and resisting forces.

In a section of channel (see Box 13.1) the tractive force operating in a body of water in the direction of flow is the downslope component of its weight:

$$T = \rho g \, Ldw \sin \alpha$$

where ρ = density of the mass, g = gravitational acceleration, L = slope length of channel, and other variables are as defined in Box 13.1.

The resisting force (F_R) is the stress per unit area (τ) multiplied by the area of the boundary over which it is applied. Thus, for a channel of rectangular cross section:

$$R = \tau(2d + w)L.$$

If there is no acceleration along the unit reach then:

$$F_\tau = F_R.$$

Thus:

$$\rho g \, Ldw \sin \alpha = \tau(2d + w) \, L.$$

Simplifying, since $A = \bar{d}w$, and $\sin \alpha = s$ for small angles:

$$\rho g As = \tau(2d + w).$$

Therefore:

$$\tau = \frac{\rho g s A}{(2d + w)}.$$

Since

$$\frac{A}{(2d + w)} = \text{hydraulic radius } (R)$$

then

$$\tau = \rho g Rs.$$

In hydraulics, the resistance to flow is related to the square of velocity, thus:

$$\text{resistance} = \tau/v^2$$

or

$$\tau = kv^2$$

where k is a constant.
By substitution:

$$\rho g Rs = kv^2.$$

Thus, by setting

$$\sqrt{\frac{\rho g}{k}} = C,$$

$$V = C\sqrt{Rs}.$$

This is known as the **Chezy equation** and expresses mean velocity as a function of hydraulic radius and channel gradient.

A similar, though more refined, version (the **Manning equation**) is more frequently employed in studies of channel flow.

$$V = \frac{1}{n} \, R^{2/3} s^{1/2}$$

where n is a measure of channel roughness. It is in part a function of the magnitude of the wetted perimeter and in part an expression of the nature of the channel boundary in terms of material calibre and the amount of vegetation. The rougher the nature of the channel margins, the greater will be the frictional resistance to flow. In natural stream channels, Manning's n varies from 0.03 in straight clean channels, to 0.10 for densely vegetated channels.

constant. The more efficient the channel the less the energy of the stream is spent in overcoming the frictional resistance of the channel margins, which permits higher velocity. Relationships between discharge and width, depth and velocity as determined in the field, are shown in Fig. 13.6.

The increased mass and velocity of the water result in a greater capacity for work over time. The rate at which energy is expended doing work is, of course, power – in this case **stream power** (Box 13.4). For these reasons, most work is carried out at times of high discharge, during floods. As discharge falls after a storm flood, so the channel variables respond in reverse sequence. Width, depth and cross-section area diminish, as does hydraulic radius, and velocity decreases. Thus the channel can be considered as an open system, in a state of dynamic equilibrium, responding instantaneously to changes in input conditions and fluctuating about a steady-state condition.

Similar relationships exist between discharge and other channel variables in the downstream direction (see Section 13.5).

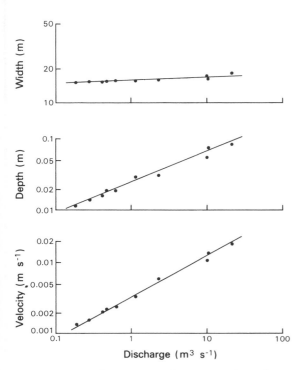

Fig. 13.6 Hydraulic geometry of a stream channel: at-a-station changes in velocity, depth and width with varying discharge (after Wolman, 1955).

13.3 Channel dynamics: erosion and transportation processes

The energy available in the channel after friction has been overcome is used in eroding the channel margin and transporting mineral debris.

Where the stream flows directly over bedrock it may erode that bedrock in various ways. The hydraulic force applied by the water may detach rock fragments, particularly if the rock is already fractured. Where a pebble lies in a hollow in the channel floor it may be swirled around mechanically, boring into the channel bed to form a pothole. The transport of a mass coarse sediment along the channel will have an abrasive effect known as **corrasion**. Clearly, all three processes depend on mechanical energy, which is more effective with the higher velocities associated with flood discharges. When the channel margins are formed from the less resistant unconsolidated materials such as alluvium, channel erosion by mechanical processes will be enhanced. In addition, the banks may become undercut at the waterline and collapse into the channel. The effect of animals trampling the river banks may hasten this process.

Debris within the channel may be set in motion as a result of the application of force by the flowing water, and thus become entrained into the mass of the stream (Box 13.5). Material can be carried in three ways. **Bed load** is material, usually of coarse calibre, rolled along the floor of the channel; **suspended load** is finer particles carried along in suspension within the water mass; and the **solution load** consists of minerals dissolved in the water.

As velocity and flow increase, so the shear stress exerted by the water on the upstream side of a sediment particle on the channel floor is increased. The force required to move a given particle (the critical tractive force) depends on the size and shape of the particle and the gradient of the bed. Once this threshold value is attained the particle is set in motion, and rolls along the channel bed in the downstream direction. Clearly, tractive force will increase with velocity, and the bed load accordingly increases at higher levels of discharge. Troake and Walling (1973) have shown that for a Devon (England) stream bedload transport is zero until a critical threshold of discharge is attained, above which it increases rapidly (Fig. 13.7a, b, c).

Finer particles (sand, silt and clay), when

Box 13.4

STREAM POWER

An important property of a river which is of great significance in relation to its ability to carry out work is that of **stream power**. This is a measure of the energy supply at the channel bed. The **gross power** (Ω) of a river is given (in watts per metre length) by

$$\Omega = \rho g \, Qs$$

where ρ = density of water (1000 kg m^{-3}), g = gravity acceleration (9.81 m s^{-2}), Q = discharge (m^3 s^{-1}), and s = slope.

An alternative approach is to define **specific power** (ω) or energy availability per unit area of stream bed thus (in watts per square metre):

$$\omega = \frac{\Omega}{W} = \tau v$$

where W = stream width and v = velocity.

Although many studies in fluvial geomorphology employ discharge, catchment area or channel length as independent variables, some authors have adopted an energy-based approach using stream power. Thus Bagnold (1977) has studied bedload transport in this way, and Ferguson (1981) examines relationships between stream power and channel morphology. Stream power is strongly dependent on gradient, and Ferguson reveals a variation of four orders of magnitude in stream power between highland and lowland streams in Britain.

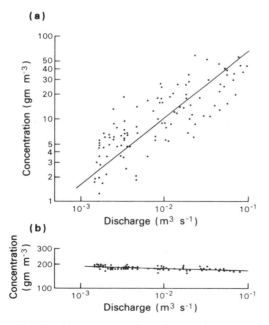

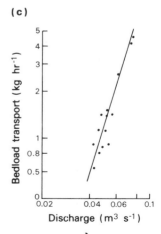

Fig. 13.7 Relationships between the various components of load and discharge for a small Devon (England) stream (after Troake and Walling, 1973). (a) sediment (b) solutes (c) bedload.

Box 13.5

ENTRAINMENT AND TRANSPORTATION OF SEDIMENT IN WATER

A fluid flowing over a bed of non-cohesive particles will apply stress to that bed and will tend to lift the particles and transport them downstream. This process of acquiring a sediment load is known as **entrainment**, and will occur when the forces applied to a particle are greater than its resistance to motion.

Five main processes contribute to entrainment. These are:

1. bed shear stress (Box 13.3), produced by the moving fluid and exerting a frictional drag (or **tractive stress**), largely a function of pressure differences on the particle between upstream and downstream faces;
2. a vertical lifting force, caused by an acceleration of flow over the obstacle represented by the particle, and directly analogous to the airflow over an aeroplane wing;
3. turbulent eddying in the fluid, creating high-velocity threads of flow oblique to the bed;
4. collisions between moving sediment particles and the bed, as **saltation** takes place, by particles hopping along the bed;
5. rolling of larger particles.

Many models of bed erosion have been put forward. One approach is to express the relationship between average bed shear stress, and average resistance along the channel margin. Thus average boundary shear stress (τ_o, in N m^{-2}) is given by

$$\tau_o = \gamma_w R \sin \beta$$

where γ_w = unit weight of water (9.81 kN m^{-3}), R = mean depth of flow (m), β = slope of water surface (in degrees).

Bed shear stress thus increases with slope, water depth and fluid density (which itself increases with sediment concentration).

Average resistance to entrainment is given by

$$n \, \gamma_{sub} \, d^3 \left(\frac{\pi}{6} \right) \tan \phi$$

where n = number of particles per bed unit area (m^{-2}), γ_{sub} = submerged unit weight of particles (kN m^{-3}), d = diameter of particles (m), and tan ϕ = coefficient of friction between particles and the bed.

Entrainment will then occur if the driving force is greater than the resisting force, i.e.

$$\text{if } \frac{\gamma_\omega R \sin \beta}{n \, \gamma_{sub} \, d^3 \left(\dfrac{\pi}{6} \right) \tan \phi} > 1$$

An alternative approach is to examine the balance between force and resistance on an individual particle (see (a)). Thus:

Drag force Moment = Drag force x Turning arm

$$= \frac{\tau_o}{n} \times \frac{D}{2} \cos \phi$$

where τ_o = bed shear stress (N m^{-2}), n = number of particles per unit area (m^{-2}), and D = grain diameter (m).

$$\begin{array}{c} \text{Submerged} \\ \text{Weight Moment} \end{array} = \begin{array}{c} \text{Submerged Weight} \\ \times \text{Turning Arm} \end{array}$$

$$= g(\rho_s - \rho) \frac{\pi}{6} D^3 \times \frac{D}{2} \sin \phi$$

where ρ = fluid density (kg m^{-3}), ρ_s = sediment density (kg m^{-3}), g = gravity constant (9.81 m s^{-2}), and D = grain diameter (m).

(a)

DRAG FORCE $= \tau_o/n$ assuming n particles in a unit area

$x = D/2 \cos \phi$

$y = D/2 \sin \phi$

SUBMERGED WEIGHT $= g(\rho_s - \rho) \pi/6 D^3$

for a sphere of diameter D

At the threshold of movement $\tau_o = \tau_{crit}$ and the applied force is equal to the resisting force, thus

$$\frac{\tau_{crit}}{n} \times \frac{D}{2} \cos \phi = g(\rho_s - \rho) \frac{\pi}{6} D^3 \times \frac{D}{2} \sin \phi$$

Therefore

$$\tau_{crit} = n\, g(\rho_s - \rho) \frac{\pi}{6} D^3 \tan \phi$$

if $\eta = nD^2$ is a measure of grain packing, then

$$\tau_{crit} = \eta\, g(\rho_s - \rho) \frac{\pi}{6} D \tan \phi$$

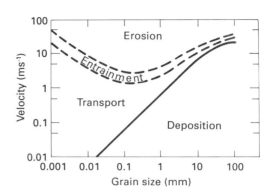

Both models indicate clearly that entrainment is strongly controlled by bed shear stress, and therefore by velocity of flow.

Threshold velocities of water required to move quartz grains have been determined by Hjülstrom (1935) as shown in (b). For particles of a given size, the velocity required for entrainment is shown to be higher than the fall velocity, at which particles settle to the bed. Thus, once entrained, particles will continue at lower ve-locities until the fall velocity is reached. For cohesive sediments in the silt and clay ranges, the velocity required for entrainment increases in order to overcome cohesion.

Particles of a given size will therefore tend to move when velocity increases appropriately, and will be deposited when velocity subsequently falls to the required level. Sediment movement in rivers is therefore an intermittent process, dependent on variations in flow velocity.

disturbed, may be lifted above the channel bed by the upward currents of water involved in turbulent flow, and carried downstream before being rede-posited. The net effect of myriad fine particles being picked up, swirled along and falling to the bed is that a mass of sediment is entrained, particularly in the lower part of the water body, as the suspended load. Provided that suitable fine-grained material is available, there will always be some material in suspension. Since, however, the amount of energy available increases with dis-charge, the amount of suspended load increases exponentially during flood (Fig. 13.8). Concen-trations of suspended sediment load typically vary over at least two orders of magnitude, in the range 10–1000 mg l⁻¹, though higher values are by no means uncommon during peak discharges.

The dissolved load of a river depends largely on the solubility of the rock-forming minerals in its catchment and the weathering regime under which it occurs. In a large drainage basin with varied geological outcrops, considerable spatial variation in solute load may exist, as in the exam-ple of the River Exe, Devon (England) (Fig. 13.9) draining areas underlain by highly soluble rocks such as limestone, which tends to have a high solute load (300–400 ppm). Elements in solution reach the river channel mainly by ground-water seepage, with additional contributions by soil throughflow and overland flow.

The concentration of solutes is greatest at low flows, when the water in the channel is derived entirely from groundwater seepage. In this situ-ation percolating water spends a greater length of time in intimate contact with the rock-forming minerals and thereby attains higher solute con-centrations. During periods of higher discharge, when the waters of throughflow and overland flow are contributing to channel flow, the water in the channel is diluted and the overall solute

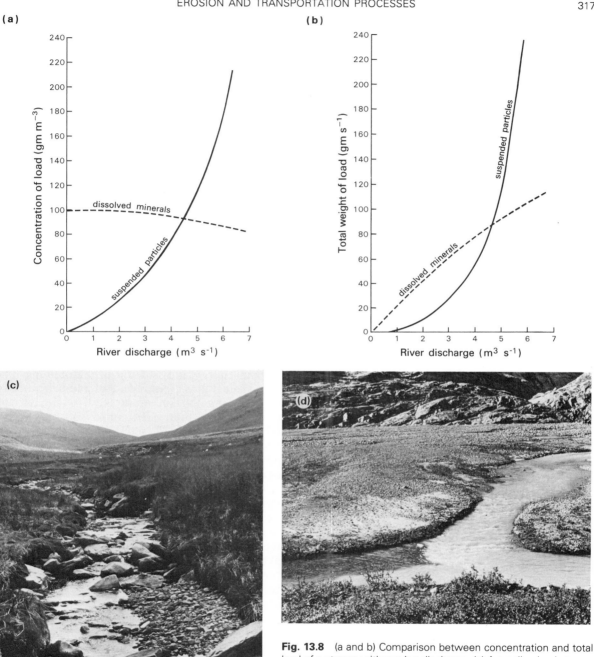

Fig. 13.8 (a and b) Comparison between concentration and total load of a stream with varying discharge. (c) A small upland stream with abundant coarse sediment resting in the channel; the coarsest sediment is probably transported only at very infrequent intervals during extreme floods (Upper Rheidol Valley, Mid-Wales) (d) Confluence between sediment-laden glacial outwash stream on the right and a sediment-free stream flowing from the left; the glacial meltwater with an abundant supply of fine sediment is very cloudy (Tunsbergdalen, Norway).

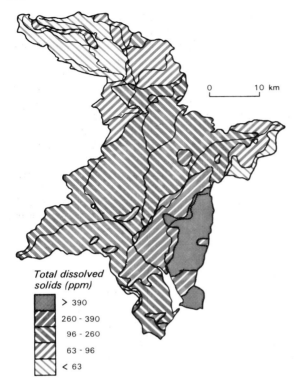

0   10 km

Total dissolved solids (ppm)

> 390

260 - 390

96 - 260

63 - 96

< 63

Fig. 13.9 Spatial variation in concentration of solutes within the catchment of the River Exe. Devon, England (after Walling and Webb, 1975).

concentration decreases (see Fig. 13.7). At times of flood, however, this dilution effect is more than offset by the increased discharge, such that the total quantity of dissolved material transported per unit time increases.

The relative importance of these three modes of transport varies from river to river, depending on the nature of the available load. In any one river channel it will vary through time, with the solute load being more important at low flows, while suspended sediment is transported in greater abundance at times of flood, and the bed load moves only once a threshold level of discharge has been attained.

13.4   Depositional processes

It has been shown that the bulk of the solid materials in a channel are set in motion at times of flood discharge. As discharge declines following a flood peak, so turbulence and tractive force

decrease and the solid particles are redeposited coarse material first, while finer particles may continue in transport at lower flow velocities. Thus while solutes, once in the channel, tend to be flushed right out of the system, the solid load is conveyed downstream at times of flood, and is subsequently redeposited.

Within the channel, abundant deposits of solid particulate material await the next flood. These may be a simple uniform spread of gravel across the channel, or be arranged in more distinctive form. Point bars, for example, represent aggradation on the insides of meander bends, and both longitudinal and transverse bars represent temporary stores of sediment in transit.

13.5   Stream power and the threshold of critical power

As this review of the dynamics of erosion, transportation and deposition has shown, streams may be regarded as sediment-transporting machines, and their behaviour analysed in terms of the availability of stream power (Box 13.4) to do work. Stream energy is dissipated in maintaining fluid flow against flow resistance, and in doing work by transporting the sediment load; where stream power is more than sufficient to transport an imposed sediment load, scour of alluvium on the bed, or of the bedrock, will occur. Where stream power is insufficient, load transport will decrease and the bed of the stream will aggrade.

A critical power threshold, therefore, separates modes of erosion and deposition in stream systems, and is dependent on the relative magnitudes of the power needed to transport the average sediment load, the critical power, and on the power available to transport the load – the available power. The threshold of critical power is defined as

$$\frac{\text{available stream power}}{\text{critical stream power}} = 1.0$$

The available stream power reflects those variables which, if increased, favour the transportation of sediment (discharge, slope); critical stream power changes with variations in sediment load and size and with hydraulic roughness, and expresses those factors which determine stream competence and capacity. Analysis of the **thres-**

hold of critical power (Bull, 1979) offers explanations for the spatial and temporal incidence of changes in fluvial systems, as can be seen in Fig. 13.10a. This depicts changes in the headwaters of a rocky drainage basin in an arid region after local thunderstorm rainfall. In reach A, available power is much greater than that needed to transport the sediment load and overcome roughness (flow resistance). The stream obtains additional sediment by vertical downcutting, so that cross-valley morphologies are characteristically V-shaped. In reach C, the available power is less than the critical power, and aggradation occurs, aided by increases in flow width, vegetation and infiltration.

In reach B, available and critical power are equal, but changing – an equilibrium condition.

Reaches near the threshold, such as B, are highly susceptible to accelerated downcutting or alluviation, because a slight change in critical power may cause a passing of the threshold. The situation is different for A, for example, since even a large increase in critical power can occur and the stream will continue to downcut. For reach, C, changes in critical power may a) accelerate alluviation; b) return the system to equilibrium; and c) initiate entrenching. Because the threshold separates modes of system operation into net deposition and net erosion, the concept can be used as the basis of a more general explanation of landform change, where the system state varies between equilibrium (no net change), threshold and rapid change states (Fig. 13.10b).

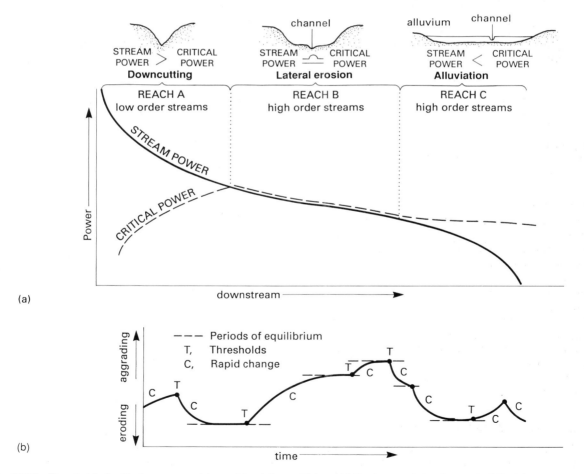

Fig. 13.10 Threshold of critical power used to explain a) the spatial and b) the temporal development of fluvial systems and fluvial landscapes. See text for further explanation (after Bull, 1979).

13.6 Channel form

The form of the river channel represents the means by which the discharge is accommodated. It can be regarded as the response by the channel to its inputs, which can be considered in terms of adjustments to the long profile, the cross profile and to the form of the channel in plan. The three are closely related in many ways, which can be explained by the interaction of hydraulic variables.

It has been recognized for some time that the long profiles of streams tend to be concave, as the gradient gradually decreases downstream. Major changes may take place in the downstream direction in drainage area and lithology. Drainage area, and therefore discharge, progressively increase, and in its lower reaches the river channel tends to be cut into fine-grained alluvial deposits, in contrast to the bedrock channel in the upper reaches. Accordingly, as channel cross-section area increases downstream in order to accommodate the increasing discharge, the channel becomes more efficient, i.e. it has a greater hydraulic radius. This tendency is reinforced by the change to a smooth channel margin in finer-grained alluvial sediments, and the stream flows along the lower gradient channel with increased efficiency. In this way the tendency to decreasing gradient downstream is a means by which channel form accommodates the increasing discharge.

Only rarely, and for a variety of reasons, is the long profile of a stream smoothly concave. First, the downstream increase of discharge is not gradual but incremental. Increase in discharge takes place mainly stepwise at tributary junctions. Secondly, increases in discharge at tributary junctions may not be accompanied by a proportional increase in sediment load. Thus the whole sediment–discharge relationship may be altered in the main channel, resulting in local adjustments of gradient by erosion or aggradation. Thirdly, variations in the resistance of bedrock to channel erosion may cause local variations in the rate of gradient adjustment. Thus an outcrop of a particularly resistant bed of rock may create a waterfall. All these circumstances are quite normal occurrences whose effects are superimposed upon the overall trend of decreasing gradient downstream.

It is clear that discharge increases in a downstream direction according to the increase in drainage area. The channel adjusts to the change in the controlling variable by adjusting its own form. Thus it becomes wider and deeper, and so possesses a greater cross-section area. In so doing, it develops a greater hydraulic radius and becomes more efficient. The downstream changes in channel variables are exemplified in Fig. 13.11, and are easy to demonstrate by simple field measurements. Not only does the size of the channel change downstream but so do its proportions. It is clear from Leopold and Maddock's (1953) data that width increases more rapidly than depth. Thus, as the width/depth ratio increases downstream, so the channel becomes proportionately wider, accommodating the increasing discharge and energy and dissipating a greater proportion of energy by friction at the channel margins as the hydraulic radius decreases.

The third major attribute of channel form is its plan. This represents yet another way in which the river adjusts to inputs of discharge and sediment. Four plan forms are normally recognized: straight, sinuous, meandering and braided channels (Fig. 13.12).

Straight natural channels are rare. Even within straight channel sections the flow of water is sinuous or helical. Even in controlled laboratory experiments straight channels cut in uniform sediment last only for a short period before non-straight flow patterns develop. It would appear,

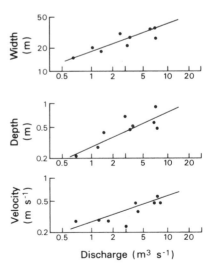

Fig. 13.11 Hydraulic geometry of a stream channel: downstream changes in width, depth and velocity with varying discharge (after Wolman, 1955).

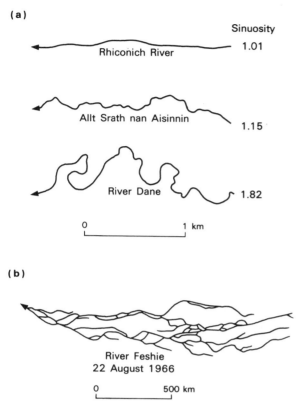

(a)

Rhiconich River   Sinuosity 1.01

Allt Srath nan Aisinnin   1.15

River Dane   1.82

0   1 km

(b)

River Feshie
22 August 1966

0   500 km

Fig. 13.12 Channel patterns: (a) single channels of varying sinuosity; (b) a braided channel.

important in influencing this characteristic. A study by Langbein and Leopold (1966) suggests that the form of the meander curve can be described by a sine function. In such curves the rate of change of curvature is minimized, i.e. there is no concentration of curvature. This can be interpreted as leading to the even distribution of stress and therefore energy loss along the channel length. Thus a meandering channel can be interpreted as being adjusted to promote evenly distributed

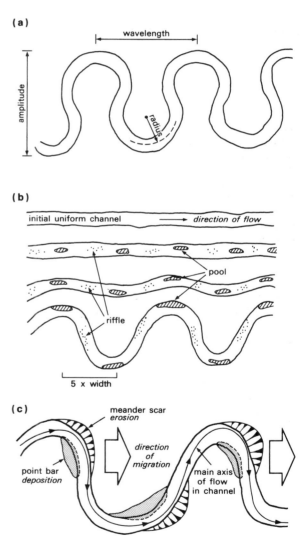

(a)

wavelength

amplitude

radius

(b)

initial uniform channel   direction of flow

pool

riffle

5 x width

(c)

meander scar erosion

direction of migration

point bar deposition

main axis of flow in channel

Fig. 13.13 Meandering channels: (a) geometric definitions; (b) development of meanders from an initial straight channel; (c) geomorphological features of a meandering channel.

therefore, that straight flow is not a normal mode of behaviour for running water.

More common is the sinuous stream pattern, in which the trend of the river channel fluctuates irregularly about its mean. Fig. 13.12a shows examples of sinuous stream patterns, in which sinuosity is measured as total channel length divided by valley length.

A particularly characteristic channel planform is the meandering habit. Meanders are well defined features possessing a high degree of geometric regularity clearly related to discharge. Fig. 13.13a shows the significant geometric features of a meandering channel. Data from meandering channels of a wide range of sizes show that meander wavelength and radius of curvature are closely and directly related to channel width, and thus to discharge. Amplitude is correlated only poorly with channel width, suggesting that local topographic or structural factors may be more

energy loss, given the discharge and load. The effect of meanders is to lengthen the stream channel through increased sinuosity, thereby decreasing its gradient.

Although the precise reason why meanders form is not well understood, their development can be reproduced in miniature form in the laboratory (Fig. 13.13b). Starting with a straight channel in non-cohesive sediments, a sequence of deeps (pools) and shallows (riffles) develops in the channel. This is an unstable form. The pools and riffles are spaced such that a channel length of 10–14 times channel width contains two pools and two riffles. These cause the flow within the channel to become sinuous, resulting in differential erosion of the channel margins. The channel form itself becomes sinuous, and ultimately meandering.

The axis of flow in a meandering channel is located at the outsides of the meander bends. It crosses over at the point of inflection. Thus the meanders develop by eroding the outside of their channel just downstream of the meander apex, and by depositing sediment on the inside of the meander in the form of a point bar. Fig. 13.13c shows the lateral migration of a meandering channel. In this way the meanders migrate down-valley, ultimately having occupied successive portions across the whole valley floor.

Braided streams (Fig. 13.12b) are associated with the deposition of bed load to form bars within the confines of the channel. Thus the thread of water within the channel bifurcates to flow around the bar. Complex braided (anastomosing) streams have an abundance of bars and a multiplicity of interconnecting channels. They appear to be related to highly variable flows and easily eroded channel margins providing an abundance of bed load. Thus, prolific transport of bed load takes place at high discharge, but it is followed by deposition as discharge subsides. In this way the bars are deposited and the dwindling water supply flows around and between them.

Channel plan form, as expressed by meandering and braiding, is clearly a complex and multivariate phenomenon. It must be considered in relation to the hydraulic variables which describe channel processes. Several empirical relationships have been put forward (Fig. 13.14) which relate channel plan form to gradient but neglect the effect of bed load calibre. Thus, for a given discharge the less efficient braided channel requires a higher gradient to do its work. Conversely, for a given gradient a decrease in discharge will convert a braided stream into a meandering one.

The ultimate cause of meandering and braiding has not been established, but it is reasonably clear that channel plan form represents a further mode of adjustment to prevailing conditions of discharge and load. In this way it cannot be dissociated from other more easily described aspects of channel form, such as gradient, cross-section and bed load calibre.

Fluvial erosion resulting from river channel processes removes only a thin section of mass directly from the landscape (Fig. 13.15). It is the operation of the tributary slope processes in opening out the valley away from the channel margin that gives fluvial landscapes their distinctive form. The fluvial system is the dominant mode of removal of mineral material from the world's land surfaces. The magnitude and frequency of fluvial events

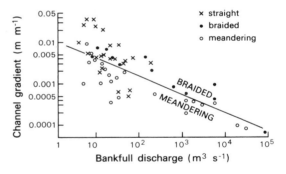

Fig. 13.14 Channel pattern relationships: bankfull discharge plotted against channel gradient for selected channels of differing type (after Leopold and Wolman, 1957).

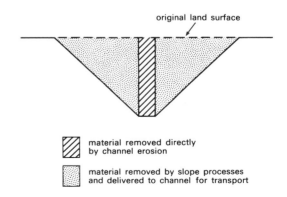

Fig. 13.15 The role of river channel processes in relation to landscape denudation.

vary widely in different environments, as do fluvial forms. However, the principles underlying their operation are fundamentally the same.

Further reading

All of the geomorphology texts referred to at the end of Chapter 10 contain sections relating to fluvial processes, and should be consulted in addition to those offered here. A useful introduction, however, remains:

Morisawa, M. (1968) *Streams – Their Dynamics and Morphology*. McGraw-Hill, New York.

For a comprehensive treatment of fluvial processes in the context of the drainage basin the reader is referred to:

Bull, W.B. (1979) Threshold of critical power in streams. *Geol. Soc. Am. Bull.*, **86**, 975–8.

Gregory, K.J. and D.E. Walling (1973) *Drainage Basin Form and Process*. Edward Arnold, London.

While still a clear introduction, the following should be consulted as a pioneering text:

Leopold, L.B., M.G. Wolman and J.P. Miller (1964) *Fluvial Processes in Geomorphology*. Freeman, San Francisco.

The following is in many ways a European equivalent:

Knighton, A.D. (1984) *Fluvial Forms and Processes*. Macmillan, London.

The following is an excellent book albeit concentrating on channel processes and bed forms:

Richards, K. (1982) *Rivers: Form and Process in Alluvial Channels*. Methuen, London.

Other useful texts include:

Burt, T.P. and D.E. Walling (eds) (1984) *Catchment Experiments in Fluvial Geomorphology*. Geobooks, Norwich.
Hadley, R.F. and D.E. Walling (eds) (1984) *Erosion and Sediment Yield: Some Methods of Measuring and Modelling*. Geobooks, Norwich.
Morisawa, M. (1985) *Rivers: Form and Process*. Longman, London.
Petts, G.E. (1983) *Rivers*. Butterworths, London.
Petts, G. and I. Foster (1985) *Rivers and Landscape*. Edward Arnold, London.
Thornes, J.B. (1979) *River Channels*. Macmillan, London.
Weyman, D. (1975) *Runoff Processes and Streamflow Modelling*. Oxford University Press, Oxford.

The glacial system

Glacial systems develop when the bulk of the precipitation received is in the form of snow, and when the total annual received solar radiation is insufficient to melt the total annual snowfall. Thus snow and ice are permitted to accumulate in large masses at the Earth's surface. Glaciers are therefore found in cold and humid environments, in polar regions and at high altitudes in temperate regions. At the present time glaciers occupy some 14.9 M km², approximately 10% of the Earth's land surface. Of this area, some 95% is accounted for by the major ice sheets of Antarctica and Greenland, while the remainder is distributed in ice bodies rarely greater than 10 000 km² in extent.

The major material input into a glacial system is precipitation in the form of snow, which once in the system becomes transformed into glacier ice. The fragile hexagonal crystals that form snowflakes become compressed into compact granular crystals. Output of mass from the glacier may be by melting (**ablation**) to form meltwater, some of which may evaporate directly to the atmosphere or, when the glacier terminates in a water body, large masses may break off as icebergs (the process of **calving**).

The glacial system can be considered to act as a store of the water at the Earth's surface, since glaciers and ice caps contain some 75% of water in the Earth's land systems. In its general structure the glacial system closely resembles other denudation systems, with the major exceptions that a) water passing through it is in the form of ice, and b) vegetation has negligible influence.

The initiation and development of glaciers is essentially a climatic phenomenon. Once significantly developed, however, glacial systems are capable of effecting striking geomorphological changes. A simplified model of a glacial denudation system (Fig. 14.1) is not dissimilar to other denudation systems. Rock debris is supplied to

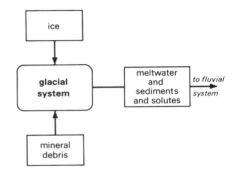

Fig. 14.1 Simplified schematic model of the glacial system.

glacial systems by erosion from the glacier bed, or by rockfalls from steep slopes overlooking the glacier surface. This is transported within the system. Ouput by release and deposition may subsequently be effected directly by glacier ice or indirectly by the action of meltwater.

14.1 Glaciers and the environment

The development of glaciers depends upon the complex interplay of several climatological, topographic and geographical factors. Fig. 14.2 is a simplified representation of the relationships between them.

Important controlling climatic factors are precipitation and temperature. For glacial development it is important that a high proportion of annual precipitation is in the solid form, as snow. This is more important than total precipitation, and usually implies winter precipitation except in the highest latitudes and the summer monsoon region of the Himalaya. The proportion of precipitation received as snow, that is, the ratio of snowfall received (in water equivalent) to total precipitation, has been defined as the nivometric coefficient. The aspect of solar radiation of greatest

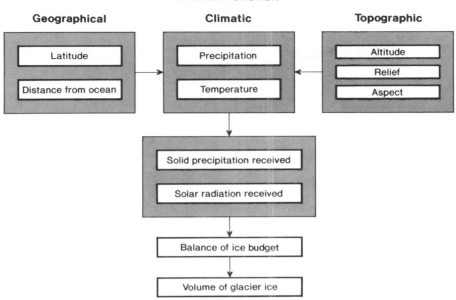

Fig. 14.2 Relationships between some major factors controlling the development and fluctuations of glacier systems.

influence in the annual snow budget is the amount of solar radiation received in the ablation season. Several studies have indicated that a useful measure of solar radiation is that of degree days, the sum of mean daily temperatures above 0°C. It is possible for glaciers to be maintained with as many as 1000 degree days per annum, provided that there is an adequate supply of snow precipitation.

These climatic requirements are more likely to be fulfilled in appropriate geographical locations. Thus latitude is an important variable in defining regions of low temperature, and distance to ocean in identifying regions of maritime influence and therefore high precipitation.

The topographic variables operate at a more local scale and will identify specific massifs, mountains or sites favourable to glacial development.

14.2 System function

In relation to climate a glacier can be viewed as a process–response system. The activity of a glacier can be described by the ice budget – the balance between ice input (**accumulation**) and output (**ablation**). The budget reflects the state of a glacier and whether it is in equilibrium or increasing or decreasing in mass. Fig. 14.3 demonstrates the

relationships between ice input and some aspects of glacier form. The morphological relationships within the glacial system exhibit a negative feedback, in that the glacier damps down the effect of increased input, in part increasing its mass and in part its output.

The balance between input and output is known as **mass balance**, and this will fluctuate on a variety of timescales – seasonal, annual, and over tens, hundreds and thousands of years. Many European glaciers have been receding over the past 250 years, and yet only 18 000 years ago they extended to cover much of northern and middle Europe (Fig. 14.4).

In the case of a European glacier, the budget year (Fig. 14.5a) begins at the first snowfall, the start of the year's accumulation. Accumulation takes place throughout the winter, tailing off towards summer. Ablation is concentrated largely in the summer period, although limited ablation may occur sporadically throughout the winter, if temperatures happen to rise above zero. The vertical axis expresses glacier mass in terms of volume (water equivalent). If the area beneath both accumulation and ablation curves is equal, then the budget is balanced and the glacier is in equilibrium. If accumulation is in excess, then the glacier has a positive budget and is growing.

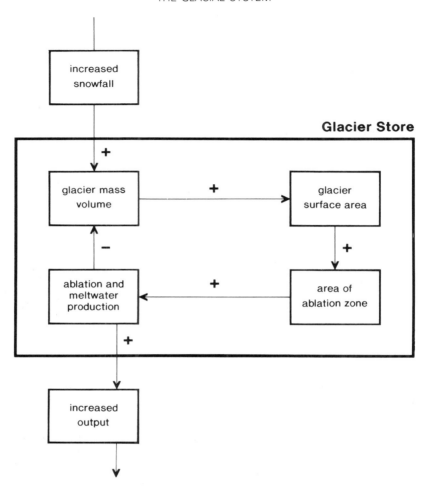

Fig. 14.3 Relationships between ice input and some aspects of glacier form. Correlations between variables are positive or negative as indicated.

If the budget is negative, then the glacier is shrinking.

It is important to note that the budget refers to the throughput of the glacier, not its volume. It is possible to distinguish high-budget glaciers characterized by high input and output, such as are found in some temperate mountains with high precipitation. These are active systems, often small in size, which transfer large volumes of ice rapidly, in contrast to the low-budget systems of polar latitudes where both precipitation and temperatures are lower.

Just as accumulation and ablation are distrib-uted unevenly through time, so also are they distributed unevenly in space over the glacier sur-face. Accumulation is greater at higher altitudes, where ablation processes are more limited due to lower temperatures. Ablation dominates over accumulation in the lower part of the glacier (Fig. 14.5b). Thus the long profile of a glacier may be divided into an upper accumulation zone and a lower ablation zone. The point at which ablation and accumulation are balanced is the equilibrium line. The maximum discharge of ice takes place through this point in the glacier channel, since down-glacier from here ablation continuously

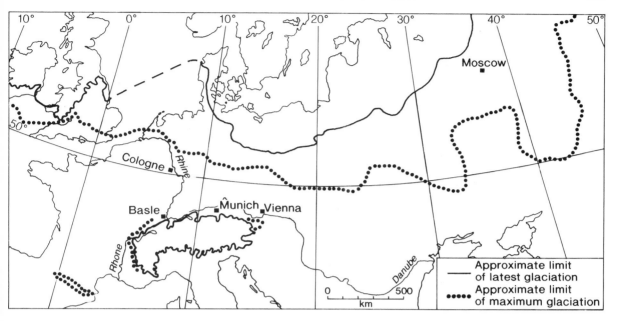

Fig. 14.4 Extent of former glaciations in Europe. Glacial landforms and deposits are present today throughout much of temperate Europe (after West, 1968).

reduces the discharge. In order to preserve a stable equilibrium form, the glacier transfers excess accumulation from the upper part by flowage, and it is only this transport of mass which sustains the glacier body within the ablation zone.

In the accumulation zone there is a downward component of motion as accumulation buries the pre-existing ice. The opposite occurs in the ablation zone as underlying ice is revealed by ablation, resulting in an upward component of motion. This is best visualized in the case of a small cirque glacier (Fig. 14.6), where both accumulation and ablation increments are shown as being wedge-shaped in long profile. Rotational movement takes place in order to redistribute the mass more evenly. In the case of longer glaciers, this model of simple rotational movement becomes attenuated and modified by changes in bed gradient, often caused by variations in lithology or the form of the underlying topography. Nevertheless, the terminal zones of most land-based glaciers where ablation is dominant show an upward component of motion.

The behaviour of a glacier is closely related to its internal temperature (Box 14.1).

14.3 The transfer of mass: glacier flow

The movement of glaciers involves two main components – internal shearing and basal sliding. Internal deformation is caused by shear stresses built up within the ice by the weight of overlying ice (overburden pressure), as a result of which the ice becomes displaced either by creep, as individual crystals deform along their cleavage planes, or by fracture. Movement takes place in the direction of the ice surface gradient (the pressure gradient) and is described by Glen's Flow Law (Box 14.2) $\varepsilon = k\tau^n$. Since

$$\tau = \rho g h \sin \alpha$$

where ρ = density of the ice, g = gravitational acceleration, h = glacier depth, α = slope of glacier surface, τ = shear stress and ε = strain rate, then

$$\varepsilon = k (\rho g h \sin \alpha)^n.$$

Internal shearing, or **intragranular creep**, is favoured when temperatures are relatively warm, as in temperate glaciers, and in polar glaciers when

(a)

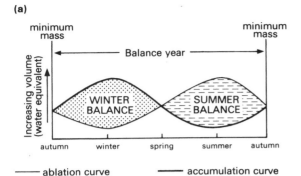

minimum
mass

minimum
mass

← Balance year →

Increasing volume
(water equivalent) →

WINTER
BALANCE

SUMMER
BALANCE

autumn   winter   spring   summer   autumn

——— ablation curve   ——— accumulation curve

Fig. 14.5 (a) The balance year of a glacier: the positive winter balance and the negative summer balance are combined to produce the annual balance. (b) Inputs and outputs of mass in a glacier system (after Sugden and John, 1976).

(b)

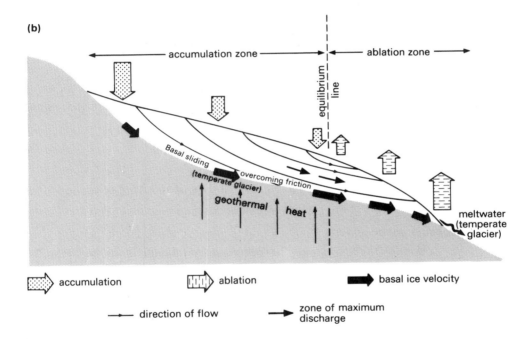

← accumulation zone →   ← ablation zone →

equilibrium
line

Basal sliding

(temperate glacier)   overcoming friction

geothermal   heat

meltwater
(temperate
glacier)

▨ accumulation   ▨ ablation   ▰ basal ice velocity

—→ direction of flow   ⟶ zone of maximum
                                   discharge

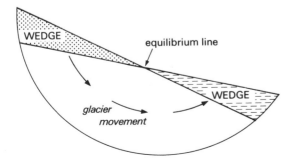

WEDGE

equilibrium line

glacier
movement

WEDGE

Fig. 14.6 Idealized diagram of a small glacier, showing wedge-shaped accumulation and ablation zone; Flow of ice is necessary to maintain an equilibrium surface profile (after Sugden and John, 1976).

Box 14.1
WARM- AND COLD-BASED GLACIERS

Heat within a glacier is derived from three main sources: solar radiation, terrestrial radiation (geothermal heat) and heat derived from the friction created by internal movement and basal sliding. The temperature of a glacier is a major control on its behaviour.

On the basis of the temperatures generated within the glacier by this heat energy, we can distinguish between **temperate ice**, which is at pressure-melting point throughout, and **cold ice** which is always at temperatures below its pressure-melting point.

Pressure melting point, the temperature at which ice melts, reduces with increasing pressure at a rate of approximately 1°C per 14 MPa. Thus the normal stress beneath a glacier 2 km deep may be calculated as

$$\tau_n = \gamma_{gh}$$

where γ = density of ice (900 kg m^{-3}), g = gravitational acceleration (9.81 m s^{-2}), and d = depth of ice (2000 m)
thus

$$\tau_n = 26.48 \text{ MPa}.$$

Under this normal stress the pressure melting point would be reduced to

$$\left(0 - \frac{26.48}{14}\right){}°C = -1.9°\dot{C}$$

The vertical distribution of temperature in a glacier at pressure melting point throughout is shown in (a). There is a slight but continuous decrease in temperature from the surface at 0°C downwards as normal pressure increases. The base of the glacier receives geothermal heat transmitted upwards through the bedrock beneath. Heat can only be conducted along a negative temperature gradient, i.e. from warm to cold. The temperature gradient up through the glacier however is positive, from cold to warm, and the geothermal heat is thus unable to follow this route. It is dissipated by melting the basal ice, and thereby transformed into latent heat. In regions of average levels of geothermal heat flow, the amount of basal melting caused in this way is approximately 5 mm a^{-1}.

In cold polar environments, surface temperatures of glacier ice are very low since they are in equilibrium with the local air temperature. Because of continuously low air temperatures, surface melting is negligible and the ice mass remains cold throughout, with the bedrock beneath commonly below 0°C. The temperature profile beneath a cold-based glacier is shown in (b). Since temperature decreases upwards from the base to the surface, terrestrial heat

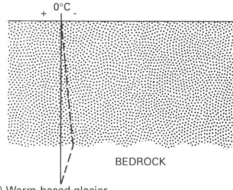

(a) Warm-based glacier

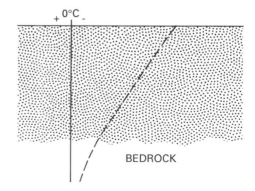

(b) Cold-based glacier

flowing from warm to cold is readily conducted through the glacier body and radiated to the atmosphere. Accordingly terrestrial energy is transmitted through the glacier without affecting its internal temperatures, and it remains frozen to its bed.

The distinction between cold- and warm-based glaciers is therefore one of thermal condition, and a particular glacier need not fall entirely into one category or the other. Conditions may vary seasonally with polar characteristics in winter and temperate ones in summer. Many glaciers are part cold- and part warm-based, conditions at any one point depending on local temperature and overburden pressure.

As a broad generalization, warm-based glaciers tend to occur in temperate environments and where ice depth is great. The cold-based condition is favoured by polar environments and thin ice. Most temperate glaciers are of the warm-based type, as also are some areas of particularly thick polar ice sheets. The geomorphological significance of thermal conditions at the glacier bed lies in the effects which they have on glacier movement and erosion processes.

overburden pressure is particularly high beneath a thick mass of ice.

Internal deformation can also take place along discrete shear planes, rather like thrust faults, within the glacier body. These planes are associated with an increase or decrease in velocity downstream, often related to a change in bed gradient. Thus, where the bed steepens the glacier is under tension and the ice cross-section thins as the glacier accelerates. Conversely, where the gradient slackens the ice is compressed and thickens as it rides up thrust planes which dip up-glacier. These phenomena have been termed extending and compressive flow (Fig. 14.7).

Basal sliding (the sliding of the glacier over its bed) takes place when the shear stress at the sole of the glacier is greater than the frictional resistance across the ice/rock interface. Clearly, resistance is much greater in the case of a polar glacier, frozen to its bed, and it is commonly assumed that movement in polar glaciers is accounted for entirely by internal deformation close to the bed, where the ice is least cold and shear stresses are highest. In the case of a temperate glacier with a film of meltwater at the ice/rock interface, the frictional resistance is much lower and basal sliding may take place freely.

Therefore, the total surface velocity of a glacier

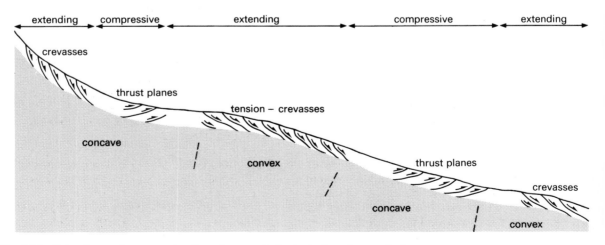

Fig. 14.7 Extending and compressive flow down a glacier bed of varying gradient.

Box 14.2
GLACIER ICE

Glacier ice is a polycrystalline substance derived from the compaction of snow crystals. In contrast to water in the liquid form, all possible hydrogen bonds are operative. The negative end of each molecule (the oxygen atom) is attracted to the positive end (a hydrogen atom) of an adjacent molecule. This constitutes a molecular bond – a form of bonding less strong than other types of bond. We may expect ice, therefore, to possess lower strength than other more strongly bonded materials, such as many rock-forming minerals.

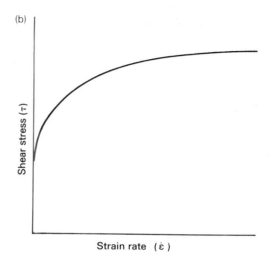

(b)

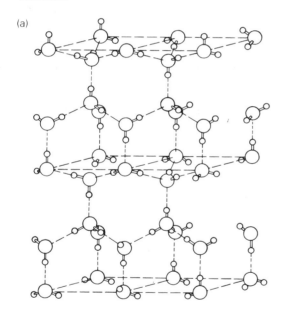

(a)

In ice each oxygen atom is enclosed tetrahedrally by four other oxygen atoms to form a regular crystal structure (a). The open nature of the ordered lattice structure requires a greater volume than the disordered liquid structure; hence water expands on freezing to form ice, which has a density of 0.9 g cm^{-3}. An ice crystal has a strong basal cleavage, determined by the layered lattice structure. As such it can be considered to resemble a pack of cards, with thin layers of card separated by cleavage planes.

The physical properties of ice can be determined by experiment under controlled conditions. The behaviour of ice under stress is illustrated in the accompanying diagram (b). It deforms slowly under low applied stresses but more rapidly under higher stresses. Therefore, its behaviour differs from both a newtonian fluid and a brittle solid. The behaviour of ice is generally described as viscoplastic. The relationship between applied stress and strain for single crystals of ice has been experimentally determined as:

$$\varepsilon = k\sigma^n$$

where ε = strain rate, σ = effective shear stress, and k and n are constants.

This expression is known as Glen's Flow Law (1954). The value of k is directly related to temperature and the value of n is generally shown to be around 3.

Ice thus deforms more readily at higher temperatures and under higher levels of stress – relationships which have important consequences in terms of glacier motion.

Deformation of ice can take place at very low stress levels, but it is strictly limited. It deforms readily at a stress level of 1 bar (100 kN m^{-2}), yielding a strain rate of 0.1 a^{-1} at temperatures close to 0°C. This severely limits the magnitude of shear stresses that the glacier is capable of transmitting through to its bed, for stresses above this level tend to be accommodated by

shearing within the ice itself. Calculated values of shear stress at the base of sliding glaciers generally yield values in the range of 0.5– 1.5 bars.

comprises the two components of basal sliding and internal shearing. The distribution of velocity in vertical profile of a temperate glacier is shown in Fig. 14.8. In the case of cold glaciers, movement may be restricted to the zone of shearing at depth within the ice, causing the ice above to be carried along. In the absence of basal sliding, surface velocity is ascribed to internal deformation alone, and in large polar ice caps, where both temperatures and ice surface gradients may be very low, surface velocities are accordingly very slow.

The pattern of glacier movement at the surface can be readily observed by recording the movement of markers in the surface (Fig. 14.8). The parabolic plan profile shows the effect of both side slip and internal shear. The latter increases toward the centre of the glacier, where the fric-

tional effect of the rough margin is least and there is a greater depth of ice, thus permitting the greatest velocity in the central part.

Variation in velocity may also be expected along the long profile of a glacier. Since the motive force at any one point is dependent on the mass of ice, assuming the surface gradient to be constant, velocity will tend to be greatest where the greatest thickness of ice is present. This will tend to be at the point of maximum discharge – the equilibrium zone. Locally, channel characteristics will influence velocity, with increasing bed gradient or decreasing cross-section causing an increase in velocity, just as in a river channel. In contrast to rivers, however, glacier velocity tends to decrease in the lower reaches as ablation reduces the glacier mass or the ice spreads out over a lowland, thus increasing its cross-sectional area.

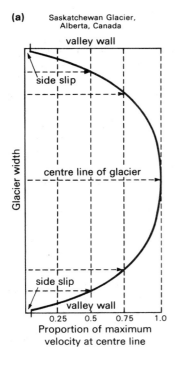

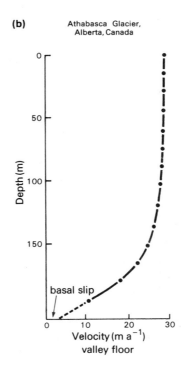

Fig. 14.8 Glacier velocity profiles: (a) surface velocity across a valley glacier: (b) vertical velocity profile of a warm-based glacier.

14.4   Glacier morphology

Morphologically, ice masses can be classified according to their relationship with relief (Table 14.1). Those dominating relief are the ice sheets and domes. The distinction between the two is one of scale. Large masses of ice of continental scale which completely submerge the bedrock topography are commonly termed **ice sheets**, while masses of a more regional scale sitting on areas of high relief are **ice caps**. On a more local scale are the glacier types constrained by relief, whose outlines are largely determined by the topography of the land surface. Thirdly, there are the **ice shelves**, which float on water bodies, usually the sea. Since these do not interact directly with the land surface, we shall not consider them further. We shall discuss the land-based glaciers in ascending order of scale.

Cirque glaciers develop in hollows in regions of accentuated relief. Snow will tend to collect in sites sheltered from solar radiation and the prevailing wind. Accordingly, wind and avalanching will concentrate snow into sheltered hollows and the snow will lie longest in sites sheltered from solar radiation. Cirque glaciers are generally small, and wide in relation to their length, with a steep surface gradient. The relative proximity of the zone of accumulation, with its downward motion, to the ablation zone with its upward flow movement, gives the incipient cirque glacier its characteristic rotational movement (see Fig. 14.6).

Trough glaciers form where one or more cirque glaciers have extended to become confluent. The well developed trough glacier constrained within a valley may attain a length of many kilometres and receive as tributaries both cirque glaciers and other trough glaciers. In plan it may develop the same kind of network form as a river system. The Aletsch Glacier in Switzerland is just one example in the Alps of a well developed trough glacier system (Fig. 14.9).

A more intense form of glaciation is the coalescent condition. Here the ice partially submerges the underlying landscape by overtopping the topographic divides. This can take place in two ways. If a trough glacier attains such a thickness that the bedrock topography is barely sufficient to contain it, then the ice will begin to spill over the divides at low points to coalesce with adjacent glaciers. This is known as **diffluent** flow (Fig. 14.10).

Table 14.1 Classification of ice masses based on their relationship with relief

Unconstrained by relief	Constrained by relief
ice sheet	icefields
ice cap	valley glaciers
	cirque glaciers
	other small glaciers

A second form of coalescence occurs when valley glaciers emerge from an upland on to a plain, where they spread laterally, fanning out until they coalesce. Here again the local relief becomes totally submerged beneath glacier ice (Fig. 14.11).

An ice mass of regional dimensions in an upland area which generated the precipitation to nourish it, is termed an ice cap. This consists of a central ice dome with marginal outlet glaciers which drain to the surrounding lower ground. Ice caps range up to a diameter of ca 250 km, and completely submerge bedrock topography (except around their margins where isolated rock masses may protrude as **nunataks** and define the outlet troughs). The ice dome tends to be symmetrical in plan and convex in profile, and it normally receives its greatest precipitation in the central summit area. Ice flow is radial outwards from the centre. Vatnajökull in southeastern Iceland is a representative example (Fig. 14.12a).

An even more extensive ice cover is termed an ice sheet, which may attain continental dimensions. Ice bodies of this magnitude become independent of the topographic surface beneath, such that the highest area of the symmetrical ice sheet may not be coincident with the highest subjacent relief. Fig. 14.12b illustrates this using the Greenland ice sheet. Since flow is radial outwards from the central and thickest part of the ice sheet, ice streams pass right through the underlying major topographic divide dissecting it deeply – a feature known as **transfluent flow**.

The various stages of glacier development have been defined in purely descriptive terms in relation to the extent of ice cover of the landscape. It seems possible, however, that there may be a generic link between them in that all ice bodies may have small beginnings – the perennial snowpatch – and subsequently develop by thickening and extending to their ultimate form. In so doing they will tend to assume, in so far as bedrock topography permits, the various forms

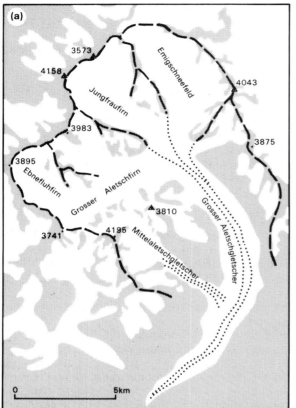

(a)

3573

4158

Jungfraufirn

Emigschneefeld

4043

3983

3875

3895

Ebnefluhfirn

Grosser Aletschfirn

3810

3741   4195   Mittelaletschgletscher

Grosser Aletschgletscher

0   5km

Fig. 14.9 Trough glacier receiving tributaries from corries. Most of the foreground is covered by snow, which, down-valley, has ablated away to reveal glacier ice beneath. The firn line, delimiting the zones of ablation and accumulation, lies across the lower middle of the frame. Down-glacier, convergent tributary glaciers form medial moraines. (Aletschgletscher, Switzerland.)

bedrock

ice – spot heights in metres

ice catchment boundary

medial moraines

(b)

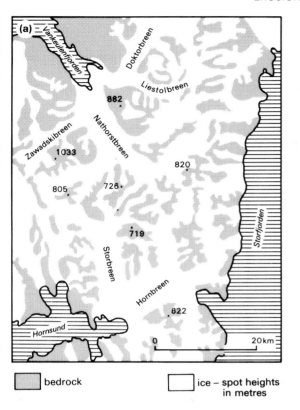

bedrock

ice – spot heights
in metres

described above, from cirque glacier, through the trough glacier stage, coalescing and perhaps developing as far as the ice cap.

14.5 Erosion processes

At the contact between the glacier and its bed the glacier applies various stresses to the bedrock, and it is here that glacial erosion takes place. Glacial erosion is commonly recognized to embrace three major groups of process (Sugden and John, 1976; Drewry, 1986), and we shall consider erosion under the headings of *rock fracturing*, *abrasion* and *meltwater*, dealing with entrainment separately.

14.5.1 Rock fracturing
It is doubtless the case that preglacial weathering plays an important role in creating a pre-existing regolith upon which glacial action takes place. This

Fig. 14.10 Diffluent glaciers. (a) Interconnecting ice streams formed by diffluent glacier flow dissect the landscape (Spitsbergen). (b) Most of the landscape is submerged beneath glacier ice and only isolated peaks of bedrock protrude above the glacial surface (Icefield Ranges, Alaska, USA).

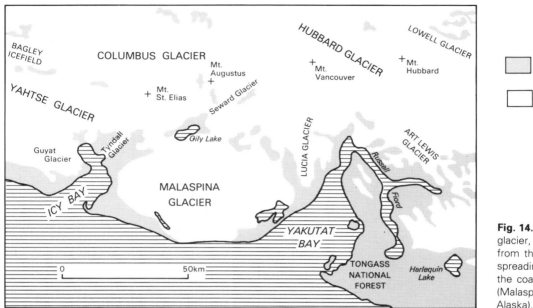

Fig. 14.11 Piedmont glacier, spilling out from the upland and spreading out across the coastal plain (Malaspina Glacier, Alaska).

regolith may take the form of a periglacial weathering profile comprising mechanically fractured rock, or a temperate profile combining coarse fractured blocks in its lower part together with finer material including resynthesized clay minerals (see Chapter 11, Section 11.2). Indeed the nature of glacial deposits in formerly glaciated temperate lands, which contain an often high proportion of clay minerals (*boulder clay*) suggests that they may be derived in large part from temperate weathering in a preceding interglacial period. Two lines of argument, however, suggest that glaciers themselves are capable of eroding fresh unweathered bedrock directly. First, the amount of material evidently removed by glacial action, particularly in glaciated uplands, far exceeds the possible depth of any pre-existing weathering profile. Secondly, till emerging from, and recently deposited by, contemporary glaciers in hard rock terrains commonly consists of fresh blocks and clasts of rock set in a matrix of sand and silt. This implies contemporary erosion by glaciers of fresh unweathered rock, in contrast to the tills in temperate lands.

The key to understanding glacial erosion of fresh rock lies in the nature and magnitude of the stresses applied by glacier ice, in relation to the resistance of rock. The deepest glaciers and ice sheets may attain a thickness of *ca.* 5000 m, and thereby create a compressive normal stress at the bed of about 45 MPa. Many indurated rocks, however, have a compressive strength in small samples considerably greater than this (see Table 10.3), although rock mass strength may be substantially reduced by fractures (Section 10.3.2). Under warm-based glaciers the fractures will contain water, exerting a pore water pressure and further reducing rock strength in the glacial environment.

Under these circumstances it may be that ice itself is capable of creating sufficient normal stress to cause bedrock to fail in compression and develop fractures. A widespread phenomenon is the local enhancement of normal stress due to the presence of boulders at the glacier bed. The normal stress is transmitted through the boulder and will become concentrated at any surfaces or asperities on which the boulder is compressed against the bed. In this way flakes can be detached from the bedrock to form features such as **chatter marks** and **crescentic gouges**.

Glacier beds are rarely plane surfaces of bedrock, but are often highly irregular with a complex pattern of protuberances and hollows. The normal stress exerted by the glacier and its contained debris is thus transmuted over much of the bed into shear stresses. A shear stress applied across the surface of a large block may be sufficient for

(a)

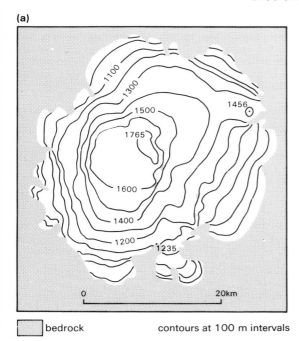

☐ bedrock   contours at 100 m intervals

(b)

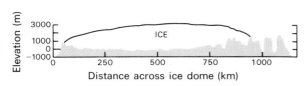

Distance across ice dome (km)

Fig. 14.12 (a) Small ice cap, Hofsjökull, Iceland. (b) Off-central ice dome, Greenland. Note the discordance between the glacier surface and the bedrock topography beneath.

that block effectively to concentrate the energy to a level sufficient to exceed the shear strength of solid rock, and thereby shear off a smaller rock protuberance in the glacier bed. McCall (1960) has calculated that the downstream stress on a 1 m³ block of granite will permit that block to shear off a rectangular protuberance of bedrock of 160 cm² in area in contact with it (Fig. 14.13a). This process multiplied across the total bed of the glacier therefore has considerable potential for erosion.

A second process of rock fracture likely to contribute to glacial erosion is the development of dilatation joints (Fig. 14.14, Chapter 11, Section 11.1.2). Waters (1954) noted the frequency of sheet joints and slabs of rock parallel to the surfaces of

glacial erosion in cirques and glacial troughs. It seems that once glacial erosion has removed a significant depth of overburden, which is replaced by a less dense body of ice, the underlying rock will respond to stress relief by expanding and fracturing. In this way fractured rock becomes readily available for transport at the glacier bed.

A third process of rock fracture is the possible effect of freeze and thaw caused by subglacial weathering of bedrock. Measurements of air temperatures in subglacial cavities have, however, suggested that freeze–thaw cycles are of very limited occurrence. If the basal ice is close to pressure melting point, it is possible that pressure variations may cause thawing and freezing. As ice flows over an upstanding mass of bedrock, basal pressure is increased, leading to pressure melting. The water thus released flows round the protrusion and refreezes on the lee side, where pressure is lower. This process is known as **regelation**, and where rock is already jointed, it may be effective in detaching small rock fragments.

Irregularities in the rock surface at the glacier bed thus play a number of important roles. In addition to the regelation effect just described, they influence the distribution of applied stresses; confining (compressive stress is increased on the upstream side, shear stress generated by normal confining stress, and dilatation, are at a maximum in the zone of stress relief in the lee side subglacial cavity, encouraging fracture in this zone (Fig. 14.15a).

In these various ways fractured rock may become available at the glacier bed for entrainment and transportation within the glacier (Fig. 14.15b).

14.5.2 Abrasion

Abrasion takes place when the base (or **sole**) of the glacier slides over the bedrock surface, a process widespread but confined to warm-based glaciers. Several experiments (Hope *et al.*, 1972; Budd *et al.*, 1979; Matthews, 1979) have demonstrated under controlled laboratory conditions the ability of clear ice to abrade bedrock. The process is enhanced by the presence of debris, both coarse and fine in calibre, at the ice/rock interface. Single clasts with conical asperities concentrate stresses, and when dragged along the bed cut scratches (striations). Fine particles form an abrasive mixture, analogous to a lapidarist's rock-polishing powder. In these ways the glacier may act as a

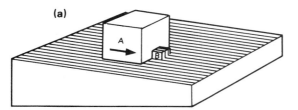

(a)

A

B

Fig. 14.13 (a) Shearing of bedrock by basal debris: shearing stress transmitted through block A in the direction arrowed is sufficient to shear bedrock protuberance B (after Embleton and Thornes, 1979). (b) The sole of a glacier: coarse clasts and fine sediment are contained within the ice; this debris acts as an abrasive on the bedrock surface beneath; as ablation proceeds the sediment is released and deposited (Tunsbergdalen, Norway).

(b)

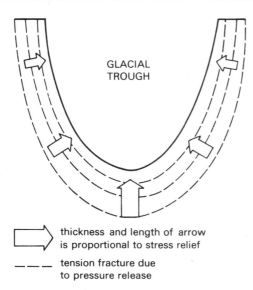

GLACIAL
TROUGH

thickness and length of arrow
is proportional to stress relief

tension fracture due
to pressure release

Fig. 14.14 Dilatation joints formed as a result of glacier trough erosion (after Sugden and John, 1976).

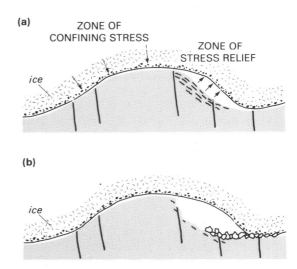

(a)

ZONE OF
CONFINING STRESS

ZONE OF
STRESS RELIEF

ice

(b)

ice

Fig. 14.15 Stress relief causing failure in bedrock beneath a subglacial cavity (after Derbyshire *et al.*, 1979).

giant sander, grinding and scratching and polishing the rock surface to produce fine debris (**rock flour**). Theoretical models suggest that abrasion rates are controlled by normal pressure, sliding velocity, clast shape and hardness, and debris concentration.

14.5.3 Meltwater

The contribution of meltwater to glacial erosion, comprising both mechanical and chemical effects, has been increasingly recognized in recent years.

The operation of meltwater streams is governed by the same fluvial principles as other rivers, but they do possess distinctive characteristics by virtue of their environment, and these render them particularly effective as agents of denudation. First they possess a wide variation in discharge, commonly of two orders of magnitude. The discharge is concentrated in summer, and by diurnal peaks of melting. Short periods of high discharge generate more energy and therefore mechanical power than the same total discharge distributed more evenly. As a consequence of strong pressure gradients and high gradient subglacial channels, the high flood peaks are associated with high velocities and turbulent flow. The temperature of meltwater is commonly <2°C, at which temperature viscosity is increased to 1.8 mN s m^{-2} (compared with 0.8 mN s m^{-2} at 30°C). Such high viscosity reduces the fall velocity of suspended sediment and means that higher concentrations of suspended sediment are characteristic. The principal mechanical effects of meltwater are abrasion of the channel margins by sediment-, ice-, and bedrock-laden waters. Erosion is enhanced in the high turbulence of these streams by the process of cavitation, in which bubbles of gas entrained in the water implode, generating shockwaves of high magnitude. These impinge on rock surfaces and lead to the development of cavities as mineral grains are loosened and detached. Cavitation is a potent process, well known from the field of fluid hydraulic engineering.

The principal chemical effects of meltwater on bedrock are the removal of cations in solution. Glacier ice, when formed, has the chemical composition of rainwater. Studies of ice and meltwater composition within glaciers show that meltwaters become enriched with cations, by solution and cation exchange, when they come into contact with the glacier bed. Freshly fractured rock surfaces and finely comminuted till are rich sources of cations and readily yield them up to passing meltwater. Collins (1979, 1981) has demonstrated strong diurnal and seasonal fluctuations in solute concentration, and has used this parameter to discriminate between meltwater of subglacial and englacial origin.

14.5.4 Rates of glacial erosion

Much remains to be learned about the relative contribution of abrasion, fracturing and meltwater processes in glacial erosion. This is due largely to the difficulty of access to the ice/rock interface. A number of boreholes through deep glaciers have exposed samples of glacier bed, though these are necessarily limited in area; direct observations in subglacial cavities and shallow tunnels have provided quantitative evidence of erosional processes, although this is likely to be unrepresentative by virtue of the marginal location beneath shallow ice.

Attempts to measure glacial erosion can adopt either direct methods, in which the process itself is observed, or indirect methods in which the total net mineral output is assessed by monitoring its discharge in glacial ice and meltwater. The latter method in effect integrates the effects of individual erosional processes over the entire glacier bed (see Chapter 17).

In respect of individual processes, subglacial abrasion rates in the range of 1–4 mm a^{-1} have been measured by Boulton (1974) beneath sliding ice of velocity 10–20 m a^{-1}. Rock fracturing has thus far eluded any effective attempts at measurement; limited measurements of mechanical erosion by meltwater exist (Vivian, 1975), and suggest that local lowering values of 20 mm a^{-1} in hard rock may occur.

Drewry (1986) provides a speculative estimate of the relative significance of the major erosional processes under differing bedrock and glacier-bed conditions (Table 14.2). These must await confirmation by further research and monitoring experiments in this intriguing field of study.

14.6 Transfer of materials: transportation by glaciers

14.6.1 Entrainment

Once debris has been made available by abrasion and fracturing processes, it can then be entrained by and incorporated into the glacier and subsequently transported. Fine particles are readily

Table 14.2 Ranking of glacial erosional mechanisms under different thermal and bedrock conditions (after Drewry, 1986)

Mode of erosion	Cold-based	Warm-based	
		Hard rock	Soft rock
abrasion	2	2	1
crushing/fracture	1	1	2
meltwater mechanical		3	3
meltwater chemical		4	4

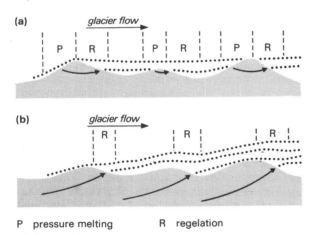

P   pressure melting   R   regelation

Fig. 14.16 Incorporation of debris into the base of a glacier by regelation. (a) In a warm-based glacier the regelation layer remains thin, repeatedly melting and reforming as the ice moves across an uneven bed. (b) In a cold-based glacier the successive freezing on of regelation layers builds up a thin layer of regelation ice (after Barlton, 1972).

incorporated by the process of regelation (14.5.1) as the basal ice experiences variations in normal pressure in flowing over an uneven bed. Under these circumstances where melting and refreezing are continuous processes, a layer of regelation ice containing sand-sized and finer particles may develop to a depth of several centimetres (Fig. 14.16). Blocks and clasts of rock may be incorporated directly into the basal ice by pressure-melting of the ice, or by fluctuations due to varying flow velocity in the size of subglacial cavities, and hence the point at which the glacier touches its bed downstream (Fig. 14.13b). Meltwater, as we have already mentioned, transports fine sediment in addition to cobbles and boulders. Although a portion of the load may be a consequence of

erosion by the meltwater itself, the bulk of the material entrained is probably material initially derived by direct glacial action.

14.6.2   Glacier power and transportation

We have considered glacial erosion processes in a systematic manner on the small scale, examining the stress-resistance relationships of the individual rock particles. It should be emphasized, however, that the processes described may operate around the entire perimeter of the glacier/bedrock contact and along a considerable portion of the glacier's long profile. The individual processes, multiplied across the area of the glacier bed, can add up to considerable amounts of erosion over a period of time.

We can assess the ability of a glacier to perform erosional work in terms of power – that is, energy per unit time ($J s^{-1}$). Power per unit area (W_T) can be defined as the product of bed shear stress (τ which $= \rho g h \sin \alpha$) and mean velocity ($\overline{U}$) as

$$W_T = \tau \overline{U} \qquad \text{(Andrews, 1972),}$$

where W_T is in $J s^{-1} m^{-2}$, τ is in $N m^{-2} = J m^{-1}$, since $1 Nm = 1 J$ (see Chapter 2), and $\overline{U}$ is in $m s^{-1}$.

The major controlling variables of glacier power at a point are therefore, mass, gradient and basal sliding velocity.

There is, therefore, a considerable distinction to be drawn between cold- and warm-based glaciers in terms of the amount of erosional work of which they are capable. With cold-based glaciers basal sliding is inhibited and erosional processes are severely limited, restricted probably to the shearing of masses of unconsolidated rock frozen to the base of the glacier. Warm-based glaciers, on the other hand, with a film of meltwater at their base, slide readily over the bedrock beneath. This permits them to apply considerable shearing forces to the bed, thereby encouraging erosion.

Once entrained, debris is carried along within the glacier towards the snout. The distribution of the load throughout the glacier's mass is important in influencing the mode of deposition at the end of its journey.

Debris (usually coarse rock fragments), fed directly on to the glacier surface by rockfall from bedrock slopes above, may simply be carried along on the glacier surface. Some debris is always present in the base and margins of the glacier, entrained in the regelation layer. A considerable portion of the total load may become embodied deep within the glacier. There are two principal ways in which this comes about. First, when two glaciers form a confluence, the lateral moraine present along their adjacent margins becomes absorbed in the confluent flow downstream as medial moraine. Clearly, where a trunk glacier is served by many tributaries, a considerable amount of moraine may be incorporated in this way. Secondly, shear planes developed under compressive flow, which often takes place towards the terminal zone of a glacier, may carry debris-rich basal ice upwards. In this way a succession of debris-rich bands of ice may outcrop at the surface of the glacier near its snout (Fig. 14.17).

At the snout, then, debris may be distributed throughout the glacier's cross-section. Concentrations may occur in the sole and at the surface, while vertical concentration of medial moraine and horizontal concentrations of thrust-plane debris may occur within the body of the glacier.

14.7 Deposition processes

Deposition takes place at the glacier margin. It embraces a variety of processes, which involve first the release of debris as the ice ablates and then the distribution or reworking of that debris by falling, flowage, glacial overriding or running water. The mode of deposition will be strongly influenced by the location in which the debris is released from the ice.

14.7.1 Deposition by ice
Beneath an active sliding glacier, pressure melting of basal ice leads to the release of debris particles carried in the sole of the glacier. These particles become plastered onto the bedrock floor as lodgement till. Where the glacier is stagnant, subglacial melting takes place caused by geothermal heat, and mineral debris accumulates by accretion on the glacier bed. On the glacier surface, meltout of debris takes place as a result of heating by solar radiation. In the case of a steep glacier front, debris so released may fall or roll directly from the glacier margin. Where the terminal slope is more gentle, the debris accumulates as a carpet on the glacier surface. It may then become saturated by water released by the melting of ice, and flow down the glacier surface to the foreland areas as a flow till. All materials that accumulate at the glacier margin may be subject to reworking by glacial pushing or overriding as the margin fluctuates in position (Fig. 14.18).

It is clear, therefore, that glacial deposition involves a variety of complex processes, and debris may be reworked several times before it finally comes to rest. It follows, therefore, that the characteristics of tills may vary considerably. However, tills have a number of common sediment characteristics:

1. they are generally unstratified;
2. they are generally poorly sorted, and contain clasts of all sizes set in a matrix of fine material;
3. they are composed of a variety of minerals and rock types, many fresh and unweathered;
4. the clasts are variable in shape, sometimes subangular or, when further transported, with smoothed and rounded facets due to abrasion;
5. they may have a preferred orientation of the elongated particles, in which a high proportion of particles lie with their long axes in a restricted azimuthal range.

14.7.2 Meltwater deposition
In contrast to the unstratified till deposits are the sediments laid down by meltwater. Mineral sediments may be picked up by meltwater streams flowing in a **supraglacial** (on the glacier surface), **englacial** (within the glacier) or **subglacial** (beneath the glacier) course and deposited subglacially or, more widely, in a proglacial location. Meltwater deposition follows the same general principles as other fluvial deposition. Its distinctive characteristics are caused by a) the widely fluctuating nature of meltwater discharge, which depends on daily variation in temperature and consequent ablation; and b) the physical constraints on the meltwater channel, which may be constricted within a subglacial tunnel or may braid widely across the glacier foreland.

Meltwater deposits, then, are sorted and stratified. Yet they may vary considerably in calibre over a short distance, due to the variations in discharge and velocity. Where laid down originally in contact with the glacier, they may be affected by subsequent faulting and slumping structures as the supporting ice melts. Alternatively, they too, like till, may be overridden and reworked by the glacier itself.

14.7.3 Conclusion
In its fundamental effect (the transportation of water and erosion of mineral sediment from the land surface) the glacial system resembles the fluvial system. Yet the forms adopted by glacier ice flowing across the land surface, and its erosive

Fig. 14.17 Glacial transport (a and b) Debris-rich layers of ice-carrying sediment to the surface of the terminal zone of a glacier (Tunsbergdalen, Norway). (c) Cones of fine debris emerging from the glacier surface and a large isolated boulder. The latter stands on a plinth of ice which it has protected from ablation (Breidamerkurjökull, Iceland). (d) Englacial bands of fine sediment are exposed at the ice cliff forming the terminus of the glacier (Kviarjökull, Iceland).

(b)

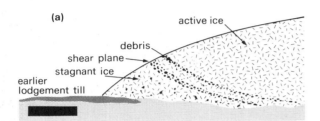

(a)

active ice

debris

shear plane

stagnant ice

earlier lodgement till

(c)

(d)

Fig. 14.18 Glacier terminus. Contorted dirt bands are visible at the glacier surface. The snout zone is completely covered by debris released by ablation. Meltwater released by ablation is stored in proglacial lakes drained by meltwater streams which carry fine sediment to the sea. (Breidamerkuriökull, Iceland.)

power, lead to a variety of distinctive landforms of both erosional and depositional origin. Where glacial occupance of a landscape has been long continued, distinctive and large-scale landforms may be formed – the often spectacular landscapes of glacial erosion. The more subtle lowland landscapes of glacial deposition are probably even more extensive, yet equally distinctive.

For more detailed discussion of glacial modifications of landscapes the reader is referred to the sources listed below.

Further reading

For a thorough treatment of glaciers and their effects, the reader is referred to:

Andrews, J.T. (1975) *Glacial Systems*. Duxbury, N. Scituate.
Sugden, D.E. and B.S. John (1976) *Glaciers and Landscape: a Geomorphological Approach*. Edward Arnold, London.

The next two books provide a broad cover of glacial and periglacial processes and landforms:

Embleton, C.E. and C.A.M. King (1975a) *Glacial Geomorphology*. Edward Arnold, London.
Embleton, C.E. and C.A.M. King (1975b) *Periglacial Geomorphology*. Edward Arnold, London.

An interesting collection of key historical studies is brought together in.

Embleton, C.E. (ed) (1972) *Glaciers and Glacial Erosion*. Macmillan, London.
Drewry, D. (1986) *Glacial Geologic Processes*. Edward Arnold, London.
Eyles, N. (ed) (1983) *Glacial Geology*. Pergamon Press, Oxford.
Paterson, W.S.B. (1981) *The Physics of Glaciers*. Pergamon Press, Oxford.

For a range of specialist papers on glacier-bed processes see:

Symposium on glacier beds: the ice–rock interface. (1979)

Journal of Glaciology 23, No. 89.

The aeolian system

15.1 Introduction

Aeolian systems in denudation (named after Aeolus the God of winds in Greek mythology) are those in which wind plays a dominant role in sculpting the landscape.

In blowing across the land surface the wind is a part of the general circulation of the atmosphere. The mass of atmospheric gas in motion behaves as a dynamic fluid, governed by the same fundamental physical principles as the motion of water, although it has a much lower fluid density (only 1.22×10^{-3} of that of water).

Wind applies mechanical stress to surface materials and is capable of entraining, transporting and depositing sediment. In this way it creates and modifies landforms, and since it is not confined to channels it may operate across the entire land surface.

The circulation of the atmosphere is driven by heat imbalances at the Earth's surface (Chapter 4), which are in turn the consequence of differential surface heating by solar radiation. It therefore follows that solar energy (in combination with the rotational and gravitational energy of the Earth itself) is the ultimate energy source that powers aeolian systems.

Wind as a geomorphological agent assumes greater significance in environments in which there is an abundant supply of surface sediment of suitable grade, that is exposed due to a lack of vegetation cover. These general conditions are most commonly met in arid environments, glacier marginal environments, and on coasts.

Arid environments, defined as regions in which potential evapotranspiration exceeds precipitation, are regions with a continual soil moisture deficit. In reality, active aeolian systems occur in regions which are circumscribed by the 150 mm rainfall isohyet. Thus defined they occupy approximately 20% of the land surface of the Earth, evenly divided between hot deserts and the cold deserts of middle and polar latitudes.

In these arid environments rock breakdown is dominated by processes of brittle fracture, and in the absence of significant chemical weathering and resynthesis of clay minerals, regoliths tend to be dominated by stony, sand and silt-grade material.

Glacier marginal environments are characterized by spreads of sediment in the glacier foreland. These include substantial volumes of pulverized rock flour in the form of silt, deposited either directly by the glacier itself, or after sorting, by the action of proglacial rivers. During Quaternary times, when continental ice sheets were extensive, considerable areas of mid-latitude continents presented land surfaces of this kind.

In the case of coastal environments, particularly on exposed oceanic coasts, large deposits of beach sediment may accumulate as a consequence of coastal processes. This sediment usually consists of well sorted sands, which at low tide are even more extensively exposed.

All three types of environment are, for different reasons, characterized by a paucity of vegetation cover – aridity in the case of deserts, the recency of the surface in glacier forelands, and exposure and salinity on coasts. Under these conditions, then, wind may become a very effective agent of geomorphological change, and it is in these environments that aeolian systems develop to greatest effect.

15.2 Aeolian sediment movement

Air in motion exerts an applied force against a flat surface ranging from 2 N m^{-2} at a velocity of 5 km h^{-1} to 100 N m^{-2} at 50 km h^{-1} and 600 N m^{-2} at 120 km h^{-1}. Its ability to carry out geomorphological work, however, depends additionally on the nature of the fluid motion of the air and also the character of the ground surface and the materials which compose it.

As air flows above the ground surface it is retarded by friction progressively downwards through its lower layers (Fig. 15.1), similar to the retarding effect of friction on a river bed (Chapter 13, Section 13.2). At the base of the airflow there is a thin layer in which the velocity is zero. On a granular surface the depth of this layer is approximately 1/30 of the diameter of particles in the surface. This layer of 'dead air' attains greater depths over a vegetated surface. Above this layer the velocity intensifies progressively in logarithmic fashion.

The rate of change of velocity with height may be described in terms of velocity difference between the height k where velocity is zero, and the height z where velocity is u, then

$$U_* = \frac{Ku}{\log\dfrac{z}{k}}$$

where K = constant. The term U_* is defined as the drag (or shear) velocity of the wind; it is a measure of the retardation of the wind by friction at the ground surface. It is related to the shear stress exerted by the wind on surface particles lying on the surface, thus

$$U_* = \frac{\sigma}{\rho}$$

where σ = frictional stress (shear), and ρ = density of the air. It is therefore the drag velocity which is the critical factor in determining surface sediment movement (Box 15.1). The threshold at which sediment grains begin to move is termed the fluid threshold, and may be expressed in values of drag velocity. As wind velocity increases, so does the drag velocity and hence the applied shear stress. The threshold drag velocity for a particular particle will depend upon the mass of that particle and its density and shape. It has been demonstrated experimentally that, for particles of similar density and shape, threshold drag velocity increases linearly for particles 0.06 mm in diameter and above. For particles smaller than this the threshold velocity for fluid motion increases again, due to their greater interparticle cohesion and therefore resistance to movement (Fig. 15.2). For the majority of desert sands the critical threshold wind velocity for sand movement is 4.4 m s⁻¹. Once grain movement is under way, then the process of saltation (Box 15.2) immediately develops. This further enhances sand movement through the impacting effect of landing grains. Fig. 15.2 also distinguishes an impact threshold velocity, which is the drag velocity required to move particles by grain impact. The

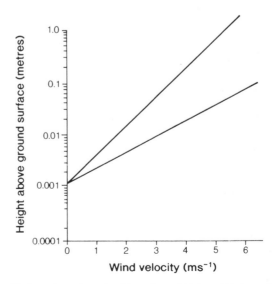

Fig. 15.1 Velocity profiles showing retardation of wind by surface friction, plotted on a) arithmetic and b) logarithmic scales.

Box 15.1
ENTRAINMENT OF PARTICLES BY WIND

Whether soil or sediment particles on a land surface become entrained under the force applied by moving air in the form of wind depends on the balance between impelling and resisting forces.

The resisting forces may be identified as:

1. inertia: the inert mass of any particle subject to gravitational acceleration;
2. friction: the resistance to sliding past adjacent particles;
3. cohesion: the tendency of adjacent grains to cling together, caused by interparticle adhesion in the case of clays, or a bond of moisture in the case of sands.

The main impelling forces are:

1. drag force: the direct impact on the particle by the fluid motion of the wind, creating a substantial difference in applied stress between the upstream and downstream sides of the particle;
2. lift force: the tendency to lift the particle as the wind accelerates over it, resulting in a decrease in normal pressure (a).
3. ballistic impact: the impact of flying particles as they return to its surface, transferring kinetic energy to static particles resting there (b).

Fluid entrainment in air is similar in fundamental principle to that in water, with notable differences due to the much lower density of air. Thus the stress applied by moving air is much lower than that of water flowing at the same velocity, and sediment particles are much less buoyant in air. On the other hand, the effect of ballistic impact in air is much greater, as the bouncing particles meet a much lower fluid resistance in flight.

Particles disturbed from the surface by wind vary enormously in the height to which they may rise, and the duration of the time they spend aloft. The magnitude of the disturbance will depend initially on the strength of the wind, but the amount of lofting into the atmosphere will be enhanced when the air is turbulent, with strong convective updraughts. Gravitational attraction of lofted particles will tend to pull them back to the surface, and the strength of this is directly related to particle size. Fig. (c) indicates the relationship between particle size and fall velocity. Larger (sand-sized) particles fall readily back to ground, small (clay-sized) particles remain aloft for long periods.

Thus sand moves across the surface in a series of intermittent short hops, in which the process of **saltation** (Box 15.2) becomes im-

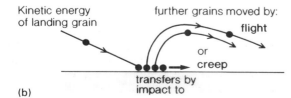

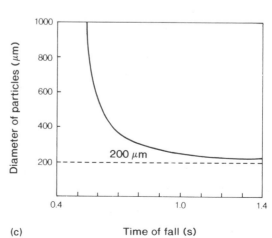

portant. Finer particles may be carried up to a height of 2 km in the atmosphere and may remain aloft for months or years, travelling thousands of kilometres before returning to the surface.

This has the further consequence that in the case of fine particles, there is a much clearer separation between the processes of entrainment and deposition than is the case with hopping sand particles.

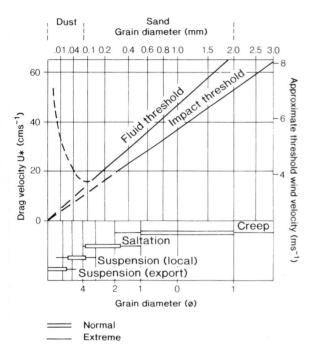

Fig. 15.2 Threshold velocities for sand transport by fluid motion and ballistic impact in relation to particle size.

impact threshold velocity is always lower than the fluid threshold velocity, which means that once grains of a given size begin to move, saltation can continue at lower levels of wind drag.

The volume of sand transported can be expressed in terms of shear velocity and grain diameter, thus

$$q = C\,D \times U'^3$$

where U' = shear velocity during saltation, q = mass of sand moved per unit time, C = constant, and D = grain diameter.

Wind velocity is thus a major determinant of the rate of sand transport, which is shown to vary as the cube of shear velocity (Fig. 15.3).

Given the relationship identified above, the

question arises as to whether high-velocity winds are overall more effective than low-velocity winds in sand transport. This, of course, depends on the respective frequency with which winds of different magnitudes blow. Assuming a realistic distribution of wind velocity through an annual cycle, the cumulative effect on sand movement can be calculated as in Fig. 15.4. It can be seen that over a 1-year period, winds of middling magnitude are most effective in sand movement.

15.3 Erosion by wind

Wind armed with a load of saltating quartz grains (Moh hardness value of 9) forms an effective agent of abrasion. This process of sand-blasting is capable of denuding solid rock at a variety of scales. The ability of desert wind to abrade was demonstrated experimentally by Sharp (1964). Vertical lucite (Moh hardness 2.5) rods were installed in the Mojave desert and showed erosion by sand-blasting of 0.3–0.9 mm over an 11-year period, with a maximum at a height of 23 cm above ground level. It is apparent that this process is also likely to have some effect on natural rock materials.

The removal of surface sand by wind action may lead to a concentration of stones and boulders at the surface of the ground which are too large to be transported by wind. A stone cover of this kind is known as **desert pavement**, or in Arabic, **reg**. Such stones resting on the surface may become abraded to develop a smoothed facet on the upwind side, separated by sharp edges or keels from the downwind face. More than one smoothed facet may develop if multidirectional winds are prevalent, or if the stone is overturned by disturbance. Wind faceted stones are termed **ventifacts**.

The attribution of larger-scale landforms to wind erosion is somewhat more conjectural. Various kinds of streamlined rock outcrop have been attributed to sand-blasting by wind, for example, isolated rock outcrops undercut at their base to create a mushroom-shaped pedestal rock. It is the case, however, that desert weathering processes

Box 15.2
SALTATION

The low fluid density of air as compared to water enables the process of saltation to become especially significant in aeolian systems. A quartz grain is only 1.6 times heavier than its equivalent volume of water, whereas it is 2000 times that of air. It thus becomes very bouncy in air.

A sand grain lifted vertically from the surface will enter higher-velocity streams of air as it elevates. It will thus be carried forward in the direction of wind. Grains larger than 0.2 mm have a fall velocity such that the wind will be unable to maintain them in suspension, and they will fall back to the surface, usually impacting at an angle of 6–12° to the horizontal. A grain landing on a sand bed will transfer its kinetic energy to the grains on which it impacts, and will cause a group of grains to be thrown up into the moving air. They too will throw further grains up on landing, and very soon there is a cloud of sand grains bouncing across the sand surface, commonly to a depth of 20 cm. This is the process of **saltation**.

A sand grain is capable of moving a grain six times its diameter. Such large grains move along the surface by intermittent creep as they are jerked along by impacts. It is estimated that during saltation some 75% of sand is moved by bouncing and 25% by creep.

The cloud of saltating grains has important feedback effects on the wind itself. They extract energy from the wind and retard it. Thus the effective surface roughness increases and the effective surface lies at *ca*. 1 cm above the ground. At this level wind velocity is found to be constant, with a value of *ca*. 4.5 m s^{-1} – the threshold velocity for movement of most dune sands.

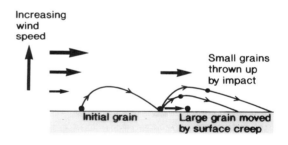

Increasing wind speed

Small grains thrown up by impact

Initial grain

Large grain moved by surface creep

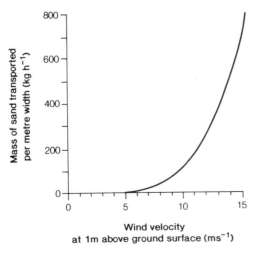

Fig. 15.3 Sand transport as a function of wind velocity.

such as salt and hydration weathering are likely to be most effective at the base of rock outcrops close to ground level, in precisely the same zone of elevation that wind erosion is most effective. It would therefore be unwise to attribute the undercutting of rock surfaces to wind action alone.

Streamlined rocks, elongated in the direction of the prevalent wind, have been described from many deserts. Up to 200 m in elevation and 1 kilometre in length, they are termed **yardangs**, and although in all probability other processes contribute to their formation, the consistent alignment with wind direction points to a significant role for wind in their formation.

At a larger spatial scale, satellite photography has revealed the development of grooved surfaces cut in bedrock over wide areas. A particularly good example occupies an area of some 90 000 km^2 near the Tibesti massif in the Sahara desert. Be-

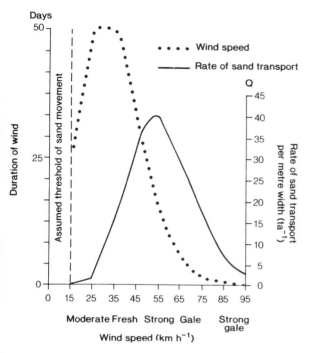

Fig. 15.4 Effectiveness in sand transportation of winds of differing magnitude and frequency (after Warren, in Embleton and Thornes, 1979).

tween half and one kilometre wide and 500 m to 2 km apart, the grooves have steep sides and flat floors. Their curved alignment is absolutely consistent with the regional wind.

Large-scale desert hollows, such as exist in the Western Desert of Egypt, have been attributed to deflation by wind. The Qattara Depression, for example, measures some 250 km by 100 km, and extends to 134 m below sea level. Whereas many geomorphological processes are likely to contribute to such a large-scale landform, the removal of sediment by wind at the regional scale is undoubtedly a major factor in its development.

At the regional scale, the volumes of sediment transported by wind are very considerable. It is estimated that winds blowing across the southwest Sahara export some 200–400 Mt a^{-1} of dust into the Atlantic, where it is deposited to be incorporated into deep-sea sediment. Windborne dust from north Africa has accumulated to form a 10 cm layer of silt on the northern shores of the Mediterranean in only 300 years. It is clear from these observations that wind is an effective agent of regional denudation of desert areas.

15.4 Aeolian deposition

The most distinctive feature of aeolian systems is the creation of dune forms by aeolian deposition, whereby distinctive landforms are shaped upon large spreads of sand. Although this is a common stereotype of desert environments, it is important to recognize that dunefields occupy only a limited proportion of the area of individual deserts commonly 20–30% (Table 15.1). Dunefields nevertheless tend to occupy large areas. It has been estimated that 99.8% of all active aeolian sand is found on sandfields of area greater than 125 km^2. The modal size of dunefields is about 190 000 km^2, with the largest known, the Rub al Khali in Arabia, at 560 000 km^2. Vast volumes of sand are present in these features.

Accumulations of sand – **ergs** – tend to collect in desert lowlands, where winds are less severe and erosive than in adjacent uplands. The sand in dunefields is believed to have originated largely from alluvial sediments deposited in desert lowlands after being eroded from nearby uplands. In several tropical deserts the local bedrock is desert sandstone laid down in earlier geological periods, and indicating climatic stability over long periods of geological time. Weathering and denudation of these rocks may thus release sediment grains originating from ancient desert conditions. Most desert sands consist of mechanically resistant quartz grains, and given their transport history, are generally well rounded in form.

The nature of the dune forms which develop depends upon a range of factors. Wind strength and direction, and the local airflow patterns which develop, are one set of factors. Sand particle size and supply form another set. In the case of coastal dunes vegetation also becomes a significant factor. The dune forms which result can be classified in terms of size, form and internal structure (Table 15.2).

The internal structure of dunes was highlighted

Table 15.1 Proportion of individual deserts occupied by dunefields and sands (after Cooke and Warren, 1973)

Australian	31.0%
Sahara	28.0%
Arabia	26.0%
Libyan	22.0%
Southwestern USA	0.6%

Table 15.2 Morphological features of some basic dune types (after Chorley *et al.*, 1984)

	Form	No. of slip faces	Mean length (km)	Mean width (km)	Mean wavelength (km)	Form of occurrence
barchan	crescentic plan form	1	0.56	0.90	0.68	isolated forms
akle	asymmetrical ridge	1	1.27	2.11	1.90	transverse to wind
parabolic	U-shaped plan form	1				coastal dunes
linear	symmetrical ridge	2	18.14	0.24	0.81	parallel to wind
star	central peak, radiating arms	>2	0.86	0.86	1.76	isolated forms

by McKee (1979) as an indicator of the nature of the wind regime responsible for particular dune types. On the lee side of a dune a slip face forms at the angle of repose of dry sand (30–34°) which produces a set of cross-stratified beds dipping downwind. These beds form the internal structure of the dune. The number of sets of such beds present in the internal structure of the dunes – one, two, three or more – is indicative of the number of dominant wind directions involved in the formation of the dune.

There appears to be a hierarchical arrangement of dune bedforms, as indicated in Fig. 15.5 (Wilson, 1972), and related to the particle size of the sand forming this feature. Based on worldwide data there is a fairly clear separation between forms at different scales – ripples, dunes and megadunes (termed **draa**) (Table 15.3).

15.5 Aeolian bedform morphology

15.5.1 Ripples
Ripples are formed during saltation. Where an irregularity exists in the sand surface, a windward-facing slope will, by virtue of simple geometry, intercept more grains than a leeward slope. Thus more grains will be thrown up from the windward slope. Assuming constant wind velocity and uni-form sand size, then the grains will tend to travel a uniform distance downwind, there to accumulate and form a further irregularity. In this way a set of small ridges, transverse to the wind direction

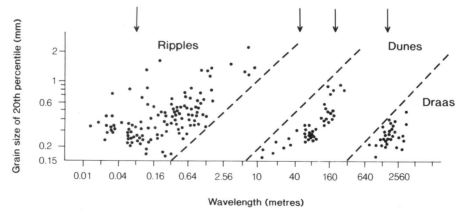

Fig. 15.5 Relationship between grain size and wavelength of dune bedforms. Arrows indicate frequency modes of different scales of bedform. The value defined by the grain size diameter, termed the 20th percentile in frequency distributions of sand grain sizes from samples of dune bed forms, is that at which 20% of the sand grains in the distribution are finer. (after Wilson, 1972).

Table 15.3 Characteristic features of the hierarchy of aeolian bedforms (after Wilson, 1972 and Selby, 1985)

	Modal wavelength	Range of wavelength	Range of height
ripples	8 cm	5.0–200 cm	0.1–5 cm
dunes (1)	40 m	3.0–600 m	0.1–15 m
dunes (2)	200 m	3.0–600 m	0.1–15 m
draa	1500 m	0.3–3 km	20.0–400 m

and equally spaced, will develop as ripples (Fig. 15.6). The ripple spacing will depend on grain size wind strength. Ripples travel in the downwind direction as the material is removed from the windward slope of each ripple, to be deposited on the next one downwind.

15.5.2 Desert dunes

Longitudinal dunes. This dune type is probably the most common form. Longitudinal dunes consist of long parallel ridges, continuous over many kilometres, sometimes joining in 'tuning fork' junctions (Fig. 15.6a). The individual ridges are often separated by sand-free stony surfaces. Formed under constant unidirectional winds, they appear to be related to horizontal roll vortices, horizontal corkscrew-like wind motions which sweep sand laterally upwards on to both flanks of the dune from the interdune surfaces.

Another type of longitudinal dune, the **seif** dune, common in Arabia, is more sinuous and less continuous in form (Fig. 15.6b). It appears to be associated with a wind climate in which there is a second major directional component, perhaps of a seasonal nature.

Fig. 15.6 (a) Longitudinal dunes, Great Australian Desert. (b) The sinuous form of a seif dune ridge, Negev Desert, Israel.

Transverse dunes. This dune type appears to develop in a unidirectional wind regime in association with wave-like surface airflow, which may develop downwind of a major obstacle such as an upland area. At the wave trough, where higher-velocity winds sweep the ground surface, sediment is entrained, to be deposited where the higher-velocity winds are aloft at the wave crests. Thus dune ridges accumulate transverse to the wind direction, and with a spacing equal to the wavelength of the windflow. Ridge growth continues until the lee (slip) face exceeds the angle of stability (30–34°) of dry sand. Stability is then restored by small-scale sand slips (Fig. 15.7), which reduce the angle of the dune face and transport sand forward towards the base of the dune. In this way the dune migrates in the downwind direction.

Fig. 15.7 Dune slip face showing slip scar and sandflow track. Dune face is *ca.* 20 m high. Ripples are evident on dune faces.

Transverse dunes may develop a sinuosity of plan form, with regularly spaced alternating downwind bulges and upwind re-entrants (Fig. 15.8). Known as **aklé dunes**, they are associated with horizontal roll vortices aligned parallel with the dominant wind and representing a smaller-scale circulation system.

Barchans. These isolated dunes are associated with unidirectional winds and a limited sand supply. They occur in fields, the individual dunes separated by sand-free surfaces (Fig. 15.9). They are characterized by a crescentic plan form, a rounded humpback upwind, and two curved horns flanking the downwind slip face.

Star dunes. This dune form is created by multi-directional winds. They have a central peak and radiating arms. They appear to grow vertically rather than laterally, and often attain elevations of tens of metres.

Field measurements of the migration of barchans indicate rates of movement of up to 50 m a^{-1}. Smaller dunes migrate more rapidly than larger ones, which leads to the general conclusion that individual dunes will tend to coalesce to form large dunes.

Coalescence may also occur when two dune patterns are superimposed on each other. Two crossing sets of linear dunes will create star dunes at their points of intersection. In this and other ways, compound dunes may be formed to create a complexity of form which reflects the local complexity of wind patterns (Fig. 15.10).

15.5.3 Coastal dunes

The development of coastal dunes is encouraged where strong onshore winds blow across a sandy beach, especially if the tidal range is large. Under these circumstances a wide expanse of sand is exposed to drying wind and becomes susceptible to aeolian onshore transport.

Coastal dunes differ in one important respect from desert dunes, and that is that they may develop in humid climatic environments in which vegetation becomes a significant factor in dune formation. The prolific development of coastal vegetation in the humid tropics, however, seems to inhibit coastal dunes in those environments.

Wind driving across a sandy foreshore will entrain and transport sand in a saltation cloud. When this sand cloud passes on to a more energy-

Fig. 15.8 Transverse dunes showing aklé form, Morocco.

absorbent surface, for example rippled sand above high-water mark, or damp sand or vegetated ground, then more energy is absorbed by the ground surface, saltation declines and deposition takes place. Thus an embryo foredune may form, which itself acts as a trap for further sand. This dune is linear in plan and parallel to the coast along the landward side of the beach. Subsequent development of the dune is enhanced by the growth of vegetation, especially maritime grasses, which further trap both energy and sand.

Coastal dune growth rates have been observed up to 1 m a^{-1} and heights over 10 m may readily be attained. Commonly, transverse coastal dunes are asymmetrical in profile, with a steeper seaward slope. As the wind passes over the crest a separation of flow occurs as high-velocity winds rise, creating a zone of sheltered air on the lee face. In this quiet zone saltation declines rapidly and

deposition takes place. The effect of erosion on the exposed face and deposition on the lee face is the inland migration of the dune mass, to be succeeded in time by the development of a further foredune.

Coastal dune systems therefore tend to comprise a series of linear transverse ridges, separated by troughs which are swept clear of sand by winds descending over the dune from seaward. The elevation of these troughs or **dune slacks** is determined by the level of the water table, which acts as a natural lower limit to wind erosion by keeping the sand moist.

Older dunes, further inland, become more irregular in their morphology and plan form. If the vegetation cover is broken, by animal or human trampling, or fire, then sand at or near the crest of the dune is again exposed to the wind and erosion ensues. Thus sand in the area of the crest

Fig. 15.9 A small barchan dune surrounded by stony desert pavement.

is blown forward and the linear plan of the crest becomes punctured by a bulge in the downwind direction. This is termed a **blow-out** (see Chapter 25, section 25.1.4, and Fig. 25.10).

Over time a linear dune may become quite irregular in plan. Blow-outs may be further extended until the modal dune form becomes parabolic in plan. The bow of the parabola faces downwind, the arms anchored by vegetation forming parallel linear ridges to windward (Fig. 15.11).

15.6 Conclusion

Aeolian systems differ from many of the other denudation systems in their external relations. Whereas weathering, slope and fluvial systems occur in a linked cascade, aeolian systems are more varied in their contacts with other systems.

Occasionally there are sharp system boundaries,

where a coastline or river intersects a desert. More commonly however, desert margins grade imperceptibly into semi-arid lands. Similarly, the outputs of aeolian systems are rarely demonstrable at a distinct system boundary. Occasional exceptions do occur, as, for example, in the case of sand dunes overrunning adjacent non-aeolian terrain.

It is, however, the large-scale and perhaps long-term export of fine aeolian sediment which has the greatest significance for other systems and terrains. Such materials return to the Earth's surface as **airfall deposits**, most notably as far-travelled silt in the form of **loess**. This material accumulated around the margins of continental ice sheets in North America and Europe in Quaternary times, and mantles parts of northern China east of the Gobi Desert to depths of more than 250 m.

In these regions loess, as the geological substrate,

Fig. 15.10 Complex pattern of crossing dunes of draa size.

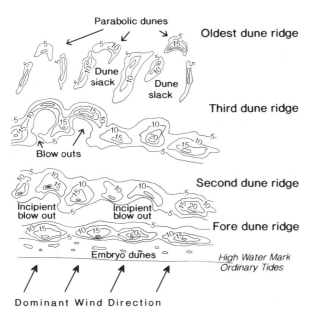

Fig. 15.11 The development of coastal dune forms; on the coast of Britain such a sequence might represent 200–300 years of accumulation.

forms a major input to fluvial denudation systems, which often produce intricately dissected landscapes in the easily eroded silt.

Further reading

The classic text which still remains the fundamental source, in spite of being dated is:

Bagnold, R.A. (1941) *The Physics of Blown Sand and Desert Dunes*. Chapman & Hall, London.

More general advanced texts on aeolian geomorphology are:

Cooke, R.U. and A. Warren (1973) *Geomorphology in Deserts*. Batsford, London.
Goudie, A.S. and A. Watson (1980) *Desert Geomorphology*, Macmillan, London.
Nickling, W.G. (ed) (1986) *Aeolian Geomorphology*. Allen & Unwin, London.
Thomas, D.S.G. (ed) (1989) *Arid Zone Geomorphology*. Belhaven, London.

The following texts extend the consideration of arid environments beyond the themes of this chapter:

Heathcote, R.L. (1983) *The Arid Lands: Their Use and Abuse*. Longman, London.
Louw, G.N. and M.K. Seely (1982) *Ecology of Desert Organisms*. Longman, London.
Spooner, B. and H.S. Mann (eds) *Desertification and Development: Dryland Ecology in Social Perspective*. Academic Press, London.
Uncod (1977) *Desertification: its Causes and Consequences*. Pergamon Press, Oxford.

Additionally, individual chapters in advanced texts are often informative, *viz*:

Chorley, R.J., S.A. Schumm and D.E. Sugden (1984) *Geomorphology*. Methuen, London.
Derbyshire, E., K.J. Gregory and J.R. Hails (1979) *Geomorphological Processes*. Dawson, Folkstone, Chapter 4.
Embleton, C.E. and J.B. Thornes (eds) (1979) *Processes in Geomorphology*. Edward Arnold, London. Chapter 10.
Selby, M.J. (1988) *Earth's Changing Surface*. Oxford University Press, Oxford.

The coastal system

Coasts occur where the world's oceans and seas lap up against the margins of land masses. It is estimated that some 440 000 km of coastline exist (Pethick, 1984), and as such represent an environmental system of considerable magnitude. Coasts are characterized by a distinctive set of environmental conditions, and are acted upon by a distinctive set of processes. They therefore merit separate consideration, and in a manner which indicates the interaction between physical and ecological systems.

It is appropriate to think in terms of a coastal zone. The direct forces of waves and tides are brought to bear in a zone defined vertically by the tidal range (plus the height of any waves superimposed on high-tide level), and laterally by the distance between high and low water marks. Yet these limits do not entirely define the scope of coastal processes, for there may be transfer of materials both onshore and offshore from this defined zone by winds, waves and currents. The coastal zone itself consists of a number of subdivisions, each associated with particular sets of processes and conditions (Fig 16.1).

Within the intertidal zone, oceanic energy in the form of tides and waves becomes very narrowly focused. This energy input is continuous through time, and yet very variable over short, diurnal and monthly intervals. Large storms may occur episodically, causing a high short-term energy input as large ocean-derived storm waves beat upon the shore. This constantly varying energy input, combined with the ready availability of mobile sediment, means that the coast is a very dynamic zone. It responds rapidly to changing inputs, and coastal change due to geomorphological action may readily be observed over short periods of time.

It is estimated that some 65% of the Earth's human population live within a few kilometres of the coast. The coastal zone thus has considerable significance for these people, who may be engaged in a variety of activities – dwelling, food production, industry, communication and recreation, for example. These activities represent a great investment both of capital and human energy. Yet environmental processes in the coastal zone create a variety of hazards in these intensively used lands, such as flooding, erosion, storm damage, and pollution. It is therefore vital to human use of the coastal zone that effective environmental management is exercised, and this in turn implies the need for a clear understanding of coastal processes.

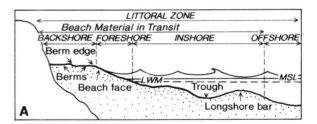

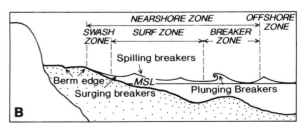

Fig. 16.1 Subdivisions of the coastal zone in relation to (a) beach profiles, and (b) waves and currents.

16.1 Sources of energy in the coastal system

The two major immediate sources of energy specific to coastal systems are those of tides and waves. Both of these, however, represent transformations of energy derived from more distant sources. Tidal energy has its origins in the gravitational energy of the Moon and Sun, in combination with the rotational energy of the Earth. (Chapter 6). Wave energy is created by wind shear across the ocean surface (see Section 6.3), thus transforming the kinetic energy of wind to the potential and kinetic energy of waves. Since the wind is a function of the general circulation of the atmosphere, itself powered by solar radiation imbalances, it can be seen that the ultimate source of wave energy is the Sun. The transformations of energy involved in the creation of tidal and wave energy are illustrated in Fig. 16.2.

16.1.1 Tides

Tidal level refers to the still-water level on a coast at a particular time, neglecting any short-term disturbance in water level due to waves. Tides are raised on a regular and predictable basis as a consequence of the relative motions of Moon and Sun in relation to the Earth (Section 6.3). In this way, regular oscillations in tide level are created, subject to the fortnightly variation in range defined by the spring/neap cycle.

The tidal variations created by these forces range from 0.5 m in the open ocean, exemplified by Hawaii, to maximum values of 10 m or more where the tide is amplified by funnelling into an estuary. Davies (1972) defines tidal environments as macrotidal (>4 m), mesotidal (2–4 m), or microtidal (<2 m) according to the tidal range at springs. A further identifiable characteristic of tides is that of **tidal duration** (Trenhaile, 1980; Carr and Graff, 1982). This defines the proportion of time the tide is present at various levels throughout its range (Fig. 16.3).

The regular and predictable oscillations of the tide are liable to modification by other episodic effects. Variations in atmospheric pressure can depress or raise sea level. A decrease in atmospheric pressure of 1 mb raises the water level by 1 cm – the inverse barometer effect. The not-infrequent occurrence of mid-latitude depressions and tropical cyclones with pressures as low as 950 mb can thus have the effect of raising sea level locally by 0.5 m.

Associated with these events is the generation

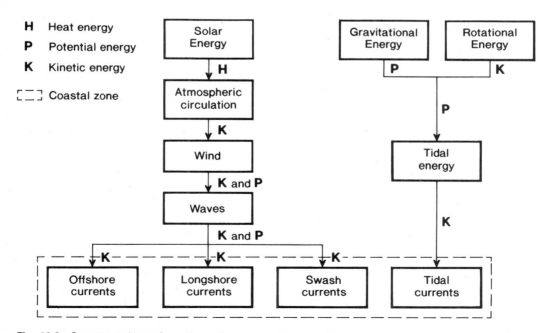

Fig. 16.2 Sources and transformations of energy in the coastal zone.

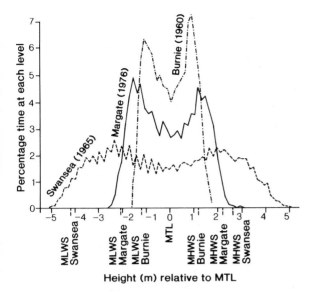

Fig. 16.3 Representative tidal duration curves for different stations (after Carr and Graff, 1982).

of strong winds. When these blow onshore they have the effect of transferring water shorewards, causing it to pile up and thereby elevating the water surface against the shore (**wind set-up**). The storm surge of January 1953 in the North Sea raised tidal levels by 2–3 m on the coasts of East Anglia and the Netherlands.

The significance of tides in relation to coastal processes is as follows:

1. They define the range of elevation (including the maximum height) in which direct wave and other marine action takes place.
2. The flow and ebb of tides around coastal irregularities (inlets, islands, shoals) create tidal currents.
3. The diurnal alternations of inundation and exposure of the intertidal zone create conditions for particular rock weathering processes (slaking, salt crystallization).
4. Tides create the intertidal habitat to which a range of plant and animal organisms is adapted, and which may have geomorphological effects.
5. With each tide an influx of nutrients is supplied to intertidal organisms.

16.1.2 Waves

Wave action is probably the most important agent of coastal change. The generation of waves at sea,

the magnitude of wave energy and its progress onshore are therefore fundamental to any consideration of coastal processes (Box 16.1).

Waves are generated at sea by the passage of wind across the water surface. A glassy calm surface will become ruffled in contact with eddies in turbulent windflow at the boundary layer. These eddies create differences in dynamic pressure normal to the water surface, and waves are initiated. Once initiated, waves will grow in height and length as the wind continues. The waves now interact with the wind, and under continuing wind shear and turbulence will continue to grow. Waves in the process of generation are known as **sea waves**, and tend to be irregular in form, with short crossing crests forming a confused pattern. They are accelerated and propagated in the direction of windflow.

The size of waves generated in open sea is related to wind duration and velocity, and the length of fetch (the length of open sea across which the wind blows). Fig. 16.4 is an example of a wave-forecasting chart based on these variables. After departing from the zone of turbulence in which they were generated, waves are propagated across the open ocean towards the continental margins, becoming smoother and more regular as they progress. Travelling across the open ocean they may encounter other wave trains of different height and period, upon which they become superimposed. The waves received by a coast at any given time may thus originate from several sources, and will demonstrate the cumulative effect of several wave trains in the form of a wave spectrum.

As they approach the coast, waves encounter shallowing water, as a result of which they undergo important changes, or **shoaling transformations**, modifying both wave sectional form and plan form. Whereas wave period remains constant, wave velocity and length both decrease, whilst wave height increases (Box 16.2).

If a train of parallel waves approaches a shallowing coast obliquely, or crosses an irregularly shallow sea bed, then the waves will be differentially retarded along their length, a process known as **wave refraction**. Fig. 16.5 shows the effects of refraction on waves approaching a straight coast obliquely, and normal waves approaching an indented coast. The distribution of wave energy can be demonstrated by constructing **wave orthogonals** – lines normal to the wave crest – which are shown to converge or diverge

Box 16.1
The nature and form of ocean waves

(a)

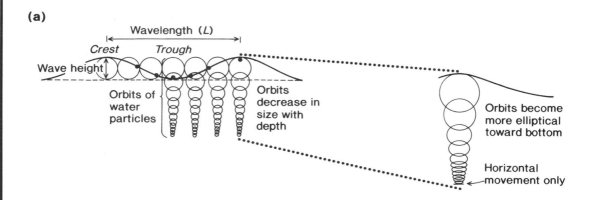

Waves form in open water and are associated with the orbital motion of individual water particles. Each particle orbits in a constant position; orbit diameter decreases with depth, becoming negligible at a depth equal to half the wavelength ($d < 0.5L$; see below and Fig. (a)). The position of the orbit, and the wave particle in it, remains in a stationary position while the wave form moves forward.

Waves can be defined with the use of four basic descriptive parameters (Fig. b). These are:

Wavelength (L) – the distance between successive crests

Still water depth (d) – water depth in the absence of surface disturbance

Wave period (T) – the time taken for consecutive crests to pass a given point

Wave height (h) – the height from crest to trough.

From these basic parameters the following are derived:

wave velocity or *celerity*, $C = \dfrac{L}{T}$ (1)

wave steepness $\quad = \dfrac{h}{L}$

Waves in deep water, defined as $d > 0.25L$, known as Airy waves, can be defined by the following equation:

(b)

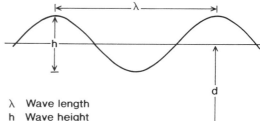

λ Wave length
h Wave height
d Still water depth
T Time taken for successive crests
    to pass given point (wave period)

$C = \dfrac{\lambda}{T}$ (wave celerity)

$$L = \frac{gT^2}{2\pi}$$

substituting for $g(9.81 \text{ m s}^{-1})$ and π gives

$$L = 1.56T^2$$
(2)

and substituting L from equation (1) in equation (2) gives

$$C = 1.56T$$

Thus longer wave period is associated with greater wave velocity ($\sim T$) and much greater wave length ($\sim T^2$). Long waves therefore travel quickly; furthermore, as they emerge from the storm zone in which they originated, they will separate out by virtue of their velocity from the less rapid smaller waves.

Coasts which face open oceans may receive a high proportion of far-travelled ocean waves, known as **swell waves**.

(e)

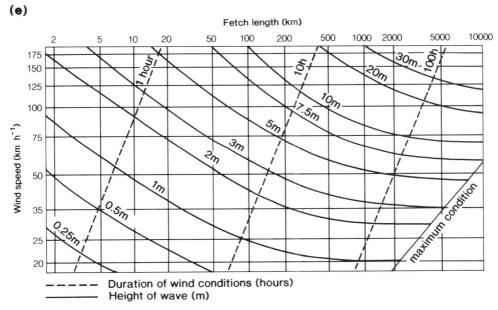

Fetch length (km)

Duration of wind conditions (hours)
Height of wave (m)

Fig. 16.4 Wave-forecasting chart illustrating the effect of windspeed, duration and fetch on wave height (after Bretschneider, 1959).

towards the coast, indicating concentration or dispersion of energy along different sections of the shore. Wave energy distribution can also be calculated for a coast with an irregular sea bed offshore, under different wave conditions. Fig. 16.6 shows a pattern of wave attack thus calculated, which is surprisingly unrelated to the outline of the coast itself (Hails, 1975).

The shoaling transformations also bring about the breaking of the waves as the water depth continues to decrease towards the shoreline. Decreasing wave length and increasing wave height result in an increase in wave steepness. When this reaches a critical value, defined by

$$\frac{h}{L} = \frac{1}{7}\,(= 0.147).$$

then the wave becomes unstable, and breaks by spilling, plunging or collapsing. In this way the mass of water moves ahead of the wave form, and the water runs up the beach in the form of **swash**, a transformation of the potential energy of wave elevation to kinetic energy.

Clearly water depth is an important factor in the breaking of waves, since it is the cause of increasing wave steepness. Thus higher waves are required to break in deeper water, while lower waves can run on into shallower water before the point of breaking is reached. There is a close relationship between wave height and water depth given by the ratio

$$\gamma = \frac{L}{d}.$$

Box 16.2
WAVES IN SHALLOW WATER

As waves approach shallowing water, the particle orbits become increasingly elliptical, elongated horizontally, until the basal water has a simple horizontal to-and-fro motion. This alternating motion is asymmetrical in duration and velocity as the orbital loops are not entirely completed in the shallowing water. This results in a net shoreward mass transport of water, an important onshore normal current.

In water of intermediate depth, defined in relation to wave length as $0.25 > d/L \geq 0.05$, the Airy equation

$$L = \left(\frac{gT^2}{2\pi}\right) \times \tan h \frac{2\pi d}{L}$$

predicts that velocity, and therefore wave length, decreases, and wave height increases. This is shown in Fig. (a), in which values of water depth and wave velocity and height are plotted as ratios relative to their initial deep-water values.

For waves in shallow water, defined as $d/L <$

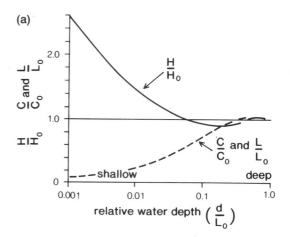

(a)

0.05, the Airy equation simplifies to $L = T\sqrt{gd}$ and since $C = L/T$, then $C = \sqrt{gd}$.

These relationships apply to conditions near the shore as waves approach the shallow water of the breaker zone, the location of greatest geomorphological activity.

The value of γ ranges from 0.6 to 1.2, with the higher values associated with steep beaches. This means that where beach slope is steep, waves progress into shallower water before breaking. This close approach of wave mass to the shore has implications also for the creation of wave-induced currents.

16.1.3　Currents
The forces generated on nearshore waters by wind, waves and tides create flows of water in the form of nearshore currents. Water is caused to concentrate locally, elevating the surface water level and thereby creating a pressure gradient which drives the current. These currents are capable of transporting sediment grains already in suspension, and may be of sufficient velocity to entrain sediment themselves (see Box 13.5). They make an im-

portant contribution to sediment movement in the coastal zone.

Wave-induced currents　The progress of waves onshore creates a concentration of water shorewards. Outside the breaker zone, the asymmetrical particle orbits and oscillations within waves result in a net shoreward transport of bottom water – **onshore mass transport** (Box 16.2). Within the breakpoint, the breaking of waves and uprush of water on to the beach elevates the mean water level toward the shore (Fig. 16.7). This is known as **wave set-up**. The height of breakers along the shore may be complicated by the presence of standing waves aligned normal to the coast, known as **edge waves**, causing alternate peaks and troughs in breaker height along the beach. The combination of wave set-up and edge waves creates a lat-

(a)

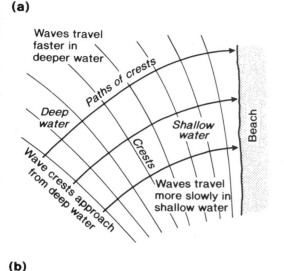

(b)

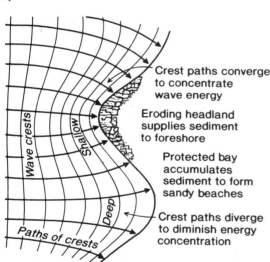

Crest paths converge to concentrate wave energy

Eroding headland supplies sediment to foreshore

Protected bay accumulates sediment to form sandy beaches

Crest paths diverge to diminish energy concentration

Fig. 16.5 Wave refraction: (a) simple refraction of waves arriving obliquely on a straight coast; (b) refraction of waves on an indented coast: wave energy is concentrated on the headlands and dispersed in the bays.

eral pressure gradient to power longshore currents in alternate directions along the shore, feeding into **rip currents** which transfer excess water seawards (Fig. 16.8). These develop at intervals along the shore to form regular circulation cells.

Longshore currents can also be generated by the incidence of waves oblique to the shoreline. The mass and energy of approaching waves can be resolved into shore-normal and longshore components (Box 16.3). Thus longshore current discharge can be calculated, and if the current cross-section is known, its velocity also. Longshore currents readily attain a velocity of 1 m s⁻¹ or more.

Tidal currents Tidal currents are created when the tidal wave encounters irregularities of the sea bed caused by sediment accumulations in shallow inshore waters, and by constriction when it enters coastal inlets. Nearshore and estuarine tidal channels are subject to alternate flood and ebb currents. These tend to reach a maximum value around mid-tide, with slack water (= zero velocity) at or around the times of high and low water. Flood and ebb currents may be of different duration and velocity, according to the height and duration of consecutive tides, and in estuaries may use different channels due to lateral deflection by the Coriolis force (see Box 4.4). The strength and direction of tidal currents in coastal and estuarine waters may therefore be quite different between flood and ebb. Tidal currents in coastal waters are quite capable of generating velocities in excess of 5 m s⁻¹.

Wind-induced currents These may occur during periods of strong onshore winds, creating wind set-up against the coast. A consequential effect is a compensatory bottom current flowing offshore to transfer the excess water mass away from the shore. This phenomenon is irregular and episodic, related to storm surges and lesser events, and is consequently difficult to evaluate in other than a qualitative sense.

Currents in inshore waters are thus both abundant and varied. They can generate transfers of water onshore, longshore and offshore, and in and out of coastal inlets. They attain significant velocities in relation to sediment movement, and are capable of substantial amounts of geomorphological work.

16.2 Materials in the coastal system

A general model of the sources and transfers of sedimentary materials of terrestrial origin in the coastal zone is shown in Fig. 16.9. The two major terrestrial sources are the erosion of coastal cliffs, and the delivery into the coastal zone of sediment transported by rivers from inland catchment basins.

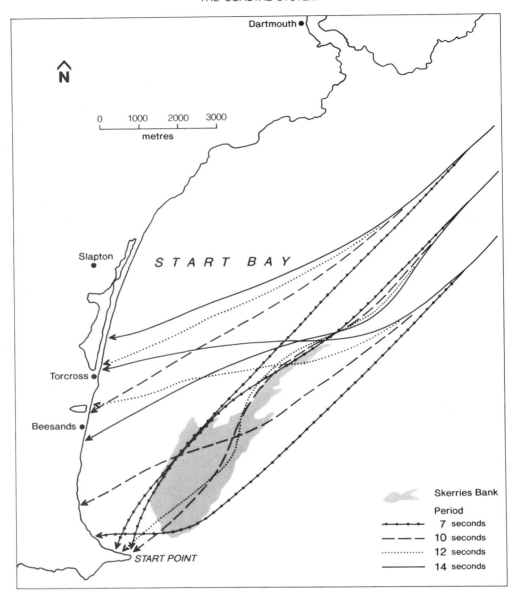

Fig. 16.6 Wave refraction in Start Bay, south Devon, caused by the submerged shoal of the Skerries Bank. The initially uniformly spaced crest paths become concentrated at particular points on the coastline as a result of refraction (after Hails, 1975).

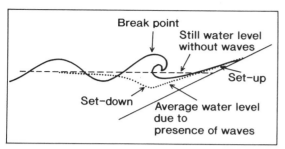

Fig. 16.7 Wave set-up: the rise in mean water level shoreward of the breaker zone due to the action of waves.

Coastal cliffs will feed a variety of material to the shore, according to the composition of the cliffs, their geological structure, and the subaerial processes operating upon them. Massive lithified rocks are likely to release a high proportion of coarse blocky debris, but only in limited quantities, since the rate of cliff recession is likely to be slow. Unlithified sediments, such as unconsolidated clays or glacial till will release fines plus any constituent material of coarse calibre, but at a high rate of delivery.

Sediments introduced into the coastal zone by rivers are generally well sorted and of fine calibre. Whilst the finest materials, clays and silts, are likely to be carried out to sea, the material most likely to be retained in the coastal zone is sand, usually composed of resistant quartz grains which, on account of their hardness, are able to withstand transportation. Occasionally, when a high-energy stream reaches the shore, coarse bouldery sediment may be delivered directly to the coast, as for example the fan at the mouth of the River Lyn, Devon. More commonly, fluvial sediments are deposited at the coast in the form of shoals or banks at the mouth of a river or estuary.

There is some transfer of material between the coast and the sea bed offshore, usually in the form of sandy sediments, by onshore and offshore currents. Where the coastline is breached by a deep river channel, or a submarine canyon exists close inshore, then coastal sediment may be transported into these deeper waters and lost to the coastal zone.

Within the intertidal zone sediment may be stored in a variety of locations, and be transferred between these shores by coastal geomorphological

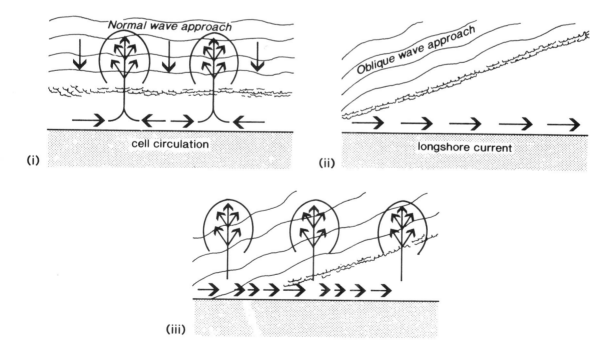

Fig. 16.8 (i) Normal onshore wave approach results in rip cell circulation. (ii) Oblique wave approach creates a unidirectional longshore current, which may develop into the rip cell circulation pattern in (iii).

Box 16.3
WAVE ENERGY

Wave energy is present in the form of potential energy due to the elevation of the wave above still-water level, and kinetic energy due to the orbital motion of the component particles.

Airy wave theory predicts that these two components of energy are of equal magnitude, and that total energy per unit length of crest is expressed by

$$E = \tfrac{1}{8}\rho g H^2$$

Where E = wave energy (J m^{-2}), ρ = water density (1025 kg m^{-2}), g = gravitational acceleration (9.81 m s^{-1}), and H = wave height (m).

The power (P) of a wave – that is energy per unit time – is given by

$$P = EC_n$$

where n = wave group velocity.

This comes about because the velocity of a group of waves is less than that of the individual waves within it. In deep water, group velocity is half that of individual waves; thus in this case $n = \tfrac{1}{2}$.

The incidence of wave power on a shore will vary according to the angle of approach of waves. Thus incident power can be resolved into shore-normal and longshore components by simple geometry. Fig (a) shows a unit width of wave crest allowed through a hypothetical breakwater to approach the shore. (i) The shore-normal normal component (ii) is given by

$$BC = \frac{\cos\alpha}{AB} = \cos\alpha$$

where α = angle of wave incidence and AB = unit length of crest = 1.

The longshore component (ii) is then given by

$$AC = \frac{\sin\alpha}{AB} = \sin\alpha$$

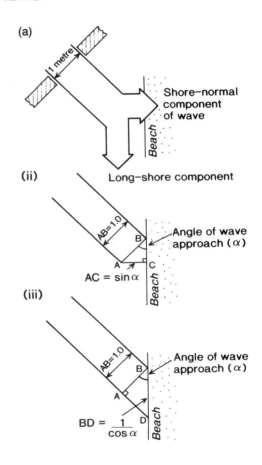

These components are both distributed along the length of shore BD (iii), calculated as $\frac{1}{\cos\alpha}$.

The components of wave power per unit length of shore are thus modified as follows:

shore-normal: $\dfrac{\cos\alpha}{\dfrac{1}{\cos\alpha}} = \cos^2\alpha$

longshore: $\dfrac{\sin\alpha}{\dfrac{1}{\cos\alpha}} = \sin\alpha\cos\alpha$

Incident wave power per unit length of wave crest therefore resolves as follows per unit length of beach:

Shore normal wave power, $P_n = EC_n \cos^2 \alpha$
Longshore wave power $P_l = EC_n \sin \alpha \cos \alpha$

For increasing values of α, therefore, shore-normal power decreases and longshore power increases. The angular resolution of wave power in this way has important consequences for the relative movement onshore and longshore of both water and entrained sediment.

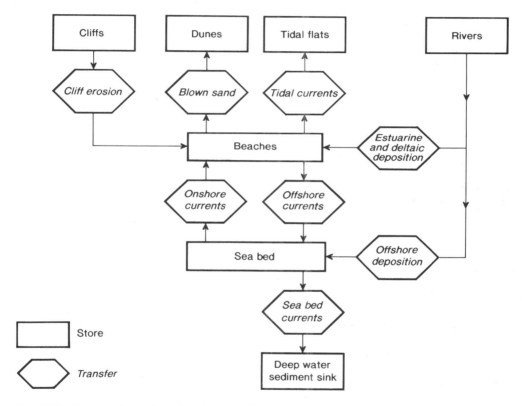

Fig. 16.9 Stores and transfers of sediment in the coastal zone.

processes. **Beaches** occur on coasts exposed to wave action. They commonly consist of shingle or quartz sand. Where cliffs make a significant direct contribution, then boulders are often present. Shell fragments may be a significant component of beach sediment, especially in tropical environments. Estuaries, deltas and saltmarshes are other inter-tidal sediment stores, usually less exposed to wave action than beaches. Onshore, coastal dunes may be considered a component store of the coastal system.

The application of erosional energy by coastal

wave, current and wind action, gives rise to processes of erosion and deposition. This dynamic interaction transfers the readily mobile sediment between the various stores and may create rapid change in the form of the coast.

16.3 Nearshore dynamics: the interaction between water and coastal sediments

At the crux of coastal processes is the interaction between the supply of wave energy and the abundance of readily mobile sediment. Waves themselves move sediments around the nearshore zone, and by their turbulence bring sediment into suspension, making it available for transport by nearshore currents. We will now examine some of the more frequent processes characteristic of the nearshore zone.

Waves progressing shorewards towards the breaker zone disturb bottom sediment and transport it shoreward as the bottom current associated with onshore mass transport begins to take effect. This transport occurs as far onshore as the breaker zone, where the current is terminated by the breaking of waves.

At the breakpoint, particularly in the case of steep plunging breakers, much energy is applied to the sea bed, disturbing the sediment and bringing it into suspension. After breaking, a bore of high-velocity water is projected up the beach, its leading edge a film of water <6 mm deep – the **swash**. This uprush of water is retarded by the gradient of the beach and by the frictional resistance of the beach surface. The water then drains back down the beach under gravity as **backwash**. The flow of the backwash may be reduced on a beach with a high permeability due to coarse texture or unsaturated condition.

The swash/backwash sequence is often asymmetrical and may be expressed in the form of a ratio of time to uprush (t), to breaking wave period (T), such that if

$$\frac{t}{T} < 0.3$$

swash and backwash are completed before the next wave breaks;

$$0.3 < \frac{t}{T} < 1.0$$

swash and backwash are not completed before the next wave breaks and the swash is diminished by the preceding backwash;

$$\frac{t}{T} > 1.0$$

continuous surf drives onshore.

The swash bore may disturb sand to a depth of several centimetres, moving sediment onshore as both bed load and suspended load. The net movement of suspended material will depend on the settling time (f), which is dependent on particle size, in relation to wave period (T). Thus when

$$\frac{f}{T} < 0.5$$

there will be a net onshore movement of sediment, and when

$$\frac{f}{T} > 0.5$$

material remains in suspension for a sufficient time for offshore transport to occur. Hence the finer fractions will be lost to the beach, whilst the coarser material remains.

In addition to their direct effect on swash and backwash processes, breaking waves have the important effect of disturbing sediment throughout the nearshore zone, and putting it into suspension (Fig. 16.10a). Once suspended it is readily transported by a variety of longshore (Fig. 16.10b) and offshore currents, as a result of which it may be redeposited along the coast, offshore, or lost entirely from the coastal system.

These wave processes, then, transport sediment around the nearshore zone and create the particular features of the beach profile. Beach profiles characteristic of differing wave conditions may be identified.

Swell waves, of low steepness, occur on humid temperate coasts, typically in the calmer conditions of summer. These longer-period waves permit the full development of swash and backwash. This promotes onshore sediment movement and aggradation of the beach. The profile associated with swell conditions (Fig. 16.11) is smoothly concave with a wide ridge at the backshore – the **berm** (Fig. 16.12). This tends to accumulate coarser beach material, preferentially moved onshore due to its shorter settling time in the backwash.

Storm waves, characterized by greater steepness, are more typical of winter conditions. At the breakpoint plunging breakers throw up sediment to form a breakpoint bar, and also limit the shoreward transport of sediment by the onshore bottom current. Shoreward of the bar is a trough which may become a channel for longshore cur-

Fig. 16.10 (a) The cloudy inshore water is indicative of high suspended sediment concentration, even under very moderate wave conditions. (b) The effects of longshore drift are visible in the impounding of beach sediment by groynes. Sediment moves laterally along the beach face away from the viewer.

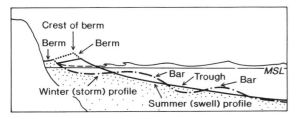

Fig. 16.11 Characteristic beach profiles associated with differing seasonal wave conditions.

rents. The uprush of the swash in these short-period waves is limited by the preceding backwash. Thus onshore transport of sediment up the beach is strictly limited, and the net effect is a combing of sediment down the beach face, either to form an inshore bar, or to be available for transport by waves and currents.

These are idealized models of two specific sets of conditions. In reality, beach profiles are a consequence of complex interactions between wave type, sediment characteristics, beach gradient and tidal range. The beach profile itself responds rapidly to changing energy inputs, as demonstrated by both controlled experiments in the laboratory, and even casual observations in the field.

16.4 Coastal erosion

If the rate of sediment removal exceeds the rate of supply, then no beach will be present to absorb the impact of wave energy, and this energy will

Fig. 16.12 A series of berm ridges is clearly visible on the face of the shingle barrier beach (Porlock Bay, Somerset, UK).

(a)

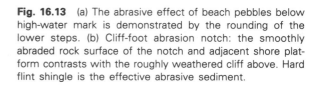

(b)

Fig. 16.13 (a) The abrasive effect of beach pebbles below high-water mark is demonstrated by the rounding of the lower steps. (b) Cliff-foot abrasion notch: the smoothly abraded rock surface of the notch and adjacent shore platform contrasts with the roughly weathered cliff above. Hard flint shingle is the effective abrasive sediment.

be applied directly to the coastal bedrock to cause erosion.

Where a cliffed coast exists, then the application of wave energy in the intertidal zone will tend to undercut and steepen the cliff. Waves breaking directly at the cliff foot will exert a percussive effect on the rock, and a compressive force on air trapped in open joints within it, loosening and detaching particles and blocks from the cliff. Where a supply of coarse abrasive sediment is present in the intertidal zone, this may be swilled to and fro to create a smooth notch of abrasion to undercut the cliff (Fig. 16.13). Denudation of bedrock may be enhanced within the intertidal zone by salt weathering (see Section 11.1.2) and by the action of marine organisms, known as **lithophages**, which directly remove surface rock by boring and rasping. In tropical seas blue-green algae may cause limestone solution around high-water level.

These processes cause a recession of the coastline around high-water mark, which in turn steepens the coastal cliff, thereby increasing the overburden stress at the base and decreasing its factor of safety (Box 12.3). Coastal cliffs above high-water mark behave as other subaerial slopes (Chapter 12). Massive indurated rocks will respond by toppling failure, rockfall or landslide, according to their jointing structure. Less lithified argillaceous rocks may respond by a variety of slide and flow processes according to their composition and attitude. Material thus released is delivered to the shore below, where it becomes incorporated into the store of beach material.

Under continued marine erosion at the cliff foot, a gently inclined plane will form normally of gradient 1–5° and steepening towards the cliff-foot notch (Fig. 16.14). The development and extension of this platform will then limit the access of large waves to the cliff foot, and through the operation of negative feedback, become self-limiting. Though the seaward extent of the platform will vary according to the available tidal range, it seems unlikely that widths of more than a few hundred metres are attainable with a constant sea level.

On tropical coasts, which tend to be characterized by a low tidal range, sandy sediments and an abundance of marine organisms, a different combination of platform cutting processes

Fig. 16.14 Cliff, beach and shore platform. Large recently fallen blocks rest on a stony beach. Shore platform extends to the left.

operates. Here **water layer weathering**, chemical weathering of rock by water in shallow surface pools, and biological erosion combine to create a near-horizontal platform at high-water level.

The erosion of coastlines by cliff retreat may be very variable in rate. The rate of recession will be controlled by the development of the cliff-foot notch, which in hard rock may be a fraction of a millimetre per year. Where coastal cliffs are formed in unconsolidated materials, then rates of recession measured in metres per year are not uncommon, creating both rapid changes in the position of the coast and also feeding an abundance of new sediment into the coastal system.

16.5 Coastal deposition

Deposition by coastal processes takes place at locations in the littoral zone, where sediment traps exist in the form of beaches, sand dunes, estuaries and tidal flats. Beaches are probably the most abundant feature of coastal deposition. Pethick (1984) suggests that beach planforms can be classified as follows.

16.5.1 Shoreline beaches
Shoreline beaches are accumulations of beach sediment adjacent to a bedrock coast. A distinction is made between beaches isolated from longshore sediment supply by bounding headlands, and therefore closed systems in relation to sediment supply (termed pocket beaches), and those which are open to a supply of sediment.

Pocket beaches (Fig. 16.16a) tend to develop an alignment normal to the approach of wave crests. As the waves become refracted in shallow water, their curvature is adopted by the beach.

This can readily be explained by reference to the longshore transport equation

$$i_{\text{longshore}} = ECn \sin\alpha \cos\alpha$$

If $\alpha > 0$, and the obliquely approaching waves have sufficient power to transport sediment, then it will be moved to areas where $\alpha \to 0$, where transport ceases and deposition occurs. This process will cause erosion of wave-oblique sections on the shore, therefore modifying the plan form of the beach until $\alpha = 0$ throughout its length, at which point no further transport takes place and the plan form is parallel to wave approach and stable in form. This is termed a swash-aligned beach (Fig. 16.15a).

In the case where a sediment supply throughput is available to a beach, in which a headland permits sediment passage alongshore, or there is an abundant supply of sediment locally from a river outlet for example, then a balance is struck between the forces tending to produce swash alignment, and those responsible for longshore drift. Now the orientation of the beach becomes adjusted to permit sufficient longshore transport to maintain an equilibrium plan form – termed **drift alignment** (Fig 16.15b).

Zeta-form beaches represent a compromise between swash-alignment and drift-alignment beaches, and occur when headlands partially block sediment supply. Here the beach adjacent to and sheltered by the headland adopts a swash alignment, whereas that section receiving a sediment supply tends to drift alignment (Fig. 16.15c).

16.5.2 Detached beaches

This category of beaches contains a variety of forms which have an existence somewhat independent of the form of the bedrock coast. It includes spits (Fig. 16.15b), which form at points where the coastline turns sharply inland, a line not followed by the sediment-bearing littoral current. The supply of sediment continues but the longshore littoral current may lose power on passing the headland, thereby causing deposition. Cuspate forelands are triangular accumulations of coastal sediment attached to the bedrock coastline only at their base, their two beach faces meeting at an acute angle. Possible explanations are the refraction of waves offshore by submerged shoals, or the development of two beach faces, each of whose orientation is determined by its

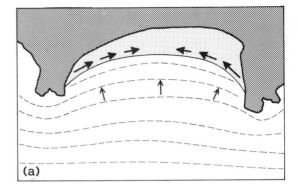

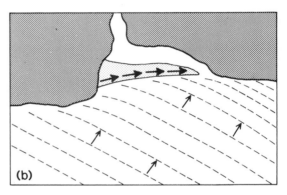

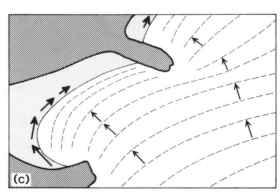

Fig. 16.15 Beach alignment in relation to sediment supply and transport. (a) Swash alignment. (b) Drift alignment. (c) Zeta-form shape.

own particular longshore sediment equilibrium in relation to sediment supply and wave climate. A related feature is the tombolo (Fig. 16.16c).

Barrier islands, a line of offshore islands up to several hundred kilometres long, appear to develop on high-energy coasts with an abundant sediment supply. A variety of origins has been postulated for these features, but their maintenance appears to be related to the ability of high-energy events to carry sediment over to the landward side by washover processes. Onshore movement of barrier islands may produce a barrier beach impounding a coastal lagoon (Fig. 16.16d).

16.5.3 Rhythmic beach morphology

Rhythmic beach forms are smaller-scale features which develop on the surface of existing beaches, modifying their form in detail, rather than forming the beach itself.

Beach cusps are intriguing features which form on the face of a beach in mixed sediment materials. They are regularly spaced, commonly 1–10 m but ranging up to 60 m in size, crescentic in plan form, with intervening horns pointing seawards. Coarser sediment is swept by the swash toward the horns, where the swash divides, swirls and returns down the trough facing the crescentic embayment. They may form quite rapidly, and with each new tide, although the reasons for their initiation are not well understood. They are a common feature of British beaches.

Cell circulation topography is the nearshore topography created by the cell circulation of water. When wave crests approach parallel to the beach, a discontinuous bar may form, breached by regularly spaced rip channels. Where two

Fig. 16.16 (a) Pocket beaches on a rocky cliffed coastline. Large landslip scars indicate former cliff falls (North Devon, UK). (b) Spit, created by longshore sediment drift toward the viewer (Borth, Wales). (c) Tombolo connecting nearshore island to the shore (Brittany). (d) Barrier beach impounding coastal lagoon, the nearer section of which has been infilled by sediment (Slapton, South Devon, UK).

opposed longshore currents meet to form the rip, then accumulation of sediment may accumulate as giant cusps. When the wave-crest approach is oblique, rip currents may be responsible for forming a series of bars attached to the shore and parallel to the rips themselves.

16.5.4 Sand dunes

Coastal sand dunes may develop where strong onshore winds blow across sandy beaches with a large tidal range. Under these conditions a wide expanse of sand is exposed to drying winds and becomes susceptible to aeolian transport onshore. Sand transport per unit width of beach is described by the following equation

$$q = C(V_{100} - V_t)^3$$

where q = weight of sand moved per unit time, C = a constant, V_{100} = velocity at 1 m above the surface, and V_t = critical threshold velocity for a given grain size. Sand transport is defined by the cube of wind speed and is therefore extremely sensitive to variation in this parameter, and to particle size (which determines the critical threshold velocity). Sand moves landwards by the process of saltation, in which grains are accelerated by being lifted into wind which increases in velocity with height above the retarding shear of the beach surface. Thus the kinetic engery within the saltating cloud is a balance between the gains from the wind energy, and the losses by impact at the beach surface. When the sand cloud passes on to a more energy-absorbent surface, for example dry rippled sand above high-water mark, then more energy is absorbed, saltation declines and deposition takes place. Thus an embryo foredune is initiated. Subsequent development of the dune is enhanced by the growth of vegetation which traps both energy and sand. The dune system may subsequently evolve by the landward migration of dunes as erosion of the seaward face is accompanied by deposition to landward. Dune morphology may subsequently be modified by blow-outs of sections of the ridge if the vegetation cover is breached to expose bare sand to the wind. See Chapter 25.

16.5.5 Estuaries

Sediment within estuaries is subject to the flow and ebb of tidal currents commingled with the discharge of fluvial waters. This creates a continuous complex and shifting pattern of water and sediment movement.

As the tidal wave progresses up the estuary it becomes increasing asymmetrical, since the crest in deep water moves more rapidly than the trough in shallow water. This means that the crest tends to catch up with the preceding trough, leading to an asymmetry in which the flood tide in the upper estuary may be reduced in duration to 2–3 hours, in contrast with the 9–10 hour ebb which completes the 12.5 hour cycle. This shorter duration of the flood means that its velocity and therefore its ability to transport sediment is much greater. Thus sediment tends to be transported up estuaries during the flood tide and deposited at slack high water, whilst the weak ebb current is insufficient to carry it back again.

The upper reaches of estuaries are thus zones of substantial sediment deposition. The middle and lower reaches experience a complex interplay of tidal and fluvial flows, each with their different temporal rhythms. The combination of these currents with an abundant supply of mobile sediment creates a complex and ever-changing morphology of sandbanks and channels, often extending seawards in the form of a delta.

16.5.6 Tidal flats

Tidal flats, comprising mudflats and saltmarshes, form in sheltered coasts, estuaries and other partially enclosed re-entrants in the coastline. Here the accumulation of fine-grained sediment forms broad flats at intertidal elevations, dissected by branching creek systems draining seawards. At each flood fine sediment is mobilized and transported shorewards, up the confining creeks at higher velocities on the rising tide, and up on to the flats, where current velocity slackens towards high tide.

A converse movement takes place on the ebb. However, higher velocities are required to entrain suspended sediment than deposit it (Box 13.4), leading to a net shoreward transport of fine sediment with each tide. Settling is enhanced by flocculation of silt and clay particles in the saline water, and by the mechanical trapping effect of vegetation once saltmarsh plants have colonized.

16.6 Conclusion

Coastal processes tend to be dominated by the mechanical energy transfers associated with the processes of fluid dynamics. An emphasis on fluid dynamics is therefore essential in explaining the interaction between coastal waters and the form of land mass margins.

The continuously varying energy inputs, at a variety of temporal scales – waves on a timescale of seconds, tides measured in hours and months, and storms episodically, combine with mobile sediment to create constant change at the coastline. Coasts are therefore rapid-response systems, which, when mobile sediment is present, are capable of rapidly adapting to new equilibrium conditions. Coastal systems are thus seen to embrace a wide range of distinctive environments, quite different in kind from other terrestrial environments. These consequences of the ocean/land interaction create the opportunity for the development of unique coastal ecosystems.

It should be remembered, however, that the sea only attained its present level at the end of the Flandrian (postglacial) transgression some 6000 years ago. Additionally, many coastlines at the present time are experiencing isostatic uplift or depression (Box 5.4). It is therefore unwise to interpret coastal forms in general as a consequence of contemporary processes, except in cases of coastal features whose nature has permitted rapid response and evolution. Thus landforms in many hardrock coasts may bear the impact of very different environmental conditions associated with past and different sea levels.

Further reading

An excellent broad perspective on coastal systems is provided by:

Pethick, J. (1984) *An Introduction to Coastal Geomorphology.* Edward Arnold, London.

An even broader perspective encompassing coastal geomorphology, coastal ecology, and management is provided by:

Carter, R.W.G. (1988) *Coastal Environments: an Introduction to the Physical, Ecological, and Cultural Systems of Coastlines.* Academic Press, London.

However, the best general text on coastal geomorphology is probably:

Davies, R. (ed) (1978) *Coastal Sedimentary Environments.* Springer-Verlag, Berlin.

A wealth of information on coastal processes, and abundant examples of forms is present in:

King, C.A.M. (1972) *Beaches and Coasts*, (2nd edn) Edward Arnold, London.

A broad survey of coastal types and processes at the global scale is provided by:

Bird, E.C.F. (1984) *Coasts.* Blackwell, Oxford.
Davies, J.L. (1972) *Geographical Variation in Coastal Development.* Oliver & Boyd, Edinburgh.
Schwarz, M.L. and E.C.F. Bird (eds) (1985) *The World's Coastlines.* Van Nostrand Reinhold, New York.

Lastly a reference to a text introducing the ecology of coastlines:

Mann K.H. (1982) *Ecology of Coastal Waters: a Systems Approach.* (Studies in Ecology, vol. 8) Blackwell, Oxford.

Accounts of the current understanding of coastal zone processes can be found in the following advanced geomorphology texts:

Chorley, R.J., S.A. Schumm and D.E. Sugden (1984) *Geomorphology.* Methuen, London. Chapter 15.
Derbyshire, E., K.J. Gregory and J.R. Hails (1979) *Geomorphological Processes.* Dawson, Folkstone. Chapter 3.
Embleton, C.E. and J.B. Thornes (eds) (1979) *Processes in Geomorphology.* Edward Arnold, London. Chapter 11.
Selby, M.J. (1988) *Earth's Changing Surface.* Oxford University Press, Oxford. Chapter 13.

Spatial variations in the denudation system

As we have seen in Chapters 5 and 10, the major energy inputs into denudation systems are precipitation and the potential energy of relief. These forces of denudation operate on the materials exposed at the surface of the lithosphere. Since precipitation, relief and surface materials vary widely across the Earth's land surface, one may expect consequent spatial variations in denudation systems. This spatial variation is expressed both in the functioning of denudation systems, and also the land surface form which they produce. We have examined the various components of the denudation system in terms of thier internal operation and have seen how they respond to inputs. This chapter, therefore, attempts a comparison of denudation systems operating under different environmental conditions with different inputs.

The output from a catchment basin can be established by a monitoring programme involving measurement of runoff, together with sediment and solute concentrations. From these data rating relationships (Figs. 13.7, 13.8) can be established. The total annual volumetric loss of minerals can then be calculated. In the case of minerals in solution, there is an additional input to the land surface via precipitation and other atmospheric fallout, and this also must be monitored. The net loss of solutes by denudation from a catchment is therefore calculated as gross output less atmospheric input.

The loss from the land surface may be expressed in various units. If determined in units of mass, these are normally standardized in respect of catchment area as tons per square kilometre per year ($t\ km^{-2}\ a^{-1}$). Mass may be converted to volume by dividing by density (normally 2.65 t m^{-3}) to produce a volumetric loss, commonly expressed as volume per square kilometre per year

($m^3\ km^{-2}\ a^{-1}$). This in turn may be expressed as a surface lowering rate by dividing by the land surface area from which the value is derived, expressed in millimetres lowering per year (mm a^{-1}). Clearly this represents a generalized measure of surface lowering and does not imply a uniform value throughout the source area, within which local variations in lithology, gradient and land use can suffer wide variations in erosion rate. It does, however, serve as a useful standard for comparison of spatial variations in the denudation system.

As indicated in Chapter 10, the major controlling variables in the operation of the catchment basin system are climate, relief, lithology and land use (or vegetation type and cover). It is possible to gain insights into the effects of these variables by comparing denudation system outputs at various spatial scales. The effect of individual controls on denudation can thus be examined by employing studies of restricted areas in which variations in other variables are limited, and can thus be controlled. Having considered variations in denudation system output, we shall make some observations on denudation system form, and the nature of the relationships between landscape form and process.

17.1 Spatial variations in denudation output

17.1.1 The global scale

Estimates of the amount of mineral material currently being denuded from the continents are presented in Table 17.1. Data of this kind necessarily present a very generalized picture, since a great variety of climatic, topographic and geological conditions may exist within the bounds of

Table 17.1 Estimates of denudation losses of suspended sediment and solutes from the continents (adapted after Meybeck, 1979; Milliman and Meade, 1983)

	Solute loss $t \times 10^{-6} a^{-1}$	Sediment loss $t \times 10^{-6} a^{-1}$	Total $t \times 10^{-6} a^{-1}$	Ratio Sediment: solute	Surface lowering $mm\, a^{-1}$
Africa	201	530	731	2.64	0.0093
N. America	758	1462	2220	1.93	0.0408
S. America	603	1788	2391	2.97	0.0504
Asia	1592	6433	8025	4.04	0.0718
Europe	425	230	655	0.54	0.0257
Oceania	293	3062	3355	10.45	0.0627

a single continent. The data only include losses in the form of solutes and suspended sediment, and they omit bed load, which is very difficult to measure in major rivers. As such, they represent an underestimate of total denudation.

Significant differences between the continents are, however, apparent. Total denudation, as mean surface lowering, differs by almost an order of magnitude between Asia, a continent with some very high relief and a number of major rivers traversing sediment-rich alluvial lowlands, and Africa, a largely lowland continent embracing considerable arid regions.

The relative proportions of solute and sediment losses also vary widely. In Europe, a continent of modest relief and humid conditions, solute loss exceeds that of sediment by a factor of almost 2. In other words, chemical denudation in this case is more effective than mechanical denudation. In all the other continents mechanical denudation is dominant, accounting for over 90% of denudation loss in Oceania, which embraces the arid interior of Australia, and the high-relief humid terrains of the oceanic islands such as New Zealand, and New Guinea.

At this scale of generalization, then, the major controls on denudation appear to be climate and relief. Indeed, Garrels and Mackenzie (1971) suggest that while solution losses per unit area do not vary widely between continents, sediment losses appear to be positively related by a power function to mean continental elevation. Various attempts have been made to map denudation at the global scale. Fig. 17.1, based on data form 1500 stations, represents the sediment yields from basins 1000–10 000 km² in area (Walling and Webb, 1983). In this way the high values expected from small high-relief basins, and the over-generalized values from very large basins, are avoided. The

highest sediment yields (values in excess of 10 000 t km^{-2} a^{-1} have been reported) are found in the loess regions of China, and the Cenozoic mountain terrains around the Pacific margin. Other high values are found in high mountain terrains and regions with seasonally arid and humid climates. Based on these data, a global average sediment yield of *ca.* 150 t km^{-2} a^{-1} can be derived, representing a surface lowering rate of 0.057 mm a^{-1}.

17.1.2 The regional scale

Denudation at the regional scale can be studied by observation of individual catchment basins. Table 17.2 lists a selection at both world and British scales, with units of stream load, representing denudation loss, standardized to permit comparison. The major world rivers tend to exhibit high values of suspended load, as they drain mountainous terrains (Brahmaputra, Orinoco) or large alluvial lowlands (Ganges, Mississippi). Only in the humid lowland basins (Zaire, Lena) does solutional denudation predominate.

The British rivers, in contrast, reveal a predominance of solutional denudation. Only in the upland streams (Dart, Esk) is sediment loss dominant. High relative and absolute dissolved loads are found in streams whose catchments comprise a high proportion of soluble rocks (Avon, Cuckmere).

Data currently available on the sediment loss from British rivers suggest a mean denudation rate of around 50 t km^{-2} a^{-1}, low by world standards (see Section 17.1.1). The pattern of data is not entirely representative and contains many significant gaps, and a clear picture of the distribution of sediment loss has yet to emerge. The pattern of chemical denudation in Britain however is beginning to become apparent (Walling and Webb, 1986). Fig. 17.2 represents an attempt to

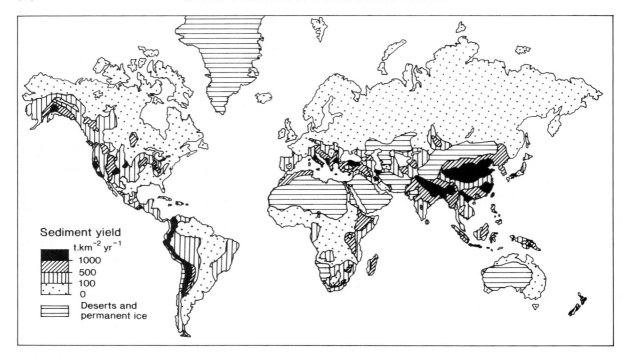

Fig. 17.1 Global pattern of sediment yields of medium-sized drainage basins (after Walling and Webb, 1983).

Table 17.2 Magnitude of dissolved and suspended loads for a selection of (a) world and (b) British rivers (after Walling and Webb, 1981, 1986)

(a)

	Area km² × 10⁶	Load (t km⁻² a⁻¹)		Ratio
		Dissolved	Suspended	Suspended: Dissolved
Ganges/Brahmaputra	1.48	102	1128	11.06
Orinoco	0.99	51	212	4.16
Zambesi	1.20	13	40	3.08
Mississippi	3.27	40	64	1.60
Indus	0.97	70	103	1.47
Zaire	3.82	12	11	0.91
Lena	2.50	34	5	0.14
(b)				
Dart	46	61	91	1.50
Esk	310	51	57	1.12
Creedy	262	61	53	0.87
Severn	6850	109	65	0.60
Usk	912	129	46	0.55
Avon	666	148	27	0.18
Cuckmere	133	74	9.2	0.12

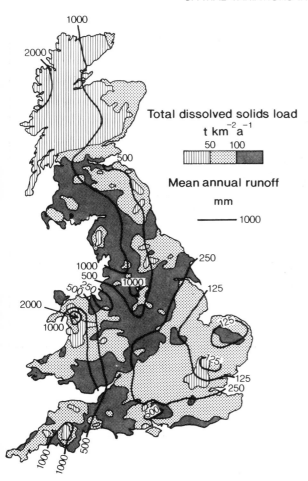

Fig. 17.2 Solute loads of British rivers. Isolines indicated annual discharge values (after Walling and Webb, 1981).

map annual dissolved solid loads for Britain based on water quality data for unpolluted small streams. Minimum rates occur in upland areas formed by chemically resistant rocks, such as the Scottish Highlands, upland Wales and Dartmoor, and regions of low annual runoff such as southeast England. Maximum rates are identified with regions where moderate levels of runoff combine with availability of soluble minerals, and especially identify regions of soluble limestone.

Denudation rates under glacial conditions may be significantly higher than in fluvial environments. Drewry (1986) reviews data on subglacial erosion rates from a variety of sources. Annual sediment yields from meltwater drainage of five glacier basins in Norway over a number of years yield

mean erosion rates in the range 0.073–0.610 mm a^{-1}. Corbel (1964) made a detailed study of the St. Sorlin glacier in the French Alps, embracing material discharged both in meltwater and as morainic debris. This yielded a bedrock lowering rate of 2.2 mm a^{-1}, much higher than adjacent unglacierized terrain. Drewry quotes a range of values of glacial denudation derived by several authors using varied methods; the majority lie in the range of 0.5–5 mm a^{-1}. All of these values represent the total effect of erosion under glacierized conditions, and treat the glacier erosion system as a black box in that they do not discriminate between the effects of the separate processes of abrasion, block removal and meltwater erosion. Nevertheless, these results prompt the conclusion that, under active temperate glaciers at least, denudation rates tend to be at least an order of magnitude higher than in un-glacierized terrain.

17.1.3 The small catchment basin scale

A study by Arnett (1979) of a series of contiguous basins in a humid temperate environment permits comparisons to be made at a more local scale. Fifteen adjacent catchment basins were monitored and their outputs assessed over a 1-year period (Table 17.3). At this scale the climatic input can be regarded as being essentially uniform, and differences in denudation outputs may be considered largely to be a function of catchment basin form, lithology and land use. Several significant conclusions may be drawn from these data. There is, even within a small region, a fourfold variation in the rate of denudation. By far the greater losses are in the form of dissolved minerals, in line with the data for the humid continent of Europe as a whole. There is, furthermore, a twentyfold variation in the proportion of sediment loss between catchments.

The major variables controlling denudation losses at this scale were found to be land use, lithology and drainage density.

17.1.4 The local scale

At this scale, which embraces both actively eroding gully systems and experimental plots, controlling variables are different again and rates of denudation may be much higher. Erosion from localized plots can take place at rates two to three orders of magnitude higher than from catchment basins as a whole (Table 17.4). In all cases the

Table 17.3 Denudation losses for the year 1975 from adjacent catchments in a humid temperate region,. North York Moors, England (after Arnett, 1979)

Catchment basin	Area (km²)	Suspended load (%)	Dissolved load (%)	Total yield (t km⁻²)	Surface lowering (mm a⁻¹)
1	11.1	1.5	98.5	115.0	0.043
2	24.2	7.3	92.7	61.3	0.023
3	85.0	5.1	94.9	52.4	0.019
4	11.6	5.1	94.9	37.0	0.014
5	46.2	9.0	91.0	46.5	0.017
6	37.2	5.7	94.3	36.7	0.013
7	22.0	2.0	98.0	40.7	0.015
8	130.7	7.0	93.0	57.1	0.021
9	18.8	3.2	96.8	77.9	0.029
10	13.6	2.7	97.3	62.8	0.023
11	9.7	7.1	92.9	30.0	0.011
12	15.1	16.8	83.2	75.5	0.028
13	19.7	30.8	69.2	104.0	0.038
14	299.4	12.4	87.6	55.6	0.020
15	155.8	3.1	96.9	71.5	0.026

Table 17.4 Denudation losses from small sample areas

	Range (mm a⁻¹)	Mean (mm a⁻¹)	Source
mid-Wales*	0–75	15	Slaymaker (1972)
Pennines[†]	–	14.8	Harvey (1974)
South Wales*	2–24	10.6	Bridges and Harding (1971)
North York Moors (vegetated) site	–	+1.35 (gain)[‡]	Imeson (1974)
(unvegetated)	–	38.10[§]	

* Sample plots.
[†] Area of gully erosion.
[‡] Mean of 7 values
[§] Mean of 3 values

process of denudation is either sheetwash over an unvegetated soil surface or active gully development. All show rates greater than 10 mm a⁻¹. This emphasizes the protective role of vegetation as a control on surface sediment removal. It also points to the conclusion that sediment sources may be highly localized within catchment basins, and that these very high local denudation rates are offset by much lower rates over wide areas.

It is clear from this brief survey of denudation rates at different scales, that many of the functions controlling denudation are themselves closely interrelated. Relief, for instance, is positively correlated with precipitation in causing orographic

rainfall. Zones of exceptionally high relief are commonly young fold mountains, often comprised of relatively erodible sedimentary rocks of Cenozoic age, thus effecting a correlation between relief and lithology. At the global scale there is a correlation between precipitation and vegetation type. These correlations between the major controlling factors in denudation mean that it is often difficult to isolate the effect of individual factors.

17.2 Individual factors controlling denudation

The effects of individual factors on denudation losses, however, can be deduced by comparing similar regions or catchments which differ in their inputs in just one factor. We shall investigate first the effects of the dominant external factor of climate, and then examine the effects of more local factors such as lithology and relief.

17.2.1 Climate
Working on a sample of drainage basins from the USA, Langbein and Schumm (1958) established a relationship between precipitation and denudation (Fig. 17.3). Referring to sediment removal alone, and considering a region with a mean annual temperature of 10°C, the maximum denudation takes place with an annual effective precipitation (i.e. runoff) of 300 mm. Other authors have shown a similar peak in sediment yield in semi-arid en-

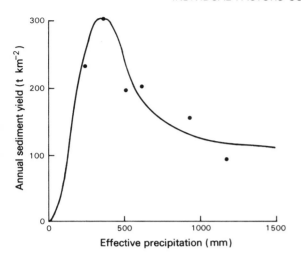

Fig. 17.3 Sediment yield in relation to effective precipitation (after Langbein and Schumm, 1958).

denudation of a particular rock type in different environments. Limestone, a rock which is denuded almost exclusively by solution, is one such rock which has been studied in different environments. Table 17.5 is compiled from rates of solutional lowering obtained from different climatic environments. The values of regional denudation are modest in relation to some already quoted. There is, however, an apparent increase in denudation rate from the tropical to the arctic–alpine environment, suggesting that limestone denudation is in general inversely proportional to mean annual temperature. The distinction, however, is far from clearcut, for there is considerable overlap between climatic environments, and these data mask the effects of other variables such as vegetation, precipitation and type of limestone.

17.2.2   Lithology
The effects of lithology on both runoff and denudation yield of sediment can be shown by comparing catchments developed on different lithologies within one region. Table 17.6 shows such a comparison from the Cheyenne River basin (S. Dakota, USA), derived from Hadley and Schumm (1961).

Five different rock formations are examined, with widely varying surface permeability as

vironments, although differing somewhat in absolute value. Thus, Douglas (1967) demonstrates a maximum denudation with 50 mm mean annual runoff in eastern Australia, while Dunne (1979) shows a peak denudation at *ca* 100 mm annual runoff for catchment basins in Kenya. Walling and Webb (1983) cite evidence from a range of authors working in varied environments that sediment yields increase again at higher values of annual precipitation and runoff. In truly arid conditions, erosion is less because of the lower erosional energy and less frequent runoff, whereas in moister environments vegetation cover increases and inhibits surface erosion of sediments. The critical combination of increasing erosion potential and increasing surface resistance causes denudation to peak under semi-arid conditions.

The influence of climate on denudation can be examined in another way, by comparing the

Table 17.5 Estimates of erosion rates for limestone areas under different climatic conditions (simplified after Smith and Atkinson, 1976)

	Mean (mm a^{-1})	Standard deviation	Number in sample
tropical	0.017	0.0125	18
temperate	0.021	0.0158	87
arctic–alpine	0.023	0.0141	24

Table 17.6 Lithology and sediment yield (from Hadley and Schumm, 1961)

Lithological unit	Rock type	Mean infiltration (cm hr^{-1})	Sediment yield (mm a^{-1})	Drainage density (km km^{-2})
Wasatch Formation	incoherent sand	23.0	0.088	8.6
Lance Formation	sandy loam	12.5	0.337	11.4
Fort Union Formation	sandy clay loam	3.2	0.876	18.2
Pierre Shale	sandy clay loam	2.5	0.943	25.8
White River Group	silt and clay loams	0.4	1.213	413

indicated by the mean infiltration rate. Infiltration rate is seen to be inversely related to drainage density. The higher density, reflecting higher surface runoff, is closely related to sediment yield. Clearly then, rock types that encourage greater surface runoff denude more rapidly.

17.2.3   Relief

The role of relief as a control on denudation rate is of paramount importance. High relief is associated with steeper slopes and therefore more active erosion, particularly the mechanical removal of material in sediment form. Fig. 17.4 shows the relationship between sediment loss and relief (as expressed by the relief ratio) for small drainage basins in the western USA, and Fig. 17.5 shows a similar relationship from a study in Papua New Guinea (Ruxton and McDougall, 1967). Young (1972) collated experimental data on surface lowering from a variety of environments (Fig. 17.6), which strongly suggest a significant distinction between areas of steep relief (mountains and individual steep slopes) and normal relief (plains, moderately dissected areas and gentle to moderate slopes). There appears to be an order of magnitude difference between the two, with normal relief showing a median denudation rate of 0.046 mm a^{-1} and steep relief 0.5 mm a^{-1}, although there is a degree of overlap in the range of values in the two categories.

These results have two important implications in spatial and temporal terms. First, there is a tendency for small drainage basins to exhibit higher gradients, since they usually form the headwater sections of larger catchments. Thus higher denudation rates tend to be associated with smaller catchments, rather than larger catchments which incorporate significant areas of lowland. Secondly, on the temporal scale, it is implied that a recently uplifted land mass will be denuded more rapidly, and that the rate of denudation will slow down as the relief is lowered. The form of the relationship indicates that, in the absence of further crustal uplift, the rate of erosion will decline exponentially through time.

It is clear that significant empirical relationships exist between catchment parameters (climatic, morphological and lithological) and rates of denudation. Spatial variations in these parameters are reflected in denudation rates. However, the relationships between gross climatic parameters and rates of denudation are still not clearly under-

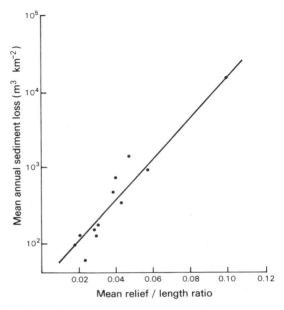

Fig. 17.4 Sediment loss as a function of basin relief (after Hadley and Schumm, 1961).

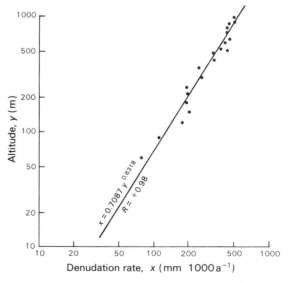

Fig. 17.5 Denudation rate as a function of altitude, eastern Papua New Guinea (after Ruxton and McDougall, 1967).

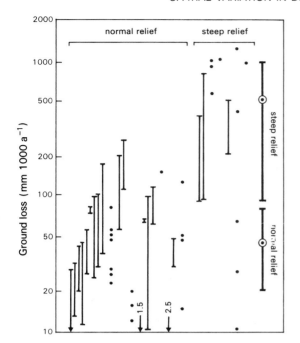

Fig. 17.6 Rates of lowering of the ground surface according to type of relief (after Young, 1969).

stood. The relationship between mean annual rainfall, for example, and denudation rate is far from clear. In theory, for a given rainfall input, the amount of runoff available for erosion is related to mean temperatures, since these will govern evaporative losses. Some authors (e.g. Fournier, 1960) have argued that the seasonal concentration of rainfall is more important than the annual total, since a dry season will permit drying and fragmentation of the soil, and the concentrated heavy storms of the wet season will bring more kinetic energy to bear on the land surface than more evenly distributed rainfall. The search for a realistic explanation of denudation patterns on a worldwide scale will depend on a more thorough understanding of such relationships, which can come about only as the result of a more representative spread of monitored catchment basins.

A final point to understand is that many of the denudation rates so far obtained cannot be considered to be the results of 'natural' processes. Many of the catchments monitored have been affected by alteration of the ground surface, by stripping forests and imposing agricultural practices, thereby exposing bare soil surfaces and modifying the hydrological balance. A more detailed examination of human impact on the denudation system follows in Chapter 26.

17.3 Spatial variation in denudation system form

Since Chapter 10, we have largely concentrated on processes in denudation systems. We shall now examine the effects of the operation of these processes in producing the form of the landscape. Landscape form can be regarded as a function of climatic and lithological inputs, and over a period of time it may reasonably be expected that a landscape will be produced which is adjusted to these major controls, operating through the individual and collective processes of denudation.

The status of landscape morphology depends very much on the scale at which is considered. In the terms outlined above, at a broad spatial and temporal scale it is clearly a dependent entity. At smaller spatial and temporal scales, however, at which the operation of specific denudation processes are considered, the landscape form may become an independent variable. At this scale, for example, drainage density may become an important control on hydrograph form, and relief and gradient are major controls on slope processes. In this section, however, we shall concern ourselves with spatial variations in landscape patterns, dependent on the major climatic and geological inputs.

The form of a landscape can be assessed by applying morphometric techniques (see Chapter 10). In this way we can analyse and compare landscapes formed under different environmental conditions and observe the consequences of a variation in the major controls. Two examples follow, which demonstrate the effects of lithology and climate upon fluvially eroded terrains.

The influence of lithology alone can be demonstrated by comparing drainage basins developed upon different rocks within the same climatic region. Brunsden (1968) offers morphometric comparison between three sets of drainage basins on and around Dartmoor, England. Each set is developed on a different lithological base, the three types being granite, Devonian slates and grits, and Culm Measures which include significant quantities of shales. The drainage basins are standardized by considering fourth-order basins only, and values are tabulated for the characteristics of basin area, total stream length, mean bifurcation ratio and drainage density (Table 17.7). Three of these basins are illustrated in Fig. 17.7. Although the size of the sample is small, there are clear differences between the basins developed on the various rock types in respect of two of the four descriptive parameters – basin area and drainage density. Total stream length and mean bifurcation ratio also show some differences between groups, although there is some overlap in these two parameters. The granite terrain is characterized by large drainage basins with a lower drainage density, while the impermeable shales of the Culm Measures produce small basins with a high drainage density. It can be concluded, therefore, that the morphometric characteristics of these basins are closely influenced by lithology.

The influence of climate on landscape form can be illustrated by comparing landscapes developed under different climatic inputs. The study by Gregory (1976) of variations in one morphometric characteristic (drainage density) in Britain in relation to precipitation is useful in this respect (Fig. 17.8). Drainage density is assessed for 13 sample areas and plotted against mean annual rainfall. Although there is a considerable range demonstrated in some of the sample areas, there is nevertheless a clear tendency for drainage density to increase in direct proportion to rainfall. Furthermore, there is a distinction between permeable and impermeable rocks, with higher drainage densities occurring on the latter.

These two illustrations suffice to show that the

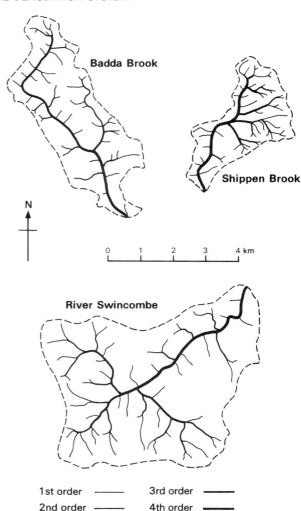

Fig. 17.7 Drainage basin morphometry on different lithologies: (a) Badda Brook (Devonian), (b) Shippen Brook (Culm), (c) River Swincombe (granite) (after Brunsden, 1968).

characteristics of landscape form can be related to the major environmental inputs. There is clearly scope for further study in this field, particularly in respect of the relationships between landscape and climate. Indeed, on a worldwide scale, Gregory (1976) demonstrates relationships between drainage density and annual precipitation, and also a more refined measure of rainfall input – that of rainfall intensity. Such studies, and closely controlled morphometric studies of landscapes developed on similar rock types but under different

Table 17.7 Morphometric properties of drainage basins on Dartmoor and adjacent areas (after Brunsden, 1968)

	Basin area (km²)	Total stream length (km)	Mean bifurcation ratio	Drainage density (km km⁻²)
Granite				
Swincombe	18.8	38.4	3.9	2.0
Cherry Brook	13.2	29.3	3.5	2.2
Walla Brook	13.7	22.4	3.6	1.6
Devonian				
Gatcombe	9.3	27.4	3.8	2.9
Badda Brook	8.0	24.3	4.2	3.0
Dittisham Creek	12.4	35.6	3.9	2.8
Culm				
Yeo River	5.4	20.3	3.6	3.7
Shippen Brook	4.1	20.0	3.7	4.8
Dorna Brook	7.5	24.3	3.4	3.2

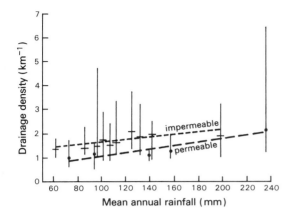

Fig. 17.8 Drainage density in relation to mean annual precipitation in Britain (after Gregory, 1976).

climatic regimes, point the way to a fuller understanding of the relationships between landscape, lithology and climate. In particular, there is a paucity of morphometric information on the morphology of glacially eroded terrain, although individual landforms, such as cirque basins, have been studied in some detail.

17.4 Relations between denudation processes and landscape form

In this chapter we have treated landscape form as a product of, or response to, denudation processes. The status of landscape morphology however, and the variables used to describe it, depend very much on the temporal and spatial scale at which it is considered.

In the terms outlined in Section 17.3, landscape form as a consequence of denudational activity is clearly a dependent entity. At smaller spatial and temporal scales however, when the operation of individual denudation processes are considered, landscape form may become an independent variable. Thus drainage density, which is presented as a dependent variable in Tables 17.6 and 17.7, and Fig. 17.8, may become an important independent variable, for example, in controlling hydrograph form. In Table 17.6, drainage density may be regarded as dependent on infiltration rate, yet independent in relation to sediment yield. The

variables of relief and gradient, themselves dependent on processes of uplift or incision, become independent variables when considered as controls on slope processes.

At a particular point in time, then, it is reasonable to expect that landscape form will attain an equilibrium in relation to major environmental inputs, providing that these are constant for a reasonable period of time. Yet as denudation processes continue to reduce the relief, those aspects of landscape form which control denudation processes will themselves be modified, and in turn affect the denudation processes themselves. Over longer periods of time, as relief is considerably reduced, the decline in potential energy, and perhaps also in orographic precipitation, will significantly affect the major inputs to denudation systems. In this way there are important feedbacks, at various scales, between landscape form and denudation process (Section 10.7).

Thus it may be appropriate over a short timescale to regard landscape form as a denudational output. Over longer timescales however, its state at a given time is better considered as one point in a continually changing response to external inputs.

Further reading

For a fuller discussion of denudation rates the reader is referred to:

Degens, E.T., S. Kempe and J.E. Richey (eds) (1990) *Biogeochemistry of Major World Rivers*. John Wiley, New York.

Drever, J.I. (1988) *The Geochemistry of Natural Waters*. Prentice-Hall, Englewood Cliffs.

Gibbs, R.J. (1970) Mechanisms controlling world water chemistry. *Science*, **170**, 1088–1090.

Walling, D.E. and B.W. Webb (1981) Water quality, in, *British Rivers*, (ed. J. Lewin). Allen & Unwin, pp. 126–169.

Walling, D.E. and B.W. Webb (1983) Patterns of sediment yield, in, *Background to Palaeohydrology: a Perspective*, (ed. K.J. Gregory). John Wiley, Chichester, pp. 69–100.

Walling, D.E. (1987) Rainfall, runoff and erosion of the land: a global view, in, *Energetics of Physical Environment: Energetic Approaches to Physical Geography*, (ed. K.J. Gregory). John Wiley, Chichester.

Walling, D.E. and B.W. Webb (1986) Solutes in river systems, in, *Solute Processes*, (ed. S.T. Trudgill). John Wiley, Chichester, pp. 251–327.

VI   ECOLOGICAL SYSTEMS

The ecosystem 18

18.1 The ecosystem concept

The deciduous woodland depicted in Fig. 18.1 is an oak/beech wood in southern England. It is, of course, a segment of the global ecosphere considered in Chapter 7, although it is doubtful whether this is the first reaction it evokes. This is simply because here we are encountering the ecosphere at a more familiar and accessible scale. We have taken a step down from the abstraction of a global model to something real and comprehensible. For a moment, let us imagine that we can walk into this woodland.

Initially we are aware of a reduction in light, though our eyes soon compensate and look down as dead wood and dry leaf litter crack under our feet. A disturbed fallen branch reveals woodlice beneath it and on its underside the lacework of fungal threads stands out white against the rotting wood. Here and there beech seedlings rise through the mantle of fallen leaves, while near the boles of the trees, green cushions of moss contrast with the browns of the litter. Elsewhere, in shade, wood ruff, dog's mercury, wood anemone and Solomon's seal compete for space.

From the massive roots of beech and oak reaching down like flared fingers into the earth, the eyes follow the towering trunks as they reach up in a colonnade to support the vaulting tracery of the branches. Here the leaves lie in a mosaic against the Sun, a mosaic more complete above the beech with few gaps and little overlap, but more open above the oak. Epiphytic mosses and lichens cover these trunks and branches, while a dead bough is resplendent with the colourful, if grotesque, fruiting bodies of fungi.

Across a glade, a deer stops browsing and, raising an antlered head to read the air, silently merges into the sunflecked birch scrub at the margin of the clearing. Startled by the resonant drumming of a woodpecker somewhere close at hand, we become aware, in the relative silence that follows, of the songs and calls of other birds, high in the canopy. The subtle sounds of myriads of insects impinge on our consciousness, and looking closely we see further evidence of their existence. Leaves are trimmed by wintermoth, dunbar, or green tortrix larvae while some have neat discs removed by leaf-cutting bees. Oak apples formed by gall wasps hang amongst the foliage, abandoned by their former tenants.

As we emerge from this woodland, let us take stock and attempt to impose some order on these images. First, the woodland consists partly of a large number of different organisms, both plants and animals, each being represented by a population of individuals. Each population has at any moment in time a particular spatial distribution so that the total three-dimensional space is partitioned between the organisms present. Although these organisms form a complex woodland community, our image of it is of more than a collection of organisms growing and living together. The still, sun-shafted air, cool within the trunkspace, the soil which tethers the tall trunks, the smooth stream with trees on either side, the still pool and clean gravel and the rotting remains of once freshly fallen leaves are all just as important to our image of the woodland as the living plants and animals. In other words, it is the community and its immediate environment which together make up our perception of the woodland, for both are inextricably linked, not only in our mind's eye but also functionally. In Chapter 7 this unity was recognized at the global scale, where the concept of the ecosphere was used to encompass the living systems of the biosphere and those parts of the atmosphere, lithosphere and hydrosphere with which they exchange matter and energy. At the scale of the plant and animal community, the term **ecosystem** (ecological system) is used to embody the same concept, and what we have been describing is, therefore, a woodland ecosystem (Box 18.1).

18.2 The structural organization of the ecosystem

18.2.1 The elements of the ecosystem and their attributes

From this initial discussion of a woodland ecosystem, it is clear that the structure of the ecosystem has two components: the living component (the organisms themselves) and the non-living (environmental) component. The structural organization of the environment is dealt with in other chapters of this book. In this chapter, therefore, we shall be concerned largely with the way in which the living organisms of the system are disposed. In the context of an ecosystem model these organisms are the elements of the system. How, then, do we describe or categorize these elements; what are their attributes? As living systems, all organisms in an ecosystem can be regarded initially as having two sets of attributes. First, they possess a set of genetic attributes enshrined in the DNA molecules in the nuclei of their cells (see Box 2.3). This is their **genotype**. It controls their development and activity and it is manifest as the second set of attributes: morphological, physiological and behavioural attributes – their **phenotype**. There are three ways in which we can choose to use these attributes to categorize the organisms of an ecosystem: taxonomically, structurally and functionally. As we shall see, the choice is largely conditioned by the type of relationship between the elements (organisms) of the system (ecosystem) that we wish to stress.

Both genotype and phenotype can be used to define the types of organism present in an ecosystem by identifying them as **species**, **subspecies** or **varieties**, and placing them in a natural hierarchical taxonomic classification (Table 18.1). The underlying assumption is that such a classification reflects the evolutionary relationships between the organisms of the biosphere and it should, therefore, be based on genetic criteria. Nevertheless, inconsistency and ambiguity can become apparent (Box 18.2), particularly when phenotypic characters are used as a basis for classification. Although the 'species list' is the common starting point for the categorization of the organisms present in an ecosystem, it is not the only approach, nor necessarily the most useful.

The phenotypic attributes of organisms, particularly the architecture of their form, or their morphology, can be used as an alternative to taxonomy to order the 'elements of the system'. Before considering such schemes, however, it will be instructive to distinguish between what have been called

---

Box 18.1

THE ECOSYSTEM CONCEPT

The concept of the ecosystem as an ecological unit composed of living and non-living components interacting to produce a stable system is not new. It has a long history in the biological and ecological literature, sometimes explicitly stated under a different terminology (e.g. biocoenosis), sometimes merely implied. In 1935 Sir Arthur Tansley, an important figure in the development of ecology in the British Isles, proposed that the term should be used to describe 'not only the organism complex but also the whole complex of physical factors forming what we call the environment' (Tansley, 1935). However, the main theoretical development of the concept, and the implementation of research associated with that development, occurred in the period since 1940, the major impetus taking place in the 1950s. This followed Lindeman's formulation of the trophic–dynamic view of the ecosystem, which came to provide a conceptual framework and stimulation for research in ecology (Lindeman, 1942).

Fig. 18.1 Images of a temperate deciduous forest eco-system.

Table 18.1 Taxonomic classification. An example of the Linnaean classification scheme applied to an animal – the red deer. Carolus Linnaeus (1707–78) was a Swedish physician and naturalist. He was Professor of Medicine at the University of Uppsala, but in 1758 he published the 10th edition of his *Systema naturae* which, for the first time, embodied a uniform system of classification for plants and animals alike, and from which all other classifications have stemmed.

kingdom	Animalia	animals
subkingdom	Metazoa	multicelled animals
phylum	Chordata	animals with notochords
subphylum	Vertebrata	animals with backbones
class	Mammalia	mammals
subclass	Theria	non-egg laying mammals
infraclass	Entheria	placental mammals
order	Artiodactyla	even-toed hoofed mammals
family	Cervidae	deer – 40 species
genus	*Cervus*	
species	*Cervus elaphus*	red deer

unitary and modular organisms (Jackson *et al.*, 1985; Harper *et al.*, 1986) (Fig. 18.2).

Unitary organisms are those like us human beings, and indeed most animals, which as embryos develop a definite form (our bodies) which remains largely unaltered until death. Certainly the bodies of unitary organisms pass through recognizable phases in their lifecycle, but apart from increases and decreases in size and weight and some effects associated with the onset of sexual maturity, no fundamental change occurs in the form of the individual. Modular organisms on the other hand develop by the repeated production of a **unit of construction**, or a **module**. There are several important groups of modular animals (e.g. corals, bryozoans, sponges), but the most familiar examples of the group are of course plants, most of which are modular. In the higher plants the unit of construction is the leaf, its axillary bud and the length of associated stem (the internode). The architecture of a plant's form is therefore determined by the number and the pattern of repetition of the modules. Modules may sever their

---

Box 18.2

THE SPECIES CONCEPT

The concept of the species as a unit in the classification of plants and animals was introduced in Chapter 7. Its use is far from simple and the term is used to describe some quite different entities. Ideally, a species is defined as a group of organisms that interbreed to produce fertile offspring. Members of different species do not normally interbreed and, if they do, the progeny are sterile. Not all species have been tested in the context of such a definition. Furthermore, it is inapplicable to species that habitually self-fertilize or reproduce only asexually. Species are perhaps best designated in accordance with the criteria used to define them:

Taxonomic species meets all criteria and satisfies the rules of international nomenclature.
Biospecies fulfils the breeding requirements of

the definition and is therefore restricted to sexual reproduction and cross-fertilization.
Ecospecies is a group of ecotypes (variants or races of a species adapted to different environments which can still cross, even though some of their adaptations may be hereditable) within a species which can cross with one another to produce fertile offspring.
Coenospecies are species which belong to a group which can intercross to form hybrids which are sometimes fertile.
Morphospecies are named only on morphological evidence – often these are modified in the light of new evidence.
Agamospecies are species with asexual reproduction only, therefore treated as morphospecies.
Palaeospecies are extinct organisms, therefore the only evidence is derived from fossils, again subject to modification.

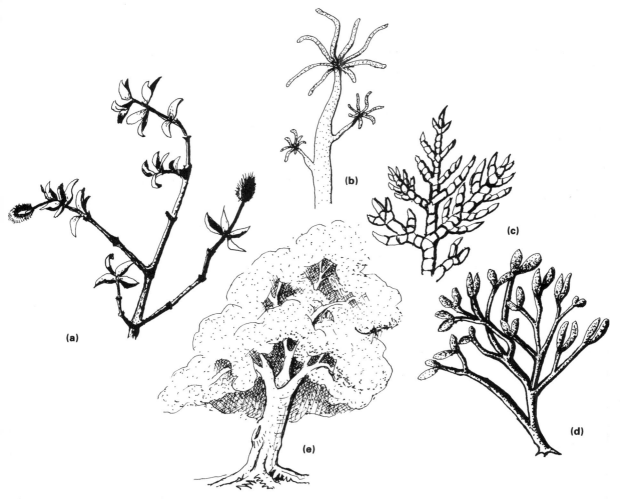

Fig. 18.2 Modular organisms. (a) *Larrea divaricata* (Creosote bush). (b) *Hydra* sp. (Hydrozoa). (c) *Corallina* (Alga). (d) *Pelvetia* (Alga). (e) *Quercus robur* (Oak).

connections and exist as separate vegetative individuals (though genetically identical members of a clone), or extend the organism through accumulation and branching. This accumulation can essentially take place as either horizontal extension and branching across or through the substrate (the soil in terrestrial plants), or vertical extension and branching towards the light. Moreover, dead modules may remain to provide skeletal or support functions (e.g. wood in trees and shrubs), while some modules may senesce, die and be shed and replaced periodically, thereby effecting temporal change in the plant's morphology (e.g. leaf shedding in deciduous species).

Clearly, therefore, the structure of vegetation in space (Section 18.2.2) is conditioned by the modular architecture of the constituent plants, or their **physiognomy**. Indeed, plants are often grouped into classes based on this architecture, or their growth habit or **lifeform** (Box 18.3), an approach which is intuitively adopted and recorded in the vernacular by the use of such terms as trees, shrubs, bushes and herbs. It is also an approach which lends itself readily to the analysis of the

Box 18.3
LIFEFORM CLASSIFICATIONS

Physiognomy when applied to **vegetation** is its outward or superficial appearance, without any necessary reference to structure or function, and even less to floristic composition. Of course an important element in this appearance is the physiognomy of the individual plants that comprise that vegetation. In this sense the physiognomy of the plant is the vegetative form of the plant body, that is, those morphological features that determine the general architecture of the plant, or its **lifeform**.

However, plants have ecological amplitudes, and the morphology of successful plants represents an expression of their adaptation to the environment. It is this implied correlation between lifeform and environment that has led to

Life-form	Size	Function	Leaf shape and size	Leaf texture
T ⚲ trees	t tall (T: minimum 25m.) (F: 2-8m.) (H: minimum 2m.)	d ☐ deciduous	n ⌒ needle or spine	f ⧄ filmy
F ⚲ shrubs	m medium (T: 10-25m.)	s ‖‖ semideciduous	g ◊ graminoid	z ☐ membranous
H ▽ herbs	(F, H: 0.5-2m.) (M: minimum 10cm.)	e ⊞ evergreen	a ◇ medium or small	x ■ sclerophyll
M ⌒ bryoids	l Low (T: 8-10m.) (F,H: maximum 50cm.)	j ※ evergreen-succulent; or evergreen-leafless	h △ broad	k ⠿ succulent;
E ✳ epiphytes	(M: maximum 10cm.)		v ⩔ compound	
L ⌔ lianas			q ◯ thalloid	

Cover			
b barren or very sparse		i discontinuous	
p in tufts or groups		c continuous	

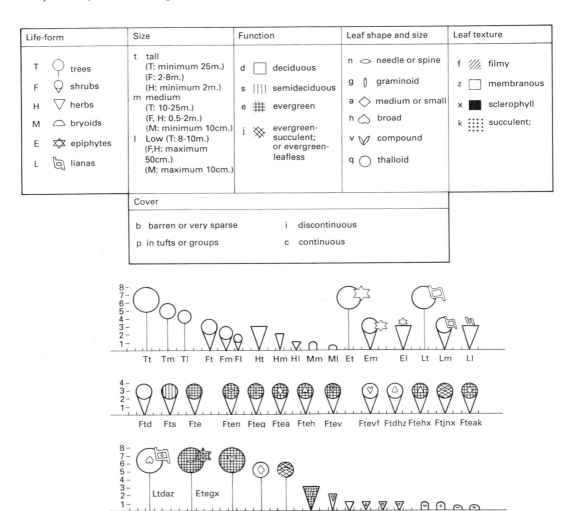

the primary division between the many life- or growth form classification schemes that have been proposed. Those schemes that are purely physiognomic, with no implied functional significance, are traditionally, though perhaps a little misleadingly referred to as **structural**. Those that deploy physiognomic characters to imply some adaptation to environment are referred to as **functional**.

Structural Lifeform Classifications

Sometimes such classification schemes are of individual physiognomic characters of the plant, such as the bark or the leaves. Size is often the principal criterion, but other characteristics of a single organ, such as a leaf, may be combined so that size may be augmented by information on shape, orientation or duration. More usually, several physiognomic characters are linked to present a comprehensive description of the morphology or lifeform of the individual plant, and, in combination, of the physiognomy of the vegetation they form. The French Canadian ecologist/biogeographer Pierre Dansereau's scheme, expounded in 1951 and later in his book published in 1957, is illustrated.

Dansereau not only provides a scheme for describing and classifying the lifeform of plants, but also by incorporating information on stratification and cover provides a means of recording the **structure** of the vegetation as a whole, both as a formula and as a scheme of graphic symbols that represent the vegetation visually in such a way that outstanding differences and similarities between stands can be recognized.

Functional Lifeform Classifications

The individual physiognomic characters of plants often also lend themselves to functional as well as purely structural interpretation. Between about 1903 and 1916, Christen Raunkiaer, a Danish botanist, developed the main functional classification of plant lifeforms (although his main publication on the topic occurred in 1934) and it remains the only one to have wide currency, albeit sometimes in a modified form.

Raunkiaer searched for a character of the plant that fulfilled three criteria. First, it must be of fundamental importance in relation to climate.

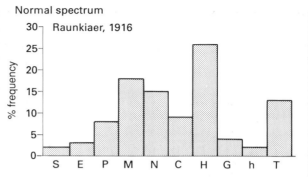

Normal spectrum
Raunkiaer, 1916

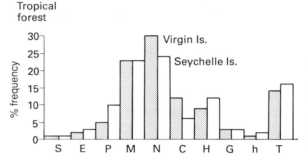

Tropical forest

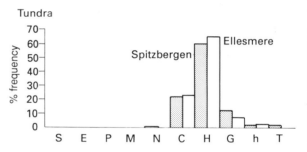

Tundra

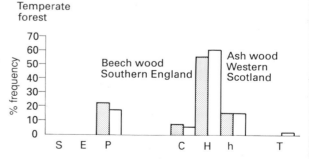

Temperate forest

S = succulents, E = epiphytes, P = megaphanerophytes and macrophanerophytes, M = mesophanerophytes, N = nanophanerophytes, C = chamæphytes, H = hemicryptophytes, G = geophytes, h = helophytes and hydrophytes, T = therophytes.

Secondly, it had to be easily recognizable and recorded, and thirdly it had to be a single attribute and susceptible to statistical treatment. He finally chose the height relative to the ground surface at which the perenating tissue (perenating bud, vegetative bud or apical meristem) of the plant was borne. The application of this relatively simple criterion, sometimes with some further qualification such as the presence or absence of protective bud scales, produced the classes of lifeform shown in Fig. 18.3.

The lifeform series was based on the assumption that the more primitive lifeforms of plants were associated with early evolutionary phases, under a more genial climate that was both more uniformly moist and more uniformly warm than that which prevails over most of the world today. Following such an assumption the most primitive lifeforms are those that dominate the humid tropics today, while the more highly evolved lifeforms, displaying progressive adaptation, are found under the non-tropical climates of the globe, associated with greater seasonality, aridity and/or cold. Now although such a functional interpretation of lifeform classes has to be a gross oversimplification, nevertheless it can be useful.

Using Raunkiaer's scheme and what are known as **lifeform spectra** (the relative proportions of the lifeform classes in the flora of a region or vegetation type), it is possible to compare areas which may be widely separated geographically and which do not contain species in common to both, i.e. the areas are floristically or taxonomically distinct. Such spectra can also be compared with his so-called **Normal Spectrum**, that he produced in 1916 by taking 1000 randomly selected plants from the flora of the world.

What this approach does is to emphasize the correlation of vegetation with climate and indicate the existence of bioclimatic zones with characteristic lifeforms. The zonation progresses from a predominantly Phanerophytic zone in the tropics, to a Therophytic one in the subtropical semi-arid to desert latitudes, a Hemicryptophytic one in the cool/cold temperate mid-latitudes, to a Chamæphytic one in the cold high latitudes. Although the approach is generally thought not to be of great value over smaller areas where one is dealing with physiognomically similar vegetation, this is probably something of an underestimate of its usefulness, and where vegetation is changing over relatively steep environmental gradients, as with altitude, lifeform analysis can often provide significant insight into the nature and causes of such change.

structural relationships of plants (particularly in space) because the lifeform of a plant is the vegetative form of the plant body. Plants that show the same general vegetative features belong to the same lifeform class irrespective of their genetic and taxonomic relationships. The lifeform, however, is also considered to be a hereditary adjustment to environment, and it can therefore reflect functional relationships, particularly those with climate (Fig. 18.3).

The use of the attributes of organisms to define structural entities is more usual and more developed in plant ecology. In animal ecology the alternative to taxonomy most often adopted is the use of functional entities, particularly those based on the trophic relationships of animals (herbivore, carnivore, omnivore, detritivore and so on). Some functional categories have also been employed for plants, though with less precision. Apart from the trophic counterparts, such as producer (or autotroph) and saprophyte, examples of functional entities from the plant kingdom usually concern adaptation to environment – as in the case of hydrophyte, xerophyte, halophyte, epiphyte and succulent – and they tend to overlap with lifeform categories.

18.2.2 Structural relationships in the ecosystem: structure in space

Once we know what organisms are present in an ecosystem and have defined them in relation to their attributes, we need to be able to describe the relationships between these elements and between their attributes. Structurally, the most obvious organizational relationships are spatial. Here the plants and the vegetation they form are paramount. In most (and certainly in terrestrial) ecosystems, it is the sedentary plant kingdom that forms the spatial framework of ecosystem structure. Plant biomass represents not only the **standing**

Phanerophytes (trees and shrubs)

perenating buds or shoot apices
borne on aerial shoots

evergreen, with or without bud scales
deciduous

nanophanerophytes	< 2 m high
micro	2 – 8 m
meso	8 – 30 m
mega	> 30 m

Myrica gale   nanophanerophyte

Chamaephytes

perenating buds or shoot apices
close to the ground:

(a) on lower portion of erect stems
     which die back during
     unfavourable season

(b) passive prostrate stems

(c) persistently prostrate stems

(d) cushion plants – reduced and compact vegetative growth

Silene aucaulis   cushion chamaephyte

Hemicryptophytes

perenating bud at ground level

some possess stolons or runners

some are rosette or partial rosette
plants

also includes grasses and sedges

Saxifraga sp.   rosette hemicryptophyte

Cryptophytes

perenating bud below ground
or below water

geophytes with bulbs, tubers or rhizomes

marsh plants (helophytes) with perenating
bud below water, but shoot above

hydrophytes – water plants with perenating
bud below water leaves submerged or
floating

Menyanthes trifoliata   helophyte

Therophytes

annual plants – life history complete in the favourable season, 'seed is perenating bud'

Succulents and Epiphytes

usually function as phanerophytes because perenating buds are often borne aloft
frequently used as special categories in life form analysis

Fig. 18.3 Lifeform classification systems. Note that *Myrica gale*, growing here as a nanophanerophyte, can attain heights greater than 2 m under more optimum conditions.

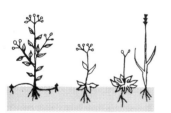

crop of the primary producers, but also living space, forming (as in our deciduous woodland, Fig. 18.1) the **habitat** of most consuming organisms. The way in which the vegetation partitions the three-dimensional space it occupies is traditionally approached by considering separately the vertical structure (**stratification**) and the horizontal structure (**areal distribution**) of the organisms concerned.

The deciduous woodland described above, with its herb layer and tree canopy, and here and there a discontinuous shrub layer, has a vertical structure consisting of two or three strata. In more complex forest types, such as those of the humid tropical lowlands, the stratification can become much more complex. Nevertheless, as Robert Louis Stevenson's tin soldier could attest lying 'in the forests of the grass', this same concept can be applied to vegetation of smaller stature; to grassland, dwarf shrub heath and aquatic communities (Fig. 18.4). However, stratification not only describes structural relationships but has functional implications as well, because what we are considering is the stratification of the photosynthesizing tissue of the primary producers. It can be seen, therefore, as a solution optimizing the utilization of available light energy. Notice that the term '*optimizing*' is used, not '*maximizing*', because a price must be paid, at least in terrestrial ecosystems, for the creation of such an elaborate energy filter as, for example, a forest canopy. This price is exacted in the form of the energy diverted to the maintenance of non-photosynthesizing support tissue, represented in our woodland ecosystem by the massive biomass of wood. Also associated with a plant's place in the stratification is a host of essentially functional adaptive features related to the microclimatic gradients that the stratification itself creates (see Chapter 8). Indeed, gradients of environmental parameters are not restricted to the lower layers of the atmosphere, and the below-ground biomass of the ecosystem also shows a distinct stratification. The stratification of organisms inhabiting litter and humus and of the roots of higher plants all reflect gradients in physicochemical conditions and in the values of water and nutrient storage parameters in the soil (see Chapters 11 and 22).

The aim of the description of the horizontal structure of vegetation is the location of each individual of all species in a two-dimensional framework. Such exact mapping is attempted only for small areas or along short transects in structurally simple communities, and usually to monitor short-term change. The more usual approach is to describe statistically the horizontal **pattern** of species populations or of the vegetation, particularly in terms of its departure from a theoretical random distribution that would occur in the absence of intrinsic or extrinsic causal factors (Fig. 18.5). The statistical detection of pattern, therefore, implies that there is an underlying cause. Some pattern occurs in response to variations in environmental conditions, such as soil properties or surface microrelief; some is sociological in the sense that it is caused by interactions, often but not always competitive, between different species, or even between individuals of the same species. Yet other instances of pattern reflect the growth morphology of the species – especially in vegetatively reproducing plants – or its seed dispersal capacity. The investigation of pattern is further complicated by the fact that individual species, or the vegetation as a whole, may display pattern at several different spatial scales, and at each scale the causes may be different. At a small scale we may be dealing with morphological pattern, at an intermediate scale with some sociological interaction, and at a larger scale with an imposed environmental pattern.

A third element in the description of spatial structure remains, and that is the **degree of presence** of each species, or a measure of its actual contribution to the community (Fig. 18.6). The most objective and quantitative measure of this contribution is the number of individuals of each species per unit area – the **density**. However, there are practical difficulties associated with density counts, and subjective estimates of abundance are often employed as an alternative. To accommodate differences in size and growth form between different plant species, measurements or estimates of a **species cover** can be made. Cover is an area equivalent to the vertical projection of the above-ground parts of the plant on to the ground surface, and is expressed as a percentage of the total area; it is related to the share that any species has of the total incoming light energy. Because the vertical separation implicit in stratified communities allows overlapping areas, the total cover of all species will normally exceed 100%. Quantitative measurements of cover are difficult to obtain in all but simple communities, but estimates of cover can be combined with those of abundance to give synthetic scales (Table 18.2). Closely

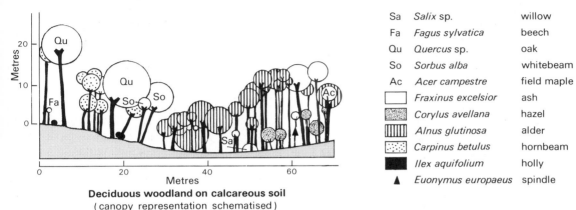

Sa *Salix* sp.   willow
Fa *Fagus sylvatica*   beech
Qu *Quercus* sp.   oak
So *Sorbus alba*   whitebeam
Ac *Acer campestre*   field maple
▢   *Fraxinus excelsior*   ash
▨   *Corylus avellana*   hazel
▥   *Alnus glutinosa*   alder
▨   *Carpinus betulus*   hornbeam
■   *Ilex aquifolium*   holly
▲   *Euonymus europaeus*   spindle

Deciduous woodland on calcareous soil
(canopy representation schematised)

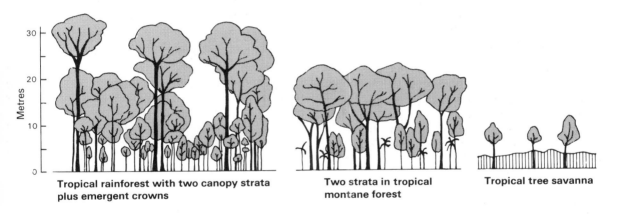

Tropical rainforest with two canopy strata plus emergent crowns

Two strata in tropical montane forest

Tropical tree savanna

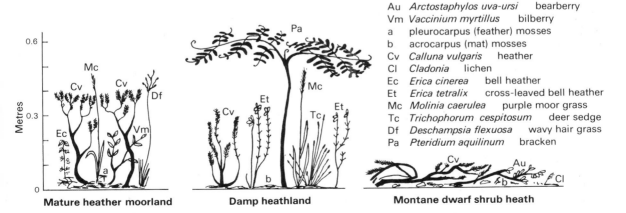

Au *Arctostaphylos uva-ursi*   bearberry
Vm *Vaccinium myrtillus*   bilberry
a   pleurocarpus (feather) mosses
b   acrocarpus (mat) mosses
Cv *Calluna vulgaris*   heather
Cl *Cladonia*   lichen
Ec *Erica cinerea*   bell heather
Et *Erica tetralix*   cross-leaved bell heather
Mc *Molinia caerulea*   purple moor grass
Tc *Trichophorum cespitosum*   deer sedge
Df *Deschampsia flexuosa*   wavy hair grass
Pa *Pteridium aquilinum*   bracken

Mature heather moorland

Damp heathland

Montane dwarf shrub heath

Fig. 18.4 Vegetation stratification.

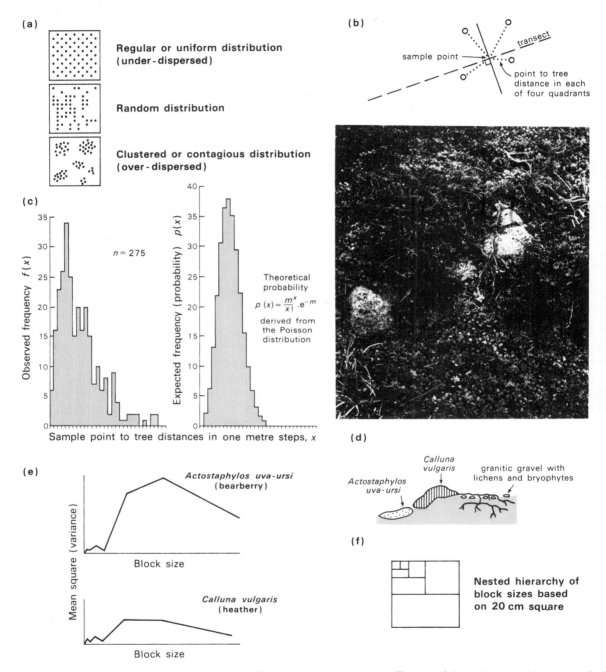

(a)

Regular or uniform distribution
(under-dispersed)

Random distribution

Clustered or contagious distribution
(over-dispersed)

(b)

transect

sample point

point to tree
distance in each
of four quadrants

(c)

$n = 275$

Observed frequency $f(x)$

Sample point to tree distances in one metre steps, x

Expected frequency (probability) $p(x)$

Theoretical
probability

$$p(x) = \frac{m^x}{x!} \cdot e^{-m}$$

derived from
the Poisson
distribution

(d)

(e)

Actostaphylos uva-ursi
(bearberry)

Mean square (variance)

Block size

Calluna vulgaris
(heather)

Block size

*Actostaphylos
uva-ursi*

*Calluna
vulgaris*

granitic gravel with
lichens and bryophytes

(f)

Nested hierarchy of
block sizes based
on 20 cm square

Fig. 18.5 The detection and analysis of pattern. (a) Types of spatial pattern (b, c) The use of the point-centred quarter method in a pine wood to test the pattern of tree distribution against that expected if the trees were distributed at random. Note that the sample frequency distribution in (c) is overdispersed relative to the expected (Poisson) distribution, indicating that the trees some degree of clustering. (d, e and f) The application of pattern analysis to a *Calluna/Arctostaphylos* montane dwarf shrub heath to detect the occurrence of pattern at different spatial scales (block sizes).

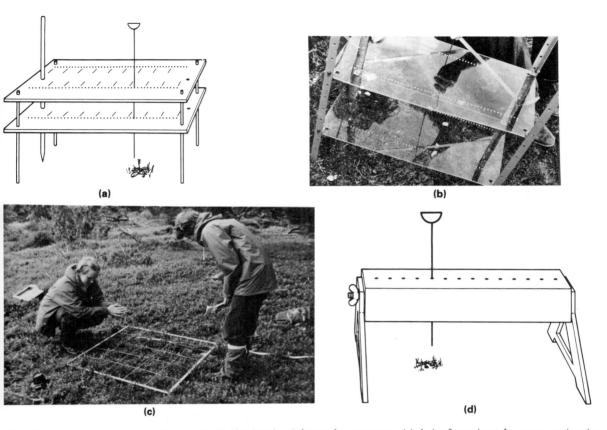

Fig. 18.6 The measurement of cover. (a, b) Kershaw's pinframe for transects. (c) A 1 m² quadrant for cover estimation. (d) A simple pinframe.

related to cover is the use of above-ground biomass measurements to determine the contribution of each species, on the assumption that those with the largest biomass have acquired the greatest share of space, energy and other resources.

Animal communities also exhibit a spatial structure, vertically and horizontally, but it tends to reflect that of the vegetation itself, or the disposition of gradients of habitat conditions the vegetation creates. The bird populations in our deciduous woodland, for example, occupy distinct strata in the woods. This vertical segregation partly reflects their feeding habits and food availability, but also the occurrence of suitable cover, nesting sites and roosting perches. The wren, for example, is restricted in nesting sites to thick understorey scrub, while the great and blue tits and perhaps the starling may occupy holes bored and previously occupied by woodpeckers. The strength of the branches of the trees of the main canopy supports the large nests of carrion crow, and beneath these the uneven extended canopy

Table 18.2 Cover/abundance scales. When the Domin scale is examined with respect to either cover or frequency data alone, it is found to be non-linear. Domin values have therefore been transformed using the equation $y = 0.0428x$ to provide a scale with the required linear characteristics (Bannister, 1966). The Braun–Blanquet scale has the same deficiency and Moore (1962) has provided a suitable transform.

	Braun–Blanquet	Domin	Transformed Domin (Bannister, 1966)
cover about 100%	5	10	8.4
cover about 75%		9	7.4
cover 50–75%	4	8	5.9
cover 33–50%		7	4.6
cover 25–33%	3	6	3.9
abundant; cover about 20%		5	3.0
abundant; cover about 5%	2	4	2.6
scattered; cover small		3	0.9
very scattered; cover small	1	2	0.4
scarce; cover small		1	0.2
isolated; cover small	+	+	0.04

of oak provides nesting sites for wood pigeon, doves and some thrushes. MacArthur (1958) describes similar stratification in different species of insectivorous warblers in evergreen forest in New England which feed exclusively within a narrow vertical zone (Fig. 18.7). It is in tropical forests, however, that the stratification of animal communities is best exemplified, where, for example in Guyana, 61% of the forest mammals are arboreal and each species is restricted to a particular level in the canopy. Aquatic ecosystems also show stratification.

Habitat. This consideration of the spatial organization of animals, of course, has been concerned with the relationship of those animals with the *places* in which they live, i.e. their **habitat**. There have been some attempts of varied success to define and classify the habitats which organisms occupy. In animal ecology one of the most influential is that proposed by Elton and Miller (1954), where the role of vegetation as a physical descriptor of animal habitats is apparent. They divide the environment into a number of habitat systems, which are then subdivided by vegetation physiognomy and stratification, both closely paralleling the lifeform classification of Raunkier (Box 18.3), and with a small group of factors called *qualifiers* which deal with minor habitats not accounted for already (Table 18.3, and Fig. 18.8).

However, the notion of the habitat is more complex than such schemes and attempts to define it beyond bland statements of the kind: *the place an organism lives*, imply. Habitat is a holistic, synthetic concept embracing both the inorganic and organic dimensions of some compartment of space, and as such immediately begins to involve the functional interaction of an organism with its environment (see for example Fig. 18.7). Environment and habitat are not, however, totally interchangeable terms. An organism may occupy different habitats in different parts of its geographical range, but the environment offered for the growth of the organism in question by each habitat may be to all intents and purposes within the organism's limits of tolerance, the same. For example the wood sage *Teucrium scorodonia* growing in the deep cleft afforded by a gryke in a limestone pavement in the Burren, in County Clare, Ireland, will experience virtually the same microclimatic environment of humidity, shelter and shade as it does growing in the field layer, under a closed canopy in a nearby woodland. The two habitats, however, are different.

As the habitat has both an *abiotic* (non-living), and a *biotic* (living) component to its definition, the latter leads to a consideration of interactions with other organisms, and becomes of necessity concerned with functional relationships such as competition for resources, herbivory and preda-

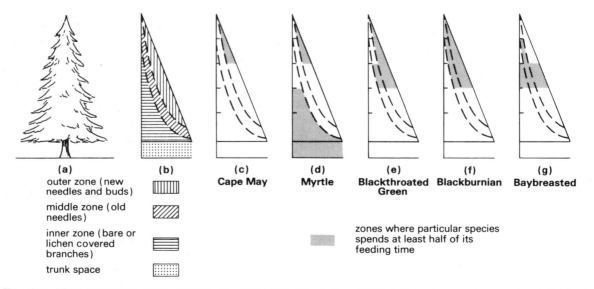

(a)
outer zone (new needles and buds)

middle zone (old needles)

inner zone (bare or lichen covered branches)

trunk space

(c) Cape May

(d) Myrtle

(e) Blackthroated Green

(f) Blackburnian

(g) Baybreasted

zones where particular species spends at least half of its feeding time

Fig. 18.7 Niche separation among species of warbler (after MacArthur, 1958). (a & b) Canopy structure; (c–g) distribution.

Table 18.3 A classification of animal habitats after (Elton and Miller, 1954)

Systems

T Terrestrial System; A Aquatic System; A-T Transition System [includes waterlogged soil, marsh, and shore habitats]; D Domestic System [associated with human activity]; G General System [small habitats associated with dead and dying material]; S Subterranean System

Terrestrial system; Formations

OGT Open Ground Type Vegetation <15 cm in height; FT Field Type Vegetation <2 m but may be grasses, herbs, or shrubs; ST Scrub Type Vegetation < 5 m; WT Woodland Type Trees are the dominant life form.

Terrestrial System

Layers: AA Air Above occurs over all formations; C High Canopy Vegetation >5 m. SL Scrub (Low Canopy) Zone Between 2–5 m; FL Field Layer Between 15 cm–2 m; GZ Ground Zone Up to 15 cm, includes carcases and fallen timber, even if higher than 15 cm; Topsoil; Subsoil.

Terrestrial system: Qualifiers

Edge: Interface habitat between two formations (ecotone) Soil: Acid pH <7. Non-acid pH >7 , Calcareous, Maritime, Arable, Woodland: Deciduous, Evergreen, Mixed.

Various minor habitats within wild and domestic systems
General Habitats

Dead herbaceous stems [Dhs], Dead and decaying wood [Dd wd], Macrofungi [Mf], Sapflows [Sf], Dung [Dg], Carrion [Cn], Animal artefacts [An-a], eg. nests [n], Human artefacts [H-a], eg. walls [wl].

Living parts of plants

Flowers [Fl], Fruit [Fr], Leaf [Lf], Twig [Tw], Branch [Br], Trunk [Tr], Root [Rt], Woody stem [Ws], Live herbaceous stem [Lhs], Galls [G1].

Other

Moss [Ms], Lichen [Ln].

Ground zone

Natural litter [N1], Bare soil [Bs], Under stones [Us], Bare mud [Bm].

tion, and as such overlaps with the concept of **niche**, which will be developed in Section 18.2.3. Even the abiotic component of an organism's habitat is not just what has been called a physicochemical *stage set* on which the organism acts out its existence. Such a view is misleading, for the physicochemical conditions are at the same time attributes of an organism's habitat *and* the conditions under which it exists, and to which it (or rather the species to which it belongs) has adjusted

through the evolutionary development of adaptation and tolerance.

Moreover, organisms respond to the habitat, and this response regulates the activity of the population in some way. Indeed, neither the physicochemical, nor the biological components of the habitat are constant. Habitats show significant variation in their favourableness for an organism, both in space and in time. In 1977, Southwood, in a presidential address to the British Ecological Society, looked at habitat variability from the perspective of the organism itself and classified habitats as continuous, patchy or isolated in space, and as constant, seasonal, unpredictable and ephemeral in time. To survive, species have to evolve strategies to cope with particular combinations of spatial and temporal variation in habitat factors. Therefore, the idea of the stage set is misleading in the sense that organism/ habitat relations are interactive, and these interactions are the principal determinant of survival. Indeed, the habitat can be regarded as an **ecological template** for the development of successful **ecological strategies**; that is, species can be regarded as evolving those strategies which maximize the numbers of their descendants which survive in their habitats (see discussion in Chapter 25, and (Southwood, 1977)).

Phenology. However, the spatial organization of animals is generally much more fluid than that of plants. Their location and pattern of distribution may change throughout the day or the year, depending on the particular activity in which they are engaged (Fig. 18.9a). Hunting and browsing, nesting and mating, or simply resting, are all activities which may produce different patterns of distribution in the same animal species. Even explicitly spatial phenomena, such as the **territorial behaviour** of many animals and birds (Fig. 18.9b) may only be established for part of the breeding cycle and may change radically from year to year, in response to complex interactions involving, for example, food availability or quality, population numbers, hormonal control of behavioural traits such as aggression, and variations in habitat diversity (see for example the red grouse: Jenkins *et al.*, 1967; and Moss, 1969).

It would be wrong, however, to think of the vegetation as being static. Some plants may be annuals, some perennial, but all have a lifecycle from germination and establishment to death. Each

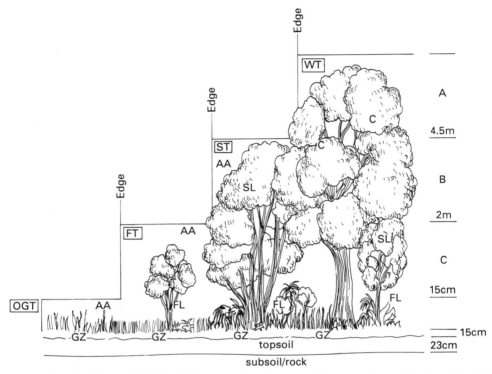

Fig. 18.8 Diagram showing Elton and Miller's habitat types for the terrestrial system. For explanation of symbols see Table 18.3.

exhibits temporal variation in growth, flowering, fruiting and other functional activities (its **phenological characteristics**) and, as it does so, its place changes in the spatial structure of the community (Fig. 18.10). Also, as each generation succeeds its predecessors, so the spatial pattern of each species population changes with time. So too do the absolute numbers of each species, whether plant or animal, for populations increase and decline through time, perhaps with random fluctuations, perhaps with an identifiable periodicity, perhaps with a recognizable long-term trend. We shall consider all of these aspects of the ecosystem's behaviour with time in more detail later (Chapter 23), but in the present context they can be seen to have an expression in terms of changes in spatial organization. The continuity of an ecosystem's structure, like the solid permanence of the woodland in Fig. 18.1, is as much a figment of our imagination as is the myth of the everlasting hills. We are looking at a structural mosaic not only in space, but also in time.

18.2.3 Structural organization of the ecosystem: functional relationships

The organization of the ecosystem is fundamentally functional. As well as spatial relationships between the organisms of an ecosystem there are also functional relationships, for they partition not only the available space but also the energy and resources that sustain life. These functional relationships facilitate the transfer of energy and matter, both between the living components of the system and between them and their non-living environment. The key to this functional structure is the trophic organization of producer, consumer and decomposer levels, introduced in Chapter 7. However, the trophic structure of real ecosystems is often extremely complex, as in the case of the deciduous forest considered at the beginning of this chapter (Fig. 18.11a). Nevertheless, these patterns of functional organization provide the framework that makes it possible both to understand the role and significance of each organism in the ecosystem's function, and to

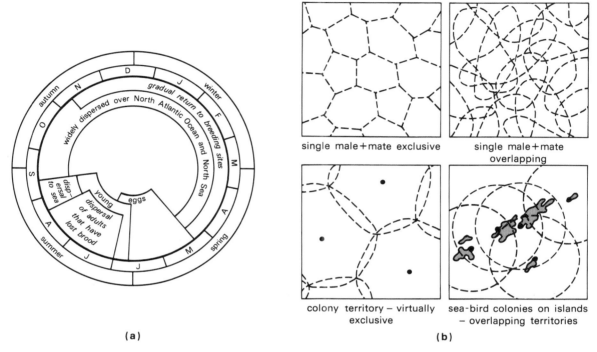

Fig. 18.9 (a) The lifecycle of the fulmar (after Fisher, 1954). (b) Territorial behaviour in animals and birds (after Wynne-Edwards).

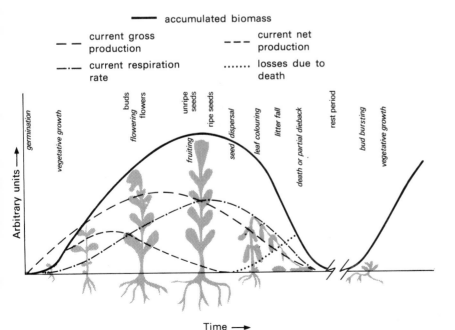

Fig. 18.10 Lifecycle and plant phenology (after Lieth, 1970).

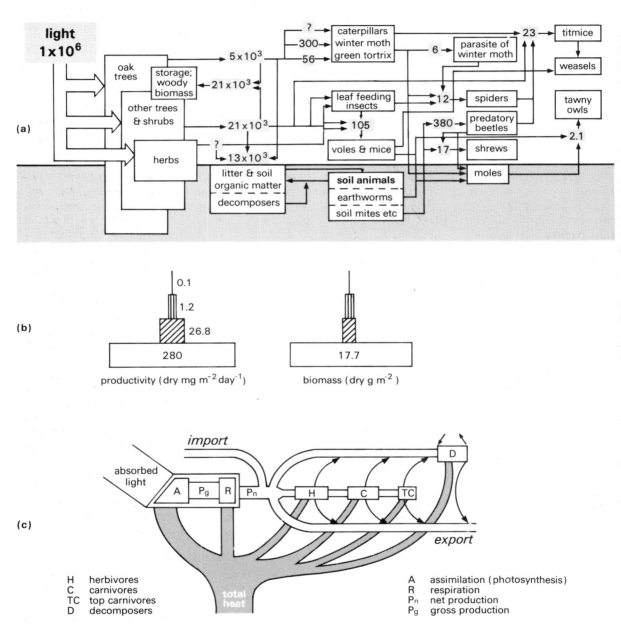

Fig. 18.11 (a) Trophic model of an oak woodland (after Varley, 1970). (b) Pyramids of productivity and biomass (after Whittaker, 1975). (c) Energy flow model for Silver Springs, Florida (after Odum, 1957).

begin to interpret and monitor the pathways of energy and matter transfer. The trophic model (Box 7.4) has been regarded as fundamental to the analysis of function and is presented as applicable in one form or another to all ecosystems. Of equal generality is the fact that the distribution of biomass between trophic levels apparently conforms to the model first introduced, in terms of numbers of animals in different size classes in consumer trophic levels, by Charles Elton in 1927, as the **pyramid of numbers**. In the **pyramid of biomass** there is a decline in standing biomass from the producer level (t_1) to the top carnivore level (t_n) as the energy available to support it diminishes (Fig. 18.11b). By combining this description of the ecosystem in terms of the distribution of biomass (pyramid of biomass) with the functional relationships of trophic organization and the pathways to energy and matter transfer, we arrive at the **trophic–dynamic models** inspired by Lindeman (1942) and developed by Howard and Eugene Odum (e.g. H.T. Odum, 1957, 1960; E.P. Odum, 1964, 1971) and others (Fig. 18.11c).

As we observed in Chapter 7, this trophic view of the ecosystem's functional organization lumps together the organisms in a trophic level, thereby reducing their significance merely to the contribution which they make to the biomass of that trophic level. Not only are the differences between organisms played down in favour of overall balance sheets of relative energy and biomass accumulation, but it is assumed that all organisms are capable of being partitioned into one trophic level or another. For these and other reasons the trophic dynamic approach to ecosystem function has limitations, particularly in relation to practical field and experimental ecology, and as we shall see in Chapter 20, is undergoing some reassessment (Cousins, 1980). Nevertheless, the pyramid of biomass and the energy flow paradigm undoubtedly have validity as general thermodynamic models of ecosystem function, and will be treated thus in Section 18.3.

The place of each organism in the food economy or trophic organization of the ecosystem is referred to as its **trophic niche**, a term also introduced by Elton (1927) who originally defined the niche as 'an animal's place in the biotic environment; its relation to food and enemies'. The concept of the **ecological niche** (attributed to Joseph Grinnell, 1917–29) as the functional role occupied by an organism has been extended and is now used in a broader sense to encompass not just an organism's trophic relationships, but also its position in the community in terms of all kinds of relationships with its habitat and with other organisms (Giller, 1984). Eugene Odum (1971) expresses this idea simply as a statement for an organism, species or population of *where it lives, what it does*, and *how it is constrained by other organisms*. To begin to understand the ecological niche of an organism is therefore to begin to able to explain its structural, physiological and behavioural adaptations. Looked at in another way, the ecological niche of an organism is a description of the extent to which it is functionally specialized. This is the **species niche**, what Paul Colinvaux (1986) has called a *Class II niche* (his *Class I niche* is the Eltonian functional or trophic niche), and it considers the niche as reflecting a specific set of capabilities relative to a specific set of needs or requirements possessed by an organism for exploiting resources, and for competing and surviving successfully. The species niche, therefore, reflects the **ecological strategies** (see Chapter 25) which organisms have evolved to survive and compete in their particular niche. In a similar way the number of distinct niches in a community can be taken as a measure of the functional complexity (and at least in some circumstances the specialization) of the community.

However, as a construct the concept of the ecological niche remains elusive (and perhaps rightly so) in terms of strict definition, and is best captured by Hutchinson's (1965) notion of **niche space**. This he describes as an abstract **multidimensional space** defined by axes that represent a large number of biological and environmental variables affecting species interactions (Fig. 18.12). This space he refers to as a **hypervolume**. In the literature Hutchinson's niche space, or hypervolume, is often called a *resource space* and the axes defining it as *resource gradients*. As long as *resource* is interpreted to include in a broad sense the organism's conditions of existence, involving both abiotic and biotic dimensions of the niche, the difficulty is merely semantic. Nevertheless, perhaps this usage should be avoided as it unnecessarily and misleadingly tends to limit the perception of the concept. According to the Hutchinsonian view, therefore, each organism can be conceived as occupying part of this total niche space to which it is adapted (this is Colinvaux's *Class III niche*). Here, however, Hutchinson makes

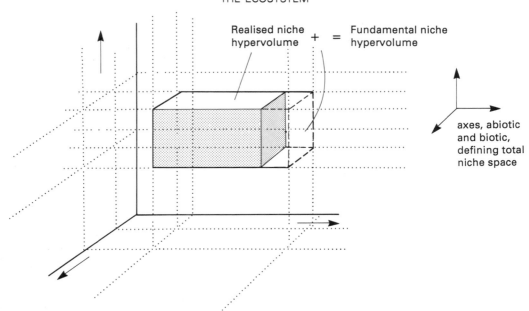

Realised niche hypervolume + = Fundamental niche hypervolume

axes, abiotic and biotic, defining total niche space

Fig. 18.12 An attempt to visualize Hutchinson's niche space as a multidimensional hypervolume, in this instance represented in three dimensions only.

a distinction between a theoretical or abstract portion of the hypervolume which would be occupied by a species when unconstrained by other organisms. This is the **fundamental niche**, and here the niche space is justifiably a resource space as it is defined by positions along resource axes corresponding to the tolerance range of the species conditioned by adaptation (see Box 19.2). In other words it is physicochemical conditions and food which define the fundamental niche. In real ecosystems organisms live as members of populations and communities, and interact and compete with each other, thereby limiting the proportion of the theoretical fundamental niche space that they can occupy under such constraints. The proportion of the niche volume which organisms can actually *realize*, Hutchinson called the **realized niche** (see Fig. 18.12). The extent to which these portions of the total niche space are mutually exclusive is a measure of the degree of **niche segregation** displayed by the community.

The concepts of ecological niche and **species diversity** are closely linked, for as the number of species in the community increases, so the niche space is divided between more and more organisms. This is because although some competition

is to be expected, large-scale overlap of niche space is not compatible with survival (see the *competitive exclusion principle*, Chapter 25). Consequently, if the ecological niche of each species is not to overlap with that of another, it must become smaller and more narrowly defined, leading to greater niche segregation. The ecological niche is, therefore, a dynamic concept and niches change as the relationships of species with their environment and their community change through the natural selection of adaptation. So niches evolve just as the species occupying them evolve, and niche segregation involves the divergent evoluion of the ecological niche paralleling the divergent evolution of the species (Fig 20.1). These are not simple relationships, however, for increases in species diversity can themselves create new and unexploited niche space. Indeed, an understanding of changes in niche space is fundamental to the appreciation of the changing functional structure of communities through time involved in models of succession. Such complications will be considered more fully in Chapter 25.

In geographically separated areas experiencing similar environmental conditions, similar ecological niches exist within structurally and functionally similar communities. Such niches, however, will be filled or occupied by organisms displaying similar adaptations but genetically unrelated. Both these organisms and the niches they occupy are said to show **ecological equivalence**, and the unrelated organisms are examples of **convergent evolution** under the stimulus of similar niche opportunities and pressures (Fig. 18.13).

18.3 Functional activity of the ecosystem: the transfer of energy and matter

The fundamental functional activities of living systems have already been identified, whether at the level of the cell, the organism or indeed the plant and animal community, as the assimilation and utilization of energy (food) and respiration (i.e. eating and breathing), and all living systems must perform these activities to live, grow and reproduce. This one-way flow of energy powers and is coupled to the closed circulation of the elements essential to life within the ecosystem (Fig. 18.14) and, because it is an open system, between it and adjacent systems. The trophic model of functional organization is, therefore, an appropriate starting place, but it needs to be extended to incorporate those parts of the inorganic environment that are also involved in the functional activities of energy transfer and nutrient circulation.

In Fig. 18.15 the model has been divided into a number of subsystems through which energy and nutrients cascade. The first of these, the **primary production system**, contains as its principal compartment the living plant, rooted in the soil and with its aerial shoots held aloft in the lower atmosphere. The plant is itself an open system, diverting part of the solar energy flux to the manufacture of organic compounds from carbon dioxide, water and mineral nutrients. In terrestrial ecosystems this is possible only because of the transport of water and essential elements from the soil via the transporting tissues of the plant to the sites of photosynthesis in the leaves. To model this subsystem, therefore, we must also incorporate the soil adjacent to the roots and the air in contact with the leaf surfaces, as well as the plant that connects them, for together they form a soil–plant–air continuum. This continuum is the key to our understanding of the way in which water and nutrient transfers take place between the plant and its environment.

The second subsystem maps the transfer of energy and materials (organic and inorganic) through the **grazing–predation pathway**. The understanding of the mechanisms involved in these transfers, and of the strategies adopted by herbivores and carnivores, is of considerable importance, partly because we ourselves fall into one of these two compartments and derive a direct harvest from both producers and consumers, but also because today we often seek to manage and conserve natural populations of animals.

The third subsystem, the **detrital system**, sees the final release of energy originally trapped by the primary producers, and provides the potential for the circulation of nutrient elements to be completed by root uptake, as organic compounds undergo the complex processes of decomposition.

Decomposition in terrestrial ecosystems takes place in the soil. The final subsystem, the **soil system**, is therefore the functional connection that completes the ecosystem model depicted in Fig. 18.11. However, it is more than a link in the circulation of nutrients, for it couples the functional activities of the ecosystem to the cascades of water and debris in the denudation system, and particularly to the processes operating in the weathering and slope systems. The soil system, therefore, must be the focus of attention in any attempt to integrate ecosystem models with those of the denudation system, and this remains as true for managed agricultural systems as it is for natural ecosystems.

18.4 The ecological organization of space: the ecology of landscape

So far we have been considering the structural and functional organization of the ecosystem itself, culminating in the model depicted in Fig. 18.15. However, there is a level of organization at a larger scale than that of the individual ecosystem. Here individual ecosystems become components in a spatial system whose dimensions are those of an entire landscape: a truly geographical system. Our perspective in this chapter remains an ecological one, even at this larger scale. The focus of our attention is still the plant and animal communities which, together with their habitats, make up the constituent ecosystems of a landscape. Nonetheless, as we have often had cause to remark, as systems are integrated into models at larger scales so the whole becomes greater than the sum of the parts, and organizational attributes

Fig. 18.13 Examples of convergent evolution in taxonomically unrelated and geographically separated animals. Similar convergence occurs in plants as for example with the Old and New World stem succulents, belonging to the Euphorbiaceae and Caetaceae respectively.

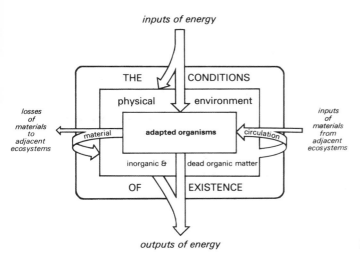

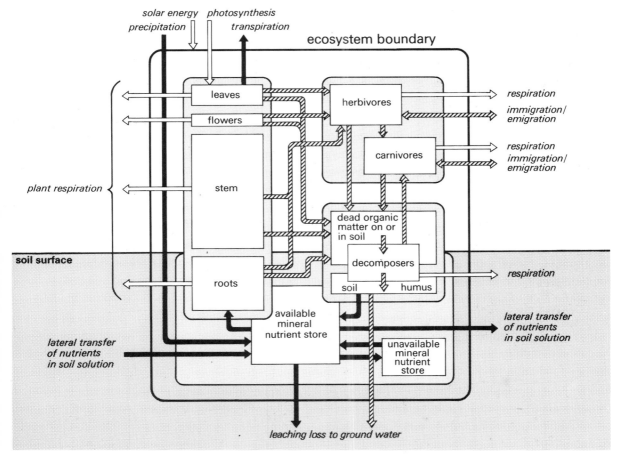

Fig. 18.14 Generalized ecosystem model.

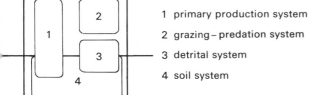

1 primary production system

2 grazing–predation system

3 detrital system

4 soil system

Fig. 18.15 Refined ecosystem model showing the primary production, grazing–predation, detrital and soil subsystems.

emerge which are unique to the model at the higher level. So here we can speak of **landscape structure** and **landscape function** in ways which, as well as subsuming the structural and functional attributes of the constituent ecosystems, also encompass those attributes of denudation, soil and atmospheric systems which are also constituents of that landscape.

In recent years, the study of ecological models of the structure, function and development of landscapes has been termed **landscape ecology** (Box 18.4) (Tjallingii and de Veer, 1982; Naveh and Lieberman, 1984; Forman and Godron, 1986). The structural components or landscape elements are **patches** of several origins, **corridors** of four types, and a **matrix**. The functional aspects of landscape involve fluxes of energy, materials and organisms (or parts thereof, e.g. seeds and spores) between landscape elements. Landscape elements, therefore, are linked by fluxes with both surrounding landscape elements and with the matrix. The model of the landscape which emerges, therefore, embodies the spatial expression of the concept of cascading environment systems as they are manifest in the real world.

18.4.1 Landscape elements: the structural components

Although landscapes vary from a few kilometres to several hundred kilometres in extent, they nevertheless represent a single level of scale in space – the landscape scale. No matter what this extent is, each landscape is an **ecological mosaic** composed of patches which vary in size, shape, origin, distinctiveness, number and configuration. Most mosaics contain narrow strips or lines which may function, in part at least, as corridors. Similarly most mosaics contain a background type in which the patches and the corridors are embedded, and which may be considered the matrix. These are the three fundamental landscape elements – the structural components of landscape: **patches**, **corridors** and **matrix**.

Patches. Four types of patch are recognized, and are widespread at the landscape scale. These are

---

Box 18.4

LANDSCAPE ECOLOGY

In recent years an apparently 'new' branch of ecology, or to some a 'new' discipline in its own right, has arisen. It has its origins in those two **holistic paradigms**, the **landscape/ landschaft** tradition in geography (Troll, 1939; 1971), and the **biocoenosis/ecosystem** tradition in biology; Tansley, 1935). To anyone steeped in the British tradition of natural history and physical geography, the declarations of continental landscape ecologists have a sense of *déja vu*. Equally, the claims of devotees from eastern Europe, who have until very recently seen it as a doctrine through which to pursue the planning of landscapes in rigidly planned societies, seem both somewhat exaggerated and dishonest. Indeed, there is still some uncertainty as to whether it is the science of the *ecology of landscape*, in which case as implied above it has a long history outside biology, or whether it is merely the discipline of *ecology at the landscape scale*. To say that it is concerned with the structure and functioning of landscapes and the patterns and processes in landscapes is to beg the question. As yet, the major impetus to development of the discipline has come from two not unrelated directions. First, nature conservation, environmental management and land-use planning have generated a need for the formulation of theory and the development of methodologies in areas espoused by landscape ecology. Secondly, the rapid development of remote sensing and image processing, and their coupling with geographic(al) information systems (spatial databases), has provided a powerful tool for analysing information at the landscape scale. The potential for landscape ecology is considerable, and its development should be. followed with interest.

listed in Table 18.4(a). The size and shape of patches also differ greatly and have major implications for the nature of the ecosystems within (Fig. 18.16). In this context patch area and the **interior/edge ratio** are two extremely important characteristics of patches. Small patches are all edge, intermediate patches may contain appreciable proportions of both edge and interior, while large patches, though having somewhat more edge, are mostly interior. Interior/edge ratio is useful in

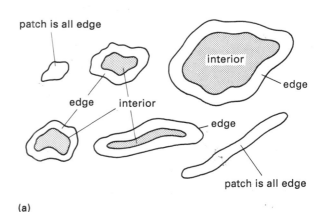

(a)

Table 18.4 Types of landscape element: (a) Patches, (b) Corridors

(a) Patches

Environmental resource patches: originate as a result of the uneven distribution of what can be regarded as relatively permanent environmental properties in space. As these properties can be considered to be resources from a biological point of view we hence have the concept of an environmental resource patch.

Disturbance patch type I. Spot disturbance patch: results from the disturbance, modification or alteration of the properties of a small area of the landscape, so that surrounding patches, corridors and matrix remain largely undisturbed.

Disturbance patch type II. Remnant patch: here the bulk of the landscape has undergone modification or disturbance and this surrounds small undisturbed (remnant) patches in the landscape.

Ephemeral patches: these are patches of very short duration associated with transient events in the landscape, and are often important mainly at fine levels of resolution.

(b) Corridors

Line corridors: such as hedgerows and roadside verges, are narrow and provide migration routes and habitats primarily for edge species.

Strip corridors: such as fire breaks, or cuts for power cables through forest, are wider and have an interior environment down the centre that is capable of providing migration routes and habitats for interior species.

Networks: are interconnected and anastomosing line or strip corridors which contain loops, and therefore provide alternative pathways for migration, predator avoidance, foraging etc.

Stream corridors: border waterways and where they are wide enough to include a significant proportion of interior environment on well drained soil may double as a strip corridor.

(b)

Fig. 18.16 (a) Hypothetical woodland remnants patches showing the relationship between the size and the shape of patches, and the patch edge to interior ratio. (b) Positive and negative influence of patch characteristics on species diversity.

understanding the species diversity as well as other ecological properties of the patch. Typically, species are predominant in, or limited to, either the edge or the interior environment of the patch. This same ratio underlies the importance of the shape of patches. Hence, narrow rectangular, strip and ring patches contain considerably less interior environment than the same area in isodiametric form.

Corridors. There are four types of corridor, partly determined by structure, but also differentiated on functional grounds (see Table 18.4(b)). Although these corridor types differ greatly in their structural characteristics as well as their functional roles, all are major integrators of landscape structure, and perform significant roles in the functioning of landscape processes. That is, patches are linked by corridors, and the matrix is permeated by and interconnected by corridors.

Matrix. The matrix is the background type of a landscape, and as such it is usually, though not always, areally the most extensive of the landscape elements. It is also usually highly interconnected, and of course it exerts a major influence on the ecology and particularly the developmental ecology of the patches and corridors embedded in it. Although it is possible to provide a typology of matrices, such attempts are less satisfactory than typologies of patches and corridors. So the general matrix concept of a relatively homogeneous background element of the landscape will suffice here.

18.4.2   Landscape ecology: the functional organization of landscape

We have already identified functional models of landscape as the spatial expression of the operation of energy and mass cascades through interconnected environmental open systems. In Chapters 10–17 this notion was pursued in the context of the subsystems of the denudation system. In much the same way we shall consider the subsystems of the ecosystem in the chapters which follow this. However, it is necessary to augment the model of the catchment basin as the fundamental functional spatial unit of landscape, in order to fully understand the ecological processes at work in that landscape.

Certainly the concept of the catchment ecosystem encountered implicitly in Chapter 10 (Fig.

10.9) allows an integration of ecosystem processes with those of the weathering, soil, slope and channel systems. This is particularly true for an understanding of the pathways by which elements are transferred in biogeochemical and debris cascades through a landscape. This approach works best when the catchment is occupied by a relatively homogeneous ecosystem, as was the case in the pioneer studies in the northeastern hardwood forests of the USA. Most catchments, however, contain a mosaic of different ecosystems: a heterogeneity which can be usefully described in terms of the ecological landscape elements outlined above. By so doing we can refine our spatial understanding of the energy and mass transfers taking place within the catchment. Here, such fluxes can be interpreted as patch-to-patch, patch-to-matrix, or within-matrix transfers, enhanced or hindered by matrix connectivity. Corridors can be seen to be operating as either conduits facilitating biomass transfer through the landscape, or as filters or barriers restricting such movement (Fig. 18.17).

However, the ecological processes operating in a landscape transcend the confines of its constituent catchment basins. The passive or active movement of organisms, involving hunting, foraging, dispersal and migration, can all extend beyond the bounding interfluves of a catchment basin. Indeed, such interfluves may themselves form important corridors facilitating movement. Here, too, the use of the concept of ecological landscape elements allows the spatial expression of such processes to be fully realized. Furthermore, it allows other important and essentially spatial ecological concepts to be integrated into an ecological understanding at the landscape scale. Concepts such as habitat, home range and territory obviously fall into such a category, but notions of niche space, competiton, species diversity, and insular biogeography (MacArthur and Wilson, 1967) all also possess an important spatial dimension (Fig. 18.16a). Even the ecology of speciation is strongly spatially dependent, at least in terms of allopatric models of speciation, requiring as they do the relative spatial (ecological/geographical) isolation between local breeding populations (**gamodemes**).

One interesting and relatively recent recruit to these spatial ecological concepts ought to have considerable significance for the development of functional units in landscape ecology. This is the

Hedge bank/permanent pasture

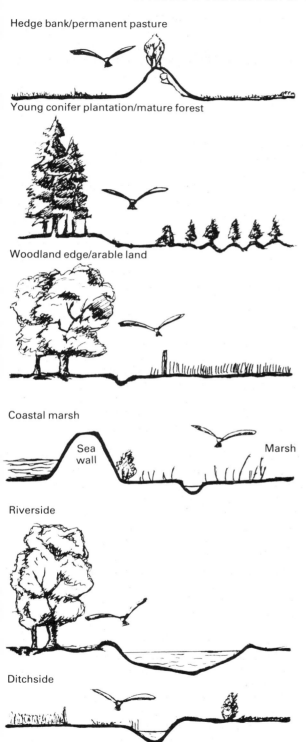

Young conifer plantation/mature forest

Woodland edge/arable land

Coastal marsh

Sea
wall

Marsh

Riverside

Ditchside

contention that the largest naturally occurring functional ecological entity is the group of organisms which constitute the food-web of the social group of the top carnivore (the largest predator) in an area. Cousins (1987) has termed this unit the **ecosystem trophic (ecotrophic) module (ETM)**. He goes on to point out that the ETM is spatially constrained to the foraging area of the social group of the largest predator by behavioural mechanisms, that it is a distinct entity in time limited by the emergence and by extinction of the social group in question, and that it is a dynamic entity in several important ways. Here it is necessary realize that over time both the composition of the social group and of the entire food-chain entity is only capable of definition in probabilistic terms, and changes according to the population dynamics of each constituent population. This means that the spatial limits of the ETM will show some plasticity, but because the lifetime of the social group is longer than that of the individuals which comprise it, the area occupied by the ETM will preserve some spatial continuity. This last observation, according to Cousins, indicates another way in which these entities are dynamic – that is, they are lineages and are capable of evolution through time.

We shall have need to return to the formalization of landscape structure and function, as embodied in modern landscape ecology, at several points as we consider in more detail the operation of the constituent subsystems of the ecosystem. Perhaps the most important application of these ideas can be appreciated, however, when we reflect that it is through the landscapes we inhabit that we have exerted our most profound influence on the operation of environmental systems. The modification, exploitation and destruction of our environment is essentially manifest as the changes we have wrought in those landscapes. It is interesting that Cousins (1987) draws attention to the fact that human intervention in the

Fig. 18.17 Linear landscape features – corridors – are important to barn owls for two reasons: (i) as hunting habitats because the damp grassland associated with these 'edges' has higher small mammal densities than either open fields or closed woodland; while the edge features of trees fences provide hunting perches, and (ii) they provide prey-rich dispersal corridors for young barn owls, allowing young birds to successfully disperse to potential nest sites (after Shawyer, 1987).

ecosystem tends to eliminate the top predators, replacing their ETMs in the same area by numerous smaller ETMs based on those predators which can survive in disturbed landscapes. Indeed, he suggests that the ratio between the size and number of predators in an area, and the size and number typical of the same area in an undisturbed condition, might act as an index of interference. Furthermore, our attempts to manage and conserve our environment must also be seen as essentially exercises in the management of entire landscapes, for only with such a holistic approach are we likely to succeed. In such a context the methodology of landscape ecology provides not just a management tool of great utility, but also a conceptual framework at an appropriate scale. However, for the moment we shall return to the model of the ecosystem represented in Fig. 18.15.

Because the flow of energy is linked with a cyclic transfer of matter between these subsystems, it does not really matter where we break into the model, for ultimately, if we follow it through, we shall have considered all four subsystems and be able to appreciate the function of the ecosystem as a whole. Nevertheless, it is logical to start with the primary production system and with the initial input of energy, which will be considered in detail in Chapter 19, while the remaining subsystems of Fig. 18.15 will be considered in turn in Chapters 20, 21 and 22.

Further reading

There is a growing number of good basic undergraduate texts available covering ecology, and all will to a greater or lesser extent serve the subject of this chapter and Chapters 19–22. Only a selection can be cited here:

Begon, M., J.L. Harper and C.R. Townsend (1990) *Ecology: Individuals, Populations and Communities*, (2nd edn) Blackwell, Oxford.
Colinvaux, P. (1986) *Ecology*. John Wiley, New York.
Krebs, C.J. (1972) *Ecology: the Experimental Analysis of Distribution and Abundance*. Harper & Row, New York.
Krebs, C.J. (1988) *The Message of Ecology*. Harper & Row, New York.
Odum, E.P. (1971) *Fundamentals of Ecology*, (3rd edn). Saunders, Philadelphia.
Pianka, E.R. (1983) *Evolutionary Ecology*, (3rd edn). Harper & Row, New York.
Ricklefs, R.E. (1990) *Ecology* (3rd edn). Freeman, New York.

The structural aspects of ecosystem description are well treated in:

Digby, P.G.N. and R.A. Kempton (1987) *Multivariate Analysis of Ecological Communities*. Chapman & Hall, London.
Goldsmith, F.B. (ed) (1991) *Monitoring for Conservation and Ecology*. Chapman & Hall, London.
Goldsmith, F.B., C.M. Harrison and A.J. Morton (1986) Description and analysis of vegetation, in *Methods in Plant Ecology*, (2nd edn), (eds P.D. Moore and S.B. Chapman), Blackwell, Oxford.
Greig-Smith, P. (1983) *Quantitative Plant Ecology*, (3rd edn) Butterworths, London.
Kershaw, K.A. and J.H. Looney (1985) *Quantitative and Dynamic Ecology*. (3rd edn) Edward Arnold, London.
Spellerberg, I.F. (1991) *Monitoring Ecological Change*. Cambridge University Press, Cambridge.

Apart from the treatment in the standard texts, the following sources shed more light on the trophic–dynamic approach to ecosystem function:

Cousins, S.M. (1985) Ecologists build pyramids again. *New Scientist*, 4th July.
Elton, C.S. (1927) *Animal Ecology*. Macmillan, New York.
Pimm, S.L. (1982) *Foodwebs*. Chapman & Hall, London.

The following text will serve as an accessible introduction to the niche concept

Giller P.S. (1984) *Community Structure and the Niche*. Chapman & Hall, London.

Landscape ecology is dealt with in the following texts, of which the first is the most accessible:

Forman, R.T.T. and M. Godron (1986) *Landscape Ecology*. John Wiley, New York.
Gorman, M. (1979) *Island Ecology*. Chapman & Hall, London.
Haines-Young, R. *et al.* (1992) *Landscape Ecology and GIS*. Taylor and Francis, London.

Moss, M. (ed) (1988) *Landscape Ecology and Management*. Polyscience, Montreal.
Naveh, Z. and A.S. Lieberman (1984) *Landscape Ecology: Theory and Application*. Springer Verlag, New York.
Peters, R.H. (1991) *A Critique for Ecology*. Cambridge University Press, Cambridge.
Pickett, S.T.A. and P.S. White (1985) *The Ecology of Natural Disturbance and Patch Dynamics*. Academic Press, London.
Tjallingii, S.P. and A.A. de Veer (eds) (1982) *Perspectives in Landscape Ecology*. Centre for Agricultural Publications, Wageningen.

Finally many of the concepts introduced in this chapter come in for critical evaluation in a stimulating and thought provoking recent critique.

The primary production system

19.1 Functional organization and activity of the green plant

Before we can begin to examine the transfer of matter and energy through the plant we need to consider more fully its functional organization. The terrestrial plant can be divided into a root system, a photosynthesizing system consisting normally (but not always) of a leaf canopy, and, linking them, a support and transporting system (Fig. 19.1).

Leaves are a system of cells organized to expose the chloroplasts to light in the most efficient manner, to ensure an adequate supply of water, nutrients and carbon dioxide, and finally to remove the surplus products of photosynthesis. A leaf resembles a sandwich whose filling consists of large (often columnar) cells occupied by streaming cytoplasm packed with chloroplasts. These are **mesophyll** cells, closely arranged at the top of the filling but with extensive air space between them lower down. Running through them are bundles of conducting tissue – the veins – so spaced that all parts of the mesophyll are near them. The top and bottom of the sandwich are layers one cell thick, transparent and waterproof, called the **epidermis**. The outer surface of the epidermis has a wax coating (**cuticle**). The continuity of the epidermis is broken in the leaves of all plants from ferns to broad-leaved flowering plants, by slit-like pores (**stomata**) between pairs of guard cells which behave as valves. Stomata are usually restricted to the lower surface of the leaf, but in some plants may be sparsely or even equally distributed on the upper surface.

The conducting tissue of the leaf veins is continuous with that of the stem. Stems, whether they are those of forest trees of small herbaceous plants, are essentially **fibrovascular** systems, with a dual function of support and transport. The cylindrical stem is bounded by an epidermis, which in trees is replaced by dead cork cells forming the bark. The vascular tissue is of two types, the **xylem** tubes and the **phloem**. The xylem occupies the inner part of the vascular bundles and consists of the lignified (cellulose and lignin in the combination we call wood) walls of dead elongated cells or vessels which connect longitudinally by large perforations, and laterally by pores and diaphragms called pits. The lignified xylem elements have a supporting function, especially in the more primitive plants such as the gymnosperms, but in more advanced xylem systems some become specialized to this function alone as wood fibres.

The phloem, in contrast, consists of living thin-walled cells which are elongated also but connected by strands of cytoplasm passing through the small perforations in the sieve plates separating them. Phloem cells or sieve tubes are unusual in that they lack a nucleus, and the cytoplasm is constantly streaming and circulating. The sieve tubes are always associated with nucleated companion cells and very long thick-walled cells, known as phloem fibres, which add support. The remainder of the stem tissue, which we well regard simply as the matrix, has two main functions: storage (**parenchyma**) and support (**collenchyma**). This matrix forms the **cortex** and the fibrovascular tissue the **stele** enclosed in an **endodermis**, though this may not be discernible.

The structure of roots, the absorption system, is essentially similar to that of the stem: an outer epidermis, a cortex and a stele containing the vascular tissue. At the root extremity, however, these distinctions are less clear. The root tip is the area of rapid cell division and growth and it is usually protected by a **root cap**. Behind the tip are regions of cell elongation and differentiation as cells develop specialized functions appropriate to their position in the root cylinder. Further behind

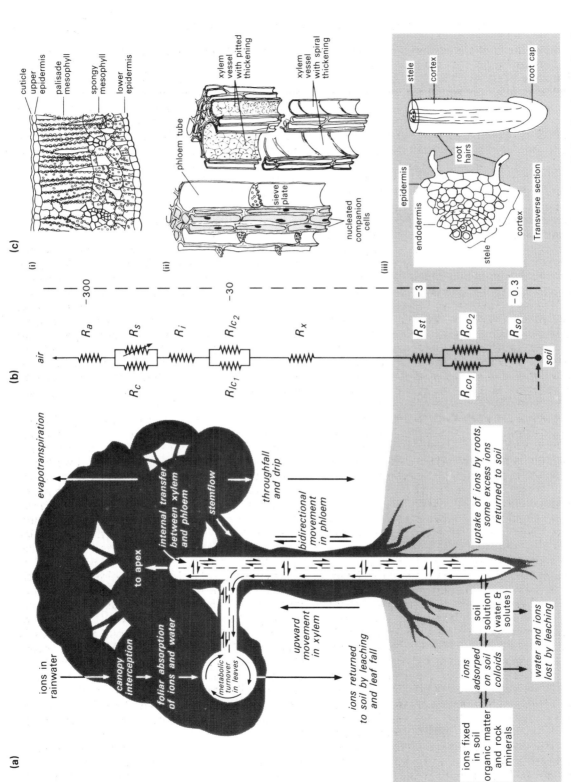

Fig. 19.1 (a) Transfer pathways in the plant. (b) An electrical analogue model of resistances to water movement in the plant, where R_{so} is soil resistance, R_{co1}, R_{co2} are root cortex free and non-free space resistances R_{st} is root stele resistance, R_x is xylem resistance, R_l is leaf cellular free and non-free space resistances, R_i is leaf intercellular space resistance, R_c is leaf cuticular resistance, R_s is stomatal resistance, and R_a is atmospheric resistance. Note that only the stomata provide a variable resistance to water movement. (c) Cell structure of the (i) photosynthetic (leaf), (ii) transport (xylem and phloem) and (iii) absorption (root) systems of the

the tip, where the root tissues have become distinguished, is an important (but not the only) absorptive zone of the root where the epidermal cells develop slender outgrowths known as **root hairs**.

There are many variations, of course, on the expression of this theme of root, stem and leaf among terrestrial plants, and there is often some implied adaptive significance to these variations thrown up by evolution in different environments (Fig. 19.2). Leaf size, shape and detailed morphology show enormous variety. The large obvate – lanceolate leaves of tropical rainforest trees have a thick waxy cuticle and elongate drip tips; the slender leaves of the grasses and sedges are sometimes fine or in-rolled with protected stomata, sometimes wide and pubescent; the needle leaves of the conifers have sunken stomata; while the leaves of heather shoots are tiny and scale-like. Some plants even lack true leaves or have leaves that have taken on an entirely different function, such as the spines of barberry and perhaps stem succulents such as the Cactaceae and Euphorbiaceae. In such cases as these the photosynthetic function may be taken over by other organs, such as the leaf petiole (stalk) or 'phyllode' in some Australian species of *Acacia*. Stem systems show variations too. At one extreme are the reduced hemicryptophyte stems of the grasses from which the leaf sheaths arise; at the other the massive trunks of forest trees reaching heights of 100 m or more. In between, every conceivable intermediate seems to exist, such as the prostrate stems of woody chamaephytes such as the dwarf willow (*Salix herbacea*) or the pliable but thickened stems of tropical lianas, and to be made more confusing by the variety of growth patterns and branching habits. Roots, however, show fewer examples of detailed refinements in their cellular structure, perhaps because conditions are more uniform below ground, although the velamen of the aerial roots of tropical epiphytic orchids forming a sponge-like tissue absorbing water from the air is one example of root specialization. Their cortex is photosynthetic, so the root appears white when the velamen is full of air but green when full of water. Nevertheless, although individual roots are less variable, the complete root systems do show variety, from dense adventitious systems to deep tap roots, not to mention peculiarities such as the stilt roots and breathing roots (**pneumatophores**) so prominent in descriptions of tropical rain forests.

In spite of this fascinating but perplexing diversity, for our present purposes we shall consider the root–stem–leaf system in terms of a generalized model of transfer pathways through the plant. These pathways can be separated into a **free-space pathway** through intercellular voids, through the lumina of xylem vessels and through the microporous structure of cell-wall material, and a **non-free space pathway** involving the crossing of biological membranes and transfer through the cytoplasm of living cells (Fig. 19.3).

19.1.1 Throughput of water and mineral nutrients

Water held in the soil by surface tension (capillary water) and by adsorption on soil colloids, will enter the root and move across the cortex to the conducting elements of the xylem as long as a water potential gradient exists across the root. Such movement occurs in response to negative hydrostatic pressure potentials developed in the xylem by the transpiration (Figs 19.1, 19.3) of water from the leaves and transmitted to the roots, and to an osmotic potential difference caused by higher solute concentration in the xylem sap than in the external soil solution.

Nutrient ions enter the root passively with the mass flow of water but, because the concentrations of free ions in the soil water are generally low, other uptake mechanisms must exist. Two hypotheses are normally invoked. Both are processes of ion exchange (see Chapter 11) and both probably occur in many soils. The first hypothesis involves close contact between the root hairs and the soil colloids – so close in fact that the oscillation volume of ions adsorbed on the root surface (particularly hydrogen ions) overlaps with that of ions adsorbed on the soil colloids. Under such conditions ion exchange occurs without the ion appearing free in the soil solution, and, not surprisingly, the process is referred to as contact exchange (Fig. 19.4a). The second hypothesis is really hydrolysis of the soil colloids, and it occurs when hydrogen ions (H^+) produced by the dissociation of carbonic acid in the soil water exchange with metallic cations adsorbed on the colloid surface. These released ions then diffuse to the root surfaces. The carbonic acid involved in this process – called the 'carbonic acid exchange hypothesis' – is derived from the dissolution of respiratory carbon dioxide (Fig. 19.4b).

The transfer of water takes place across the root by two alternative pathways. The first is a passive

Agave sp.

Zeltova sp.

Betula nana

Gynkgo biloba

Juniperus communis

Victoria amazonica

Platanus orientalis   *Pinus*

Figure 17.2   Variations in plant growth morphology, illustrated here by variations in leaf shape.

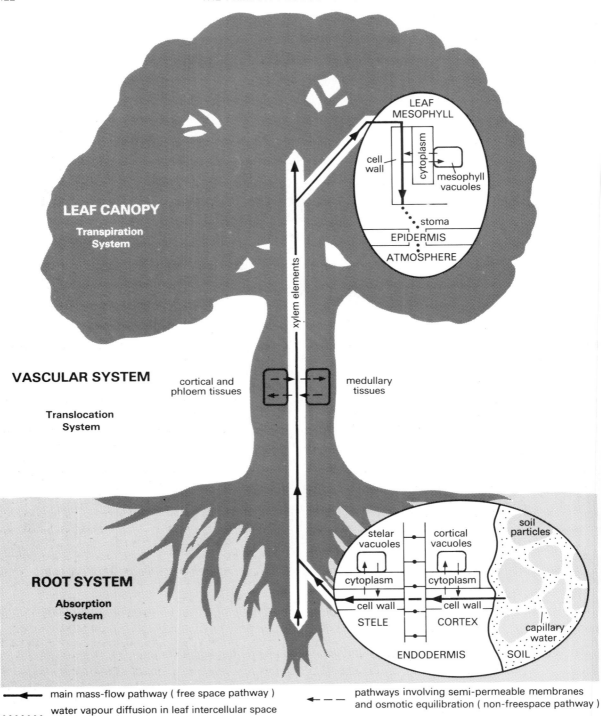

Fig. 19.3 Mass-flow and diffusion pathways of water movement in the plant (after Weatherly, 1969).

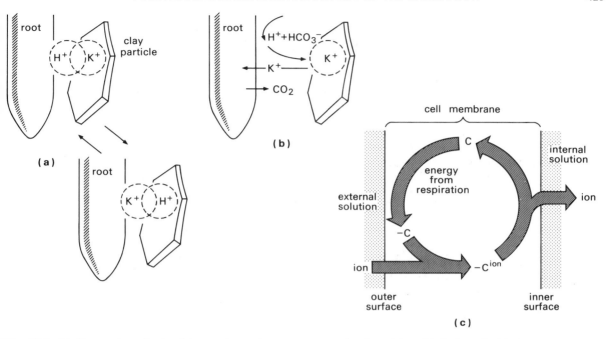

Fig. 19.4 (a) Contact hypothesis of ion exchange between root and soil particles. (b) Carbon dioxide hypothesis of ion exchange. (c) Carrier hypothesis of ion movement across cell membranes.

mass flow pathway through the freely permeable intercellular spaces and the microporous structure of the cell walls, i.e. the **root free space**, largely in response to the pressure potential gradient (Fig. 19.3). Secondly, water enters the cells and moves through the cytoplasm and vacuoles and across the cell-wall membranes (**non-free space**) by diffusion. Although the entry of some nutrient ions may be a mass flow phenomenon in the free space, most are adsorbed on the cell walls of the cortex or cross the cell membranes to join the non-free space pathway. The transfer of water and nutrient ions through the cortical cells is a passive response to diffusion gradients, particularly those across differentially permeable cell membranes. These are electrochemical gradients associated with the charges on the ions, the fixed charges on the membrane and the relative concentrations of ions, and the mechanism of transfer is **electro-osmosis**. There is a great deal of evidence to suggest that ions can also move against such diffusion gradients by means of active ion pumps. These are now known to involve enzyme carrier molecules in the cell membrane which harness metabolic ATP energy (see Chapter 2) to accomplish the work of

transporting ions against the resistance of electrochemical potentials (Fig. 19.4c).

The free-space and non-free space pathways through the cortex join in order to cross the endodermis, which represents a barrier to free-space transfer, for the partial thickening (**suberization**) of endodermal cell walls renders them impermeable. Diffusion, however, can still take place from cortical cells to the stele by the non-free space pathway, from which both water and nutrient ions are released into the xylem elements.

As water evaporates from the leaf mesophyll cells during transpiration, their solute concentration increases and lowers their osmotic potential (ψ_π) causing the movement of water from adjacent cells. In this way an osmotic potential gradient is established across the leaf and draws water from the leaf xylem elements. The resulting pull on the water in the xylem, placing it under tension or negative pressure, draws water up the stem in response to the pressure potential difference (Fig. 19.5). This is believed to be possible because the cohesive and adhesive forces, due to intermolecular attraction between water molecules and between them and the sides of the xylem capillaries,

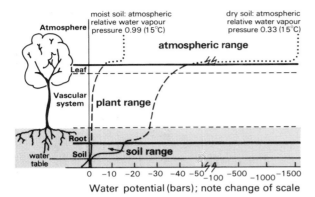

Fig. 19.5 Water potential gradient in the soil–plant–air continuum (after Etherington, 1975).

allow the maintenance of unbroken water columns from root to leaf. Most nutrient ions move passively with this transpiration stream in the same form as they were absorbed from the soil. Others, notably nitrogen, phosphorus and sulphur, are translocated as organic derivatives which must have formed in the root, while such elements as iron, which tend to precipitate in high-valency forms (see Chapter 11), may be aided in their mobility by forming complexes with organic chelating agents. Some ions do not make the complete trip to the leaves, for during upward translocation they may pass selectively across cell membranes to enter other cells, by either passive diffusion or active transfer.

19.1.2 Photosynthesis and plant primary production

The concentration of chloroplasts in the mesophyll of leaves is the site of photosynthesis. These cells are supplied with water and mineral nutrients via the transpiration stream in the xylem of stems and leaves, and by alternative pathways across the mesophyll itself, similar to those described for the root cortex. In the case of the leaf, movement of water in the free space around the leaves seems to be the principal pathway. Only a very small percentage of the water absorbed by roots is used for the vital functions of the plant in which water is involved, and of that a fraction of 1% is used directly in photosynthesis. The remainder evaporated from the micropores of the mesophyll and cuticular cell walls, and escapes to the atmosphere as water vapour by diffusion across the

intercellular air space to the stomata, and thence by either molecular diffusion in calm conditions, or by turbulent diffusion over the leaf surface in moving air.

Carbon dioxide, the remaining raw material for photosynthesis, is apparently freely available in the atmosphere surrounding the leaf, but in practice CO_2 concentrations can become rate-limiting for photosynthesis. This is because it is the concentrations at the chloroplasts that are important, and to reach them CO_2 has to diffuse through the stomata and the mesophyll air spaces, across the cell walls and through the cytoplasm. So CO_2 supply to the chloroplasts is determined by the rate of these diffusion processes, and this depends on the maintenance of CO_2 concentration gradients between the air outside and the chloroplasts (Box 19.1).

Finally, the photosynthetic process requires the energy input of solar radiation. Fig. 19.6 illustrates the complex energy balance of a single leaf, at night and during the day (see Chapter 3). Of the incident visible light, normally less than 20% is reflected (higher in the presence of waxy or light-coloured layers), but reflectivity rises sharply for longwave radiation with 40–60% of infrared radiation being reflected. Little transmission of energy takes place through the leaf, although again it is higher for infrared (*ca.* 30%) than visible light (<10%), and of course it varies with leaf thickness. The remaining radiation input to the leaf is absorbed (50% of total radiation), but differentially, with *ca.* 80% of visible wavelengths absorbed as opposed to 10% of infrared. High absorption by such thin structures as leaves is due to multiple reflection by cell/water interfaces in the mesophyll. Only *ca.* 2% of absorbed energy (in the photosynthetically active wavelengths) is used in photosynthesis, the remaining energy, which may be as high as *ca.* 80% of total incoming energy (*ca.* 26×10^6 J m^{-2} day^{-1} in midsummer), is available to raise the temperature of the leaf.

To avoid cell death by high temperatures, much of this heat load must be dissipated by the leaf. Cooling of the leaf takes place by convective cooling, particularly forced convection, but also free convection, when there is little or no air movement. Evaporative cooling also occurs as long as the stomata are open and transpiration is not limited by soil or plant moisture deficits. These cooling mechanisms have already been discussed in some detail in Chapter 4. Finally, the leaves

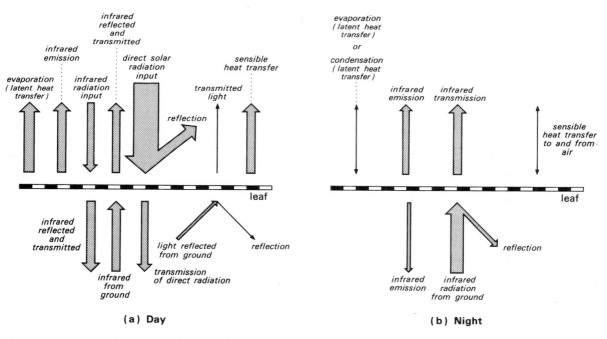

Fig. 19.6 The energy budget of a leaf, by (a) day and (b) night (after Meidner and Sheriff, 1976).

will emit longwave radiation, particularly at night (see Chapter 3).

We have seen how the reactants and the energy necessary for photosynthesis to proceed are brought together in the leaf canopy. However, a further consideration remains, for high rates of photosynthesis can occur only if the products (organic assimilates) of photosynthesis are removed from their source in the leaves. These are the fuel molecules on which the metabolism of the rest of the plant and its growth depend and, of course, they are translocated through the plant by the bidirectional diffusion in the phloem sieve tubes (see Fig. 19.1). Nevertheless, they must not be allowed to accumulate at their 'source' and, when produced in greater quantities, the plant's immediate metabolic needs would require that they be removed to a 'sink' in some storage tissue.

The process of photosynthesis itself has been treated at the biochemical level, when we considered the autotrophic cell model in Chapter 7. Because it is essentially a process of carbon fixation or assimilation, the rate of photosynthesis is usually expressed as the rate of CO_2 uptake by the photosynthesizing system (g CO_2 m^{-2} s^{-1}) irrespective of whether we are considering a single leaf, a plant or an entire canopy. The net photosynthetic rate is the difference between gross photosynthetic rate and respiration. The net photosynthesis (i.e. the net photosynthetic rate integrated over time) which accumulates in the plant (or stand of vegetation) with the addition of nutrient elements in protein synthesis represents the **net production** of the plant or stand. Similarly, the accumulation of gross photosynthesis is **gross production** (see Chapter 7).

19.2 Ecosystem primary production

The net primary production over a period of time of an ecosystem such as the woodland seen in Fig. 18.1 is, therefore, partly represented by the change in the biomass during that period $(t_1 - t_2)$ (Fig. 19.7)

$$P_n = \Delta B$$

Such a view is, of course, an oversimplification. If we refer to the woodland ecosystem, we can represent the biomass of all the primary producers at the beginning of the time period (i.e. at t_1) as B_1. At the end of the period t_2, some of this initial biomass will have died and been shed as litter (L_o); some will have been consumed by the herbivore population (whether by defoliating insects or the browsing of deer is unimportant) and can be represented as a grazing loss (G_o). Finally, a proportion of this initial biomass will have been leached from the canopy (S_o). During the same period of time $(t_1 - t_2)$ an increment of new biomass will have been added. This of course is the net production (P_n), but it too will have suffered grazing (G_N) and some new tissue will have died, perhaps as a result of late frosts or wind damage to young shoots, and been shed as litter (L_N). In addition, some of the constituents of the new biomass will have been leached (S_N). So at the end of the time $t_1 - t_2$, not only will the initial biomass have been reduced but the new biomass remaining will be less than that created by net production. The change in biomass will fall short of net production by the same amount as the sum of the old and new biomass lost to litter, grazing and leaching. A more accurate view of net primary production, in terms of changes in accumulated biomass is therefore:

$$P_n = \Delta B + (L_o + L_N) + (G_o + G_N) + (S_o + S_n)$$
$$= \Delta B + L_{total} + G_{total} + S_{total}.$$

(see Fig. 19.7)

Bearing this relationship in mind, we might make some observations concerning our woodland ecosystem. The thick carpet of leaf litter, together with the inconspicuous nature of the herbivore element and the existence of a lush green canopy with no obvious evidence of heavy grazing pressure, suggests that $L + S$ in the above equation is probably more important than G. Is this the case in all ecosystems? Furthermore we might observe, particularly if we were regular visitors to the woodland, that the biomass of the woodland, especially of the trees, represents a massive accumulation of organic matter, but apart from the predictable seasonal changes in appearance it would seem not to change greatly through time. Why should this be so? This might further prompt the observation that nevertheless, at some time there has been an investment of net production in accumulating the biomass that we see now. When was this and why is biomass apparently no longer accumulating?

These are profound observations and the answers to the questions they pose provide some of

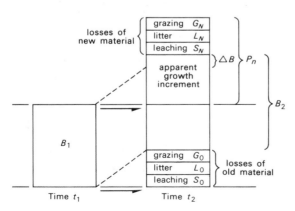

Fig. 19.7 The components of biomass change with time, and their relationship to production (after Chapman, 1976).

the most important theoretical statements concerning production. Fig. 19.8 shows the energy flow through the primary producer trophic level of several ecosystems. Here we can see the answer to the first question. There appears to be a distinction between those ecosystems with a high proportion of plant-derived dead organic matter and a low herbivore element, and those with little plant derived dead organic matter and a high herbivore element. Under natural conditions most terrestrial ecosystems, including our woodland, fall into the former category and most of the animal biomass is to be found in the detrital system. Because they are not being destroyed by grazing, the plants (particularly the perennial ones) accumulate a proportion of their net production over time in support tissue and structures such as the wood of the forest trees, a necessary feature of adaptation to terrestrial environments. This accumulation is capital and it must be paid for (see Chapter 18). Much of this tissue in trees is dead wood, but it also contains the living cells of the transport system and **cambial** cells that permit further growth. These non-photosynthesizing cells are a respiratory drain on the total assimilated energy or gross production. The greater the biomass the greater the maintenance cost the plant sustains in terms of respiratory heat loss.

In our woodland the law of diminishing returns is operating because, for every increment of photosynthesizing surface, and even greater maintenance respiration is required to support it. Initially, the increase in **leaf-area index** (ratio of leaf surface to ground area) will more than compensate, but, as total biomass increases the respiratory load increases faster than the gross production, and net production falls off (Fig. 19.9). Here, then, is part of the answer to the remaining questions, namely that maintenance costs increase in a mature ecosystem, leaving a small proportion of assimilated energy as net production. In addition, there is evidence that the photosynthetic efficiency) (CO_2 fixed per unit of energy input, g CO_2 J^{-1}) of old plants declines so that both net and gross production are reduced. Finally, we can say that if the woodland is to maintain a steady state (i.e. $\Delta B/\Delta t = 0$), then the net production must be just sufficient to offset total losses by grazing, leaching and litterfall, i.e.

$$P_n = G + L + \bar{S},$$
$$\text{so } \Delta B = P_n - (G + L + S) = 0.$$

In terrestrial ecosystems, then, a large proportion of the energy flow through the primary producer level is trapped or stored in non-productive

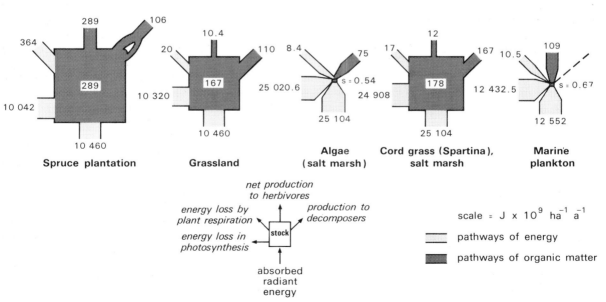

Fig. 19.8 Primary production in different ecosystems (note that production to herbivores in the case of the spruce plantation is misleadingly high, for it includes a timber crop extracted by man) (after Macfadyen, 1964).

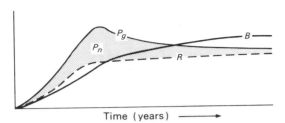

Fig. 19.9 Change in ecosystem gross primary production (P_g), plant respiration (R), net primary production (P_n) and plant biomass (B) with time (from Odum, 1971, after Kira and Shidei, 1967).

tissues, destined to be released after a considerable time lag by decomposition. Up to a point, the same is true of the nutrient elements immobilized in the organic compounds of the woodland biomass, although nutrient availability is ensured by other strategies, as we shall see. In the aquatic ecosystems represented in Fig. 19.8 by the marine phytoplankton, all the living tissue is productive and hence there is no respiratory sink. They are small organisms with high surface-area/volume ratios, a high metabolic rate and high productivity. They may be grazed many times over or, to put it another way, several generations may be grazed in a year, so that annual production figures are cumulative. Both the energy input and energy output to herbivores are high and because lifecycles are short there is little or no accumulation of biomass, even if the algal cells escape grazing, as in a sewage pond, death and decomposition are also rapid. In contrast, in the woodland most of the plant material remains free of herbivore attack over the same period. The grassland represents an intermediate situation, though closer to the woodland, for the ratio of net production to biomass (P_n:B) is low (0.06), compared with the high figure for the phytoplankton (162.7). This P_n:B ratio is a useful index of the amount of non-productive tissue in the plants.

19.3 Regulation and limits to photosynthesis and primary production

Primary production is, of course, directly linked to photosynthesis, so the mechanisms regulating and the factors limiting the photosynthetic rate will ultimately affect the primary productivity of the ecosystem. These can be divided into endo-

genous – **genotypic** and **phenotypic** – variables which affect the photosynthetic rate per unit of photosynthetic tissue, and exogenous – **environmental** – variables which regulate the size of the photosynthesizing system by their effect on the plant's metabolism and growth. The first category can be divided further into morphological, physiological and biochemical factors.

17.3.1 Genotypic and phenotypic variables

Morphological variables. Morphological variables are manifested at two scales. At the scale of the individual leaf certain variables become important, such as leaf thickness and volume, number and distribution of chloroplasts, morphology of the mesophyll, and number and position of the stomata. The variation in leaf size, form and structure mentioned earlier in this chapter will influence the photosynthetic rate and may represent the evolutionary selection of leaf characteristics of adaptive significance. In many species distinct **sun** and **shade** leaves can be distinguished. The former are exposed to full sunlight, often being smaller but thicker than the latter, which show reciprocal adaptations to low light intensities. At the scale of the whole plant the number, arrangement and length of life of the leaf canopy are important variables. **Leaf-area index** (LAI) (the ratio of leaf area to ground area) interacts with leaf position and angle of leaf insertion (Fig. 19.10) to govern light penetration into the canopy. High values of leaf-area index inevitably mean the existence of shelf shading. Diffuse radiation is reduced exponentially from the surface of the canopy towards the ground, but the effect on photosynthesis is complicated by changes in the spectrum of radiation inside the canopy as the photosynthetically active wavelengths (red and blue) are selectively absorbed, and by the fact that the shade leaves may show physiological and biochemical adaptations to low light intensities. Steep angles of insertion of leaves, as in grasses, will favour more effective penetration of light than will leaves held horizontally, but it must be remembered that the leaves of many species can be orientated in relation to light (**phyllotaxy**). The duration of maximum leaf-area index is also an important factor, which may compensate for other variables.

Physiological variables. Physiological controls are many and complex, but among the most import-

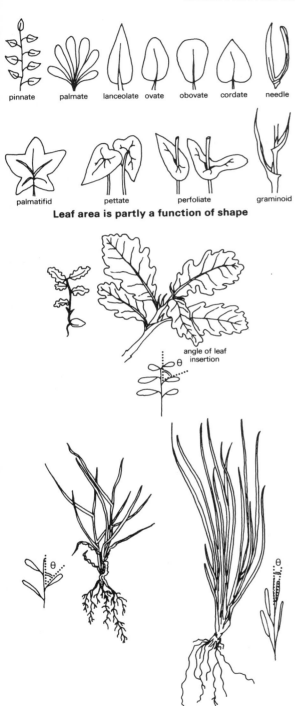

pinnate   palmate   lanceolate   ovate   obovate   cordate   needle

palmatifid   pettate   perfoliate   graminoid

Leaf area is partly a function of shape

angle of leaf
insertion
θ

θ

θ

Fig. 19.10 Leaf area and angle of leaf insertion, which contribute to the leaf-area index.

ant are the plant's resistances to diffusion and flow of water and to the diffusion of CO_2 in the leaf (see Figs 19.1, and 19.4). One of the most important resistances is that of the stomata, and here physiological factors become significant, such as the existence of endogenous rhythms of stomatal opening and closing. Although the concentration of CO_2 and light and temperature are environmental variables, the fact that they have a physiological expression justifies considering them under this heading. Photosynthesis ceases under conditions of too much or too little heat, and it is optimal under a particular temperature range for each species. The high and low temperature extremes are known as **temperature compensation points**. Similarly, there is a low light intensity, the **light compensation point**, below which there is a net loss of CO_2 by respiration which is not compensated for by carbon fixation in photosynthesis. Indeed, leaves low in vegetation canopies may be close to or below their compensation points and represent a respiratory energy sink. One way in which plants overcome this problem is to shed leaves and branches from the lower part of the canopy as they grow. The upper extreme of light intensity above which photosynthesis ceases is the **light saturation point**. Light and temperature are not independent, however, for at high light intensities the optimum temperature for photosynthesis seems to be higher and the temperature range is increased (Pisek *et al.*, 1969). Atmospheric carbon dioxide concentrations are not normally limiting, except perhaps in full sunlight at the top of leaf canopies, but often the diffusion gradient from the air to the chloroplasts is limiting. In this context the **carbon dioxide compensation point** (see Box 19.1), which is a measure of the plant's ability to reuse CO_2 released during respiration and to take up CO_2 from the air, is an important physiological property.

Biochemical variables. The principal biochemical controls on photosynthesis, and hence on productivity, concern the carbon fixation pathways present in the plants of the ecosystem. There are three such possible biochemical pathways involved in photosynthesis, the **C-3**, (**Benson-Calvin**, or normal) pathway, the **C-4** (after **Kortshchak, Hatch** and **Slack**), and the **CAM** (**Crassulacean acid metabolism**) pathway. The utilization of the light energy made available as ATP (see Box 7.2) during photosynthesis for the assimilation of CO_2

and the regeneration of ATP, takes place through a 13-reaction cycle called the Benson–Calvin cycle. The immediate end product of this cycle is a three carbon compound (C_3) and all green plants use this cycle in carbon assimilation. Those that depend on it alone are therefore called C-3 plants. However, some plants make use of an additional series of reactions which yield a four-carbon compound as the first product of CO_2 fixation, which then fuels the Calvin cycle with the release of CO_2. These fall into two groups depending on the details of the pathways involved, that is the C-4, and the CAM plants. In the C-4 plants the two sequences of reactions which typify the C-4 pathway are spatially separated. That is, the success of the process depends in these plants partly on the modified biochemistry and partly on a modified leaf anatomy. Unlike C-3 plants, where the mesophyll containing the chloroplasts is distributed more or less uniformly through the leaf, in typical C-4 plants the chloroplasts are restricted to two successive layers of photosynthetic tissue surrounding the leaf veins. These layers are an outer mesophyll and an inner (vascular) bundle sheath, and the two parts of the C-4 reaction pathway are segregated, with the C-4 compound generation in the mesophyll and the Calvin C-3 cycle in the **bundle sheath**. In plants with the CAM pathway the reactions are essentially similar to C-4 pathway plants, but the separation of the two parts of the reaction sequence is temporal, not spatial. This is because the CAM pathway is utilized mainly by *succulent* or *semi-succulent* plants (though not all members of the Crassulacease) growing in arid areas, which normally close their stomatal apertures during the day to restrict moisture loss. So, the stomata open at night, CO_2 enters and is fixed to a four-carbon compound, and the next day, behind closed stomata, this is broken down to a three-carbon compound and CO_2, which feeds the Calvin cycle.

C-4 plants are characterized by low CO_2 compensation points, a lack of photorespiration and – usually – high light saturation points, all of which are responsible for a photosynthetic rate two or three times greater than in C-3 plants. These high-capacity producers are all plants of the tropics and subtropics and they include several important crop species such as maize (*Zea mais*) and sugar cane (*Saccharum officinarum*). In contrast, the presence of photorespiration causes high CO_2 compensation points (see Box 19.1) and reduces the efficiency of carbon fixation in Calvin plants by liberating CO_2 into the leaf intercellular space at the same time as photosynthesis is removing it. The combination of morphological, physiological and biochemical factors which characterizes high- and low-capacity producers respectively is shown in Table 19.1.

17.3.2 Environmental variables

All of the endogenous variables that influence the plant's performance as a primary producer are characters that have appeared during the course of evolution, and have been selected, as far as we can judge, because they have adaptive significance (Box 19.2). The existence of these characters not only sets limits to the plant's photosynthesis and production, but, by their relation to and interaction with exogenous or environment variables, they control and regulate the plant's response to its environment, as expressed in its growth and productivity.

Light. The prime environmental variable is the radiation flux density or net radiation receipt per unit area of surface, and this is under the control of climate. The input in the photosynthetically active wavelengths ($0.4–0.7\ \mu m$) at any point on the Earth's surface is the maximum energy potentially available for photosynthesis and production. The variation in energy receipt over the Earth's surface has already been considered at length in Chapter 3, and although there is a correlation between net radiation and productivity, it is complicated by other relationships, particularly with temperature and water availability (see Fig. 7.9). However, light has effects other than direct energy input, for adapted plants are able to respond to variations in light intensity both spatially and temporally. For example, plants adapted to shade have light saturation points and intensities only a small fraction of full sunlight, while *sun plants* have higher light saturation points. In the case of some tropical plants with the C-4 carbon fixation pathway, no light saturation appears to exist and they can photosynthesize at very high light intensities. The horizontal and vertical disposition of plants in an ecosystem can be viewed, at least in part, as a response of species adapted in these ways to the three-dimensional distribution of light intensities within the canopy microclimate.

Light availability and intensity vary with time as well as with space. There are, of course, diurnal

Table 19.1 The characteristics of plants with high and low production capacities (modified from Black, 1971)

	High-capacity producers (C-4 plants)	Low-capacity producers (C-3 plants)
General type of plant	herbaceous and mostly grasses or sedges	herbs, shrubs or trees from all plant families
Morphology		
(1) leaf characters	bundle sheath cells around vascular bundles packed with chloroplasts	no chloroplasts in the bundle sheath cells
Physiology		
(2) rate of photosynthesis	40–80 mg CO_2 dm^{-1} h^{-1} in full sunlight; no light saturation	10–35 mg CO_2 dm^{-1} h^{-1} in full sunlight; light saturation at 10–25% full sunlight
(3) response to temperature	growth and photosynthesis optimal at 30–45°C	growth and photosynthesis optimal at 10–25°C
(4) CO_2 compensation point	0–10 ppm CO_2	30–70 ppm CO_2
(5) sugar transport out of leaves	rapid and efficient: 60–80% in 2–4 h (at high temperatures)	slower and less efficient: 20–60% in 2–4 h
(6) water requirements (g water needed to produce 1 g dry matter)	260–350	400–900
Biochemistry		
(7) carbon fixation	Hatch–Slack (or C-4) and Calvin cycle (or C-3) pathways	Calvin cycle (C-3) pathway only
(8) photorespiration	not detected	present

variations in the radiation received (see Fig. 19.6) and these are mirrored in diurnal patterns of photosynthesis, respiration and production. Such diurnal rhythms may be very important and predominant in tropical ecosystems which lack any marked seasonality. In seasonal environments the growth response of many plants, such as the ground flora herbs in our woodland ecosystem, is regulated by the seasonal periodicity of radiation input, which defines the ecosystem light climate (Fig. 19.11). The lifecycles of different species (phenology) are adapted to different phases of the light climate, and they attain optimal leaf-area indices and high daily rates of production at different times. Most natural ecosystems in seasonal climates will, therefore, show a succession of small peaks of production attributable to different species. This is in strong contrast to agricultural crops, which rise from zero leaf-area index to a high value and a high single peak of daily production – at which point they may be harvested. In the natural system, however, the production is not all present at one time (which means it is difficult to

harvest) but there is a more consistent, prolonged but lower daily rate of production.

The regulatory role of light is to elicit photoperiodic responses from adapted species which result from **circadian rhythms** of light and dark, or variations in day length (see Box 25.1). Their significance lies not only in their control of ecosystem productivity over the annual cycle, but also in the regulation they exert on reproductive processes and hence on the dynamics of species populations. In plants this is usually expressed in the time of flowering, and the existence of **short-day** and **long-day** plants is well documented and of economic significance in plant breeding, horticulture and agriculture. There are also similar responses in animal populations, but these will receive some attention in the next section.

Temperature. So far we have considered radiation input in terms of the visible waveband, but the limiting and regulating effects extend to the exchange of longer wavelengths and involve the plant's heat balance (Fig. 19.12). Air, leaf and soil

Box 19.2
ADAPTATION AND TOLERANCE

The relationships of organisms with their abiotic environment have already been explored in Chapter 18 under the guise of habitat and niche. In this chapter these relationships are being explored as limiting, controlling or regulating factors in the growth and productivity of green plants. However, whether plant or animal, at the level of the individual organism the principal relationships with the environment are primarily physiological. Physiological processes proceed at different rates under different conditions, and for any organism there is a limited range of conditions under which it can survive. Furthermore, there is an even more restricted range of conditions in which it can perform at its optimum. So it is possible to draw, as in (a), a curve of the performance of a physiological process against a gradient of physicochemical conditions. Such curves are tolerance curves. From such curves for individual physiological processes,

(a) The performance of a physiological process (e.g. photosynthesis) against a gradient of an environmrental factor (e.g. temperature) — i.e. **a tolerance curve**

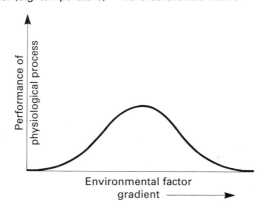

(b) Overall tolerance curve for organism

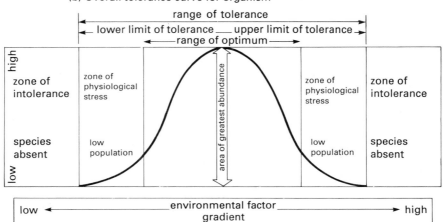

overall tolerance curves for the *whole* organism can be derived, and a series of nested limits of different severity can be identified, (b). Although the effects of environmental variables can be treated in isolation as limiting factors, there is considerable interaction between them. Hence, deficiencies or excesses of one often affects tolerance to all, e.g. the interrelationships of temperature and water availability, of light and temperature, and of nitrogen and drought (see text). The considerations about the effects of environmental factors are captured by two well-worn laws of ecology: **Liebig's Law of the Minimum**, and **Shelford's Law of Tolerance** (Box 19.4)

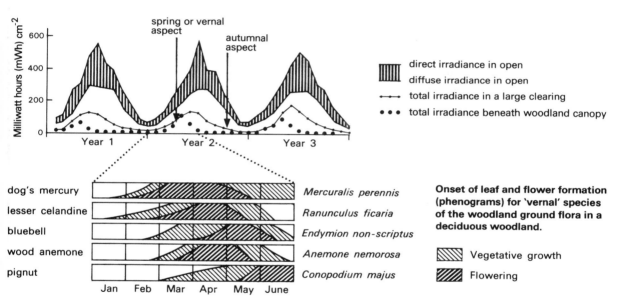

Fig. 19.11 Plant phenology and woodland light climate (Anderson, 1964).

temperatures are, therefore, important environmental variables. High and low temperature extremes are lethal and result in cell and plant death. At the physiological level the mechanism in both cases involves damage (**denaturing**) to enzyme and structural proteins, particularly those associated with membranes, as a result of mechanical stress. At high temperatures this is a consequence of increased kinetic energy at the molecular level, but at low temperatures it is due to dehydration as water is withdrawn from cells by extracellular freezing. The effect seems to be the loss of cellu-

lar organization as biological membranes break down and become permeable. Between these extremes, temperature affects all metabolic processes and hence growth and production, because for every 10°C rise in temperature the rate of chemical reaction roughly doubles (the **Arrhenius relationship**). Biochemical reactions conform to this relationship but, because they are catalysed by enzymes, biochemical reaction rates progressively decline above 40°C, simply because many of the enzymes are damaged and become inactive above this temperature.

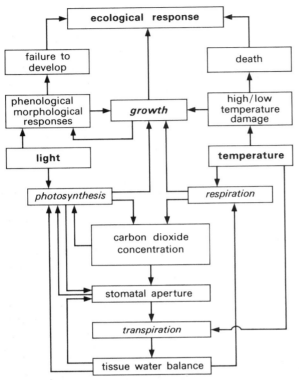

Fig. 19.12 Plant response to light and temperature (after Bannister, 1976).

At the macroscale the photosynthetic rate will reflect this interaction of light and temperature as expressed in the macroclimate, but perhaps more important, within a vegetation canopy sun and shade leaves will experience different microclimates, and hence combinations of light and temperature, and display different patterns of photosynthesis. Secondly, under conditions of high temperature but low light intensities, the respiration rate can exceed photosynthesis and the leaves will be below their compensation points. These interactions of light and temperature within vegetation canopies are not constant of course; they will change throughout the day and from season to season, and affect the overall net production of the plant community (Fig. 19.13). As with light, temperature is not just a limiting variable, but acts to regulate metabolic activity, growth and production by the response of plants to temperature fluctuations through inherited adaptive characters. However, perhaps one of the most

The net production of an ecosystem is, however, a function of both photosynthesis and respiration, and because these processes respond in slightly different ways to temperature its differential effect can greatly influence net production. Respiration rate is affected by temperature in the way outlined above, rising up to 35–40°C, increasing initially at higher temperatures, but then falling rapidly. As was pointed out when discussing temperature compensation points, photosynthesis responds rather differently. This is because in reality it is two reactions: light and dark (see Chapter 7). The light reaction rate is little influenced by temperature (i.e. it has a low temperature coefficient), whereas the dark phase responds to temperature in a way analogous to respiration. Therefore, under low light intensities when light is limiting, the dark reaction rate will not respond appreciably to temperature, as it is rate-limited by the supply of reactants from the light reaction. Photosynthesis will increase only with rise in temperature when light is not limiting. The implications are twofold.

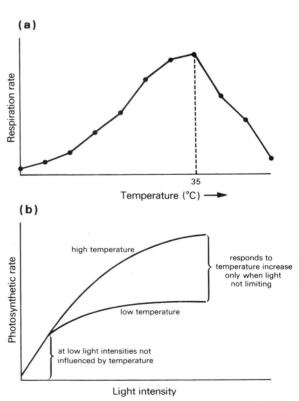

Fig. 19.13 The effects of temperature on (a) respiration and (b) photosynthesis.

important aspects of radiation (long and short wavelength) is its indirect effect on the plant and on production through its regulation of the surface water balance (see Chapter 6).

Water. The development of a plant water deficit has far-reaching physiological implications affecting many of the plant's metabolic activities and retarding growth and production (Box 19.3). Quite small water deficits are now known to reduce photosynthesis, even though a very small proportion of the absorbed water is used directly in photosynthesis. The reasons are complex, but they include the effect on CO_2 diffusion of stomatal closure under water stress, the reduced permeability of membranes and hence a slower rate of removal of photosynthetic products, some adverse effects on the photosynthesis process *per se*, and the reduction of leaf-area index because of loss of cell **turgor** and reduced tissue expansion. Respiration is also reduced by water deficit, but not at the same rate as the reduction of photosynthesis. In some circumstances respiration rate may even increase. In either case, however, a situation can arise where weight loss occurs as respiration rate exceeds that of photosynthesis.

The plant water balance is maintained by the transpiration stream in response to the water potential gradient from soil to free atmosphere (see Fig. 19.4). The rate of uptake from the soil will therefore depend on the rate of water loss by transpiration and the availability of soil water. As we saw in Chapter 6, the transpiration rate will depend on the same environmental factors that determine evaporation from any surface, namely the heat energy supply, the vapour pressure (or pressure potential) difference between the surface and the air, and air movement across that surface, in this case the leaf. Transpiration therefore increases with increases in vapour pressure difference, which itself increases with temperature. Again in Chapter 6 it was pointed out that evaporation rate would also depend on the properties of the evaporating surface. It is in this connection that the characteristics of plant structure affect transpiration rates, and some of the most important of these characters are listed in Table 19.2. The regulation of transpiration is governed mainly by the stomatal resistance (see Fig. 19.1) and its relationship to the resistance of the air above the leaf. In fact, in some circumstances it is the air resistance (R_a) which is predominant (Fig. 19.14). Xerophytes lose as much water in still air as mesophytes, for under these conditions transpiration is controlled by R_a. In a wind, however, the anatomical features of xerophyte leaves (such as hairiness, thick cuticle and sunken stomata) retain a layer of moist air in proximity to the stomata and R_s becomes the regulating resistance (Fig. 19.1).

Box 19.3 (see also Boxes 12.1 and 12.2)
PLANT WATER POTENTIAL

Water in the plant is held by forces of retention, and the potential free energy of that water can be expressed in terms of water potential. Water potential in living cells results from a balance between two forces, one due to the presence of solutes in the cell sap and one due to the multidirectional hydrostatic pressure within the cell, the turgor pressure.

$$\psi_{cell} = \psi_{osmotic} + \psi_{pressure}$$
$$\quad c \qquad\quad \pi \qquad\qquad (turgor)$$
$$\qquad\qquad\qquad\qquad\qquad\quad p$$

In living cells turgor pressure is positive, i.e. it enhances cell water potential operating in the direction opposite to osmotic pressure. Plant water potential (ψ_{plant}) is therefore a mean value of the cell water potential in all tissues of the plant, plus the matric potential of the plant matrix, i.e. the intercellular spaces and microporous structure of the cell walls as well as perhaps some xylem vessels (Meidner and Sheriff, 1976). ψ_{plant} also includes the contribution of negative hydrostatic pressure in the non-living matrix (the free space), especially in the xylem.

Table 19.2 The relationship between certain structural characters of plants and water loss by transpiration

leaf area	Plants with large foliage area transpire more rapidly; plants with small foliage area transpire less rapidly. Shedding of leaf area and reduction of leaf area/plant are characteristics of xerophytes.
stomata	The number, distribution and size of stomata affect transpiration rate. Stomatal transpiration is proportional to number and size of stomata and to the linear perimeter of the pores, not to surface area.
leaf intercellular space	Transpiration is higher in leaves with large surface areas of mesophyll cells exposed, i.e. open spongy mesophyll promotes high rates. Contrasting mesophyll structures may occur on the same plant in sun and shade leaves.
cuticle	Nature of cuticle important, particularly at night when stomata are closed. In some shade plants up to 30% of total water loss may occur via cuticle. In xerophytes this may be reduced to zero.
leaf hairs, scales, leaf rolling and furrows	All these features help retain a layer of moist air in contact with leaf and reduce transpiration. Some have additional effects, such as increasing reflection or radiation.
root/shoot ratio	Plants of dry situations, particularly desert plants, have high root area to shoot area ratios to ensure adequate water absorption.

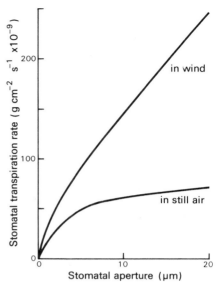

Fig. 19.14 Stomatal water loss and air movement (after Sutcliffe, 1968).

Mineral nutrition. Of course, water can be considered to be a nutrient because it is essential to plant growth, though it is usually treated separately. Mineral nutrition is, however, a further exogenous variable, the effect of which may be to limit production (Box 19.4) (Fig. 19.15). The principal nutrients are listed in Table 19.3, together with their function and sources, and the broad outline of nutrient circulation as global biogeochemical cycles was discussed in Chapter 7. These nutrients are essential requisites for plant growth, either as components of major structural compounds or as critical constituents of enzymes and other compounds active in biochemical re-

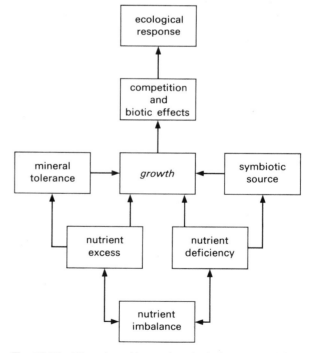

Fig. 19.15 Mineral nutrition and ecological response (after Bannister, 1976).

Box 19.4
LIEBIG'S LAW OF THE MINIMUM

In 1840 Justus Liebig recognized explicitly for the first time that an organism is constrained by the essential mineral nutrient available to it in amounts which most nearly approach the minimum requirements for the survival of that organism. Today this *law of the minimum* is applied to requirements and conditions other than mineral nutrients, though perhaps it is easier to justify the more restricted meaning. Closely related to Liebig's law is that accredited to the

American ecologist V.E. Shelford, in 1913, who recognized that excesses even of essential requirements, as well as deficiencies, could be limiting for the growth and success of organisms. This is **Shelford's Law of Tolerance**, which formalized the notion of a range of tolerance and explored the interaction between factors in setting limits of tolerance for particular organisms.

---

actions. Photosynthesis would be impossible without the magnesium in chlorophyll. Similarly, there would be no nitrogen fixation without molybdenum. If we discount carbon, hydrogen and oxygen then with the partial exception of mineral nitrogen, and to some extent sulphur, the availability of mineral nutrients to primary producers depends on the microbial decomposition of organic matter and the detrital system (see Chapter 21), with rock weathering and precipitation input representing secondary sources. If these sources are inadequate, nutrient deficiency will occur, and growth and productivity will be limited even when other factors are favourable. Most natural soils show some nutrient deficiency and respond to artificial applications of nutrients with increased production. However, the situation is far from simple, for it is often the case that additions of nutrients lead to a change in species composition, a fact which would suggest that the original species were genetically adapted to grow in nutrient-deficient soils. Sometimes such an effect is indirect, acting through competition. Such a case is reported by Jeffrey and Piggott (1973) where the survival of *Kobresia simpliciuscula*, a rare sedge in the relict flora of Upper Teesdale (England), is due not to a positive tolerance of soil phosphorous deficiency but to the fact that this restricts competition.

The reasons for nutrient deficiency are very complex, but soil type, soil texture, ion exchange capacity, weathering and decomposition rates are

all important controlling variables. As in the case cited above, phosphorus is perhaps the most widespread soil mineral nutrient deficiency, and it is in short supply in almost all soils. Nitrogen too is often limiting, particularly where decomposition is slow, as for example in cool, acid or waterlogged soil environments where C:N ratios are high (Fig. 19.16). Many plants are adapted to such environments by the presence of nitrogen-fixing bacterial or fungal symbionts associated with their roots, or perhaps more spectacularly by insectivorous habits which enhance their nitrogen supply. In cases such as these, and more particularly that of phosphorus, where an element is limited in its availability in the soil and the rate of replenishment by weathering and precipitation is low, the amount of the element in the living and dead biomass of an ecosystem becomes critical. Furthermore, the circulation of the element between them becomes a negative-feedback loop to which the productivity of the vegetation must become adjusted by the conservation of that element . The circulation of nutrients such as these, where the bulk of the ecosystem's stock is held in the organisms themselves and the fraction in the soil is subject to rapid turnover between release from decaying litter and uptake again by roots, can be said to be tight.

In order to conserve such **tight circulation** elements against loss there is a redistribution and reuse of mobile ions of these elements within the plant prior to any senescence. Ions of elements

Table 19.3 Mineral nutrition (from Collinson, 1977)

Nutrient	Physiological function	Sources	Environmental effects
oxygen	respiratory metabolism	photosynthesis of green plants	Important determinant of plant distribution in relation to soil aeration. Where soil is badly aerated or permanently water-logged anaerobic conditions prevent root growth in plants without special adaptations and toxic materials, e.g. H_2S may be generated.
carbon dioxide	source of carbon in photosynthesis	decay, respiration and the oceans release some 10^{12} tons/year	Some slight natural variations of atmospheric concentration but with little effect on plant growth and distribution. Important ecological effects on soil acidity. Total concentration of CO_2 (320 ppm) sets ultimate limit on photosynthesis. Plants can increase photosynthesis up to 3 × normal with increasing concentration.
nitrogen	essential element of proteins. Can only be absorbed in fixed form (NH_4, NO_2, No_3)	drawn from the atmosphere by a host of microbes, and by lightning	In most well aerated soils of intermediate acidity fixed N usually freely available. Deficiencies associated with cold, wet soils, very porous soils and tropical soils where vegetation cover is cleared. Where destruction of organic matter is slow or dead material highly lignified, acid 'mor' peat may accumulate as humification may be inhibited.
sulphur	essential for protein synthesis and vitamin synthesis	sulphates in well aerated soils, pyrites and gypsum in arid lands, H_2S and reduced sulphur in airless soils	Cycled rapidly by micro-organisms similarly to nitrogen. 'Downhill' losses replaced by weathering, airborne dust, salt spray and volcanic gases. In arid regions strong concentrations of SO_4^{2-} ions exist which select for tolerance. Pollutant sources – some 146 000 000 tons annually of SO_2 – increasingly added to biosphere.
phosphorus	incorporated into many organic molecules, essential for metabolic energy use	Fe, Al and Ca phosphates; free anions in solution (H_2PO_4 in acid, HPO_4 in alkaline conditions)	Great differences in demand between species and hoarded tenaciously in most ecosystems. Cycled on a world scale with downhill losses replaced similarly to sulphur above. Oceanic reservoir returns deepwater reserve along cold currents via plankton, fish and guano of fish-eating birds.

Element	Function	Source	Notes
calcium	essential to metabolism but not incorporated into fabric molecules of living matter	feldspars, augite, hornblende, limestone, and sulphates and phosphates in arid lands	Strong selective effects in all habitats – lakes, marshes, grasslands, forests, rock outcrops. Important determinant of prime physicochemical characteristics of soil. Antagonistic to toxic effects of K, Mg and Na. Retention of ions by colloids closely related to climate, especially rainfall.
potassium	essential to many metabolic reactions, especially protein-building and transphosphorylation	feldspars, micas, clay minerals	Deficiency has marked effects on carbon assimilation thus lowering production and biomass. Certain crops – beet, cotton, vine, legumes – are very sensitive.
magnesium	vital constituent of chlorophyll	biotite, olivine, hornblende, augite, dolomite, and clays of the montmorillonite group	Excess produces serpentine barrens, e.g. in California, Spain, New Jersey, southern Urals, Japan, New Zealand. Natural climax is replaced on these by impoverished, often scrubby vegetation commonly with specialized, e.g. *Quercus durata* in California.
iron	iron oxidation and reducing reactions in respiration	iron silicates, iron sulphates, free ions chelated with organic molecules	Calcareous or alkaline soils may be deficient as iron may be precipitated as insoluble hydroxides. May also be deficient where copper or manganese is present to excess. Vines and fruit crops may be easily affected by iron deficiency.
manganese	minute amounts needed for certain enzymatic reactions	ferromagnesian minerals; absorption dependent on other metallic cations	Deficiencies noted in mid-latitudes especially. Tropical soils, especially feralites, may have excess manganese which has toxic effects.
zinc	enzymatic metabolism	zinc-bearing vein minerals	Often leached out of the soil profile in acid soils. May be insoluble in alkaline soils. Certain species, e.g. *Viola calaminaria* of the Harz Mountains in Germany, are endemic to zinc-rich soils.
copper	essential for respiratory metabolism	copper-bearing vein minerals	Deficiency frequent in alkaline soils. Any excess has strong selective effects, e.g. in Katanga, the 'copper flower', *Haumaniastrum robertii*, has 50 × the normal copper content in its leaves; also *Becium homblei* cannot germinate without 50 p.p.m. of copper at least in the soil. The latter is a reliable prospecting index for mineral veins.
boron	necessary for successful cell division during growth	soluble borates are the only assimilable form	May be leached out in acid soils. Some crop plants – beet, potato, cauliflower – show considerable sensitivity to any deficiency.
molybdenum	essential for nitrogen fixation and assimilation	vein minerals	Deficiency in acid soils frequent and also in certain tropical soils on ancient land surfaces.

such as nitrogen, phosphorus and potassium move readily from older tissues to more metabolically active sites. Phosphorus is particularly mobile, and an atom of phosphorus may be incorporated and released continuously and make several complete circuits of the plant in one day (Biddulph, 1959). As the leaves of deciduous perennial plants senesce and die, considerable quantities of nutrients (of which N, P, K, S, Cl and perhaps Fe and Mg are most important) are withdrawn from them to minimize losses. Calcium, silicon, boron and manganese are translocated to the leaves before abscission, and little fall occurs, enabling the plant to shed excess nutrients not used in metabolic processes. Key nutrients may also be preferentially accumulated in storage organs such as rhizomes to be available for rapid vegetative growth in perennials, while these same elements are concentrated in seeds in both annual and perennial plants, to be remobilized and transported into the growing embryo during germination. In this way even the death of an individual plant need not represent a complete loss of the nutrient elements it contained, for some will have be passed on to the next generation.

Other elements which are present in the environment, and which are released by weathering or supplied by precipitation in appreciable quantities, can be said to be in fairly **loose circulation**. In such cases the bulk of the ecosystem's store of the element is usually in the soil, not in the plant biomass, and hence availability is not so critically affected by turnover rates and decom-position. The ecosystem can afford to be more careless in their use, and it does not actively conserve against loss to the hydrological cascade through the denudation system. Analysis of the solute loads of streams, such as the figures quoted for the Hubbard Brook catchment in Chapter 26 (see Fig. 26.7 and Table 26.1) often shows a net loss of such elements (e.g. Na, K, Ca) when precipitation input and stream discharge output are compared. If the ecosystem is actively growing and accumulating biomass, or just maintaining a steady-state biomass, the difference must be made good by weathering.

These concepts of tight and loose nutrient circulations are important, for not only may the circulation type be different for different elements in the same ecosystem, but ecosystems in contrasting environments may be characterized by distinct types of nutrient circulation as a whole. For example, tropical forest ecosystems have tight circulations for all nutrients when compared with temperate forests (Fig. 19.17 and Table 19.4).

It is not only the macronutrients that may be in short supply. Deficiencies of trace elements also occur and limit growth and production, but perhaps the most spectacular effects of deviations from normal trace concentrations by these elements are when they are present in abnormally high concentrations. Most of these trace elements are metals with high atomic numbers and they give rise to **heavy metal toxicity**, of which many examples are documented. Some are entirely natural, like the sparse unproductive

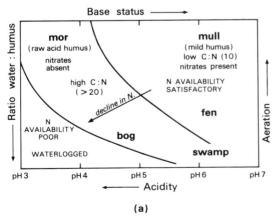

Fig. 19.16 (a) Carbon/nitrogen ratios in relation to soil water and aeration status. (b) The sundew (*Drosera*), an insectivorous plant.

vegetation of the copper-rich soils of Zambia and the nickel-chromium toxicity and calcium–magnesium imbalance of serpentine soils (see Chapter 5), with their unique and largely unproductive flora. Other sites are the result of environmental pollution by toxic mining and smelting wastes. In all cases, however, some species or ecotypes have evolved tolerance mechanisms to

specific metal toxicity which appear to involve the immobilization of the metallic ion by complexing at cell walls (by chelating agents), thereby keeping it away from metabolically active sites where it would block enzyme activity. Such adapted species have often been used as indicator species in prospecting and, more recently, have received renewed attention as indicators, this time of

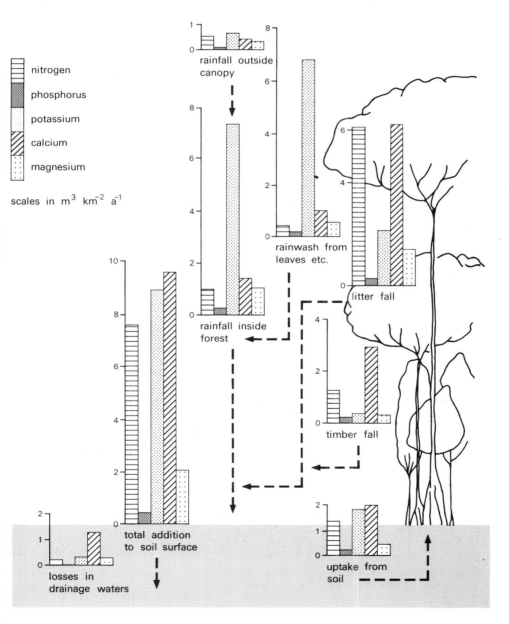

Fig. 19.17 Nutrient circulation for tropical rainforest.

Table 19.4 Distribution of all major nutrient elements in store in three forest types, (a) by weight and (b) as a proportion of total biomass.

Nutrients	Pine forest		Beech forest		Tropical rainforest	
(kg ha⁻¹)	a	b	a	b	a	b
forest biomass	1112	1.0	4196	1.0	11 081	1.0
annual litter	40	0.036	352	0.084	1540	0.138
soil and partly decomposed litter	649	0.58	1000	0.28	178	0.0016

sources of water-supply pollution by metals such as lead, as awareness of potential health hazards has increased.

There are, however, other examples of soil/nutrient imbalance which are more common. The most familiar is the acid–alkaline continuum in soil reaction, or soil chemical environment. At both extremes nutrient imbalances occur to which particular groups of plants have evolved tolerance, the so-called **calcifuge** and **calcicole** species respectively (Fig. 19.18). This topic has been the subject of a considerable volume of research and the mechanisms involved are not simply tolerance of high acidity at one end and high calcium carbonate levels at the other. Acid soils are generally leached and nutrient-deficient but, more important, at low pH aluminium is more mobile so that adaptation to acid environments involves tolerance of aluminium toxicity as well as of low nutrient status. On the other hand, basic soils with high pH and well drained oxidizing environments lead to iron deficiency, as it is present in an unavailable Fe III state, in addition to the excess of calcium.

The variation in mineral nutrient levels, therefore, not only sets limits to plant growth and primary production but also regulates the distribution, composition and productivity of the community through the ecological response of adapted or tolerant species. Such response, as with water availability, will depend on the **phenotypic plasticity** of the species concerned, or the extent to which they can accommodate variations in nutrition without genetic change. It will also depend on the genetic flexibility of the species and the speed with which new races or ecotypes are produced and selected for their tolerance of particular nutrient regimes.

19.3.3 Primary Production and plant competition

So far in this chapter we have been considering the structural organization and functional activity of the primary production system of the ecosystem. We have seen the way in which these features reflect the interplay between the endogenous variables which characterize the system itself (i.e. the inherited genetic and phenotypic attributes of the plants themselves) and the exogenous, or environmental, variables that define the conditions under which the primary production system operates. We have seen how this interplay controls and regulates the primary productivity of the system, with the result that any ecosystem operating under a given set of conditions will arrive at an optimum state for a sustained production under those constraints. The optimum state is achieved, therefore by competition between the available species for light, space, water and nutrients over the duration of their lifecycles. Such competition is for the primary producer niche in the ecosystem, and it leads to niche differentiation and segregation, not only in space but also in time, so that a diverse but integrated, efficient and stable community emerges.

19.4 Geographical variation and comparison of ecosystem primary production

In Chapter 7, global figures for the net primary production (NPP) of the land and oceans were given, together with a warning that many such figures are mere estimates, some almost guesses. Today the potential exists to provide more reliable regional and global estimates of NPP, based on the methodology of **satellite remote sensing**. As a result of the absorption of visible light by chlorophyll, vegetation shows a ratio between the thematic mapper bands 4 and 3 of the LANDSAT satellite of >>1. This ratio has been shown by ground truth sampling to correlate with leaf-area index (LAI), and as there is a fairly direct relationship between LAI and NPP, the way is clear for regional (Fig. 19.19) and global extrapolations. Additionally, data from the very high-resolution radiometer (AVHRR) on the NOAA-7

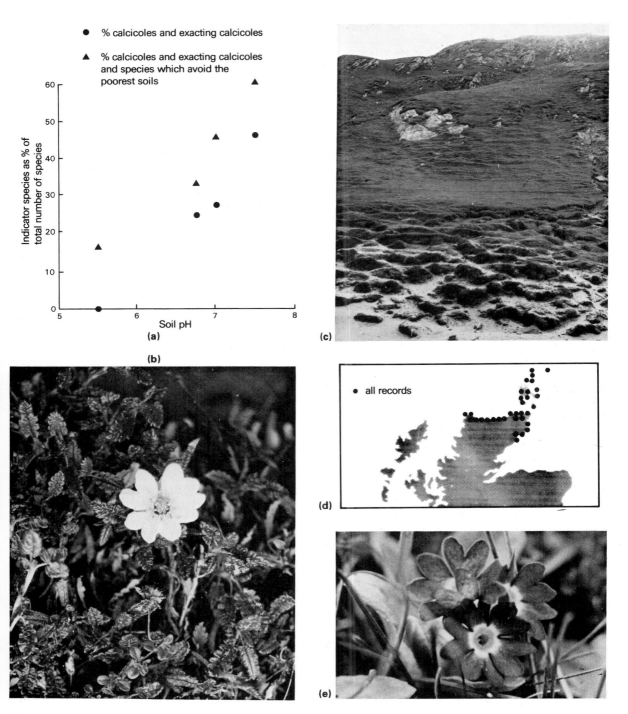

Fig. 19.18 Soil base status and the distribution of calcicolous species. (a) The increase in the representation of calcicoles as soil pH increases at sites showing a progressive increase in the effect of wind-blown shell sand on soil calcium (b) Wind-blown shell sand banked against the headland and colonized by *Dryas octapetala*, forming distinct hummocks. (c) *Dryas octapetala*, the mountain avens, a calcicole. (d) The distribution of an exacting and rare calcicole, *Primula scotica* (e), correlating with coastal sites enriched with shell sand. (All at Bettyhill, Sutherland, Scotland.)

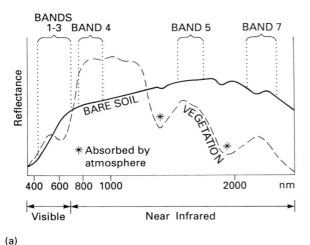

(a)

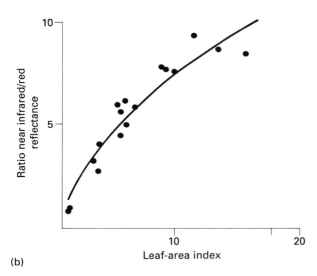

(b)

Fig. 19.19 (a) The proportions of the reflectance from bare soil and leaf canopies which are measured by the LANDSAT satellite thematic mapper bands. (b) Relationship between reflectance in the LANDSAT thematic mapper bands 4 and 3 (near IR/R) and the leaf-area index (LAI m^2 m^{-2}) for forest stands in western North America (after Peterson *et al.*, 1987).

satellite indicates that what has been called a 'greenness' index can be calculated from the ratio between quantities based on the relationship between the near-infrared reflectance and reflectance in the visible wavebands. This index is termed the normalized difference vegetation index (NDVI):

$$NDVI = \frac{(\text{near IR} - \text{visible})}{(\text{near IR} + \text{visible})}$$

Again, by utilizing a relationship between NDVI, LAI, and NPP, and with suitable calibration, estimates at regional and global scales are possible. So, we shall consider below to what extent generalizations can be drawn from the pattern of primary production in different ecosystems at the surface of the planet.

From the figures shown in Table 19.5 the annual production of dry matter can be divided into four main groups. At the high end of the range (2.0 kg m^{-2} a^{-1}) are some tropical forests, some developmental or successional communities (see Chapter 25) and some swamp and marsh communities (annual values for *Spartina* and *Phragmites* reach 150–180 kJ m^{-2} a^{-1}) and intensive year-round tropical agriculture with such crops as sugar cane, which holds the record for the highest annual production (Java), and maize, which holds the daily rate record (in subtropical Israel). However, the highest transient productivities in natural ecosystems occur in shallow marine waters over coral reefs, estuaries and perhaps some shallow freshwater springs. These ecosystems with dry matter production 1–2 kg m^{-2} a^{-1} include the remaining tropical forests, most temperate forests, some grasslands, wetland habitats and highly productive agriculture, mainly the energy-subsidized (fuel, fertilizers, etc.) agriculture of temperate zone. Below this there is a middle range of the order 0.25–1.0 kg m^{-2} a^{-1}, which contains a variety of communities, grasslands, shrublands, some woodlands and the continental shelf and, interestingly enough, most cereal crops. Finally, there is a group of habitats and plant communities with low net annual production figures up to 0.25 kg m^{-2} a^{-1}. These are the extreme habitats of aridity, and high or low temperatures, desert, semi-desert and tundra on land, and also much of the open ocean.

These groups reflect fairly clearly the integrated effects of the limiting and regulating factors that we have already discussed as controlling photosynthesis and productivity. Light energy is probably not the major variable here, for the variations in net photosynthetic rate or carbon fixation per unit of photosynthesizing surface do not vary enough to account for variations in production figures. Total photosynthesizing surface, however, does vary and it is reflected in the leaf-area index. Both this and the variations in

Table 19.5 Net primary production of the Earth (modified after Whittaker and Woodwell, 1971 and Whittaker and Likens, 1975)

	Average P_n (kg m^{-2} a^{-1})	Average P_n (10 J m^{-2} a^{-1})	Area (10^6 km^2)	World P_n (10^{19} J a^{-1})	Average B (kg dry wt m^{-2})	Ratio P:B
tropical rainforest	2.2	3780	17	75.6	45	84
temperate deciduous forest	1.2	2475	7	44.6	30	92
boreal coniferous forest	0.8	1512	12	18.1	20	76
woodland and shrubs	0.7	1134	8.5	7.9	6	189
tropical savanna	0.9	1323	15	19.8	4	331
temperate grassland	0.6	945	9	8.5	1.6	630
high-latitude and alpine tundra	0.14	265	8	2.1	0.6	442
desert and semi-desert shrub	0.09	132	18	2.4	0.7	189
rock, sand and ice desert	0.003	6	24	0.14	0.002	300
agriculture	0.65	1229	14	17.2	1.0	1229
marsh and swamp	2.0	3780	2	7.6	15	315
Total or average (land)	0.77	1380	149	205.6	12.3	110
open ocean	0.125	242	332	80.3	0.003	80 667
continental shelf	0.36	662	27	17.9	0.01	6620
estuaries and littoral	2.5	3780	2	7.6	1.0	3780
Total or average (ocean)	0.15	293	361	105.8	0.009	32 556

productivity reflect gradients of temperature, moisture and nutrient availability. The high production categories above are all associated with non-limiting conditions of temperature, water and nutrients, and all carry large leaf areas; and, in the case of sugar cane, maize and reedswamp, the efficiency of light penetration is enhanced by steep angles of leaf insertion. In conifers and evergreen forests the high leaf area indices, and particularly the leaf duration, probably compensate for less favourable climatic and environmental conditions. The steep angle of leaf insertion may also partly account for the very high production of some grass and grass-like ecosystems.

These are structural considerations and their effect is emphasized if we look at the production to biomass ratio (P_n:B). Immediately the picture changes, with aquatic ecosystems having the highest ratios, then agriculture, then grassland, wetland, desert and tundra communities, savanna and finally forests (with the lowest ratios). As we have seen earlier, this ratio reflects the extent to which the ecosystem carries a large surplus of non-photosynthesizing and respiring tissue, which explains the low ratios for forests. Xeromorphic communities with their sclerophytic (thickened) or succulent habits also suffer from this handicap, and are therefore inherently inefficient.

The interpretation of production figures will,

in the last analysis, depend on the point of view adopted. For example, if we are concerned with the production per unit area of the Earth's surface (P_n m^{-2}), tropical forests, reedswamps and marshes head the list. If, however, it is production per unit biomass that interests us, then it is the oceans, or rather their productive margins, that come out on top, for reasons already discussed. Alternatively, we may be interested in production of the total ecosystem area relative to the total area of the Earth's surface. Tropical rainforest again scores here, for although the oceans occupy two-thirds of the surface they contribute only one-third of global primary production, because the large area of deep ocean has low values due to nutrient limitations. Indeed, terrestrial forests contribute something of the order of 45% of total world primary production, a fact of some significance when we reflect on the extent of past deforestation and its currently accelerating rate (see Chapter 26).

We shall continue to explore some of the implications of production data in Chapter 20, but first we must consider the secondary production of the animals of the grazing and predation chain. For the moment, suffice it to observe that from the human standpoint, moisture and nutrients appear to be the principal limits to natural primary production, other things being equal; that herbaceous

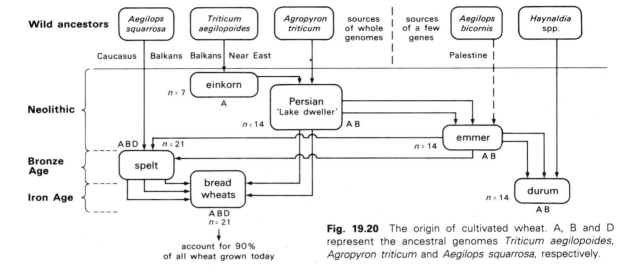

Fig. 19.20 The origin of cultivated wheat. A, B and D represent the ancestral genomes *Triticum aegilopoides*, *Agropyron triticum* and *Aegilops squarrosa*, respectively.

communities are more productive than forests under non-limiting conditions; and that even with massive technological investment and energy subsidies, argiculture is only slightly more productive than natural communities, and much agriculture is actually less productive. However, in this context it must be said that crop plants are more useful and more easily harvested, facts which reflect centuries of cultural selection of genotypes and programmes of crop breeding (Fig. 19.20).

Further reading

The ecology of primary production is well covered by the general texts cited at the end of Chapter 18. However, some specific topics can be pursued further with reference to the following sources:

Photosynthesis and Productivity:

Bjorkman, O. and J. Berry (1973) High energy photosynthesis. *Scientific American*, **229**, 80–93.

Berger, W.H., V.H. Smetack and G. Wefer (eds) (1989) *Productivity of the Oceans: Present and Past*. Wiley, New York.

Cooper, J.P. (ed) (1975) *Photosynthesis and Productivity in Different Environments*. Cambridge University Press, Cambridge.

Fitter, A.H. and R.K.M. Hay (1987) *Environmental Physiology of Plants*, (2nd edn) Academic Press, London.

Grace, J. (1983) *Plant–Atmosphere Relationships. (Outline Studies in Ecology Series)*. Chapman & Hall, London.

Hall, D.O. and K.K. Rao (1972) *Photosynthesis*. Edward Arnold, London.

Hobbs, R.J. and H.A. Mooney (1990) *Remote Sensing of Biosphere Functioning*. Academic Press, San Diego.

Meidner, H. and D.W. Sheriff (1976) *Water and Plants*. Blackie, London.

Walker, D. (1979) *Energy, Plants and Man*. Packard, Funtington.

Webb, W.L., W.K. Lauenroth, S.R. Szarek and R.S. Kinerson (1983) Primary production and abiotic controls in forests, grasslands, and desert ecosystems in the United States. *Ecology*, **64**, 134–151.

Woodwell, G.M. (ed) (1984) *The Role of Terrestrial Vegetation in the Global Carbon Cycle: Measurement by Remote Sensing*. John Wiley, New York.

The grazing–predation system

20.1 The heterotroph

Part of the net primary production of an ecosystem represents the energy and material input to the heterotrophic or consuming organisms of that system – the animal community. As we saw in Chapter 7 when discussing the heterotrophic cell model and heterotrophic level, animals rely either directly (herbivore) or indirectly (carnivore) for all their carbohydrates on those originally produced by plants. Much of the plant carbohydrate material is, however, unavailable to animals, in the sense that although they may eat it they cannot use it. The cellulose of plant cell walls cannot be digested by most animals except by those (such as cattle) which possess **cellulase**-secreting bacteria living symbiotically in their digestive tracts. The lignin of thickened woody plant tissue is virtually completely indigestible, though here again a few animals have symbiotic gut fungi which can degrade lignin. In both cases the animal only digests the simple products of degradation. Of course, the unavailability of much plant production for utilization by herbivores underlines the importance of the alternative energy flow pathway: the detrital system, which we will examine in Chapter 21.

All animals can use their digested carbohydrate to manufacture other related compounds, for structural tissue, for energy-yielding reactions or for storage, usually as fats. They can also convert their carbohydrates to organic acids which can react with ammonia to form amino acids (the building blocks of protein) with the elimination of oxygen. However, the extent to which animals produce amino acids in this way is negligible when compared with plants, mainly because they have very small quantities of the enzyme systems necessary to do so; but given a few amino acids to start with they can manufacture many others. Some exceptions remain – those amino acids which require the synthesis of hydrocarbon rings – and these they must acquire in their food, from plants. Some amino acids, therefore, are essential components of the animal diet, as too are vitamins of which about 16 are needed by animals.

Therefore, animals are totally dependent on plants. They cannot manufacture carbohydrates or proteins (except to a very limited extent) from any materials except other carbohydrates and proteins, and such substances as vitamins they may not be able to manufacture at all. So the transfer of food from producer to consumer is not only a transfer of an energy source – a respiratory substrate – but also the transfer of essential nutrient elements, as components of essential compounds. Although animals can and do obtain mineral nutrients directly from the inorganic environment, as for example in drinking water or the 'salt licks' put out by the farmer for domestic herbivores, it is usually the nutrient content of their plant food which is critical. Moss (1969) has shown conclusively the effects of variation in the nutrient status of heather shoots in the diet of the hen red grouse (*Lagopus lagopus scoticus*) and ptarmigan (*L. mutus*) on the properties of the egg and chicks. Fluctuations in lemming (*Lemmus* sp.) populations in Alaska have been partly explained in terms of nutrient deficiencies in the vegetation, and hence in the milk of the females during lactation, which affects the survival rate of the offspring (Shultz, 1964). The production of the red deer (*Cervus elaphus*) in the Scottish Highlands is higher when grazing over basic rock types than over acid rocks.

20.2 Modelling ecosystem animal production

The framework for the discussion of the functional activities of the ecosystem in this section has been the model of trophic structure. In Chapter 19 it was relatively easy to consider the organization of the individual plant and of the primary producer trophic level from the point of view of energy flow and throughput of nutrients. As we shall also appreciate with trophic models of the detrital system in Chapter 21, the modelling of consuming systems is much more difficult. In the first place, the consumer, or grazing–predation system, is concerned with more than one trophic level. This would be unimportant if each organism could be classified unequivocally as a herbivore or carnivore. Even when this is possible, it is difficult to define the diet, for very few species are restricted to a single food type (**stenophagous**), as for example the koala, which feeds exclusively on *Eucalyptus* leaves. In such species, which are limited to one food source, the distribution of that food presents a geographical barrier limiting dispersal and speciation. The juniper moth, whose larvae are restricted (in the field) to the juniper as a food plant, have become discontinuous in their distribution in the British Isles (southeast England, Lake District, north Pennines and Scotland) as the distribution of the shrub has contracted (Huxley, 1942).

Most animals, however, have varied diets even when food is abundant, and they can show considerable variations in food taken at different times or from place to place. Such behaviour is referred to as **eurphagous** or **polyphagous**. Indeed, many species will regularly, occasionally, or under the pressure of limited food, behave as **omnivores** or **diversivores**. In other words, it is often extremely difficult to assign organisms to particular trophic levels and food-web relationships, and hence the pathways of energy and nutrient transfer become both diverse and complex. Many of the small animals in the woodland ecosystem described in Chapter 18 (see Fig. 18.1) such as mice and voles, will, though mainly herbivores, also eat insects, particularly in summer. Blackbirds will feed on worms and insects in the summer because they are plentiful, but in winter they eat fruit, berries and seeds. Even the red grouse mentioned above, with a diet consisting almost exclusively of young heather shoots, will take appreciable quantities of insects during the summer. Now, not only may the food taken vary from season to season, as in these examples, but it may also change through the lifecycle of an organism, as in the common frog, which for a brief interval functions as a herbivore while still a tadpole, before leaving the pond to exist on a carnivorous diet of slugs, worms and insects. Unfortunately, the confusion of trophic levels does not stop with the herbivore and carnivore levels, for as you may have noticed in some of these examples the detritivore level (represented here by earthworms) was being tapped. Other examples of similar confusion arise with animals such as hyenas and vultures which take dead flesh as food. Are they carnivores or detritivores? Indeed, perhaps there should be a separate level – the scavengers – for many carnivores will behave in this fashion, especially under food stress, even the so-called king of beasts, the lion.

In spite of these observations, however, most animals show to a greater or lesser degree some food preference under normal conditions, and this is, of course, an important factor in the definition of their ecological niche, particularly the trophic niche (Fig. 20.1). Nevertheless, it remains true that although the concept of trophic levels is helpful, any study of energy flow through the grazing–predation system will require due attention to be paid to food sources, preferences, feeding habits and life histories. However, there is one further difficulty in modelling consumer production and that is the mobility of many animals. As with primary production, the biomass and production of animal communities is expressed as mass or

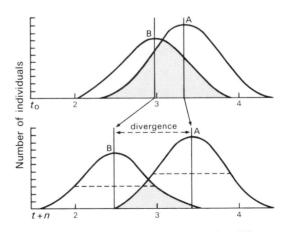

Fig. 20.1 Prey size and niche differentiation. When two species overlap in mean prey size, competition favours those individuals taking prey of a size where there is no overlap, so that mean prey sizes for the two species diverge, over the time interval $t_o - (t + n)$, as indicated by the arrows. (After Whittaker, 1975.)

energy per unit area of the Earth's surface (and in the case of production, per unit time). Unfortunately, unlike plants, animals will not necessarily wait to be measured: they will enter and leave the area of study. To a small extent such movements do affect measurements of primary production. This is especially true in aquatic ecosystems, where input and output in running water may be important, but it also occurs in terrestrial ecosystems as, for example, when leaf litter is carried out of a woodland by the wind. In studies of animal production these factors may be very significant indeed, but, as in the case of migratory species, their effect may be difficult to assess. For example, in the arctic tundra the relationship between the primary production of a relatively small area of vegetation, say 1 km², and the production of a herd of caribou or reindeer, may be very misleading (Fig. 20.2). In reality these migratory herbivores will range over perhaps hundreds of square kilometres, and be dependent on the production of a vast area of tundra vegetation, but in winter they may also crop the forest–tundra zone.

Because of these difficulties we shall not attempt to model the trophic structure of real grazing–predation chains too closely, but concentrate on some of the more important theoretical relationships of such models. For similar reasons it is not easy to present a model of the transfer pathways of energy and nutrients through the individual animal, as was the case with plants. The great diversity of animal trophic niches, food type and availability means that a large variety of morphophysiological mechanisms have evolved in the animal kingdom, affecting feeding mechanisms, structure and shape of the digestive system, and the nature of the digestive process, including the types of enzymes involved. In very general terms, however, it might be possible to draw functional parallels between the digestive and respiratory systems of animals and the xylem and stomatal pathways in plants, and the blood vascular systems of animals and the phloem pathway in plants, at least as far as the movement of materials is concerned.

20.2.1 Secondary production: the individual

The kinds of difficulty discussed above were highlighted when the scientific community began to digest the results of the International Biological Programme throughout the 1970s. It became apparent that although studies of primary productivity had been reasonably successful at the level of the entire primary producer community, those of secondary production were much less so. Indeed, most studies of animal production had not progressed beyond the species level, let alone successfully addressed the productivities of entire consumer trophic levels (Heal and Maclean, 1975; Rigler, 1975). Throughout the following decade some attempt has been made to reassess the utility

Fig. 20.2 Reindeer, relatively conspicuous herbivores in an open landscape.

of the Lindeman model of ecosystem energetics and to return to Charles Elton's emphasis on number and body size in seeking to understand the feeding relationships of animals (Platt and Denman, 1977; Cousins, 1980, 1985).

So, to begin with, we shall consider the conversion of primary production to secondary, or animal production, in relation to the individual animal, because there are additional problems involved if we attempt to discuss the production of the total animal community of an ecosystem. The term conversion is used in preference to consumers, as we are not really dealing with production. This distinction is perhaps clarified by an economic analogy: winning a ton of coal is primary production in the economic sense and is comparable, therefore, to photosynthesis (indeed photosynthesis *is* economic primary production in the case of crop plants); the ouput of coal is said to be consumed by the economy, but this consumption will involve waste, for some coal will be left at the pit unexploited or in tip heaps or lost as coal dust; the use of the coal to smelt iron ore, to generate electricity and to manufacture a motor car is strictly conversion, some of it useful conversion of one material to another, some of it waste, such as the heat lost to cooling water (unless used for domestic heating) or the ash from the power station and slag from the blast furnace.

While feeding, the animal will destroy (D) a certain amount of the biomass of the previous trophic level, but it will not necessarily consume (C) this amount. Herbivores often damage or destroy more vegetation than they actually eat. Sometimes this may be directly related to feeding, as when the squirrel scatters shells and frass (W) when eating buds and seeds, or in other cases it may be indirect, such as trampling by large herbivores. After the kill, many carnivores do not consume their prey in its entirety. Part of the carcass may be left to scavengers, and even these may leave the skin, hair, teeth and parts of the skeleton (W). We can summarize these activities as

$$D = C + W \quad \text{so} \quad C = D - W.$$

Not all of the energy in the food eaten or consumed by the animal will be absorbed or assimilated. Some food will pass through the digestive system without any chemical change, while other food substances, though experiencing chemical breakdown in the gut, are nevertheless not absorbed into the animal's cells. In both cases these materials and the chemical energy they represent are discarded as faeces (F). The analysis of faecal pellets can be the only evidence as to the food of consumers, and it is particularly useful in reconstructing the diet of polyphagous carnivores from the fragments of fur, bone, feathers and the chitin exoskeletons of insects they may contain.

The absorbed food is used for growth, for the repair, maintenance and turnover of cell constituents, and also for the breakdown of fuel molecules and cell components in respiration, thereby releasing the energy necessary for vital functions, which in animals, of course, include their movement (see Chapter 7). Compounds absorbed in food in excess of the animal's immediate metabolic needs are stored in the body, as glycogen in the case of carbohydrates. However, excess protein broken down into its constituent amino acids is not stored in this way. The amino acids are deaminated to form fatty acids which are stored as body fat, and relatively simple nitrogenous compounds which are excreted as urine (U). The compounds in urine, therefore, are not available as sources of energy to the consumer and are usually added to the energy loss in faeces. The remaining food absorbed by the animal is assimilated energy (A):

$$A = C - F - U \quad \text{and} \quad C = A + F + U.$$

But, as we have just seen, some of the assimilated food is oxidized and lost during respiration (R) (catabolic heat loss), while the remainder adds to the biomass of the animal and is production (P). So

$$A = P + R$$

where A (assimilation) is the equivalent of gross primary production by plants and P (production) is the equivalent of net primary production. Therefore

$$C = P \quad + R \quad + U \quad + F$$
consumption = production + respiration + excretion + egestion

and

$$P = C - R - U - F.$$

In animals, however, neither R nor P is a simple parameter. In warm-blooded animals R is a complex parameter involving heat used in the maintenance of body temperature (**thermoregulation**) (Box 20.1, Fig. 20.3). In these animals, R varies proportionally with the surface-area/volume ratio of the body and with age, tending to be higher in the young than in adults. R is also influenced by activity and behaviour. For example, the congregation of animals will reduce energy consumption, as will the ability to maintain normal life activity while increasing the economy of energy expenditure (by lowered metabolic rate), as in the native species of the tundra such as the

lemming. Many tundra natives, however, have an enlarged heart relative to related species and are hence endowed with the ability to increase their motor activity (muscular activity, movement) when needed and yet cope with the associated increase in oxygen demand.

Production is also a composite term, for some of the assimilated energy will contribute to the growth of the individual, while some will be diverted to the development of sexual functions and products (eggs or sperms), which are then lost to the individual but represent the initial energy input into the next generation. So

$$P = \underset{\text{growth + reproduction}}{P_g + P_r.}$$

The laying of eggs or the growth of young in the uterus, and even after birth during lactation, represent huge investments of energy in P_r for the female. For example, many birds lay the equivalent of their own body weight of eggs over short periods of time. Reproduction, therefore, is very closely correlated with food availability and quality, and this goes some way to explain the territorial and social behaviour of many species during the breeding season. It also brings us to the relationship between animals or their life-cycles and production.

In young animals assimilation will exceed respiration ($A > R$), leaving a positive value of P to be channelled into growth of the individual (P_g). At sexual maturity and during breeding, A may still be greater than R but, as we have just seen, the excess production will be accounted for not by P_g (the growth of the individual) but by P_r (reproduction). The rearing of young, particularly in birds, will also involve the expenditure of energy by the parents in food-gathering, energy ultimately derived from their own food consumption. So maturity and breeding tend to lead to a reduction or even cessation of growth and, in some species, the exploitation of food reserves laid down in body fat earlier. Although growth may cease with maturity (mammals and birds cease skeletal growth) the number of breeding events and number of offspring will affect the calorific value of the body, and it may fluctuate with the storage or dissipation of fat and other food reserves. The value of P, therefore, may be either positive or negative at any particular time in the adult.

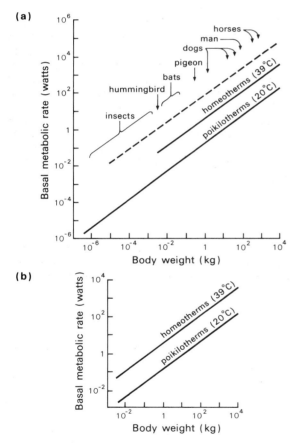

(a)

(b)

Fig. 20.3 (a) Relation between basal metabolic rate of homeotherms, maximum metabolic rate for sustained work by homeotherms (pecked line) and basal metabolic rate for poikilotherms at 20°C. (b) Relation between basal metabolic rates for homeotherms and poikilotherms. (Both after Hemmingsen, 1960.)

20.2.2 Secondary production: the animal community

When we come to apply the relationships considered above for individual animals to complete animal communities, the task of estimating or measuring secondary production or conversion becomes even more difficult. In the first place, there are additional variables to consider. The first of these is the need to measure population size, and this is far from easy, except in the case of large and conspicuous animals. Although it may be relatively easy to count polar bear or caribou from the air in an open tundra landscape, such situations are comparatively rare, and in theory at least we need such census material on all species in the community. Number alone, however, is not sufficient, for the production of any population will also depend on its age structure, each age class displaying a different productivity relationship. Furthermore, it will be necessary to know the rate at which each population is being added to by both birth and immigration, and diminished by both death and emigration. To measure all of these parameters for populations of all animal species in the community is a daunting task, rarely undertaken. In practice, effort is concentrated on a single or small number of species which are quantitatively most important in functional terms, or we rely on theoretical relationships to estimate the values of variables such as birth rate, using constants based on limited empirical data. Finally, even when these additional population parameters have been measured or estimated, they are used to multiply up or extrapolate production figures obtained from a small number of individuals under controlled conditions, and there is no guarantee that such figures will always be valid under the variety of conditions experienced by populations in the field.

It is not surprising, therefore, for all of these reasons, that our knowledge of secondary production on a global scale is still very far from complete, and in most ecosystems it is very sketchy indeed, as the results of the IBP suggested (see

---

Box 20.1

WARM-BLOODED AND COLD-BLOODED ANIMALS

The nature of the respiratory heat loss is different in warm- and cold-blooded animals. Warm-blooded animals (**endotherms** or **homeotherms**) which maintain a body temperature (35°–42°C) that is usually higher than ambient environmental temperatures, need to produce heat in amounts that will partly depend on the temperature difference between their body and that of the environment, and partly on their insulation and **thermoregulation** mechanisms. Heat generated during the performance of vital metabolic activities will include a proportion used to maintain body temperature, even when the animal is at rest. The production of heat by resting animals is known as their **basal metabolic rate** (BMR). Heat is also produced, however, by what is known as the **specific dynamic action** of food (SDA) and is due to the metabolic processes which follow the absorption of digested food. The SDA is not restricted to warm-blooded animals, for it also occurs in cold-blooded creatures (**ectotherms or poikilotherms**) but represents wasted energy as it is unavailable for growth or movement, and is therefore either neglected or included in a general term for respiration. In warm-blooded animals, however, the heat generated by SDA of the food assimilated is useful heat and can be used to maintain body temperature, and therefore reduces the demand for heat from basal metabolism (BMR) or from regulatory movements such as shivering. Ultimately, heat from both sources is lost from the animal, but the rate of loss will be controlled by insulation and other adaptations for thermoregulation. So, in warm-blooded animals the respiratory heat loss (R) can be split into two parts, R_1 (the SDA of assimilated food) + R_2 (the energy loss during vital activities, including movement).

beginning of Section 20.2.1). Because of this we shall make no attempt to summarize world secondary production as we have done for primary production. We will consider instead what is known of the efficiency with which energy fixed by photosynthesis is utilized by grazing–predation systems as it is passed to successively higher trophic levels.

20.2.3 Secondary production: the efficiency of energy transfer

Ratios within trophic levels (see Box 20.2). At the level of the individual population, or of single trophic levels, the **ecological** (or **Lindeman** see below) **efficiency** is an expression of the effectiveness with which organisms exploit their food resources and convert them into biomass available to the next higher trophic level. Ecological efficiency depends on (a) the proportion of consumed energy assimilated, and (b) the proportion of assimilated energy incorporated in biomass, and reproduction. The first proportion is called the **assimilation efficiency** (E_a) and the second, the **net production efficiency** (E_t) or as it was originally called the **tissue growth efficiency**.

Net production efficiency for the producer level is the ratio of net to gross production, and apparently varies between 30 and 80%, depending on plant growth form and environment. Rapidly growing plants in the temperate zone have uniformly high net production efficiencies (70–80%). Similar vegetation types in the tropics exhibit lower net production efficiencies of 40–60%, because as we saw in Chapter 19, tropical forest vegetation carries a large respiratory load of non-productive support biomass.

Net production efficiencies of plants and the factors affecting them have been discussed at length in Chapter 19. In warm-blooded animals, particularly active homeotherms, where thermoregulation and activity are an important energy demand, small amounts of available energy may go towards growth and net production production efficiencies are low. In poikilotherms, however, often more sedentary, net production efficiencies may be as high as 75%.

The palatability and nutritional value of plant foods depends upon the **digestibility reducers** present. These are principally the complex plant polymers cellulose, hemicellulose, pectins, lignins, cutins and tannins, as well as silica in crystalline form, all of which can dramatically reduce digestibility and hence nutritional value. Suitably adapted herbivores, with modified digestive tracts and mutualistic gut bacteria and protozoans can assimilate as much as 60% or so of the energy in young vegetation. However, most grazers and browsers assimilate 20–40% of the energy in their food. **Detritivores** such as mites and millipedes (see Chapter 21), which eat decaying organic matter, are particularly inefficient and assimilate only 15% or so of material ingested.

Assimilation efficiencies of predatory species vary from 50 to 90% because of the greater digestibility of their diet and its greater biochemical similarity to their own tissues. However, the exoskeletons of insect prey cause problems for digestion, so that assimilation efficiencies of insectivores vary between 70 and 80%, whereas those of most carnivores are about 90%. The percentage of energy lost in respiration, however, tends to increase along the food chain from plant to carnivore.

Ratios between trophic levels (Box 20.2). Also of importance are the ecological efficiencies which consider ratios *between* trophic levels. Here the most important relationships are the **Lindeman efficiencies** (E_L) which express the ratio of the rates of ingestion (consumption) or energy intake of pairs of trophic levels. These ratios are what Lindeman termed the progressive efficiencies of the food chain. The Lindeman efficiencies of energy transfer between trophic levels are the fundamental efficiencies that set the energy flux to the next trophic levels. All other efficiencies result from processes that also affect Lindeman efficiency. All efficiencies, therefore, are related to the fundamental Lindeman efficiency (Colinvaux, 1986). (*NB. there is considerable confusion in the definition of Lindeman efficiencies in textbooks. This is because the energy intake to the primary producer level can be equated with gross photosynthesis, or gross production. At consumer trophic levels, however, the rate of energy intake or consumption is not the same as assimilation or secondary gross production as is sometimes erroneously assumed.*)

There are two other useful ratios. The first is that between the rates of production of pairs of trophic levels, i.e. the net productivity of the higher trophic level of the pair expressed as a proportion of the lower trophic level. This is known as the **trophic level production ratio** or **efficiency** (E_p).

Box 20.2
EFFICIENCIES OF ENERGY TRANSFER

E_u = utilization efficiency

$$= \frac{I_2}{P_1} = \frac{I_3}{P_2} = \frac{I_4}{P_3} = \frac{I_5}{P_4}$$

E_a = assimilation efficiency

$$= \frac{A_2}{I_2} = \frac{A_3}{I_3} = \frac{A_4}{I_4} = \frac{A_5}{I_5}$$

E_t = tissue growth efficiency (net production)

$$= \frac{P_2}{A_2} = \frac{P_3}{A_3} = \frac{P_4}{A_4} = \frac{P_5}{A_5}$$

E_e = ecological growth efficiency

$$= \frac{P_2}{I_2} = (E_t)(E_a)$$

E_1 = Lindeman efficiency
(ratio of intakes of trophic levels)

$$= \frac{I_2}{I_1} = (E_e)(E_u)$$

E_p = trophic level production ratios

$$= \frac{P_2}{P_1} = (E_u)(E_e) = (E_u)(E_t)(E_a)$$

where:

P = production rate (rate of net organic synthesis in the form of the species of the trophic level or in storage products)

A = rate of assimilation

I = rate of ingestion (consumption) or energy intake

R = rate of respiration

after Lindeman (1942)

Secondly, the **exploitation** or **utilization efficiency** (E_u) expresses the energy intake or consumption of one trophic level as a proportion of the available energy, that is, of the net productivity of the trophic level below it. The first of these, the trophic level production ratio, provides some insight into the efficiency of energy transfer and is considered briefly here.

We already know that the ratio of gross primary production to incident light is very low indeed, being of the order of 1–4%. The ratio of herbivore production to green plant production is often less than 10%, as far as the available figures from natural ecosystems are concerned. The reasons would appear to be twofold. First, the efficiency with which plant tissue is assimilated by grazing animals is reduced by the biochemical differences between plants and animals, as we have noted above. Secondly, consumption by herbivores acts directly to reduce the size of the photosynthetic system, thereby decreasing the primary production rate. The 10% figure may therefore represent an agreed compromise between a sustained food yield to the herbivore and irreparable damage to the vegetation and seriously reduced primary productivity.

Higher trophic levels fare rather better and the ratio of predator production to prey production is usually higher than 10%. The percentage of energy lost in respiration, however, tends to increase along the food chain from plant to carnivore.

The length of food chains is limited by the efficiencies of energy transfer between trophic levels and by the Second Law of Thermodynamics. Ecological efficiencies are usually lower in terrestrial habitats; as a rule the top carnivores in terrestrial communities can feed no higher than the third trophic level on average. Aquatic carnivores, on the other hand, seem to feed as high as the fourth or fifth levels (Fenchel, 1988). Indeed, calculations suggest the average number of trophic levels to be about 7 for marine plankton-based systems, 5 for inshore aquatic communities, 4 for grasslands, and 3 for wet tropical forests.

20.3 Energy flow and population regulation

In Chapter 17 and in this chapter we have been tracing the way the original light energy is transferred and partitioned between the trophic levels of the ecosystem, thereby sustaining the populations of plants and animals which make up the living community. These populations are not static, as we have noted on several occasions: they respond to variations in their energy supply or to variations in other conditions, such as light and temperature. The organisms of the detrital pathway, which we shall consider in detail in Chapter 21, increase their population size in response to the supply of energy as dead organic detritus. Therefore, they tend to consume their food or energy source almost as fast as it is supplied to them, although the picture may be complicated by seasonal time lags. The result is that under most circumstances undecomposed litter does not accumulate excessively at the soil surface. The only exceptions occur where some other environmental factor such as waterlogging restricts their numbers and rate of increase, so restricting decomposition and promoting the accumulation of peat. Herbivores, however, cannot expand their population size in direct response to the green plant tissue available to them as food. If they did so, the surface of the planet would be denuded of its 'verdant pastures' and 'majestic forests'!

We are familiar with the consequences of the occasional ascendancy of the herbivore, if only from awe-inspiring tales of the desert locust. Outbreaks of plagues of defoliating insects such as the green tortrix moth caterpillar (*Tortrix viridiana*) on our native oaks in Britain can have similar devastating effects. In years (*lamas* years) when such outbreaks occur the oak grows a new, second leaf canopy on a fresh crop of young shoots the (*lamas* shoots). Usually, however, in natural ecosystems, for some reason, herbivores settle for about 10% or less of the primary production, and expand their populations only to a size at which this level of cropping is sustained. In other words, some factor other than food or energy supply would appear to be regulating their populations.

There is a considerable amount of evidence, however, that predator populations do increase in response to increased availability of energy as production by prey species. In some cases maximum predator numbers may be limited by direct competition for this food or for living space, but in the higher animals the situation is complicated by social interactions and conventions. Nevertheless, it is broadly true to say that in terrestrial ecosystems decomposer and carnivore numbers are directly limited or regulated by food resources. The same is true, of course, for the primary producers, but in the case of plants we have to read food resources to mean solar energy, water and mineral nutrients. Herbivore numbers, however, are not directly linked to the food resources available to them as vegetation. In contrast, their numbers appear to be mainly regulated by predation. Now, the significance of these relationships can perhaps be realized when it is appreciated that a community where herbivores are held down to population levels which are not damaging to the vegetation (itself resource-limited) is most likely to persist or remain stable through time. Conversely, a community where space is continually being made available by excessive grazing pressure from high herbivore populations is most likely to run the risk of replacement, and suffer invasion and change in plant composition, i.e. it is potentially unstable.

These views have emerged from population ecology and from models of population dynamics and the theories underlying them. However, work on the ecology of herbivory is increasingly modifying, and regards the plant as much more than a passive player in this game. The apparently limitless supply of food represented above as the verdant pastures and majestic forests of the natural world of vegetation is *not* free for the taking. Much of it is protected mechanically, has limited digestibility, or contains toxins and is genuinely poisonous to most potential grazers. Indeed most species of plant have complex combinations of these kinds of **mechanical** (spines etc), **structural** (digestibility reducers), or **allelochemical** defences (in this context allelochemicals are plant toxins such as *alkaloids* (many of which are now used as cytotoxic drugs in chemotherapy), which kill or repel herbivores at very low concentrations). The first two categories are usually permanent (**constitutive** defences), but the toxins are often inductive responses of individuals to tissue damage (**inducible** defences). It is probable that the balance between plant production and herbivore numbers is due in no small way to the coevolution of these defences and the complex array of adaptations which herbivores have evolved to overcome them, at least sufficiently to survive, reproduce and compete. The role of density-dependent regulation by predators may be less significant, therefore, than has hitherto been thought. However, the relationship between pathways of energy and matter transfer, the mechanisms of population regulation and the maintenance of a stable community structure are complex subjects, and we shall return to them in Chapter 25.

Further reading

Once again the material covered in this chapter is well covered in the general texts cited at the end of Chapter 18. However the following texts will exemplify particular aspects of the herbivore and carnivore role in ecosystems, particularly in relation to production:

Crawley, M.J. (1982) *Herbivory: the Dynamics of Animal–Plant Interactions. (Studies in Ecology, vol 10)*. Blackwell, Oxford.

Harbourne, J.C. (1977) *Introduction to Ecological Biochemistry*. Academic Press, New York.

Heal, O.W. and S.F. MacLean (1975) Comparative productivity in ecosystems: secondary productivity, in, (eds W.H. Dobben and R.H. Lowe-McConnell) *Unifying Concepts in Ecology*, Junk, The Hague, pp. 89–108.

Howe, H.F. and L.C. Westley (1988) *Ecological Relationships of Plants and Animals*. Oxford University Press, Oxford.

Humphreys, W.F. (1979) Production and respiration in animal populations. *J. Animal Ecol.*, **48**, 427–454.

Moss, R., A. Watson and J. Ollaoon (1982) *Animal Population Dynamics*. (Outline Series in Ecology). Chapman & Hall, London.

Robbins, C.T. (1983) *Wildlife Feeding and Nutrition*. Academic Press, New York.

Taylor, R.J. (1984) *Predation*. Chapman & Hall, London.

See also references to animal population dynamics at the end of Chapter 25.

The detrital system

21.1 Decomposition, weathering and the soil

After death or excretion, organic compounds cease to be components of living systems and undergo the processes of alteration and adjustment collectively known as decomposition. In terrestrial ecosystems these processes take place largely on or within the soil. Here too an analogous sequence of processes affects the rocks of the lithosphere and their constituent minerals (see Chapter 11). Both these processes of weathering and those of decomposition proceed towards the establishment of new equilibrium states for materials of inorganic and organic origin respectively. These states are weathered residues and secondary minerals on the one hand, and humus on the other. However, the mobile materials released by weathering and by decomposition are of equal importance for they are the nutrients upon which plant growth depends. In Chapter 7 it became evident that both rock weathering and decomposition, therefore, are critical and potentially rate-limiting steps in the circulation of materials in the ecosphere.

The detrital system of the soil can be modelled in the same way as the weathering system in Chapter 11, by distinguishing between the processes and conditions that promote or activate decomposition and the primary mechanisms of decay. Similarly, these primary mechanisms can be viewed as forming a sequence from the initial breakdown of fresh organic detritus to the production of relatively stable decomposition products. Before considering such a model, however, we shall begin by looking at the nature of the input; that is, at the supply of organic matter.

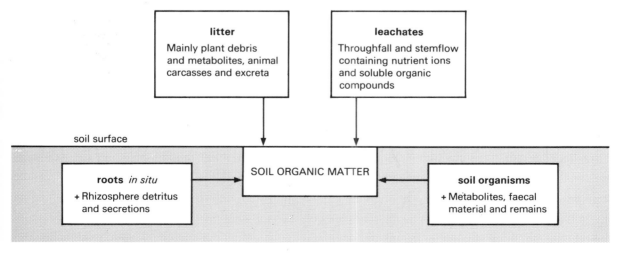

Fig. 21.1 The sources of soil organic matter.

21.2   The input to the detrital system: the supply of organic matter

The sources of organic matter to the decomposition system of the soil are shown schematically in Fig. 21.1. The soil surface in terrestrial ecosystems receives a supply of litter, the major proportion of which is plant debris of varying sizes from leaves, inflorescences, twigs, bark and bud scales, to the so-called macrolitter of branches and fallen tree trunks. It also includes the carcases of animals as well as their metabolites and excreta, though this 'animal litter' is normally quantitatively less significant. The exceptions are those ecosystems with a very large herbivore element, such as managed grazing land where the detrital input from dung can be of great importance. Apart from reproductive structures such as seeds and spores, all of this material is potentially available for decomposition. The litter present at any particular time can be expressed as the biomass of litter in kilograms per square metre or in terms of its energy equivalent in joules per square metre. The amount of litter shed by an ecosystem in a period of time such as a year is the litter production in kilograms per square metre per unit time. Of course, this is not the total increment of dead material, for some will remain above the soil surface (for example the dead limbs of trees) and this is known as the **standing dead biomass**.

The quantity and type of litter will vary from one ecosystem to another (Table 21.1) and, although generally proportional to the biomass of

Table 21.1 Variations in annual litter production under different vegetation types (Mason, 1970, after various sources)

Trees	Location	Litter production kg/m²/a
Norway spruce	Norway	150
oak	England	300
temperate oakwood	Netherlands	354
	–	440
beech beechwood	England	580
evergreen oak	S. France	380–700
tropical forests	Ghana	1055
	Thailand	2330
blanket bog	England	3 tonnes ha⁻¹a⁻¹

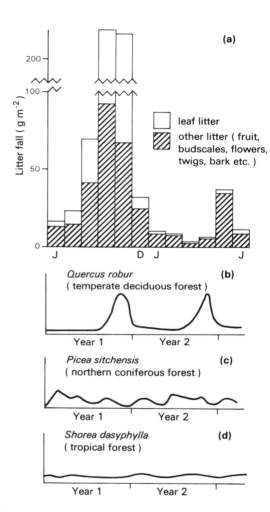

Fig. 21.2 Temporal variations in litter supply. (a) Variations in both quantity and type of litter received by the soil surface over an annual cycle under beech woodland (after Mason, 1977). (b), (c), and (d) The annual pattern of litter supply under temperate deciduous, northern coniferous and tropical forest, respectively. Note the autumnal bias in the temperate environment, the regularity of litter supply and the lack of seasonality in the tropics.

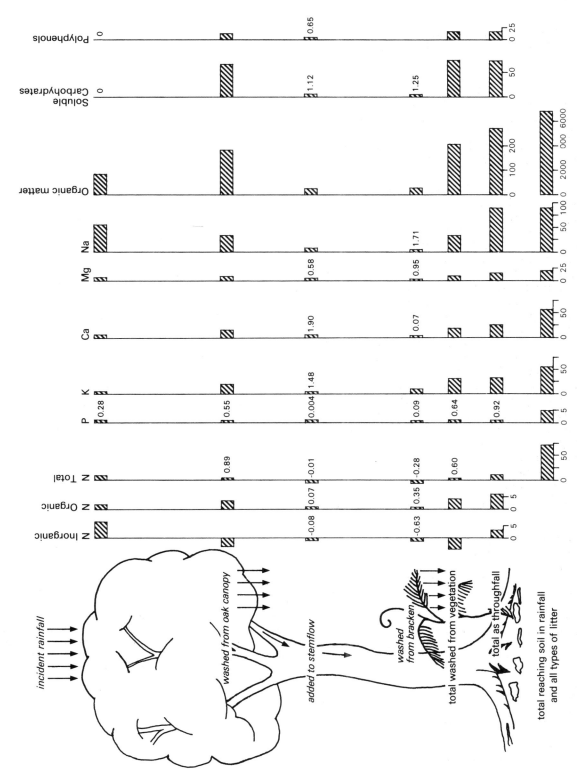

Fig. 21.3 Nutrients washed from vegetation and leaching the soil surface in rainwater and litter in a sessile oakwood (*Quercus petraea*) (kg ha^{-1} a^{-1}) (after Carlisle *et al.*, 1967).

the vegetation, it also responds to variations in environment. Within any one ecosystem the litter supply will vary, not only spatially with variations in canopy density and composition but also temporally both in type and quantity in accordance with the life cycle of the community (Fig. 21.2).

In addition to the litter components, the soil surface also receives organic compounds dissolved or suspended in water that has been intercepted by the vegetation canopy. This rain is, of course, not pure water: it will already contain inorganic material. That portion of the rainfall which reaches the ground directly through gaps in the canopy and as drip from leaves and stems is known as **throughfall**, while the portion that converges to run down the stems is termed **stemflow**. Both components are involved in a complex exchange system between the canopy and rainfall. Although the canopy may gain water and mineral nutrients through foliar absorption, there is often a net loss of nutrients leached from the leaves, as well as substantial losses of soluble carbohydrates. The rain reaching the soil, therefore, is considerably enriched (Fig. 21.3) and differs from pre-interception rainwater in relative composition.

Beneath the soil surface the death of roots represents a supply of organic matter to the soil detrital system *in situ*. In theory, it should be possible to measure both the biomass and production of these dead roots, but in practice it proves extremely difficult. Nevertheless, it seems probable that large amounts of root tissue become available for decomposition. Even before death, however, live roots are a source of organic compounds in the form of root exudates, root-cap cells, moribund root hairs and epidermal and cortical cells. In some soils, perhaps half the dry weight of lateral roots might be shed in this way.

Although all of this material represents the primary substrate for the production of humus, it is also the food supply or energy source that sustains the final contributor to the organic matter in soils: the organisms of the detrital food chain (Box 21.1, Fig. 21.4). On their death these soil organisms represent, like the roots of plants, a direct *in-situ* supply of organic detritus. The same is true of the metabolites and faeces they release into the soil during life. Their overwhelming importance to the functioning of the system, however, is due more to their role as the principal activating mechanism of the decomposition process. Without soil organisms the breakdown of

organic detritus would be very much slower and far less complete. The importance of this is clearly evident in environments unfavourable to large and diverse populations of soil organisms, such as the arctic tundra, where largely undecomposed litter accumulates on the soil surface.

21.3 Decomposition: a process–response model

Although the model depicted in Fig. 21.5 is similar to that of the weathering system (Fig. 11.2), and in both the active processes are those of fragmentation and the breakdown of chemical structure, the decomposition system is much more complex. The reasons are twofold. In the first instance, the input of dead organic tissue – i.e. the initial state of the system (see Fig. 2.3) – is chemically and structurally more complicated than the equivalent rock and mineral input to the weathering system. Secondly, the role of soil organisms makes decomposition more difficult to model, for in addition to the inorganic processes of weathering, decay involves extremely complex biochemical reactions, many of which are still little understood. Furthermore, and even more so than was the case with the weathering system, the sequential arrangement of these processes is merely an artificial device to aid the clarity of explanation. In the real world they may be proceeding simultaneously; the interactions of populations of the various soil organisms involved and the complementary nature of their roles in decomposition can be very intricate.

21.3.1 Fragmentation: the mechanical comminution of litter

The fragmentation of litter is activated in part by purely physical processes such as raindrop impact, expansion and contraction and freeze/thaw. Indeed, some of these processes may have started before the litter reached the soil surface. With the onset of senescence, for example, the moisture content of leaves decreases and they become dry, brittle and susceptible to abrasion as a result of movement by wind. Once on the soil surface, trampling can further break up large fragments of litter. By far the most important process activating fragmentation, however, is the maceration and comminution of litter by soil detritivores. These are soil organisms which consume particulate organic detritus, and include members of several

Box 21.1
SOIL ORGANISMS

There are several bases for the classification of soil organisms, ranging from their habitat preferences and activity to their trophic relationships, but perhaps the most often used schema is based on size.

micro	20–200 μm
meso	200 μm–1 cm
macro	>1 cm

The soil microfauna is made up mainly of **protozoa** – uni- or non-cellular animals, including many motile forms with cilia or flagella. There are also many **amoeboid** forms, sometimes with chitin or silica sheaths, and some of these can encyst and survive desiccation for long periods. A third group of protozoa includes the **rotifers**, which have a more differentiated cell structure and which characteristically construct mucous nests from soil particles and faeces. Protozoa are very numerous in soil, with figures of tens of thousands per gram of soil being quite common.

The mesofauna contains the smaller types of worms such as the **nematodes** (eelworms) and **turbellarians**, both unsegmented worms, 1 mm long and 2 μm in diameter, which inhabit soil water. They are very numerous in certain soils (several million per hectare) and feed on decaying organic matter and fungi, but many are parasitic and infect plant roots. Slightly larger are the **enchytraeids**, or potworms. These are small segmented worms common in acid and organic soils. Figures of 200 000 m^{-2} have been recorded from heathland and coniferous woodland soils. They feed on algae, fungi, bacteria and organic matter in various states of decay. They are particularly significant in acid soils, where they replace earthworms.

Two important groups of small arthropods occur in the mesofauna. These are the **Acari** or mites, common in acid litter and organic matter, where they can constitute up to 80% of the soil fauna. Their diet varies greatly, but though some consume litter, many feed on fungal hyphae and spores. Similar food preferences are shown by the **Collembola** or springtails, also arthropods, though some are carnivorous. The **myriapods** include two important groups of soil mesofauna, the vegetarian millipedes (**Diplopoda**) and the active predatory centipedes (**Chilopoda**). Smaller forms of several groups of animals such as ants, beetles, molluscs and particularly the larval stages of many insects also fall into the category of mesofauna. The main role of the group as a whole is to function as activating mechanisms in the fragmentation of litter by consuming plant detritus and its attached bacteria and fungi, reducing it to colloidal dimensions, providing a substrate for humification and moving it deeper into the soil.

On size grounds alone some of the larger members of groups considered above, such as the beetles, are placed in the soil macrofauna. However, the most important members of the macrofauna are the earthworms. These large segmented or **annelid** worms belonging to the Lumbricidae are extremely significant both in the decomposition of litter and in the mixing of mineral and organic components of the soil. There are 25 British species, of which only ten are common and, although some species can tolerate mild acidity, earthworms are rare in soils with a pH value below 4.5 and under anaerobic conditions. They ingest both litter and soil, so that the comminuted organic fragments and the soil minerals are intimately mixed in their gut, promoting the formation of organomineral complexes. In addition, by secreting calcium from a calciferous gland, the pH of these complexes is raised and they are egested as water-stable crumbs forming worm casts.

The only other members of macrofauna that will be mentioned here are the **Isopoda** or woodlice. These are particularly common in dry and acid litter and soil organic horizons, where they replace earthworms as litter destroyers.

The soil microflora is dominated by **bacteria**

and **fungi**, though their inclusion in the flora is more convention than anything else, for perhaps neither group should be described as plants. The bacteria are 1–5 μm across, and they occur in various shapes, while some are active forms with cilia or tufts of flagella. Many secrete polysaccharide gums, perhaps to form a protective sheath which binds and holds small clay and iron oxide particles, these gums are of great significance in soil aggregation. It is their activity that determines their grouping into aerobes and anaerobes, autotrophic, chemosynthetic and heterotrophic groups, while many play important roles in the cycling of elements such as nitrogen and sulphur, some developing symbiotic roles with the roots of higher plants. However, in decomposition their importance lies in their role as efficient decomposers and in the promotion of humification and the resynthesis of humic polymers.

The fungi, which can be very numerous, have two important effects. One is mechanical, as fungal hyphae and mycelia develop and push through decaying litter, aiding fragmentation. The other consists of complex chemical effects, as they secrete enzymes which digest organic matter. In acid soils they are often more important and more effective than bacteria, but even on the better soils they may be as important as the bacteria. Some fungi are symbiotic, forming **mycorrhizae** with the roots of species such as the Scots pine. Closely related to both the fungi and the bacteria are the **Actinomycetes** with their fine, branching filaments similar to, but finer than, fungal mycelia. They are aerobic organisms and can survive in dry soils. Most are saprophytic and some, such as **Streptomycetes**, produce antibiotic substances.

The final group of soil microflora is the **algae**. Blue-green algae are most numerous in neutral soils, green algae in more acid environments. Both occur as single-celled microorganisms, as colonies or as filaments. Because they contain chlorophyll and photosynthesize, most occur near or at the surface and, of course, some are nitrogen fixers of great significance.

groups of invertebrate animals. Perhaps the most important are the segmented worms (Annelida), including earthworms and enchytraeid worms, millipedes (Myriapoda), insect larvae (Insecta), mites (Acari), springtails (Collembola) and woodlice (Isopoda) (see Box 21.1). Of course, this digestion by detritivores also involves biochemical change, for the organism digests and assimilates some constituents of the detrital tissue. These are usually only the simpler constituents, for the enzyme systems of these soil invertebrates are apparently incapable of attacking more complex structural molecules such as cellulose, lignin and phenolic complexes. As a consequence, the assimilation efficiencies (food assimilated as a percentage of that consumed) of these organisms are low: 1–3% in earthworms, 6–15% in millipedes, 10% + in mites, and 15–30% in woodlice. The net effect, therefore, is the physical comminution of the detritus until much of its original character is lost, it is reduced to colloidal dimensions and its total surface area is vastly increased. In this form it is returned to the soil incorporated in faecal pellets, where it may be intimately mixed with inorganic soil particles. These pellets have a better moisture/aeration status than the original litter and a higher nitrogen content as a result of the digestive release of nitrogen (mainly as ammonia) from organic compounds. Such faecal material forms an ideal substrate for microbial activity.

21.3.2 Structural breakdown and microbial decomposition

The initial processes of chemical change are the straightforward inorganic reactions of oxidation, hydration, hydrolysis and solvation, as the dead organic material is exposed to the atmosphere, to light and to water. Such reactions and the release of readily volatilizable and soluble products (*ca.* 70% of the K and Na in litter is leached into the soil during the first 2–3 months after deposition) are aided by the initial mechanical break-up of the litter. The majority of the litter and detritus, however, consists of complex structural molecules, and the unaided progression of these inorganic reactions would be slow and many of the original compounds would persist. This does not happen, largely because of microbial decomposition which may precede, run parallel with or follow, the initial

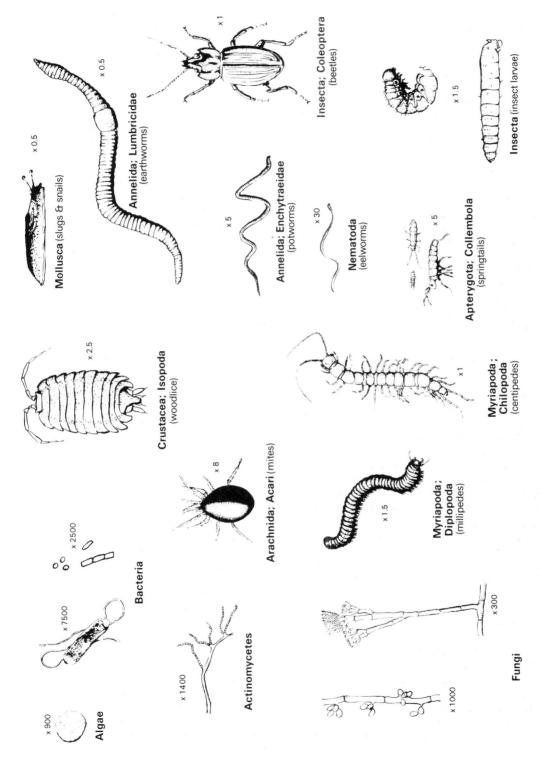

Fig. 21.4 The major groups of soil organisms (mostly detritivores), and decomposers and some of the carnivorous groups dependent on them.

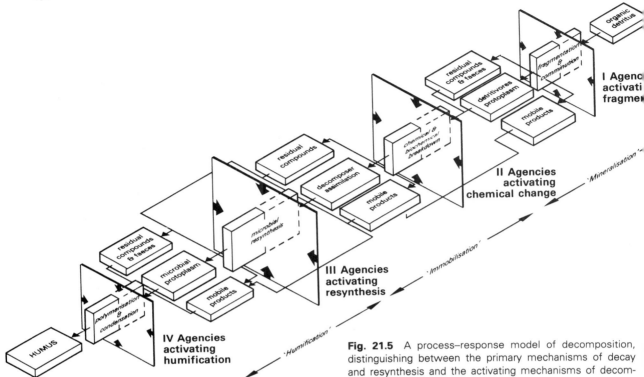

Fig. 21.5 A process–response model of decomposition, distinguishing between the primary mechanisms of decay and resynthesis and the activating mechanisms of decomposition (compare with Fig. 11.2).

weathering, leaching and fragmentation by detritivores. It is these decomposer microorganisms (mainly bacteria and fungi) which permit the detrital substrate to be reduced to a small but highly resistant residue, or to be completely decomposed.

The primary mechanism of microbial decomposition is the enzymatic cleavage of the structural macromolecules of organic detritus. These molecules are long polymers with high molecular weights and are large relative to the size of microbial cells. To utilize the molecules or their components as food, the microorganisms secrete extracellular degradative enzymes, either on the surface of their cell walls or 'free' into the soil. Here they may be adsorbed onto the organic substrate, or on the surfaces of clay minerals.

These extracellular enzymes are highly specific, not just to the substrate on which they act but to particular linkages. They are often sensitive to the presence of particular functional groups on the molecule, and some can even distinguish between different isomers of the same substance. Many of the enzymes are inductive – that is, they are produced only in the presence of particular substrates. The products of these enzymatically catalysed reactions are the constituent monomers which made up the original polymer, or groups of such units. These are now available to be transported across the microbial cell wall, usually by other carrier enzymes, again highly specific, before they can form a respiratory substrate for the organism. Such products of enzymatic reactions are also available to other soil organisms which do not possess the enzymes necessary to release them and, of course, they are also available to be leached by percolating water (see Chapter 11).

The complexity of these processes has certain important implications. First, any one organism is unlikely to produce the many different enzymes necessary to degrade the large variety of molecules present in the natural substrate. Complete breakdown, therefore, will depend on a succession of microorganisms, each contributing particular enzymes and catalysing particular reactions. Secondly, the susceptibility of an organic polymer to degradation is inversely proportional to its heterogeneity. In other words, the more diverse its composition and the more diverse the bonds and linkages involved, then the less likely it is to be successfully degraded by microbial enzymes.

Changes in the biochemistry of decomposing leaf litter suggest that components disappear in the following order: soluble sugars, hemicellulose, cellulose and finally lignin (Burges, 1958) (Fig. 21.6). This supports the view that a succession of microorganisms colonizes the substrate in waves, each wave capable of decomposing a particular component. So sugar fungi and other organisms utilizing simple carbohydrates are replaced in time by decomposers of cellulose or comparable polysaccharides, and finally by decomposers of lignin. This successional concept is not entirely simple, for it is related not only to the physical and nutritional status of the substrate but also to differential growth rates of the species concerned. Nevertheless, such schemes are useful (Table 21.2) and the soil animals can be incorporated in them, for not only does their comminution of detritus pave the way for microbial colonization but some microbial decomposition of, for example, leaf cuticles may be necessary for detritivores to gain access to ingestible tissue. Furthermore, some detritivores possess bacterial gut floras which secrete cellulase, the degradative enzyme of cellulose, and therefore cannot be clearly separated from other microbial activity (see Chapter 20).

Table 21.2 Sequence of utilization of substrate and waves of decomposers in the breakdown of plant litter (after Garrett, 1981 and Frankland, 1966)

Implied succession	Decomposer colonists	Substrate utilization
1st stage (0–1 year)	parasites and primary colonizers including fungi and prokaryotes*	sugars utilized and other simple carbohydrates (incorporated into fungal mycelia)
2nd stage (1–2 to 3 years)	cellulolytic soil saprophytes including soil fungi, protistans, soil invertebrates and prokaryotes*	cellulose decomposition (excess simple carbohydrates produced are used by non-cellulolytic fungi)
3rd stage (2 to 3–5 to 6 years)	lignin-decomposing saprophytes, fungi and prokaryotes*	lignins, tannins and other complex compounds, later chitin of fungal cell walls utilized

* Bacteria and/or actinomycetes

21.3.3 Resynthesis: the final state of the system, humification and the overall effects of decomposition

The overall effects of decomposition can be summarized as the disappearance of litter and detritus and the associated release of CO_2, H_2O and mineral nutrients (**mineralization**). These phenomena are accompanied by the appearance of decomposer (particularly microbial) protoplasm and the **immobilization** of some of these nutrients. Also associated with these events is the appearance of a residuum of organic compounds largely resistant to further breakdown. This is the process of **humification**, and the residual material is **humus**. Humus is a mixture of complex compounds which are variable in composition and amorphous, lacking crystallinity. These properties have ensured that, in spite of continuing research efforts extending over 100 years, the precise nature of soil humus is still obscure. Indeed, it has been suggested that no two molecules of humus may be exactly alike (Fledgmann and George, 1975).

We can, however, arrive at some broad conclusions regarding the origin, composition and properties of humus. It appears to consist of

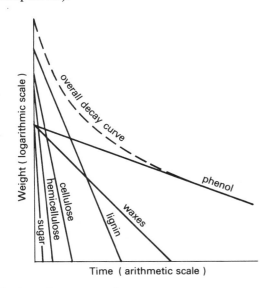

Fig. 21.6 Decay curves for the principal constituents of plant litter (after Minderman, 1968).

insoluble heterogeneous polymers which have been produced within a microbial cell, either when alive or from its protoplasm on its death. These polymers form particles of colloidal dimensions ($<2\ \mu m$) which are relatively stable and comparatively resistant to acid hydrolysis and to further microbial decomposition because of their heterogeneity relative to microbial enzymes. Humic polymers are believed now to result from reactions between polyphenolic compounds of both plant and/or microbial (especially fungal) origin with amino acids, peptides and protein; also from plant and/or microbial sources. Humus may also contain some of the more resistant polysaccharides from litter, but much of this carbohydrate material in humus is derived from microbial cell walls (e.g. chitin from fungal cell walls) and from bacterial extracellular gums.

Colloidal humus particles have a vast surface-area/volume ratio and considerable ion exchange properties. Like clay minerals, they carry a net negative charge, but because of the dissociation of hydrogen ions from the OH (hydroxyls) of the carboxylic and phenolic active groups under different conditions of soil acidity this charge is pH-dependent (Fig. 21.7). These humic particles

SYSTEM

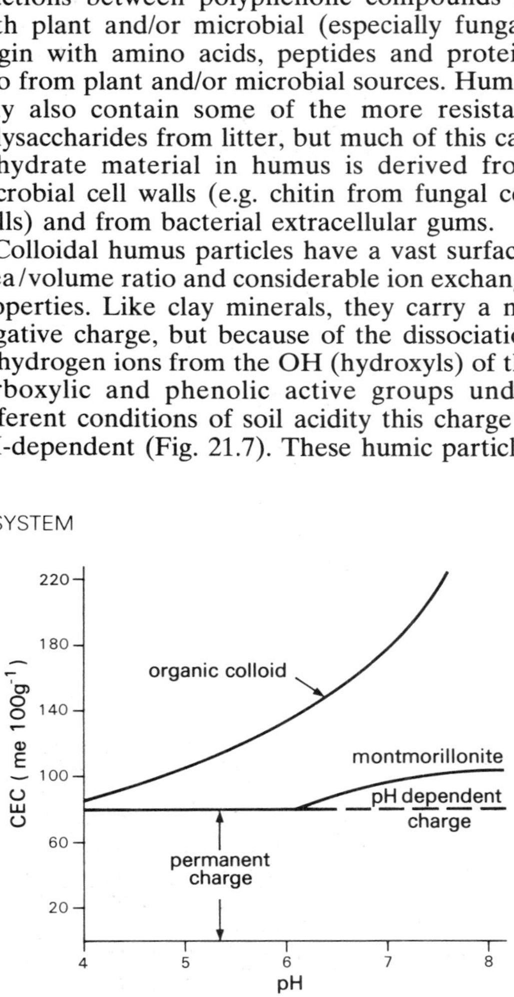

Fig. 21.7 Cation exchange capacity (CEC) plotted against pH for an organic colloid and the 2:1 lattice clay mineral, montmorillonite. The CEC that reflects the number of negatively charged adsorption sites on the organic colloid is pH-dependent throughout the soil pH range. (Note CEC units, me = milliequivalent – a quantity chemically equivalent to 1 mg of hydrogen) (after Buckman and Brady, 1974).

and clay minerals together make up the major part of the colloidal size fraction of the soil, although under most circumstances they exist not as separate particles but as intimately associated **organomineral complexes**. Here the humic colloid is adsorbed on the surface of the clay mineral, and perhaps within its lattice, by various forces (such as hydrogen bonds, coordination to exchangeable cations, or van der Waal's forces) which overcome the inherent electrostatic repulsion of the net negative charge carried by each particle. The importance of these organomineral complexes to the retention of water and mineral nutrients in the soil cannot be overstated. Together with immobilization in microbial tissue, ion adsorption by colloids constitutes the most important regulator of leaching loss to the solute throughput of the denudation system and of nutrient availability to higher plants.

21.4 Decomposition: a trophic model and the pathway of energy flow

The detritivores and decomposers considered above as activating mechanisms in the process–response model of decomposition obtain the energy for their metabolism from dead organic matter. They either ingest this dead material whole or absorb food molecules and nutrients previously made available by the decomposer activity of microorganisms, a process that may take place outside or inside the digestive system. There are, however, other soil organisms that prey on living detritivores and decomposers, and those that feed on their dead remains. These constitute secondary grazing–predation and secondary decomposer systems respectively.

Many groups of soil animals – for example the mites – graze either selectively or indiscriminately on bacterial cells and fungal hyphae. Some of the larger soil invertebrates, such as the centipedes, are voracious carnivores, preying on mites, springtails and nematodes. These trophic relationships are further complicated by antagonistic and parasitic relationships, particularly those involving fungi. Although it is possible to construct trophic dynamic models of these detritivore–decomposer–carnivore food webs (Fig. 21.8), they prove to be extremely complex to unravel and as yet we know very little in detail of the way energy is partitioned and transferred between the different trophic

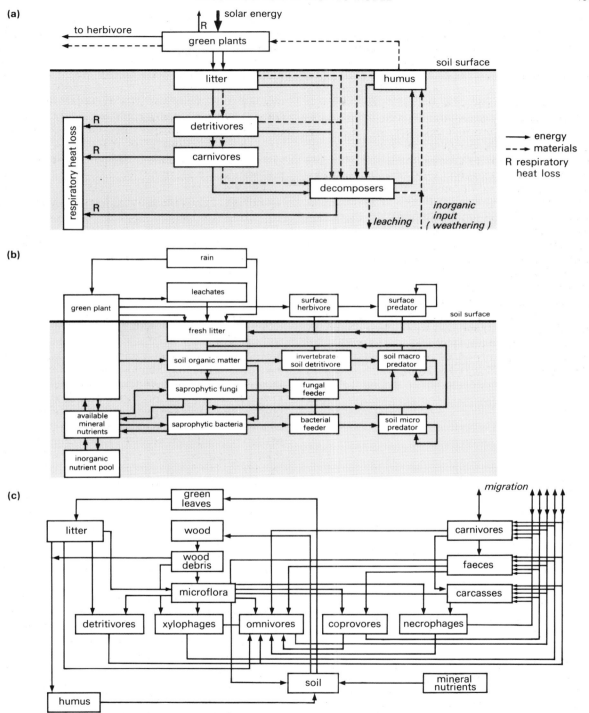

Fig. 21.8 Three attempts to depict the trophic relationship of soil organisms and the pathways of matter and energy transfer in the detrital system: (a) after Wallwork, 1970, (b) after Fortescue and Martin, 1970, and (c) after Edwards *et al.*, 1970.

levels. Nevertheless, some important conclusions can be made.

From the point of view of the process–response model of decomposition developed above, decay is a process of mineralization or release of the chemical elements present in the detritus, and their conversion to a more oxidized state, ultimately as simple inorganic compounds such as carbon dioxide. The detritivores and decomposers, however, and the carnivores dependent on them, assimilate organic compounds originally derived from detritus and take in mineral nutrients, both as constituents of these compounds and from soil water. Using respiratory energy they synthesize from these components the constituents of their own protoplasm. So, from the point of view of the trophic model of decomposition, decay is a process of fixation or immobilization of these same chemical elements in the protoplasm of the organisms of the detrital food web. (Indeed some of these organisms can use their respiratory energy to fix dinitrogen gas from the soil atmosphere.) All of these elements are, therefore, effectively immobilized, albeit temporarily, and are unavailable either to the roots of higher plants or to be lost by leaching. They only become available with the demise of the decomposers concerned and the digestion or breakdown of their protoplasm. Here, then, is a further way in which the decomposers are seen to act as a regulator on the rate of turnover of nutrient elements.

In thermodynamic terms, the organisms of the detrital food web decrease their internal entropy as they develop, grow and reproduce, sustained by the flow of chemical energy and materials from the litter. In accordance with the Second Law of Thermodynamics, however, the decomposing litter increases in entropy as it is degraded to simpler constituents and energy is dissipated as heat – think of the high temperatures in the centre of a compost heap – during the exothermic reactions of decay and by the respiration of the detrital population.

21.5 Decomposition: a pedological model

In Chapter 11 the weathering system was modelled as a profile through the weathered mantle or regolith. Similarly, decomposition can be seen in the same terms, with different soil layers or horizons corresponding to different stages in decomposition. Surface horizons in closer proximity to the input of litter will be dominated by relatively fresh organic remains. Conversely, lower horizons will show progressively more advanced alteration, more humified material and more mixing of humus with mineral soil and regolith. Following this gradient in the state of the detritus are parallel gradients in physicochemical conditions, in populations of soil organisms and in their habitat conditions. It must be noted, however, that in certain circumstances such distinctions may be

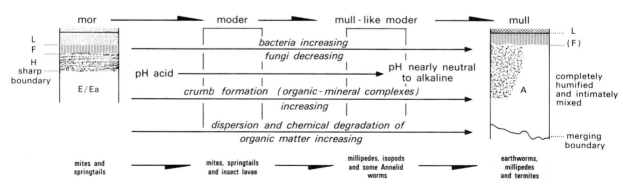

Fig. 21.9 The pedogenic model of the detrital system and its expression in the soil profile as a series from mull to mor organic horizons shown in relation to the controlling gradients of soil conditions and changes in the dominant groups of soil organisms (partly after Wallwork, 1970).

obliterated by the mixing activities of the larger soil animals.

The extent to which this horizonation is developed will depend on the nature of the initial input of litter, on the properties of regolith and mineral soil and, above all, on the climatic environment of soil formation. These factors will regulate the number, type and activity of the soil population, which in turn will regulate the rate of decomposition and nutrient turnover. Together all of these factors will influence the state of the soil profile (Fig. 21.9, see Chapter 22).

An understanding of the detrital system is important to mankind in several ways, apart from its obvious contribution to soil fertility. Even here, however, modern farming has opted to reject so-called 'organic farming' in favour of massive applications of fertilizer. Loss of organic matter affects soil structure detrimentally and reduces its capacity to retain these fertilizers. The result – a vicious circle. Society also uses the decomposition process directly in the treatment of sewage and increasingly in the composting of domestic refuse. The effluents released from sewage works, however, are still rich in nutrients, particularly nitrates and phosphates. These, together with fertilizers leached from agricultural soils, have been responsible for many enrichment or eutrophication problems in rivers and lakes, with their associated algal blooms. Conversely, the overloading of natural detrital systems with untreated waste allows the oxygen demand of decomposer organisms to depress oxygen levels to a point where vegetation and fish die of asphyxiation.

Further reading

Accessible introductory texts are:

Jackson, R.M. and F. Raw (1966) *Life in the Soil*. Edward Arnold, London.
Mason, C.F. (1977) *Decomposition*. Edward Arnold, London.
Richards, B.N. (1974) *Introduction to the Soil Ecosystem*. Longman, Harlow.

More advanced treatments will be found in:

Burges, A. (1958) *Micro-organisms in the Soil*. Hutchinson, London.
Burges, A. and F. Raw (1967) *Soil Biology*. Academic Press, London.
Dickinson, C. and G. Pugh (1974) *Biology of Plant Litter Decomposition*. Academic Press, London.
Kononova, M.M. (1966) *Soil Organic Matter*, (2nd edn). Pergamon, Oxford.
Swift, M.J., O.W. Heal and J.M. Anderson (1979) *Decomposition in Terrestrial Ecosystems*. (*Studies in Ecology, vol 5*). Blackwell, Oxford.
Wild, A. (ed) (1988) *Russell's Soil Conditions and Plant Growth*, (11th edn). Longman, London. (especially Chapters 14–19).

Particular groups of soil organisms are dealt with in:

Griffin, D.M. (1972) *Ecology of Soil Fungi*. Chapman & Hall, London.
Wallwork, J.A. (1970) *The Ecology of Soil Animals*. McGraw-Hill, Maidenhead.

Two International Biological Programme handbooks are useful introductions to techniques used in studying soil organisms:

Parkinson, D. *et al.* (1971) *Ecology of Soil Micro-organisms*. IBP Handbook No. 19. Blackwell, Oxford.
Phillipson, J. (1971) *Quantitative Soil Ecology*. IBP Handbook No. 18. Blackwell, Oxford.

The soil system

22.1 Defining the soil system

We have so far encountered the soil in several different contexts. For example, its relationship with the regolith was discussed under the heading of the weathering system in Chapter 11. The decomposition of organic matter was seen to take place largely on or within the soil, which was also seen to form the habitat of the organisms of the detrital system in Chapter 21. We also recognized, in Chapter 12, that the soil is intimately associated with the movement of water and debris in the operation of denudation processes on slopes, but at the same time it was seen in Chapter 19 to function as a reservoir of water and mineral nutrients exploited by plant roots. Clearly, the soil has been a part – either explicit or implicit – of many environmental systems that we have considered, and has been involved in the cascades of matter and energy with which they are concerned (Fig. 22.1). But can we define the soil as a component system of such cascades in its own right?

22.1.1 Modelling the soil: its three-dimensional organization

From our experience of the soil, we know that it is a three-phase system: solid, liquid and gaseous. The solid phase is in part made up of inorganic material, ranging from clastic fragments of largely unaltered rock, through disaggregated mineral grains actively undergoing weathering, to the secondary minerals derived from weathering products. As we saw in Chapter 11, this inorganic fraction occurs in a range of size classes and the

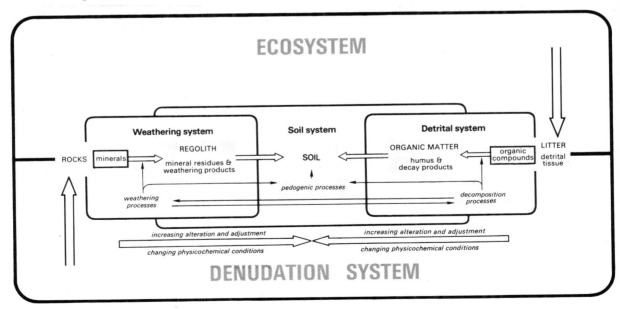

Fig. 22.1 The relationships between the soil system and other systems in the environment.

relative proportions of these size categories define the texture of the soil (Fig. 22.2). The second component of the solid phase consists of organic matter in various states of decay, the humic substances resynthesized from the products of decay (and perhaps also the populations of soil organisms directly or indirectly involved in decomposition). The solid phase of the soil, however, does not usually consist of discrete particles, whether organic or inorganic. More usually these particles are intimately associated to form **soil aggregates** or **peds**. They result from the forces of attraction between soil particles and involve the formation of hydrogen and ionic bonds, and may be further promoted by the wetting and drying of the soil, by the pressures exerted by developing roots and by the presence of polysaccharide gums secreted by the soil microflora (Fig. 22.3). Between these soil aggregates and within them there are voids of various kinds. A network of micropores

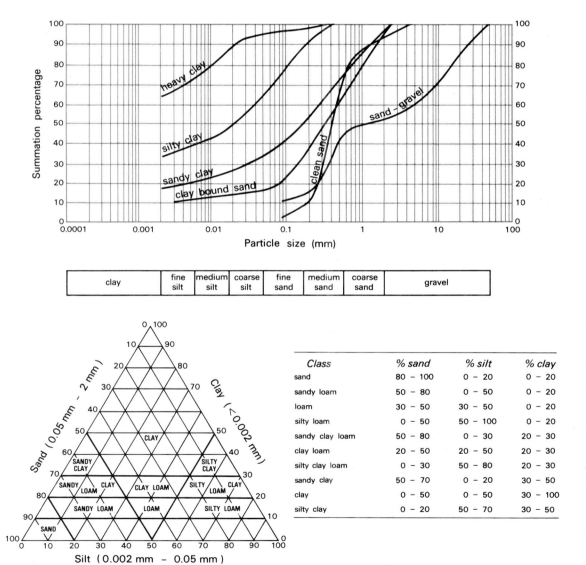

Fig. 22.2 Soil texture.

Class	% sand	% silt	% clay
sand	80 – 100	0 – 20	0 – 20
sandy loam	50 – 80	0 – 50	0 – 20
loam	30 – 50	30 – 50	0 – 20
silty loam	0 – 50	50 – 100	0 – 20
sandy clay loam	50 – 80	0 – 30	20 – 30
clay loam	20 – 50	20 – 50	20 – 30
silty clay loam	0 – 30	50 – 80	20 – 30
sandy clay	50 – 70	0 – 20	30 – 50
clay	0 – 50	0 – 50	30 – 100
silty clay	0 – 20	50 – 70	30 – 50

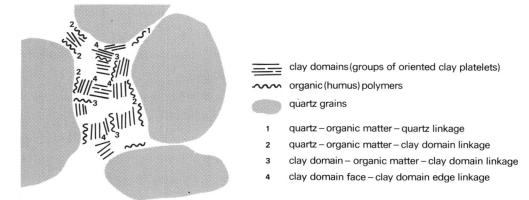

clay domains (groups of oriented clay platelets)

organic (humus) polymers

quartz grains

1   quartz – organic matter – quartz linkage

2   quartz – organic matter – clay domain linkage

3   clay domain – organic matter – clay domain linkage

4   clay domain face – clay domain edge linkage

Fig. 22.3 The Emerson model of soil aggregation.

2–20 μm in diameter) lies between individual particles or between microaggregates within the peds, while in clay soils these micropores may be as narrow as 10 nm. Between major soil aggregates, macropores (7200 μm in diameter) occur as well as larger voids and cracks (Fig. 22.4). Sometimes interconnecting tubular passages, probably biotic in origin, may also occur and their significance has already been alluded to in Chapter 12, where they were referred to as soil pipes. Together, the kind, size and distribution of soil aggregates and soil voids and pores determine the structure or fabric of the soil.

The liquid and gas phases are represented by the soil water and air which occupy the pores and voids. We have already seen that water entering the soil by infiltration is subject to certain forces which determine the soil water potential (ψ_s) (Box 12.2). Some components of soil water potential such as the matrix (ψ_m) and osmotic potentials (ψ_π) are due to retention forces and are responsible for holding water in the partially air-filled pores above the water table. Immediately following precipitation and below the water table, the pores may be saturated, being entirely water-filled. Soil water, however, is not pure water. As we know, it contains a variety of substances in solution, particularly various cations and anions which are potentially available as plant nutrients. These solutes are derived from a number of sources. Some are already present in rainwater, some originate in the vegetation canopy, while others have passed into solution in the soil as a result of weathering, decomposition and ion-exchange

processes. Indeed, soil water is often referred to as the soil solution. However, some of the soil constituents themselves, although not in true solution, are so small (<2 μm) that they may occur as a colloidal suspension. From the point of view of soil chemistry, it is common to regard the whole of the solid and liquid phases as behaving as a soil–water suspension. Most prominent amongst the colloidal fraction are the hydrated aluminium silicate clay minerals, the humic polymers and the hydrated amorphous oxides and hydroxides of iron, manganese, aluminium and titanium (see Chapter 11, Box 11.2 and Chapter 21).

When the pores are not full of water they are partly or wholly air-filled, but the composition of this soil atmosphere may differ significantly from the free atmosphere above. Furthermore, these differences may increase with depth and fluctuate with time. The main difference is that oxygen concentrations are somewhat lower than in the atmosphere, and carbon dioxide concentrations are significantly higher (Table 22.1). Both fluctuate, however, and depend on the respiratory demand for oxygen and the rate of carbon dioxide evolution by roots and soil organisms. The relationship is complicated by the very different diffusion coefficients of the two gases in air and water, and by their solubilities in water.

The soil is a three-dimensional natural body, occurring at the surface of the Earth, reaching vertically to about the lower limit of root penetration and extending laterally as a component of the landscape (Fig. 22.5). However, solid, liquid

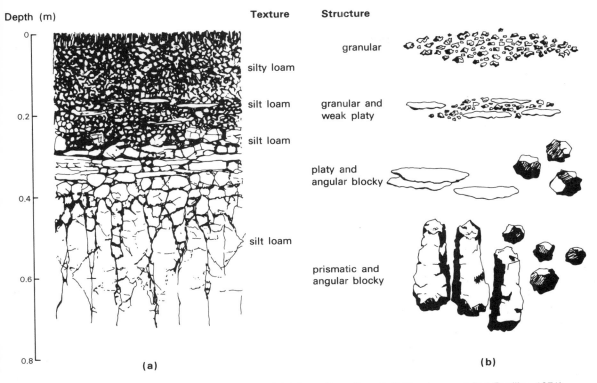

Fig. 22.4 Soil structure (modified from Agricultural Advisory Council on Soil Structure and Soil Fertility, 1971).

Table 22.1 The oxygen and carbon dioxide composition as percentage by volume of the gas phase of well aerated soils, compared with that of dry air (modified from Russell, 1973)

	Oxygen (%)	Carbon dioxide (%)
dry air	20.95	0.03
arable soil		
fallow	20.7	0.1
unmanured	20.4	0.2
manured	20.3	0.4
manured sandy soil cropped with potatoes	20.3	0.6
pasture	18–20	0.5–1.5

and gas phases do not form a random mosaic in these three dimensions but are more or less organized to impart a definite vertical and lateral structure to the system. The properties of the soil solids, and of the voids and their associated soil water and soil air, vary both vertically and laterally. Layers are differentiated vertically within the soil body and differ in the relative proportions of the soil-forming materials and in their characteristics, so that each layer presents a different set of physical, chemical and biological attributes. These sets of attributes reflect the processes operating, or which have operated, in the particular soil layer concerned, but they also form the conditions, or environment, under which contemporary processes take place. This vertical organization of the soil is referred to as the **soil profile** and the layers as **soil horizons** (Fig. 22.6). Laterally, too, there is some semblance of organization, for the pattern of horizonation and the characteristics of the horizons vary laterally in a largely predictable manner in response to the soil's position in a landscape,

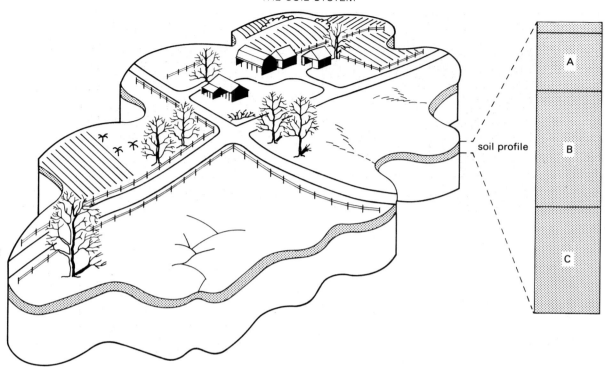

soil profile

A

B

C

Fig. 22.5 Soil as a component of landscape (adapted from R.W. Simonson, *Soil Sci. Soc. Am. Proc.* **23**, 152–156 by permission of the publishers).

particularly its position relative to slope (see Fig. 22.13).

Indeed the extension of the concept of the soil profile into a three-dimensional unit at a number of scales has been formalized in the literature of soil science by notions such as the **pedon**, the **polypedon**, and the **soilscape** (Hole, 1978; Hole and Campbell, 1986). The former is simply an extension of the two-dimensional soil profile into a volumetric unit at the same scale and usually of fixed dimensions (e.g. 1 m^2 × 2 m, Van Wambeke, 1966). Individual soil bodies such as that represented in Fig. 22.5 are, therefore, **polypedons**, and it is the combined pattern formed by the clustering of these larger spatial units which constitutes the **soilscape** (multipolypedonic units) at the landscape scale. There are of course great similarities in these pedological approaches to landscape organization with those associated with *landscape ecology* as discussed in Chapter 18. Size, shape and orientation are terms applied to polypedons as they are to landscape patches. The

soil scientists also talk of a landscape matrix (*L*-matrix) in a way analogous to landscape ecologists, though some of the terminology used by soil scientists derives from that employed in soil micromorphology (e.g. landscape plasma, see Fig. 22.8) (Brewer, 1975). The concept of the soilscape also has functional significance and facilitates the consideration of the operation of soil processes in the landscape (Huggett, 1975), and in conjunction with landscape ecology forms a valuable framework for studies of soil erosion in agricultural landscapes (Farres *et al.*, 1991) (Fig. 22.7).

22.1.2 A process model of the soil

The three-dimensional organization of the soil system, and particularly the presence of horizons, is the result of a complex suite of processes. Some of these processes have been treated explicitly in other chapters, but their role within the soil system means that they can all be considered as pedogenic processes and contribute to pedogenesis or soil formation. In detail the pattern, type and degree

fresh litter deposited during previous litter fall L

partly decomposing litter remaining from earlier periods of litter fall, original plant structures visible F

well decomposed litter, original structures cannot be seen H

mineral horizon with incorporation of humidified organic matter as a result of biological activity Ah where mixed by cultivation – known as Ap and where gleyed as Ag

subsurface horizon with less organic matter and/or clay and/or iron than immediately underlying horizon – a result of leaching and translocation prominent in podsols as bleached Ea horizon absent in chernozems and brown earth soils

transitional horizons with characteristics of both A or E and B horizons

mineral subsurface horizon characterised by in situ alteration (weathering) of original material – decalcification; formation, or residual accumulation of clay or oxides; formation of large scale peds (structure) Bw or (B) and/or the illuvial concentration of clay Bt, sesquioxides Bs, humus Bh, or the formation of a cemented pan by the accumulation of iron, aluminium and organic matter Bf. B horizons may be gleyed Bg

weakly consolidated little altered mineral horizon that retains rock structure or otherwise lacks properties characteristic of A, E and B horizons

C horizons Ck (Cca) and Cy (Ccs) are layers of accumulated calcium carbonate and sulphate found in some soils

C horizons may be periodically gleyed, Cg or intensely gleyed, G as in ground water gley soils

very hard bedrock is known as R horizon

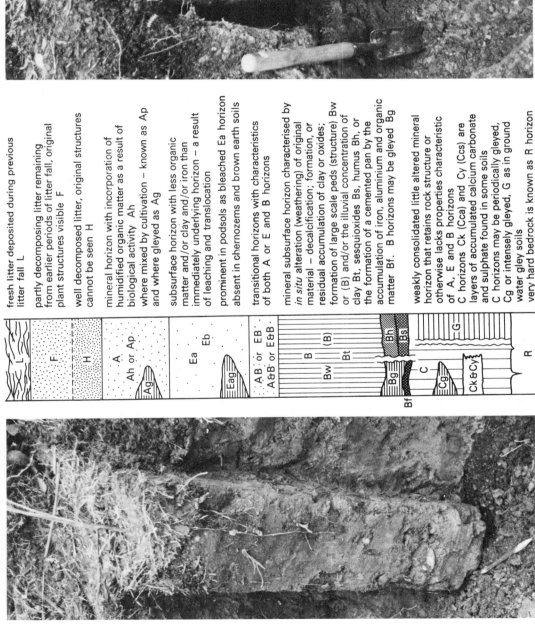

Fig. 22.6 The soil profile and soil horizons. Note that the recognition and description of horizons is not as straightforward as the schematic diagram would suggest, for, as shown in the photographs, the boundaries often merge. Note also that all of the horizons shown in the diagram would never occur together in the real world.

A   Run-off generation area
B   Ephemeral gully feature
C   Colluvial build-up of material and concentration of flow
D   Break through point at hedge bank,
     flow directed down to dry valley
E   Deep meandering gully form
F   Remnants of old hedge boundary
G   Widening of gully form, major flow lines dictated
     by linearities of tractor wheelings
H   Severe head cut forms at cascade into sunken lane
I   Major sunken lane feature
J   Flow of water and sediment potential off site drainage

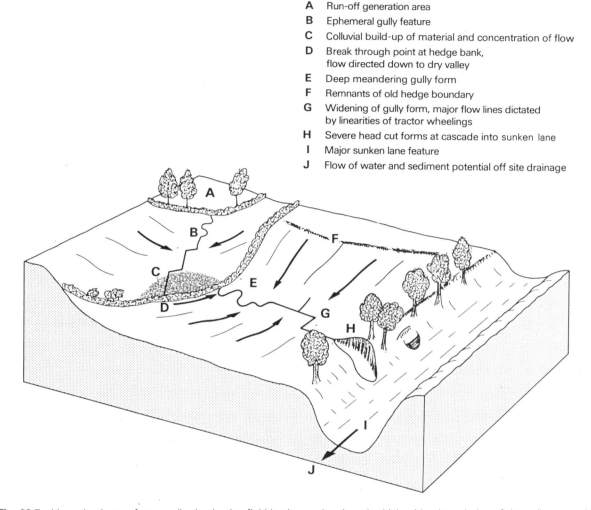

Fig. 22.7 Linear landscape features (hedge banks, field banks, sunken lanes) which with a knowledge of the soilscape units improve our understanding of the expression of soil erosion at the landscape scale (after Farres, 1991).

of expression of horizonation displayed by the soils of the world are extremely varied. Nevertheless, the presence of horizons in all genetically developed soils (i.e. where sufficient time has elapsed to allow soil development) suggests that certain processes are common to the development of all soils and that each soil is not, therefore, the product of a set of processes unique to it. Indeed, Simonson (1959) states that a soil forms 'as a result of an aggregate of many physical, chemical and biological processes, all *potential* contributors to the development of *every* soil'. Before we consider

the implications of this statement, we need to consider what these pedogenic processes are in the context of a systems model of the soil (Fig. 22.8).

In terms of an open-system model, the characteristics of a soil body – its elements, their attributes and their relationships – represent the state of the system. This will depend on the inputs of matter and energy to the soil system – the **input processes**. For example, these include the addition of organic matter from the vegetation above, the addition of rainwater by infiltration, the input of

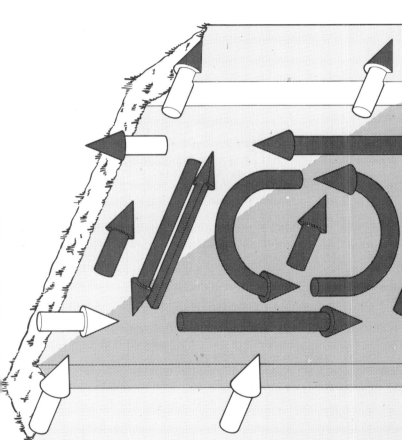

TRANSFORMATION PROCESSES

Internal reorganisation of matter and redistribution of energy, but *in situ*, e.g. decay of organic matter, weathering of primary and secondary minerals. Net loss of mass and free energy accompanies such processes

TRANSFER PROCESSES

Internal reorganisation of matter and redistribution of energy, but involving movement e.g. translocation of iron, clay, humus, and hydrated ions, diffusion of gases, ion exchange, mass-movement and through flow, capillary rise, mixing by soil fauna, cryoturbation

INPUT PROCESSES

Inputs of mass, (e.g. organic matter as litter, rainwater, respiratory; CO_2, regolith by weathering, mass movement and through flow from upslope) and energy as chemical, kinetic, radiant, or mechanical energy or as some combination

OUTPUT PROCESSES

Many similar to input processes e.g. downslope mass movement and through flow, deep percolation and leaching. Some uniquely output processes such as water and nutrient uptake by plant roots

Fig. 22.8 Soil processes.

regolith by the weathering of bedrock, and the increment of materials supplied from upslope by mass movement and throughflow. All of these materials represent inputs of chemical energy, while materials in motion, such as percolating water, will represent kinetic energy inputs capable of doing work on the system. Solar radiation will be absorbed by the soil and heat energy transferred, particularly by conduction. Even the height of the soil body endowing each soil particle with gravitational potential energy can be thought of as an energy input inherited from the geomorphological event responsible for the elevation of the land surface.

The state of the system (the soil) will also depend on processes operating within it either to maintain or to change its state. These processes are of two types. The **transformation processes** involve the reorganization of matter and redistribution of energy, but largely *in situ* and within particular horizons. The **transfer processes** are associated with the pathways of throughput within the system and involve both vertical and lateral transfer between different stores of matter and energy, usually between different horizons. Perhaps the best examples of *in-situ* reorganization are the transformations of organic matter during decomposition and of primary minerals during weathering, and secondary mineral and humus formation within the soil. Both involve the structural reorganization of matter but they do so with a net loss of mass and dissipation of free energy, as heat, for as spontaneous processes they proceed in accordance with the Second Law of Thermodynamics. The lost mass is accounted for by the products of weathering and decomposition which are removed and become involved in transfer processes.

Indeed, one major class of transfer processes within the soil system is concerned with the translocation of the products of pedogenic processes. For example, the movement of hydrated ions in solution through the soil, their exchange between the soil solution, the surfaces of colloids and plant roots, the diffusion of respiratory carbon dioxide, the mechanical translocation of clay minerals through soil pores (Fig. 22.9), or the translocation of iron as hydrated hydroxides in colloidal suspension or as complexed ions by cheluviation – are all transfer processes involving the mobile products of weathering and decomposition. Gravity is the major control of such transfers and

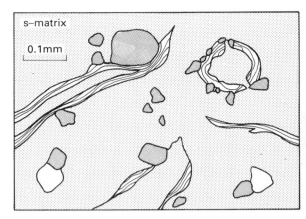

Fig. 22.9 Clay skins (cutans) formed by the mechanical translocation of clay minerals.

the usual result is movement to lower horizons, or laterally downslope, as with the movement of water by matrix throughflow. Nevertheless, other forces may outweigh the attraction due to gravity and, under certain circumstances, an upward movement can occur. Such a situation is involved in the capillary rise of water in soil micropores, which under suitable conditions may result in the enrichment of horizons near the surface with precipitated soluble salts derived from depth.

All of these transfer processes are relatively specific, but there are others which have a more general effect on the state of the system. While many transfer processes are selective and result in the differentiation of specific horizons, others operate in a largely non-selective manner and militate against horizonation; they may even destroy what horizons exist. Mixing by the soil fauna, particularly the macrofauna (especially earthworms in temperate soils and termites in the tropics), brings organic matter, particularly humified material, into intimate contact with the mineral soil and eliminates any clear distinction between organic and mineral horizons. In appropriate environments, **cryoturbation** (frost heave) will have a similar effect in that it will disrupt the development of horizons. However, the latter group of processes may give rise to distinct patterns within the soil. These are a result of the differential responses of each size fraction to freeze/thaw (frost sorting) and, although genetic features, they are not true soil horizons (Fig. 22.10). Finally, any mass movement process, particularly if rapid, will tend to blur, if not

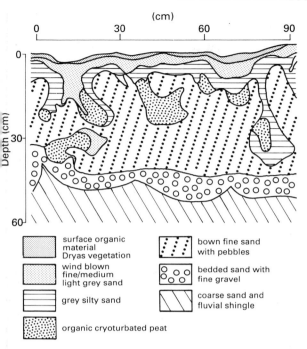

(cm)

Depth (cm)

surface organic
material
Dryas vegetation

wind blown
fine/medium
light grey sand

grey silty sand

organic cryoturbated peat

bown fine sand
with pebbles

bedded sand with
fine gravel

coarse sand and
fluvial shingle

Fig. 22.10 Cryoturbation effects, upper Kellett River basin Banks Island, Northwest Territories, Canada.

obliterate, the differentiation of vertical horizons as the whole soil matrix is subject to lateral movement downslope.

The last group of processes which will determine the nature of a soil or the state of the system are the **output processes**, by which matter and energy are transferred across the boundary of the three-dimensional soil system. As transfer processes, many are similar to those responsible for input to or translocation within the system; indeed many are the same. The downslope mass movement and throughflow processes that deliver material to a soil body from upslope are also the processes that transfer that material through the soil and across its boundary in a downslope direction. The leaching and translocation processes that transfer elements between soil horizons are also responsible for loss to drainage water at depth or by throughflow downslope. However, some processes, such as nutrient and water uptake by plant roots, are uniquely output pathways and processes.

If we now return to Simonson's contention that all of these groups of processes are potential contributors to the development of any soil, it becomes clear that a soil developed under tropical rainforest in Africa, a prairie soil in Saskatchewan, and a soil under an old permanent pasture in the English Midlands, are not in any true sense unique. Though the differences between them are real, they merely reflect differences in the magnitude, in the rate of operation, and in the relative combination of the processes outlined above. As any of these properties can change over relatively short distances, it is not surprising that real soils show such bewildering variation and have largely defied attempts at successful classification. The state of each of these soils – indeed of any soil – will represent to some extent the balance or the steady state attained by the system in response to the particular set of processes it experiences. However, the questions arise: what controls the way specific processes are combined to form a set of input, transformation, transfer and output processes; what controls the magnitude and rate of these four suites of processes; and what therefore controls the nature of the steady-state soil?

22.2 The control and regulation of pedogenesis

From the beginnings of modern soil science and pedology, soils have been described as a function of five factors of soil formation (Jenny, 1941). These are inorganic, organic, climatic, relief or topographic, and time factors, and they can be integrated with the open-system model of the soil by regarding them as sets of exogenous variables which define the operating conditions of the system. It is this whole complex of environmental conditions that determines the characteristics of any suite of processes and of the equilibrium state attained by the soil.

The nature of the parent material (the inorganic factor) controls the properties of the inorganic input to the soil and largely determines the initial textural state of the soil and, by so doing, regulates many of the properties and processes operating in a soil. The mineralogy of the parent material controls in part the initial chemical state of the soil – whether free $CaCO_3$ is present, for example – and this influences many soil properties and processes, such as weathering, the dominant ions in the exchange complex and the presence of earthworms, to quote only a few examples. The chemistry of the parent material also represents part of the initial potential chemical energy of the system.

Climate, through the soil radiation balance, controls the input of radiant energy and, through the surface water balance, the input of water. Through the magnitude of these inputs climate also largely determines the weathering regime, for they function as activating processes in weathering. Through its integrated effect on plant growth, climate also indirectly regulates the supply of organic matter from the vegetation, and in part the type and activity of soil organisms. Climate not only controls the input of water to the soil but also partly regulates its movement in the soil, again through the surface water balance. Discounting runoff, change in soil moisture (ΔS) and groundwater storage (ΔG) will depend on the precipitation/evapotranspiration ratio ($P{:}E_T$). Where $P > E_T$ there will be a net downward movement of water in the soil profile, promoting a whole complex of transfer processes. Where $P < E_T$ water infiltrating into the soil will soon be evaporated and soluble salts will be precipitated at a percolation or evaporation front within the profile.

The organic factor acting through the vegetation regulates the amount and composition of organic matter supplied to the soil in the form of dead litter, and hence part of the input of matter and chemical energy to the system. Within the system the number and type of soil organisms, themselves partly determined by the quantity and kind of litter input, largely control the pathways by which this matter and energy is transformed and transferred during decomposition. In addition, through its interception of precipitation and its absorption, transmission, reflection and redistribution of radiant energy, the microclimate of the vegetation canopy regulates the effects of the climatic factor. Finally, the root systems of plants apply mechanical stress, which may help to control the development of soil aggregates as well as promoting weathering by fracture. Roots also affect the distribution of voids and soil aeration and, certainly in the volume of soil in close contact with them (the **rhizosphere**), can control the soil chemical environment.

The development and maintenance of a steady-state (or equilibrium) soil depend on the long-term stability of the land surface on which it occurs. In other words, it will depend on the balance between on the one hand the pedogenic processes promoting the existence of an ordered three-dimensional soil system and the processes of erosion tending to remove it and on the other those of deposition tending to bury and obliterate it. This balance is controlled, therefore, by the overall effect of the geomorphological factor and, as stated earlier, it is closely related to the soil's position in a landscape – notably its relationship to slope and to the processes operating on slopes. Slope will also control the hydrological characteristics of the site, the drainage relationships and the position of the water table. These factors will in turn influence the operation of pedogenic processes such as leaching and waterlogging. Indeed, the control exercised by slope has long been recognized in the literature in the **catena concept** of Milne (1947) and the toposequences and hydrological sequences of workers such as Fitzpatrick (1971) and Glentworth and Dion (1949) (Fig. 22.11). These models are closely related to those of slope form, such as the nine-facet model of Dalrymple *et al.* (1968), and of slope process (see Chapter 12).

Finally, the time factor can be built into this factorial approach, for soils are dynamic bodies and the state of the system will also depend on their stage of development, or on how far any soil has travelled along the path to equilibrium. However, the dependence of many properties of the system on elapsed time is not straightforward; neither is their equilibrium state a simple function of time (Fig. 22.12). Some properties adjust slowly, and although their initial rate of change may be rapid, it gradually becomes asymptotic and they become virtually time-independent over relatively short time intervals. Such properties only change very slowly, reaching a steady state after perhaps 10^3 or 10^4 years. The development of the textural properties of a soil and soil fabric (Kubiena, 1938; Brewer, 1964) would fall into this category. In contrast, other properties adjust very rapidly in their response to the external conditions of existence. The pH of the soil solution is such a property, which, because of the buffer capacity of the soil solid phase, can be adjusted rapidly to re-establish an equilibrium after a disturbance such as a rainfall event. The occurrence of reversible redox reactions allows rapid adjustments to fluctuations in the redox environment of the soil associated particularly with waterlogging (Fig. 22.13). The soil population of decomposers can also respond rapidly to changes in the amount or type of litter supplied to the soil by increasing

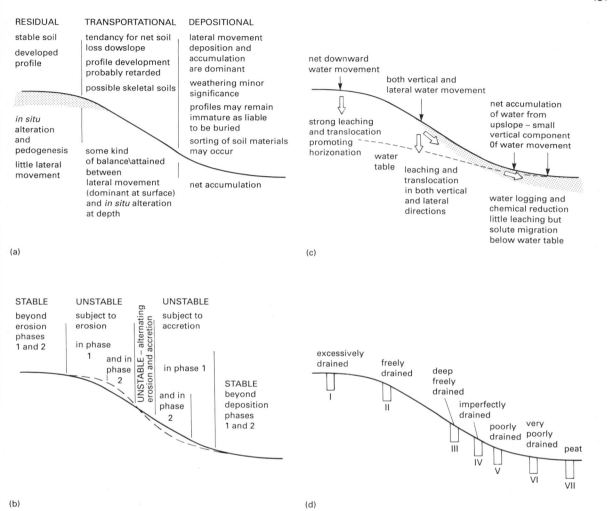

RESIDUAL | TRANSPORTATIONAL | DEPOSITIONAL

stable soil

developed
profile

in situ
alteration
and
pedogenesis

little lateral
movement

tendancy for net soil
loss dowslope

profile development
probably retarded

possible skeletal soils

some kind
of balance\attained
between
lateral movement
(dominant at surface)
and *in situ* alteration
at depth

lateral movement
deposition and
accumulation
are dominant

weathering minor
significance

profiles may remain
immature as liable
to be buried

sorting of soil materials
may occur

net accumulation

(a)

net downward
water movement

both vertical and
lateral water movement

net accumulation
of water from
upslope – small
vertical component
Of water movement

strong leaching
and translocation
promoting
horizonation   water
table

leaching and
translocation
in both vertical
and lateral
directions

water logging and
chemical reduction
little leaching but
solute migration
below water table

(c)

STABLE | UNSTABLE | UNSTABLE

beyond
erosion
phases
1 and 2

subject to
erosion

in phase
1

and in
phase
2

UNSTABLE – alternating
erosion and accretion

subject to
accretion

in phase 1

and in
phase
2

STABLE
beyond
deposition
phases
1 and 2

excessively
drained

freely
drained

deep
freely
drained

imperfectly
drained

poorly   very
drained poorly
         drained peat

I

II

III

IV

V

VI

VII

(b)

(d)

Fig. 22.11 Soil–slope relationships (partly after Gentworth and Dion 1949 and Butler 1959). (a) Stressing the erosional and depositional relationships of soil profiles developing at different positions on a slope, but in (b) introducing two phases of erosion and deposition. (c) The drainage relationships of soil profiles developing at different positions on a slope. (d) Such a hydrological sequence of profiles used as a basis of soil classification.

their number to approach the limits of their food supply. These rapidly adjusting features, and the processes involved, are seen to be maintaining a steady state even over a short interval of time. However, some processes are irreversible and self-terminating, and are dominated by positive feedback. The weathering of primary silicate minerals to clay minerals is this kind of process, for when the reserve of unaltered primary minerals is exhausted the process stops, and the clay minerals will not spontaneously reconstitute to

form feldspars, for example. The development of an iron or clay pan is in a similar way irreversible, and it terminates when all of the iron or clay has been translocated from higher horizons. The equilibrium these properties attain is thermodynamic. Often these irreversible processes convey the soil system as a whole through a threshold which initiates a new system state. For example, the production of an impermeable pan at depth restricts the percolation of water, produces a perched water table within the profile,

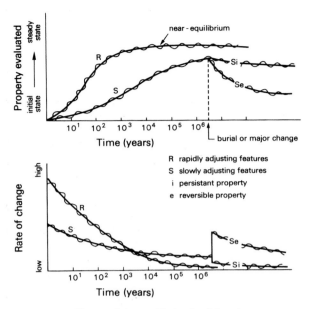

Fig. 22.12 Time and the equilibria of soil-forming processes (after Yaalon, 1971).

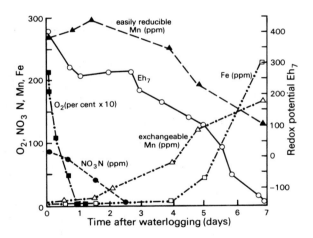

Fig. 22.13 Changes in oxygen, nitrate, manganese, iron and redox potential of a silt clay after waterlogging (Armstrong, 1975, after Patrick and Turner, 1968).

alters the chemical environment, restricts the effectiveness of leaching and translocation processes to that part of the profile above this water table, and initiates reducing conditions below it. The new state is very different from the freely draining soil which encouraged the leaching and translocation that led to pan development in the first place.

The response of the system with time, therefore, will reflect a complex combination of the behaviour of all three kinds of process with time (see Chapter 23).

22.3 Formal processes of pedogenesis

In the real world, pedogenic processes are not combined at random. They occur consistently in particular combinations determined and controlled by reoccurring sets of conditions. Each of these commonly recurrent combinations is known collectively as a **formal process** of soil formation. Since the ultimate factor regulating pedogenesis is often climate, particular formal processes are broadly correlated with particular climates. We shall not consider these formal processes in great detail here, but they are treated in several of the texts recommended as further reading at the end of the chapter. In the brief treatment that follows, the formal processes have been grouped to provide models of pedogenesis in temperate, tropical and arid environments. The nomenclature used will mainly follow that of the **FAO/UNESCO** (FAO, 1988) soil classification, although alternative terminology will also be used wherever it is thought helpful (Box 22.1).

22.3.1 A model of temperate pedogenesis
In a humid temperate environment, on parent materials with a moderate to rich reserve of weatherable minerals, and supporting a woodland or grassland vegetation supplying a litter rich in base cations and maintaining an efficient nutrient circulation, a soil develops in which many of the soil constituents have become **stabilized** (Fig. 22.14a). In such brown earth soils, or **Cambisols**, distinguished by the presence of a diagnostic **cambic horizon** of *alteration*, the dominant process is leaching, particularly of calcium, which tends to be the most important cation saturating the exchange complex. However, this leaching tendency is normally very slight for it is offset by the *in-situ* weathering (*alteration*) of the parent material and the release of calcium and other cations from the decomposition of litter. Such soils support a large and diverse soil population, with bacteria as the principal microbial decomposers. They are characterized by efficient humification, deep and intimate mixing of humus and mineral soil by the soil fauna and the formation of a deep mull horizon (see Fig. 21.9). Structurally they show a

Box 22.1

FAO-UNESCO SOIL CLASSIFICATION

Over a 20-year period between 1961 and 1981, at the recommendation of the International Society of Soil Science, and under the auspices of the Food and Agricultural Organization of the United Nations, and the United Nations Educational, Scientific and Cultural Organization, a world soil map was completed at a scale of 1:5 000 000. The soil classification scheme used here was developed and refined for use as the legend for this world map and as the field mapping units used in its compilation. Originally the soils of the world were divided into 26 major groups at the first level in the classification, and into 106 soil units at the second, more detailed, level. The latest revised legend (FAO, 1988) increases the number of categories to 28 major soil groups and 153 soil units. The FAO-UNESCO scheme

has now become accepted as one of two internationally established standards for soil description, classification and mapping, the other being the United States Soil Conservation Service's Soil Taxonomy (1975).

The FAO-UNESCO classification is based on the use of observable or measurable attributes of the soils themselves. These properties are taken to be *diagnostic* of certain soil types, and may in conjunction define recognizable **diagnostic horizons**, again *diagnostic* of certain soil types. Both diagnostic properties, and more particularly diagnostic horizons, reflect certain processes or suites of processes (formal processes) of soil genesis, although the soil units are not themselves defined directly in terms of process *per se*. For further guidance see Further reading.

---

well developed **crumb structure** promoted by the flocculation effects of divalent cations, such as calcium, on the clays, and by the aggregation effects of bacterial polysaccharide gums.

Where the parent material has poorer reserves of weatherable minerals, the movement of water through the profile is greater, or the rate of cation replacement from litter is inadequate to offset leaching, the exchange complex will become at least partially dominated by hydrogen ions. Under these conditions aggregate stability will decrease, clay will deflocculate and in the dispersed state will be subject to mechanical translocation through the profile to accumulate as a finer textured Bt horizon at depth (the diagnostic **argic B** horizon). Such a suite of processes is the formal process of **lessivage** (Fig. 22.14b) and the soils are known as **luvisols** (acid brown earths, sols lessivées), characterized by the possession of an argic B, which when formed by clay illuviation may have oriented clay skins, clay coatings, **argillans**, or clay **cutans** (see Fig. 22.9) on the sides of soil pores, in fissures and channels, and on the faces of the peds.

On siliceous parent materials with few weather-

able minerals, which are freely drained and support a vegetation which supplies a base-deficient litter and promotes the formation of raw acid mor organic horizons, the formal process of **podsolization** occurs (Fig. 22.14c). Here the base cations are leached rapidly, particularly under a cool temperate climate. These cations are not replaced either by *in-situ* weathering or by decomposing litter. The exchange complex is dominated by hydrogen, and clay minerals – never present in great quantities in such parent material – become unstable and are broken down. In addition, the soil population, dominated by arthropods and fungi, is less efficient than that inhabiting cambisols (brown earth soils), with the result that decomposition and humification of the acid litter remains incomplete. Many organic compounds, particularly fulvic acids and the polyphenols, are not incorporated in humic polymers and are leached from the organic horizons and become available as complexing (chelating) agents. In the mineral soil they preferentially complex polyvalent metal cations such as iron (Fe^{3+}) and aluminium (Al^{3+}) which pass into solution as metal ion

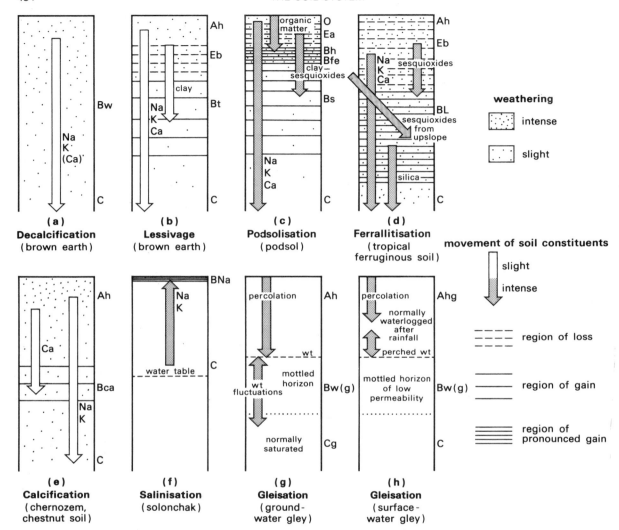

Fig. 22.14 The formal processes of pedogenesis (after Knapp, 1979).

chelates and are translocated down the profile, leaving a bleached **albic E** horizon (E$_a$) of uncoated quartz particles. Progressively the ratio between the organic anion part of these complexes and the complexed metallic cations alters, until they become saturated by iron and aluminium as the complexes move down through the soil. This occurs partly as a result of a relative increase in the iron and aluminium as more is taken up by the complex, and/or a relative decrease in the organic anion as it continues to degrade through microbial decomposition. Ultimately the saturated

complex becomes unstable and is precipitated to give rise to the diagnostic horizon, the **spodic B**. This is a horizon of accumulation enriched with certain high valency forms of iron and aluminium and sometimes organic carbon, below the albic E when present. Where organic carbon is precipitated as well as the oxides of iron and aluminium it is a B$_h$ horizon, but the organic matter may not be in evidence and diffuse horizons (B$_w$) of iron accumulation or thin discrete horizons with an indurated surface creating iron pans (B$_{Fe}$) may occur.

Some workers have placed cambisols, luvisols and these **podsol** soils in a developmental sequence. However, in most situations the controlling variable is not time but the initial characteristics of the parent material and the type of vegetation, which act as the primary regulators operating within the overall control of a cool humid temperate climate.

22.3.2   A model of tropical pedogenesis

The dominant formal process in the humid tropics is **ferralitization** (Fig. 22.14d) which is the very intense and advanced hydrolysis of weatherable minerals. Where leaching is not excessive, and intense *in-situ* weathering under a hot and humid climate leads to the breakdown of clay minerals, soil profiles are characterized by the accumulation of 1:1 kaolinite clay and hydrated oxides of iron and aluminium. The types of iron compound are different from those of temperate environments and impart a rich red colour to many of these soils. Hematite, and in the subtropics goethite, are normally important minerals, but there may be other iron compounds involved. The reddening is due to the dehydration of iron compounds and is known as **rubefaction**.

With greater leaching intensity the stability of ferruginous soils is disrupted and lessivage of clay minerals in particular and some hydrated oxides occurs leading to a textural B, or **argic B**, horizon. These soils differ in the details of their profile characteristics, partly because of differences in the base status of their parent materials, or in the extent to which weathering and ferralitization have proceeded. **Nitosols** are associated with basic parent materials and distinguished not only by clay translocation but by shiny ped faces (*L. nitidus* = shiny). **Alisols** and **acrisols** are comparable soils, both with argic B horizons, but on poorer parent materials than nitisols. In acrisols weathering and ferralitization have proceeded further than in alisols or nitisols.

However, where both leaching and weathering are intense, the lattice breakdown of clay minerals proceeds further and deeply weathered soil profiles develop. These soils have a bimodal texture dominated by residual quartz of sand size and hydrated oxides of iron and aluminium of clay size, together with some kaolinite. Such soils (the **ferralsols**), become highly acidic (pH 4) and under these conditions amorphous silica is removed from the profile, thereby enhancing the relative accumulation of the iron and aluminium oxides (B_L horizon). However, in many such soils the amount of iron present cannot be explained by *in-situ* weathering of the parent material, or by translocation from the upper horizons (E_b), and their occurrence on footslopes points to the reprecipitation of iron gathered by laterally moving soil water. Indeed, ferralsols seem to have developed best in areas of fluctuating water table under alternately wet and dry tropical seasonal climates. These conditions allow iron III oxides to pass into solution as reduced iron II ions at times of high water table, and to migrate laterally downslope to reprecipitate in an oxidizing environment when the water table falls. With desiccation (perhaps associated with climatic change) or with erosion the profile may undergo irreversible dehydration leading to the formation of indurated **plinthite** (laterite). Soils with significant amounts of plinthite or plinthitic horizons are known as **plinthosols**.

22.3.3   A model of arid pedogenesis

In semi-arid and arid environments where effective precipitation is insufficient, leaching of solutes from the profile is incomplete. In mid-continental grassland environments of semi-aridity, precipitation is usually sufficient for sodium and potassium to be removed from the profile. However, calcium is precipitated as calcium carbonate (**calcification**, Fig. 22.14e) forming a B_{Ca} or B_K horizon as carbon dioxide concentrations fall below the rooting zone, or as percolating water begins to evaporate. These soils are **calcisols** and they are characterized by a **calcic** horizon, a **petrocalcic** (hardened $CaCO_3$) horizon, or concentrations of powdery lime. Under more severe aridity, intense evaporation means that only the most mobile ions – sodium and potassium – pass into solution, but even these are soon precipitated at a percolation front. In extreme cases, where texture permits and the water table depth is not great, saline groundwater may be drawn up in response to strong surface evaporation, leading to the precipitation of sodium and potassium salts (**salinization**, Fig. 22.14f). These soils are **solonchaks** and they are common in internally draining depressions such as *playa* lakes and inland basins. Often associated with solonchaks are **solenetz** soils. These are characterized by diagnostic **natric B** horizons with high exchangeable sodium percentages and a columnar structure. These columnar peds have rounded tops due to clay destruction (*deflocculation*). Indeed, where saline ground water is too close to the surface, **solonization** occurs, producing saline soils of very high pH, where high concentrations of exchangeable sodium have the effect of deflocculating and dispersing the clay minerals, allowing their translocation and the loss of soil aggregation.

22.3.4   Gleisation
Certain formal processes are controlled less by climate than by other dominant sets of conditions. **Gleisation** is such a process, where the drainage relationships of the soil and the type and position of the water table are the paramount controls (Figs. 22.14g, h). Because appropriate conditions can occur under a variety of climatic regimes, gleying can occur in conjunction with several other formal processes. Gleying occurs in soils which experience either periodic or permanent waterlogging. It is associated with the presence of high groundwater tables in the true **gleysols** or **groundwater gleys** which exhibit **gleyic properties**. In these soils the water table at shallow depth lies within the soil profile and oxidized horizons lie *on top of* a fully reduced subsoil. Where the impedance of soil drainage is due to the existence of horizons of low permeability, and perhaps the formation of perched water tables, a different pattern of oxidation and reduction occurs. With the reduced horizon associated with perched water above, a slowly permeable subsurface horizon occurs on top of an oxidized subsurface horizon. The true water table lies at greater depth. These conditions are termed **stagnic soil properties** and soils possessing them are not gleysols but **stagnic** variants of other soil groups, or **surface water gleys**. However in both groundwater and surface water gleys anaerobic conditions are established rapidly as available oxygen is depleted by the respiratory demand of aerobic soil organisms and plant roots. These are replaced by anaerobic organisms and as the redox potential falls and chemically reducing conditions are established, many redox couples – both inorganic and organic – are converted to their reduced forms (Fig. 22.13). The most obvious sign of these changes in soil chemical environment is the conversion of reddish-brown iron III or ferric compounds to greyish iron II or ferrous compounds. These pass into solution as mobile ferrous ions and migrate both within soil

aggregates, forming coatings on the ped faces, and out of the profile with flowing groundwater. Where oxygen penetrates the gley horizons, secondary oxidation takes place, producing characteristic reddish mottling. This happens particularly where the water table fluctuates, perhaps seasonally, or where radial oxygen loss from live roots occurs.

Further reading

Useful introductory texts are:

Courtney, F.M. and S.T. Trudgill (1984) *The Soil: an Introduction to Soil Study in Britain*, (2nd edn). Edward Arnold, London.
Fenwick, I.M. and B.J. Knapp (1982) *Soil Processes and Response*. Duckworth, London.
Knapp, B.J. (1979) *Soil Processes*. Allen & Unwin, London.
Paton, T.R. (1978) *The Formation of Soil Materials*. Allen & Unwin, London.
Pitty, A.F. (1979) *Geography and Soil Properties*. Methuen, London.

At a more advanced level the following texts all have their own but different strengths:

Birkeland, P.W. (1984) *Soils and Geomorphology*. Oxford University Press, Oxford.
Duchaufour, P. (translated by T.R. Paton) (1982) *Pedology: Pedogenesis and Classification*. Allen & Unwin, London.
Gerrard, A.J. (1981) *Soils and Landforms*. Allen & Unwin, London.
White, R.E. (1979) *Introduction to the Principles and Practice of Soil Science*. Blackwell, Oxford.
Wild, A. (ed) (1988) *Russell's Soil Conditions and Plant Growth*, (11th edn). Longman, London.

The following texts address the soilscape and the spatial aspects of soils, and could also include the book by Gerrard above:

Farres, P.J., J. Poesen and S. Wood (1992) *Some Characteristic Soil Erosion Landscapes of N.W. Europe*. Catena.
Hole, F.D. and J.B. Campbell (1986) *Soil Landscape Analysis*. Routledge & Kegan Paul, London.
Northcliffe, S. (1984) Spatial analysis of soil. *Progress in Physical Geography*, **8**, 261–269.
Trudgill, S.T. (1983) Soil geography: spatial techniques and geomorphic relationships. *Progress in Physical Geography*, **7**, 345–360.

The FAO Soil Classification and soil genesis in different environments are covered by the following:

Driessen, P.M. and R. Dudal (eds) (1989) *Geography, Formation, Properties, and Use of the Major Soils of the World*. Agricultural University, Wageningen, & Catholic University, Leuven, Wageningen/Leuven.
FAO-UNESCO (1989) *Soil Map of the World – Revised Legend 1988*. ISRIC, Wageningen.
Farres, P.J. (1991) Soil classification: the updated FAO system and soils in the UK. *Geog. Rev.* **5** (1), 27–31.

Part D
Systems and change

In modelling the organization and operation of Earth and environmental systems we have chosen to ignore the explicit expression of change and have not attempted to isolate those mechanisms and processes leading to changes in systems states, except insofar as they are implicit in the normal operation of these systems. In Chapter 1 it was stressed that one of the important properties of any model, including a systems model, is its generality. The models we use to represent environmental systems should, therefore, be applicable to any broadly similar situation. The drainage basin, the ecosystem, the pressure cell – indeed all of the models we have used – are generalizations that are valid at different points in space and time to systems of matter which share the same general form, and which function in the same way. Such generality is possible because our models are abstractions from the real world. In the real world the systems with which we have been concerned vary enormously in the details of their structural and functional organization, and system processes vary in their magnitude, relative significance and rates of operation. There are two directions that such variation can take, one in space and one in time. To a certain extent we have examined variations in space in all of the models used so far. Geographical or areal variations have been stressed repeatedly, in the disposition of systems elements and their attributes, and in the magnitude of inputs, or processes. We have considered spatial variations in energy and water balances, and in the operation of the denudation system, compared geographical variations in ecosystem primary productivities, and considered the way in which the soil system responds to spatial variations in its operating conditions. However, in each case, we have treated each system as if it were in some broadly stable equilibrium state, and although our attention has been directed to the dynamics of systems operation, apart from a few exceptions, we have not considered explicitly the dimension of time. In Part D, therefore, the behaviour of environmental systems with time will be explored at some length.

VII THE NATURAL WORLD

Change in environmental systems

23.1 Equilibrium concepts and natural systems

The equilibrium state of thermodynamic open systems (and that is the way that we have modelled environmental systems throughout this book) is a steady state. However, the notion of equilibrium is more complex than this tacit assumption would suggest, and furthermore it is intimately associated with ideas of change, or indeed lack of change, through time. What then have we meant by the concept of a steady-state equilibrium as we have

used it in earlier chapters? Central to the concept is the maintenance of an average condition of the system, the trajectory of which remains unchanged in time (*homeostasis*) (Fig. 1.4a; Fig. 23.1a). This average condition can be defined in terms of both the disposition of the morphological components of the system and the flows of mass and energy. At any instant, the actual condition of the system will only approximate to the average state and over a period of time will be seen to fluctuate about, but may never actually accord with, this average state. One immediate implication of this view is that the

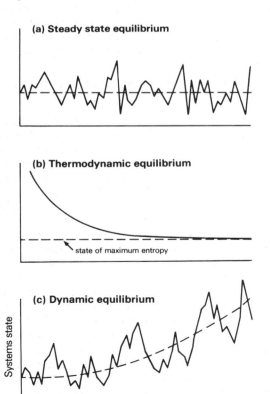

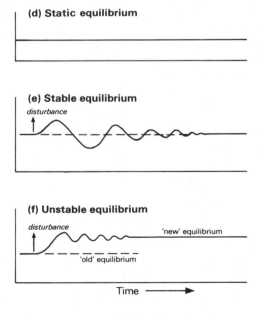

Fig. 23.1 Types of equilibrium.

recognition that a system is maintaining a steady state is dependent on the timescale over which it is considered. Indeed, the importance of both timescale and spatial scale in the interpretation of equilibrium conditions in environmental systems is a subject to which we shall return shortly.

Also central to the concept of steady-state equilibrium is the significance of self-regulation and the ability of the system, consequent on the operation of negative feedback loops and the action of regulators, to damp the amplitude of the cyclic departures from an average state. Therefore, a dynamic open system is able to maintain an orderly state within controlled limits determined by the precision and efficacy of its capacity for self-regulation. Furthermore, this orderly state is maintained by, and in the face of, often massive throughputs of mass and energy.

There are, however, other conceptions of equilibrium which prove useful in understanding the relationships between equilibrium and change in environmental systems. For example, embodied in the classic concepts of isolated system thermodynamics is the tendency towards maximum entropy which we have already encountered in Chapter 2. When a system attains such a distribution of mass and energy it is said to be in thermodynamic equilibrium (Fig. 23.1b). Implicit in this concept is irreversible change marked by a decline in free energy capable of doing work on the system, and a corresponding increase in the entropy of the system. Also emphasizing progressive changes of state, but this time explicitly, is the concept of dynamic equilibrium. Here, as in a steady-state equilibrium, controlled fluctuations occur about average systems states. However, in this case they are unrepeated average states through time (homeorhesis) (Fig. 1.4b; Fig. 23.1c). The existence of dynamic equilibrium is not always easy to recognize because directional change in the average condition of the system is often obscured by the greater magnitude and rate of change of the fluctuations about it. Over short time periods, therefore, the system appears to maintain a steady state.

Three further definitions of equilibrium remain that are best applied to relatively simple mechanical and chemical systems, and they are defined by the system's response to limited external forces. The first is static equilibrium, where force and reaction are balanced and no resultant force exists, and the properties of the system to which the concept is applied remain unchanged or static through time. The second and third definitions are stable and unstable equilibrium, respectively. Here the system displays tendencies either to return to (stable) or to be displaced further from (unstable) the initial equilibrium state in response to an externally applied disturbance (Figs 23.1d, e, f). The notion of balance as applied in static equilibrium is familiar in a number of situations, but one example of its application that we have encountered is the balance between the reactants and products when a weathering reaction reaches equilibrium. It can be applied also to systems where a balance pertains between inputs and outputs, as in the slope system. Stable and unstable equilibria are familiar both in relation to systems of gases when considering atmospheric stability, and in relation to simple mechanical systems concerned with the erosion and transport of rock materials. Related to the last two of these definitions is the concept of metastability. This is the tendency of a system to move from one equilibrium state to another when an external force causes it to cross a threshold beyond which the probability of recovery becomes remote. These concepts of stable, unstable and metastable equilibria have been used to some effect by Trudgill (1977) to model the response of soil–vegetation systems to disturbance (Fig. 23.2).

23.2 Thermodynamics, equilibrium and change

In the above discussion it will have become obvious that almost all views of equilibrium demand an appreciation of the manifestations and mechanisms of change in the state of the system through time. However, before we turn our attention to change, it will be useful to remember that environmental systems are energy systems and are subject to the laws of thermodynamics. In accordance with the Second Law (Chapter 2), the operation of natural irreversible processes leads to a decrease in the free energy and an increase in the entropy of an isolated system. It was the Austrian physicist Ludwig Boltzmann, who recognized, in 1896, that the entropy of a system described in this way is a measure of the way in which the total energy of a system is distributed among its constituent elements (atoms or molecules). More precisely, he demonstrated that the entropy of a system (S) is dependent on the number of statistically independent ways of

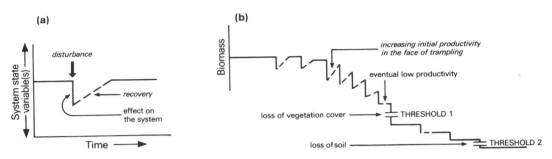

Fig. 23.2   (a) General disturbance and recovery model; (b) the effect of trampling on vegetation biomass (both after Trudgill, 1977).

distributing the elements amongst a number of energy levels (or quantum states, P):

$$S = k \log_e P$$

where k = Boltzmann's constant = 1.38054×10^{-23} J K^{-1}. At thermodynamic equilibrium the entropy of the system is at a maximum and the distribution of elements among the energy levels conforms to the most probable state – that is, at random.

This statistical or probabilistic view of entropy has been extended beyond Boltzmann's original formulation and linked with Shannon's work on communication and information theory (Jaynes, 1957) so that the entropy (S) of a system can be expressed as the sum of the probabilities of possible states or properties of the system, p_1, p_2, p_3 ... p_N:

$$S = - k \sum_{i=1}^{N} p_i \log_e p_i$$

where $\sum_{i=1}^{N} p_i = 1$.

The entropy of the system is at a maximum when all of the states or properties have an equal probability ($p = 1/N$) or are distributed at random in time and space. Conversely, entropy is at a minimum when only one state is possible and all others have a probability of zero, i.e. the most organized state. Therefore, entropy is seen to embrace concepts of order and disorder and, as we saw in Chapter 7, the information content of the system. (Shannon showed that the total information content of a system could be given

by log N, and that if base 2 is used instead of 10, or e, then the smallest possible unit of information is the 'bit'.) The Second Law, therefore, gives us a direction to natural processes: from high to low energy levels, from organized (ordered) to random (disordered) configurations, and from high to low information content.

However, these concepts were developed originally to apply to isolated systems. Environmental systems are open systems. In such systems the input of mass and energy – precipitation and kinetic and potential energy in channel systems, food molecules and chemical energy in animals, water vapour and thermal energy in atmospheric systems – increases or maintains the free energy, the organization and information content of the system, and in so doing reduces its internal entropy. How, then, can these facts be equated with the view of equilibrium, in thermodynamic terms as a state of maximum entropy and, in probability terms as the most probable state?

First, the creation and maintenance of order and information in an open system is dependent on the transformation of free energy by the system. Nevertheless, this consumption of free energy in doing work on the system by irreversible processes will lead inevitably to the production of entropy. So secondly, individual processes will proceed in the direction of increasing entropy given the constraints imposed by, for example, the energy environment under which the system operates. However, under open-system conditions a true thermodynamic equilibrium cannot be reached. Therefore, the key to understanding the equilibrium state of an open system is to grasp the system's need to minimize the rate of entropy increase per unit of preserved structure, where

rate of
increase in
the entropy
of the
system

=

rate of
internal
generation
of entropy
by the
system

−

rate of outflow
of entropy
to the
surroundings.

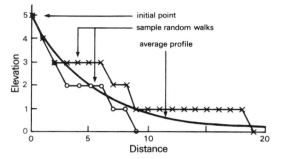

Fig. 23.3 Sample random walks generating an average stream long profile (after Leopold and Langbein, 1962).

In theory, the equilibrium state of an open system should be one in which the rate of increase in entropy is zero. That is, the rate of internal generation of entropy by the processes maintaining an ordered or low entropy state is balanced by the outflow of entropy to the surroundings. The most probable state to satisfy these equilibrium conditions is one where the rate of internal entropy generation is minimized. Such a state can be equated with a configuration of the system which requires the minimum work and minimum expenditure of free energy to maintain its structure.

This concept of minimum energy expenditure and minimum internal production of entropy has been applied to generate the most probable long profile of a stream. Here adjustments in the hydraulic geometry of the channel along its length are seen to establish the most efficient form of long profile for the prevailing conditions of energy and mass flow, such that the expenditure of energy on work is minimized. In such a state there is a uniform rate of internal entropy generation per unit stream length, and this rate per unit discharge rate equals the rate of outflow of entropy as heat. This form, therefore, represents the most efficient and probable long profile and accords well with natural stream profiles. The fact that such a profile can be simulated by purely random processes, apart from the constraint introduced by the need for the stream to flow downhill, confirms that it is the outcome of the minimum-work tendency of the system (Leopold and Langbein, 1962, 1964; Fig. 23.3).

This example, and similar work such as Yang's (1970, 1971) attempt to use the same work-minimizing reasoning partially to explain meander form, concern relatively simple physical systems. However, perhaps the most profound insight we gain from these entropy considerations is the beginnings of an explanation of the capacity of living systems to maintain complex information-rich structures. If we return to the cellular level of Chapter 7 we can reinterpret the complexity of organic macromolecules and the cellular structures they build in the light of an understanding of irreversible open-system thermodynamics.

We can see now that the apparently highly improbable complexity of the living cell is the automatic outcome of the chance occurrence of the information-rich programming system represented by the DNA molecule. We have seen already that DNA (Box 2.3) programmes the amino acid sequence of protein polypeptide chains, and that protein enzyme systems programme the synthesis of other molecules. The enzyme proteins depend on their three-dimensional structure for their function, and this structure can be seen as the result of the tendency displayed by all systems to seek that state which possesses the least free energy, i.e. an entropy-maximizing state. In this case the amino acid side chains tend to arrange themselves in relation to their neighbours in such a way as to minimize the energy content of the structure. The result is that the whole polypeptide chain bends and coils so as to arrive at the most stable arrangement with the lowest energy content under the constraining conditions of pH, temperature and ionic composition that pertain in the cell. The DNA, therefore, has not only coded the amino acid sequence, but by so doing has automatically determined the three-dimensional geometry of the chain and ultimately of the protein molecule (Fig. 23.4). The same reasoning can be applied to supramolecular structures such as enzyme systems, or biological membranes (Fig. 23.4). Again, because of the large negative free-energy change associated with certain arrangements, it is these configurations that are the most stable, most probable, and therefore the automatic outcome of cellular reactions. Even more importantly, they are the configurations that require the least expenditure of energy to maintain under open-

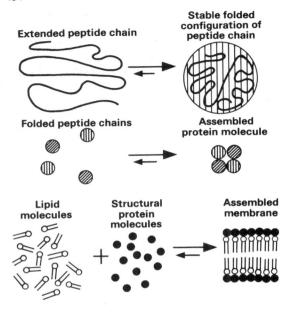

Fig. 23.4 The natural tendency of supramolecular organization (modified from Lehninger, 1965).

system conditions, and are, therefore, the most likely to persist. There is no reason why these thermodynamic considerations may not be valid, in a broad way, at the level of the individual organism or, as we shall see, at the level of the entire ecosystem.

However, the flow of mass and energy through an open system, although maintaining its organization, also has a disruptive effect on the existence of a constant equilibrium state. This is partly because the inputs to environmental open systems are never uniformly distributed in either time or space. Furthermore, the complexity of the pathways of throughput, the residence times of mass and energy in the various stores and the lags experienced in the response to processes all mean that at any instant the rate of entropy increase is unlikely to be zero. Over a period of time the system wll fluctuate about this statistical average, or most probable state. In other words, on thermodynamic grounds the equilibrium state of an open system is a *steady state of minimum work, minimum internal generation of entropy, maximum preserved order per unit of energy flow and with an average increase in the entropy of the system equal to zero.*

23.3 Manifestation of change

The above discussion of the equilibrium state attained by environmental open systems allows us to recognize and describe the types of change that we should look for in the following chapters. Chorley and Kennedy (1971) recognize four classes of change: first, changes in the energy content and energy distribution within the system; secondly, those due to changing inputs or input/output relationships of both mass and energy; thirdly, those associated with shifts in the internal organization or integration of the system itself; and fourthly, those associated with the development of energy and mass stores which introduce time lags in the operation of systems processes, usually acting as buffers. However, it is important to realize that the operation of all environmental processes involves change in the system; indeed, in Chapter 1 we defined a process as merely the method of operation by means of which a change in state is effected. Furthermore, in Chapter 2 and again in this chapter we have recognized that all processes occurring in open systems are time-dependent, for open systems maintain a steady state by the transfer of mass and energy by real or irreversible processes with the surroundings. Time and rate are critical variables in the energetics of such systems, while entropy production itself is time-dependent (see Chapter 2).

However, as we saw in Section 23.2, the maintenance of a steady state implies that the net effect of the operation of these irreversible processes is to return the system to the most probable state, or at least keep it fluctuating about such a state. As we know, this is accomplished by self-regulation under the control of negative feedback mechanisms. Therefore, the first kind of change we should look for is fluctuation about the steady state, and indeed much of Parts B and C has done just that. These fluctuations may be associated with all four of the causes of change classified by Chorley and Kennedy. Both the atmospheric energy balance and the atmospheric circulation discussed in Chapters 3 and 4 reflect changes in the energy content and energy distribution of the atmosphere and the feedback mechanisms that redress imbalances. The hydraulic geometry of a stream channel changes as it adjusts in response to fluctuating water and sediment discharge characteristics, themselves a reflection of changing input/output relationships. The steady-state condition may in many cases be that which is adjusted to bankfull discharge, but in others the mean annual flood, or even base flow may be the condition to which the channel returns. Variations in hydrograph characteristics reflecting differences in the contributing areas of a drainage basin under different rainfall conditions is an example of fluctuation reflecting changes in the organization of internal functional linkages within the basin. The immobilization of nutrient elements in litter and soil organic matter represents the development of a storage compartment with a lagged output which influences the rate of nutrient turnover.

These causes of variation in the state of the system may themselves possess recognizable periodicities to which the system can become adjusted, and for which it develops **memory**. Variations in input, in patterns of energy distribution, in functional linkages or in changes in storage, may all reflect predictable diurnal or seasonal fluctuations. Sometimes the periodicity may be longer and simply be termed cyclic, while in other cases variation may be apparently random. The magnitude of the response of the system, and the length of the **relaxation time** (time over which adjustments take place) will vary from system to system, but will reflect the memory and efficiency of negative feedback control that the system has developed.

However, not all systems which we chose to define in the natural environment will have attained a steady-state equilibrium. We must expect to encounter systems which are immature, which represent merely transient states in a sequence of states that form part of a developmental sequence. Here we must be prepared to recognize directional change – specifically evolutionary change – as the system seeks to attain a steady state under the prevailing conditions. Such change will be dominated by the cumulative effect of positive feedback propelling the system towards its most probable state. The growth of an immature organism, the attainment of a 'graded' stream long profile, the development of a mature ecosystem on an initially virgin surface, and the increasing integration of a developing drainage network, are all changes which can be viewed in these terms. Nevertheless, they are all the inevitable result of inherent properties of the systems concerned, and represent the diversion of surplus energy throughput towards the development of more integrated and efficient functional organizations, and may be manifest as

any combination of the four classes of change recognized by Chorley and Kennedy. As such changes in systems states approach the most probable steady-state condition, we would expect negative feedback self-regulatory mechanisms to become dominant, and directional change to be replaced by fluctuation about a mean condition.

Finally, we must expect that the capacity of any defined environmental system exhibiting a

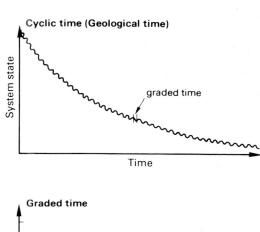

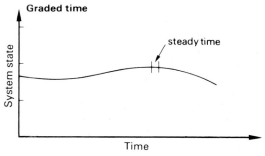

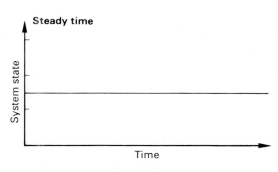

Fig. 23.5 Time, space and causality in geomorphic systems (after Schumm and Lichty, 1965). Cyclic, graded and steady time as reflected in changes in stream gradient over time.

steady state to resist or accommodate changes in the energy inputs to it, or changes in the energy environment of its surroundings, to be limited and dependent on the efficiency and resilience of its self-regulatory mechanisms. When this capacity is exceeded, we must anticipate that the system would again show directional change with a sequence of transitory states replacing each other through time as the system moves towards a new equilibrium, determined by the new level of inputs or new energy environment it experiences. Again the dominant control of change will be positive feedback, but this time externally induced. Adjustments in the energy and mass balance of the atmosphere resulting from changes in planetary inputs, or changes in denudation and ecosystem function reflecting climatic change, are examples of such externally induced change (to the system as defined).

All of these manifestations of change in systems states are examined in detail in the chapters that follow, but their recognition depends on a clear definition of the system being considered and of the timescale concerned. This timescale must be appropriate to the scale at which the system is defined, and in the case of environmental systems particularly, its spatial scale. This important point was first explicitly recognized in geomorphology in the now classic paper by Schumm and Lichty (1965) (Fig. 23.5).

Appropriate further reading for this chapter can be found in the references at the end of chapter 1 addressing the nature, behaviour and modelling of systems. Reading providing exemplification of the general principles expounded in this chapter will be found at the end of the chapters that follow.

Change in physical systems

24.1 Change in climatic systems

Climate is defined in terms of statistical averages of, for example, air temperature, hours of bright sunshine and precipitation. Within any climate so defined there are periodic changes in these and other parameters over diurnal and annual time scales, which are an integral part of the operation of the Earth–atmosphere system, and do not involve it in long-term structural changes. However, it is possible to identify longer-term changes in climate occurring over periods of the order of 10^1–10^5 years. Some of these changes represent considerable modifications to the mode of operation of Earth-surface and atmospheric systems.

The Earth's surface reveals evidence of periods of extreme cold, when large areas were under thick sheets of ice. Areas which are deserts today contain landforms that are clearly fluvial in origin, suggesting periods of wetter climatic conditions. Measurements of climatic parameters (which have been available only since the early 18th century) reveal, for example, general increases in global mean air temperatures over the period from the late 19th century. More recently, the greater frequency of failure of monsoon rains to reach north-west India, and the extension of desert conditions along the south side of the Sahara in the Sahel, indicate that major changes in the Earth–atmosphere system are currently taking place.

These changes can be viewed alternatively as a result of the changing operation of factors external to the Earth–atmosphere system, or as modifications of the disposition of energy and matter within it. For example, changes in the temperature of the lower troposphere may be due to a changing radiant energy exchange between the Earth–atmosphere system and its surroundings (space). Alternatively, it may be that structural modification within the system has resulted in more or less energy being stored as sensible heat within the atmosphere.

If we view the Earth–atmosphere system as a black box (Chapter 3), then our first alternative concerns us with a discussion of relationships between input and output and we may make general statements regarding the throughput. Changes within the system concern its response to energy inputs, in which case we must consider the pathways along which energy and matter are transferred, the nature of the stores and the rate at which energy and matter are transferred between them. Critical in this respect are the regulators which are the control points within the system.

24.1.1 External changes

If the Earth and its atmosphere were full radiators (Chapter 3), we should expect that any change in the amount of radiation received from the Sun would be matched by increased absorption and emission. A net radiative balance or dynamic equilibrium would thus be maintained. This assumes that the Earth–atmosphere system is capable of establishing such a radiative equilibrium with its surroundings. However, as radiant energy passes through a complex energy cascade, such changes give rise to a temporary state of disorder within the system which results in changes in surface temperatures.

If we were to deal only with radiative exchanges, then a change in the solar constant (ΔR_0) would produce a change in Earth-surface temperature (ΔR_s) given by:

$$\Delta T_s = \frac{1}{4}\left\{\frac{R_0(1-r)}{4\sigma}\right\}^{1/4}\left\{\frac{\Delta R_0}{R_0}\right\}$$

where r = planetary albedo (≈ 0.3) and σ = Stefan's constant = 5.57×10^{-8} W m^{-2} K^{-4}

The information available on solar radiation received by the Earth is limited in quality and quantity for mainly technical reasons. Only in the past three decades has it been possible to monitor incoming solar radiation above the filtering effects of the atmosphere, by using satellites. The alternative is to use high alpine observatories which are still subject to problems of atmospheric transparency. Surface observations of direct solar radiation made at mountain observatories between 30° and 60°N (Fig. 24.1a) show variations within 10% of an overall mean value, while mean daily hours of bright sunshine in Southampton, at only 20 m above sea level, show a haphazard scatter about the mean which is due in large part to a much greater influence of atmospheric transparency (Fig. 24.1b).

In the absence of an atmosphere, long-term fluctuations in solar radiation arriving at the Earth's surface may be due to changes either in emission from the Sun or in the characteristics of the Earth's orbit. The gaseous structure of the Sun undergoes continuous modification through time, which produces changes in total emission. Variations of this nature are relatively small in relation to the solar constant. Sunspots are the centres of localized disturbances in the centre of which the Sun's surface temperature is reduced by as much as 2000 K. Around them are brighter zones, referred to as faculae, from which there is

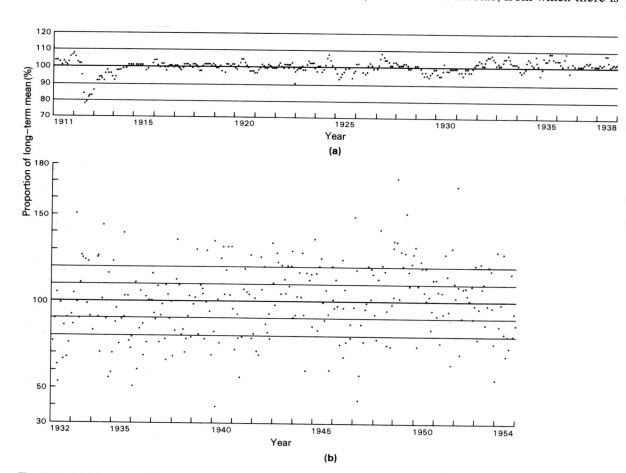

Fig. 24.1 (a) Mean monthly values of the strength of direct solar radiation from observations at mountain observatories between 30°N and 60°N in America, Europe, Africa, and India from 1883 to 1938 as a percentage of the long-term average (after Lamb, 1972). (b) Mean monthly hours of bright sunshine at Southampton (East Park) expressed as percentage of the long-term (1932–54) mean.

intensified radiation. Associated with the occurrence of these disturbances are increases in the emission of both ultraviolet and infrared radiation, and of solar particles, which directly affect the upper atmosphere. The increase in high-energy radiation increases the probability of dissociation of oxygen molecules in the upper atmosphere, which has the effect of increasing the rate of production of the unstable ozone (O_3). The presence of more ozone and the absorption of the extra ultraviolet radiation produces a slight increase in temperatures in the stratosphere. The number of sunspots on the solar surface exhibits a well marked periodicity, the time between successive maxima being about 11 years. Several attempts have been made to relate these sunspot cycles to surface weather patterns, but these have been largely inconclusive, or at best tentative.

There are three characteristics of the Earth's orbit which vary over periods of time measured in tens of thousands of years. These affect mainly the seasonal and geographical distribution of solar radiation over the Earth's surface and not the total energy received. The tilt of the Earth's axis relative to its plane of orbit changes over a period of approximately 40 000 years. The probable range of variation lies between 21°48' and 24°24'. It is currently 23°27' and is decreasing from a maximum value some 10 000 years ago. The significance of this variation lies in the latitudes of tropical and polar circles. An increase in the angle of tilt reduces the latitude of the polar circles and raises the latitude of the tropical circles (Fig. 24.2). With regard to the former, this implies an increase in the area of the Earth's surface which experiences day-long polar nights and days. The effect of changes in tilt of the Earth's axis is, therefore, to alter the seasonal variation in solar radiation over its surface. The shape of the ellipse described by the Earth's orbit around the Sun changes over periods of approximately 100 000 years. The more elliptical the orbit becomes, the more marked are the differences between perihelion and aphelion (Chapter 3). If the Earth described a circular orbit, this difference would be zero. At its most elliptical there is a ± 15% variation either side of an annual mean amount of solar radiation received, while current variation is ± 3.5%. The Earth's orbit itself rotates around the Sun, which causes a change in the timing of perihelion and aphelion, relative to the Earth's seasons. Perihelion currently occurs on 3 January, during the southern hemisphere

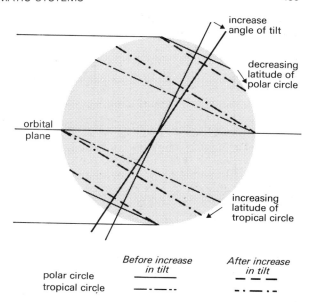

Fig. 24.2 The effect of changing the tilt of the Earth's axis on the polar and tropical circles.

summer. It takes approximately 21 000 years for the cycle to be completed, which means that perihelion in the Earth's orbit occurs one day later every 58 years.

Milankovitch (1930) has estimated the net effect of these three changes in the Earth's orbit (Fig. 24.3), which shows that they have resulted in long-term fluctuations in solar radiation received at the Earth's surface. These variations were linked tentatively with the periods of expansion of the polar ice caps during the ice ages. In excluding the transfers of heat by conduction and convection, and by the complex radiational exchanges that occur at or near the Earth's surface, Milankovitch's estimates of temperature changes resulting from changes in radiation input tend to overestimate fluctuations that would take place. However, the analysis of the isotope ^{18}O in cores taken from ocean core sediments (Shackleton and Opdyke, 1973) has indicated some degree of coincidence with the chronology of climatic changes suggested by the Milankovitch model.

24.1.2 Internal changes

The response of the Earth–atmosphere system to inputs of solar energy is determined by the operation of regulators (Fig. 24.4). The most important of these are the transparency of the atmosphere

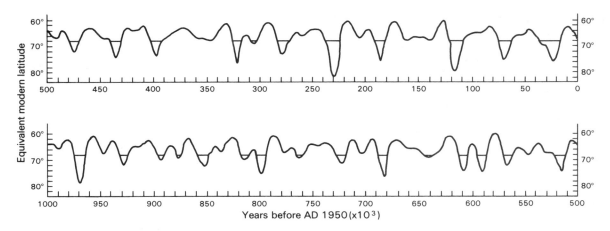

Fig. 24.3 Amounts of solar radiation available in the summer half of the year at latitude 65°N over the past million years, expressed as equivalent to the radiation available at various latitudes at the present day (after Milankovitch, 1930).

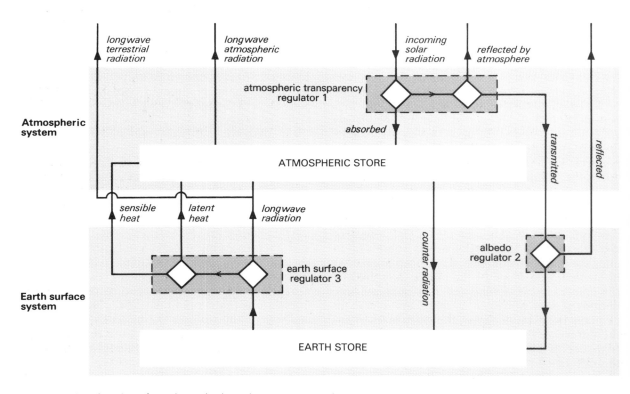

Fig. 24.4 The function of regulators in the solar energy cascade.

to solar radiation, the albedo of the surface, the physical properties of the subsurface, and the action of greenhouse gases. Climatic changes may be a result of modifications to one or all of these.

The operations of regulators, while directly affecting energy distribution through the Earth–atmosphere system, are themselves inextricably linked to the disposition of matter in the stores within the system. For example, an increase in water storage in the form of ice increases the albedo of the Earth's surface. The materials that have the greatest bearing upon regulators are solids and water, together with greenhouse gases (see Section 3.2.2).

Solid materials. Small particles in the atmosphere are derived principally from three major sources: volcanic eruptions, surface dust and the byproducts of combustion (Chapter 4). The force of ejection and the strong thermal convection associated with a volcanic eruption send large quantities of dust and debris into the atmosphere. Most of this falls back to the surface, but some rises to 25 km or more into the stratosphere. For example, the Mount St Helens eruption in May 1980 in the northwest USA sent 1.3×10^5 kg of particles of diameter less than 2 μm into the stratosphere (Hobbs *et al.*, 1982). Dust from volcanic sources is distributed around the world by the upper winds. The dust veil directly affects the transparency of the atmosphere by scattering and absorbing incident solar radiation. Following the Bali eruption in 1963, observations of solar radiation revealed a marked decrease in direct and increase in diffuse components (Fig. 24.5). The reduction in solar heating of the Earth's surface results in a decrease in daytime temperatures, but the dust also acts to reduce longwave radiant energy loss from the surface at night. Immediately after the Mount St Helens eruption, daytime near-surface air temperatures fell by as much as 8°C, while night temperatures increased by almost the same value (Mass and Robock, 1982). The magnitude of temperature changes falls away quickly as much of the dust settles back to the surface. Effects on mean air temperatures over longer timescales tend to be relatively small, of the order of 0.5°C, but may persist for 2 years or more (Fig. 24.6). Many of the cool summers in western Europe during the past 200 years appear to have followed major periods of volcanic activity elsewhere in the world. Similar effects could arise from

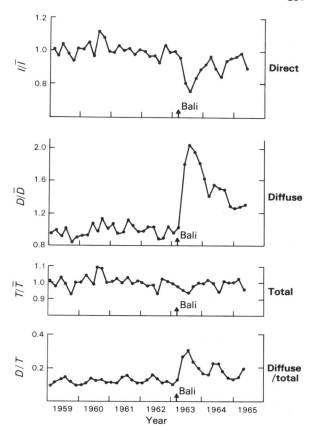

Fig. 24.5 Direct (*I*), diffuse (*D*) and total (*T*) solar radiation, and the contribution of the diffuse (sky) radiation to the total at Aspendale, Melbourne (38°S 145°E) after the Bali eruption in 1963; fractions of the 1959–62 averages (after Dyer and Hicks, 1965).

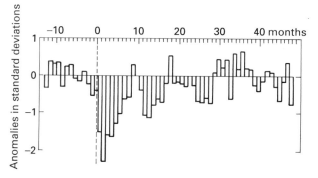

Fig. 24.6 Changes in mean surface air temperature after volcanic eruptions (from Budyko *et al.*, 1988).

a large-scale nuclear conflict, which would produce a protracted period of significantly lower temperatures, referred to as a *nuclear winter*.

Although ice-cores from the polar ice-caps have revealed evidence of the world-wide dispersal and deposition of volcanic dust, it is questionable whether volcanic activity, on its own, would cause a sufficiently persistent reduction in surface temperature to initiate major growth of ice sheets. In order to produce cooling on a large enough scale the dust veil would have to be replenished by major eruptions every few years. On the basis of the available evidence, it seems improbable that they have occurred with sufficient frequency to maintain a persistent dust veil.

Dust may also be lifted from the land surface by the action of wind, and may be carried aloft by turbulent mixing of the atmosphere. Much of this is derived from the deserts but some may be a product of soil erosion. Poor land management may lead to rapid erosion, such as that which occurred in the 1930s in the American 'dust-bowl'. Bryson and Baerreis (1967) have suggested that dust from dry land surfaces in northwest India directly enhances subsidence in the lower atmosphere effectively limiting precipitation from the advancing monsoon. As most terrestrial dust is confined to the troposphere, it is likely to be washed out by precipitation. While it may produce localized and short-term modifications to atmospheric transparency, its long-term effects upon climate may be negligible.

Water. Water is stored within the Earth–atmosphere system in oceans, ice deposits and the atmosphere and on the land surfaces (Chapter 4). Changes may occur in the amount stored in any one of these and also in the rate at which water is transferred between them. The amount of water in the form of ice has fluctuated during the Earth's history, reaching several distinct maxima during ice ages, when polar ice caps and valley glaciers expanded to cover large areas (see Section 24.4). These periods were certainly accompanied by a general lowering of surface and air temperatures and possibly may have experienced greater amounts of precipitation in the form of ice crystals. Climatic data for the latter part of the period 1650–1850, during which there were advances of valley glaciers, suggest that these were accompanied by reduced temperatures and increased precipitation.

The melting of large quantities of ice requires a supply of latent heat at the rate of 3.33×10^5 J kg^{-1}. During the last ice advances, some 10 000 years ago, there were approximately 72×10^6 km^3 of ice over the Earth's surface, compared with the 33×10^6 km^3 of today. Melting of the 39×10^6 km^3 would have consumed approximately 10^{27} J, which, in the absence of an external source of energy, would have to be found from elsewhere in the Earth–atmosphere system.

The increase in albedo, as ice covers the ground and ocean surfaces, reduces the amount of solar energy they absorb. The cold surface reduces the temperature of the air in contact with it which, under conditions of weak atmospheric circulation, encourages stability in the lower atmosphere and the formation of a shallow anticyclone. The divergent nature of airflow prevents the advection of heat over the ice surface and also spreads cold air beyond the ice limits. The advent of a more vigorous atmospheric circulation breaks down the weak anticyclone and initiates ice melt. Under these conditions, air temperatures around the ice limits undergo very marked increases. During the first half of the 20th century, a decrease in the extent of north-polar ice produced amelioration of winter mean temperatures in Spitzbergen of the order of 6°C over less than 20 years.

Changes in atmospheric moisture may arise from a general increase or decrease in surface and air temperatures which enhance or restrict not only evaporation but also the moisture-holding capacity of the air. The atmosphere represents only a very small store of water (Chapter 4), in the form of water vapour, but changes in its contents have far-reaching effects upon surface climates. Increases in atmospheric moisture content are associated with greater amounts of cloud and possibly also of precipitation. As the major absorber of terrestrial radiation, atmospheric moisture also has a considerable bearing upon the **greenhouse effect**.

Atmospheric moisture content is only one of the factors affecting rates of precipitation. The intensity of dynamic cooling of the atmosphere, particularly by cyclonic mechanisms, and the presence of conditions conducive to the growth of cloud droplets are important considerations.

The evidence for long-term changes in precipitation in the record from individual rain-gauges is inconclusive. This is due to the effect of local controls on precipitation mechanisms and on gauge performance. However, aggregated totals for large areas of the Earth's surface do reveal marked changes, such as those which have been taking place since the early 1950s (Fig. 24.7). While rainfall amounts appear to be increasing in middle and higher latitudes, they are decreasing in subtropical latitudes. The recent very wet winters in northern Britain contrast with the persistent droughts in, for example, the Sahel and Ethiopia.

The hydrological system undergoes changes through time which are complex and are a result of internal modifications to the stores and the rate of exchange of water between them. The system is intimately linked to the solar energy cascade not only in the form of radiative, sensible and latent heat transfers, but also in the dynamics of thermally driven atmospheric motion. For example, the changes in rainfall that have already been identified are linked to the dynamics of the polar front convergence and intertropical convergence zones within the atmospheric circulatory system (Chapter 3).

Greenhouse gases. Although water vapour and carbon dioxide are the principal greenhouse gases, others, including nitrous oxide (N_2O), methane (NH_4), tropospheric ozone (O_3) and chlorofluorocarbons (particularly CFCs 11 and 12) are making an increasingly significant contribution to the greenhouse effect (Section 3.2.2). The absorption of radiation by these gases retains heat within the troposphere. An observed increase in global mean temperature of approximately 0.5°C during the 20th century (Fig. 24.8), culminating in the exceptionally warmer years of the mid to late 1980s, may be attributable to increases in the amounts of these greenhouse gases in the atmosphere.

The burning of fossil fuels, deforestation and the burning of vegetation have contributed to an increase in the *carbon dioxide* in the atmosphere. In addition to such anthropogenic sources, the amount of CO_2 in the atmosphere is determined by the dynamic balance between production by living organisms, removal by photosynthesis in plants, and uptake into, or release from, stores such as the oceans. This balance varies periodically over both diurnal and annual timescales. In the oceans, for example, uptake of CO_2 is dominant in winter and release is dominant in summer. Fluxes from and to the ocean surface dominate the global carbon cycle (Fig. 24.9a). Monthly

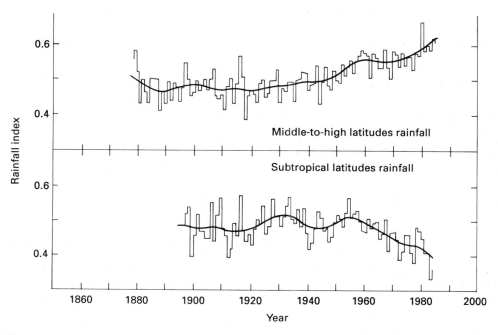

Fig. 24.7 Changes in annual rainfall index for two latitudinal zones (after Bradley *et al.*, 1987).

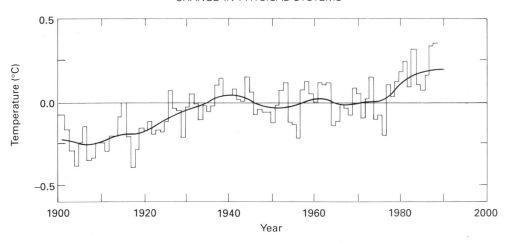

Fig. 24.8 Globally averaged near-surface air temperature in relation to the 1950–79 average (after Jones *et al.*, 1988).

concentrations of CO_2 reveal not only an annual periodicity but also a long-term increase over the period 1958–1988, from approximately 315 ppm to 352 ppm (Fig. 24.10). This amounts to an increase of 12%. As concentrations in the middle of the 19th century were approximately 280 ppm, there has clearly been a recent acceleration in the rate of increase in atmospheric CO_2, which can be attributed to anthropogenic sources (Fig. 24.11). Current predictions are that concentrations will have exceeded 600 ppm by 2050 if no steps are taken to control emissions from such sources. The role of the oceans as the major store of carbon is as yet not fully understood (Fig. 24.9b) but they will certainly exert a dominant control over future atmospheric CO_2 levels.

Methane is derived from the bacterial breakdown of organic matter in swamps, paddy fields and waste tips, and in cattle, together with direct leakage during fossil fuel extraction. It is currently increasing at the rate of approximately 1% per year.

CFCs are exclusively an industrial product, being used as coolants in fridges and air-conditioning systems, propellants in aerosol cans, and in the production of plastic foams. International agreements such as the Montreal Protocol of 1987 may slow down the annual rate of increase in atmospheric CFC, which is currently 6% per year.

Nitrous oxide is a byproduct of fertilizer manu-

facture and use, and of fossil fuel and biomass burning. It is also emitted from vehicle exhausts and is, therefore, a particular problem in large cities such as Los Angeles and Mexico City, where strong solar radiation results in its photochemical conversion to a noxious urban haze. Atmospheric N_2O is currently increasing at approximately 0.4% per year.

Tropospheric ozone (as distinct from stratospheric) results from the photochemical transformation of the carbon monoxide, oxides of nitrogen and hydrocarbons produced by the internal combustion engine. It is increasing at the rate of approximately 2% per year.

The clearance of forests for cattle grazing, the extension of paddy-field agriculture, the increase in the use of nitrogenous fertilizers, the growth of population, especially urban population and the associated amount of road traffic, and the continuing use of organic fuels as a primary source of energy all contribute to a general increase in atmospheric constituents which enhance the heating of the troposphere. If we ignore water vapour, CO_2 accounted for about half of the greenhouse effect in the 1980s, but the increase in, for example, CFCs, which have a very long lifespan and are as up to 20 000 times more effective than CO_2 as greenhouse agents (Table 24.1), will inevitably change the relative contributions by the 21st century unless emission controls are implemented. If current rates of CFC increase are sustained,

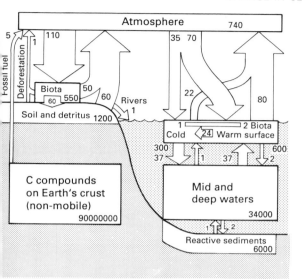

Estimates of the major reservoirs (in 10^9 tonnes of carbon) and fluxes in 10^9 tonnes carbon per year) involved in the global carbon cycle.

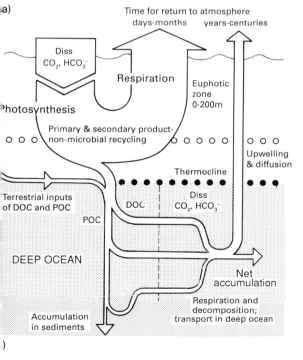

Fig. 24.9 The effect of the oceans on atmospheric CO_2: (a) Oceans and the global carbon cycle; (b) CO_2 uptake and release (from NERC, 1989).

by the year 2000 there will be 90% more in the atmosphere than at present, compared with an estimated 15% increase in CO_2.

Predictions of future changes in climate resulting from increases in greenhouse gases have been based on a range of physical models. The simplest of these are based on two-dimensional heat energy transfers, as represented in Fig. 24.4, in which we can operate on the regulators such as albedo and atmospheric transparency to calculate potential redistributions of thermal energy. However, as we have seen in the preceding chapters, the solar energy cascade, hydrological system and atmospheric and ocean circulatory systems are all interlinked, so it is important that such *coupling* is incorporated into our models. Thus global circulation models (GCMs) deal with a spatially and temporally variable Earth–atmosphere system. However, in addition to the unpredictable course of human behaviour, the role of the oceans in the carbon cycle and in the distribution of heat energy in surface currents and near-surface mixing layer (Chapter 6), the role of clouds in modifying (and being modified by) the heat energy balance of the troposphere, and the dynamics of ice-cap growth and decay, all represent substantial uncertainties in even the most sophisticated of predictive models.

Predictions of future climatic change, or climate-forced environmental changes, are thus going to be in the form of ranges of possibilities. For example, if we assume that atmospheric CO_2 will double by 2050, the globally averaged near-surface air temperature may increase by between 0.5°C and 3.0°C, with larger changes occurring in higher latitudes due to decreases in surface albedo caused by ice-melt. Changes in the hydrological and atmosphere–ocean circulatory systems are predicted with considerably less confidence. Any increase in available heat energy in the troposphere has implications for surface evaporation and the water-holding capacity of the warmer air (Chapter 4), which may in turn lead to increased cloudiness and precipitation. Current predictions indicate a 5–15% increase in globally averaged rainfall within which there will be areas with reductions (e.g. continental interiors, subtropics) and increases (e.g. middle-latitude maritime areas).

An increase in surface temperature will lead to a release of more water from the ice stores within the hydrological system and the expansion of the ocean volume by direct energy uptake from the

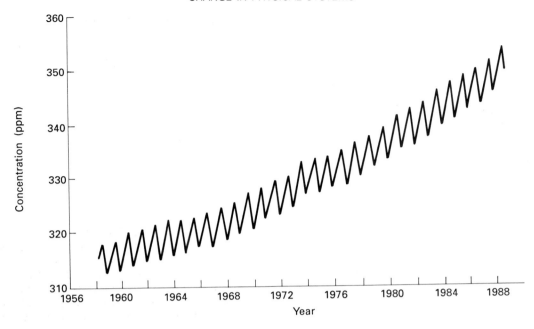

Fig. 24.10   Carbon dioxide in the atmosphere 1958–1989 (from Gribbin, 1989).

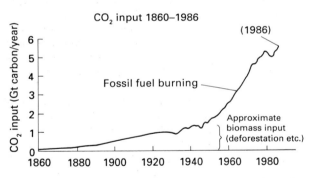

Fig. 24.11   Evolution of CO_2 inputs to the atmosphere due to human activity (from Rowntree, 1990).

solar energy cascade, resulting in a net increase in ocean levels relative to the land masses. The observed increase in global mean temperature in Fig. 24.8 has been accompanied by an increase in mean sea level (Fig. 24.12a) at an average rate of 1.6 mm per year since 1890. Although the cause-effect relationships involved are very complex, the implication is that global warming has already led to rises in sea level. The predictions of future global sea-level rise by 2050 (Fig. 24.12b) indicate the large range of uncertainty that exists, but expectations are that rises will lie within the range 25–40 cm. Within this range, thermal expansion

Table 24.1   Principal greenhouse constituents of the atmosphere (Association for the Conservation of Energy, 1989)

	Global mean increase %	Life	Strength relative to CO_2	Contribution to greenhouse effect in 1980s % in dry air
carbon dioxide	0.5	7 y	1	50
methane	1.0	10 y	30	18
CFCs	6.0	70–20000 y	<20 000	14
nitrous oxide	0.4	170 y	150	6
tropospheric ozone	2.0 approx	weeks	2000	12

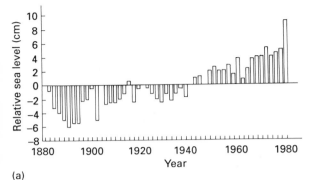

(a)

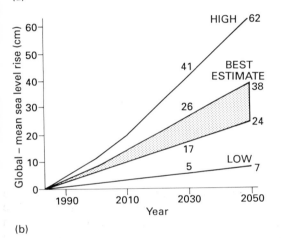

(b)

Fig. 24.12 Changing sea levels: (a) Sea levels since 1880 (Doornkamp, 1990). (b) Predicted sea-level changes to 2050 (after Warrick *et al.*, 1989).

Table 24.2 Estimated contribution to sea-level rise 1980–2050 (Doornkamp, 1989)

	Percentage contribution	
	Highest scenario	*Lowest scenario*
thermal expansion	47	58
Alpine snow/ice melting	25	28
Greenland ice-cap melting	7	6
Antarctic ice-sheet breakup	20	8

servations over the Antarctic revealed dramatic decreases. During the Antarctic spring (September to November) most of the ozone at a height of 15–20 km, usually the level of greatest concentration, now undergoes rapid depletion, almost to zero (Fig. 24.13). The photochemical processes which take place at this height in the atmosphere are complex, but samples taken from an aircraft during the Airborne Antarctic Ozone Experiment in 1987 found unexpectedly high concentrations of chlorine monoxide (ClO) in the areas where ozone was most depleted. Satellite measurements have since shown that the depletion extends over much of the Antarctic stratosphere, and may also be developing in the northern hemisphere. Although stratospheric ClO can originate from the natural source of oceanic methyl chloride, the recent dramatic changes can be attributed directly to the ultraviolet photolysis of industrial CFCs. These are extremely stable and are virtually indestructible in the troposphere. When lifted into the stratosphere by the general circulation they are transformed by the photolytic reactions

$$Cl + O_3 \rightarrow ClO + O_2$$
$$ClO + O \rightarrow Cl + O_2$$

This results in the depletion of O_3 and the formation of an *ozone hole*. Although international efforts are being made to reduce CFC emissions, their longevity may result in yet further reductions in stratospheric ozone. The implications for the Earth's climate remain uncertain, but the reduction in atmospheric stability in the stratosphere may increase the mixing potential of the troposphere, which has implications for vertical heat and moisture transfers.

drives about half the rise, with the remaining half dependent on the dynamics of ice deposits (Table 24.2). The implications of such sea-level changes for coastal systems are considerable (see Chapter 17).

Stratospheric ozone. The absorption of ultraviolet radiation by stratospheric ozone (Section 4.1) fulfils two vital functions in the Earth–atmosphere system. Not only are surface lifeforms protected from the potentially lethal effects of ultraviolet excess, but the absorption from the solar beam also creates a temperature inversion in the stratosphere. This acts as a stable upper limit to convective mixing in the atmosphere. Until recent years, average ozone concentrations remained relatively constant, but in the 1980s ob-

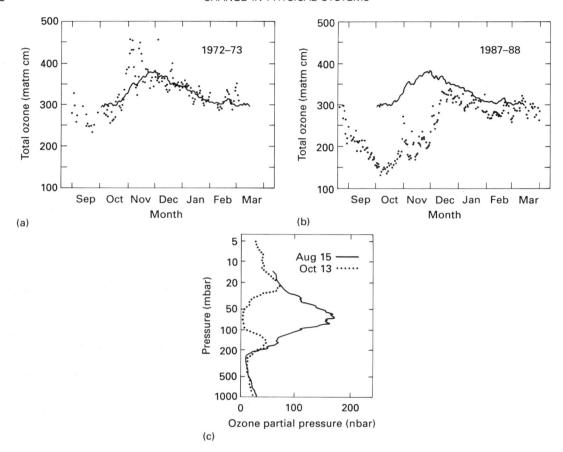

Fig. 24.13 Ozone measurements at Halley Bay: (a) 1972–73; (b) 1987–88; (c) Vertical profile in 1987 (from Gardiner, 1989).

24.2 Feedback mechanisms

External and internal changes to the Earth–atmosphere system are inextricably linked and it is difficult to define singular cause-and-effect relationships. Mitchell (1976) has referred to a resonance between external and internal forces of change. Each observed effect is the end-product of a complex chain of causes and effects. Some of these chains are assembled in such a way as to produce a compensating mechanism for any initial changes imposed on the Earth–atmosphere system. This negative feedback provides for long-term system stability. In others, an initial impetus may produce increasing degrees of change within the system as positive feedback occurs.

If we consider a change in the solar constant, or energy input into the system, it could be argued that the resulting increase in surface heating would increase rates of evaporation. The consequently larger amounts of moisture in the atmosphere would result in increasing cloud cover and hence a decrease in solar radiation reaching the surface (Fig. 24.14a). The Earth–atmosphere system thus compensates for change in a negative feedback loop, which may be joined at any point.

If there were an increase in carbon dioxide, this would cause an increase in the absorption of terrestrial heat energy in the lower atmosphere which would inhibit the cooling of the Earth's surface. The resulting increase in surface temperature could encourage evaporation, increase the water vapour content of the atmosphere and consequent cloudiness, and thereby reduce the

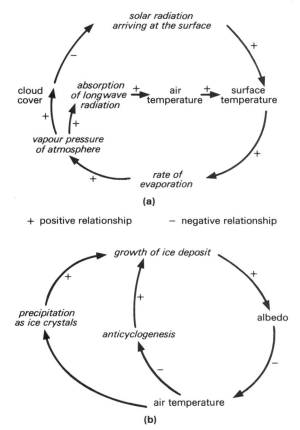

+ positive relationship   − negative relationship

Fig. 24.14 (a) Feedback in the Earth–atmosphere system. (b) Positive feedback in the growth of an ice sheet.

amount of radiation reaching the surface. Alternatively, we can see the effect of changes in solar radiation as initiating positive feedback. For example, the increases in surface temperature would greatly enhance the greenhouse effect. This would cause further increases in surface temperature through inhibiting heat loss and reducing global albedo (Fig. 24.14a).

The covering of a surface by ice increases albedo, thereby reducing solar heating. The air above is cooled and this generates surface divergence of airflow, which restricts the advectional inflow of heat. Therefore, the ice sheet would extend its area not by further inputs but by restricting ablation, thereby exhibiting positive feedback (Fig. 24.14b). An alternative view is that cooling of the air increases the likelihood of pre-

cipitation in the form of ice crystals, thereby adding to the ice deposit.

These feedback loops are clearly oversimplifications of the very complex interrelationships that exist between meteorological variables within the Earth–atmosphere system. They do, however, illustrate that the system is, to a large extent, self-regulating in that it is capable of modifying its mode of operation to accommodate changes in inputs or its own regulators.

24.3 The form of change

Changes in climate are normally expressed in terms of trends. Over short periods, measured in decades, monotonic trends may be identified, such as the increase in air temperatures which took place during the early part of the 20th century. However, most analyses of long-term climatic changes have sought to identify oscillations in the data. Such oscillatory changes take place about a mean long-term value. Most climatic data may not exhibit a single periodicity similar to that of the sunspot cycle, but may be compounded of several interlocking periodic changes through time. For example, Milankovitch's calculations of solar radiation represent a superimposition of three periodicities. An analysis of ice cores from the Greenland ice sheets has identified two possible periodicities in global temperature of 80 and 180 years, over the past 800 years.

In the absence of recognizable periodicities there is the possibility that climatic change is irregular, or possibly random, as suggested by Curry (1962). In the established presence of regular changes in the orbital characteristics of the Earth and of the gaseous content of its atmosphere it seems unlikely that climatic changes could be random. However, it may prove difficult to assign causes to all observed variations in, for example, air temperature and precipitation.

24.4 Zonal and dynamic change in the denudation system

In considering the operation of geomorphological systems, Thornes (1987) expresses the view that steady-state equilibrium may in reality be unusual, and that changes in system behaviour through time may be the norm. He employs the term **dynamical system** to describe situations in which the dynamics of a system change through time (Box

24.1). Change in a system will take place when that system is subject to a perturbation, or disturbance. This may be due to a change in external (boundary) conditions, or to an internal structural instability which results in a critical threshold being crossed. The response of a system to such a change may follow a variety of paths, as a plot of system variables through time demonstrates (Fig. 24.15). Thus a damped response is one in which the variation diminishes with time, whereas an explosive response demonstrates progressive change (see Chapter 23). A periodic response shows cyclic change and an unsystematic response shows chaotic or random behaviour. During all of these types of response the system is by definition in a state of change (see also Chapter 25), and as a consequence the characteristic form associated with the steady state of inputs and processes is replaced by an adjusting form associated with the change. This may apply both to the operation of the system and its physical form in the landscape.

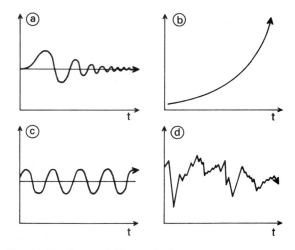

Fig. 24.15 Types of Change in Systems.
(a) damped response, (b) progressive response, (c) cyclic change, (d) random change.

Box 24.1
DYNAMICAL (MODELS) SYSTEMS

As we saw in Chapter 1, and in Box 1.3, one has to be aware of how the term 'system' is being used to avoid confusion. Here, Thornes is using it to refer to both a system of equations constituting a mathematical model, and a system of relationships constituting the conceptual model upon which the mathematical model is predicated. The real thermodynamic system – the denudation or geomorphic system – is conceptualized and modelled as one where its behaviour through time can be described from a knowledge of its initial state, the length of the time interval involved, and information on the inputs to it and outputs from it over that time interval. This conceptual model is translated into its mathematical equivalent, where both the output from the system and its change of state over a time interval, are *functions* of the initial state and the inputs. All systems which can be modelled as having both a **state**

transition function and an **output function** can be regarded as **dynamical** systems. A moment's reflection should indicate that virtually all environmental energy systems can be so modelled. However, Thornes' contention goes beyond modelling to stress that the **dynamical systems approach** moves the intellectual focus away from long-term steady or stable states towards an understanding of transient states, multiple equilibria and the recognition of bifurcation points and inherent instability in the behaviour of systems. Furthermore, the dynamical systems approach allows the integration of conceptual developments such as **catastrophe theory** and **chaos theory** and permits the behaviour of environmental systems to be viewed in the mathematical landscapes they create (Thornes, 1987) (see also similar approaches to modelling change in Ecosystems, Section 2.1.4).

Change in systems due to disturbance can take place at all scales, from the global to the local. Examples of changes in boundary conditions at the global and regional scales are fluctuations in sea level and climatic change. Events at these scales affect entire denudation systems at the scale of the continent and the large drainage basin. At a smaller physical and temporal scale, a pertinent example is provided by the cessation of basal removal from a slope system (Brunsden and Kesel, 1973), which results in a progressive change in the form of the slope system as it adjusts to the change in its boundary conditions. Change in smaller geomorphological systems may be brought about by a single denudational event. Thus Anderson and Calver (1977) study the effect of an exceptional river flood event with a large recurrence interval, and evaluate the persistence in the landscape of the changes it created.

It is implicit in the previous paragraphs that landscapes will comprise both forms associated with a steady-state system and forms resulting from change. Brunsden and Thornes (1979) have attempted to embrace these by developing the concept of **landscape sensitivity to change**. This is evaluated in terms of the recurrence interval of perturbing (or formative) events, in combination with the relaxation (or recovery) time of the landscape itself. Landscape sensitivity to change is then expressed in terms of a transient form ratio:

$$TF_r = \frac{\bar{R}}{\bar{D}}$$

where $\bar{R}$ = mean relaxation time and $\bar{D}$ = mean recurrence interval of the formative event.

If $\dfrac{\bar{R}}{\bar{D}} < 1.0$, then stability returns;

If $\dfrac{\bar{R}}{\bar{D}} > 1.0$, transient behaviour occurs.

Thus following a perturbation, a characteristic form resulting from steady-state conditions will be replaced by an adjusting condition (Fig. 24.16). If the relaxation time is less than the recurrence interval of the event, then a new characteristic form will be attained. If, however, perturbing events are frequent and their recurrence interval is less than the recovery time, then transient forms indicative of progressive change will ensue.

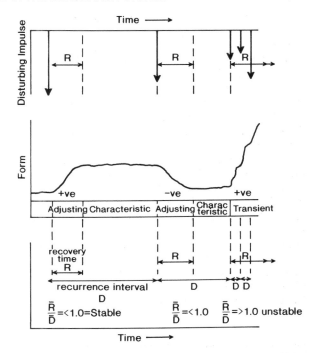

Fig. 24.16 Schematic representation of landscape sensitivity to change, indicating relations between perturbing events, recovery time, and adjusting, characteristic and transient forms (after Brunsden and Thornes, 1979).

The sensitivity of landscapes to change will vary according to their resistance to particular perturbing events, and will vary spatially according to landform and material variations. It therefore follows that any given landscape will comprise zones of sensitivity and short relaxation time, in which systems are in equilibrium with current conditions, and less sensitive zones which retain forms created by former and different conditions. In the case of major environmental changes, such as climatic change, these are known as **relict landforms**.

In recent years evidence has accumulated to suggest that the major external controls on geomorphological systems have been subject to cyclic change continuously over the past 3 million years. Evidence from cores of deep ocean sediments reveals changes in the isotopic composition of seawater, as recorded in the tests (shells) of calcareous organisms. During periods of accumulation of ice sheets on the continents, ocean water becomes depleted in the lighter ^{16}O isotope of oxygen in relation to the heavier ^{18}O isotope. Thus the ratio of ^{16}O:^{18}O provides a measure of the extent of glacial ice accumulation on land, and the associated lowering of global (eustatic) sea level as the ocean store of water is depleted (Fig. 24.17). The evidence suggests that there have been some 30 significant episodes of glacier ice expansion during the last 3.2 Ma (Shackleton and Opdyke, 1973, 1977), with a regular periodicity of 10^5 years. There appears to be a regular pattern of slow accumulation of ice over 90 000 years, followed by rapid disintegration and a 10 000-year interglacial episode. The oceanic record is supplemented by the stratigraphic record provided by fine-grained terrestrial sediments accumulated in unglaciated terrains. Thus accumulations of loess, and the record contained within it of weathering cycles in the form of soils, demonstrates a sequence

of 17 glacial episodes during the past 1.7 Ma in localities as widespread as middle Europe (Fink and Kukla, 1977) and China (Heller and Liu Tungsheng, 1982).

These recurrent major climatic and sea-level changes at the global scale have profound implications for environmental systems, in that change appears to have been the norm for over 3 million years. Thus the potential energy for denudation of the continents has waxed and waned with each fall and rise of sea level. The growth and decay of ice sheets would have been accompanied by the march of climatic belts back and forth across the continents. A particular point on land would therefore have experienced a sequence of climatic changes, with their associated perturbations of local environmental systems. As a consequence, evidence of environmental change is widely manifest in contemporary landscapes. Thus arid forms are present in the humid tropics of the Amazon basin (Tricart, 1985), pluvial forms are present in arid lands (Goudie, 1977; Street-Perrott *et al.*, 1985), and polar forms are widespread in temperate lands. the following section briefly draws attention to some selected examples of evidence of change.

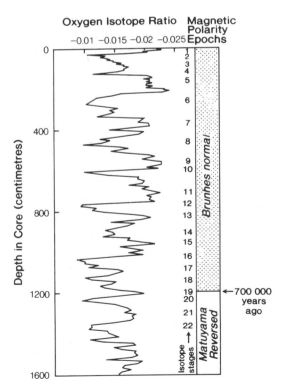

Fig. 24.17 Oxygen isotope ratios derived from Pacific ocean core V28–238. Values are expressed as variation in parts per thousand of ^{18}O from a standard (after Shackleton and Opdyke, 1973).

24.4.1 Dynamic change

Changes in the relative levels of sea and land masses can be brought about by changes in either factor. Land-level changes (isostatic) are outlined in Chapter 5. The chief causes are orogenic uplift of fold mountain belts, uplift or depression of fault blocks as a result of rifting, epeirogenic deformation of broad continental areas, and downwarping of continental margins adjacent to basins of sedimentation. Locally, the development of an ice sheet can increase the load on a continent and depress its elevation, which returns to its original level subsequent to the glacial episode. While in the long term this can be viewed as a temporary effect, there are several areas which are at the present time recovering from glacial loading and being uplifted at a rate of several millimetres per year. These processes operate independently of sea level and they affect land masses unequally across their area. Accordingly, the changes in potential energy are distributed unevenly, and within one land mass some areas may be increasing and others decreasing in potential energy. (See Figure 5.10)

Sea-level changes (eustatic) operate on a worldwide scale, affecting all land masses equally and simultaneously. There are two main causes:

1. The mobility of the oceanic crust renders it inherently unlikely that the ocean basins can retain a constant form. There is the probability of change in the form and volume of the ocean basins, particularly due to the growth and decay of ocean ridge systems. Assuming a constant mass of water in the ocean basins, any change in basin volume would result in a rise or fall of water level relative to the continents.
2. The transfer of large masses of water during glaciation, from storage in ocean basins to storage in ice sheets, has resulted in repeated changes in the volume of water in the ocean basins, lowering sea level by more than 100 m.

The relative effects of eustatic and isostatic change can be determined from tide-gauge records in areas of continental crust of contrasting stability. Stable coastlines will record only eustatic changes, while in unstable areas local effects will be superimposed on the worldwide process.

Glacially controlled changes in sea level may attain a range of at least 100 m between phases of interglacial high sea level, when the water is in the oceans, and phases of glacial low sea level, when much more of the water is in glacial storage on the land masses.

Minor oscillations occurred on a smaller time-scale. This can be illustrated with reference to the last major (postglacial) rise in sea level, in Britain termed the **Flandrian transgression** (Fig. 24.18). More detailed evidence is available for this event, and shows a sea level rising with marked oscillations towards the present level at 6000 BP (before present).

On any one land mass the overall change of elevation in relation to sea level will be the result of the combined effect of eustatic and isostatic changes, superimposed upon each other. The relative rates of change generated by these different mechanisms varies considerably (Table 24.3). The

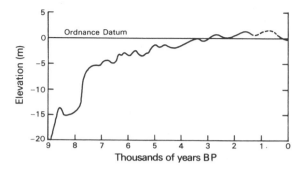

Fig. 24.18 Flandrian sea-level change in northwest Britain (after Tooley, 1974).

most rapid changes appear to be associated with glacial eustatic sea levels. In a crustally stable region such as southern Britain, the last major event of this kind, with the most recent effects, was the Flandrian transgression, which terminated as recently as 6000 years ago, although the current rate of eustatic rise is by no means negligible.

The effect of a change in base level is felt most immediately in the coastal zone. Thus the coastline will tend to advance or retreat, and when a sea level is stabilized again coastal erosion and deposition processes will make their mark at the new stable level.

The effects of increased potential energy caused by uplift are transmitted through the denudation system. Initially, river channel gradients downstream are locally steepened, leading to increased velocity and erosion and the incision of the river channel. Fig. 24.19 shows how this can occur at both coastal and inland locations. Incision of the channels has the effect of both lengthening and steepening valleyside slopes, thus imposing greater erosional stresses on them. Accordingly, there will be a greater volume of eroded material from the slopes delivered to the river channel, and the rate of operation of the denudation system, i.e. the rate of erosion, increases. Downstream, a large river will tend to become incised into its former floodplain, which is left upstanding as a dissected river terrace, while the river forms a new floodplain at a lower level. Repeated incision in

Table 24.3 Representative rates of relative land and sea level changes (after Carson and Kirkby, 1972, from various sources)

	mm a⁻¹
glacial eustatic	up to 25
current eustatic rise	1.2
orogenic uplift	
California	3.9–12.6
Japan	0.8–7.5
Persian Gulf	3.0–9.9
epeirogenic uplift	0.1–3.6
isostatic	
Fennoscandia	10.8
Southern Ontario	4.8

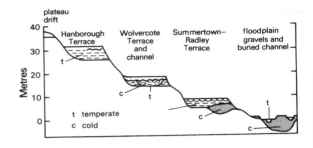

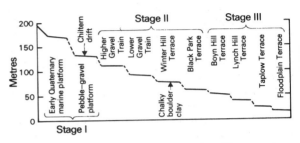

Fig. 24.20 Terrace sequences of the River Thames; above in the upper Thames (after Sandford, 1954) and below in the middle and lower Thames (after Wooldridge, 1960).

(a) Interglacial conditions (humid–temperate)

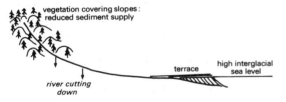

(b) Periglacial condition (frost climate)

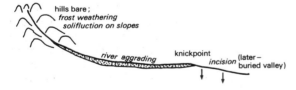

Fig. 24.19 The incision of river channels in both inland and coastal locations and under different climatic controls (after Clayton, 1977).

A rise in sea level relative to land will tend to affect denudation systems more locally. First, the seaward positions of river valleys may become drowned, particularly if such valleys are deeply incised. Thus all the major rivers of southern Britain have deep buried channels, formed in response to low glacial sea levels. Such buried valleys become sediment traps, and as the river's velocity is reduced as it flows into the lower part of its valley being drowned by the rising sea level, sediment accumulates in the submerging valley. However, such effects are not transmitted upstream in the same way as incision, and the upstream operation of the denudation system may continue as before.

24.4.2   Zonal change
Although the overall world pattern is still far from being completely understood, it is clear that con-

this way may result in a flight of terraces, each one bearing the deposits of a former floodplain. Fig. 24.20 shows the terrace sequence of the Thames formed in this way.

siderable zonal changes have taken place during the Quaternary era. On the geomorphological timescale these changes are sufficiently recent to have formed features in the landscape that are clearly visible today. Currently temperate environments may have experienced phases of glacial or periglacial conditions during this time. In subtropical latitudes there were fluctuations between arid and moist (pluvial) phases, although they were not necessarily synchronous throughout the tropics, nor in step with the changes in temperate latitudes. The Quaternary fluctuations in the world circulation pattern produced effects that would vary from place to place, and which remain to be worked out. It would appear, however, that the zonal changes were most pronounced in midlatitude and marginal environments, while the core of the equatorial zone remained largely unaffected by climatic change. This general equatorward shift of climatic belts means that a particular point on the Earth's surface would experience a change in climatic inputs. Depending on the existing regime and the nature of the changed conditions, a wide variety of responses took place in denudation systems.

A trend toward greater humidity in a cold environment may lead to the development of glaciation, if winter snow accumulation exceeds summer ablation. In humid and arid environments greater river runoff will tend to occur, with increased frequency of floods. Increased humidity would tend to allow the development of more luxuriant vegetation and accelerate soil and weathering processes. In arid and semi-arid environments the increased precipitation and runoff would lead to an increase in the volume of inland lakes. Trends towards aridity result in a decrease in vegetation cover and an increased susceptibility of surface material to transport. Under these more arid conditions the transporting agent is likely to be wind, and this results in the loess accumulations of the tundra zone and sand-dune activity in the low-latitude arid zones.

Under the changed environmental conditions a variety of new features may develop. Altered denudation processes may produce new landforms, and new agents and environments of deposition may produce new kinds of sedimentary deposit. Changed weathering processes may modify weathering mantles and soils. Features may remain in the landscape which are the product of previous conditions. These are **relict landforms** or,

in the case of fossil soils, **palaeosols**. Together they constitute evidence from which the history of climatic change can be reconstructed. They will tend to persist in the landscape until they become obliterated under the new conditions. Fortunately, such major changes in environmental inputs occurred throughout the Quaternary, and so little time has elapsed since, that abundant relict forms currently exist in many different environments.

It is marginal areas, at the boundary between climatic zones or systems, that are most sensitive to change, and in such environments it is possible to monitor zonal change even on a short timescale. Glacier fluctuations and changes in desert margins represent two environments where change is sufficiently rapid to do this.

Within the timescale available to modern methods of scientific measurement of environmental variables, changes can be demonstrated in the arid zone boundary in west Africa south of the Sahara. Rainfall records over the period indicate a progressive decrease in precipitation. In such a marginal environment a desiccation of this magnitude brings about a change from semi-arid condition to desert. It is estimated that the desert boundary has shifted southwards and extended the desert in this region by up to 100 km during the period 1968–76. Within this marginal zone, therefore, there has been a change from semi-arid to arid environmental conditions.

Another type of system in which zonal change can be identified readily over a contemporary and historic timescale is the glacial system. This is simplified by the fact that the system boundary is easily identified, i.e. the glacier margin. Fig. 24.21 shows the changes in the margin of the valley glacier Tunsbergdalsbreen, in southern Norway (Mottershead and Collin, 1976). These have been identified on the basis of historical and botanical evidence, notably the use of lichens as indicators of ground-surface age (Mottershead and White, 1972, 1973). The evidence indicates that the glacier has receded some 2.4 km since the mid-18th century, at an ever-increasing rate.

The foregoing are just two examples of environmental change currently taking place in the vicinity of system boundaries. During Quaternary times and earlier, changes of much greater magnitude took place, causing zonal shifts of hundreds of kilometres in some cases. The operation of denudation systems no longer active in these regions formed suites of landforms which are out

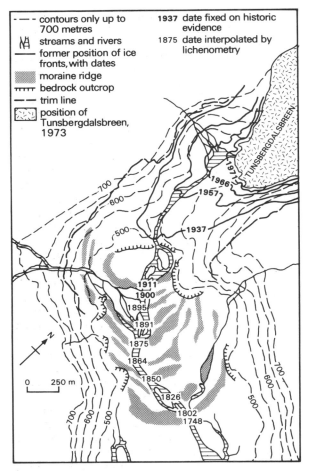

- - - contours only up to 700 metres
- streams and rivers
- former position of ice fronts, with dates
- moraine ridge
- bedrock outcrop
- trim line
- position of Tunsbergdalsbreen, 1973

1937 date fixed on historic evidence
1875 date interpolated by lichenometry

Fig. 24.21 Tunsbergdalen, southern Norway, showing recessional moraines and lichenometric dates (after Mottershead and White, 1973).

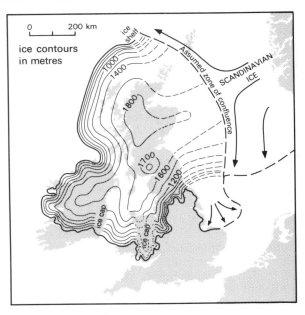

Fig. 24.22 The Devensian ice sheet in the British Isles (after Boulton *et al.*, 1977).

of phase with contemporary conditions – fossil or relict forms. We shall examine two cases which show the contrasting effect of environmental change. In Britain, where glacial and periglacial systems were formerly active, contemporary temperate conditions of denudation have permitted the survival of glacial and periglacial forms. In arid regions of Africa, landforms created under pluvial conditions persist in the landscape, preserved by the limited effects of arid zone denudation.

Glacial features in the temperature zone. Fig. 24.22 shows a reconstruction of conditions in the

British Isles during the late Devensian glacial maximum – *ca.* 18 000 BP (Boulton *et al.*, 1977). Glacial conditions are seen to cover much of Britain north of a line from the Severn to the Trent. Periglacial conditions existed to the south of the ice margin. In these two zones relict glacial and periglacial landscapes can be detected readily today, as evidence of former conditions.

The late Devensian ice sheet measured 1000 km from north to south, and 600 km west to east, where it became confluent with the Scandinavian ice cap. It accumulated to a maximum depth of over 1800 m in the Scottish Highlands, with subsidiary highs extending across northwest Ireland and southwards from the Southern Uplands towards Wales. This reconstruction of the ice sheet can be matched with patterns of glacial erosion (Fig. 24.23), using a scale devised by Clayton (1974) and summarized in Table 24.4.

There is a general gradient from the landscapes most intensely affected by glacial erosion close to the centre of the ice sheet, with the least affected landscapes towards the margins where the effective pressure of ice on the bed is least. It should be remembered, however, that this ice-sheet model applies only to the maximum glacial extent. Glacial conditions will have lasted longer in the core

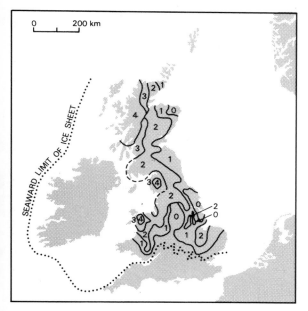

Fig. 24.23 Zones of glacial erosion in the British Isles, using the scale devised by Clayton (1974) and summarized in Table 24.2.

regions of the highlands, both preceding and postdating the glacial maximum. Furthermore, during the early and late stages of glaciation the ice mass was of a different form – i.e. as cirques and valley glaciers, often in discrete bodies – in these regions. These may explain the intensely glaciated landscapes in Snowdonia and the Lake District.

In regions of intense glacial erosion, spectacular assemblages of glacial erosional landforms remain. Since such regions in Britain are invariably highlands developed on resistant bedrock, the glacial landforms have survived the 10 000 years since deglaciation in almost pristine condition (Fig. 24.24).

Periglacial features in the temperate zone. Beyond the limits of glaciation in southern England periglacial conditions existed, probably at several different periods and periglacial forms are well developed and remain preserved unmolested by the ice sheets which would have destroyed any such forms further north. Evidence of permafrost is present in the form of fossil tundra polygons and ice wedges. More widespread, however, and

with a more significant morphological impact on the landscape, are the effects of accelerated mass movement. These are particularly prevalent on the chalklands and in southwest England. This can be illustrated with reference to the landscape of the south Devon coast to the west of Start Point (Mottershead, 1971). Here periglacial conditions brought about a considerable metamorphosis in the landscape. Mass movement caused the transfer of regolith down bedrock slopes of 20–30° to form sediment accumulations at the base. These periglacial sediments are typically unsorted, with angular frost shattered clasts, and are very stony. Consisting of locally derived debris, the deposits are in part bedded, suggesting that the material was transported in sheets or layers, as in gelifluction terraces. (See Figure 25.20).

The accumulation of slope-foot sediments forms gently sloping terraces, concave in profile and thinning towards the sea. Where sediments have accumulated at the base of two opposed slopes confined in a valley, a wedge-shaped fill of material has accumulated on the valley floor, as in some of the small valleys running down towards the sea. The removal of regolith from the upper parts of the slopes has resulted in the exposure of salients of unweathered bedrock to form tors. Their craggy outlines, defined by the fracture patterns in the schists of which they are composed, dominate the crestlines. Periglacial frost shattering of exposed rock outcrops resulted in their partial modification and the production of angular rubble which strews the slopes beneath to form blockfields. This assemblage of tor, blockfield and gelifluction deposit represents a periglacial landscape in microcosm (Fig. 24.25). The degree of

Table 24.4 Zones of intensity of glacial erosion (after Clayton, 1974)

0   no erosion

1   ice erosion confined to detailed or subordinate modification

2   extensive excavation along main lines of flow, glacial troughs common

3   comprehensive modification of preglacial forms, scoured landscapes common in lowlands, interconnecting systems, troughs extensive in uplands

4   ice-moulded streamlined forms dominant in the landscape

Fig. 24.24 Landscape of glacial ice-sheet erosion. The lowland shows evidence of widespread glacial scouring with isolated knolls of rock swept clean of regolith, and enclosed rock basins containing lakes. Elongated remnants of high land were streamlined by the ice, which flowed from right to left across this landscape (Sutherland, Scotland).

modification of the landscape by periglacial conditions can be estimated from the depth of accumulated sediments, in places up to 25 m. Thus the elevation of the pre-periglacial valley floors would have been lower by such an amount, while the crests may well have carried a mantle of regolith, increasing their elevation slightly. The degree of modification of relative relief under periglacial conditions, then, may locally be in excess of 25 m.

That the periglacial forms are not in equilibrium with contemporary conditions is shown by the fact that they are currently being destroyed by erosion. Marine erosion is trimming back cliffs cut in the coastal gelifluction deposits, while contemporary streams are incising themselves into the valley fills to form valleyside terraces.

The legacy of periglacial conditions in this small area is therefore considerable, and indeed well preserved. However, this example is by no means untypical. Similar landscapes are preserved in many locations in southwest England, while the chalklands of southern England have their own distinctive periglacial features (French, 1973).

Pluvial features in the arid zone. Geomorphological evidence of former conditions both more and less arid than the present is to be found in the Sahel region to the south of the Sahara (Fig. 24.26).

Dune systems indicate the extent of truly arid conditions. Under such conditions a perennial cover of closed vegetation cannot exist and loose surface sediment is distributed readily by wind to create dune forms. The present limit of active dunes is shown in Fig. 24.26, while extending over 500 km to the south is a zone of old vegetation-covered dunes. In this zone clearly arid conditions existed, permitting the extension of the desert. A subsequent change to more humid conditions permitted the development of vegetation which covered the old dunes, stabilizing them and creating a fossil dune landscape.

Evidence of former pluvial conditions is preserved around Lake Chad, a basin of inland drainage (Fig. 24.27). This is a system in which inputs are derived mainly from surface runoff from adjacent highland regions augmented by direct

Fig. 24.25 Relict periglacial landscape. The vegetated cliffs represent the accumulation of periglacial gelifluction deposits derived from the slopes above, which are crowned with salients of bedrock (tors). A dissected raised-shore platform is indicative of a former high sea level, predating the periglacial conditions (Devon, England).

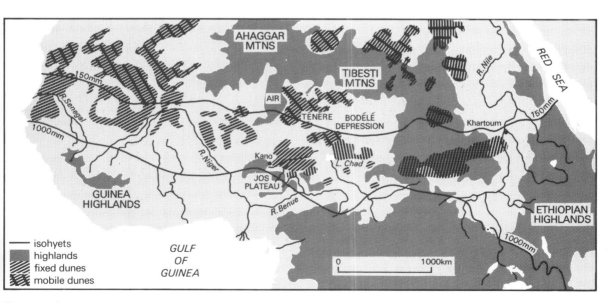

Fig. 24.26 The distribution of mobile and fixed dunes in relation to present rainfall in the Sahel and southern Sahara (adapted from Grove and Warren, 1968, *Geog. J.* **134**, 194–208, by permission of A.T. Grove).

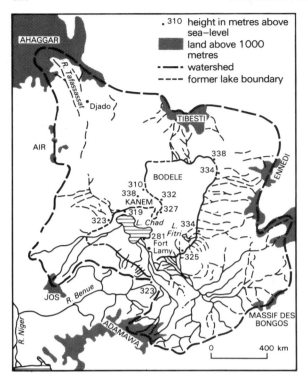

Fig. 24.27 The present extent of Lake Chad compared with the old shore lines of the former 'Mega-Chad' at about 320 m (after Grove, 1967).

precipitation, and are balanced by outputs in the form of direct evaporation from the water surface. The lake, therefore, acts as a store, sensitive to climatic changes which affect both precipitation and evaporation. The morphology of the lake basin is such that small changes in the depth of water present results in large variations in surface area. Its present mean depth ranges between 3 m and 7 m, and its area varies from 10 000 km² to 25 000 km². The shallowness of the basin means that its area is very sensitive to changes in lake volume.

Past pluvial conditions are indicated by elevated strandlines, sand bars and deltaic features. A strandline at 50 m above the present lake would have been associated with a former lake with an area of 400 000 km². Such a lake would have suffered evaporative losses of about 16 times the present rate, and would be balanced by surface runoff of the rivers into it. Clearly the maintenance of the lake store at this high level requires

conditions which are considerably more humid than those prevailing at the present, and Grove (1967) suggests that such conditions existed as recently as 10 000 or even 5000 years ago.

It should be stressed that these isolated case studies at different scales and from disparate regions are illustrative of a widespread phenomenon. Most of the world's landscapes bear at least some inherited features, indicating different zonal conditions in the past. The range of possible inherited forms is quite considerable, from fossil soils through sediments, which may be of sufficient volume to have morphological expression, to the spectacular landforms of glacial erosion.

Further reading

Various aspects of climate change are treated in the following:

Bach, W. (1983) *Our Threatened Climate*. D. Reidel, Dordrecht.
Cannell, M.G.R. and M.D. Hooper (1990) *The Greenhouse Effect and Terrestrial Ecosystems of the UK*. ITE Res. Pub. No. 4. HMSO, London.
Denton, G.H. and T.J. Hughes (1983) Milankovitch theory of Ice Ages: hypothesis of ice sheet linkage between regional insolation and global climate.
Flohn, H. and R. Fantechi (eds) *The Climate of Europe: Past, Present and Future*. D. Reidel, Dordrecht.
Houghton, J.T., G.T. Jenkins and J.J. Ephraums (eds) (1990) *Climate Change: the IPCC Scientific Assessment*. Cambridge University Press, Cambridge.
Lamb, H.H. (1977) *Climate: Present, Past, and Future*. Methuen, London.
Lamb, H.H. (1982) *Climate, History and the Modern World*. Methuen, London.

Acid rain is the subject of an increasing number of publications, just three of which are given here:

Pearce, F. (1987) *Acid Rain*. Penguin, London.
Reuss, J.O. and D.W. Johnson (1986) *Acid Deposition and the Acidification of Soils and Waters*. Springer Verlag, New York.
UK Terrestrial Effects Review Group (1988) *The Effects of Acid Deposition on the Terrestrial Environment in the UK*. HMSO, London.

Some theoretical aspects of landscape change are considered in the following:

Brunsden, D. and J.B. Thornes (1979) Landscape sensitivity and change. *Trans. Inst. Brit Geog. NS*, **4**, 463–484.
Thorn, C.E. (ed) (1982) *Space and Time in Geomorphology*. Allen & Unwin, London.
Thorn, C.E. (1988) *Introduction to Theoretical Geomorphology*. Unwin Hyman, London.
Thornes, J.B. (1983) Evolutionary geomorphology. *Geography*, **68**, 225–235.
Thornes, J.B. (1987) Environmental systems – patterns, processes, and evolution, part 1.2 in *Horizons in Physical Geography*. (eds M.J. Clark, K.J. Gregory, and A.M. Gurnell) Macmillan, Basingstoke.

Thornes, J.B. and D. Brunsden (1977) *Geomorphology and Time*. Methuen, London.

Quaternary environmental change is well covered by the following texts:

Catt, J.A. (1988) *Quaternary Geology for Scientists and Engineers*. Ellis Horwood, Chichester.

Goudie, A.S. (1977) *Environmental Change*. Clarendon, Oxford.

Imbrie, J. and K.P. Imbrie (1979) *Ice Ages: Solving the Mystery*. Macmillan, London.

Lowe, J.J. and M.J.C. Walker (1984) *Reconstructing Quaternary Environments*. Longman, London.

Recent developments in our understanding of ice ages are covered in:

John, B.S. (1979), *The Winters of the World*. David & Charles, Newton Abbot Ch. 7.

Selby, M.J. (1985) *Earth's Changing Surface*. Clarendon Press, Oxford, Ch. 16.

Change in living systems

25.1 Inherent change in the ecosystem

Biological (living) systems are characterized by a capacity for progressive directional change – the cumulative manifestation of positive feedback. The functional activities of cells are concerned not only with the maintenance of a steady state but also with the investment of surplus matter and energy in cell division and growth. As we saw in Chapter 7, the accumulation of mass, the construction of complex ordered structures, such as organic macromolecules, and their intricate and precise arrangement to form the living cell, are all the expression of the inherent capacity of living systems to decrease their *internal entropy* and to increase in complexity and order as the result of controlled growth. In the following sections we shall examine these process at the level of the population, community and ecosystem, but first we shall consider the individual organism. Many of the features of the growth model developed for the organism can still be recognized when applied to higher levels.

25.1.1 Growth and development of the organism

In multicellular organisms organized cell growth and division take place rapidly at first and lead to the appearance of specialized regions of differentiated cells. For example, in flowering plants terminal or axillary buds and root tips are growing masses of cells, sources for the production of initially unspecialized cells from which multiple structures develop successively. As long as these shoot and root **meristems** continue this function, new stem, new leaves and new roots will be added to the growing plant. Such growth, however, is controlled. In this example, a combination of external regulating factors, such as temperature and photoperiod, and internal control

mechanisms which include the effect of the **auxins** (plant growth hormones) and particularly the **phytochrome P_r and P_{fr} system**, will lead at some time to the production of a flower and the cessation of growth in that system (Box 25.1).

In animals the developing embryo gradually produces specialized regions of differentiated cells which assume different functions, and it progressively develops towards an identifiable individual organism. Growth continues until sexual maturity is reached (Chapter 20), but in many animals (for example most vertebrates) at this point there is no further increase in size. In the adult animal, the organism as a living system is simply maintaining a steady state. However, as it passes into a post-mature phase this steady state takes on a downward trend as production per unit biomass falls and ageing becomes apparent (Fig. 25.1). The time scale of individual growth masks the activity going on within the organism at the molecular and cellular levels. Tissue growth occurs on a longer timescale than cell growth, so that the tissues of an organism outlive individual cells. (Note that there are important exceptions, such as the cells of the central nervous system and the human retina, which are not replaced through the life of the individual.) For example, in humans the 4.5×10^6 red blood cells present in every cubic millimetre of blood have a life of only 120 days or so, but are being replaced continuously by the 8×10^6 to 10×10^6 that are produced in the bone marrow every hour (see Chapter 7), while as many are removed by the spleen and liver. As in the case of plants, growth in animals is under the control of hormone systems that regulate the direction and rate of growth and coordinate growth in different tissues. In mammals the anterior lobe of the pituitary gland both secretes general growth hormones and acts indirectly by stimulating other endocrine systems.

Box 25.1

PHYTOCHROME P_r AND P_{fr}

Phytochrome is a highly reactive protein molecule capable of changing its shape to exist in two distinct forms (isomers), each with a characteristic absorption spectrum. A blue form P_r responds to red light (wavelength 650–660 nm) and changes to a blue–green form P_{fr}. The reverse change from P_{fr} to P_r occurs on illumination with far-red light (wavelength 725–730 nm) or spontaneously but slowly in the dark. Phytochrome is now known to be the fundamental mechanism that switches on and off many biological responses to light in higher plants. These include many rhythmic responses associated with the alternation of light and dark in the 24-hour cycle and with variations in daylength over the annual period. Phytochrome plays a part in the phasing of many diurnal rhythms, such as stomatal movements, photosynthesis, respiration, ion uptake and cell division. Photoperiodic responses involving phytochrome systems are at least partially responsible for seasonal changes such as bud dormancy, internode elongation and stem growth, and of course the initiation of flowering (Kendrick and Frankland, 1976).

From the very start (even before birth in mammals) the potential for ageing is present and is associated always with growth and development in the organism. If growth is viewed as an accumulation of excess cell multiplication over cell death, then in maturity this ratio must level off to a steady-state relationship maintained by a continual turnover at the cellular level. As ageing becomes apparent this turnover slows down: repair and replacement slow down; skin wounds, for example, heal less readily; tissues and organs lose weight; cell death rate increases and blood volume decreases. This inherent ageing process is evident in the decline (by 50%) in the physical capacity of humans during their lifetime. The human heart beats 140 times per minute at birth but this rate declines to 70 at the age of 25 years. At 20 years the blood takes up an average 4 litres of O_2 min⁻¹ but by 75 years this is reduced to 1.5 litres of O_2 min⁻¹. The brain decreases in weight by 10% between the ages of 30 years and 90 years. The inescapable conclusion of the cumulative effect of ageing is ultimately a breakdown of bodily function that leads either directly or indirectly to the death of the individual.

This view of the life and death of an individual organism can be simplified into three stages, each characterized by different patterns of control exerted by both external and internal feedback mechanisms. First, the growth stage where the balance of control is dominated by positive feedback reinforcing the upward trend of growth; secondly, a stage where negative feedback or conservative mechanisms gain the upper hand, slowing the rate of growth and establishing a controlled steady state – the mature organism. Ultimately, conservative control breaks down and once more a stage dominated by positive feedback takes

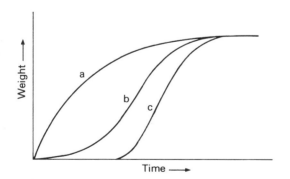

Fig. 25.1 Three different theoretical curves, based on different mathematical assumptions, that have been used to describe biological growth. Curve b is the logistic growth curve that will be encountered later in this chapter, a is the monomolecular curve assuming constant rates of addition and subtraction, and c is the more complex Gompertz curve.

over as ageing progresses. This time it is a downward accelerating trend and finally it causes the system to cross a threshold and take on an entirely new state: death. Although this model is true for the individual organism, the steady state of its adult stage can be envisaged as being perpetuated in its offspring. So although the individual organism does not escape the consequences of the Second Law of Thermodynamics (i.e. death) the capacity for self-replication and reproduction allows the steady state to be maintained, but at the level of the entire species population.

25.1.2 Population dynamics

One of the most important properties of any species population is its size – the number of individuals in the population. At any moment these individuals will be in different stages of their lifecycles, so that the population as a whole will have an age structure composed of several overlapping generations of individuals. (Sometimes, however, particularly among insects, the life span of the species is the same as its period of reproductive maturity and the population is composed of generations that replace each other and scarcely overlap in time. In mammals, for each age group

the rate of addition by birth and subtraction by death will be age-specific. Birth rate and death rate will change with age. In humans, for example, the death rate is highest in the first year and in old age, and the birth rate is highest at about 20 years, but lower before and after, while the death rate is lowest at approximately 11 or 12 years. If these age-specific birth and death rates remain constant, or relatively so, then the population will exhibit a stationary age structure, and if overall the total birth rate and total death rate are the same, then the population size remains constant. Such a situation implies that each individual is replaced only once in its lifetime. Fig. 25.2 illustrates this theoretical state for an idealized bird population. In reality the input and output never quite balance and the population size fluctuates within narrow limits, maintaining a steady state.

From the time of Darwin it has been recognized that most species possess an impressive ability to increase in number. Evolution has provided most species populations with reproductive strategies endowing them with the capacity to produce many more offspring than would be necessary to replace the population. Chapman (1928) called this multiplicative tendency inherent in the reproductive

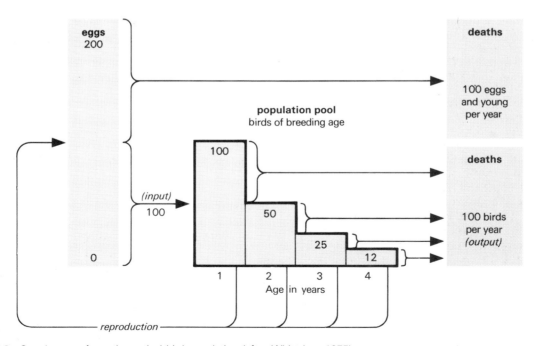

Fig. 25.2 Steady state for a theoretical bird population (after Whittaker, 1975).

process the **biotic potential** of the species. This kind of growth can be described by the following equation:

$$dN/dt \propto N$$

where the change in the number of individuals with time dN/dt (i.e. the increase in the population) is proportional to the number of individuals already there (N). This is a classic self-reinforcing positive feedback loop. We can remove the proportionality sign and replace it with an equals sign if we insert a proportionality constant in the equation to give

$$dN/dt = rN$$

where the rate of population growth is given (r), which is known as the **intrinsic** or **instantaneous rate of increase**. The results of such growth are graphed in Fig. 25.3 on both an arithmetic and a logarithmic scale. In the former this **exponential growth** is seen as a characteristic U-shaped curve as population increase accelerates upwards, but in the latter it appears as a straight line (notice the similarity to graphs of stream order against number of streams in Chapter 10).

There are many amusing theoretical calculations in the literature on population dynamics which demonstrate the consequences of applying this concept of exponential growth to populations of various kinds, but of course it rarely occurs in the real world, and where it does it is short-lived. Exponential growth is valid only in the absence of crowding, competition or resource depletion. The importance of exponential growth is that it is a model of the capacity for increase possessed by most species. 'Real' growth curves represent the result of a conflict between this capacity – Chapman's biotic potential – and what he called the **environmental resistance**. This term embraces the limits set by resources such as light, space, nutrients, food and by intraspecific (usually called crowding in this context) and interspecific competition which any species encounters in the real world. In combination these limits help to define the maximum **carrying capacity** of the environment for any particular species population. So the population growth equation becomes

$$dN/dt = rN\left(\frac{K-N}{K}\right)$$

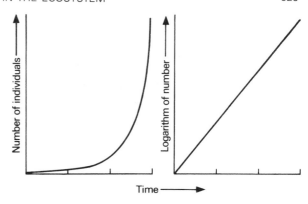

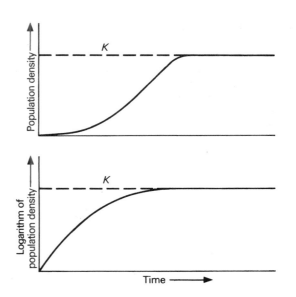

Fig. 25.4 Curves showing sigmoidal (logistic) population growth on both arithmetic and logarithmic scales.

where K is the upper limit of population growth or carrying capacity. When N is small, $(K-N)/K$ approaches 1 and the equation reduces to the equation for exponential growth. As N increases and approaches the value of K, $(K-N)/K$ becomes a smaller and smaller fraction and hence the rate of increase slows down. When graphed, this model gives an S-shaped (sigmoidal) curve (Fig. 25.4).

When the growth curves of natural populations are compared with this theoretical model, although some accord reasonably well with it, many do not. This is because the feedback mechanisms that operate on real populations do not act as instantaneous switching mechanisms but as components of complex feedback loops. This introduces delay and lag effects such as those that occur when a population, having reached its carrying capacity K, continues to increase until the cumulative effect of resource depletion (for example) produces a decline in population numbers. Overshooting in this way is very common in the real world (Fig. 25.5) and the subsequent decline sometimes continues to low population densities before recovery begins and another overshoot–recovery cycle starts. In other cases,

after the initial overshoot the population numbers drop to and oscillate about K with decreasing amplitude until stabilized. Furthermore, there is some evidence, particularly among the higher animals with well developed social organization, that population densities are stabilized at a level somewhat below the maximum carrying capacity. Such a situation could be interpreted as a tendency to maximize the quality of life (i.e. share of space, resources, etc.) per individual, rather than to maximize the quantity of individuals per unit area.

In species that display relatively unstable growth curves with irregular or periodic fluctuations, the capacity for reproductive success, for wide dispersal and for rapid population growth are all of adaptive significance in ensuring their survival. Consequently, such species are grouped on the basis of 'r' selection. On the other hand, species that exhibit stable populations at or near their K values, and which coexist with other species in stable ecosystems where they occupy specialized defined niches to which evolution has made them well suited, are said to be 'K'-selected species (see Section 25.1.3).

If population fluctuations were random, and increases and decreases occurring in response to change in some environmental factor were entirely independent of density, then sooner or later numbers would fluctuate down to a level where reproduction would fail or the last remaining individuals would be eliminated, leading to extinction. Whittaker (1975) expresses this conclusion succinctly when he maintains that 'in principle, a population that randomly walks in time, without some density-dependent limitation, must walk randomly to extinction. . . . Influences limiting fluctuations are necessary to the long-term survival of populations.' So again, as we saw in Chapter 23, random tendencies are constrained and feedback interactions maintain an organized state: this time a population density regulated within limits. Density-dependent mechanisms must regulate the upper and lower limits of populations by

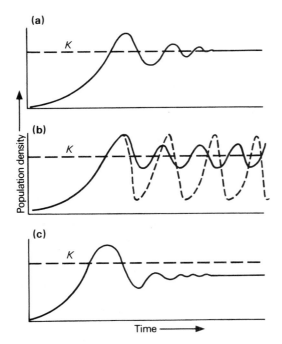

Fig. 25.5 Models of population growth in relation to the carrying capacity (K). (a) Initial overshoot followed by damped oscillations of declining amplitude finally stabilizing at an upper limit determined by the value of K. (b) Initial overshoot followed by regular cyclic fluctuation; in some cases (pecked line) the recovery limb of each cycle may approximate to exponential growth. (c) Initial overshoot followed by damped oscillations of decreasing amplitude but stabilizing at an upper limit below the theoretical value of K, i.e. selection for quality of existence.

1. increasing the mortality of individuals proportionally or decreasing the birth rate per individual as the population grows;
2. decreasing the mortality of individuals proportionally or increasing the birth rate per individual as the population declines.

All natural populations show some degree of regulation or control and the principal interactions involved are shown in Fig. 25.6. Competitive or crowding interactions between individuals in a population often result in density-dependent regulation as resource limits are approached. However, this regulation is rarely evident, as starvation and death, and even undernourishment and stunted growth, which may occur in plants, are not usually observed in natural populations of animals. In the case of animals, Wyne-Edwards (1965) cites two strategies of regulation. Both are mediated through physiological/behavioural changes and are under the control of social feedback mechanisms.

These strategies are, first, the limitation on the number of individuals allowed to breed and, secondly, influences on the number of young each pair are permitted to produce. They are implemented by social behaviour such as the establishment of territories (particularly among birds) which impose a ceiling density for a particular habitat and thereby exclude a non-breeding surplus population that may perish, but in any event does not replace itself by successful mating. Other mechanisms include the adoption of limited acceptable or traditional breeding sites (as with colony-nesting birds or seal breeding grounds) where again a non-breeding surplus is excluded. The development of social hierarchies inhibits the onset of sexual maturity in young non-dominant adults in extended family groups, while harem systems will exclude unsuccessful males from

mating. Pressure from crowding may increase socially induced mortality through, for example, a decline in the maternal care given to the eggs or to the young, and in rarer cases may be manifest as cannibalism. However, all of these mechanisms are more complex than they appear at first sight and they usually involve regulation at the biochemical level as hormonal systems control physiological and behavioural responses.

Interspecific competition, on the other hand, appears initially to have a destabilizing effect, at least on one of the competing populations. If we first modify the equation for sigmoidal growth to incorporate competitive interactions between two species, this effect will become apparent (Lotka–Volterra competition equation, Hutchinson, 1965):

$$\frac{dN_1}{dt} = r_1 N_1 \frac{(K_1 - N_1 - aN_2)}{K_1}$$

$$\frac{dN_2}{dt} = r_2 N_2 \frac{(K_2 - N_2 - bN_1)}{K_2}.$$

Here, N_1 and N_2 are the numbers of species 1 and 2 respectively, r_1 and r_2 their relative growth rates, and K_1 and K_2 the carrying capacity or saturation densities. Expressing through aN_2 and bN_1, the effects of population change of one species on the population of the other are the competition coefficients a and b. The critical parameters in these equations are these competition coefficients and the carrying capacity densities, and the model predicts that all relationships between the values of these parameters, but one, will lead to the extinction of one or other of the competitors. This prediction has become known as the **competitive exclusion principle**. It implies that if two species are direct competitors utilizing the same resource at the same time, at the same location i.e. occupying the same niche – then as their populations reach equilibrium in a stable community one of the species will become extinct. Therefore, competition tends to bring about the ecological separation of closely related or otherwise similar species. That is, it promotes niche differentiation to avoid extinction.

There is, however, one condition where the two species can coexist. In 1962 Slobodkin showed that if the values of the coefficients were small in relation to the ratios of the carrying capacity densities $a < K_1/K_2$ and $b < K_2/K_1$ both species

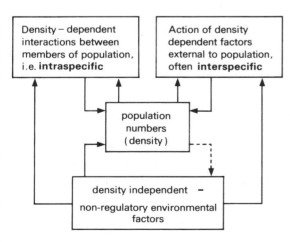

Fig. 25.6 Density-dependent, density-independent and environmental interactions in the regulation of population size (after Solomon, 1969).

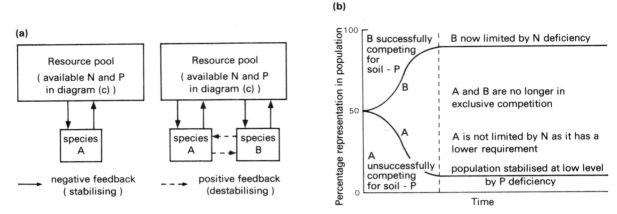

Fig. 25.7 (a) The establishment of positive and negative feedback relationships between competing species, and between them and a mutual resource (after Margalef, 1968). (b) The interpretation of Margalef's model in terms of the coexistence of two plant species in relation to limitation by two nutrient elements. Species B is a better competitor for soil phosphorus than species A, but has a higher nitrogen requirement and so cannot entirely exclude species A by competitive stress (i.e. the development of intraspecific density dependence in B before interspecific competition eliminates A) (after Etherington, 1978).

could survive. In effect each species is inhibiting its own population growth by intraspecific density-dependent regulation more than that of the other species by interspecific competition. Therefore, each population is kept at a level below that which would culminate in an exclusive struggle for survival. These relationships have also been modelled more explicitly in terms of feedback by Margalef (1968; Fig. 25.7). Here, intraspecific density-dependent regulation is shown by negative feedback loops established between each population and a mutual resource, and interspecific competition by a positive feedback loop between the two species populations. This positive feedback is therefore potentially a self-reinforcing tendency towards the exclusion of one species, unless the negative feedback succeeds in regulating the densities of each species in the way outlined above.

There are many other kinds of interspecific interaction apart from competition, and all can involve a degree of density-dependent regulation, but equally under many conditions they may not (Table 25.1). For example, in the case of predation, stability is possible if the proportional loss to predation increases as the prey population grows, but if a lag occurs in the predators' response to feedback from prey numbers, then such a predator–prey system is likely to be inherently unstable. However, to consider a single predator–

Table 25.1 Types of interspecific population interaction (after Odum, 1971)

Type of interaction	General nature of interaction
neutralism	neither population affects the other
competition: direct interference type	direct inhibition of each species by the other
competition: resource use type	indirect inhibition when common resource is in short supply
parasitism	the parasite generally smaller than the host
predation	the predator generally larger than the prey
commensalism	the commensal population benefits while the host is not affected
protocooperation	interaction favourable to both but not obligatory
mutualism	interaction favourable to both and obligatory

prey system in isolation is, in most circumstances, unreal. The prey, if a herbivore, may be limited directly by food resources or indirectly by breeding sites or territory size; it may even be the prey of more than one predator. Therefore, the two-part predator–prey system is really one aspect of a larger network of interactions and it may indeed exhibit stability as a composite result of all interactions, some of which are density dependent. In fact, many density-dependent mechanisms involve combinations of different kinds of effects. For example, the kind of intraspecific competition that may result in the exclusion of surplus individuals (e.g. territorial behaviour) initially results in emigration and then dispersal of the surplus, but its ultimate effect may be increased mortality of these individuals by predation.

Rarely is a single mechanism responsible for the regulation of natural populations and, although the notion of the steady-state population is useful, it is commonly not the case. It is true that many species populations are maintained at or near their upper limits by self-regulatory mechanisms, but many others show wide fluctuations in numbers. However, that they survive is due again to some density-dependent factor which operates at the lower limits to prevent complete extinction.

25.1.3 Ecological strategies and population regulation

In earlier chapters we encountered the concept of the **ecological strategy** adopted by an organism to endow it, or more accurately, selected by evolution as endowing it, with an enhanced fitness in capturing and utilizing resources and in competing successfully. This idea was evident in Chapter 18, when discussing an organism's relationship with its habitat and in attempting to define its ecological niche. Here we shall concentrate on an approach to the strategy displayed by an organism through its life history for survival; an approach which grew out of a debate among population ecologists.

The first step was to identify that the population regulation mechanisms, which we have just discussed, correlate with distinctive sets of environmental conditions. So, density-independent regulation was seen to be paramount in relatively extreme (from the point of view of the organism) environments where environmental variables could fluctuate widely. On the other hand, relatively stable genial environments with minimal

or predictable fluctuations were seen to correlate with density-dependent regulation. Both density-independent/density-dependent mechanisms and marginal/equable environments represent the extreme ends of continua which run parallel in the real world.

The next step in this logic was to recognize that these continua present organisms with different combinations of biotic (particularly regulatory) and abiotic interactions. Therefore, to survive, to reproduce, and to coexist in communities, they must possess appropriate strategies. Then came the recognition of a continuum of groups of species, each group sharing a common life-history strategy. The groups at the ends of the spectrum of strategies became known as r and K strategists.

These groups were first proposed by MacArthur and Wilson (1967) as r-selected and K-selected species, as they were concerned with the selection of different strategies from within the species. Specifically they were researching into the ecological biogeography of islands, and identified strains which were successful colonizers – r selection – of remote habitats. From these populations arose successful competitors – K selection – which subsequently survived in equilibrium communities supplanting the pioneering or colonizing strains. The interpretation of the r and K groups, however, was broadened and developed by Pianka in 1970, and subsequently by others (e.g. Southwood, 1977) so that it has come to pervade much of current ecological thinking. In the context of this chapter on change in ecological systems the r and K paradigm is important in aiding the understanding of changes in community organization through time, particularly during succession (Section 25.1.4).

The r and K, of course, are derived from the population growth equation, because r-selected species are selected for their ability to reproduce rapidly, i.e. for their intrinsic rate of population increase, or r; K-selected species are selected because of their ability to maintain stable populations at their carrying capacity, densities (K). These are paralleled, as we saw above, by two distinct categories of environment: r-selecting environments – ephemeral, extreme, unpredictable – and K-selecting environments – equable, predictable and stable.

The r strategists are often referred to as opportunists and are organisms whose life histories are adapted to rapid population growth and high fecundity. As their name suggests, they are capable of exploiting any chance opportunity, wherever and whenever favourable conditions occur. They have effective and efficient dispersal mechanisms enabling them both to colonize remote places and to exploit ephemeral habitats (the weed or ruderal strategy). In both kinds of habitat the niche space is unsaturated. They often do not compete well with other species, particularly where growing or living under non-limiting conditions. Indeed this observation led Hutchinson to call them fugitive species, in the sense that their prevalence in unfavourable habitats is because it is there that they take refuge, fleeing as it were, from more vigorous competitors. This is a reminder that the opportunist strategy cannot be defined solely in terms of abiotic factors. Furthermore, opportunist or fugitive species are usually short-lived, attain only temporary or localized population equilibria, with numbers fluctuating widely, perhaps becoming locally extinct. All of these features reflect the strong influence of density-independent factors.

The K strategists are usually termed equilibrium species, are strongly competitive and adapted to persist in closed communities with a saturated niche space. They are the opposite of opportunists, with high reproductive rates and dispersal being of less importance than continuity, persistence in the face of resource stress, and adaptation to intense interspecific competition in closed communities. The appropriate strategy is one of diverting production or energy to defence (including allelopathic adaptations), endurance and competition.

Although the r and K scheme has become part of ecological orthodoxy, it is clearly not adequate to cope with the range of subtle variants of ecological strategy encountered between the two end members of the continuum described above. As yet, however, there have been few real attempts to replace or supplant it, with the possible exception of the C-S-R scheme proposed by Grime (1977, 1979; Box 25.2).

25.1.4 Succession and climax

The complex, apparently stable and persistent ecosystems described in Chapter 18 and on a global scale in Chapter 7, do not 'arise as it were ready made'. Such mature communities with the highly developed interdependence of their constituent species and their complex network of interaction with the environment are the result of inherent processes of change – directional change akin to the growth and development of the organism. This sequence of changes has been known traditionally as **succession**, and it is not caused by any change in the external environment (Cowles, 1901; Clements, 1916). As an inherent, or **autogenic** process of change, succession was for many years thought to consist of a number of successive developmental stages – **seral stages** – which together form an identifiable sequence, a **sere**, from the initial colonization of a vacant habitat to the fully developed ecosystem. This view is now known to be an oversimplification and has been subject to modification.

During the initial stages of colonization the environment will be the dominant influence on the success or failure of the plants and animals of the pioneer phase. To a greater or lesser extent this environment will be 'extreme', in that these pioneers will have to cope with a new mineral regolith, bare rock, not a *true* soil, with the full range of temperature fluctuations, with the full impact of precipitation and perhaps also desiccation. They are usually species with wide and effective dispersal mechanisms. Among the plants, most pioneers of terrestrial environments have light wind-borne seeds or spores which are produced in great numbers, as for example *Epilobium* in Fig. 25.8, which produces *ca.* 80 000 viable seeds in one year. Once established, however, many show impressive rates of population increase often associated with vegetative or asexual reproduction, all characteristics of r selection species. Although such vegetative clones are at an evolutionary disadvantage in that they consist of genetically identical individuals, they possess two immediate advantages when colonizing new environments (Davis and Heywood, 1963). First, they can spread rapidly over an unsaturated habitat where competition is lacking, and secondly, they avoid the risks of elimination that accompany fertilization, seed dispersal, germination and establishment in sexually reproducing plants. Some species that occur during the pioneer stage, however, are ephemerals (usually annuals) which avoid the worst extremes of the environment by passing the unfavourable season as dormant seeds. Such is the case with many of the driftline plants which are the initial colonists of the sand-dune succession (Fig. 25.9). Adaptation amongst pioneer species, therefore, is primarily adaptation

Box 25.2
ECOLOGICAL STRATEGIES

The scheme of ecological strategies developed by McArthur and Wilson (1967) and introduced in Section 25.1.3 is expressed in terms of two contrasting groups of species, the *r*-selected opportunist, fugitive, species, and the *K*-selected or equilibrium species. It has long been appreciated that these two groups actually represent the extreme ends of what is in reality a continuum of ecological strategies. Indeed, in recognition of this situation Grime (1977, 1979) recognizes three basic ecological strategies which he applies specifically to plants, although conceptually they are equally applicable to animals. These became known as the *C–S–R* strategies, and they originate from a classification of external factors which affect plants and the vegetation they form. These factors Grime divided into two broad categories:

Stress: factors which restrict or inhibit growth, i.e. limiting resources and suboptimal conditions (in plants specifically, factors which affect photosynthesis, such as shortages of light, mineral nutrients and/or water, or suboptimal temperatures).
Disturbance: the total or partial destruction of plant biomass by the activity of herbivores, pathogens, anthropogenic factors and factors of the inorganic environment including natural hazards, fire, wind, drought, erosion, etc.

Using this fundamental division Grime then constructs a matrix using two intensity levels, high and low stress, and high and low disturbance.

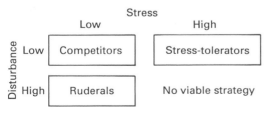

(a)

Using the three viable strategies which he derives from this matrix as the apices, Grime then constructs a triangular ordination, which allows the further recognition of intermediate strategies. The *R* or ruderal strategy is the equivalent of the opportunist or fugitive *r* strategist, with most of the energy and resources allocated to dispersal, reproduction and fecundity, and where disturbance and lack of competition are characteristic of the habitat.

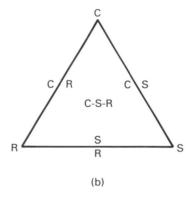

(b)

The *C* or Competitor strategy is one where disturbance or stress are of little significance relative to competitive interactions, so that most resource allocation is to competition.

The *S* or stress tolerator, and in the *S–C* or stress–competitor strategies the species divides its allocation of energy and resources between competition and stress resistance. Stress in this context is a kind of competition, but it refers to that imposed by absolute constraints affecting physical and chemical resources.

The advantage of Grime's scheme is that it allows a more sophisticated analysis to be made of the changing relationships in communities approaching or at the climax state, than was possible merely regarding them as being composed of equilibrium species (*K*-selected).

(a)

(b)

(c)

(d)

(e)

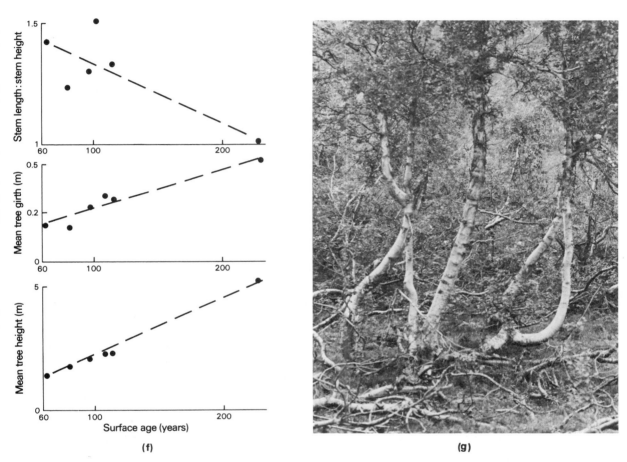

(f) (g)

Fig. 25.8 Plant succession in front of the Tunsbergdalsbreen Glacier, Norway. (a) The glacial foreland and the moraine sequence (see Fig. 24.22). (b) The initial surface available for colonization. (c) Three pioneer species: *Epilobium anagallidifolium, Oxyria digyna,* and *Saxifraga groenlandica.* (d) Increase in percentage cover of the vegetation with time. (e) Dwarf shrubs dominate and trees begin to invade as the vegetation closes. Tree growth is at first tortuous and prostrate, but the ratio of stem length to height approaches 1 as a more erect growth habit is assumed, as here on the 1743 moraine, (f) and (g). (White, 1973.)

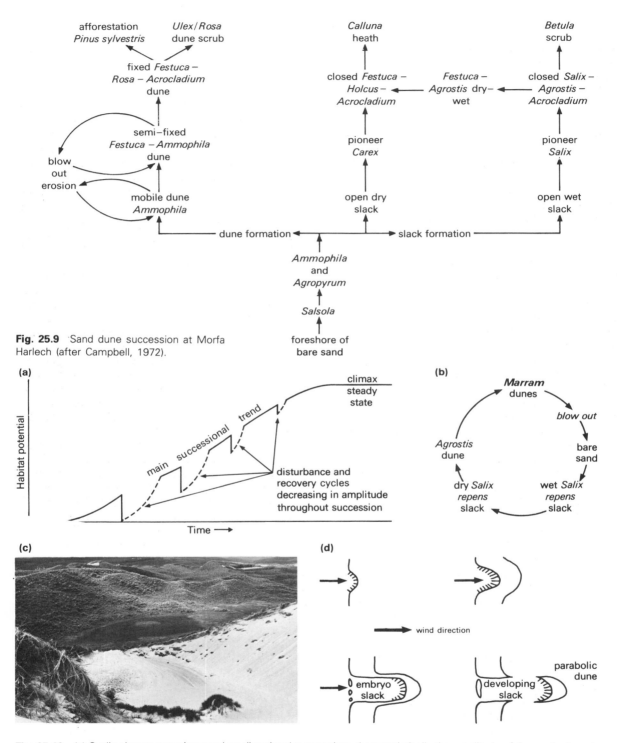

Fig. 25.9 Sand dune succession at Morfa Harlech (after Campbell, 1972).

Fig. 25.10 (a) Cyclic change superimposed on directional succession; characteristically the amplitude of the cyclic departures declines as the community approaches the climax state. (b) Sand dune erosion–recolonization cycle. (c) Dune erosion at Balnakiel, Sutherland. (d) Formation and development of a parabolic dune (after Ranwell, 1972).

to survive in an unfavourable environment, rather than to confer an adaptive advantage in competition with other species.

Although pioneers characteristically display wide tolerance limits for external environmental parameters, they are not normally tolerant of competition. Most are 'open habitat' species that do not tolerate shade, while equally many require good root aeration with low levels of soil carbon dioxide and little root competition (for example, marram grass (*Ammophila arenaria*) in Fig. 25.9). Indeed *intraspecific* competition is more important during this phase than *interspecific* competition while the habitat remains unsaturated. In terms of their ecological strategy these pioneer, opportunist of fugitive species are called 'ruderals' by Grime (1979) (Box 25.2).

The structural simplicity of these early communities means that the extent to which they can control fluctuations in the environment by any sort of negative feedback is severely limited. Consequently, it is often the case that the magnitude of such fluctuations exceeds the survival capacity of the pioneer community, which is destroyed, and colonization begins again. Often the destruction is not total, and, for example, roots and humus and perhaps propagules remain and survive in the substrate to initiate growth once more. In time, the instability of the environment is modified by the developing community and its susceptibility to disruption by extreme events is reduced. Nevertheless, the superimposition of such cyclic processes on the general successional tend remains a feature of the early part of succession, although the amplitude of the oscillations decreases with time (Fig. 25.10).

As the pioneer species produce a more closed vegetation cover, they alter the nature of the surface. They provide some shelter and conserve moisture for the germination and establishment of other plant species and create microhabitats for animals, particularly insects, arthropods and small herbivores. Below ground their roots and the decay of the litter they produce begin to create a true soil (Fig. 25.11). They have begun to modify their environment, but in such a way as to allow other species to enter the community. This is the **facilitation model** of succession, a positive feedback process reinforcing change, and it leads to an increase in species diversity, slowly at first and then more rapidly. Although the newcomers compete more successfully in the modified

environment than the species of the pioneer stage, some pioneer species, particularly those with the wider tolerance ranges, will persist for some time, though their population numbers decline and often show more marked fluctuations.

As species diversity increases rapidly during the middle phase of succession, the tolerance range shown by each species entering the community becomes narrower. This is because, for a species to survive in competition with those already there, it must be able to occupy an unexploited functional niche, or to partition an existing niche by being more specialized in part of that role than the current occupant. So this middle phase of succession is one of intense interspecific competition and of increase in diversification and segregation of ecological niche, and it affects plants and animals alike. It is marked by a development of negative feedback loops, both between the species of the community and between them and the environment, so that the level of integration and the degree of self-regulation increase.

Ultimately a more stable community emerges, especially among the plants, when the structural complexity of the vegetation reaches its maximum. This is associated with the development of the most complex stratification of the plant community possible under the prevailing environment. However, with shrubs and trees providing a range of new arboreal habitats, a renewed phase of diversification among the animals ensues, as they compete for new niche space in the developing canopy. Furthermore, the existence of wood and dead timber, and increases in the amount as well as changes in the type of leaf litter, all provide additional niches. These are exploited by an increasing number of specialized microorganisms, saprophytes and saprovores. This late increase in the potential for invasion, niche segregation and competition gives a final spurt to species diversity that thereafter tends to decline slightly, as some of the species of the earlier stages are finally eliminated. Although the diversity remains high at the end of succession, because of intense competition the population density of each species is relatively low. These successful species, that contribute to the establishment of a relatively stable community, tend to be genetically adapted to be tolerant of competition and specialization, which in conjunction with resources define the carrying capacity for them; in other words they show K selection. The rates of invasion of r and

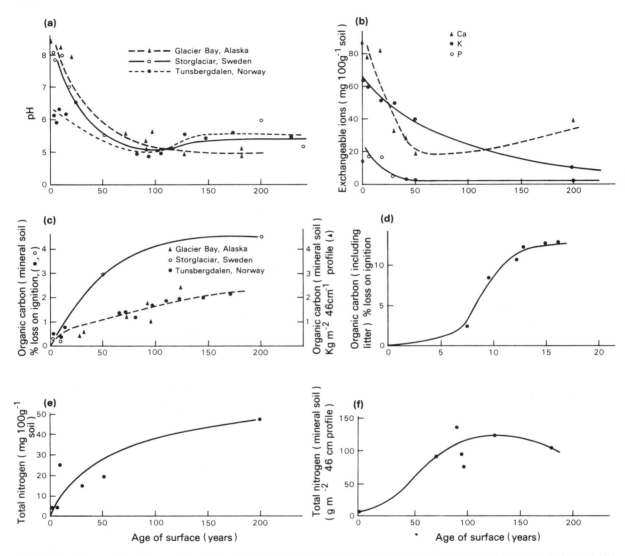

Fig. 25.11 Changes in soil parameters through succession. (a and b) Decline in pH and exchangeable ions in soils from glacial foreland succession reflecting progressive leaching of originally fresh morainic material (after Crocker and Major, 1955; Stork, 1963; White, 1973). Increases in soil organic matter in (c) primary and (d) secondary succession (after Crocker and Major, 1955; Stork, 1963; White, 1973; Maris, 1980), and in total soil nitrogen, (e and f) (after Crocker and Major, 1955; Olson, 1958).

K species through succession are summarized in Fig. 25.12.

Although succession can be seen in terms of the interplay between groups of organisms with different strategies summarized by MacArthur and Wilson's distinction between *r* and *K*-selected groups, the realities of the process are better served by the use of Grimes' three categories of **ruderals**, **competitors** and **stress tolerators**. Here

the middle phase of succession is seen as a phase when organisms which allocate resources to competitive strategies reach an ascendancy over the opportunists. However, towards the final phase, with the establishment of **climax**, organisms which are adapted to tolerate stress of various forms associated with resource availability, as well as continuing to allocate resources to competitive strategies, become more important (Fig. 25.13).

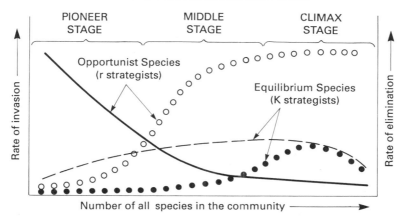

Fig. 25.12 Succession modelled in terms of rates of invasion and subsequent elimination of species with different ecological strategies (derived from Colinvaux, 1986)

Throughout succession the kind and intensity of the interactions between the community and its non-living environment change. The soil develops under changing conditions as the degree of control and regulation exerted by the community increases. Organic matter content increases throughout succession (Fig. 25.11c, d.). The development of soil organic horizons and the complexity of the soil population and decomposition processes parallel changes in the plant and animal communities. The balance of weathering processes changes in response to increases in carbon dioxide from soil respiration, and to increased availability of organic acids and complexing or chelating agents. Soil water and drainage regimes adjust to the increased demand from transpiration. The chemistry of the soil reaches a balance that reflects the nutrient demand of the vegetation and return of nutrients by litterfall and decomposition (a

classic case of a negative feedback loop). Surface and subsurface erosional processes are affected also, particularly by the increased stabilization of the soil and regolith effected by the developing root system and organic horizons, and the reduction in the soluble loss to streams as elements are immobilized or conserved in the community (again negative feedback). Also the effects of climate are modified progressively by the developing community until the canopy microclimate is controlled by the characteristics of the community itself. Such an effect can be seen in Fig. 25.14, where diurnal and seasonal temperature and humidity ranges are clearly damped down.

There has been a tacit assumption throughout this discussion that the process of succession has been proceeding from the initial colonization of a bare virgin surface. In other words we have been following the course of a **primary succession**. **Secondary successions** are those where the process proceeds from the cessation of a major disturbance to the pre-existing ecosystem, but where that ecosystem is not totally destroyed. Research into secondary successions, but also some empirical studies of primary successions, have called into question the nature of succession. In the first instance, the simple sequential model of seral stages forming a more or less smooth progression as they replaced each other through time has been criticized. This model requires the species of each stage to modify the conditions of the habitat in such a way as to **facilitate** the entry of the species of the next stage, and to hasten their own demise.

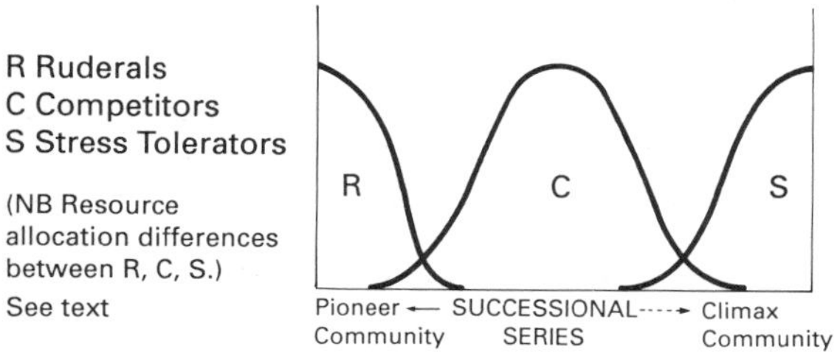

R Ruderals
C Competitors
S Stress Tolerators

(NB Resource
allocation differences
between R, C, S.)
See text

R   C   S

Pioneer ◄— SUCCESSIONAL ----► Climax
Community   SERIES   Community

Fig. 25.13 The proposed relationship between Grime's *R-C-S* strategy model and succession (after Grime, 1979).

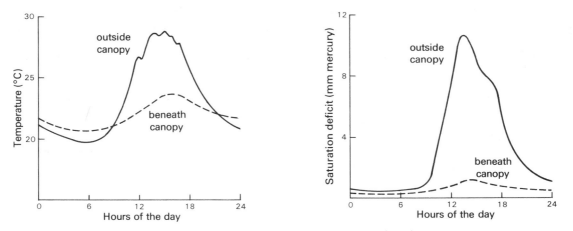

Fig. 25.14 The effect of a climax forest canopy on microclimatic parameters (after Hopkins, 1962).

The validity of what Egler (1954) called the **relay floristics model** (Fig. 25.15) with such waves of colonization has been widely questioned. The alternative model is one where, except in the very early stages of primary successions, at least for the plants, the propagules and reproductive structures (in vegetatively reproducing plants) are present from quite early on in the process. In secondary successions this is because they have survived, as part of the soil seed bank for example, and persist from a time before the disturbance. Therefore, in this model, the apparent waves of species usually regarded as seral stages arise

because different groups of species reach maturity and rise to prominence at different rates. These rates, in turn, reflect the life-history strategies of the species concerned. Egler terms this alternative the **initial floristic composition model** (Fig. 25.15). This model is now almost universally accepted as typifying secondary successions after disturbance (Drury and Nesbit, 1973; Horn, 1974, 1981).

There remain several aspects of succession, both empirical and conceptual, which either have been or are currently being re-evaluated. For example, many ecologists would not accept the trends in species diversity throughout succession as outlined

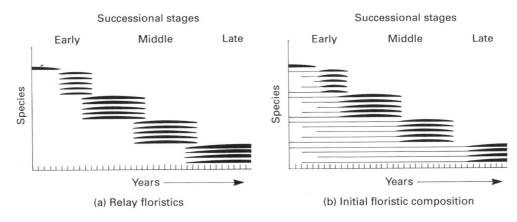

(a) Relay floristics   (b) Initial floristic composition

Fig. 25.15 Two models of succession: (a) relay floristics model with discrete waves of colonization, expansion and decline; b) initial floristic composition model where time of expansion is due to life history strategy of species, not to time of immigration or colonization. (b) Is the preferred model for secondary successions.

above. Such debates, however, are beyond the scope of this chapter and the reader is referred to the excellent review in Burrows' book (1990; see also Further reading) on the processes of vegetation change.

The end-point of succession is a mature self-regulating ecosystem which maintains a steady state 'equilibrium' through time. It is characterized by a complex, highly integrated community structure with high species diversity but relatively low individual population densities, not subject to serious fluctuation. The alternative pathways by which energy and matter are transferred through the system are also complex and diverse, and not easily disrupted. Furthermore, this community has largely created its own environment and is buffered against reasonable environmental fluctuations. This steady-state ecosystem is a **climax ecosystem**. It is often stated that the relative stability of such a climax ecosystem is due to its complexity (Elton, 1958; Hutchinson, 1959). However, May (1971a,b, 1975), by the use of mathematical models based on Volterra's equations, but varying the coefficients of interaction randomly, was able to show that complexity, in terms of either number of species, or number of interactions, tends to produce instability in the system as a whole. The implication is that if complex natural ecosystems are as stable as many observations would suggest, then the interactions occurring in them are highly non-random (Maynard-Smith, 1974).

Nonetheless, a degree of controversy still surrounds attempts to define the notion of the climax ecosystem. Much of the debate is concerned with the interpretation of the idea of **stability** as it is applied to ecosystems. In the conventional use of stability by mathematicians and engineers, the word denotes conditions at or very near to equilibrium or most-probable states. In the recent past ecologists have broadened the concept of stability to include **topological models** and notions of multiple stable states and metastability in interpreting the meaning of climax in non-equilibrium (in a mathematical sense) ecosystems. This trend is similar to, and preceded that in geomorphology, outlined in Section 24.4.

The culmination of this trend is the application of **chaos theory** as a formal mathematical approach to the problems posed for linear and deterministic mathematical models by non equilibrium systems. This approach would for example allow May's (1974) equilibrium-centred modelling of stability and diversity or complexity referred to above, to be resolved. For if there is more than one region within which stability occurs (more than one domain of attraction in terms of topological models), then increases in complexity can be seen as merely moving the system from one stable state (domain) to another.

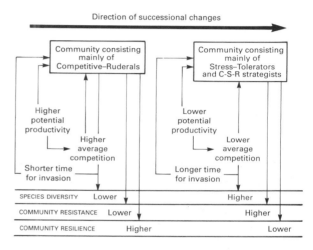

Fig. 25.16 An attempt to relate life history, productivity and ecological strategy of species to changes in species diversity, community resistance and community resilience through succession.

Holling (1973, 1976), made an important contribution to this debate when, using topological, non-equilibrium models of ecosystems, he introduced the term **resilience**. In the wider ecological literature this has now come to mean the capacity of the ecosystem to absorb or respond to changes, fluctuations and disturbances; to return to, and to persist in, a stable state. However, that state may be a dynamic stable state (homeorhesis) rather than a static state (homeostasis) (Waddington, 1975). Subsequently the concept of **resistance** to change has been introduced to represent the extent to which the system is buffered against or resists change. Highly resilient communities therefore correlate with both early successional, and apparently climax, situations, but ones where there is inherent instability, e.g. from recurrent forest fires or flooding. Although resilient, such communities are not resistant. Resistant communities are integrated climax communities with well developed feedback, such as a tropical rainforest.

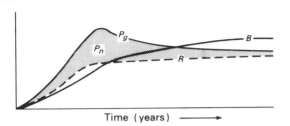

Fig. 25.17 Changes in gross (P_g) and net (P_n) production, respiration (R) and biomass (B) through succession.

Resistant communities, however, are not necessarily resilient in the face of perturbations and disturbances. On the contrary, they may prove to be **fragile** when disturbed and their conditions of existence are altered (see Chapter 26), and may be propelled towards a new but different 'stable' state. *Resistance, resilience* and *fragility* are all concepts which can be related to the life history and ecological strategies of the individual species comprising communities. Furthermore, they are concepts which have great utility in the interpretation of successional changes, and of anthropogenic disturbance to natural ecosystems (Fig. 25.16).

25.1.5 *The bioenergetic model of succession*

So far, the process of succession has been discussed in terms of changes in the structural and functional organization of the ecosystem, and the climax has been defined in the same terms. However, our discussion of succession would not be complete without examining changes in the dynamics of the system through succession, and particularly changes in the pattern of energy flow. This bioenergetic approach is central to the approach adopted in this book. However, as with the concepts of succession and climax it does not find universal acceptance, and some of its broader generalizations are open to question.

In the early stages of succession, production exceeds respiration ($P > R$) and the excess (net production) is channelled into growth and the accumulation of biomass through time (Fig. 25.17). Therefore, a large proportion of the energy input to the system is being diverted into store, and the output as respiratory heat loss is relatively small. The processes by which species interactions improve the habitat potential and promote increased

species diversity has already been recognized as positive feedback. Its effect is to increase both the capacity and the complexity of the energy storage compartments (as the total biomass of all species and trophic levels), as well as the complexity of energy transfer pathways. Some of this increasing energy store will be represented by increments in photosynthesizing tissue and will feed back to increase the energy input. However, the rate of this increase will decline through succession as the vegetation reaches its maximum potential. Additional biomass or stored energy will continue to accumulate, but much of it will consist of non-photosynthetic plant tissue (particularly wood) and the biomass of heterotrophic organisms. A maintenance must be paid for this additional biomass as respiration, which therefore increases through succession. However, as long as $P > R$, biomass will continue to accumulate, but this ratio progressively decreases through succession and approaches a value of 1, i.e. $P = R$. By this time, competition, the stabilization of niche structure and of population densities as negative feedback processes, have established and continue to maintain a community steady state, with little or no change in store (biomass) through time, but in equilibrium with the available energy flow.

The implications of this view of succession are very far-reaching. In the first place, the amount of biomass supported per unit of energy flow $B:E$ (where $E = P + R$), increases to a maximum in climax ecosystems. Secondly, the ratio of total respiration (or maintenance cost) to biomass (or structure), $R:B$, decreases and is at a minimum in climax ecosystems. Both of these observations would suggest that the climax ecosystem is energetically efficient. Margalef (1968) couched these same observations in the language of information theory and entropy. He viewed succession as a progressive accumulation of information, which can be taken to mean order or organization. In these terms, R represents the tendency of the system to produce entropy, because the respiratory heat loss is unusable energy unavailable to do work on the system, and B represents the ability of the system to preserve an ordered structure, because the biomass represents the complex structural and functional organization of the living systems of the ecosystem. The second ratio therefore tells us that the production of entropy per unit of preserved and transmitted information is at a minimum. The ratio $B:E$ tells us that it takes less energy to maintain a complex information-rich system than is necessary for the relatively simple systems characteristic of the earlier stages of succession.

25.2 Cliseral change

In the life of an individual organism, its growth and development is set against an environment subject to fluctuation and change. External environmental events vary in magnitude and frequency of occurrence, but most are predictable changes and conform to an established diurnal and seasonal pattern. Indeed, such patterns would have been 'learned' long ago and accommodated in the evolution of adaptation and tolerance by the species that experienced them. This accommodation of predictable rhythmic patterns of change in the external environment is equally true at the level of the community and ecosystem (Fig. 25.18). For this reason they have been largely ignored in the preceding sections and we have assumed that all the manifestations of the inherent processes of change in populations, communities and ecosystems have occurred against a background of a 'constant' external environment through time.

However, we know that such an assumption is an oversimplification. Regulation and control mechanisms in physicochemical systems are generally less sophisticated, less precise and less able to maintain a stable condition than their counterparts in living systems. Although the state of climatic, hydrological and erosional systems can be defined by statistical averages, or by statements of statistical probability, they are nevertheless subject to considerable variation about any steady-state condition. Events of high magnitude but low frequency, such as large-scale flooding, periods of pronounced drought, particularly cold springs, or the sudden occurrence of slope failure, are all examples of the extremes of such variation. However, the effects they have on the ecosystem vary widely, as changes in the inputs to it or to its conditions of existence.

In a climax seasonal forest ecosystem, for example, fluctuations in the extent of the summer water deficit or in the air temperature during the growing season may be recorded merely as fluctuations in the radial growth increment as reflected in the annual ring widths of the tree species (Fig. 25.19). Alternatively, such fluctuations may modify the community directly or indirectly by the elimination of one or more species. In extreme cases, fluctuations in external environmental conditions can lead to a total, although usually localized, destruction of the ecosystem, as for example with the occurrence of a natural forest fire or a period of rapid mass movement (Fig. 25.20). However, when such changes in the conditions of existence of the ecosystem are fluctuations – albeit sometimes violent – about some steady state, their effect tends to be short-lived. The resilience of living systems soon re-establishes the *status quo*.

As we have already seen in Chapter 24, events of the kind considered above may be more than

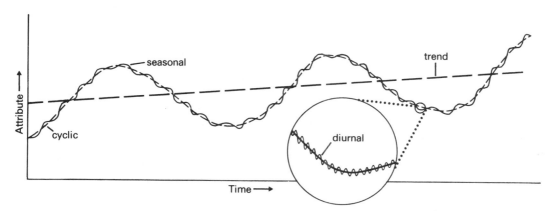

Fig. 25.18 Predictable rhythmic change in environmental parameters accommodated by the community adaptation of ecosystems, but often, as here, masking an underlying trend.

Fig. 25.19 (a) Taking an increment core from which (b) a plot of annual growth increments against time can be constructed. Unusually narrow or wide rings represent environmental variation as reflected in the pattern of tree growth.

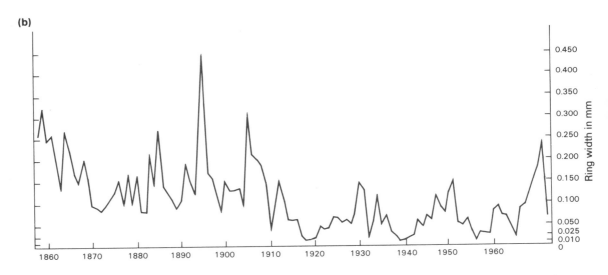

fluctuations about some average steady state. They may instead mask some long-term trend of climatic and environmental change (see Fig. 25.18). Temporary setbacks to the growth and vitality of individual organisms, or minor modifications to species composition, give way to wholesale change in the ecosystem as the underlying trend of environmental change asserts itself. The effects of such trends of climatic and environmental change are well illustrated by the Quaternary period, to which we alluded several times in Chapter 24.

During the Quaternary, any point on the land surface of northern Europe and northern North

America will have passed through a sequence of vegetation change, perhaps from glacial conditions devoid of vegetation, through tundra vegetation under a periglacial environment, then through a pretemperate stage of birch woodland, through an interglacial deciduous forest, to the boreal coniferous forest of a post-temperate stage, back to open tundra and perhaps again to ice. This broad sequence can be regarded as externally induced change in response to broad categories of climatic change. We may therefore envisage the climatic zones of the northern hemisphere moving south, then north and then south again,

Fig. 25.20 Localized destruction of vegetation and burying of soil following rapid mass movement; gelifluction lobe on Ben Arkle, Sutherland (after White and Mottershead, 1972).

with the climax ecosystems or biomes – tundra, boreal coniferous forest and summer deciduous forest – following in sequence (Figs 25.21, 25.22). This concept of zonal ecosystem migration in response to changed climatic inputs is known as **cliseral change** (climatic sere), but its attractive simplicity is complicated by several factors.

It is true that climatic shifts result in shifts of vegetation and animal populations under stress from climatic change, but in the case of plants, migration is dependent on the production of propagules, their quantity and their dispersal capacity. Some species are prolific seed-producers and these seeds may be beautifully adapted to wide dispersal. Other species produce fewer seeds and may not set viable seed every year, particularly near the limits of their geographical range, while the seeds of some species may be heavy and not easily dispersed over large distances. As a result, even within the same climax ecosystem, species will possess diferrent rates of potential migration. However, the possibility of migration under climatic stress depends, for both plants and animals, on the disposition of barriers in relation to potential migration routes (Fig. 25.23).

The absence of the spruce from the native British flora in the postglacial (Flandrian) and the impoverished Irish flora relative to the rest of the British Isles are ascribed to the barriers presented by the drowning of the English Channel – southern North Sea and Irish Sea respectively by the eustatic rise in sea level after the last glaciation (Devensian). The differential distribution of bar-

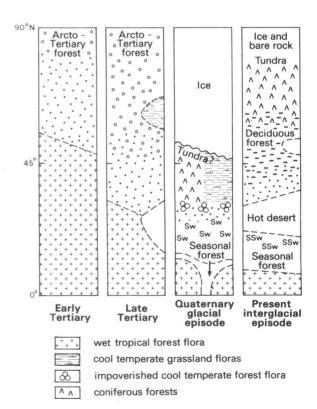

Fig. 25.21 Zonal or cliseral change in the northern hemisphere during the Tertiary and Quaternary (partly after Collinson, 1978).

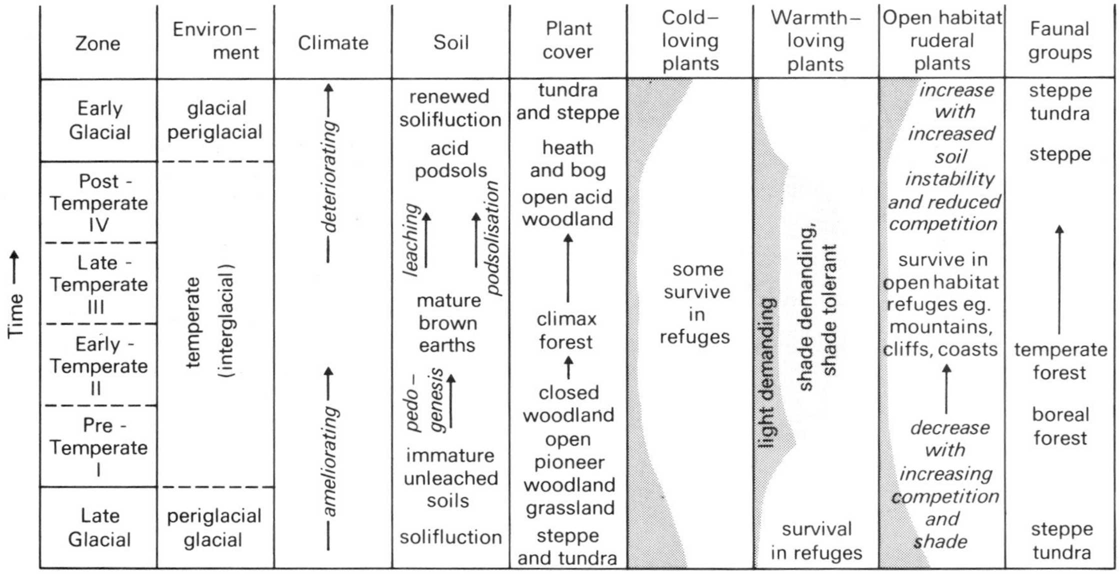

Fig. 25.22 Floral, faunal and environmental changes associated with Quaternary climatic oscillations (modified from Sparks and West, 1972).

Zone	Environment	Climate	Soil	Plant cover	Cold-loving plants	Warmth-loving plants	Open habitat ruderal plants	Faunal groups
Early Glacial	glacial periglacial	*deteriorating* ↑	renewed solifluction / acid podsols	tundra and steppe / heath and bog			*increase with increased soil instability and reduced competition*	steppe tundra / steppe
Post-Temperate IV	temperate (interglacial)		*leaching* ↑ / *podsolisation* ↑	open acid woodland	some survive in refuges	light demanding — shade demanding, shade tolerant	survive in open habitat refuges eg. mountains, cliffs, coasts	
Late-Temperate III			mature brown earths	climax forest ↑				temperate forest
Early-Temperate II				closed woodland				
Pre-Temperate I		*ameliorating* ↑	*pedo-genesis* ↑ / immature unleached soils	open pioneer woodland / grassland			*decrease with increasing competition and shade*	boreal forest
Late Glacial	periglacial glacial		solifluction	steppe and tundra		survival in refuges		steppe tundra

Time ↑

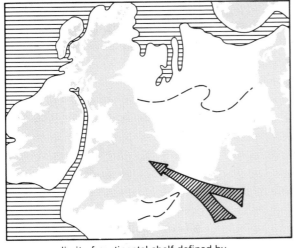

limit of continental shelf defined by 100m contour

- - - approximate shore line in English Channel and southern North Sea c. 9000 BP

migration route of forest genera from continental Europe during early Flandrian

Fig. 25.23 The English Channel, Irish Sea, and southern North Sea as barriers to floral migration during Quaternary interglacials. In the Flandrian at 9000 years BP Ireland would already have been cut off, while a land connection still existed in the southern North Sea. By 7500 years BP something approaching the present coastline would have been established.

riers can also be invoked, for example, to explain the differences in the floristic diversity of the temperate deciduous forests of North America, Europe and eastern Asia. The key here is the differences in the alignment of the potential topographic barriers in relation to the direction of migration (Fig. 25.24). Even when migration routes are available, they rarely offer the same opportunities to all species and in practice often act as differential filters.

For plants it is not only dispersal of the seed but successful germination, establishment and propagation that are also necessary before migration can be said to have taken place. So the rate of establishment of vegetation becomes another important consideration. In the Flandrian, Iversen (1964) has shown that forest communities actually became established a long time after climatic conditions had become favourable for forest growth. This is partly because of the time lag before propagules become available (through dispersal and migration), but mainly because the development of a climax forest ecosystem in a postglacial environment requires not only a favourable climate but also favourable soil conditions. Yet these soil conditions can be developed only under a succession of vegetation types. In such cases it may be difficult to disentangle the effects of autogenic succession from those of cliseral change.

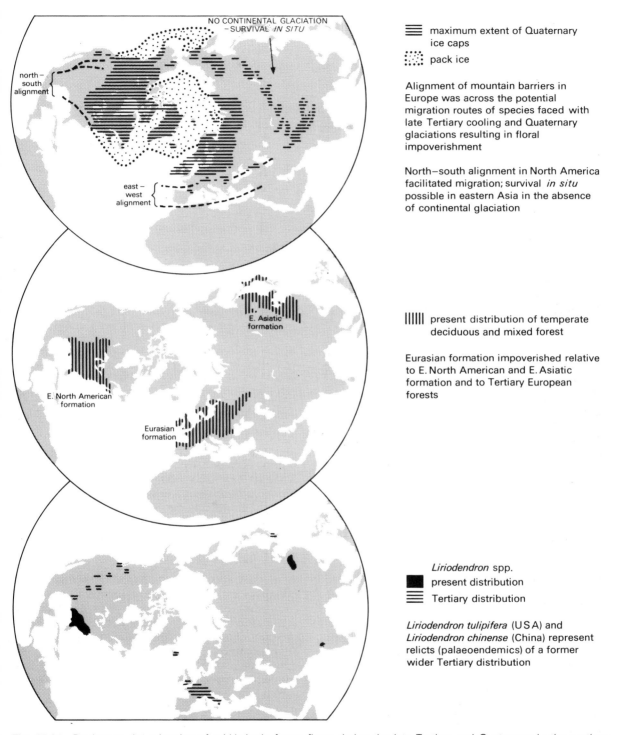

Fig. 25.24 Barriers to the migration of mid-latitude forest floras during the late Tertiary and Quaternary in the northern hemisphere. The differential disposition of the barriers in part explains the differential patterns of elimination in the three continents.

In a similar way, animal migrants are only vagrants until they become successfully established breeding populations, and this will depend on the existence of the correct habitat conditions, and hence on vegetation development.

Since extremes and not average conditions are critical to vegetation, broad climatic variations are often of less importance than specific features (such as degree of slope, aspect exposure or proximity to the sea or fresh water) that may modify regional climate. Therefore, during an overall climatic change local conditions could either hasten or prevent a related vegetation change, depending on whether they cause a local or microclimate to cross a threshold of vegetation change.

In addition, closed climax ecosystems, particularly forest ecosystems, offer little scope for invaders as the niche space is fully occupied. They therefore possess an enormous inertia (Smith, 1961). For example, a climatic change may prevent regeneration but not kill the existing population, so that the effect of the change is delayed for the lifespan of those individuals, which in the case of some tree genera could be more than 100 years. In certain circumstances such an inertia may have an effect longer than the life of the individuals concerned. The effect of a climatic change is not uniform, and in some regions local ecological conditions may compensate for it, as we have seen. So although under some stress, relict communities may survive long after the initial climatic change. In other regions the threshold for survival may have been crossed and that community (or species) may have disappeared (Fig. 25.25).

There is one final complication to the simple view of cliseral change: the plants and animals involved in these migrations are not static entities. They are subject to change in their genotype, largely by mutation and gene recombination. Before genetic change has culminated in the differentiation of a new species, however, small changes in genotype may be sufficient to alter the tolerance of the population and thereby affect its subsequent behaviour under environmental stress. For example, it has been suggested that the tolerance of the hazel (*Corylus avellana*) changed during the Quaternary, for it re-enters Britain progressively earlier with each successive interglacial (Fig. 25.26). Deacon (1974) suggests that this is because it was able to survive the glacial stages in refuges progressively closer to the ice

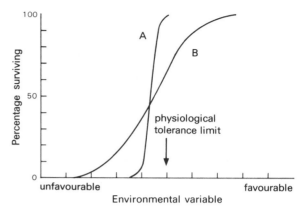

Fig. 25.25 Unbuffered and buffered population response to an unfavourable environment. A represents an experimental genetically homogeneous population for which an environmental variable (such as temperature) becomes increasingly unfavourable past a physiological tolerance limit or threshold. Past this limit the population declines rapidly to extinction. B represents a buffered gradual population decline in which some individuals are more vulnerable and others less so, because of genetic differences among them or differences in the microenvironments they occupy, or both. Genetic or microenvironmental heterogeneity can permit the population to survive environmental fluctuations that would make the population extinct in a homogeneous environment. (After Whittaker, 1975.)

limits as its cold tolerance evolved throughout the 2 Ma of the Quaternary. This last example introduces the final manifestation of change in the ecosystem, namely evolutionary change.

25.3 Evolutionary change

As we saw at the end of Chapter 7, the immediate precursor of life on Earth was the appearance in the 'primordial soup' of molecules or molecular concentrations which were autocatalytic in some manner. This self-reproducing tendency is not an uncommon chemical property. However, the appearance of a molecule that would produce chance variations among its 'offspring' and pass such variations on to the next 'generation' must have been a very rare event, even in the complex chemical mix of the early seas. Nevertheless, once such a system had arisen by chance, natural selection could operate on this variation; its consequence – adaptation – would appear and the

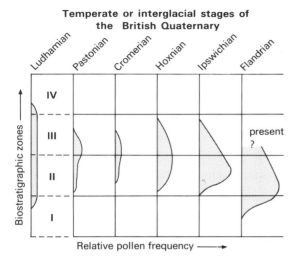

Temperate or interglacial stages of the British Quaternary

Ludhamian Pastonian Cromerian Hoxnian Ipswichian Flandrian

IV

III   present ?

II

I

Biostratigraphic zones →

Relative pollen frequency ⟶

Fig. 25.26 The changing behaviour of the hazel *Corylus avellana* in the temperate stages of the Quaternary. The hazel, an understorey shrub of mixed oak forest, arrives in the British Isles progressively earlier, finally arriving in the Flandrian well before the tree dominants of the forest of which it is normally a part. There is a massive expansion in its representation in the pollen diagrams as it spreads successfully on fresh unleached soils and flowers freely in the absence of competition. The explanation of this behaviour appears to reflect genetic change in the species itself; see text. (After Walker and West, 1970; Deacon, 1970.)

Earth would harbour life for the first time. Accepting this view of the origin of life emphasizes that from the very beginning three features have characterized living systems:

1. hereditable variation arising at random;
2. natural selection;
3. the appearance of adaptation.

Today, the sources of variations in genetic information are the segregation of chromosomes derived from different parents, the recombination of chromosome segments by crossing over, the spontaneous multiplication of chromosome sets and the occurrence of mutations. It is important to realize that in principle all of these mechanisms appear to operate at random. During the formation of the male and female gametes, chromosomes that originated from the two parents and were pooled at fertilization are separated into two groups. This separation or segregation is at random so that for n pairs of chromosomes there are 2^n combinations of differently endowed gametes possible. However, before segregation the chromosomes come together in pairs, one from each of the parents, and exchange segments of the DNA-encoded information that they carry, so giving rise to new combinations of genetic information. There is effectively no limit to the number of possible recombinations that may be produced in this way. Polyploidy (the production of multiple chromosome sets) is a complex group of processes involving some form of failure of cell division, apparently at random, during the production of the gametes. It is a very important and abrupt source of speciation in plants. These three sources of variation all involve the rearrangement of existing genetic information in the offspring. Mutations, on the other hand, are random alterations of this information itself. More specifically, they are changes in the base sequence of the DNA molecule.

In practice, this capacity for random genetic variation is constrained and regulated by natural selection, which determines which genotypes survive and by so doing determines the rate and direction of evolution. We have seen already in Chapter 7 that in a stable environment it is only the best-suited genotypes that are preserved. If such an environment remains unchanged for a long period and the ecosystem is a stable climax with no available niche space, then **stabilizing selection** dominates and inhibits evolutionary change in the population, even though mutation and genetic change will still occur among its individual members. Under most conditions these variants would be eliminated. However, Federov (1966) argues that, where the environment is genial and there is a degree of reproductive isolation experienced by individuals of a population in a diverse community (as in tropical rainforest), then mutations might persist and accumulate at random in the population. He calls this phenomenon **genetic drift** and uses it to explain the variety of physiognomic features of tropical rainforests. Many workers, notably Ashton (1969), would not agree with this view. Nevertheless, Simpson (1953) maintained that evolutionary stabilization is characteristic of organisms that occupy continuously available, generalized or homogeneous environments, such as the major forests and oceans, where the range of variation in the environment does not greatly affect the capacity of the available genotypes to exploit it. Of course, in theory, the reverse should also be true. That is, in such generalized environments a greater range of genotypes would be expected to survive as selective pressures are also more general, and hence some genetic drift might be expected. However, many of these hypotheses are rather speculative because of the nature of the evidence and of the timescales involved. Even so, it is safe to say that, in general, stable environments tend to promote evolutionary stabilization of populations, whether or not one accepts the possibility of random drift round this stabilized genotype.

Again in Chapter 7, we noted that in contrast, environmental variability, whether in time or space, allows natural selection to promote the differentiation of populations in response to environmental changes. Environmental change in time or variation in space will lead, by the selection from random recombination of existing genotypes, to adaptive shifts in the population to occupy either new or formally unoccupied niche space. If a stable environment becomes diversified, or migration to a new and diverse environment occurs, diversifying selection leads to adaptive radiation. If the trajectory of environmental change is maintained for a long enough period of time (millions of years), then sucessive adaptive shifts can produce what Simpson (1953) called **directional** selection. Mutations contribute to these changes largely indirectly, by their interaction with other genes and by producing new genetic combinations which may enhance the adaptive value of the whole genotype. In plants, polyploidy has been a very effective and rapid contributor to adaptive shifts under selective pressure from environmental change.

In a lucid review of evolution Stebbins (1974) arranges adaptive shifts in what he terms a spectrum, from those with a simple genetic base such as shifts in the frequency of particular genes which promote survival in an altered habitat, citing the case of industrial melanism in moths. The spectrum is followed through single phenotypic characters controlled by several genes, to the differentiation of subspecies and ecotypes with more complex alterations affecting many characters. Ultimately, if such shifts affect reproductive isolating mechanisms, restricting gene flow, then new evolutionary lines emerge at the species level and above. Therefore, there is a unity in the evolutionary process which differs not in kind but in magnitude of effect, at different taxonomic levels.

Further reading

The relevant chapters on population dynamics, on succession and on evolution in the general ecology texts cited at the end of Chapter 18 are an appropriate starting point. Additional insight into population regulation will be provided by:

Bergon, M. and Mortimer M. (1986) (2nd ed) *Population Ecology: a unified study of plants and animals*. Blackwell, Oxford.

Krebs, J.R. and N.B. Davies (eds) (1984) *Behavioural Ecology: an Evolutionary Approach*, (2nd edn) Blackwell, Oxford. (particularly part 2)

Moss, R., A. Watson and J. Ollason (1982) *Animal Population Dynamics*. (*Outline Series in Ecology*). Chapman & Hall, London.

Silvertown, J.W. (1987) (2nd ed) *Introduction to Plant Population Ecology*. Longman, London.

Solomon, M.E. (1969) *Population Dynamics*. Edward Arnold, London.

The concept of r and K strategies is well treated in the standard ecology texts already cited, but further information on the R-C-S system can be gained from:

Grime, J.P. (1979) *Plant Strategies and Vegetation Processes*. John Wiley, Chichester.

Grime, J.P., J.G. Hodgson and R. Hunt (1990) *(The Abridged) Comparative Plant Ecology*. Unwin Hyman, London.

The development of the concepts of ecological succession and the climax ecosystem are covered in:

Burrows, C.J. (1990) *Processes of Vegetation Change*. Unwin Hyman, London.

Connell, J.H. and R.O. Slayter (1977) Mechanisms of succession in natural communities, and their role in community stability and organisation. *American Naturalist*, **111**, 1119–1144.

Golley, F.B. (ed) (1977) *Ecological Succession*. (*Benchmark Series in Ecology*). Dowden, Stroudsburg. (contains reprints of most of the classic papers.)

Gray, A.J. *et al.* (eds) (1987) *Colonisation, Succession and Stability*. Blackwell, Oxford.

Miles, J. (1978) *Vegetation Dynamics*. (*Outline Series in Ecology*). Chapman & Hall, London.

Ecological change in response to recent Quaternary climatic and environmental change is exemplified in:

Godwin, H. (1975) *History of the British Flora: a Factual Basis for Phytogeography*, (2nd edn) Cambridge University Press, Cambridge.

Goudie, A.S. (1983) *Environmental Change*, (2nd edn) Clarendon Press, Oxford.

Lowe, J.J. and M.J.C. Walker (1984) *Reconstructing Quaternary Environments*. Longman, London.

Roberts, N. (1989) *The Holocene: an Environmental History*. Blackwell, Oxford.

Change on an evolutionary timescale, but set in an ecological context, is the theme of an excellent recent book:

Cockburn, A. (1991) *An Introduction to Evolutionary Ecology*. Blackwell, Oxford.

Access to recent debates in evolutionary biology can be gained from:

Dawkins, R. (1986) *The Blind Watchmaker* Longman, London.

Edwards, K.J.R. (1977) *Evolution in Modern Biology*. Edward Arnold, London.

Eldridge, N. (1985) *Time Frames*. Simon & Schuster, New York.

Several of the paradigms of ecological and evolutionary change come in for scrutiny in

Peters, R.H. (1991) *A Critique for Ecology*. Cambridge University Press, Cambridge.

VIII THE HUMAN'S IMPACT

Human modification of environmental systems

The interactions of humans with the environment are enormously complex. At one level *Homo sapiens* is just one kind of organism, a primate distinguished by an upright posture, sparse body hair and a highly developed brain. As such we may regard him as fitting naturally into the structural and functional organization of the ecosystems in which he occurs. Indeed, such an approach is perfectly valid and the treatment of relatively simple human societies in this way has helped to improve our understanding of human–enrivonment relationships. Fig. 26.1 shows the tropic relationships of Mesolithic Man in the forests of the British Isles during the climatic optimum of the Flandrian; similar relationships involving *Homo sapiens* at the end of the Devensian glaciation are depicted in Fig. 26.2. This application of systems modelling to complete human ecosystems is not restricted to primitive societies and is equally valid with advanced technological societies, often shedding new light on the energy relationships and transfers that pertain in these complex systems (Fig. 26.3).

However, in the introductior to this book it was stressed that humans are not merely omnivorous animals occupying a trophic niche in the food web of natural ecosystems: they are conditioned also by a complex and heterogeneous cultural inheritance. We stated that humans' interactions with the natural environment can be understood only by considering their perception of it and their behavioural responses to it, both of which are conditioned by their cultural environment. The implication is that to do justice to human–environment interactions we have to be able to mesh models of social and economic systems with our models of natural systems, so that the linkages between the activities of these systems become clear. We must be able to accommodate flows of investment, technology, decision-making, generated energy and other elements of what have been called human information and control sytems in a comprehensive human–environment system model (Chorley, 1973). Perhaps this is one of the most important academic challenges that exist today. There have

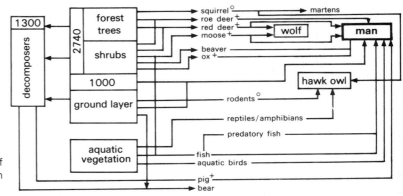

+ total *c*. 6 kg/ha
o total *c*. 5 kg/ha

Fig. 26.1 The trophic relationships of Mesolithic Man in the early Flandrian forests of Britain (after Simmons, 1973).

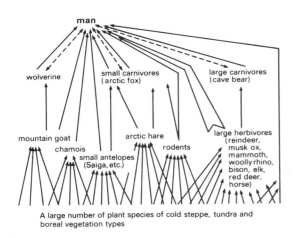

Fig. 26.2 The trophic relationships of humans at the end of the Devensian glaciation, inferred from accumulations of animal bones (after Cox *et al.*, 1976).

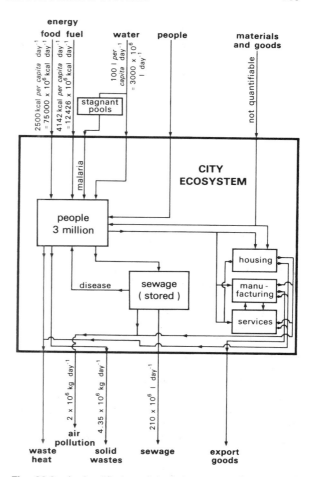

Fig. 26.3 A simplified model of the metropolitan area of greater Calcutta viewed as an ecosystem (after Learmonth, 1977).

been attempts at such complete models, for example the world models of Forrester (1971) and Meadows and the 'limits to growth' team (Meadows *et al.*, 1971), but this kind of comprehensive approach is beyond the scope of this chapter, as is the detailed consideration of information-control systems (Fig. 26.4). What these observations make clear, however, is that many of the models we have used to explain and understand natural environmental systems do not, as they stand, accommodate the cultural activities of humans. Such activities must be considered therefore as exogenous variables, as inputs to our models of environmental systems, and this approach will predominate in our treatment of humans and their interactions with their environment.

The effect of human activities as inputs to environmental systems is to redistribute matter and energy between the stores of the system and to alter both the magnitude of, and the pathways by which, energy and matter are transferred. Such interference with transfer pathways applies not only to those within a particular system, but also between systems. Because natural systems are open, and are organized (no matter at what scale we may view them) as energy and mass cascades, the consequences of human intervention in the operation of a system will almost inevitably have

ramifications beyond the boundaries of that system.

Even the planting of human feet on the ground causes localized compression of the soil and a consequent reduction in the amount of rainfall that will infiltrate it. For example, multiplied manyfold, the trampling of footpaths creates localized high surface runoff and hence a more consequential modification of the hydrological system. Indeed, human activities can change the nature of the land surface in ways that are quite fundamental to the operation of natural systems, by, for example, the removal or modification of the accumulated living and dead biomass of

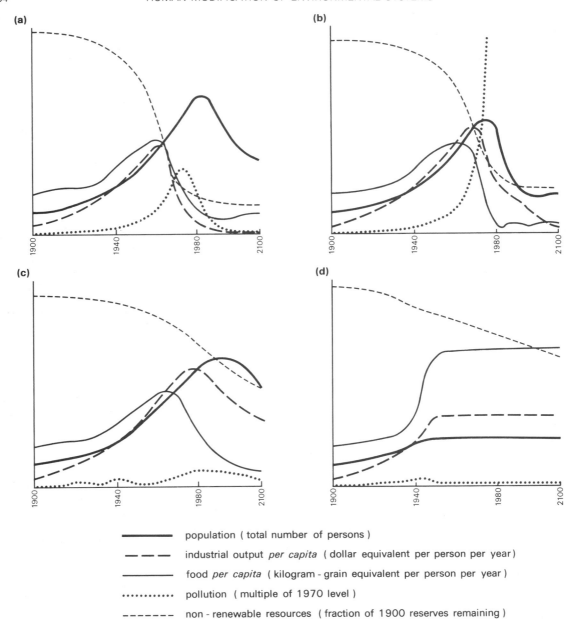

(a)

(b)

(c)

(d)

——————— population (total number of persons)

– – – – – industrial output *per capita* (dollar equivalent per person per year)

——————— food *per capita* (kilogram - grain equivalent per person per year)

•••••••••••• pollution (multiple of 1970 level)

– – – – – – non - renewable resources (fraction of 1900 reserves remaining)

Fig. 26.4 The Forrester–Meadows world model. (a) A standard run of the computer simulation model based on 1970 values. (b) Pollution-induced collapse predicted even when known natural resource reserves are doubled. (c) Collapse due to population growth even though resources are set as unlimited and pollution controls are assumed. (d) Stabilized model producing an equilibrium state sustainable into the future. However, assumptions include birth rate set equal to death rate, capital investment equal to capital depreciation, and that a range of technological policies are implemented, including resource recycling, pollution control, and restoration of eroded and infertile soils. (After Meadows, 1971.)

the natural vegetation and the disruption of the ecosystem's functional organization. As a result, natural regulators to the flow of water through the system are fundamentally modified, such as vegetation and the ability of the land surface to store water. However, the extent to which the hydrological properties of the land surface are altered will reflect a sequence of changes in land use, initially by deforestation, then by the creation of arable land, with its exposed soil surfaces, and in more recent times the spread of urban and industrial areas with paved surfaces.

These modifications lead to changes in the water balance of catchment areas, usually increasing the proportion of surface runoff. In addition, the distribution of runoff through time is also changed, normally producing concentration of runoff into higher peak flows. These hydrological changes in turn permit greater removal of mineral material from the land surface in the form of increased rates of erosion. In order to accommodate these increased flows of water and minerals, river channel adjustments take place. Such changes in process are normally progressive, as the negative feedback processes prevalent in natural systems are replaced by positive feedback mechanisms.

We have referred already in our discussion in the preceding chapters to some of the modifications humans may make to atmospheric systems. Their ability to effect major long-term changes is limited by the ability of the Earth–atmosphere system to compensate for the relatively small temporal and spatial scales of change which arise from their activities. Most of these are localized and are usually short-lived. The nuclear bomb represents an event of the highest order of magnitude that humans are capable of generating. Although the localized effect is total devastation, modifications to the whole Earth–atmosphere system of a single explosion are small. Most of the changes that humans make are not, in fact, a result of such high-magnitude events but of continuous and cumulative modifications to the operation of regulators within the Earth–atmosphere system.

The effects of human activities, some of which we have just considered, are brought about either as an inadvertent byproduct or through conscious efforts to manipulate the environment, though the latter are often accompanied by the former. In both cases, as we have seen already, the mechanisms by means of which change is effected are the system regulators – the parameters that dominate the feedback relationships controlling the operation of the system. Nevertheless, the impact of human intervention in the dynamics of natural systems, whether purposeful or unwitting, is so widespread that to present a comprehensive picture of its extent would require us to retrace our steps through this book, identifying at almost every turn the effect that humans have had, are having, or could have in the future. In the sections that follow, therefore, we can provide only a glimpse of the profound and pervasive influence of these strange, erect primates with the highly developed brains.

We shall begin by looking at some of the accidental changes in the operation of environmental systems that have resulted from and accompanied different levels of modification and alteration that humans have exerted on the environment. For convenience we shall consider such levels of interaction under the headings of deforestation, agriculture and urbanization, although each of these terms will be interpreted in a rather broad way.

26.1 Deforestation

The potential natural climax of most of the land areas of the globe that today support high human populations is some type of forest ecosystem, or at least an ecosystem in which tree cover forms a significant component. Therefore, it is justifiable to begin by considering the effects that a long history of human modification, exploitation and clearance of these forests has had, both on the ecosystem itself and on its functional relationships with the denudation system.

The biomass of the living vegetation of the ecosystem and the dead biomass store represented by the litter and organic matter of the soil are critical regulators of catchment hydrology, both directly and indirectly, through their control of interception, surface and soil moisture storage, infiltration and evapotranspiration. Therefore, under natural conditions mature forest ecosystems regulate the two-way transfer of energy and water between the soil and the atmosphere. They regulate weathering and slope processes and conserve elements, both in the biomass of the forest itself and in its massive closed nutrient circulation, against leaching and erosion loss. Only in recent decades have controlled scientific experiments permitted an assessment of the magnitude of change in system operation following deforestation. In this way, flows of water and nutrients through a forested catchment can be measured during a calibration period, and then the forest is cut and the ensuing effect on runoff monitored. Alternatively, a pair of similar catchments may be monitored simultaneously. One of them can be deforested and the ensuing effects can be compared against its still-forested neighbour.

The primary effect of deforestation is to decrease the loss of water by evapotranspiration as the biomass is reduced. The fundamental change in water balance leads to an increase in surface runoff. This effect is illustrated dramatically by results from the Coweeta catchment in the Appalachian Mountains (Fig. 26.5), operated by the US Forest Service. With an annual precipitation of approximately 2000 mm, clear felling of the hardwood forest produced an increased surface runoff equivalent to 373 mm depth of precipitation. As the forest grew again, the excess water yield declined exponentially with time over a period of 20 years, until a second clear felling operation produced a similar increase in surface runoff.

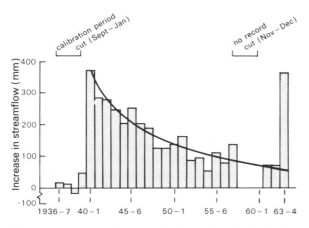

Fig. 26.5 Increase in water yield after clear felling in the Coweeta catchment (after Hibbert, 1967).

A similar effect was observed at the Hubbard's Brook catchment in New Hampshire, USA (Bormann and Likens, 1970). An overall increase in surface runoff of 40% was recorded after deforestation, with the increased stream discharge particularly marked in summer, when evapo-

transpiration would have been at its highest. The summer period showed a fourfold increase in surface runoff, which would be achieved in large measure by increased flood peaks and increased velocity of streamflow.

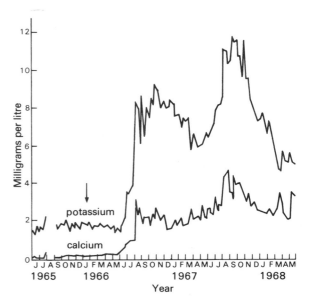

Fig. 26.6 Increased output of potassium and calcium in stream water from watershed no. 2 following deforestation (arrowed), Hubbard Book, New Hampshire, USA (modified from *The nutrient cycles of an ecosystem* by F.H. Bormann and G.E. Likens, copyright © 1970 by *Scientific American*, Inc., all rights reserved).

Removal of solutes was also monitored in the Hubbard's Brook experiment. Fig. 26.6 shows the increase in concentrations of calcium and potassium in stream water consequent upon felling. On average, concentrations of dissolved elements increased fourfold to produce a net output 14.6 times greater than when forested. The increased availability of unprotected surface soil and the greater energy of the channel runoff led in addition to a fourfold increase in sediment output from the basin. Table 26.1 shows the budget for several elements for both the forested and felled catchments at Hubbard Brook, and emphasizes the control the forest exerts over nutrient loss to drainage under undisturbed conditions. In the forested catchment, there is a net input of nitrogen from total nitrogen fixation in the ecosystem, and a conservation of the element in the nitrogen cycle.

Table 26.1 Nutrient budgets for catchments at Hubbard Brook (after Likens and Bormann, 1972)

	Precipitation input (kg ha^{-1} a^{-1})	Streamflow – precipitation net output (kg ha^{-1} a^{-1})	
		Forested catchment	Cutover catchment
calcium	2.6	9.1	77.9
sodium	1.5	5.3	15.4
magnesium	0.7	2.1	15.6
potassium	1.1	0.6	30.4
NH$_4$ nitrogen	2.1	−1.8	1.6
NO$_3$ nitrogen	3.7	−1.7	114.0

In the cleared catchment, the nitrogen cycle is broken and organic nitrogen and ammonia are oxidized rapidly to nitrate-nitrogen and lost to the stream. The increased concentration of the nitrate anion affects the ionic balance of soil and stream water and increases their ability to transport metallic cations.

Such clear felling of an entire catchment illustrates the magnitude of the effect of a forest cover on denudation and ecosystem processes in terms of nutrient cycling, surface runoff and rate of erosion. Studies such as these have prompted a re-evaluation of the long-term benefits of conservation and intelligent management of natural ecosystems by humans. We are learning to look on and conserve ecosystems not from the point of view of productivity and economic return alone, but in terms of the control that their highly complex regulatory mechanisms exert over environmental processes. The original undisturbed natural forests of the world were cleared gradually and partially, and often allowed to regenerate (Fig. 26.7). Forest ecosystems display a remarkable resilience in the face of human intervention as long as the degree of exploitation is limited and does not exceed the threshold beyond which recovery is impossible. Their species diversity, the complexity of their functional organization and the concentration of resources and information in the living forest all mean that they are to an extent buffered against change; they possess the capacity for self-regulation and regeneration. Therefore, they can coexist with the limited demands of shifting cultivation, limited grazing pressure, limited timber exploitation and settlement. Indeed, much of the floral and faunal diversity and aesthetic wealth of woodlands as perceived in the

temperate zone of the northern hemisphere – the small grassy clearing, the rather open canopy, the variety of woodland birds and the flush of spring flowers – is a reflection of centuries of low-level exploitation and structural modification (Streeter, 1974; Rackham, 1971, 1986) (Fig. 26.7). Wholesale commercial timber exploitation and clear felling of virgin forest is an entirely different matter.

As we have seen in Chapter 25, there are considerable theoretical and empirical arguments indicating that complex ecosystems are dynamically fragile. They may well possess great stability in the relatively narrow range of environments within which they have evolved, but faced with disturbances wrought by humans they are inherently unstable. Such a view applies particularly to primary tropical rainforest, which has proved much less resistant to intervention by humans than have simpler and more robust temperate ecosystems. The reasons again lie with regulatory mechanisms. May (1975) has argued that evolution has selected organisms which show genetic, morphological and physiological characters that regulate population densities at or near equilibrium values, i.e. $N \simeq K$ (where N is the population number and K is the carrying capacity). Animals and birds produce fewer progeny than their temperate counterparts (Southwood, 1974) but occupy more precisely defined niches and are more successful competitors in the sense that they therefore avoid direct competition. Plants are selected also for competitive ability in the face of extreme interspecific competition for space and resources in a genial environment. In the face of habitat disturbance, such organisms are poorly equipped to respond: they are not characterized by opportunism, by wide tolerance limits, or by a potential for rapid population growth. They are adapted instead to survive and compete in a niche precisely defined by the structural and functional complexity of the forest itself; destroy that and you destroy the organism's capacity to survive. The seeds of many species, for example, have little or no dormancy period and are adjusted to the moist cool shade of the forest and the stable microclimate of the forest soil (23–26°C). Soil heating on clearance therefore results in seedling death and an inability to recolonize, once sufficiently large areas have been cleared. These characteristics led Gomez Pompa (1972) to refer to tropical rainforest as a non-renewable resource, and one which is in fact

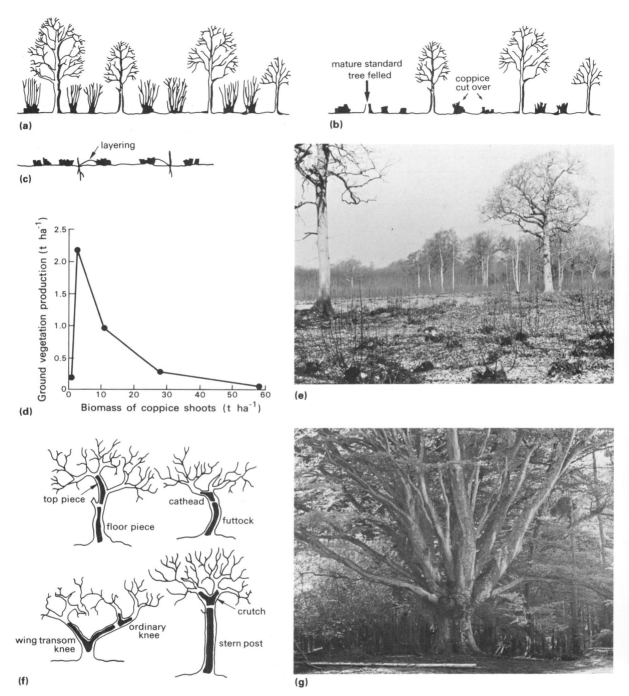

Fig. 26.7 Coppice with standards system of woodland management, (a) before and (b) after cropping – note that one standard tree has been felled. (c) Layering of coppice to produce new stools (all after Ovington, 1965). (d) Relationship between annual production in the herb layer and biomass of coppice shoots at different stages in the coppice cycle (after Ford and Newbould, 1977). (e) Coppice with standards, Stansted Forest, Hampshire, England. (f) Training of oak growth to produce timber suitable for wooden ships and for timber frame housing (after Albion, 1926). (g) Old pollarded (cut back to the top of the bole) beech, New Forest, England.

(a)

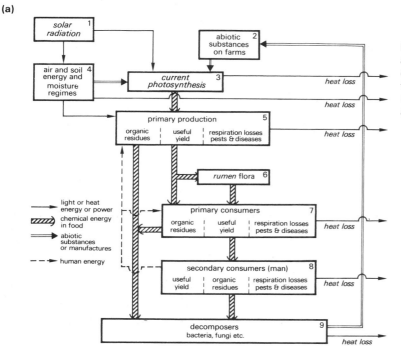

Fig. 26.8 Flows of mass and energy in (a) a simple farming system in Uganda cropping domesticated herbivores, and (b) a complex farming system in the UK, linked to the urban–industrial economy (after Duckham and Masefield, 1970).

(b)

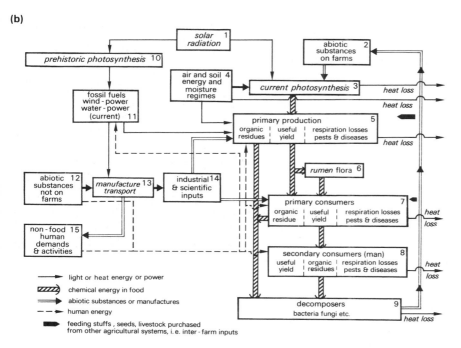

disappearing as a result of timber operations in southeast Asia and of the opening of Amazonia to commercial ranching following the road-building programme.

The plight of the rainforests of the humid tropics is perhaps an extreme and urgent contemporary case, but to exploit any natural or seminatural forest ecosystem, particularly for timber, is to decrease its complexity. Unwanted trees are eliminated; access is improved by reducing understorey growth; other herbivores competing with humans are culled, or excluded by fencing; uniformity is increased by planting policies; and animal diversity declines further as the range of available habitats is reduced. Managing and cropping such forests for a sustained timber yield has been likened to the management of permanent pasture for grazing by domesticated herbivores. Afforestation, on the other hand, is the equivalent of arable agriculture in the sense that seedlings raised in nurseries from selected seeds are planted and maintained with the aid of fertilizers, pesticides and a considerable manpower and management input. Like field crops, and for the same economic reasons, these new forests are often monocultures, although mixed planting is employed sometimes, with, for example, a slow-growing hardwood species such as beech (*Fagus sylvatica*) planted with a faster-maturing conifer such as the larch (*Larix decidua*).

However, there are problems associated with all intensively managed forests, especially with new commercial afforestation, which indicate their long-term ecological instability. First, the uniformity of dense stands of a single tree species makes them particularly susceptible to explosive increases in pest populations and to outbreaks of disease. Combating such problems in the absence of natural regulation mechanisms is a costly operation often involving the aerial spraying of pesticides, sometimes with unforeseen side effects. Secondly, and again because of their uniformity, commercial plantations and managed seminatural forests are more prone to fire hazard than are most natural forest ecosystems. Finally, the sustained cropping of timber represents a break in

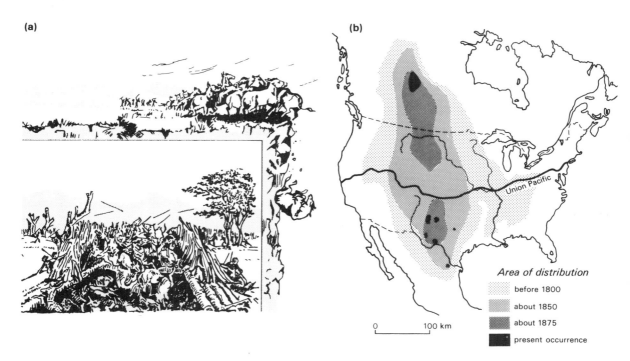

(a)   **(b)**

Area of distribution

before 1800

about 1850

about 1875

present occurrence

0   100 km

Fig. 26.9 (a) Cooperative hunting by driving and pitfall trapping and by driving over a cliff. (b) The dramatic contraction in the distribution of the bison in North America following the opening of the Union Pacific railway and the extension of commercial hunting (from Illes, 1974, after Ziswiler, 1965).

the natural nutrient circulation that, together with the accelerating effects of timber extraction operations on runoff and erosion, is a progressive drain on the fund of available plant nutrients. This is especially true of softwood forests in the northern hemisphere, with short cropping cycles of 30–40 years, established on soils of low inherent fertility, and ultimately yields may decline unless the nutrient status is maintained by chemical fertilization.

26.2 Agriculture

All agricultural activity can be seen in the widest sense as a deliberate attempt on the part of humans to modify and manipulate the trophic relationships that pertain in nature (Fig. 26.8). In almost all cases the effect is to simplify the complexity of natural ecosystems in order to increase the direct harvest to humans. By so doing, the ecological balance of the natural community

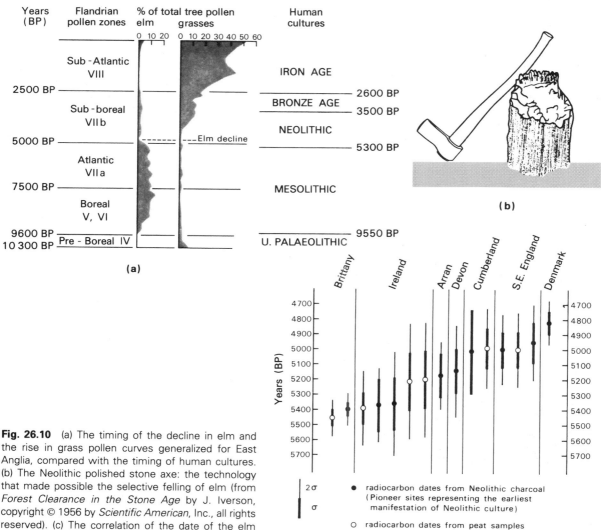

Fig. 26.10 (a) The timing of the decline in elm and the rise in grass pollen curves generalized for East Anglia, compared with the timing of human cultures. (b) The Neolithic polished stone axe: the technology that made possible the selective felling of elm (from *Forest Clearance in the Stone Age* by J. Iverson, copyright © 1956 by *Scientific American*, Inc., all rights reserved). (c) The correlation of the date of the elm decline and the data of the earliest appearance of Neolithic culture in western Europe (Seddon, 1967).

is modified or destroyed as the complex but stable network of interactions is broken within that community and between its organisms and their abiotic environment.

As hunter–gatherers, humans were (and in some parts of the world still are) an integrated part of the dynamic balance of the steady-state climax ecosystem. However, with increasing social organization and cooperative capacity, and with the development of tools and weapons, the magnitude of their effect on natural herbivores and carnivores increased as they upset the regulation of population densities inherent in many natural grazing–predation systems. The extinction of many large Quaternary mammals in North America, Africa and, to a lesser extent, Eurasia has been attributed to the cooperative hunting techniques of Palaeolithic humans (Fig. 26.9). Although there is some debate as to whether environmental change could account adequately for these extinctions, it remains true (as we shall see) that the extinction of some and the reduction in the numbers of many animal species are due directly to hunting by humans.

The relationships between humans and their fellow heterotrophs change fundamentally with the domestication of grazing herbivores and the development of pastoral economies. The stocking densities of domesticated herbivores are usually higher – often very much higher – than those of natural herbivores. The consequences of such densities are to reduce the standing biomass and to alter the structure and composition of the primary production system. As we have already seen, vegetation change may initially be promoted directly by human action. The selective lopping of elm branches for fodder by Neolithic people is thought partially to explain the decline in elm pollen in the Flandrian (Fig. 26.10). However, the clearance of forest, often using fire, is the ultimate expression of such modification motivated by a desire to extend pasture, and at least some parts of the major grassland biomes of savanna and prairie were probably created in this way. Indeed, as we have just seen, it is still continuing on a large scale in the case of the Amazonian cattle ranches. Nevertheless, the large domesticated herbivore population itself brings about more

(a)

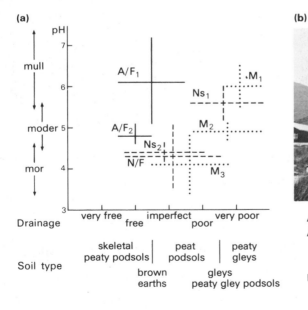

(b)

A/F$_1$	species rich	*Agrostis / Festuca*
A/F$_2$	species poor	*Agrostis / Festuca*
Ns$_1$	species rich	*Nardus stricta* (sub - alpine)
Ns$_2$	species poor	*Nardus stricta* (sub - alpine)
N/F	*Nardus – Festuca* (with *Deschampsia*)	
M$_1$	*Molinia – Agrostis*	
M$_2$	*Molinia – Festuca / Agrostis*	
M$_3$	*Molinia – Festuca / Deschampsia*	

Fig. 26.11 (a) Habitat ranges in terms of pH, humus type, soil type, and drainage class for neutral and acid grassland communities in the British uplands (after Burnett, 1964). (b) Extensive sheep grazing in the Scottish Highlands. Sheep stocking rates and grazing pressure, acting in conjunction with the edaphic gradients depicted in (a), largely determine the sward types present.

The twayblade, *Listera ovata* (L.) R. Br.

The presence and size of these ant hills at Noar Hill, Hampshire indicates that the pasture has remained undisturbed by the plough for a considerable time.

The sites of former chalk pits at Noar Hill.

The milkwort, *Polygala vulgaris* L.

rmander speedwell, *Veronica drys* L.

The pyramidal orchid, *Anacamptis pyramidalis*.

The cowslip, *Primula veris* L.

6.12 The grassland sward of the English chalklands, with its rich complement of flowering herbs and short springy a classic example of a community whose character has been largely produced and maintained by grazing.

subtle changes in the structure and composition of the sward.

Free-range or extensive grazing of unimproved or rough pasture allows the grazing preference of the herbivore to be expressed. For example, the extensive studies of Hunter (1962a,b) and others have shown that in the neutral and acid grasslands of upland Britain sheep have recognizable preferences which change seasonally and influence the way in which they interact with a largely edaphically controlled vegetation mosaic (Fig. 26.11). Where socially determined feeding ranges cover a variety of sward types, selective grazing pressure, particularly if combined with high stocking densities, can lead to an extension of the less palatable herbage species such as mat grass (*Nardus stricta*) and purple moor grass (*Molinia caerulea*). In addition, every grazing animal has a distinctive way of obtaining herbage, and sheep can graze very close indeed. Combined with selective grazing, such intense defoliation results in lower yield, a reduction in the recovery capacity of the sward, and ultimately in overgrazing. The existence of an unsaturated niche then allows opportunities for invasion by coarser and less palatable species. Virtually all of the dwarf shrub moor and grassland communities so characteristic of the British upland landscape originated as byproducts of their exploitation for grazing after forest clearance, and from the use of fire as a management tool. They represent a balance between edaphic and climatic factors on the one hand and the differential effect of the particular kind and intensity of grazing, both now and in the past, on the other.

The complete plant assemblage, as well as many of the characteristics of the soils of the short-cropped downland turf of southeast and south-central England, are also accidental products of centuries of grazing, particularly by sheep and rabbits (Fig. 26.12). The latter were introduced by the Normans as a semidomesticated herbivore, but subsequently they spread naturally. The reduction in grazing intensity and the temporary virtual annihilation of rabbit populations by myxomatosis in 1954, have made this status clear as areas of downland turf have reverted through tall grass communities to scrub and woodland (Thomas, 1963). Comparable examples of the effect of grazing can be seen in many other parts of the world. The destruction of the sclerophyllous evergreen oak forests of the Mediterranean by goats and its replacement by maquis and garrigue communities, if not by virtual desert, is a well documented example (Fig. 26.13). In New Zealand the natural tussock grasses of South Island were invaded and replaced by introduced species of grass under grazing pressure from sheep and introduced wild herbivores. Thorsteinsson (1971) recounts the deforestation of Iceland by sheep and the impoverishment of the vegetation and erosion of up to 40% of the soil cover that resulted from it.

The effect of grazing on largely unmanaged pastures is to convert the ecosystem to something very similar to the early stages of succession. Apart

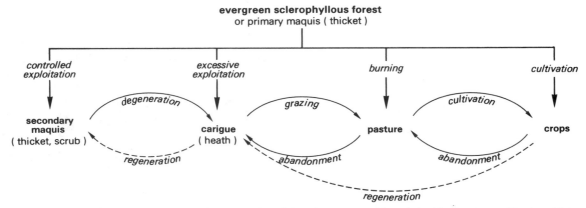

Fig. 26.13 Stages in the degeneration and regeneration of the plant communities of the Mediterranean. The long history of modification by man from at least classical times to the present has replaced almost all of the natural evergreen sclerophyllous forests by secondary communities with associated soil erosion and environmental deterioration. (After Polunin and Huxley, 1967.)

Table 26.2 Lifeforms found in chalk grassland, acid grassland and neutral grassland

Lifeform	Position of overwintering bud	Chalk grass land (%)	Acid grass land (%)	Neutral grass land (%)
chamaephytes	soil surface to 25 cm	7.6	20.8	7.2
hemicryptophytes	at soil surface	67.6	54.6	70.3
geophytes	below soil surface	16.5	5.6	13.5
theraphytes	as seeds	8.2	18.8	8.6

from the reduction in above-ground producer biomass, many of the features of the plants themselves, such as the position of the perenating bud (Table 26.2), the presence of protective spines, and the prevalence of annuals or plants with vegetative means of reproduction, equip them to withstand grazing. The last-named adaptation is associated often with below-ground food storage organs such as rhizomes endowing the plant with the capacity for rapid growth and recovery. Indeed, the below-ground production of pasture grassland is often several times higher than the above-ground production available to herbivores – for example, there is twice as much annual production below ground in the Missouri prairies (Kucera *et al.*, 1967). Most, if not all, of the adaptations that prove to be advantageous in the face of grazing are also those that would be important in the relatively extreme environments

experienced by pioneer plant species in the early stages of natural succession.

There are parallel changes in soil conditions accompanying grazing. Exposure under a thin grass cover may lead to a greatly changed soil microclimate resulting in greater soil aridity. Alternatively, under a wetter climate reduced interception and transpiration can lead to localized waterlogging. This tendency will be promoted further where burning is practised, for the induration of the soil surface that results will increase the runoff component by changing the infiltration capacity regulator. The area occupied by gley and peaty gley soils in depressions and along drainage lines in upland Britain has probably been extended in this way. Even in lowland Britain under a drier climate the same process can be invoked to explain partially the valley mires of the New Forest (Fig. 26.14). Perhaps the most important soil modification, however, is the drain of grazing on its inherent fertility. The removal o' the most of the above-ground primary production and hence the nutrient elements that it contains, as a crop first to domesticated herbivores and eventually to humans, represents a loss to the detrital and soil systems. Although the deposition of dung and urine can to some extent offset this loss, it can lead to alterations in the carbon/ nitrogen ratio and may lead to the breakdown of effective nutrient cycles. Furthermore, the dense adventitious root system of the pasture plants is shallower than that of the woodland or forest it may have replaced, and it cannot draw on the nutrient reserves of the subsoil. Again the result

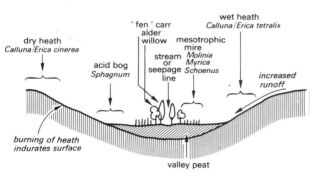

Fig. 26.14 A New Forest valley 'bog'. Strictly this community is a soligenous mire maintained by impeded downvalley seepage of water. Its nutrition is minerotrophic (soil water) and of moderate base status. The removal of forest from the interfluves and the maintenance of heath communities under a grazing and burning regime has led to intense podsolization, induration of the soil surface, reduced infiltration and accelerated runoff. As a result marginal acid bog communities have developed, characterized by ombrotrophic (rain water) nutrition, low base status and low pH.

is an alteration in the pattern of nutrient cycling in the soil.

Animals are also affected by changes in the habitat associated with extensive grazing. For example, with the creation of bare areas, perhaps by localized overgrazing or by excessive trampling, the habitat becomes unsaturated, and available niches appear which can be exploited by burrowing animals, such as the rabbit in Britain, or the gopher in the American Midwest. Carnivores are also affected, partly by changes in the habitat, but also by hunting and trapping, as humans try to control the loss of domestic herbivores to predators (and, in the case of Britain particularly, of wild but partially managed game species, e.g. pheasant, grouse, deer). The wolf and the golden eagle in Britain, the lion and leopard in Africa, the jaguar, puma and coyote in North America, and a marsupial 'wolf' in Tasmania are all examples of predators that have been exterminated or drastically reduced in number (except perhaps in national parks). Natural herbivores such as many species of deer, the bison and some of the larger species of kangaroo have also been culled to reduce competition with domestic herbivores, or to reduce damage to arable crops. Ironically it is now realized that, because they are adapted to utilize a larger proportion of the herbage, many of these natural herbivores have a higher productivity than domesticated herbivores, especially where the latter have been introduced.

Apart from the alterations to natural ecosystems that result from grazing, it is important to realize that the kind of pastoral grazing system we have been discussing is also inherently inefficient. The animals are too selective in their intake, too much energy is dissipated in their metabolism and locomotion, and lost as faeces, and as a result the conversion of plant production to animal protein is too wasteful. However, the need for some animal protein in the human diet has led to the development of more efficient 'pastoral' systems which are best considered under the heading of intensive agricultural systems.

If extensive grazing management leads to the modification of natural ecosystems, then intensive grazing, including store-fed and factory livestock systems, and arable agriculture lead to the establishment of almost entirely artificial systems (Table 26.3). In fact, the modification of the landscape to suit agricultural practice can involve not only the formation of artificial ecosystems of great simplicity, but also profound alteration of the land surface itself. Even in the early days of agriculture, quite prodigious feats of civil engineering were carried out using human energy aided by simple tools. Many primitive agricultural civilizations adopted hillside terracing for cultivation or irrigation of steep hillslopes. The conversion of even a slope of modest gradient (15°) to terraces 2 m in width is illustrated in Fig. 26.15. Each terrace is formed by the displacement by excavation and depositing of 0.134 m³ of soil per metre width of slope; for a completely terraced hillside this represents a transfer of soil of 67 000 m³ km⁻². The widespread extent of such terracing in hilly country in such regions as the Himalayas, the Andes and Japan is an indication of the total volume of material transferred, volumes which are even greater on steeper slopes. The extensive ridge-and-furrow agricultural systems characteristic of the English Midlands involve modifications of similar magnitude. A wavelength crest to crest of 9 m and an average amplitude of 0.5 m requires a transfer of earth of 62 500 m³ km⁻². In north Buckinghamshire alone some 35 km² of ridge-and-furrow forms exist, representing the uplift of well over 3.5 million tons of earth by an average of half a metre.

The modifications to the operation of environmental systems that result from bringing land into arable cultivation are many, but a major characteristic of arable land is that the biomass present in cereals, root crops or grass is considerably less than in natural ecosystems. Accordingly, there will be a lower rate of evapotranspiration and greater runoff. Often this is facilitated by the creation of artificial drainage ditches, which in effect extend the natural drainage network. Thus drainage density, which in a humid temperate environment such as Britain is normally in the range 1.5–3.5 km km⁻², may be raised to between 5 and 10 km km⁻². Drainage may be effected further by tile drains beneath the soil, which are spaced at 1–3 m, depending on the soil texture. These drains may raise the drainage density further at the times when water is flowing in them to between 100 and 350 km km⁻². The effect of this increasingly efficient evacuation of water from the soil surface is to transfer it more rapidly to the main river channels and thus to increase peak flows.

The conversion of forest to arable land also destroys the forest litter and lays bare the surface of the soil. This is particularly the case when the land is fallow or before the crop has fully matured.

Table 26.3 The modifications and manipulation of natural ecosystem aimed for by the arable and pastoral farmer in modern agricultural systems

If cropping plant production	If cropping animal production
ESTABLISH almost entirely ARTIFICIAL ECOSYSTEM (perhaps also landscape)	SELECT SINGLE HERBIVORE – Manage for meat and animal products – breed to increase yields

SELECTION, BREEDING, PROPAGATION of SPECIALIZED CROP PLANTS:
  High yield of storable material
  satisfying dietry needs for
  energy (Carbohydrates) and
  growth (protein)
  Disease and pest resistant
  Genetic and physiognomic uniformity – 'machinable'
	EXCLUDE competing herbivores
	INCREASE Population Density/Stocking rates
	
	MANAGE HERBAGE BY:
	  Regulation and Control of Stocking Density
	  Exploitation of social behaviour,
	  Selective grazing preferences
	  Eradicate unpalatable species including shrubs and trees
	  Extend area of herbage by clearance

CULTIVATION: tillage, planting, fertilizing, drainage, irrigation, weeding, harvesting, i.e. increased inputs.	MAINTAIN and ENCOURAGE HERBAGE PRODUCTION BY:
	  Direct intervention – cutting and burning to encourage vegetative regrowth
	  Increase Inputs: redistributing manure, apply fertilizer, drain, re-seed.

REDUCTION in STANDING BIOMASS relative to natural ecosystem BUT increase in ABOVE: BELOW	
INCREASE in SEASONAL VARIATION Single peak of productivity – L.A.I	NB INTENSIVE GRAZING SYSTEMS STORE-FED LIVESTOCK and FACTORY FARMING SYSTEMS best considered with ARABLE

Even then, with the exception of pasture, most crops still leave a proportion of soil directly exposed, and rainsplash erosion, rill action and overland flow can readily detach soil particles and carry them towards the river channel, increasing sediment concentration and rates of erosion. Thus Evans and Morgan (1974) observed soil erosion under immature crops in an area of Cambridgeshire of 3.3 t ha^{-1}, representing a surface lowering of 0.25 mm in a few days. Douglas (1967) quotes an increase in sediment yield in Java from 900 m^3 km^{-2} yr^{-1} in 1911 up to 1900 m^{-3} km^{-2} yr^{-1} in 1934, as a result of increasing deforestation and an extension of cultivation. Arable agriculture, therefore, can lead to increased total runoff, increased peak flows and increased sediment yield.

The loss of organic matter from the agricultural soil will further promote this soil loss by erosion as the structure of the soil is lost. This loss of structure results from the reduction of aggregating forces within soil crumbs as organic matter is lost. No longer water-stable, soil aggregates disintegrate on raindrop impact and the now discrete soil particles accumulate in the surface pores, forming a soil crust, reducing the infiltration capacity and initiating overland flow (see Fig. 12.4c). The breakdown of organomineral complexes and loss

of structure reduce the nutrient- and water-retention capacity of the soil and, together with the progressive drain on soil nutrient reserves that cropping represents, lead inevitably to reduced fertility. The addition of manure, chemical fertilizers and perhaps, in periods of summer water deficit, irrigation water, then becomes a necessity for sustained crop yields. However, low soil-storage capacity means that leaching losses of fertilizer and liquid manure can be high, and when they reach drainage channels, this can result in pollution and eutrophication problems in the aquatic ecosystems of streams and lakes. Soil drainage and soil aeration can also be affected adversely by agricultural practice, particularly where heavy machinery is used consistently and produces soil compaction. Repeated ploughing to the same depth has a similar effect with the development of a plough pan at the depth of the furrow, leading to impeded drainage (Fig. 26.16).

From the ecological point of view, the single most important effect of cultivation is the destruction of habitats. This reduction in habitat and species diversity in the rural landscape increases alongside the trend for greater mechanization, reduced labour costs and greater intensification of production. It is well illustrated by the removal

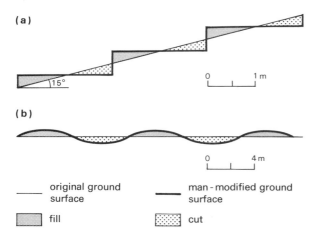

(a)

15°

0   1 m

(b)

0   4 m

—— original ground
     surface

—— man-modified ground
     surface

fill

cut

Fig. 26.15 Human modification of the ground surface by (a) hillslope terracing and (b) ridge and furrow. (c) An entire landscape artificially modified by terracing (Nepal).

(c)

of hedges in parts of rural England (Fig. 26.17). Between 1950 and 1970, 4000–7000 miles of hedge were lost in Britain, representing between 3000 and 5000 acres of habitat. Ecologically, the hedge and its associated timber simulate the woodland edge or clearing habitat, in both structure and species composition. Hedgerow removal, therefore, can mean a reduction in the species diversity of an average rural 10-km grid square in Britain of about 30%, affecting not only the plants but the insects, birds and small mammals dependent on them (Hooper, 1970).

Again in Britain, the use of mechanical excavators has accelerated the trend towards the drainage of potential agricultural land, as well as improving the drainage of that already under cultivation. The introduction of herbicides, providing a cost-effective method of controlling ditch bank

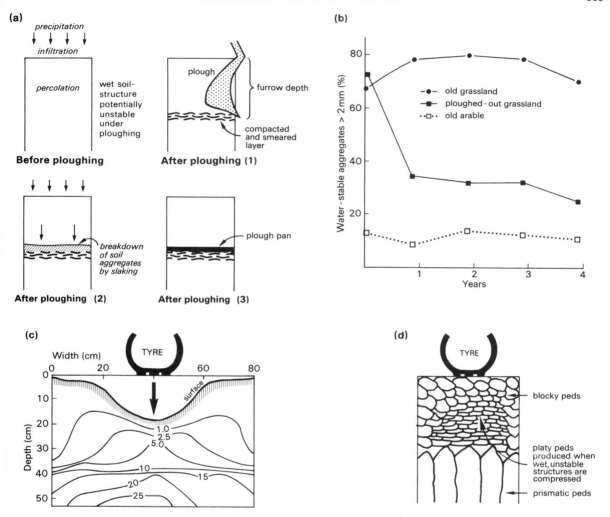

Fig. 26.16 (a) The formation of a plough pan. (b) Changes in the degree of soil aggregation under different land uses (after Low, 1972). (c) Distribution of cone resistance (bar) after passage of a test tyre; contours show compression of the soil under tyre track and sideways displacement of surface soil (after Soane, 1973). (d) Alteration of soil structure by the passage of a tractor wheel.

vegetation and aquatic weeds, has helped to reduce labour costs. Even so, mechanized farming with increased field sizes, requiring the removal of hedges and the filling-in of ditches, has removed drainage water to mole and tile drain systems below the surface. Finally, the misuse of herbicides and pesticides, some of which persist in the habitat, has further contributed to an impoverishment of the wild flora and fauna of agricultural landscapes. Some of the characteristics of modern farming systems in the UK viewed both as **agroecosystems**

and as **agribusinesses** are shown in Table 26.4. Also shown, in the third panel of the table, are considerations which suggest that modern farming may be entering a period of adjustment against a changing sociopolitical backdrop.

26.3 Urbanization and industrialization

The agglomeration of industry and housing in large urban complexes creates local modifications to the Earth's surface and the atmosphere above it which

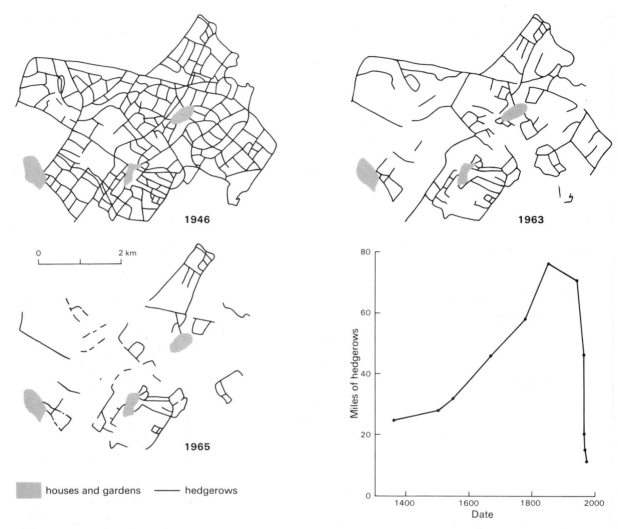

Fig. 26.17 The reduction in the estimated mileage, and change in the pattern of hedges, in part of Huntingdonshire 1364–1965 (after Moore *et al.*, 1967).

may extend beyond the immediate urban area. The main characteristics of these urban climates are outlined in Table 26.5. The reason for these effects lies in fundamental modification to systems regulators. The amount of suspended particles in the urban atmosphere is much higher than in the surrounding suburban and rural areas. In the solar energy cascade this directly affects atmospheric transparency, reducing the amount of direct radiation reaching the ground surface. Much is lost by reflection, scattering and absorption in the urban atmosphere.

Another regulator of processes in the atmosphere is the number of hygroscopic nuclei in the air, directly affecting the amount of cloud and possibly also of precipitation. The most notable form of condensation in urban areas has been the urban fog, for which London was well known until the 1956 Clean Air Act gradually brought a reduction in air pollution concentrations. During the London smog of December 1952, the atmosphere contained approximately 276 gm km^{-3} of smoke and sulphur dioxide and 124 200 gm km^{-3} of condensed water (Fig. 26.18). Fig. 26.19 shows the

Table 26.4 Some perspectives on modern farming in the UK (R & D: research & development, CAP = common agricultural policy (EC), WCA = wildlife conservation agreements, ESA = environmentally sensitive areas scheme, Agric Imp Schs = Agricultural improvement scheme)

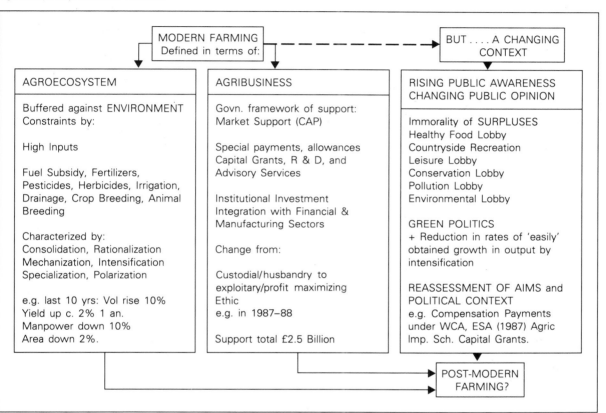

smoke concentrations in central London in 1958. Los Angeles (California, USA) still has a major urban fog problem, but this is due largely to the photochemical effects of solar radiation upon the gases emitted from vehicle exhausts, and not to the action of hygroscopic nuclei.

The effects of increased numbers of condensation nuclei in Rochdale (Lancashire, England) were noted by Ashworth (1929), who suggested that lower average rainfall on Sundays was a direct consequence of cleaner air while local mills were closed. More recently, Atkinson (1975) and others have suggested that free and forced convection over the aerodynamically rough and relatively warmer urban surface may be instrumental in producing higher urban rainfall totals.

The surfaces of urban areas have physical properties that contrast sharply with their rural surroundings. Albedo may be slightly higher, due to the large amounts of highly reflective concrete and glass used in construction (Table 26.6). More important, however, is the fact that the urban surface has a highly variable morphology, and that among the buildings there are multiple reflections from their vertical faces and shading of the streets below, which result in a low albedo.

The relatively high thermal capacity of the fabric of older urban structures and the large amounts of heat lost from heating systems operating within buildings combine to create an urban '**heat island**' effect. In most urban areas, air temperatures recorded near their centres of activity, where building density and height are usually greatest, are frequently higher than in the surrounding suburban areas. This temperature anomaly is well developed during the late evening, when solar heat

Table 26.5 Modifications in local climate produced by urban areas (from Landsberg, 1981)

Element	Compared to surrounding rural areas
Contaminants	
condensation nuclei	10x
particulates	10x
gaseous admixtures	5–25x
Radiation	
total on a horizontal surface	0–20% less
ultraviolet:	
winter	30% less
summer	5% less
sunshine duration	5–15% less
Cloudiness	
clouds	5–10% more
fog:	
winter	100% more
summer	30% more
Precipitation	
amounts	5–15% more
snowfall:	
inner city	5–10% less
lee of city	10% more
thunderstorms	10–15% more
Temperature	
annual mean	0.5–3.0°C greater
winter maximum (average)	1–2 °C greater
summer maximum	1–3°C greater
number of heating days	10% less
Relative humidity	
annual mean	6% less
winter	2% less
summer	8% less
Windspeed	
annual mean windspeed	20–30% less
extreme gusts	10–20% less
calms	5–20% more

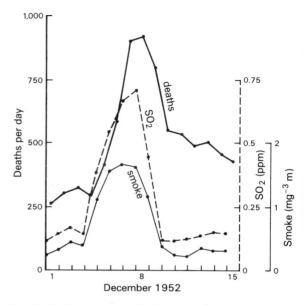

Fig. 26.18 Deaths and pollution levels for smoke and sulphur dioxide during the London fog of December 1952 (after the Royal College of Physicians, 1970).

stored in the urban fabric during the day is released and heat is still being supplied to the urban atmosphere from heating systems. Indeed, as Table 26.6 shows, the latter may constitute a major proportion of heat inputs into the urban environment. Chandler (1965) has shown that, under calm anticyclonic weather conditions, London develops a well defined heat island (Fig. 26.20) in the centre of which air temperatures at street level are more than 6°C higher than in surrounding areas of lower building density and stature.

Therefore, in constructing large urban agglomerations, humans have modified the character of both the Earth's surface and the atmosphere above it. The distinctive local climates generated are a product of consequential changes to regulation within the Earth–atmosphere system. The alteration of the Earth's surface mentioned above represents the direct reshaping of landform by urbanization, if not the creation of entirely artificial landforms. Constructions associated with prehistoric settlement sites can involve substantial accumulations of material. Even using primitive techniques, human landforms of the magnitude of the pyramids of Egypt and Silbury Hill (Wiltshire) were constructed. The latter, some 10 km west of Marlborough, is 40 m high and covers a basal area of 2.1 ha. It is estimated that 350 000 m³ of material were raised during its construction. Even the more modest artefacts such as hill forts required the transfer of substantial volumes of earth in the creation of their rampart and ditch systems. It is estimated that the surface of central London has been raised by the accumulation of construction materials and waste by an average of 3.5 m. If we assume that London is 2000 years old, then made ground has accumulated at an average rate of 1750 m³ km⁻² a⁻¹.

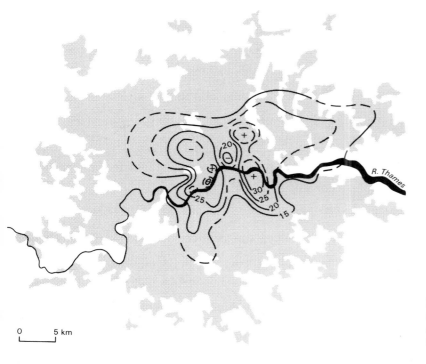

Fig. 26.19 Average distribution of smoke concentrations in London, April 1957–March 1958 in mg 100 m^{-3} (after Chandler, 1965).

0   5 km

Table 26.6 Radiative properties of typical urban materials and areas (from Oke, 1978, after Threlkeld, 1962; Sellers, 1965, van Straaten, 1967; Oke, 1974)

Surface	α Albedo	ε Emissivity	Surface	α Albedo	ε Emissivity
roads			windows		
asphalt	0.05–0.20	0.95	clear glass		
			zenith angle less than 40°	0.08	0.87–0.94
walls			zenith angle 40–80°	0.09–0.52	0.87–0.92
concrete	0.10–0.35	0.71–0.90			
brick	0.20–0.40	0.90–0.92			
stone	0.20–0.35	0.85–0.95	paints		
wood		0.90	white, whitewash	0.50–0.90	0.85–0.95
			red, brown, green	0.20–0.35	0.85–0.95
roofs			black	0.02–0.15	0.90–0.98
tar and gravel	0.08–0.18	0.92			
tile	0.10–0.35	0.90			
slate	0.10	0.90	urban areas*		
thatch	0.15–0.20		range	0.10–0.27	0.85–0.95
corrugated iron	0.10–0.16	0.13–0.28	average	0.15	?

* Based on mid-latitude cities in snow-free conditions

In more recent times, the development of extensive transport systems and the expansion of urban areas associated with advancing technology have led to large-scale exercises in land levelling. In this way deep cuttings are made through areas of positive relief and the material removed is used to form embankments across low-lying areas. Railway and motorway construction, requiring the maintenance of modest gradients, are examples of this kind of modification. The Panama Canal can be considered as one of the more spectacular modern examples of such engineering.

Restricted urban sites can be extended by building land out into the sea. Material can be acquired for this purpose either by quarrying from higher ground or by dredging from river or nearshore channels. Some 11% of the land area of the county borough of Belfast was created by humans, and Hong Kong Airport is constructed entirely on reclaimed land. Spectacular extension of land surface area can be achieved also by building dykes in order to extend the coastline seawards, as in the classic case of the Dutch Polders.

Extractive industries also modify land surface form. Opencast working of building stone, coal, sand and gravel, and raw materials such as lime and iron ore produces pits which scar the landscape. Many of these activities are associated with adjacent accumulations of spoil, the unwanted byproduct. In prehistoric times the extensive cave system of Grimes Graves was created by flint mining. In mediaeval times the extraction of chalk for agricultural purposes in East Anglia created 30 000 pits and ponds, and the extraction of peat led to the formation of the Norfolk Broads. The magnitude of contemporary extractive processes in Britain is illustrated by the facts that some 8 km^2 of land is excavated annually for sand and gravel, 4 km^2 for chalk and limestone and 1.8 km^2 for clay, to provide for the needs of the brick industry. In addition to the pits directly produced by the extraction process, the accumulation of adjacent spoil heaps can produce landforms of considerable magnitude, as illustrated by the china clay mining landscape on the southwestern fringe of Dartmoor and around St Austell in Cornwall (Fig. 26.21). The underground extraction of salt has led to surface subsidence and the formation of water-filled hollows – the 'flashes' of Cheshire. A salutary reminder of the magnitude of extractive processes is the quarrying of limestone in the Mendip Hills of Somerset. Some 10×10^6 t a^{-1} are currently

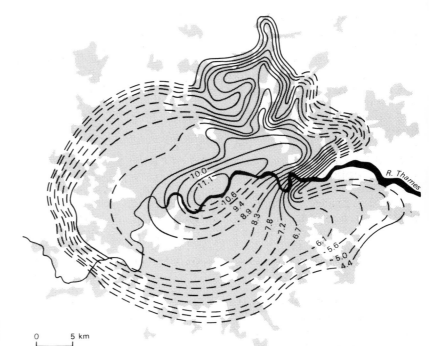

Fig. 26.20 Distribution of minimum air temperature (°C) in London, 14 May 1959 (after Chandler, 1965).

0   5 km

Fig. 26.21 A human landform: spoil from the extraction of china clay. Note the extensive modification by subaerial denudation resulting in gullying (Dartmoor, England).

being removed, representing a loss of 800 m³ km⁻² a⁻¹ over the limestone outcrop as a whole. Losses of rock by erosional processes are in the range of 50–100 m³ km⁻² a⁻¹, indicating that the extractive industry is eroding the landscape at a rate 8–16 times greater than natural erosion (Smith and Newson, 1974).

These are mostly highly localized examples of human activity in directly creating landforms, and they are of limited area and volume in relation to the land surface as a whole. The greatest contrasts in artificially created relief are usually the juxtaposition of a pit and a spoil heap. Of more lasting significance, and affecting a much broader surface area, is the effect of humans on the dynamics of surface processes by altering the nature of the land surface. In this way hydrological and denudation processes may be changed permanently, and at the scale of the drainage basin.

The process of urban development usually involves the encroachment of the built area on to former agricultural land. Initially there is a period of construction, during which there is much mech-

anical disturbance of the ground surface as drains, roads and foundations are prepared. Ultimately, the hydrological surface of the urban area becomes increasingly impervious, consisting of a high proportion of paved surfaces – roads, footpaths and roofs – which drain via a system of gutters and storm sewers into pre-existing natural drainage channels. Such paved surfaces (with the exception of leaking roofs) have an infiltration capacity of virtually zero and therefore zero storage. A high proportion of precipitation leaves as direct runoff via the storm drainage system, and storage and evaporation are limited to the soil and vegetation surfaces of parks and gardens.

Under urban conditions, therefore, the water balance will be modified again, leading to increased surface runoff and more rapid response of runoff to storm inputs. These effects have been confirmed by Gregory (1974) from observation of a small catchment (0.26 km²) on the urban fringe of Exeter. In four years (1968–72) urbanization extended to cover 12.2% of the catchment area. It was found that total runoff increased between two- and threefold and that peak discharge

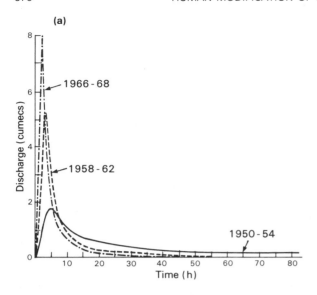

Fig. 26.22 (a) Mean unit hydrographs for three stages in the urbanization of the Canon's Brook (after Hollis, 1974), (b) River channel artificially straightened and confined in training walls to prevent lateral channel erosion and diminish overbank flooding (Afon Wyre, Llanrhystud, Dyfed, Wales).

creased in the same proportion, while the lag time of floods decrease from an average 70–80 minutes to 35 minutes. Similar results were recorded by Hollis (1974) from the growth of Harlow New Town in Essex to cover 21.4% of the catchment of Canon's Brook. Over an 18-year period the mean hydrograph progressively increased in dis-charge, while time to rise decreased. The catchment thus became increasingly flashy in character (Fig. 26.22).

A further effect of urbanization may be the addition of imported water. Water pumped from groundwater storage, or transferred to the urban area from an adjacent catchment, may be discharged as effluent into urban rivers. This in effect increases the natural catchment area or leads to a further increase in total discharge in the urban river channel.

Important effects on water quality may also accompany urban development. During the construction phase the disturbance of the soil surface by machinery may contribute sediment to the river channel, and values of sediment concentration in excess of 3000 ppm have been recorded. Once construction has ceased and paved surfaces are extensive, the scope for sediment erosion is much reduced and sediment concentration in streams may be lower than in the initial natural forested state. Water quality may also be affected by industrial processes. Effluent from industry may be discharged into river channels, increasing the concentration of dissolved solids.

The effects of this sequence of land use change have been summarized by Wolman (1967) who studied its effect on sediment yield (Fig. 26.23) in the eastern USA. Initially under forest, yields are estimated at 250 t km^{-2} a^{-1}. As forest is replaced by arable farming, this rises to 2000 t km^{-2} a^{-1}, declining as the urban fringe approaches and farmland is degraded to grazing and forest. The massive disturbance associated with construction may produce erosion rates of 250 000 t km^{-2} a^{-1} over small areas, falling to an estimated 125 t km^{-2} a^{-1} when urbanization is complete.

Thus profound changes in the water balance can be brought about as a result of human activities changing the land use and consequently the hydrological characteristics of catchment areas. These frequently lead to increased channel runoff, increased peak discharges and increased sediment and solute loads. In the case of urban development, normal sediment loads, in contrast, are reduced. Since the form of river channels is adjusted naturally to the output of runoff and sediment, it follows that river channel changes can

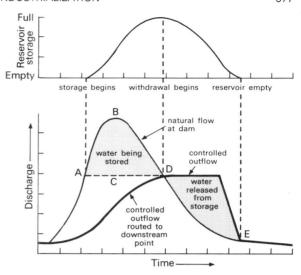

Fig. 26.24 The use of a reservoir as a regulator in flood control (after Linsley et al., 1949).

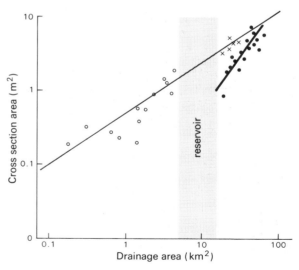

o   bankfull cross section above reservoir

x   former cross section below reservoir

•   present cross section below reservoir

Fig. 26.25 Relationship between channel capacity and drainage area above and below a reservoir, River Tone, Somerset, England (after Gregory and Park, 1974).

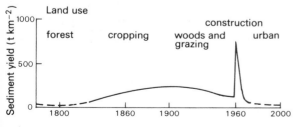

Fig. 26.23 Land-use change and sediment yield in the Piedmont region of Maryland (after Wolman, 1967).

be expected to accompany changes in land use and flow regime.

Since channel form is regarded as a function of the hydrological characteristic of the catchment area, it is possible to establish a relationship between channel capacity and drainage area. Thus channel capacity is shown to be closely related to drainage area, increasing in direct proportion in the downstream direction as discharge increases. The precise nature of this relationship will, of course, vary from river to river, but the same general tendency holds true. It is then possible to use the relationship established for a particular river in order to predict expected channel capacity further downstream. If actual channel capacity is measured downstream, below a point encompassing a major change in land use within the catchment area, then the effect of the land use change can be assessed by comparing the predicted channel capacity against that measured. In this way Gregory and Park (1974) show that stream channel capacity below a reservoir on the River Tone, Somerset, England has been significantly reduced. On the basis of the channel capacity–drainage area relationship established above the reservoir, former bankfull capacity measured below the reservoir is shown to fit in well. The current channel capacity below the reservoir is reduced to 50–80% of its former value and this reduction persists downstream until that part of the catchment area controlled by the reservoir comprises an insignificant proportion of the total drainage area. (Fig. 26.25)

A similar approach can be adopted to demonstrate natural channel adjustments below an urban area. In the same region as the previous example, the River Swale shows an increase in channel capacity downstream of the urbanized area of Catterick. Within this urban channel capacities average 1.66 times that predicted from rural channels upstream. Downstream from the urbanized area, channel capacity averages 2.62 times that expected.

On a smaller scale, Gregory and Park (1976b) demonstrate the dramatic effect of diverting runoff from a paved road surface via a storm sewer into a small natural channel in Devon. While not increasing the drainage area of 0.55 km^2, the paved road locally decreases infiltration and storage capacity and contributes high peakflow discharges of sediment-free water into the existing natural channel. This has led to rapid channel enlargement over a period of 29 years, from an estimated original average channel cross-section of 0.39 m^2 to a value of 2.07 m^2 over a length of 500 m of channel (see Fig. 26.30).

26.4 Control of environmental systems by humans: some examples

Most human activities and the consequent decision-making processes are sensitive to environment. Choice of home, place of work and recreation, in addition to the choice and success of agricultural enterprise, are affected to varying degrees by the environment. Kates (1970) has represented this relationship between human and their environment as an interaction between a 'human-use system' and a 'natural-events system' (Fig. 26.26). An event in the natural-events system which directly disrupts human activities is commonly referred to as 'a natural hazard'.

The definition of a hazard involves the identification of critical magnitudes and frequencies of events. The snow and low temperatures which disrupted transport throughout western Europe during the winters 1962–63 and 1978–79 clearly constituted a hazard in these areas. However, if we examine the climatic data for northern Canada or Siberia, we would expect low air temperatures and attendant snowfall during most winters, but considerably less disruption of normal patterns of human activity. In considering the interaction between Kates' human-use and natural-events systems therefore, we must consider expectations of events in the latter, and human adjustment to them. In the preceding example the expectation of severe cold in western Europe is relatively low, yet its impact in terms of disruption of human activity is great.

If we consider the alternatives in Fig. 26.26, it appears that there is a range of human adjustments to events such as snowfall, floods, drought or gale. In the short term, the effects of the event may be tolerated as a short-term inconvenience, perhaps due to the infrequency of their occurrence or their minimal long-term impact upon human activity. However, should there be a fund of experience of the event, then human adjustment is likely to take a relatively well organized form. The impact of the event is evaluated and an adjustment policy is formulated. For example, in the case of urban snowfall this may incorporate a range of adjustments, from greater investment in snow-moving

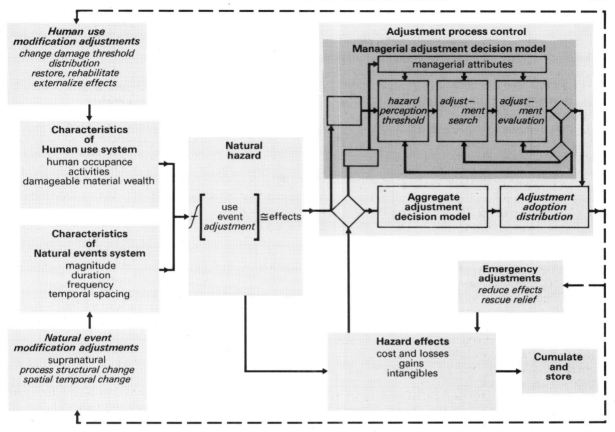

Fig. 26.26 Natural hazards represented as an interaction between a human-use system and a natural-events system (after Kates, 1970).

machinery to long-term modifications to the design of buildings and modes of transport. The outcome of decision-making inevitably results in the adoption of one of two alternatives – either to modify the 'human-use system' to reduce the impact of the hazardous event, or to modify the 'natural-events system'.

If we consider agriculture, the recurrence of, for example, damaging frosts and droughts may lead to changes in the types of crops grown. To some extent this may be avoided by the development of less sensitive hybrid crop varieties, an approach which is being adopted in many Third World countries. The alternative is to seek to exert a degree of control over Earth-surface and atmospheric systems to create a more favourable environment for crop and animal husbandry. The basic principles adopted in such control are illustrated in the following examples.

Flooding. For many reasons the concentration of human activity along river valleys and on floodplains has inevitably increased their susceptibility to flood damage. During the first 6 months of 1978, for example, floods caused widespread loss of human life and livelihood throughout the world (Table 26.7). A number of possible human adjustments to this flood hazard are listed in Table 26.8, one of which is to 'modify the flood'.

The essence of flood control is to reduce the peak discharge of the river to a level where it does not exceed the maximum capacity of its channel (see Chapter 13). In the simplified flood hydrograph in Fig. 26.27, the river banks would be overtopped between X and Y. A more desirable form shows a delayed rise, a protracted recession and, most important of all, a reduced peak discharge. If we examine the hydrographs for the River Derwent in Derbyshire (Fig. 26.28),

Table 26.7 Major flood events over the first 6 months of 1978

January	9–20	Severe floods in southern Iran; 142 villages affected, roads washed away; 20 000 families homeless; ten dead.
	13–16	Rain and flash floods in Brazil; 1400 homeless; 26 dead.
February	10–13	Heavy rain, up to 300 mm in Uruguay causes floods.
	21–26	Widespread flood in southwest England; roads and homes flooded; one death.
	24–25	Severe rainstorm causes serious flooding to 2000 km² in Western Australia, livestock losses; roads cut.
March	1–4	Floods in Tijuana City, Mexico, leave 15 dead; 20 000 homeless.
	15	Worst flood for 21 years in Brazil; 22 000 homeless.
	15–24	Northern Argentina floods leave 11 000 homeless; extensive crop damage; eight dead.
	20	Floods, the worst for 40 years, in Omaha and Nebraska, USA; 2000 homeless; one dead.
	21–31	Heavy rains and floods along the River Zambesi in Mozambique; 250 000 homeless; 45 dead; 56 000 hectares of agricultural land destroyed; damage estimated $70 million.
	24	Floods up to 2 m deep in Indiana, USA; damage estimated $10 million.
April	2	Worst floods for 23 years in Paris.
May	3	225 mm rain falls in 5 h in New Orleans, USA; flood 1.5 m deep in the city; two dead.
	22–26	Heavy rain and serious floods in southwest Germany; roads blocked; damage estimated at 100 million D-marks.
	26–27	125 mm rain in West Texas; flash flood up to 4 m deep; homes, cars, campsites washed out; three dead.
June	26	Heavy rain in Japan; 259 mm rain in 24 h; eight dead.

for example, we can see that the presence of the Ladybower Reservoir clearly introduces a damping of the flood peak. The construction of reservoirs, which increase the storage capacity of a drainage system, thus results in a degree of control over river discharge. In the case of the River Severn, adjustments to frequent flooding have included both the development of a flood-warning scheme and the employment of flood-control measures. In the flood-warning scheme, information on exceptional falls of rain and critical river levels is fed into a well organized network for its dissemination (Fig. 26.29). A fund of experience in Shrewsbury has apparently led to a well organized form of human response.

The effects of such reservoir and dam construction are to modify the pattern of channel discharge so that total discharge and peak discharge may be either increased or decreased by deliberate regulation. Deliberate acts of manipulation inevitably generate side effects. The increase in open-water surface area in reservoirs results in greater loss by evaporation, and hence discharge downstream from the reservoir is reduced in the long term. Furthermore, the control of discharge will precipitate changes in channel form as a response to human-induced changes in the surface characteristics of the basin. We have already seen the response where natural channels adjust themselves freely to human-induced changes in the surface of the contributing catchment area. However, artificial channel modification can also take place as a deliberate policy, as part of urban drainage or flood relief. In this way new channels may be cut or existing channels may be artificially enlarged and constrained by the construction of fixed artificial channels. Thus both channel cross-section and plan form are modified deliberately in order to permit high discharges and consequently reduce overbank flows and flooding. This may be accomplished in a number of ways. The roughness of the cross-section may be decreased by clearing the banks of vegetation and other obstacles, and by lining the channel with a hydraulically smooth surface such as concrete. The channel capacity may be increased by widening and deepening, or by the construction of marginal embankments. Alternatively, the gradient can be increased, shortening the channel by straightening bends and cutting out meanders. Many river channels in and downstream of urban areas and

Table 26.8 Adjustments to the flood hazard (after Beyer, 1974, adapted from Sewell, 1964 and Sheaffer *et al.*, 1970)

Modify the flood	Modify the damage susceptibility	Modify the loss burden	Do nothing
flood protection (channel phase)	land-use regulation and changes statutes	flood insurance	bear the loss
dykes	zoning ordinances	tax write-offs	
flood walls	building codes	disaster relief	
channel improvement	urban renewal	volunteer	
reservoirs	subdivision regulations	private activities	
river diversions	government purchase of lands and	government aid	
watershed treatment (land phase)	property	emergency measures	
modification of cropping practices	subsidized relocation	removal of persons and property	
terracing	floodproofing	flood fighting	
gully control	permanent closure of low-level windows and other openings	rescheduling of operations	
bank stabilization	waterproofing interiors		
forest fire control	mounting store counters on wheels		
revegetation	installation of removable covers		
weather modification	closing of sewer valves		
	covering machinery with plastic		
	structural change		
	use impervious material for basements and walls		
	seepage control		
	sewer adjustment		
	anchoring machinery		
	underpinning buildings		
	land elevation and fill		

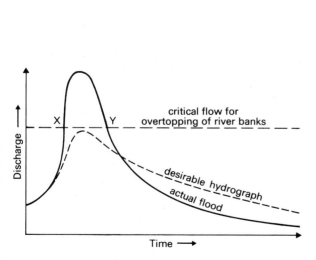

Fig. 26.27 Hypothetical flood hydrograph and a desirable modified flood hydrograph.

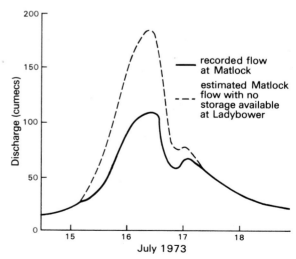

Fig. 26.28 The effect of storage in the Ladybower reservoir on the flood hydrograph of the River Derwent at Matlock (after Richards and Wood, 1977) (see Fig. 26.24).

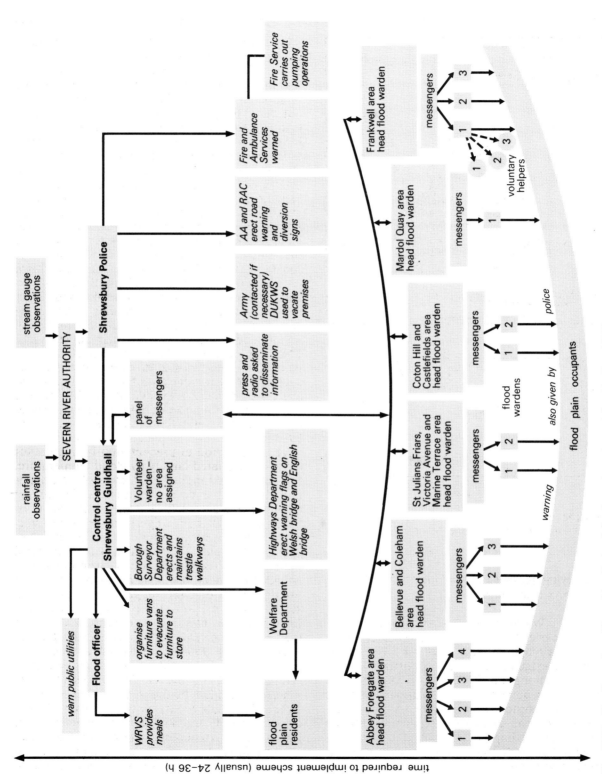

Fig. 26.29 The Shrewsbury (Shropshire, England) flood-warning scheme (after Harding, 1974).

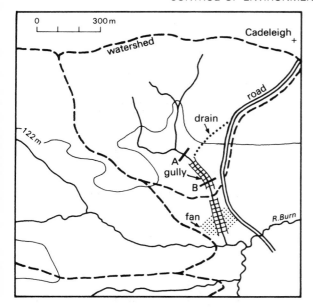

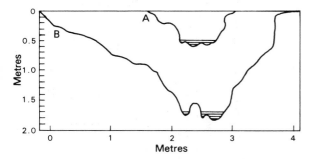

Fig. 26.30 Sections in a gulley in the valley of the River Burn, Devon, England (after Gregory and Park, 1976b).

in low-lying agricultural land may be modified in these ways (Fig. 26.30).

Drought. Water shortages have serious implications for agriculture and, as in the case of floods, there are a range of human adjustments to them. Humans engage in a number of control activities, principal among which are the construction of reservoirs to manipulate the seasonal discharge of drainage systems, which we have just considered, and the enhancement of precipitation through the seeding of clouds.

The initiation of precipitation from clouds was explored first between 1940 and 1950 and most at-

tempts were based upon the Bergeron–Findersen ice-crystal process (see Chapter 4). In the USA in 1946, Schaefer discovered that, at temperatures less than −39°C, there was spontaneous freezing of supercooled water in the free atmosphere. If objects at temperatures less than −39°C were introduced to air in the laboratory, ice crystals were formed. In clouds these could initiate the process of ice-crystal growth and precipitation. Altostratus clouds at −20°C were seeded at an elevation of 5000 m using dry ice (solid carbon dioxide), which resulted in the precipitation of snow, which re-evaporated before reaching the ground. Vonnegut, also in the USA, experimentally introduced silver iodide smoke to supercooled clouds and observed that this caused snowflakes to form. The silver iodide provided freezing nuclei upon which ice crystals formed within the cloud. Successful experiments were conducted in Australia in the 1950s in the seeding of selected cumulus clouds using silver iodide ejected from aircraft. Of clouds with temperatures below −5°C in their upper levels, 72% produced precipitation within 20–25 minutes of seeding, and 21% evaporated. Similar seeding of randomly selected clouds was less successful. Recent research in the USSR into the seeding of cumulus clouds has suggested increases of up to 30% in rainfall over the Ukraine (Battan, 1977).

The production of precipitation from cumulus clouds is regulated largely by cloud temperature, the number and rate of production of ice crystals and the number of water droplets. In seeding clouds, humans directly influence the second of these, but so far their impact has been relatively minor. The resulting amounts of precipitation are small and localized, due mainly to the physical limitations on the rates at which silver iodide may be added and to the constraints on the other regulators. The latter suggests that successful seeding, based upon the ice-crystal process, is limited at present to deep cumulus clouds. Five seeding trials carried out on Hurricane Debbie during August 1969 produced an apparent decrease in maximum wind speed of as much as 35% (Gentry, 1970; Fig. 26.31). This represents a significant reduction in the kinetic energy of the storm. Although results as yet are inconclusive, any measure of control of such an energetic system represents a considerable achievement. However, the additional possibility of steering these storms by seeding has revealed a more sinister aspect of

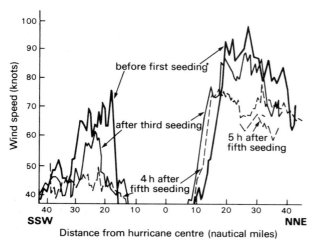

Fig. 26.31 Windspeed changes with time at 3600 m in hurricane 'Debbie' on 18th August 1969 (after Gentry, 1970).

weather control – that of weather warfare. Such has been the degree of concern that international agreements have been reached on the limitation of this aspect of environmental warfare.

Further reading

A selection of texts dealing with a range of interactions between environment and society, or with the applied aspects of physical geography and environmental science are given below:

Blunden, J. and N. Curry (1988) *A Future for our Countryside.* Blackwell, Oxford.

Briggs, D.J. and F.M. Courtney (1985) *Agriculture and Environment: the Physical Geography of Temperate Agricultural Systems.* Longman, London.

Burton, I., R.W. Kates and G.F. White (1978) *The Environment as Hazard.* Oxford University Press, New York.

Common, M. (1988) *Environmental Economics,* Longman, London.

Cooke, R.U. and J.C. Doornkamp (1974) *Geomorphology in Environmental Management.* Oxford University Press, Oxford.

Davidson, J. and R. Lloyd (eds) (1979) *Conservation and Agriculture.* John Wiley, Chichester.

Detwyler, T.R. and M.G. Marcus (eds) (1972) *Urbanisation and Environment: the Physical Geography of the City.* Duxbury, Belmont.

Douglas, I. (1983) *The Urban Environment.* Edward Arnold, London.

Garner, J.F. and B.L. Jones (1991) *Countryside Law.* Shaw and Sons, London.

Green, B. (1987) *Countryside Conservation.* (2nd ed) Unwin Hyman, London.

Haigh, N. (1989) *EEC Environmental Policy and Britain.* Longman, London.

Hails, J.R. (ed) (1977) *Applied Geomorphology.* Elsevier, Amsterdam.

Hewitt, K. (ed) (1983) *Interpretation of Calamity.* Allen & Unwin, Boston. (especially Chapter 1).

Jordan, W.R., M.E. Gilpin and Aber, J.B. (1990) *Restoration Ecology.* Cambridge, University Press, Cambridge.

Kirkby, M.J. and R.P.C. Morgan (eds) (1980) *Soil Erosion.* John Wiley, Chichester.

Maybey, R. (1980) *The Place for Nature in Britain's Future.* Hutchinson, London.

Newson, M.D. (1975) *Flooding and Flood Hazard in the United Kingdom.* Oxford University Press, Oxford.

Park, C.C. (1981) *Ecology and Environment Management.* Butterworths, London.

Parker, D.J. and E.C. Penning-Rowsell (1980) *Water Planning in Britain.* Allen & Unwin, London.

Pepper, D. (1984) *The Roots of Modern Environmentalism.* Croom Helm, London.

Perry, A.H. (1981) *Environmental Hazards in the British Isles.* Allen & Unwin, London.

Rackham, O. (1986) *The Making of the Countryside.* Dent, London.

Ramade, F. (1985) *Ecotoxicology.* Wiley, London.

Reuss, J.O. and D.W. Johnson (1986) *Acid Deposition and the Acidification of Soils and Waters.* Springer Verlag, New York.

Spellerberg, I.F., Goldsmith, F.B. and M.G. Morris, (1991) *The Scientific Management of Temperate Communities for Conservation.* Blackwell, Oxford.

Ward, R. (1978) *Floods: a Geographical Perspective.* Macmillan, London.

Warren, A. and C.M. Harrison (eds) (1983) *Conservation in Perspective.* John Wiley, Chichester.

Wathern, P. (1988) *Environmental Impact Analysis.* Unwin Hyman, London.

Whittow, J.B. (1980) *Disasters: The Anatomy of Environmental Hazards.* Allen Lane, London.

Systems retrospect and prospect

If we were now to look again on the view of the Earth from space portrayed in Fig. 1.1d, the image in our mind's eye would contain more than an evocative impression of the 'blue planet'. Our picture of a picture is enhanced and our awareness heightened by insight and understanding. It is now not only aesthetically appealing but also intellectually stimulating. Instantaneously we are aware of water molecules breaking free from the ocean's surface, cloud droplets growing, colliding and falling as raindrops. We see the enormous and inevitable evacuation from the continents of this water and the erosive load it entrains. We see the colossal and continuous gaseous exchange between the atmosphere and biosphere, as well as the energy flux that sustains it. Not only do we perceive the planet as it is, but also as it was. Like some vast time-lapse sequence we can see the extension of ice sheets, the contraction of deserts and the continents riding on their plates – colliding or cleaving – progressing like giant Noah's Arks carrying their cargoes of creatures through millions of years of evolution.

Our mind's eye is equally active at a different scale. As well as the delight that the tranquillity of the highland glen of Fig. 1.1 evokes, we recognize the valley for what it is – a relict of glacial overdeepening now drowned by eustatic rise in sea level. At the same time it interprets the tilted platform above the loch side as evidence of differential isostatic uplift. The saturated, anaerobic peat is seen as a store of organic matter and mineral nutrients imprisoned in a slowly decomposing blanket across the hillside. Simultaneously we see the turbulent pipeflow draining the peat, the tenacious trickle of water emerging along stream sides and the seemingly oily smear of the sulphide-stained surface of bog pools in terms of slope hydrology. The cow quietly cropping the grass by the loch shore is seen as part of the energy flow from producer to consumer. This same acid and neutral grassland and the dwarf shrub moor are perceived not only as an integral component in the scale and appeal of the landscape, but as communities established and maintained by the anthropogenic imposition of grazing and burning regimes that modify and control their ecology. Indeed, the treelessness of the landscape can roll back before our mind's eye and reveal the oceanic birch and oak woods which once held sway along the lochside, or the magnificent pines which, on the higher slopes, were once the Atlantic bulwark of the Caledonian Forest. Our understanding can embrace both the complex functional relationships of these natural forest ecosystems and the labyrinth of interactions with the cultures which at first modified, then exploited and finally destroyed them. Today we can recognize the conflicting pressures, assess the points of view and make informed value judgements as to the social and economic effects, as well as the environmental impact of an extension of commercial afforestation into the valley.

In short, our mental picture is a montage of images at once comprehensive in the panorama it commands and incisive in the perspective it produces. Just as the fragments of pigment in a painting, resonating individually but responding and relating to each other, are held together by the weft and warp of the supporting canvas, so our images of environment are underpinned and supported by a framework of systems and a fabric of scientific law and principle. Breadth of understanding becomes possible without superficiality, and detailed knowledge without the isolation of specialization.

As a way of looking at our world and as a framework for thought the systems approach is richly rewarding, but it is also undeniably an attitude of mind – some would say a philosophy.

Nevertheless, as a *modus operandi* it has its shortcomings. Perhaps, therefore, in conclusion we should take a brief retrospective view of its application to the natural environment before looking ahead at the prospects for its continued application and development.

In this book we have adopted a largely pragmatic approach to the application of systems thinking, opting to present models at a variety of convenient and useful scales that for the most part have been purely descriptive and explanatory. We have concentrated on the structural and functional organization of environmental systems and stressed the functional relationships within and between these systems. The flow of energy and the transfer of matter have been the unifying themes in an essentially thermodynamic perspective. To this end the underlying scientific laws and principles that condition the properties and behaviour of natural phenomena, and govern the operation of natural processes, have been repeatedly emphasized. The focus has remained the explanation of the real world by modelling it as linked, ordered systems.

However, the formal development of the systems approach requires that it be seen as more than a framework for thought and for the conveyance of explanation. Indeed, the broader applications of systems thinking across the sciences and social sciences and the claims embodied in general systems theory lead some to maintain that it represents a philosophy of science. That we have hardly mentioned this dimension is perhaps excusable, for it is at this level that the most vociferous criticism has been generated. A second omission is more significant, for ultimately the systems approach must involve reducing the description of the system, the analysis of its organization and the prediction of its behaviour to the language and rigour of mathematics; for the most part, and in spite of a largely scientific treatment, we have avoided this. Nevertheless, the full power of the qualitative models of environmental systems that we have discussed can be realized only when they are translated into quantitative mathematical models.

This process began early in some branches of the environmental sciences, such as meteorology (with its roots in physics), and soil mechanics and hydrology with their parent disciplines of civil and hydraulic engineering respectively. In other areas such as biology it came later, and most recently it has begun to transform physical geography. Mathematical models represent the simplification and abstraction of the underlying rationale of the real world in its purest form. Nevertheless, like the empirical and intuitive models, the inductive hypotheses and conceptual theories on which they build, they are not the complete answer to our future understanding of the environment. They model environmental systems in terms of sets of equations at the appropriate level of complexity. Some are equations of state, perhaps specifying dependent endogenous or state variables as vectors of state and relating these to exogenous independent variables. Processes bringing about change in state variables, or factors influencing such changes, may be modelled as transfer functions, or where not affected by the components of the system as vectors of input or forcing functions. Relationships may be expressed through the symbolism of matrix algebra and sets of difference equations, or where appropriate both linear and higher-order differential equations may be used to describe mathematically the way the system changes through time. These express the rates of change of a number of state variables as functions of each other and of other characteristics of the system. When combined differential models are written in matrix form they allow many of the properties of the model to be determined by the methods of applied mathematics. However, because of the complexity of real environmental systems these models begin to approach the limits of analytical techniques.

Most such models are **deterministic** in that they allow the prediction of the outcome of some operation performed on the system. However, others are **stochastic** in the sense that the model incorporates some element of randomness or uncertainty, and is therefore only able to predict the probability of the outcome of an operation. These models recognize that certain variables are truly random, and that in the case of others their variability is so complex that our only option is to treat their behaviour as if it occurred at random. As with deterministic models, so the sophistication of stochastic models can vary. At the simple level, an otherwise deterministic model may be improved by one or more inputs being generated at random. On another level, the 'mathematical techniques and assumptions are incorporated in the model' in order to 'provide the uncertainty from within' (Brunsden and Thornes, 1977).

However, the entropy modelling developed from statistical mechanics and considered in Chapter 21 perhaps represents the most abstract of these stochastic approaches.

As with the conceptual models that we have used, mathematical models vary in their realism, in their resolution of systems attributes and in their completeness or the number of processes and interactions they incorporate. They also vary in their generality and breadth of application. There is, however, no best fit, maximum or optimum model, although, unlike conceptual models, quantitative mathematical models can be evaluated by their precision and their numerical predictive power. To increase this power there is always a tendency for models to be expanded: that is, the independent variables of simple models may be made more realistic by recognizing their dependence on processes going on in the system and by introducing new functions. Alternatively, descriptive state variables may be refined by subdivision, or the number of variables influencing each process may be increased by the inclusion of more detail. Mechanisms which have been treated as 'black box' compartments may be opened up, again in the interests of greater realism and greater predictive precision.

In spite of the progress that has been made in mathematical systems analysis and modelling, including dynamical systems modelling and the application of catastrophe and chaos theories to atmospheric, geomorphic and ecological systems, if it is not to become a sterile exercise it must retain a continuous and constant dialogue with reality. To validate any such model requires real data, data that can be provided only by the study of the real world. That the age of the computer and of simulation has dawned over the environmental sciences does not mean that the age of exploration and discovery has set forever. Field measurement and monitoring, the description of spatial variation, laboratory experimentation and hardware modelling are just as important as they ever were. The constants (or 'parameters', as they are known) in the elegant equations of our models must be estimated with as much precision as possible to improve prediction. Although values may be arrived at by trial and error or by regression, measurement and experiment are the only real answer. Equally, the expansion of models must be based on a better understanding of what goes on in the real world.

This understanding of the real world is still far from complete and certainly far from uniform. Our detailed knowledge of the accessible and heavily populated developed world with its post-Renaissance tradition of scientific enquiry is far greater than that of the less accessible and less developed parts of the globe. This state of knowledge has two important consequences. First, many of the conceptual and mathematical models that are part of the accepted wisdom of physical geography and environmental science are based on empirical observation and the construction of inductive hypotheses that occurred particularly in Europe and North America. Secondly, the validation and testing of these models has taken place largely in these same environments. There is no *a priori* guarantee that their generality can be extended to other environments except in the most abstract way. Some of the assumptions of these models might well be questioned in, for example, the humid tropics. Indeed by their very nature, abstract mathematical models tend to obscure the rich geographical variety of real-world systems. One of the most rewarding challenges is to modify, extend and refine the model as our knowledge of the environment increases. An example of the sterility which can creep in if a fundamental model fails to develop towards greater realism is afforded by Cousin's (1985) and Hubbell's (1971) criticism of the use of Lindemann's trophic–dynamic model in productivity studies. Hubbell maintains that 'the prevalent treatment of organisms as passive agents has hindered further developments in the field of ecological bioenergetics by producing few significant questions as to what living systems are really doing with energy'. By failing to do so, 'the capacity of living organisms to regulate, within the bounds established by the laws of thermodynamics, the rates at which they accumulate and dissipate energy' is overlooked. This criticism is akin to Thornes' plea (1987) that the tendency of many geomorphic models to concentrate on the steady or equilibrium states of systems denies the importance of understanding transient states in both time and space.

Perhaps the most important development of systems modelling, however, is what Bennett and Chorley (1978) call **systems interfacing**. The interfacing of 'natural' physicoecological and 'human' socioeconomic systems involves an understanding of the links and interactions which explain the way human activities mesh with environmental systems. It is necessary to be able to model these interfaces quantitatively, because we need to project them into the future and to predict the probable outcome of alternative strategies available to civilization. The study of such interfacing, whether it be in terms of intervention or harmonious symbiosis, is the only way to generate the understanding necessary to formulate sound policies to direct and secure the future for both mankind and for the environment. The decisions to develop and implement such policies will not rest with environmental scientists, but with politicians and governments. If they can say with William Golding's Neanderthal heroine 'I have a picture ... it is a picture of a picture ... I am ... thinking', then perhaps there is hope for the future.

References

Ackerman, H., C. Lippens and M. Lechevallier (1980) Volcanic material from Mount St. Helen's in the stratosphere over Europe. *Nature* **287**, 614–616.

Addison, K. (1981) The contribution of discontinuous rockmass failure to glacier erosion. *Annals of Glaciology*, **2**, 3–10.

Agricultural Advisory Council on Soil Structure and Soil Fertility 1971. *Modern farming and the soil*. London: HMSO.

Ahnert, F. (1970) Functional relationships between denudation, relief and uplift in large mid-latitude drainage basins. *Am. J. Sci.* **268**, 243–263.

Albion, R.G. (1926) *Forests and sea power*. Cambridge. Mass.: Harvard University Press.

Anderson, M.C. (1964) Studies on the woodland light climate II seasonal variations in light climate. *J. Ecol.* **52**.

Anderson, M.G. and A. Calver (1977) On the persistence of landscape features formed by a large flood. *Trans. Inst. Brit. Geog.*, *NS*, **2.2**, 243–254.

Anderson, M.G. and T.P. Burt (1978) The role of topography in controlling throughflow generation. *Earth Surface Processes*, **3**, 331–344.

Andreae, M.O. (1986/7) The oceans as a source of biogenic gases. *Oceanus* **29** (4), 27–35.

Andrews, J.T. (1972) Glacier power, mass balance, velocities and erosional potential. *Z. Geomorph. N.F.* **13**, 1–17.

Andrews, J.T. (1975) *Glacial Systems*. N. Scituate, Duxbury.

Armstrong, W. (1975) Waterlogged soils, in *Environment and plant ecology* (eds J.R. Etherington and W. Armstrong). Wiley, Chichester.

Arnett, R.R. (1979) The use of differing scales to identify factors controlling denudation rates, in *Geographical approaches to fluvial processes* (ed A.F. Pitty). Geobooks, Norwich, 127–147.

Arya, S.P. (1988) *Introduction to Micrometeorology*. Academic Press, New York.

Ashby, J.F., D.I. Edwards, P.J. Lumb and J.L. Tring (1971) *Principles of Biological Chemistry*. Blackwell, Oxford.

Ashton, P.S. (1969) Speciation among tropical forest trees; some deductions in the light of recent evidence. *Biol J. Linn.* **1**, 155–196.

Ashworth, J.R. (1929) The influence of smoke and hot gases from factory chimneys on rainfall. *Q. J. Meteorol. Soc.* **55**, 341–350.

Association for the Conservation of Energy (1989) *Solving the Greenhouse Dilemma: A Strategy for the UK*. ACE, London.

Atkinson, B.W. (1975) *The mechanical effect of an urban area on convective precipitation*. Occasional paper 3. Department of Geography, Queen Mary College, University of London.

Atkinson, B.W. (1981) *Meso-scale Atmospheric Circulations*. Academic Press, London.

Atkinson, B.W. (ed) (1981) *Dynamical Meteorology: an Introductory Selection*. Methuen, London.

Atkinson, B.W. (1987) Atmospheric energetics, in *Energetics of Physical Environment: energetic approaches to physical geography* (ed K.J. Gregory). John Wiley, Chichester.

Atkinson, T.C. (1978) Techniques for measuring subsurface flow on hillslopes, in *Hillslope hydrology*, (ed M.J. Kirkby). Wiley, Chichester, 73–120.

Bach, W. (1983) *Our Threatened Climate*. D. Reidel, Dordrecht.

Bagnold, R.A. (1941) *The Physics of Blown Sand and Desert Dunes*. Chapman & Hall, London.

Bagnold, R.A. (1977) Bed load transport by natural waters. *Water Resources Research*, **13**, 303–312.

Bange, G.G.J. (1953) *Acta Bot. Neerl.* **2**, 225.

Bannister, P. (1966) The use of subjective estimates of cover-abundance as a basis for ordination. *J. Ecol.* **54**, 665–674.

Bannister, P. (1976a) Physiological ecology and plant nutrition, in *Methods in plant ecology* (ed S.B. Chapman). Blackwell Scientific, Oxford.

Bannister, P. (1976b) Water relations of plants, in *Introduction to Physiological Plant Ecology* (ed P. Bannister). Blackwell, Oxford.

Bannister, P.J. (1976c) *Introduction to Physiological Plant Ecology*. Blackwell, Oxford.

Barnes, R.S.K. and R.N. Hughes (1988) *An Introduction to Marine Ecology* (2nd edn). Blackwell, Oxford.

Barry, R.G. (1970) A framework for climatological research with particular reference to scale concepts. *Trans Inst. Br. Geogs* **49**, 61–70.

Barry, R.G. and R.J. Chorley (1976) *Atmosphere, weather and climate* (3rd edn). Methuen, London.

Bates, E.M. (1972) Temperature inversion and freeze protection by wind machine. *Agric. Meteorol.* **9**, 335–346.

Battan, L. (1977) Weather modification in the Soviet Union *Bull. Am. Meteorol. Soc.* **58**, 4–19.

Baumgartner, A. and E. Reichel (1975) *The world water balance*. Elsevier, Amsterdam.

Beer, T. (1983) *Environmental Oceanography*. Pergamon, Oxford.

Begon, M., J.L. Harper and C.R. Townsend (1990) *Ecology: Individuals, Populations and Communities* (2nd edn). Blackwell, Oxford.

Beishon, J. and G. Peters (1972) *Systems Behaviour*. Harper & Row, London.

Bennett, D.L., D.A. Ashley and B.D. Doss (1966) Cotton responses to black plastic mulch and irrigation. *Agron. J.* **58**, 57–60.

Bennett, R.J. and R.J. Chorley (1978) *Environmental systems philosophy, analysis and control*. Methuen, London.

Berger, W.H., V.H. Smetack and G. Wefer (eds) (1989) *Productivity of the Oceans: Present and Past*. Wiley, New York.

Berner, E.K. and R.A. Berner (1988) *The Global Water Cycle*. Prentice-Hall, Englewood Cliffs.

Beyer, J.L. (1974) Global summary of human response to natural hazards; floods. In *Natural hazards* (ed G.F. White). Oxford University Press, New York.

Biddulph, O. (1959) Translocation of inorganic solutes, in *Plant physiology – a treatise, Vol II*. (ed F.C. Steward). Academic Press, New York.

Billings, M.P. (1954) *Structural geology* (2nd edn). Prentice-Hall, Englewood Cliffs, NJ.

Bird, E.C.F. (1984) *Coasts*. Blackwell, Oxford.

Birkeland, P.W. (1984) *Soils and Geomorphology*. Oxford University Press, New York.

Bjorkman, O. and J. Berry (1973) High energy photosynthesis. *Scientific American*, **229**, 80–93.

Black, C.C. (1971) Ecological implications of dividing plants into groups with distinct photosynthetic production capacities. *Adv. Ecol Res.* **7**, 87–114.

Black, J.N., C.W. Bonyphon and J.A. Prescott (1954) Solar radiation and the duration of sunshine. *Q.J. Meteorol. Soc.* **80**, 231–235.

Bloom, A.L. (1969). *The surface of the Earth*. Prentice-Hall, Englewood Cliffs, NJ.

Bolin, B. and R.B. Cook (eds) (1983) *The Major Biogeochemical Cycles and their Interactions*. John Wiley, New York.

Borisov, A.A. (1945) *Climates of the USSR* (transl. R.A. Ledward). Oliver & Boyd, Edinburgh.

Bormann, F.H. and G.E. Likens (1970) The nutrient cycles of an ecosystem. *Scient. Am.* 92–101.

Boulding, K.E. (1966) The economics of the coming spaceship Earth, in *Environmental quality in a growing economy resources for the future*. Johns Hopkins University Press, Baltimore, MD, 3–14.

Boulton, G.S. (1972) Rôle of thermal regime in glacial sedimentation, in *Polar geomorphology* (eds R.J. Price and D.E. Sugden). Inst. Br. Geogs Sp. Publ. 4, 1–19.

Boulton, G.S. (1974) Processes and patterns of glacial erosion, in *Glacial geomorphology* (ed D.R. Coates). SUNY, New York, 41–87.

Boulton, G.S., A.S. Jones, K.M. Clayton and M.J. Kenning (1977) A British ice sheet model and patterns of glacial erosion and deposition in Britain, in *British Quaternary studies: recent advances* (ed F.W. Shotton). Oxford University Press, Oxford, 231–246.

Bowen, N.L. (1928) *The evolution of the igneous rocks*. Princeton University Press, Princeton, NJ.

Box, E. (1975) Quantitative evaluation of global primary productivity models generated by computers, in *Primary productivity of the biosphere* (eds H. Lieth and R.H. Whitaker). Springer-Verlag, New York, 266–283.

Bradley, R.S. *et al.* (1987) Precipitation fluctuations over northern hemisphere land areas since mid 19thC. *Science*, **237**, 171–175.

Bradley, W.C. (1963) Large scale exfoliation in massive sandstones of the Colorado plateau. *Geol Soc. Am. Bull.* **74**, 519–528.

Brandt, C.J. and J.B. Thornes (1987) Erosional energetics, in *Energetics of Physical Environment: Energetic Approaches to Physical Geography* (ed K.J. Gregory). John Wiley, Chichester.

Bretschneider, C.L. (1959) *Wave variability and wave spectra for wind generated gravity waves*. US Army Corps of Engineers, Beach Erosion Board Technical Memorandum 118.

Brewer, R. (1964) *Fabric and mineral analysis of soils*. Wiley, New York.

Bridges, E.M. and D.M. Harding (1971) Microerosion processes and factors affecting slope development in the Lower Swansea Valley, in *Slopes: form and process*. (ed D. Brunsden). Inst. Br. Geogs Sp. Publ. 3, 65–80.

Briggs, D.J. and F.M. Courtney (1985) *Agriculture and Environment: the Physical Geography of Temperate Agricultural Systems*. Longman, London.

Briggs, D.J. and P.A. Smithson (1985) *Fundamentals of Physical Geography*. Hutchinson.

Brimblecombe, P. & A.Y. Lein (eds) 1989. *Evolution of the Global Biogeochemical Sulphur Cycle*. John Wiley, Chichester.

Brinkman, A.W. and J. McGregor (1983) Solar radiation in dense Saharan aerosol in northern Nigeria. *QJR Met. Soc.*, **109**, 831–847.

Broecker, W.S. (1974) *Chemical Oceanography*. Harcourt Brace Jovanovich, New York.

Brown, E.H. (1970) Man shapes the earth. *Geog. J.* **136** (1), 74–85.

Brownlow, A.H. (1978) *Geochemistry*. Prentice-Hall, Englewood Cliffs.

Brunsden, D. (1968) *Dartmoor*. Geographical Association, Sheffield.

Brunsden, D. (ed) (1971) *Slopes: Form and Processes*. Inst. Br. Geog. Sp. Pub. 3.

Brunsden, D. (1979) Weathering, in *Process in geomorphology* (eds C. Embleton and J.B. Thornes). Edward Arnold, London, 73–129.

Brunsden, D. and D.B. Prior (eds) (1984) *Slope Instability*. John Wiley, Chichester.

Brunsden, D. and R.H. Kesel (1973) The evolution of a Mississippi river bluff in historic time. *J. Geol.*, **81**, 576–597.

Brunsden, D. and J.B. Thornes (1979) Landscape sensitivity and change. *Trans. Inst. Brit Geog.* NS **4**, 463–484.

Bryson, R.A. and D.A. Baerreis (1967) Possibilities of major climatic modification and their implications; north-west India: a case study. *Bull. Am. Meteorol. Soc.* **48**, 136–142.

Budd, W.F., P.L. Keage and N.A. Blundy (1979) Empirical studies of ice sliding. *J. Glaciology*, **23** (89), 157–170.

Budyko, M.I. (1958) *The heat balance of the Earth's surface* (transl. N.A. Strepanova). US Dept. of Commerce, Washington DC.

Budyko, M.I., G.S. Golitsyn and Y.A. Izrael (1988) *Global Climatic Catastrophes*. Springer-Verlag.

Bunting, A.H., M.D. Dennett, J. Elston and J.R. Milford (1976) Rainfall trends in West African Sahel. *Q.J. Meteorol. Soc.* **102**, 59–64.

Burges, A. (1958) *Micro-organisms in the soil*. Hutchinson, London.

Burges, A. and F. Raw (1967) *Soil Biology. Academic Press*, London.

Burnett, J.H. (ed) (1964) *The vegetation of Scotland*. Oliver & Boyd, Edinburgh.

Burrows, C.J. (1990) *Processes of Vegetation Change*. Unwin Hyman, London.

Burt, T. (1986) Runoff processes and solutional denudation rates on humid temperate hillslopes, in *Solute Processes* (ed S.T. Trudgill). Wiley, Chichester.

Burt, T.P. and D.E. Walling (eds) (1984) *Catchment Experiments in Fluvial Geomorphology*. Geobooks, Norwich.

Burt, T.P., R.W. Crabtree and N.A. Fielder (1984) Patterns of solutional denudation in relation to the spatial distribution of soil moisture and soil chemistry over a hillslope hollow and spur, in *Catchment experiments in fluvial geomorphology* (eds T.P. Burt and D.E. Walling). Geobooks, Norwich, 431–446.

Burton, I., R.W. Kates and G.F. White (1978) *The Environment as Hazard*. Oxford University Press, New York.

Butler, B.E. (1959) *Periodic phenomena in landscapes as a basis for soil studies*. Soil Publ. CSIRO, Australia.

Byer, H.R. (1974) *General meteorology*, 4th edn. McGraw-Hill, New York.

Campbell, D.A. (1972) *Morfa Harlech: the dune system and its vegetation*. B.A. Hons CNAA Geography dissertation, Portsmouth Polytechnic.

Campbell, I.M. (1977) *Energy and the atmosphere: a physical chemical approach*. Wiley, Chichester.

Cannell, M.G.R. and M.D. Hooper (1990) *The Greenhouse Effect and Terrestrial Ecosystems of the UK*. ITE Res. Pub. No. 4. HMSO, London.

Carlisle, A., A.H.F. Brown and E.J. White (1967) The nutrient content of tree stemflow and ground flora litter and leachates in a sessile oak (*Quercus petraea*) woodland. *J. Ecol.* **55**, 615–627.

Carr, A.P. and J. Graff (1982) The tidal immersion factor and shore platform development: discussion. *Trans. Inst. Brit. Geog.*, N.S., **7**, 240–245.

Carson, M.A. (1971) *The Mechanics of Erosion*. Pion, London

Carson, M.A. and M.J. Kirkby (1972) *Hillslope form and process*. Cambridge University Press, Cambridge.

Carter, R.W.G. (1988) *Coastal Environments: an Introduction to the Physical, Ecological and Cultural Systems of Coastlines*. Academic Press, London.

Catt, J.A. (1988) *Quaternary Geology for Scientists and Engineers*. Ellis Horwood, Chichester.

Chandler, T.J. (1965) *The climate of London*. Hutchinson, London.

Chandler, T.J. and S. Gregory (eds) (1976) *The Climate of the British Isles*. Longman, London.

Chang, S., D. DesMarais, R. Mack, S.L. Miller and G.E. Strathearn (1983) Prebiotic organic syntheses and the origin of life, in *Earth's Earliest Biosphere* (ed J.W. Schopf). Princeton University Press, Princeton.

Chapman, C.A. and R.L. Rioux (1958) Statistical study of topography, sheeting, and jointing in granite. Acadia National Park, Maine, *Am. J. Sci.* **256**, 111–127.

Chapman, R.N. (1928) The quantitative analysis of environmental factors. *Ecology* **9**, 111–122.

Chapman, S.B. (1976) *Methods in plant ecology*. Blackwell Scientific, Oxford, 229–295.

Chorley, R.J. (1973) Geography as human ecology, in *Directions in geography* (ed R.J. Chorley). Methuen, London, 155–159.

Chorley, R.J. and B. Kennedy (1971) *Physical geography: a systems approach*. Prentice-Hall, Hemel Hempstead, England.

Chorley, R.J., Schumm, S.A., and D.E. Sugden (1984) *Geomorphology*. Methuen, London.

Churchman, C.W. (1968) *The Systems Approach*. Delacorte Press, New York.

Clark, S.P. (1971) *Structure of the Earth*. Prentice-Hall, Englewood Cliffs.

Clayton, K.M. (1974) Zones of glacial erosion, in *Progress in geomorphology* (eds E.H. Brown and R.S. Waters). Inst. Br. Geogs Sp. Publ. 7, 163–176.

Clayton, K.M. (1977) River terraces, in *British Quaternary studies, recent advances* (ed F.W. Shotton). Oxford University Press, Oxford, 153–67.

Clements, F.E. (1916) *Plant Succession: an Analysis of the Development of Vegetation*. Carnegie Inst. Washington, Publ. No. 242.

Cockburn, A. (1991) *An Introduction to Evolutionary Ecology*. Blackwell, Oxford.

Cocks, L.R.M. (ed) (1981) *The Evolving Earth*. (published for the British Museum [Natural History]). Cambridge University Press, Cambridge.

Coffey, W. (1981) *Geography, towards a General Spatial Systems Approach*. Methuen, London.

Cole, L.C. (1958) The ecosphere. *Scient. Am.* **198** (4), 83–92.

Colinvaux, P. (1986) *Ecology*. John Wiley, New York.

Collins, D.N. (1979) Quantitative determination of the subglacial hydrology of two alpine glaciers. *J. Glaciology*, **23** (89), 347–363.

Collins, D.N. (1981) Seasonal variation of solute concentration in meltwaters draining from an alpine glacier. *Annals of Glaciology*, **2**, 11–16.

Connell, J.H. and R.O. Slayter (1977) Mechanisms of succession in natural communities, and their role in community stability and organisation. *American Naturalist*, **111**, 1119–1144.

Cooke, R.U. and I.J. Smalley (1968) Salt weathering in deserts. *Nature* **220** (1), 226–227.

Cooke, R.U. and A. Warren (1973) *Geomorphology in Deserts*. Batsford, London.

Cooke, R.U. and J.C. Doornkamp (1974) *Geomorphology in Environmental management*. Oxford University Press, Oxford.

Cooper, J.P. (ed) (1975) *Photosynthesis and Productivity in Different Environments*. Cambridge University Press, Cambridge.

Corbel, J. (1964) L'érosion terrestre, étude quantitative (Méthodes – techniques – résultats). *Ann. Geog.* **73**, 385–412.

Correns, C.W. (1949) *Growth and dissolution of crystals under linear pressure*. Disc. Faraday Soc. **5**, 271–297.

Courtney, F.M. and S.T. Trudgill (1984) *The Soil: an Introduction to Soil Study in Britain*, (2nd edn). Edward Arnold, London.

Cousins, S.H. (1980) A trophic continuum derived from plant structure, animal size and a detrital cascade. *J. Theor. Biol.*, **82**, 605–618.

Cousins, S.M. (1985) Ecologists build pyramids again. *New Scientist*, 4th July.

Cousins, S. (1987) The decline of the trophic level concept. *Trends in Ecology and Evolution*, **2**, 312–316.

Cowles, H.C. (1901) The physiographic ecology of Chicago and vicinity. *Botanical Gazette*, **31**, 73–108, 145–182.

Cox, B.A. and P.D. Moore (1985) *Biogeography: an Ecological and Evolutionary Approach*, (4th edn). Blackwell, Oxford.

Cox, C.B., I.N. Healey and P.D. Moore (1973) *Biogeography: an ecological and evolutionary approach*. Blackwell Scientific, Oxford

Coxon, J.M., J.E. Fergusson and L. Philips (1980). *First Year Chemistry*. Edward Arnold, London.

Crawley, M.J. (1982) *Herbivory: the Dynamics of Animal–Plant Interactions*. (*Studies in Ecology vol 10*). Blackwell, Oxford.

Crisp, P.J. (1964) *Grazing in terrestrial and marine environments*. Blackwell Scientific, Oxford.

Crocker, R.L. and B.A. Dickinson (1957) Soil development on the recessional moraines of the Herbert and Mendenhall glaciers, S.E. Alaska, *J. Ecol.* **45**, 169–185.

Crocker, R.L. and J. Major (1955) Soil development in relation to vegetation and surface age at Glacier Bay, Alaska. *J. Ecol.* **43**, 427–448.

Crompton, E. (1960) *The significance of the weathering leaching ratio in the differentiation of major soil groups.* Trans 7th Int. Cong. Soil Sci. **4**, 406–412.

Crowe, P.R. (1971) *Concepts in climatology.* Longman, London.

Curry, L. (1962) Climatic change as a random series. *Ann. Assoc. Am. Geogs* **52**, 21–31.

Curtis, C.D. (1976) Chemistry of rock weathering: fundamental reactions and controls, in *Geomorphology and climate* (ed E. Derbyshire). Wiley, Chichester.

Cushing, D.H. and J.J. Walsh (eds) (1976) *Ecology of the Seas.* Blackwell, Oxford.

Daily Telegraph (1976) USA accused of fouling Cuba's weather. 28 June.

Dalrymple, J.B., R.L. Blong and A.J. Conacher (1968) A hypothetical nine unit land surface model. *Z. Geomorph.* **12**, 60–76.

Dansereau, P. (1957) *Biogeography: an Ecological Perspective.* The Ronald Press Company, New York.

Davidson, D.A. (1978) *Science for Physical Geographers.* Edward Arnold, London.

Davidson, J. and R. Lloyd (eds) (1979) *Conservation and Agriculture.* John Wiley, Chichester.

Davies, J.L. (1972) *Geographical Variation in Coastal Development.* Oliver & Boyd, Edinburgh.

Davis, P.H. and V.H. Heywood (1963) *Principles of angiosperm taxonomy.* Oliver & Boyd, Edinburgh.

Davies, R. (ed) (1978) *Coastal Sedimentary Environments.* Springer-Verlag, Berlin.

Davis, R.A. (1972) *Principles of Oceanography* (2nd edn). Addison Wesley, Mass.

Dawkins, R. (1986) *The Blind Watchmaker.* Longman, London.

Deacon, J. (1974) The location of refugia of *Corylus avellana* L. during the Weichselian glaciation. *New Phytol.* **73**, 1055–1063.

Dearman, W.R., F.J. Baynes and T.Y. Irfan (1976) Practical aspects of periglacial effects on weathered granite. *Proc. Ussher Soc.*, **3** (3), 373–381.

Dearman, W.R., F.J. Baynes and T.Y. Irfan (1978) Engineering grading of weathered granite. *Q.J. Engng Geol.* **12**, 345–374.

Defant, F. (1951) Local winds, in *Compendium of meteorology*, (ed T.F. Malone). Am. Meteorol. Soc. Boston, Mass.

Degens, E.T., S. Kempe and J.E. Richey (eds) (1990) *Biogeochemistry of Major World Rivers.* John Wiley, New York.

Denton, G.H. and T.J. Hughes (1983) Milankovitch theory of Ice Ages: hypothesis of ice sheet linkage between regional insolation and global climate.

Derbyshire, E., K.J. Gregory and J.R. Hails (1979) *Geomorphological processes.* Butterworth, London.

Detwyler, T.R. and M.G. Marcus (eds) (1972) *Urbanisation and Environment: the Physical Geography of the City.* Belmont, Duxbury.

Dewey, J.F. (1972) Plate tectonics. *Scient. Am.* **226** (May), 56–66.

Dickinson, C. and G. Pugh (1974) *Biology of Plant Litter Decomposition.* Academic Press, London.

Dietz, R.S. and J.C. Holden (1970) The breakup of Pangaea. *Scientific American.* October.

Digby, P.G.N. and R.A. Kempton (1987) *Multivariate Analysis of Ecological Communities.* Chapman & Hall, London.

Dobben, W.H. van, and R.H. Lowe-McConnell (eds) (1975) *Unifying Concepts in Ecology.* Junk, The Hague.

Doornkamp, J.C. (ed) (1990) *The Greenhouse Effect and Rising Sea Levels in the UK.* M1 Press.

Douglas, I. (1967a) Man, vegetation, and the sediment yields of rivers. *Nature* **215**, 925–928.

Douglas, I. (1967b) Natural and manmade erosion in the humid tropics of Australia, Malaysia and Singapore, in *Symposium on river morphology.* Inst. Ass. Sci. Hydrol, 17–29.

Douglas, I. (1983) *The Urban Environment.* Edward Arnold, London.

Drever, J.I. (1988) *The Geochemistry of Natural Waters.* Prentice-Hall, Englewood Cliffs.

Drewry, D. (1986) *Glacial Geologic Processes.* Edward Arnold, London.

Driessen, P.M. and R. Dudal (eds) 1989. *Geography, Formation, Properties, and Use of the Major Soils of the World.* Wageningen/Leuven: Agricultural University, Wageningen, & Catholic University, Leuven.

Drury, W.H. and I.C. Nesbit (1973) Succession. *J. Arnold Arboretum*, **54**, 331–368.

Duchaufour, P. (translated by T.R. Paton) (1982) *Pedology: Pedogenesis and Classification.* Allen & Unwin.

Duckham, A.N. and G.B. Masefield (1970) *Farming systems of the world.* Chatto & Windus, London.

Duncan, G. (1975) *Physics for Biologists.* Blackwell, Oxford.

Duncan, N. (1969) *Engineering geology and rock mechanics.* Leonard Hill, London.

Dunne, T. (1979) Sediment yield and land use in tropical catchments. *J. Hydrol.* **42**, 281–300.

Duxbury, A.C. (1971) *The Earth and its Oceans.* Addison-Wesley, Mass.

Dyer, A.J. and B.B. Hicks (1965) Stratospheric transport of volcanic dust inferred from solar radiation measurements. *Nature* **208**, 131–133.

Edwards, C.A., D.E. Reichle and D.A. Crossley (1970) The role of soil invertebrates in turnover of organic matter and nutrients, in *Analysis of temperate forest ecosystems*, (ed D.E. Reichle). Chapman & Hall, London, Ch. 12.

Egler, F.E. (1954) Vegetation science concepts I. Initial floristic composition – a factor in old-field vegetation development. *Vegetation*, **4**, 412–417.

Egler, F.E. (1964) Pesticides in our ecosystem. *Am.. Sci.* **52**, 110–136.

Eldridge, N. (1985) *Time Frames.* Simon and Schuster, New York.

Eliassen, A. and K. Pedersen (1977) *Meteorology; an introductory course.* Vol. 1: *Physical processes and motion.* Universitetsforlaget, Oslo.

Elsom, D. (1987) *Atmospheric Pollution.* Blackwell, Oxford.

Elton, C. (1927) *Animal ecology.* Sidgwick & Jackson, London. (Paperback edn Methuen 1966.)

Elton, C.S. (1958) *The ecology of invasion by animals and plants.* Methuen, London.

Elton, C.S. and R.S. Miller (1954) The ecological survey of animal communities: with a practical system of classifying habitats by structural characters. *J. Ecol.*, **42**, 460–496.

Embleton, C. (ed) (1972) *Glaciers and glacial erosion.* Macmillan, London.

Embleton, C.E. and C.A.M. King (1975) *Periglacial Geomorphology.* Edward Arnold, London.

Embleton, C. and J.B. Thornes (eds) (1979) *Process in geomorphology.* Edward Arnold, London.

Emery, F.E. (1969) *Systems Thinking.* Penguin, London.

Engstrom, D.R. and H.E. Wright Jr. (1984) Chemical

stratigraphy of lake sediments as a record of environmental change, in *Lake sediments and environmental history* (eds E.Y. Haworth and J.W.G. Lund). Leicester University Press, Leicester, 11–67.

Etherington, J.R. (1975) *Environment and plant ecology*. Wiley, Chichester.

Etherington, J.R. (1978) *Plant physiological ecology*. Edward Arnold, London.

Evans, I.S. (1970) Salt crystallisation and rock weathering: a review. *Rev. Géomorph. Dyn.* **19**, 153–177.

Evans, R. and R.P.C. Morgan (1974) Water erosion of arable land. *Area* **6** (3), 221–225.

Eyles, N. (ed) (1983) *Glacial Geology*. Pergamon Press, Oxford.

FAO-UNESCO (1989) *Soil Map of the World* – Revised Legend 1988. ISRIC, Wageningen.

Fairbridge, R.H. (1961) Eustatic changes of sea-level, in *Physics and chemistry of the Earth*, Vol. 4 (ed L.H. Ahrens *et al.*). Pergamon, Oxford, 99–185.

Farres, P. J. (1978) The rôle of time and aggregate size in the crusting process. *Earth Surf. Proc.* **3**, 243–254.

Farres, P.J. (1991) Soil classification: the updated FAO system and soils in the UK. *Geography Rev.*, **5** (1), 27–31.

Farres, P.J., J. Poesen, and S. Wood (1992) *Some Characteristic Soil Erosion Landscapes of N.W. Europe*. Catena (in Press).

Federov, An. A. (1966) The structure of tropical rainforest, and speciation in the humid Tropics. *J. Ecol.*, **54**, 1–11.

Fenchel, T. (1988) Marine plankton food chains. *Ann. Rev. Ecol. Syst.*, **19**, 19–38.

Fenwick, I.M. and B.J. Knapp (1982) *Soil Processes and Response*. Duckworth, London.

Ferguson, R.I. (1981) Channel forms and channel changes, in *British rivers* (ed J. Lewin). Allen & Unwin, London, 90–125.

Fifield, R. (ed) (1985) *The Making of the Earth* (*New Scientist Guides*). Blackwell, Oxford. (Reprints of New Scientist reports covering the period of development of the modern theory of Plate Tectonics and related areas).

Finlayson, B. and I. Stratham (1980) *Hillslope Analysis*. Butterworth, London.

Fink, J. and G. Kukla (1977) Pleistocene climates in central Europe: at least 17 interglacials after the Olduvai event. *Quaternary Research*, **7**, 363–371.

Fisher, J. (1954) *Bird recognition I. Sea birds and waders*. Penguin, London.

Fitter, A.H. and R.K.M. Hay (1987) *Environmental Physiology of Plants* (2nd edn). Academic Press, London.

Fitzpatrick, E.A. (1971) *Pedology*. Oliver & Boyd, Edinburgh.

Flegmann, A.F. and R.A.T. George (1975) *Soils and other growth media*. Macmillan, London.

Flohn, H. and R. Fantechi (eds) *The Climate of Europe: Past Present and Future*. D. Reidel, Dordrecht.

Fookes, P.G., W.R. Dearman and J.A. Franklin (1971) Some engineering aspects of rock weathering with field examples from Dartmoor and elsewhere. *Q.J. Engng Geol.* **4**, 139–185.

Ford, E.D. and P.J. Newbould (1977) The biomass and production of ground vegetation and its relation to tree cover through a deciduous woodland cycle. *J. Ecol.* **65**, 201–212.

Forman, R.T.T. and M. Godron (1986) *Landscape Ecology*. John Wiley, New York.

Forrester, J.W. (1971) *World dynamics*. Wright Allen, Cambridge, Mass..

Fortescue, J.A.C. and G.G. Martin (1970) Micronutrients: forest ecology and systems analysis, in *Analysis of temperate forest ecosystems* (ed D.E. Reichle). Chapman & Hall, London.

Fournier, F. (1960) *Climat et érosion: la relation entre l'érosion du sol par l'eau et les précipitations atmosphériques*. Presses Univ. de France, Paris.

Frankland, J.C. (1966) Succession of fungi on decaying petioles of *Pteridium aquilinum. J. Ecol.* **54**, 41–63.

French, H.M. (1973) Cryopediments on the chalk of southern England. *Bull. Periglac.* **22**, 149–156.

Frenkiel, F.N. and D.W. Goodall (eds) (1978) *Simulation Modelling of Environmental Problems*. John Wiley/SCOPE, Chichester.

Gardiner, B.G. (1989) The Antarctic Ozone Hole. *Weather*, **44** (7), 291–297.

Gardiner, V. (1974) *Drainage basin morphometry*. Tech. Bull. No. 14, British Geomorph. Res. Group.

Garrels, R.H. and F.T. Mackenzie (1971) *Evolution of sedimentary rocks*. Norton, New York.

Garrels, R.M., F.T. MacKenzie and C. Hunt (1975) *Chemical Cycles and the Global Environment*. Kaufman, California.

Garrett, S.D. (1981) *Soil fungi and soil fertility*, 2nd edn. Pergamon, Oxford.

Gaskell, T.F. (1967) *The Earth's Mantle*. Academic Press, London.

Gass, I.G., P.J. Smith, and R.C.L. Wilson (1973) *Understanding the Earth: a Reader in the Earth Sciences*. (published for the Open University) Artemis Press, Horsham, Sussex.

Gates, D.M. (1962) *Energy exchange in the biosphere*. Harper & Row, New York.

Geiger, R. (1965) *The climate near the ground*. Cambridge, Harvard University Press, Mass.

Gentry, R.C. (1970) Hurricane Debbie modification experiments. *Science* **168**, 473–475.

Gerrard, A.J. (1981) *Soils and Landforms*. Allen & Unwin, London.

Gibbs, R.J. (1970) Mechanisms controlling world water chemistry. *Science*, **170**, 1088–1090.

Glentworth, R. and H.G. Dion (1949) The association or hydrologic sequence in certain soils of the podsolic zone of N.E. Scotland. *J. Soil Sci.* **1**, 35–49.

Glymer, R.G. (1973) *Chemistry: an Ecological Approach*. Harper and Row, New York.

Godwin, Sir H. (1975) *History of the British Flora: a Factual Basis for Phytogeography* (2nd edn). Cambridge University Press, Cambridge.

Goldich, S.S. (1938) A study in rock weathering. *J. Geol.* **46**, 17–58.

Golding, W. (1961) *The inheritors*. Faber & Faber, London.

Goldsmith, F.B., C.M. Harrison and A.J. Morton (1986) Description and analysis of vegetation, in *Methods in Plant Ecology*, (2nd edn) (eds P.D. Moore and S.B. Chapman). Blackwell, Oxford.

Golley, F.B. (ed) (1977) *Ecological Succession*. (*Benchmark Series in Ecology*). Dowden, Stroudsburg. (Contains reprints of most of the classic papers).

Gomez Pompa, A., C. Vàzquez-Yanes, S. Guevara (1972). The tropical rainforest: a non-renewable resource. *Science* **177**, 762–765.

Gorman, M. (1979) *Island Ecology*. Chapman & Hall, London

Goudie, A.S. (1977a) Sodium sulphate weathering and the disintegration of Mohenjo-daro, Pakistan. *Earth Surf. Proc.* **2**, 75–86.

Goudie, A.S. (1977b) *Environmental change*. Oxford University Press, Oxford.

Goudie, A.S. and A. Watson (1980) *Desert Geomorphology*. Macmillan, London.

Grace, J. (1983) *Plant–Atmosphere Relationships. (Outline Series in Ecology)*. Chapman & Hall, London.

Green, B. (1984) *Countryside Conservation*. Allen & Unwin, London.

Gregory, K.J. (1974) Streamflow and building activity, in *Fluvial processes in instrumented watersheds* (eds K.J. Gregory and D.E. Walling). Inst. Br. Geogs Sp. Publ. 6.

Gregory, K.J. (1976) Drainage networks and climate, in *Geomorphology and climate* (ed E. Derbyshire). Wiley, Chichester, 289–315.

Gregory, K.J. (1987) *Energetics of Physical Environment: Energetic Approaches to Physical Geography*. John Wiley, Chichester.

Gregory, K.J. and C.C. Park (1974) Adjustments of river channel capacity down stream from a reservoir. *Water Resources Res.* **10**, 870–873.

Gregory, K.J. and C.C. Park (1976a) Stream channel morphology in N.W. Yorkshire. *Rev. Géomorph. Dyn.* **25** (2), 63–72.

Gregory, K.J. and C.C. Park (1976b) The development of a Devon gully and man. *Geography*, **61**, 77–82.

Gregory, K.J. and D.E. Walling (1973) *Drainage basin form and process*. Arnold, London.

Greig-Smith, P. (1983) *Quantitative Plant Ecology*, (3rd edn). Butterworths, London.

Gribbin, J. (1989) The global greenhouse. *Scope*, Summer, 4–7.

Griffin, D.M. (1972) *Ecology of Soil Fungi*. Chapman & Hall, London.

Grime, J.P. (1977) Evidence for the existence of three primary strategies in plants and its relevance to ecological and evolutionary theory. *Amer. Naturalist*, **111**, 1169–1194.

Grime, J.P. (1979) *Plant Strategies and Vegetation Processes*. John Wiley, Chichester.

Grime, J.P., J.G. Hodgson, and R. Hunt (1990) *(The Abridged) Comparative Plant Ecology*. Unwin Hyman, London.

Gross, M. Grant (1989) *Oceanography: a View of the Earth*, (5th edn). Prentice-Hall, Englewood Cliffs.

Grove, A.T. (1967) The last 20 000 years in the tropics, in *Tropical geomorphology* (ed A.M. Harvey). BGRG Spec. Publ. 5.

Grove, A.T. and A. Warren (1968) Quaternary landforms and climate on the south side of the Sahara. *Geog. J.* **134** (2), 194–208.

Guardian, The (London) (1977) Convention bans weather war. 19 May.

Hack, J.T. (1957) *Studies of longitudinal stream profiles in Virginia and Maryland*. USGS Prof. Paper 294B.

Hadley, R.F. and D.E. Walling (eds) (1984) *Erosion and Sediment Yield: Some Methods of Measuring and Modelling*. Geobooks, Norwich.

Hadley, R.F. and S.A. Schumm (1961) Sediment sources and drainage basin characteristics in upper Cheyenne River basin, in *USGS Water Supply Paper* **1531-B**, 137–196.

Haigh, M. (1985) Geography and general systems theory philosophical homologies and current practice. *Geoforum* **16**, 191–203.

Hails, J.R. (1975) Some aspects of the Quaternary history of Start Bay, Devon. *Field Studies*, **4**, 207–222.

Hails, J.R. (ed) (1977) *Applied Geomorphology*. Elsevier, Amsterdam.

Haines-Young, R. and J. Petch (1986) *Physical Geography: its Nature and Methods*. Harper & Row, London.

Hall, D.O. and K.K. Rao (1972) *Photosynthesis*. Edward Arnold, London.

Hanwell, J. (1980) *Atmospheric Processes*. Allen & Unwin, London.

Hanwell, J.D. and M.D. Newson (1970) *The storms and floods of July 1968 on Mendip*. Wessex Cave Club Occ. Publ. Ser. 1.2.

Hanwell, J.D. and M.D. Newson (1973) *Techniques in physical geography*. Macmillan, London.

Harbourne, J.C. (1977) *Introduction to Ecological Biochemistry*. Academic Press, New York.

Harding R.J. (1979) Radiation in the British uplands. *J. App. Ecol.*, **16**, 161–170.

Harding, D.M. and D.J. Parker (1974) Flood hazard at Shrewsbury. In *Natural hazards* (ed G.F. White). Oxford University Press, New York.

Harper, J.L., R.B. Rosen and J. White (eds) (1986) The growth and form of modular organisms. *Phil Trans. R. Soc. Series B*, **313**, 1–250.

Harvey, A.M. (1974) Gully erosion and sediment yield in the Howgill Fells, Westmorland, in *Fluvial processes in instrumented watersheds* (eds K.J. Gregory and D.E. Walling). Inst. Br. Geogs Sp. Publ. 6, 45–58.

Harvey, J.G. (1976) *Atmosphere and ocean: our fluid environments*. Artemis Press, Sussex.

Heal, O.W. and S.F. MacLean (1975) Comparative productivity in ecosystems: secondary productivity, in *Unifying Concepts in Ecology* (eds W.H. van Dobben and R.H. Lowe-McConnell. Junk, The Hague.

Heathcote, R.L. (1983) *The Arid Lands: their Use and Abuse*. Longman, London.

Heller, F. and Liu Tung-sheng (1982) Magnetostratigraphic dating of loess deposits in China. *Nature*, **300**, 431–443.

Hemmingsen, A.M. (1960) *Energy metabolism as related to body size and respiratory surfaces*. Rep. Steno Meml Hosp., Copenhagen 9, part 2, 7.

Hengeveld, R. (1990) *Dynamic Biogeography. (Cambridge Studies in Ecology)*. Cambridge University Press, Cambridge.

Hewitt, K. (ed) (1983) *Interpretation of Calamity*. Allen & Unwin, Boston.

Hewson, E.W. and R.W. Longley (1944) *Meteorology, theoretical and applied*. Wiley, New York.

Heywood, V.H. (1967) *Plant taxonomy*. Edward Arnold, London.

Hide, R.·(1969) Some laboratory experiments on free thermal convection in a rotating fluid subject to a horizontal temperature gradient and their relation to the theory of global atmospheric circulation, in *The global circulation of the atmosphere* (ed G.A. Corby). R. Meteorol. Soc., London

Hjülstrom, F. (1935) Studies of the morphological activities of rivers as illustrated by the River Fyris. *Bull. Geological Inst. Univ. Uppsala*, **25**, 221–527.

Hobbs, R.J. and H.A. Mooney (1990) *Remote Sensing of Biosphere Functioning*. Academic Press, San Diego.

Hole, F.D. (1978) An approach to landscape analysis with the accent on soils. *Geoderma*, **21**, 1–23.

Hole, F.D. and J.B. Campbell (1986) *Soil Landscape Analysis*. Routledge and Kegan Paul, London.

Holland, H.D. (1978) *The Chemistry of the Atmosphere and Oceans.* John Wiley, New York.

Holling, C.S. (1973) Resilience and stability of ecological systems. *Ann. Rev. Ecol. Syst.*, **4**, 1–23.

Holling, C.S. (1976) Resilience and stability in ecosystems, in *Evolution and Consciousness: Human Systems in Transition* (eds E. Jantsch and C.H. Waddington). Addison-Wesley, Mass.

Hollis, G. 1974). The effect of urbanisation on floods in the Canon's Brook, Harlow, Essex, in *Fluvial processes in instrumented watersheds* (eds K.J. Gregory and D.E. Walling). Inst. Br. Geogs Sp. Publ. 6, 123–139.

Hooper, M. (1970) Dating hedges. *Area* **4**, 63–65.

Hope, R., H. Lister and R. Whitehouse (1972) The wear of sandstone by cold sliding ice, in *Polar Geomorphology*, (eds R. J. Price and D.E. Sugden). Inst. Brit. Geog. Spec. Pub., **4**, 21–31.

Hopkins, B. (1965) *Forest and savanna.* Heinemann, London.

Horn, H.S. (1974) The ecology of secondary succession. *Ann. Rev. Ecol. Syst.*, **5**, 25–37.

Horn, H.S. (1975) Forest Succession. *Scientific American,* (May), 91–98.

Horn, H.S. (1981) Some causes of variety in patterns of secondary succession, in *Forest Succession: Concepts and Application* (eds D.C. West, H.H. Shugart and D.B. Botkin). Springer Verlag, New York.

Houghton, J.T., G.T. Jenkins and J.J. Ephraums (eds) (1990) *Climate Change: the IPCC Scientific Assessment.* Cambridge University Press, Cambridge.

Howe, H.F. and L.C. Westley (1988) *Ecological Relationships of Plants and Animals.* Oxford University Press, Oxford.

Hubbell, S.P. (1971) Of sowbugs and systems: the ecological energetics of a terrestrial isopod, in *Systems analysis and simulation ecology.* Academic Press, New York.

Huggett, R.J. (1975) Soil landscape systems: a model of soil genesis. *Geoderma*, **13**, 1–22.

Huggett, R.J. (1980) *Systems Analysis in Geography (Contemporary Problems in Geography).* Oxford University Press, Oxford.

Hughes, R. and J.M.M. Munro (1968) Climate and soil factors in the hills of Wales in their relation to the breeding of special herbage varieties, in *Hill land productivity.* Occ. Symp. no. 4, British Grassland Soc.

Humphreys, W.F. (1979) Production and respiration in animal populations. *J. Animal Ecol.*, **48**, 427–454.

Hunter, R.F. (1962a) Hill sheep and their pasture. A study in sheep grazing in S.E. Scotland. *J. Ecol.* **50**, 651–680.

Hunter, R.F. (1962b) Home range behaviour in hill sheep, in *Grazing* (ed D.J. Crisp). Proc. 3rd Symp. on grazing, Bangor. Br. Ecol Soc. Blackwell Scientific, Oxford.

Hutchinson, G.E. (1959) Homage to Santa Rosalia, or why are there so many kinds of animals? *Am. Nat.* **93**, 145–159.

Hutchinson, G.E. (ed) (1965) *The ecological theatre and the evolutionary play.* Yale University Press, New Haven, Conn..

Hutchinson, G.E. (1970) The biosphere. *Scient. Am.* **223** (3), 45–53.

Illies, V. 1974. *Introduction to zoogeography* (transl. W.D. Williams). Macmillan, London.

Imbrie, J. and K.P. Imbrie (1979) *Ice Ages: Solving the Mystery.* Macmillan, London.

Imeson, A.C. (1974) The origin of sediment in a moorland catchment with particular reference to the rôle of vegetation, in *Fluvial processes in instrumented watersheds* (eds K.J. Gregory and D.E. Walling). Inst. Br. Geogs Spec. Publ. 6.

Iverson, J. (1956) Forest clearance in the Stone Age. *Scient. Am.* **194**, 36–41.

Iverson, J. (1964) Retrogressive vegetational succession in the post-glacial. *J. Ecol.* **52** (suppl.), 59–70.

Jackson, J.B.C., L.W. Buss and R.E. Cooke (eds) (1985) *Population Biology and Clonal Organisms.* Yale University Press, New Haven.

Jackson, R.J. (1967) The effect of slope aspects and albedo on PET from hillslopes and catchments. *N.Z.J. Hydrol.* **6**, 60–69.

Jackson, R.M. and F. Raw (1966) *Life in the Soil.* Edward Arnold, London.

Jaynes, E.T. (1957) Information theory and statistical mechanics. *Phys. Rev.* **104**.

Jeffery, D.W. and C.D. Pigott (1973) The response of grassland on sugar limestone to applications of phosphorus and nitrogen. *J. Ecol.* **61**, 85–92.

Jenkins, D., A. Watson and G.R. Miller (1967) Population fluctuations in the red grouse (*Lagopus lagopus scoticus*). *J. Anim. Ecol.*, **36**, 97–122.

Jenny, H. (1941). *Factors of soil formation.* McGraw-Hill, New York.

Jones, P.D. *et al.* (1988) Evidence for global warming in the past decade. *Nature,* **332**, 790.

Kates, R.W. (1970) *Natural hazard in human ecological perspective: hypothesis and models.* Nat. Hazards Res. Working Paper, no. 14.

Keller, W.D. (1957) *The principles of chemical weathering.* Lucas, Columbia, Miss.

Kendrick, R.E. and B. Frankland (1976) *Phytochrome and plant growth.* Edward Arnold, London.

Kepner, R.A. (1951) *Effectiveness of orchard heaters.* Bulletin 723, California Agric. Exp. Stat.

Kershaw, K.A. and J.H. Looney (1985) *Quantitative and Dynamic Ecology,* (3rd edn). Edward Arnold, London.

King, C.A.M. (1972) *Beaches and Coasts,* (2nd edn). Edward Arnold. London.

Kira, T. and T. Shidei (1967) Primary production and turnover of organic matter in different forest ecosystems of the western Pacific. *Jap. J. Ecol.* **17**, 70–87.

Kirkby, M.J. (1971) Hillslope process–response models based on the continuity equation, in *Slopes: form and process* (ed D. Brunsden). Inst. Brit. Geog. Spec. Pub., 3, 15–30.

Kirkby, M.J. (ed) (1978) *Hillslope Hydrology.* John Wiley, Chichester.

Kirkby, M.J. (1987) Models in physical geography, part 1.3, in *Horizons in Physical Geography* (eds M.J. Clark, K.J. Gregory and A.M. Gurnell). Macmillan, Basingstoke.

Kirkby, M.J. and R.P.C. Morgan (eds) (1980) *Soil Erosion.* John Wiley, Chichester.

Knapp, B.J. (1979) *Soil processes.* George Allen & Unwin, London.

Knighton, A.D. (1984) *Fluvial Forms and Processes.* Macmillan, London.

Kononova, M.M. (1966) *Soil Organic Matter,* (2nd edn). Pergamon, Oxford.

Kramer, P.J. (1969) *Plant and soil water relationships.* McGraw-Hill, New York.

Krebs, C.J. (1972) *Ecology: the Experimental Analysis of Distribution and Abundance*. Harper & Row, New York.

Krebs, C.J. (1988) *The Message of Ecology*. Harper & Row, New York.

Krebs, J.R. and N.B. Davies (eds) (1984) *Behavioural Ecology: an Evolutionary Approach* (2nd edn). Blackwell, Oxford.

Kubiena, W.L. (1938) *Micropedology*. Ames, Collegiate Press, Iowa.

Kucera, C.L., R.C. Dahlmann and M.R. Krelling (1967) Total net productivity and turnover on an energy basis for tall grass prairie. *Ecol.* **48**, 536–541.

Kurtén, B. (1969) Continental drift and evolution. *Scient. Am.* **220** (3), 54–64.

Lamb, H.H. (1972) *Climate present, past and future*. Vol. 1: *Fundamentals and climate now*. Methuen, London.

Lamb, H.H. (1977) *Climate: Present, Past and Future*. Methuen, London.

Lamb, H.H. (1982) *Climate, History and the Modern World*. Methuen, London.

Landsberg, H.E. (1960) *Physical climatology* (2nd edn). Gray Printing Co., Dubois, Penn..

Langbein, W.B. and L.B. Leopold (1964) Quasi equilibrium states in channel morphology. *Am. J. Sci.* **262**, 782–794.

Langbein, W.B. and L.B. Leopold (1966) *River meanders – the theory of minimum variance*. USGS Prof. Paper 422–H.

Langbein, W.B. and S.A. Schumm (1958) Yield of sediment in relation to mean annual precipitation. *Trans Am. Geophys. Union*, **39**, 1076–1084.

Learmonth, A. (1977) *Man–environment relationships as complex ecosystems*. Open University Press, Milton Keynes.

Lee, R. (1978) *Forest microclimatology*. Columbia University Press, New York.

Lehninger, A.L. (1965) *Bioenergetics. The molecular basis of biological energy transformations*. Benjamin, New York.

Lenihan, J. and W.W. Fletcher (1978) *The built environment*. Vol. 8: *Of environment and man*. Blackie, Glasgow.

Leopold, L.B. and W.B. Langbein (1962) *The concept of entropy in landscape evolution*, 20. USGS Prof. Paper 500–A.

Leopold, L.B. and T. Maddock (1953) *The hydraulic geometry of stream channels and some physiographic implications*. USGS Prof. Paper 252.

Leopold, L.B. and M.G. Wolman (1957) *River channel patterns: braided, meandering and straight*. USGS Prof. Paper 282–B.

Leopold, L.B., M.G. Wolman and J.P. Miller (1964) *Fluvial processes in geomorphology*. W.H. Freeman, San Francisco.

Lieth, H. (1964) Versuch einer kartographischen Darstellung der produktivität der Pflanzendecke auf der Erde, in *Geographisches Taschenbuch 1964/65*, Steiner, Wiesbaden, 72–80.

Lieth, H. (1970) Phenology in productivity studies, in *Analysis of temperate forest ecosystems* (ed D.E. Reichle). Chapman & Hall, London, Ch. 4.

Lieth, H. (1971) The net primary productivity of the Earth with special emphasis on land areas, in *Perspectives on primary productivity of the Earth* (ed R. Whittaker). Symp. AIBS 2nd Natl Congr., Miami, Florida, October.

Lieth, H. (1972) Über die primärproduktion der Pflanzdecke der Erde. Symp. Deut. Bot. Gesell. Innsbruck, Austria, September 1971. *Z. Angew, Bot.* **46**, 1–37.

Lieth, H. (1975) Modelling the primary productivity of the world, in *Primary productivity of the biosphere* (eds H. Lieth and R.H. Whitaker). Springer-Verlag, New York, 237–263.

Likens, G.E. and F.H. Bormann (1972) Nutrient cycling in ecosystems. In *Ecosystem structure and function*. J.H. Wrens (ed), 25–67. Oregon State Univ. Ann. Boil. Colloquia 31.

Lieth, H. and E. Box (1972) Evapotranspiration and primary productivity; C.W. Thornthwaite memorial model, in *Papers on selected topics in climatology* (ed J.R. Mather). C.W. Thornthwaite Associates, Elmer, NJ, 37–46.

Lilienberg, D. *et al.* (1975) L'analyse morphostructurale des mouvements verticaux actuels de la partie Européenne de l'URSS, in *Problems of recent coastal movements,* Int. Union Geol. and Geophys. Tallinn: Valgus, 57–67.

Lindeman, R.L. (1942) The trophic dynamic aspects of ecology. *Ecology*, **23**, 399–418.

Linsley, R.K., M.A. Kohler and J.L.H. Paulhus (1949) *Applied hydrology*. McGraw-Hill, New York.

Lockwood, J.G. (1962) The occurrence of Föhn winds in the British Isles. *Meteorol. Mag.* **91**, 57–65.

Lockwood, J.G. (1974) *World climatology, an environmental approach*. Edward Arnold, London.

Lockwood, J.G. (1979) *The Causes of Climate*. Edward Arnold, London.

Loughnan, F.C. (1969) *Chemical weathering of the silicate minerals*. Elsevier, New York.

Louw, G.N. and M.K. Seely (1982) *Ecology of Desert Organisms*. Longman, London.

Low, A.J. (1972) The effect of cultivation on the structure and other physical characteristics of grassland and arable soils (1945–70). *J. Soil Sci.*, **23**, 363–380.

Lowe, J.J. and M.J.C. Walker (1984) *Reconstructing Quaternary Environments*. Longman, London.

Lowman, P.D. and J.B. Garvin (1986) Planetary landforms, in *Geomorphology from space: a global overview of regional landforms* (eds N.M. Short and R.W. Blair). NASA, Washington DC.

Lukashev, K.I. (1970) *Lithology and geochemistry of the weathering crust*. Israel Program for Scientific Translation, Jerusalem.

Lutgens, F.K. and E.J. Tarbuck (1982) *The Atmosphere*. Prentice-Hall, Englewood Cliffs.

Lydolph, P.E. (1977) *Climates of the Soviet Union: world survey of climatology*, vol. 7. Elsevier, Amsterdam.

MacArthur, R.H. (1958) Population ecology of some warblers of northeastern coniferous forests. *Ecology*, **39**, 599–619.

MacArthur, R.H. and E.O. Wilson (1963) An equilibrium theory of insular zoogeography. *Evolution*, **17**, 373–387.

MacArthur, R.H. and E.O. Wilson (1967) *The Theory of Island, Biogeography*. Princeton University Press, Princeton.

Macfadyen, A. (1964) Energy flow in ecosystems and its exploitation by grazing, in *Grazing in terrestrial and marine environments* (ed D.J. Crisp). Blackwell Scientific, Oxford, 3–20.

MacIntosh, D.H. and A.S. Thom (1972) *Essentials of Meteorology*. Wykeham Pubs, London.

MacIntyre, F. (1970) Why the sea is salt. *Scientific American*, **223**, 104–115.

Mann, K.H. (1982) *Ecology of Coastal Waters: a Systems Approach. (Studies in Ecology vol 8)*. Blackwell, Oxford.

Margalef, R. (1968) *Perspectives in ecological theory*. University of Chicago Press, Chicago.

Maris, S.L. (1980) *A study of the initial stages of succession on six disused railway tracks in Warwickshire and Leicestershire*. B.A. Hons CNAA Geography dissertation, Portsmouth Polytechnic.

Mason, B. (1952) *Principles of geochemistry*. Wiley, New York.

Mason, B.J. (1975) *Clouds, Rain, and Rainmaking*. Cambridge University Press, Cambridge.

Mason. V.J. (1990) Acid rain – causes and consequences. *Weather*, **45**, 70–79.

Mason, C.F. (1977) *Decomposition*. Edward Arnold, London.

Mass, C. and A. Robock (1982) The short-term influence of the Mount St Helens volcanic eruption on surface temperature in the northeast United States. *Mon. Weath. Rev.*, **110**, 614–622.

Mathews, W.H. (1979) Simulated glacial erosion. *J. Glaciology*, **23** (89), 51–56.

Maxwell, A.E. *et al.* (1970) Deep sea drilling in the South Atlantic. *Science*, **168**, 1047–1059.

May, R.M. (1971a) Stability in model ecosystems. *Proc. Ecol. Soc. Aust.* **6**, 18–56.

May, R.M. (1971b) Stability in multispecies community models. *Bull. Math. Biophys.* **12**, 59–79.

May, R.M. (1974) *Stability and Complexity in Model Ecosystems*, (2nd edn). Princeton University Press, Princeton.

May, R.M. (1975) Will a complex system be stable? *Nature* **238**, 413–414.

Maybey, R. (1980) *The Place for Nature in Britain's Future*. Hutchinson, London

Maynard-Smith, J. (1974) *Models in ecology*. Cambridge University Press, Cambridge.

McCall, J.G. (1960) The flow characteristics of a cirque glacier and their effect on cirque formation, in *Investigations on Norwegian cirque glaciers* (eds J.G. McCall and W.V. Lewis), R. Geog. Soc. Res. Series IV, 39–62.

McKee, E.D. (ed) (1979) A study of global sand seas. *USGS Professional Paper*, 1052.

McLusky, D.S., M. Teare and A.P. Phizacklea (1980) Effects of domestic and industrial pollution on the distribution and abundance of aquatic oligochaetes in the Forth Estuary. *Helgolander Meeresunters*, **33**, 384–392.

Meadows, D.H., D.L. Meadows, J. Randers and W.W. Behrens (1971) *The limits to growth. Report of the Club of Rome*. Earth Island, London.

Meidner, H. and D.W. Sheriff (1976) *Water and plants*. London, Blackie.

Menard, H.W. and Smith (1966) Hypsometry of ocean basin provinces. *J. Geophys. Res.* **7**, 4305–25.

Meteorological Office (1972) *Tables of temperature, humidity, precipitation and sunshine for the world, Part III. Europe and the Azores*. HMSO, London.

Meybeck, M. (1979) Concentrations des eaux fluviales en éléments majeurs et apports en solution aux oceans. *Révue de Géol. Dynamique et Géog. Physique*, **21**, 215–246.

Milankovitch, M. (1930) Mathematische klimalehne und astronomische theorie der Klimaschwaukungen, in *Handbuch der Klimatologie I* (eds I.W. Köppen and R. Geiger). Borntraeger, Berlin.

Miles, J. (1978) *Vegetation Dynamics. (Outline Series in Ecology)*. Chapman & Hall, London.

Miller, D.H. (1977) *Water at the Earth's Surface: an Introduction to Ecosystem Hydrodynamics*. Academic Press, New York.

Miller, J.M. (1984) Acid Rain. *Weatherwise*, **37**, 227–251.

Miller, S.L. (1953) A production of amino acids under possible primitive Earth conditions. *Science*, **117**, 528.

Miller, S.L. (1957) The formation of organic compounds on the primitive Earth. *Annals of the New York Academy of Sciences*, **69**, 260–275.

Milliman, J.D. and R.H. Meade (1983) Worldwide delivery of sediment to the oceans. *Journal of Geology*, **91**, 1–21.

Milne, G. (1947) A soil reconnaissance journey through parts of Tanganyika territory, December 1935–February 1936. *J. Ecol.* **35**, 192–265.

Minderman, G. (1968) Addition, decomposition, and accumulation of organic matter in forests. *J. Ecol.* **56**, 355–362.

Ministry of Agriculture and Fisheries and Food (1967) *Potential Transportation*. Technical Bulletin No 16. HMSO, London

Mitchell, J.M. (1976) An overview of climatic variability and its causal mechanisms. *Quatern. Res.* **6**, 481–494.

Monteith, J.L. (1973) *Principles of environmental physics*. Edward Arnold, London.

Mooney, H.A., Vitousek, P.M. and Matson, P.A. (1987) Exchange of materials between terrestrial ecosystems and the atmosphere. *Science*, **238**, 926–932.

Moore, D.M. (1982) *Green Planet: the Story of Plant Life on Earth*. Cambridge University Press Cambridge.

Moore, J.J. (1962) The Braun-Blanquet system: a reassessment, *J. Ecol.* **50**, 761–769.

Moore, N.W., M.D. Hooper and B.N.K. Davis (1967) Hedges: I. Introduction and reconnaissance studies. *J. Appl. Ecol.* **4**, 201–220.

Morgan, R.P.C. (1979) *Soil erosion*. Longman, London.

Morisawa, M. (1968) *Streams – their dynamics and morphology*. McGraw-Hill, New York.

Morisawa, M. (1985) *Rivers: Form and Process*. Longman, London.

Moss, M. (ed) (1988) *Landscape Ecology and Management*. Polyscience, Montreal.

Moss, R. (1969) A comparison of red grouse stocks with the production and nutritive value of heather. *J. Anim. Ecol.* **38**, 103–122.

Moss, R., A. Watson and J. Ollason (1982) *Animal Population Dynamics. (Outline Series in Ecology)*. Chapman and Hall, London.

Mottershead, D.N. (1971) Coastal head deposits between Start Point and Hope Cove, Devon. *Field Studies*, **3**, 433–453.

Mottershead, D.N. (1982) Coastal spray weathering of bedrock in the supratidal zone at East Prawle, Devon. *Field Studies*, **5**, 663–684.

Mottershead, D.N. and G.E. Spraggs (1976) An introduction to the hydrology of the Portsmouth region, in *Portsmouth Geographical Essays II* (eds D.N. Mottershead and R.C. Riley). Portsmouth Polytechnic, Department of Geography, 76–93.

Mottershead, D.N. and I.D. White (1972) The lichonometric dating of glacier succession: Tunsbergdalen, southern Norway. *Geog. Ann.* **54** (A), 47–52.

Mottershead, D.N. and I.D. White Lichen growth in Tunsbergdalen – a confirmation. *Geog. Ann.* **54** (A), 3–4.

Mottershead, D.N. and R.L. Collin (1976) A study of Flandrian glacier fluctuations in Tunsbergdalen, southern Norway. *Norsk Geol. Tids.* **56**, 417–436.

Muffler, L.J. P. and D.E. White (1975) Geothermal energy, in *Perspectives on energy: issues, ideas and environmental dilemmas* (eds L.C. Ruedisili and M.W. Firebaugh). Oxford University Press, New York, 352–358.

Myers, A.A. and P.S. Giller (eds) (1988) *Analytical Biogeography: an Integrated Approach to the Study of Animal and Plant Distributions*. Chapman & Hall, London.

Natural Environment Research Council (1989) *Oceans and the Global Carbon Cycle*. Swindon, NERC.

Naveh, Z. and A.S. Lieberman (1984) *Landscape Ecology: Theory and Application*. Springer Verlag, New York.

Neiburger, M., J.G. Edinger and W.D. Bonner (1971). *Understanding our atmospheric environment*. W.H. Freeman, San Francisco.

Neumann, G. and W.J. Pierson (1966) *Principles of Physical Oceanography*. Prentice-Hall.

Newbould, P.J. (1971) Comparative production of ecosystems, in *Potential crop production* (eds P.F. Waring and J.P. Cooper). Heinemann, London, 228–238.

Newson, M.D. (1975) *Flooding and Flood Hazard in the United Kingdom*. Oxford University Press, Oxford.

Nickling, W.G. (ed) (1986) *Aeolian Geomorphology*. Allen & Unwin, London.

Nockolds, S.R., R.W.O'B. Knox and G.A. Chinner (1978) *Petrology for Students*. Cambridge University Press, Cambridge.

Northcliffe, S. (1984) Spatial analysis of soil. *Progress in Physical Geog.*, **8**, 261–269.

Odum, H.T. (1957) Trophic structure and productivity of Silver Springs, Florida. *Ecol. Monog.* **27**, 55–112.

Odum, E.P. (1964) The new ecology. *Biol Sci.* **14**, 14–16.

Odum, E.P. (1971) *Fundamentals of ecology*, (3rd edn). Saunders, Philadelphia.

Odum, H.T. (1960) Ecological potential and analogue circuits for the ecosystem. *Am. J. Sci.* **48**, 1–8.

Oke, T.R. (1974) *Review of urban climatology 1968–73*. World Met. Organisation, Tech. Note 134. WMO, Geneva.

Oke, T.R. (1978) *Boundary layer climates*. Methuen, London.

Oldfield, F. (1987) The future of the past: – a perspective on palaeoenvironmental study, in *Horizons in Physical Geography* (eds M.J. Clark, K.J. Gregory and A.M. Gurnell). Macmillan, London.

Ollier, C. (1969) *Weathering*. Oliver & Boyd, Edinburgh.

Ollier, C.D. (1981) *Tectonics and Landforms*. Longman, London.

Ollier, C.D. (1988) *Volcanoes*. Blackwell, Oxford.

O'Neill, P. (1985) *Environmental Chemistry*. Chapman & Hall, London.

O'Sullivan, P.E., M.A. Coard and D.A. Pickering (1982) The use of laminated lake sediments in the estimation and calibration of erosion rates, in *Recent Developments in the Explanation and Prediction of Erosion and Sediment Yield* (ed. D.E. Walling). Int. Ass. Hydrologie Scientifique Publication, 137, 385–396.

Ovington, J.T. (1966) *Woodlands*. English Universities Press London.

Oxburgh, E.R. (1974) *The Plain Man's Guide to Plate Tectonics*. Proceedings of the Geologists Association.

Palmen, E. (1951) The rôle of atmospheric disturbances in the general circulation. *Q.J.R. Meteorol Soc.* **77**, 337.

Parker, D.J. and E.C. Penning-Rowsell (1980) *Water Planning in Britain*. Allen & Unwin, London.

Parkinson, D. *et al.* (1971) *Ecology of Soil Micro-organisms*. IBP Handbook No. 19. Blackwell, Oxford.

Paterson, W.S.B. (1981). *The Physics of Glaciers*. Pergamon Press, Oxford.

Paton, T.R. (1978) *The formation of soil material*. George Allen & Unwin, London.

Pauling, L. (1970) *General chemistry*, (3rd edn). W.H. Freeman, San Francisco.

Pedgley, D.E. (1962) *A course in elementary meteorology*. HMSO, London.

Pegg, R.K. and R.C. Ward (1971) What happens to rain? *Weather*, **26**, 88–97.

Pepper, D. (1984) *The Roots of Modern Environmentalism*. Croom Helm, London.

Perry, A.H. (1981) *Environmental Hazards in the British Isles*. Allen & Unwin, London.

Perry, A.H. and J.M. Walker (1977) *The Ocean–Atmosphere System*. Longman, London.

Peters, S.P. (1938) *Sea breezes at Worthy Down, Winchester*. Met. Office Prof. Notes. No. 86. London: HMSO.

Peterson, D.L., M.A. Spanner, S.W. Running and K.T. Teuber (1987) Relationship of thematic mapper simulator data to leaf area index of temperate coniferous forests. *Remote Sensing of Environment*, **22**, 323–341.

Pethick, J. (1984) *An Introduction to Coastal Geomorphology*. Edward Arnold, London.

Pethick, J. (1985) *Slopes*. Geography Today Software. Manchaster, Granada TV.

Petts, G.E. (1983) *Rivers*. Butterworths, London.

Petts, G. and I. Foster (1985) *Rivers and Landscape*. Edward Arnold, London.

Phillipson, J. (1971) *Quantitative Soil Ecology*. IBP Handbook No. 18. Blackwell, Oxford.

Pianka, E.R. (1970) On r- and K-selection. *Amer. Naturalist*, **104**, 592–597.

Pianka, E.R. (1983) *Evolutionary Ecology*, (3rd edn). Harper & Row, New York.

Pickard, G. and W. Emery (1982) *Descriptive Physical Oceanography*, (4th edn). Pergamon.

Pickett, S.T.A. and P.S. White (1985) *The Ecology of Natural Disturbance and Patch Dynamics*. Academic Press, London.

Pimm, S.L. (1982) *Foodwebs*. Chapman & Hall, London.

Pisek, A., W. Larcher, W. Moser and I. Pack (1969) Kardinale temperaturbereiche des lebens der blätter verschiedener spermatophyten III temperaturabhängigkeit und optimaler temperaturbereich der nettophotosynthese. *Flora*, Jena **158**, 608–630.

Pitty, A.F. (1968) The scale and significance of solutional loss from the limestone tract of the central and southern Pennines. *Proc. Geol. Assoc.*, **40**, 601–612.

Pitty, A.F. (1979) *Geography and Soil Properties*. Methuen, London.

Platt, T. and K. Denman (1977) *Hel. Wiss. Meeresunt*, **30**, 575.

Polunin, O. and A. Huxley (1967) *Flowers of the Mediterranean*. Chatto & Windus, London.

Porter, R. and D.W. Fitzsimons (1978) *Phosphorus in the Environment: its Chemistry and Biochemistry*. Elsevier, Amsterdam.

Postgate, J.R. (1978) *Nitrogen fixation*. Edward Arnold, London.

Postgate, J.R. (ed) (1971) *The chemistry and biochemistry of nitrogen fixation*. Plenum, New York.

Press, F. and R. Siever (1978) *The Earth*. Freeman, San Francisco.

Proctor, J. (1971) The plant ecology of serpentine. *J. Ecol.* **59**, 375–410.

Rackham, O. (1971) Historical studies and woodland conservation, in *The scientific management of animal and plant communities for conservation* (eds E.Duffey and A.S. Watt). Blackwell Scientific, Oxford, 563–580.

Rackham, O. (1986) *The Making of the Countryside*. Dent, London.

Raiswell, R.W., P. Brimblecombe, D.L. Dent and P.S. Liss (1984) *Environmental Chemistry*. Edward Arnold, London.

Ranwell, D. (1972) *Ecology of salt marshes and sand dunes*. Chapman & Hall, London.

Read, H.H. (1970) *Rutley's Elements of Mineralogy*, (26th edn). Allen & Unwin, London.

Reuss, J.O. & D.W. Johnson (1986) *Acid Deposition and the Acidification of Soils and Waters*. Springer Verlag, New York.

Rice, R.J. (1977) *Fundamentals of geomorphology*. Longman, London.

Richards, B.N. (1974) *Introduction to the Soil Ecosystem*. Longman, Harlow.

Richards, K. (1982) *Rivers: Form and Process in Alluvial Channels*. Methuen, London.

Richards, K.S. and T.R. Wood (1977) Urbanisation, water redistribution and their effect on channel processes, in *River channel changes* (ed K.J. Gregory). Wiley, Chichester, 369–388.

Ricklefs, R.E. (1990) *Ecology*, (3rd edn). Freeman, New York.

Riehl, H. (1965) *Introduction to the atmosphere*, (3rd edn). McGraw-Hill, New York.

Riehl, H. (1979) *Climate and Weather in the Tropics*. Academic Press, New York.

Rigler, F.H. (1975) In *Unifying Concepts in Ecology* (eds W.H. van Dobben and R.H. Lowe-McConnell). Junk, The Hague, 15–26.

Riley, J.P. and R. Chester (1971) *Introduction to Marine Chemistry*. Academic Press, London.

Robbins, C.T. (1983) *Wildlife Feeding and Nutrition*. Academic Press, New York.

Roberts, N. (1989) *The Holocene: an Environmental History*. Blackwell, Oxford.

Rose, S. (1970) *The Chemistry of Life*. Penguin, London.

Rosenberg, N.J. (1974) *Microclimate: the Biological Environment*. John Wiley, New York.

Ross, S.M. (1987) Energetics of soil processes, in *Energetics of Physical Environment: Energetic Approaches to Physical Geography* (ed K.J. Gregory). John Wiley, Chichester.

Rowntree, P.R. (1990) Estimates of future climatic change over Britain. Part 1 Mechanisms and models. *Weather*, **45**, 38–42.

Russell, R.C.H. and D.H. Macmillan (1952) *Waves and Tides*. Hutchinson, London.

Ruxton, B.P. and I. McDougall (1967) Denudation rates in north-east Papua from K–Ar dating of lavas. *Am. J. Sci.* **265**, 545–561.

Sandford, K.S. (1954) *The Oxford region*. Oxford University Press, Oxford.

Sass, J. H. (1971) The Earth's heat and internal temperatures, in *Understanding the Earth* (eds I.G. Gass, P.J. Smith and R.C.L. Wilson). Artemis Press, Sussex, 81–87.

Schlesinger, W.H. (1991). *Biogeochemistry: an Analysis of Global change*. Academic Press, San Diego.

Schopf, J.W. (ed) (1983) *Earth's Earliest Biosphere*. Princeton University Press, Princeton.

Schultz, A.M. (1964) The nutrient-recovery hypothesis for arctic microtine cycles, in *Grazing in terrestrial and marine environments* (ed D.J. Crisp). Blackwell Scientific, Oxford, 57–68.

Schultz, A.M. (1969) The study of an ecosystem: the arctic tundra, in *The ecosystem concept in natural resource management* (ed G. Van Dyne). Academic Press, New York, 77–93.

Schumm, S.A. and R.W. Lichty (1965) Time space and causality in geomorphology. *Am. J. Sci.* **263**, 110–119.

Schwarz, M.L. and E.C.F. Bird (eds) (1985) *The World's Coastlines*. Van Nostrand Reinholt, New York.

Seddon, B. (1967) Prehistoric climate and agriculture: a review of recent palaeoecological investigations, in *Weather and agriculture* (ed J.A. Taylor). Pergamon, Oxford.

Selby, M.J. (1982a) Controls on the stability and inclinations of hillslopes formed on hard rock. *Earth Surface Processes and Landforms*, **1**, 449–467.

Selby, M.J. (1982b). *Hillslope Materials and Processes*. Oxford University Press, Oxford.

Selby, M.J. (1985) *Earth's Changing Surface*. Oxford University Press. Oxford.

Sellers, W.D. (1965) *Physical climatology*. University of Chicago Press, Chicago.

Shackleton, N.J. and N.D. Opdyke (1973) Oxygen isotope and palaeomagnetic stratigraphy of equatorial pacific core V. 28–238: oxygen isotope temperature and ice volumes on a 10^5 and 10^6 year scale. *Quatern. Res.* **3**, 39–55.

Shackleton, N.J. and N.D. Opdyke (1977) Oxygen isotope and palaeomagnetic evidence for early northern hemisphere glaciation. *Nature*, **270**, 216–219.

Sharp, R.P. (1960) *Glaciers*. Condon Lectures, Oregon State System of Higher Education, Oregon.

Sharp, R.P. (1964) Wind-driven sand in Coachella Valley, Calif. *Bull. Geol. Soc. America*, **91**, 724–730.

Shawyer, C.R. (1987) *The Barn Owl in the British Isles: its past, present and future*. The Hawk Trust.

Shreve, R.L. (1966) Statistical laws of stream numbers. *J. Geol.* **74**, 17–37.

Simmons, I.G. (1975) The ecological setting of Mesolithic man in the Highland zone, in *The effect of man on the landscape: the Highland zone* (eds J.G. Evans, S. Limbrey and H. Cleere). Res. Rep. No. 11. Council for British Archaeology.

Simonson, R.W. (1959) Outline of a generalized theory of soil genesis. *Proc. Soil Sci. Soc. Am.* **23**, 152–156.

Simpson, G.G. (1953) *The major features of evolution*. Columbia University Press, New York.

Simpson, J. (1964) Sea breeze fronts in Hampshire. *Weather*, **19**, 208–220.

Skipworth, J.P. (1974) Continental drift and the New Zealand biota. *NZ J. Geog.* **57**, 1–13.

Slaymaker, H.O. (1972) Patterns of present sub-aerial erosion and landforms in mid-Wales. *Trans Inst. Br. Geogs*, **55**, 47–68.

Smagorinsky, J. (1979) Topics in dynamical meteorology: 10, a perspective of dynamical meteorology. *Weather*, **34**, 126–135.

Smith, A.G. (1961) The Atlantic-sub-boreal transition. *Proc. Linn. Soc.* **172**, 38–49.

Smith, D.I. and M.D. Newson (1974) The dynamics of solutional and mechanical erosion in limestone catchments on the Mendip Hills, Somerset, in *Fluvial processes in instrumented watersheds* (eds K.J. Gregory and D.E. Walling). Inst. Br. Geogs Sp. Publ. 6, 155–167.

Smith, D.I. and P. Stopp (1978) *The River Basin*. Cambridge University Press.

Smith, D.I. and T.C. Atkinson (1976) Process, landform and climate in limestone regions, in *Geomorphology and climate* (ed E. Derbyshire). Wiley, Chichester, 367–409.

Smith, K. (1975) *Principles of applied climatology*. McGraw-Hill, London.

Smith, P. (1973) *Topics in Geophysics*. Open University Press, Milton Keynes.

Soane, B.D. (1973) Techniques for measuring changes in the packing state and cone resistance of soil, after the passage of wheels and tracks. *J. Soil Sci.* **24**, 311–323.

Solomon, M.E. (1969) *Population dynamics*. Edward Arnold, London.

Soper, R. (ed) (1984) *Biological Science*. Part 1. Organisms, Energy and Environment, and Part 2. Systems, Maintenance and Change. Cambridge University Press, Cambridge.

Southwood, T.R.E. (1977) Habitat, the templet for ecological strategies? *J. Animal Ecol.*, **46**, 337–365.

Southwood, T.R.E., R.M. May, M.P. Hassell, and G.R. Conway (1974) Ecological strategies and population parameters. *Am. Nat.* **108**, 791.

Sparks, B.W. and R.G. West (1972) *The ice age in Britain*. Methuen, London.

Spellerberg, I.F. (1991) *Monitoring Ecological Change*. Cambridge University Press, Cambridge.

Spooner, B. and H.S. Mann (eds) (1982) *Desertification and Develop-ment: Dryland Ecology in Social Perspective*. Academic Press, London.

Sprent, J.I. (1988) *The Ecology of the Nitrogen Cycle*. Cambridge University Press, Cambridge.

Statham, I. (1977) *Earth surface sediment transport*. Oxford University Press, Oxford.

Stebbins, G.L. (1974) Adaptive shifts and evolutionary novelty, in *Studies in the philosophy of biology*, 285–306.

Street-Perrott, F.A., N. Roberts and S. Metcalfe (1985) Geomorphic implications of late Quaternary hydrological and climatic changes in the Northern Hemisphere tropics, in *Environmental Change and Tropical Geomorphology* (eds I. Douglas and T. Spencer). Allen & Unwin, London.

Stoddart, D.R. (1969) World erosion and sedimentation, in *Water, earth and man* (ed R.J. Chorley). Methuen, London, 43–64.

Stork, A. (1963) Plant immigration in front of retreating glaciers, with examples from the Kebnekajse area, Northern Sweden. *Geog. Ann.* **XLV** (1), 1–22.

Stowe, K.S. (1984) *Principles of Ocean Science*, (2nd edn). John Wiley, Chichester.

Strahler, A.N. (1952) Hypsometric analysis of erosional topography. *Geol Soc. Am. Bull.* **63**, 923–938.

Strahler, A.N. (1972) *Planet Earth: its physical systems through geological time*. Harper & Row, New York.

Strahler, A.N. and A.H. Strahler (1973) *Environmental geoscience*. Hamilton, California.

Strakhov, N.M. (1967) *Principles of lithogenesis*. Vol. 1: *Consultants bureau*. Oliver & Boyd, New York, London.

Streeter, D.T. (1974) Ecological aspects of oak woodland conservation, in *The British oak* (eds M.G. Morris and F.H. Perring). Bot. Soc. Brit. Isles Conf. Report 14, Faringdon: Classey.

Sugden, D.E. and B.S. John (1976) *Glaciers and landscape. A geomorphological approach*. Edward Arnold, London.

Summerfield, M.A. (1991) *Global Geomorphology*. Longman, London.

Sumner, G. (1988) *Precipitation: Process and Analysis*. John Wiley, New York.

Sutcliffe, J. (1968) *Plants and water*. Edward Arnold, London.

Sutcliffe, R.C. (1948) *Meteorology for aviators*. Met. Office 432. HMSO, London.

Sverdrup, H.U., M.W. Johnson and R.H. Fleming (1942) *The Oceans: Their Physics, Chemistry and General Biology*. Prentice-Hall.

Swift, M.J., O.W. Heal and J.M. Anderson (1979) *Decomposition in Terrestrial Ecosystems. (Studies in Ecology vol 5)*. Blackwell, Oxford.

Tansley, Sir A.G. (1935) The use and abuse of vegetational concepts and terms. *Ecology*, **16**, 284–307.

Taylor, A.M. and E. Burnett (1964) Influence of soil strength on the root growth habits of plants. *Soil Sci.* **98**, 178–180.

Taylor, A.M. and L.F. Ratcliff (1969) Root growth pressure of cotton peas and peanuts. *Agron. J.*, **61**, 389–402.

Taylor, J.A. (1970) The cost of British weather, in *Weather economics* (ed J.A. Taylor). Pergamon, Oxford.

Taylor, R.J. (1984) *Predation*. Chapman & Hall, London.

Ternan, J.L. and A.G. Williams (1979) Hydrological pathways and granite weathering on Dartmoor, in *Geographical Approaches to Fluvial Processes* (ed A.F. Pitty). Geobooks, Norwich, 5–30.

Thomas, A.J. (1978) Worldwide weather disasters. *J. Meteorol.* (UK), **3**.

Thomas, A.S. (1963) Further changes in the vegetation since the advent of myxomatosis. *J. Ecol.* **51**, 151–186.

Thomas, D.S.G. (ed) (1989) *Arid Zone Geomorphology*. Belhaven, London.

Thomas, R.W. and R.J. Huggett (1980) *Modelling in Geography: a Mathematical Approach*. Harper & Row, London.

Thorarinsson, S. (1939) Observations on the drainage and rates of denudation in the Hoffelsjökull district. *Geog. Ann.* **21**, 19–215.

Thorn, C.E. (ed) (1982) *Space and Time in Geomorphology*. Allen & Unwin, London.

Thorn, C.E. (1988) *Introduction to Theoretical Geomorphology*. Unwin Hyman, London.

Thornes, J.B. (1978) The character and problems of theory in contemporary geomorphology, in *Geomorphology, present problems and future prospects* (eds C. Embleton, D. Brunsden, and D.K.C. Jones). Oxford University Press, Oxford.

Thornes, J.B. (1979) *River Channels*. Macmillan, London.

Thornes, J.B. (1983) Evolutionary geomorphology. *Geography*, **68**, 225–235.

Thornes, J.B. (1987) Environmental systems – patterns, processes, and evolution, in *Horizons in Physical Geography* (eds M.J. Clark, K.J. Gregory and A.M. Gurnell). Macmillan, Basingstoke.

Thornes, J.B. and D. Brunsden (1977) *Geomorphology and time*. Methuen, London.

Thorsteinsson, I., G. Olafsson and G.M. Van Dyne (1971) Range resources of Iceland. *J. Range Mgmt*, **24**, 86–93.

Thrush, B.A. (1977) The chemistry of the stratosphere and its pollution. *Endeavour*, **1**, 3–6.

Thurman, H.V. (1987) *Essentials of Oceanography*. Merrill, Columbus.

Tjallingii, S.P. and A.A. de Veer (eds) (1982) *Perspectives in Landscape Ecology*. Centre for Agricultural Publications, Wageningen.

Tolmazin, D. (1985) *Elements of Dynamic Oceanography*. Allen and Unwin.

Tooley, M.J. (1974) Sea-level changes during the last 9000 years in north-west England. *Geog. J.* **140**, 18–42.

Trenhaile, A.S. (1980) Shore platforms: a neglected coastal feature. *Progress in Physical Geography*, **4** (1), 1–23.

Trewartha, G.T. (1961) *The earth's problem climates*. University of Wisconsin Press, Madison.

Tricart, J. (1985) Evidence of Upper Pleistocene dry climates in northern South America, in *Environmental Change and Tropical Geomorphology* (eds I. Douglas and T. Spencer). Allen & Unwin, London, 197–217.

Troake, R.P. and D.E. Walling (1973) The natural history of Slapton Ley Nature Reserve. VII The hydrology of the Slapton Wood stream; a preliminary report. *Field Studies* 3, 719–740.

Troll, C. (1968) Landschaftsokologie, in *Planzensoziologie und Landschaftsokologie* (ed R. Tuxen). Junk The Hague, 1–21.

Troll, C. (1971) Landscape ecology (geo-ecology) and biocoenology – a terminological study. *Geoforum*, 8, 43–46.

Trudgill, S. (1977) *Soil and vegetation systems*. Oxford University Press, Oxford.

Trudgill, S.T. (1983) Soil geography: spatial techniques and geomorphic relationships. *Progress in Physical Geog.*, 7, 345–360.

Trudgill, S. *et al.* (1984) Hydrology and solute uptake in hillslope soils on Magnesian Limestone: the Whitwell Wood project, in *Catchment Experiments in Fluvial Geomorphology* (eds T.P. Burt and D.E. Walling). Geobooks, Norwich, 183–215.

Turekian, K.K. (1976) *Oceans*. Prentice-Hall, Englewood Cliffs.

UK Terrestrial Effects Review Group (1988) *The Effects of Acid Deposition on the Terrestrial Environment in the UK*. HMSO, London.

UNCOD (1977) *Desertification: its Causes and Consequences*. Pergamon Press, Oxford.

Valentine, J.W. and E.M. Moores (1970) Plate tectonics regulation of faunal diversity and sea level: a model. *Nature* 228, 657–659.

Van Dobben, W.H. and R.H. Lowe-McConnell (eds) (1975) *Unifying Concepts in Ecology*. Junk, The Hague.

Van Wambeke, A. (1966) Soil bodies and soil classification. *Soils Fert.*, 29, 507–510.

Villee, C.A. *et al.* (1989) *Biology*, (2nd edn). Saunders, Philadelphia.

Vincent, P. (1990) *The Biogeography of the British Isles*. Routledge, London.

Vivian, R. (1975) *Les Glaciers des Alpes Occidentales*. Imperimerie Allier, Grenoble.

Waddington, C.H. (1975) *Catastrophe Theory of Evolution in the Evolution of an Evolutionist*. Cornell University Press, New York.

Walker, D. (1979) *Energy Plants and Man*. Packard, Funtington.

Walker, D. and R.G. West (eds) (1970) *Studies in the vegetational history of the British Isles*. Cambridge University Press, Cambridge.

Walker, J.C.G. (1977) *Evolution of the Atmosphere*. Macmillan, New York.

Walker, J.C.G. (1984) How life affects the atmosphere. *Bioscience*, 34, 486–491.

Walling, D.E. (1987) Rainfall, runoff and erosion of the land: a global view, in *Energetics of Physical Environment: Energetic Approaches to Physical Geography* (ed K.J. Gregory). John Wiley, Chichester.

Walling, D.E. and B.W. Webb (1975) Spatial variation of river water quality: a survey of the River Exe. *Trans Inst. Br. Geogs*, 65, 155–71.

Walling, D.E. and B.W. Webb (1981) Water quality, in *British Rivers* (ed J. Lewin). Allen & Unwin, 126–169.

Walling, D.E. and B.W. Webb (1983) Patterns of sediment yield, in *Background to Palaeohydrology: a Perspective* (ed K.J. Gregory). John Wiley, Chichester, 69–100.

Walling, D.E. and B.W. Webb (1986) Solutes in river systems, in *Solute Processes* (ed S.T. Trudgill). John Wiley, Chichester, 251–327.

Wallwork, J.A. (1970) *The ecology of soil animals*. McGraw Hill, Maidenhead.

Wambeke, A. Van (1966) Soil bodies and soil classification. *Soils Fert.*, 29, 507–510.

Ward, R.C. (1975) *Principles of Hydrology*, (2nd edn). McGraw-Hill, Maidenhead.

Ward, R. (1978) *Floods: a Geographical Perspective*. Macmillan, London.

Warneck, P. (1988) *Chemistry of the Natural Atmosphere*. Academic Press, London.

Warren, A. and C.M. Harrison (eds) (1983) *Conservation in Perspective*. John Wiley, Chichester.

Warrick, R., A. Wilkinson and T.M.L. Wigley (1989) *Estimating global mean sea-level change 1982–2050*. Climate Research Unit Report, University of East Anglia.

Waters, R.S. (1954) Pseudobedding in the Dartmoor granite. *Trans R. Geol Soc.* 18, 456–462.

Watson, J.D. (1970) *The Double Helix*. Penguin, London.

Watts, A.J. (1955) Sea breeze at Thorney Island. *Meteorol. Mag.* 84, 42–48.

Weatherly, P.E. (1969) Ion movement within the plant and its integration with other physiological processes, in *Ecological aspects of the mineral nutrition of plants* (ed I.H. Rorison). Blackwell Scientific, Oxford, 323–340.

Webb, W.L., W.K. Lauenroth, S.R. Szarek and R.S. Kinerson (1983) Primary production and abiotic controls in forests, grasslands, and desert ecosystems in the United States. *Ecology*, 64, 134–151.

Wellburn, A. (1988) *Air Pollution and Acid Rain: The Biological Impact*. Longman.

Weller, G. and B. Holmgren (1974) The microclimates of the arctic tundra. *J. Appl. Meteorol.* 13, 854–862.

West, D.C., H.H. Shugart and D.B. Botkin (eds) (1981) *Forest Succession: Concepts and Application*. Springer Verlag, New York.

West, R.G. (1968) *Pleistocene Geology and Biology*. Methuen, London.

Wetherald, R.T. and S. Manabe (1975) *Run-off processes and streamflow modelling*. Oxford University Press, Oxford.

Weyman, D.R. (1975) *Run-off processes and streamflow modelling*. Oxford University Press: Oxford.

Weyman, D. (1981) *Tectonic Processes*. Allen & Unwin, London.

Whalley, W.B. (1976) *Properties of materials and geomorphological explanation*. Oxford University Press, Oxford.

Whalley, W.B. and J.P. McGreevy (1985, 1987, and 1988) Weathering. *Progress in Physical Geography*, 9, 11, and 12, 559–581, 357–369 and 130–143.

White, I.D. (1973a) *Plant succession and soil development on the recessional moraines*. Tunsbergdalen Res. Expedition Report No. 2. Dept. of Geography, Portsmouth Polytechnic.

White, I.D. (1973b) *Preliminary report on the results of dendrochronological studies*. Tunsbergdalen Res. Expedition Report No. 2. Dept. of Geography, Portsmouth Polytechnic.

White, I.D. and D.N. Mottershead (1972) Past and present

vegetation in relation to solifluction on Ben Arkle, Sutherland. *Trans Bot. Soc. Edin.* **41**, 475–489.

White, R.E. (1979) *Introduction to the principles and practice of soil science.* Blackwell Scientific, Oxford.

Whittaker. R.H. (1975) *Communities and ecosystems.* Macmillan, New York.

Whittaker, R.H. and G.E. Likens (1975) *The biosphere and man.*

Whittaker, R.H. and G.M. Woodwell (1971) Measurement of net primary production of forests, in *Productivity of forest ecosystems* (ed P. Davigneaud). Unesco, Paris, 159–175.

Whittow, J.B. (1980) *Disasters: The Anatomy of Environmental Hazards.* Allen Lane, London.

Wild, A. (ed) (1988) *Russell's Soil Conditions and Plant Growth* (11th edn) Longman, London.

Williams, A.G., J.L. Ternan and M. Kent (1984) Hydrochemical characteristics of a Dartmoor hillslope, in *Catchment Experiments in Fluvial Geomorphology* (eds T.P. Burt and D.E. Walling). Geobooks, Norwich.

Wilson, I.G. (1972) Universal discontinuities in bedforms produced by the wind. *J. Sedimentary Petrology*, **42**, 667–669.

Wilson, J.T. (1963) Continental drift. *Scient. Am.* **208** (April), 86–102.

Wilson, J. Tuzo (1976) *Continents Adrift and Continents Aground.* (Readings from Scientific American) Freeman, San Francisco.

Wilson, K. (1960) The time factor in the development of dune soils at South Haven Peninsula, Dorset. *J. Ecol.* **48**, 341–359.

Wilson, R.C.L. (1983) *Residual Deposits: Surface-Related Weathering Processes and Materials.* Geol. Soc. Sp. Pub. 11. Blackwell, Oxford.

Winkler, E.M. (1975) *Stone: Properties. Durability in Man's Environment,* (2nd edn). Springer Verlag, Berlin.

Winkler, E.M. and E.J. Wilhelm (1970) Salt burst by hydration pressures in architectural stone in urban atmosphere. *Geol Soc. Am. Bull.* **81**, 567–572.

Winstanley, D. (1978) The drought that won't go away. *New Scient.* **164**, 57.

Woldenberg, M. (ed) (1985) *Models in Geomorphology.* Allen & Unwin, London.

Wolman, M.G. (1955) *The natural channel of Brandywine Creek, Pennsylvania.* USGS Prof. Paper 271.

Wolman, M.G. (1967) A cycle of erosion and sedimentation in urban river channels. *Geog. Ann.* **49**(A), 385–395.

Woodwell, G.M. (ed) (1984) *The Role of Terrestrial Vegetation in the Global Carbon Cycle: Measurement by Remote Sensing.* John Wiley, New York.

Wooldridge, S.W. (1960) The Pleistocene sucession in the London Basin. *Proc. Geol Assoc.* **71** (2), 113–129.

World Meteorological Organisation (1956) *International cloud atlas.* WHO, Geneva.

Wyllie, P.J. (1976) *The Way the Earth Works: an Introduction to the New Global Geology and its Revolutionary Development.* John Wiley, New York.

Wynne-Edwards, V.C. (1965) Self-regulating systems in populations of animals. *Science,* **147**, 1543–1548.

Yaalon, D.H. (1971) *Palaeopedology: origin, nature and dating of palaeosols.* Israel University Press, Jerusalem.

Yang, C.T. (1970) On river meanders. *J. Hydrol.* **13**, 231–253.

Yang, C.T. (1971) Potential energy and stream morphology *Water Resources Res.* **7** (2), 311–322.

Young, A. (1969) Present rate of land erosion. *Nature* **224**, 851–852.

Young, A. (1972) *Slopes.* Longman, London.

Young, A. (1978) A twelve-year record of soil movement on a slope. *Z. Geomorph.* N.F. Suppl. **29**, 104–110.

Index